P R E N T I C E

BRIEF REVIEW IN NEW YORK

Earth Science

The Physical Setting

2006 Edition

Jeffrey C. Callister, Retired
Newburgh Free Academy
Newburgh, New York

ORDER INFORMATION

Send orders to:

PRENTICE HALL SCHOOL DIVISION
CUSTOMER SERVICE CENTER
P.O. Box 2500
Lebanon, Indiana 46052-3009

or

CALL TOLL-FREE: 1-800-848-9500
(8:00 A.M.–6:00 P.M. EST)

or

FAX TOLL-FREE: 1-877-260-2530
(24 hours a day, 7 days a week)

• Orders processed with your call.
• Your price includes all shipping and handling.

PEARSON
Prentice Hall

Boston, Massachusetts
Upper Saddle River, New Jersey

Teacher Reviewers and Contributors

Timothy Denman
Planetarium Director, Retired
Newburgh Free Academy
Newburgh, New York

Tom Elkins
John L. Miller Great Neck North High School
Great Neck, New York

David Keith
Science Chair
John L. Miller Great Neck North High School
Great Neck, New York

Peter Le Rose
Roy C. Ketcham High School
Wappingers Falls, New York

Elise Russo, Ed.D.
Administrator for Secondary Curriculum
Amityville Union Free School District
Amityville, New York

Peter Suchmann
Science Chair
Great Neck South Middle School
Great Neck, New York

STAFF CREDITS

The people who made up the Earth Science team—representing design services, editorial, editorial services, and production services—are listed below.

Lisa J. Clark, Patricia Cully, Kathleen J. Dempsey, Terence Hegarty, Kim Schmidt, Jerry Thorne

ISBN 0-13-250993-8

3 4 5 6 7 8 9 10 10 09 08 07 06

PEARSON
Prentice Hall

Brief Review in

Earth Science: The Physical Setting

About This Book

This book was designed for the students who plan to take the Regents Examination for The Physical Setting: Earth Science. Most students using this book will be taking or will have taken an investigative-approach earth science course. This book should be useful for studying to pass the end-of-year Regents Examination in earth science as well as studying for topic tests during the course. Most students taking this course will need to pass the Regents Examination to meet the requirements for high school graduation.

The Regents Examination will be based on the content and understandings addressed in the core curriculum for The Physical Setting: Earth Science (as covered in your earth science course), the *Earth Science Reference Tables,* and the laboratory skills learned during your course. This book is organized to enhance your review of the concepts, skills, and application of the core curriculum that may be tested in The Physical Setting: Earth Science Regents Examination. Since there is no preferred order for teaching concepts in the core curriculum, this book is organized according to an older New York State Syllabus in Earth Science. For those students exposed to a different order of topics in the classroom, note that each topic (chapter) of this book is independent of the order of presentation in this text except for the listing of vocabulary words (see below). The use of the table of contents and the index will enable you to find any topic that you need to review.

- *Review of Content:* This review book focuses on the basic content that will be tested on the Regents Examination. It includes more than 180 illustrations to help you visualize and understand the concepts and vocabulary of earth science. The illustrations will familiarize you with the types of drawings you will be required to interpret in Regents Examination questions. You should carefully read the captions to the illustrations and explanations within the text.

- *Vocabulary:* You will need to know the definitions of vocabulary words listed at the beginning of each topic to answer questions on the Regents Examination. These words are shown in bold type within the topic where they are first defined. These words also may appear in other topics, where they are underlined. Vocabulary words are also defined in the glossary, where they are identified by an asterisk.

- *Underlined words:* Words that are underlined in the text of a topic are words that appear in the vocabulary list and in bold type in another topic or other words that you need to know to understand basic earth science concepts. Although you are not likely to be tested on the specific definitions of these other words, they may be used in Regents Examination questions. You will need to understand the meaning of these words, which are defined in the glossary.

- *Memory Jogger and Digging Deeper:* The Memory Jogger features in the topics are designed to refresh your memory of relevant information in the topic that was covered in previous topics of this book or in previous science courses. The Digging Deeper features provide specific examples of how to do problems, describe a specific example of some subject matter, or describe some information that may expand core information content.

- *Review Questions:* Review questions appear frequently throughout each topic to help you clarify and reinforce your understanding of the content. The questions, totaling nearly 700, are similar to the types of questions that may appear on the Earth Science Regents Examination.

- *Questions for Regents Practice:* These questions, totaling about 500, appear at the end of each topic. These practice questions are written and organized in the format of the Regents Examination with Part A, Part B, and Part C questions. In many Part B and Part C questions you will notice a number in brackets. This number indicates how many points the question is worth on a Regents Examination.
  - Part A questions are entirely multiple-choice and test your knowledge of concepts in the Earth Science core curriculum.
  - Part B questions test skills and understandings of concepts outlined in the core curriculum and include both multiple-choice and constructed-response questions.
  - Part C questions often require an extended constructed response. For these questions you will often need to provide a more detailed answer, supported with applications or examples.

In the back of the book you will find several appendices and other items that will be helpful as you study the topics and review for the Earth Science Regents Examination.

- *Appendix 1:* You will need to make use of the *Earth Science Reference Tables* in most topics. When a reference table is useful, the text will refer to the *Earth Science Reference Tables* in italic type and will state the name of the specific table. A list of all of the reference tables can be found in Appendix 1 on page 281.
- *Appendix 2:* This appendix provides Strategies for Answering Test Questions. It includes strategies for answering multiple-choice and constructed-response questions and for using the *Earth Science Reference Tables*.
- *Appendix 3:* The Performance Test that will be included in Part D of the Earth Science Regents Examination is partly described in Appendix 3. The Performance Test will evaluate various laboratory and classroom procedures.
- *Regents Examinations:* Past Regents Examinations are reproduced in actual size near the end of the book to provide practice in taking a Regents Examination.
- *Landscape Regions of New York:* For easy reference, a map and descriptions of the landscape regions of New York State are found on pages 487–488. In many topics the landscape region map, which is similar to the one in the *Earth Science Reference Tables*, is referred to and is needed to answer questions.

Many individuals have helped in the completion of this book. Most of these people are acknowledged on page ii. In addition, I would like to give special thanks for help in editing the manuscript to my wife Angie Callister of the English Department of Newburgh Free Academy, Newburgh, New York. I would also like to thank Lee Cabe, Velma Garey, Barry Heil, Paul Hill, Alexandra Passas, and Alfred Romano, Jr—my former colleagues in the science department at Newburgh Free Academy, Newburgh, New York—for their research and guidance.

Jeffrey C. Callister

Introduction to Earth's Changing Environment

TOPIC 1

VOCABULARY		
classification	interface	percent deviation
cyclic change	mass	pollution
density	measurement	prediction
dynamic equilibrium	natural hazard	rate of change
inference	natural resources	universe
instrument	observation	volume

Note to student: *Much of the information presented in this topic explains process skills that you will use as you study earth science. You have probably already used some of these process skills. Most of the information in this topic will be discussed in more detail in the context of specific content in later topics. While you will not be tested on concepts in this topic independently, many of these concepts will be tested in conjunction with specific concepts. For example, you will be expected to use equations found in this topic to solve problems in percent deviation, density, and rate of change.*

Earth science is the study of Earth and its position in the universe. Often Earth science is considered a combination of four science disciplines:

- Geology is the study of the history, structure, processes, and composition of Earth's solid surface down to Earth's center.
- Oceanography is the study of all aspects of Earth's oceans.
- Meteorology is the study of Earth's atmosphere, including weather and climate.
- Astronomy is the study of the **universe**—that is all matter, time, energy, and space.

Observations

An **observation** is the perception of some aspect of the environment by one or more human senses—sight, hearing, touch, taste, or smell.

Observing with Senses and Instruments

The senses are limited in range and precision. An **instrument** is a human-made device that extends the senses beyond their normal limits, thus enabling them to make observations that would otherwise be impossible or highly inaccurate. For example, a microscope makes it possible to see objects and details too small for the unaided eye to detect, and a magnet allows one to observe something the senses do not perceive at all. We would not even know about many forms of energy—such as X-rays, radio, microwave, and ultraviolet—without certain instruments.

Inferences and Misconceptions

An **inference** is an interpretation of an observation. It is a mental process that proposes causes, conclusions, or explanations for what has been observed. For example, the observation of an impression shaped like a dog's foot leads to the inference that a dog has been present. This inference may or may not be correct. Additional observations may make the inference more likely to be true.

A prediction of a future event, such as a hurricane or snowstorm, is a type of inference. Misconceptions about environmental characteristics often result from incorrect inferences that become commonly believed. A misconception is a mistaken belief or a misunderstanding. For thousands of years, humans believed the inference that Earth was the center of the universe and didn't spin or move around the sun.

Classification

Scientists group together similar observations and inferences to make the study of objects and events in the environment more meaningful or easier to understand. This grouping is called **classification.** In later topics, you will study classifications of objects and events such as planets, stars, minerals, storms, types of energy, and natural disasters.

Measurement

A **measurement** is a means of expressing an observation with greater accuracy or precision. It provides a numerical value for an aspect of the object or event being observed. Every measurement includes at least one of the three basic dimensional quantities—length, mass, and time—which are defined as follows:

- Length may be described as the distance between two points.
- The amount of matter in an object is its **mass.** It is often determined by using a balance. Mass should not be confused with weight, which is the pull (force) of Earth's gravitation on an object. The weight of an object may vary with its location, but its mass remains the same.
- Time is often described as our sense of things happening one after another or as the duration of an event.

Some types of measurements require the mathematical combination of basic dimensional quantities. For example, a unit of volume is actually a unit of length cubed, as in 25 cm^3 (25 cubic centimeters). **Volume** is the amount of space that an object occupies. The volume of solid objects is often determined by finding the volume of water an object displaces when it is placed in water in an instrument such as a graduated cylinder. Other examples of such types of measurement are density (mass per unit volume, as in 4 g/cm^3) and speed (distance per unit time—9.8 km/sec).

All measurements of the basic quantities (length, mass, and time) are made by a direct comparison to certain accepted standard units of measurement. For example, length is measured by comparing it to a standard unit such as the centimeter, mass to a unit such as the gram, and time to a unit such as the second. Imagine you asked some friends how much time an experiment took to perform, and they answered 60. Wouldn't you have to ask 60 of what? A measurement must always state the units used, for example, 27.9 grams or 26 kilometers.

PERCENT DEVIATION OR PERCENT ERROR

No measurement is perfect because of the imperfection of the senses and of instruments. People may also introduce errors by carelessness or the improper use of an instrument. Any measurement is therefore an approximation of a true value and must be considered to contain some error.

Science has accepted values for some given qualities (for example, the 1 g/cm^3 for the density of liquid water at 4°C). It is therefore possible to determine the accuracy, or amount of error, of a given measurement by comparing it with the accepted value.

$$\textbf{percent deviation} = \frac{\textbf{difference from accepted value}}{\textbf{accepted value}} \times 100\%$$

For example, suppose a student measures the mass of an object as 127.5 grams and the accepted value is 125.0 grams. Then,

$$\text{amount of deviation} = \text{difference from accepted value}$$
$$= 127.5 \text{ grams} - 125.0 \text{ grams}$$
$$= 2.5 \text{ grams}$$

$$\text{percent deviation} = \frac{\text{amount of deviation}}{\text{accepted value}} \times 100\%$$
$$= \frac{2.5 \text{ grams}}{125.0 \text{ grams}} \times 100\%$$
$$= 2.0\%$$

When the amount of error is expressed as a percentage, it is called **percent deviation,** or percent error. The percent deviation is obtained by dividing the difference between the measured and accepted values by the accepted value and multiplying the result by 100 percent. The formula for percent deviation is found in the *Earth Science Reference Tables.*

Review Questions

1. In the classroom during a visual inspection of a rock, a student recorded four statements about the rock. Which statement about the rock is an observation? (1) The rock formed deep in Earth's interior. (2) The rock cooled very rapidly. (3) The rock dates from the Cambrian Period. (4) The rock is black and shiny.

2. To make observations, an observer must always use (1) experiments (2) the senses (3) proportions (4) mathematical calculations

3. Which statement made by a student after examining a rock specimen is an inference? (1) The rock is of igneous origin. (2) The rock has rounded edges. (3) The rock is light-colored. (4) The rock contains large crystals.

4. A classification system is based on the use of (1) the human senses to infer properties of objects (2) instruments to infer properties of objects (3) observed properties to group objects with similar characteristics (4) inferences to make observations

5. Which statement best illustrates a classification system? (1) A glacier melts at the rate of one meter per year. (2) Ocean depths are measured using sound waves. (3) Snowfall predictions for winter storms vary. (4) Stars are grouped according to color.

6. Organizing information in a meaningful way is an example of (1) prediction (2) measurement (3) observation (4) classification

7. A measurement is best defined as (1) an inference made by using the human senses (2) direct comparison to a known standard (3) an interpretation based on theory (4) a group of inferred properties

8. A student measures the velocity of the water in a stream as 2.5 meters per second. The actual velocity of the water is 3.0 meters per second. What is the approximate percent deviation of the student's measurements? (1) 0.50% (2) 17% (3) 20% (4) 50%

9. A sphere was dropped into water in a graduated cylinder as shown in the following diagram. The water level rose to the new level shown.

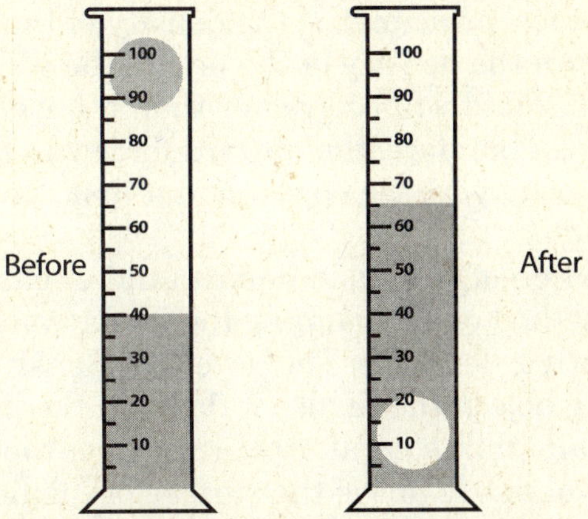

What is the volume of the sphere? (1) 15 mL (2) 25 mL (3) 40 mL (4) 65 mL

Density

The concentration of matter in an object is known as its **density.** If we model matter with students, then a full classroom with 35 students has a high density, and a classroom with only two students has a low density. The density of an object is the ratio of its mass to its volume; that is, density is the mass per each unit of volume. Density is important in the study of earth science. This is because density is a property of materials that affects the way in which materials interact in the environment.

Computing Density

To determine the density of an object, divide its mass by its volume. This formula for density is found in the *Earth Science Reference Tables:*

$$\textbf{density} = \frac{\textbf{mass}}{\textbf{volume}}$$

The density of a material does not depend on the size or shape of the sample as long as the temperature and pressure remain the same. For example, a cube of the mineral graphite with a volume of 20 cm^3 has a mass of 44 g. Its density is 44 grams divided by 20 cm^3 = 2.2 g/cm^3. A graphite ball with a volume of 40 cm^3 (twice the volume of the cube) will have a mass of 88 g (twice the mass of the cube), but its density will be the same: 88 g/40 cm^3 = 2.2 g/cm^3. The use of density in mineral identification is discussed in Topic 11.

Determining Relative Density

An object immersed in a liquid is buoyed by a force equal to the weight of liquid it displaces (Archimedes' principle). If the density of the object is less than the density of the liquid, the weight of displaced liquid will be greater than the weight of the object. Therefore, the upward force will be enough to support the object, and it will float in the liquid.

Flotation of objects in liquids (and in gases) is one method of determining their relative, or comparative, densities. The lower the density of a floating object, the higher it floats in the liquid. This means that a greater percentage of the object's volume is above the surface of the liquid. As the density of a floating object increases, it sinks relatively deeper into the liquid. If its density is greater than that of the liquid, it will sink to the bottom. If an object and a liquid have exactly the same density (for example, a fish in water), the object can remain stationary anywhere in the liquid.

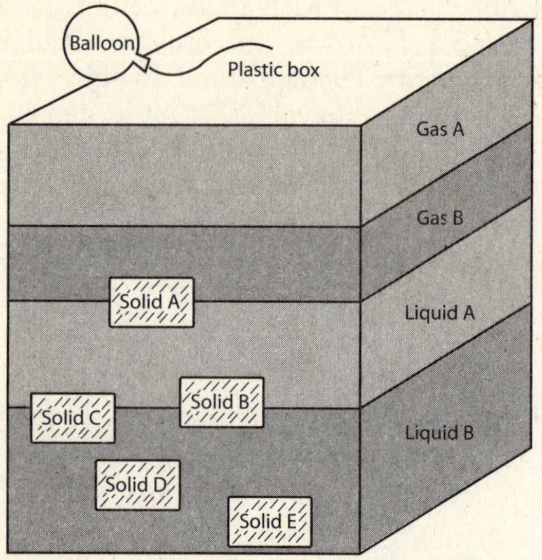

Figure 1-1. Density and height of flotation: The various substances in the plastic box are arranged by the height at which they float. Make inferences about the relative density of the substances before you continue to read. From highest to lowest, the relative densities of the substances are solid E, solid D and liquid B (about equal), solid C, solid B, liquid A, solid A, gas B, gas A, the balloon.

Physical Changes That Affect Density

Changes in temperature and pressure affect the densities of substances, especially those of gases. If the temperature of a gas increases and its pressure remains the same, its molecules move farther apart (the gas expands). The temperature rise thus results in less mass per unit volume, so that the density of the gas decreases. This explains why hot air rises when surrounded by cooler air. This rising of less-dense objects and falling of more-dense objects results in the convection currents (see Topic 5), which affect movements in many portions of Earth and help to distribute heat energy.

If the pressure on a gas increases, the molecules come closer together (the gas contracts). The pressure rise thus results in more mass per unit volume, and the density increases. The effects of pressure on liquids and solids can be significant. This is illustrated by the increasing densities of the zones of Earth's interior illustrated in the Inferred Properties of Earth's Interior in the *Earth Science Reference Tables*.

Phases of Matter and Density

Matter on Earth usually exists in three forms: solid, liquid, and gas. Each form is known as a phase, or state, of matter. The density of a substance changes with changes in its phase. The layering of the outer spheres of Earth—discussed in Topic 2—is the result of the effects of gravity pulling differently on solids, liquids, and gases due to their density differences. Almost all substances increase in density as they change from gas to liquid and from liquid to solid. They have their highest density as a solid because the atoms are closest together in that phase.

Water is an exception; its highest density is in the liquid state at a temperature of 3.98°C. Therefore, solid water (ice) floats on liquid water, while in most substances the solid sinks in the liquid. Water at 4°C will also lie below layers of water at any other temperature.

Review Questions

10. A student finds the density of an ice cube to be 0.80 g/cm³; it is actually 0.90 g/cm³. What is the percent deviation (percent error) in this calculation? (1) 6 percent (2) 11 percent (3) 13 percent (4) 88 percent

11. The amount of space a substance occupies is its (1) mass (2) volume (3) density (4) weight

12. Which is the best definition for the mass of an object? (1) the amount of space the object occupies (2) its ratio of weight to volume (3) the quantity of matter the object contains (4) the force of gravity acting on the object

13. Under the same conditions of temperature and pressure, three different samples of the same uniform substance will have the same (1) shape (2) density (3) mass (4) volume

14. As shown in the following diagram, an empty 1000 mL container has a mass of 250 g. When filled with a liquid, the container and the liquid have a combined mass of 1300 g

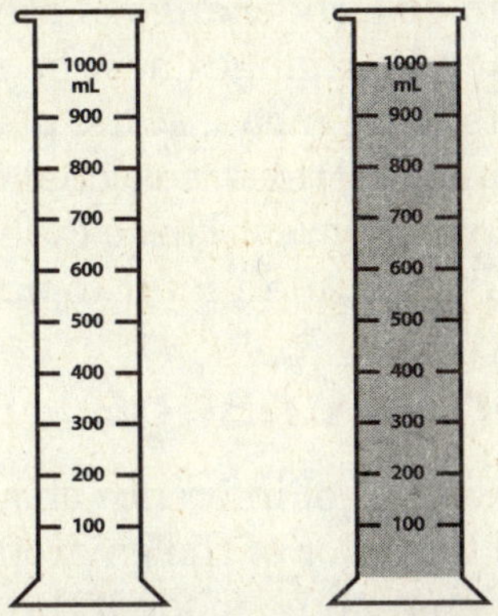

Empty Mass = 250.0 g   Filled Mass = 1300 g

What is the density of the liquid? (1) 1.00 g/mL (2) 1.05 g/mL (3) 1.30 g/mL (4) 0.95 g/mL

15. An empty 250 mL beaker has a mass of 60 g. When 100 mL of oil is added to the beaker, the total mass is 140 g. The density of the oil is approximately (1) 1.7 g/mL (2) 1.4 g/mL (3) 0.8 g/mL (4) 0.6 g/mL

16. The diagram at the right represents a cylinder containing four liquids—W, X, Y and Z—each with a different density (D), as indicated. A piece of solid quartz having a density of 2.7 g/cm³ is placed on the surface of liquid W. When the quartz is released, it will pass through

(1) W, but not X, Y, or Z
(2) W and X, but not Y or Z
(3) W, X, and Y, but not Z
(4) W, X, Y, and Z

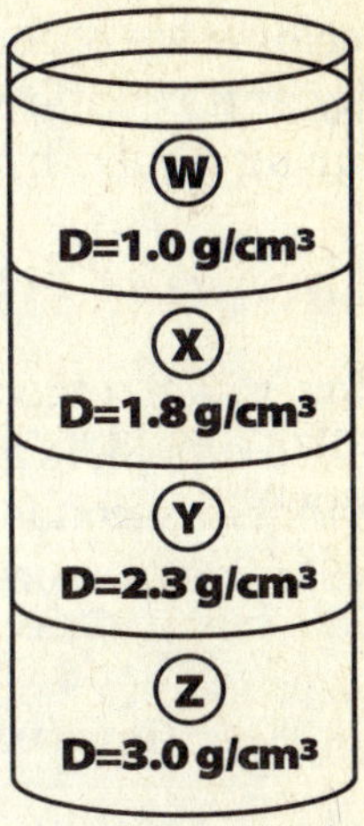

17. A student mixed several different 12 mm plastic beads together by mistake. Luckily the beads had different densities. One group of beads had a density of 0.6 g/cm³; the other beads had a density of 1.2 g/ cm³. Describe a method the student can use to sort the beads.

18. A student has a sample of a mineral that is too big to fit in a graduated cylinder. The density of the sample is known. How can the student determine the volume of the sample?

The Changing Environment

Human observations indicate that most, if not all, environments of Earth and the rest of the universe are undergoing changes. Much of the study of earth science involves analysis of these environmental changes.

What Change Is

A <u>change</u> occurs when the properties or characteristics of a part of the environment have been altered. As a result, descriptions of an environment at two different times are never exactly the same in all details. The occurrence of a change in the properties of an object or a system is called an event. Events may be almost instantaneous, as in the case of lightning and meteors, or they may occur over long periods of time, as in changes in sea level or elevation of mountains.

Change can be described with respect to time and space (location). Time and space are called the frames of reference for studying change. As an example, you might say that Earth's moon changes because we observe it in different locations in the sky and in different phases at different times during a month.

Rate of Change

How much a measurable aspect of the environment, called a field, is altered over a given time—years, hours, or seconds—is the **rate of change.** Fields are described in more detail in Topic 2.

Time (P.M.)	Rock Temperature (°C)
2:10	30
2:11	33
2:12	38
2:13	40
2:14	41
2:15	42
2:16	43
2:17	43
2:18	43

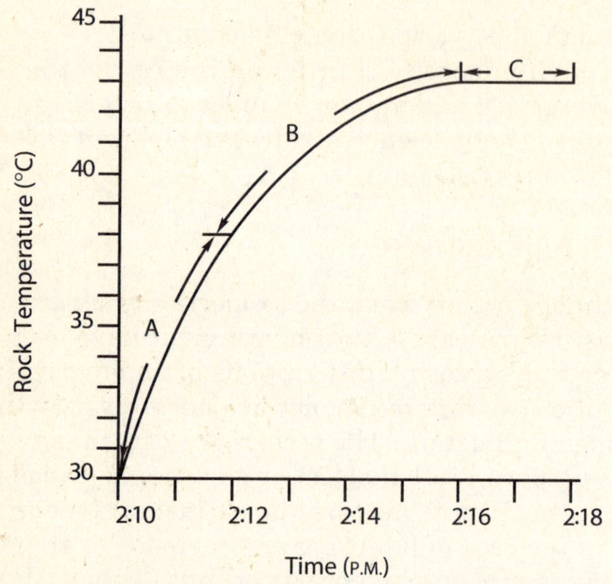

Figure 1-2. Data to be used to compute sample rate of change problems: The data represents the temperature of an area of rock that is exposed to nearby lava (liquid rock at Earth's surface) that just erupted from a volcano in Hawaii.

Consider the information in the data table and graph in Figure 1-2. The rate of change in temperature—the field—can be estimated by comparing the steepness (how close the plotted line is to a vertical) of the plotted line. In part A of the graph, the rate of temperature change is greater than that of part B. In part C, there is a zero rate of change, as shown by a horizontal line. More precision about rate of change can be obtained by using the following equation, which is also found in the *Earth Science Reference Tables.*

$$\textbf{rate of change} = \frac{\textbf{change in field value}}{\textbf{change in time}}$$

Based on the data table in Figure 1-2, what is the rate of change in rock temperature from 2:10 to 2:13 P.M.? The amount of change in the temperature (the field value) is 40°C – 30°C, or 10°C. The amount of change in time is 2:13 P.M. – 2:10 P.M., or 3 minutes.

$$\textbf{rate of change} = \frac{\textbf{10°C}}{\textbf{3 min}} = \textbf{3.3°C/min}$$

Cyclic Change

Many changes in the environment occur in some orderly fashion in which the events constantly repeat. Such an orderly change is called a **cyclic change.** Some cyclic changes are the movement of celestial objects (sun, moon, stars, planets), the numbers of sunspots, tides, seasonal events, and the water and rock cycles. These cyclic changes will be studied in detail in later topics of this book.

Prediction of Change

A **prediction** is a type of inference about the conditions and behavior of the environment in the future. Predictions can be made if the amount, type, and direction of change can be determined. For example, because eclipses are cyclic, astronomers can predict the occurrence of solar and lunar eclipses many hundreds of years into the future.

It is important in protecting against loss of life, personal injury, and loss of property that we have accurate predictions of natural hazards. A **natural hazard** is a non-human-related object, process, or situation that has the possibility of causing loss of life, personal injury, or loss of property. Natural hazards include asteroid impacts, blizzards, earthquakes, floods, hurricanes, thunderstorms, tornadoes, tsunami, and volcanic erruptions.

In the United States from 1975 to 1994, approximately 24,000 people died from natural disasters, and there were approximately 120,000 injuries. The loss of property in these same years was approximately $500 billion. Weather-related events accounted for approximately 80 percent of the deaths, injuries, and loss of property in this time period. Each of these types of natural hazards will be discussed in later topics. With accurate predictions of these hazards, people can use predetermined escape routes and follow emergency-action plans.

Energy, Interfaces, and Change

All change involves a flow of energy from one part of the environment (that loses energy) to another part (that gains it). For example, during an earthquake internal Earth energy stored in rocks is released, or lost, from the rocks and converted into sound, heat, and the mechanical energy of the Earth's shaking surface. Energy is usually exchanged across an **interface,** which is the boundary between regions with different properties. In the case of an earthquake, there is an interface between the place where the rocks break and start moving and Earth's surface where the shaking occurs. Other interfaces are where Earth's atmosphere blends into outer space or at the shoreline where Long Island, New York, borders the Atlantic Ocean.

Environmental Equilibrium

Although change occurs continuously throughout the environment, certain general characteristics tend to remain constant. For example, the wooded shore of a lake tends to look the same from one day to the next. Although it may go through a cycle of seasonal changes, it tends to look the same each year at the same season. This is the result of a natural balance among all the changes taking place, called environmental equilibrium. This equilibrium is easily upset on a small scale, for example by the burrowing of a worm in the soil, but it is normally not upset on a large scale. Human activities, however, often severely disrupt the environmental equilibrium. For example, construction equipment can rapidly change a rain forest into an urban area, thus introducing change on a large scale.

Environmental equilibrium resulting from opposing forces or actions balancing out is called a **dynamic equilibrium.** An example would be the level of a lake remaining the same even though thousands of liters of water move in and out of the lake per day. Another example would be the stable amount of oxygen in the atmosphere. Billions of molecules of oxygen are moving into and out of the atmosphere each day, but the total amount stays the same. In your lifetime, the elevation of a local mountain, the amount of gravity pulling on your body in your house, or the average distance to a star in the night sky will stay about the same. When environmental equilibrium results from little noticeable change it is static equilibrium.

Review Questions

19. Which graph represents the greatest rate of temperature change?

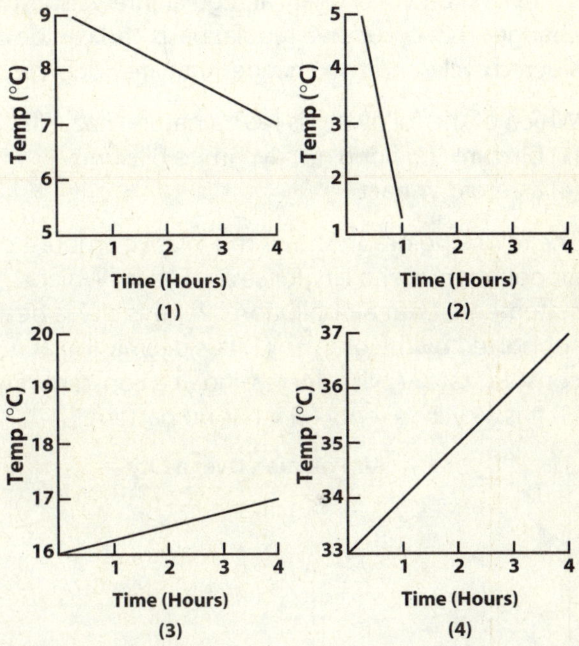

20. Refer to the following data table. At approximately what rate did the temperature rise inside the greenhouse between 8:00 A.M. and 10:00 A.M.?

Time Average	Greenhouse Temperature
6:00 A.M.	13°C
8:00 A.M.	14°C
10:00 A.M.	16°C
12:00 noon	20°C

(1) 1.0°C/hr (2) 2.0°C/hr (3) 0.5°C/hr (4) 12.0°C/hr

21. The following graph shows temperature readings for a day in April.

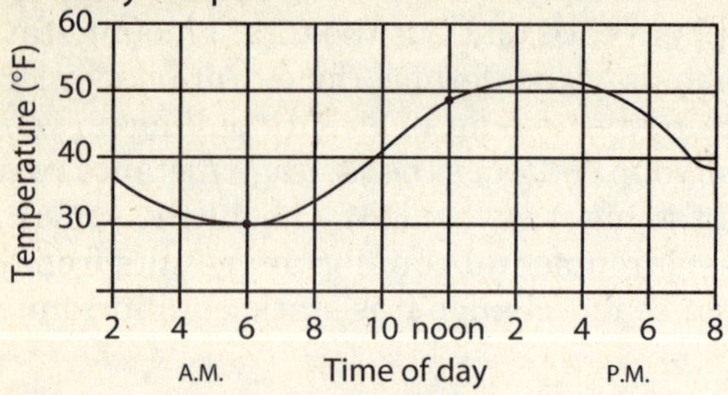

The average rate of temperature change, in Fahrenheit degrees per hour, between 6 A.M. and noon was (1) 6°/hr (2) 8°/hr (3) 3°/hr (4) 18°/hr

22. Some changes that occur on Earth are cyclic—they occur over and over again at regular intervals. List two changes that occur on a regular basis and can be considered cyclic and two changes that are not cyclic.

23. Which of the following is NOT a natural hazard? (1) blizzard (2) flood (3) dynamite blasting (4) asteroid impact

24. The following graph shows the relative amount of air pollution over a city for several years. Which statement about air pollution over this city is best supported by the graph? (1) It is decreasing at a constant rate. (2) It is increasing at a constant rate. (3) It is a cyclic event. (4) It has no pattern.

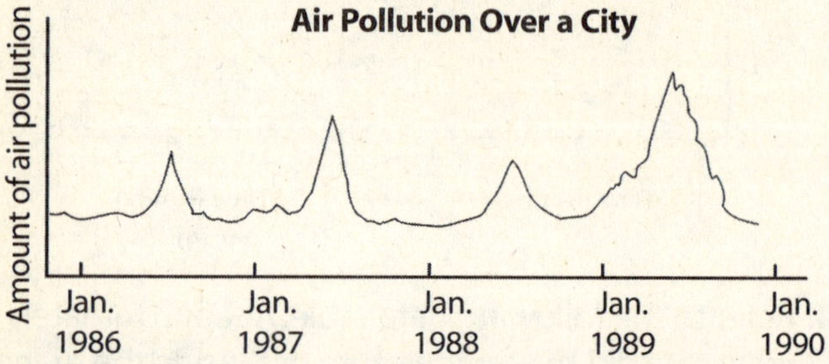

25. A student measures the distance from a bridge to a rock every day for a week. What is indicated by the following graph of the student's data?

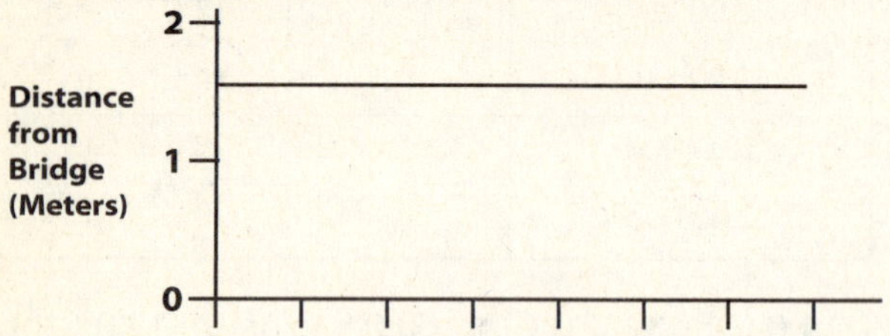

(1) No change in time or distance took place. (2) As distance decreased, time increased. (3) As distance increased, time decreased. (4) As time increased, distance remained the same.

26. What always happens when a change occurs? (1) Pollution is produced. (2) The temperature of a system increases. (3) The properties of a system are altered. (4) Dynamic equilibrium is reached.

27. Which statement is true about the environment of most of Earth? (1) It is usually greatly unbalanced. (2) It is usually in equilibrium and hard to change. (3) It is usually in equilibrium but easy to change on a small scale. (4) It is normally heavily polluted by natural occurrences.

28. Humans can cause rapid changes in the environment that may result in catastrophic events. Which statement below is the best example of this? (1) Concrete and brick buildings cause a city to absorb more heat than surrounding areas do. (2) A project to straighten out a river causes flooding at a city downstream. (3) A new housing development uses well water, causing the water table level to drop. (4) Streetlights make it more difficult to observe stars and planets at night near a city.

Human Interaction With the Environment

Natural Resources

The materials and energy sources found in the environment that humans use in their daily lives are **natural resources.** Natural resources include the air you breathe, the water you drink, the plants and animals you use for food and clothing, and the energy from the sun. Natural resources also include the fossil fuels and rock and mineral resources discussed in detail in Topic 11.

Individuals and societies are always in conflict over the ownership, supply, and use versus preservation of often-limited natural resources. There have to be trade-offs between the use of a natural resource and other uses of the land, water, or air. Often, through the use of human technology, natural-resource shortages can be solved by substituting one material for another or by recycling limited resources so that "new" resources don't have to be harvested, mined, or otherwise acquired.

Pollution of the Environment

Pollution of the environment occurs when the concentration of any substance or form of energy reaches a proportion that adversely affects people, their property, or plant or animal life. The key word in defining a polluted situation is concentration. Think of sodium chloride (NaCl), rock salt, which is essential to life. If the concentration of salt in an organism's body is too high, it can cause injury or death.

CAUSES OF POLLUTION Many forms of pollution are the result of technology, which often produces and distributes harmful concentrations of substances and forms of energy. Other forms of pollution are the result of natural events or processes and would occur without the presence of people. High concentrations of pollen in the air, volcanic ash and gases from volcanic eruptions, and X-rays from uranium and radon in rocks and soil are examples of natural pollution. Substances and forms of energy that pollute—pollutants—include solids, liquids, gases, biologic organisms, and forms of energy such as heat, sound, visible light, and nuclear radiation. Generally, the more urban an environment and the more industrial processes, the higher the pollutant levels.

Review Questions

29. Which of the following statements is an example of a way of conserving our natural resources? (1) opening new mines for valuable metals that are running out (2) using well water in desert areas for growing crops and building golf courses (3) finding alternate sources and materials to replace natural petroleum products (4) building industrial plants in national parks where they are away from populated areas

30. When organisms, sound, and radiation added to the environment reach a level that harms people, these factors are referred to as environmental (1) interfaces (2) pollutants (3) phase changes (4) equilibrium exchanges

31. List three substances or forms of energy that are common natural (NOT caused by humans) pollutants.

32. Which energy source is LEAST likely to pollute the environment? (1) solar (2) petroleum (3) coal (4) wood

33. Which type of human environment usually has the highest pollution levels?

Questions for Regents Practice

Part A

Base your answers to questions 1 through 4 on the graph at the right, which shows the masses and volumes of four Earth materials.

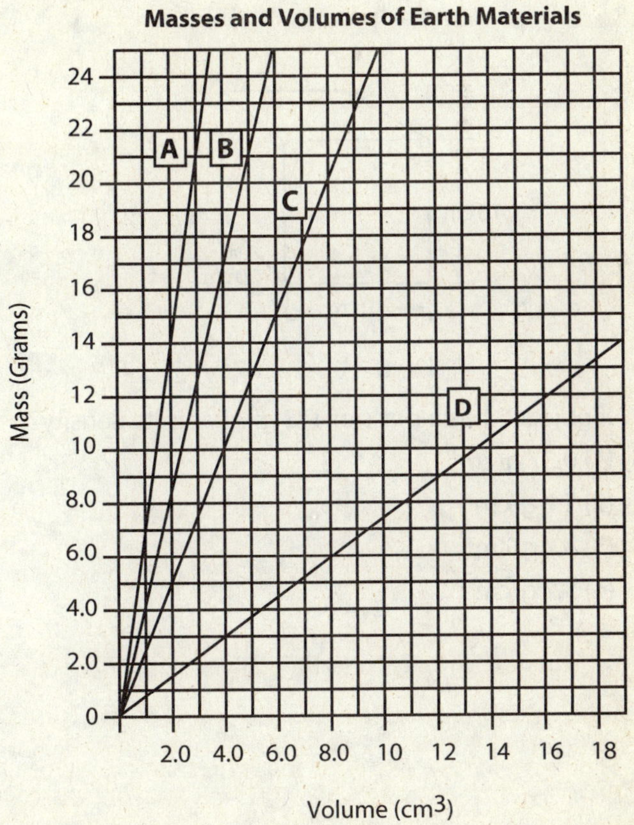

1. Which material has the greatest density?

(1) A (3) C
(2) B (4) D

2. If the density of water is 1 g/cm^3, which material will float in water?

(1) A (3) C
(2) B (4) D

3. If the volume of sample C is 3.0 cm^3, what is its mass?

(1) 7.5 g (3) 3.0 g
(2) 2.1 g (4) 21 g

4. Which material has a density of about 4.0 g/cm^3?

(1) A (3) C
(2) B (4) D

5. Substances A, B, C, and D are at rest in a container of liquid as shown by the following diagram. Which choice lists the substances in order of lowest to highest density?

(1) A, B, C, D

(2) A, D, C, B

(3) D, C, B, A

(4) C, B, A, D

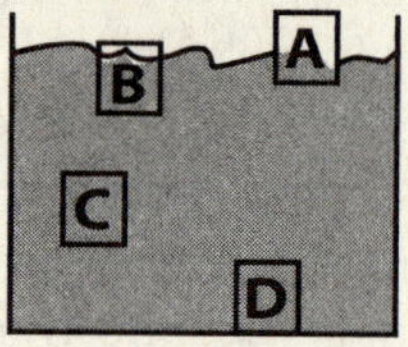

6. Substances A, B, C, and D are at rest in a container of liquid as shown in the following diagram. Which substance probably has a density closest to that of the liquid?

(1) A

(2) B

(3) C

(4) D

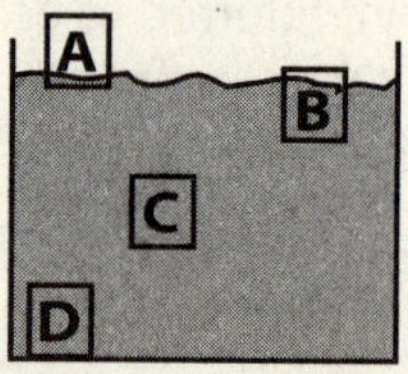

Base your answers to questions 7 through 11 on the following diagrams. Objects A and B are solid and made of the same uniform material.

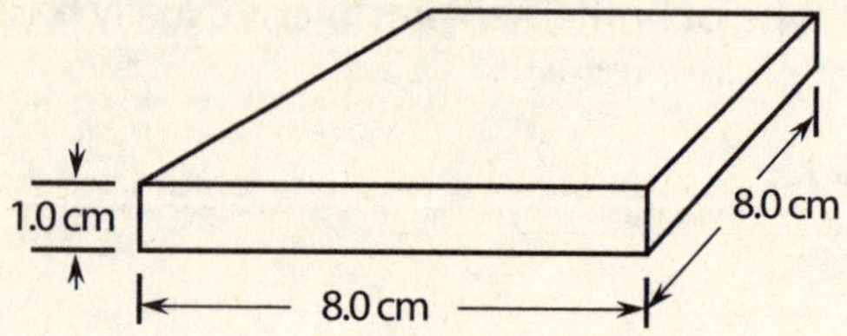

A

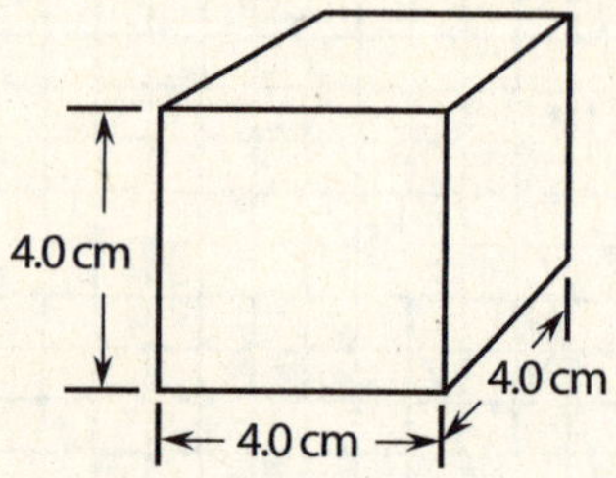

B

7. If object B has a mass of 173 g, what is its density?

(1) 0.37 g/cm^3

(2) 2.7 g/cm^3

(3) 3.7 g/cm^3

(4) 5.7 g/cm^3

8. Object A expands when it is heated. Which graph best represents the relationship between the temperature and the density of object A?

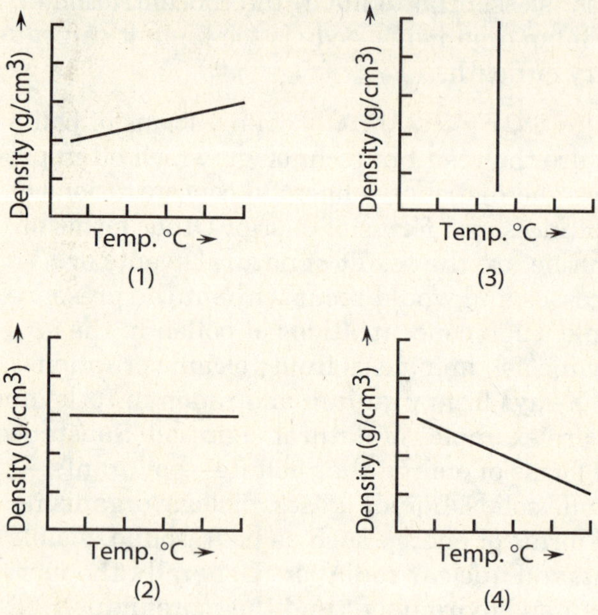

9. A student measures the mass of object B as 156 g, but the actual mass is 173 g. What is the student's approximate percent deviation (percentage of error)?

(1) 1 percent

(2) 5 percent

(3) 10 percent

(4) 20 percent

10. A third object is made of the same uniform material as object B, but it has twice the volume of object B. How does the density of this third object compare to the density of object B?

(1) It is one-half as dense as B.

(2) It has approximately the same density as B.

(3) It is twice as dense as B.

(4) It is four times as dense as B.

11. How does the mass of object B compare to the mass of object A?

(1) The mass of B is less than the mass of A.

(2) The mass of B is greater than the mass of A.

(3) The mass of B is the same as the mass of A.

---

12. If pressure is applied to a rock until its volume is reduced by one half, how does its new density compare to its original density?

(1) It is half its original density.

(2) It is twice its original density

(3) It is the same as its original density.

(4) It is one-third its original density.

13. The following diagrams show physical changes in four materials over time. The chemical composition of each material remains the same. Which material most likely changed in density?

(1) 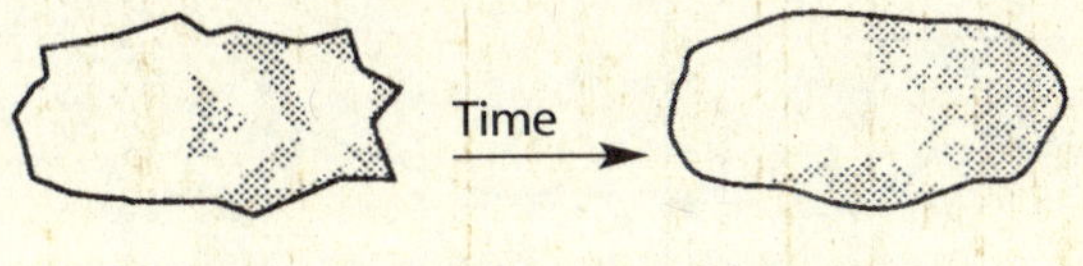

(2) 

(3) 

(4) 

14. Which graph best represents for most Earth materials, excluding water, the relationship between the density of a substance and its state of matter (phase)?

[Key: S = solid, L = liquid, G = gas]

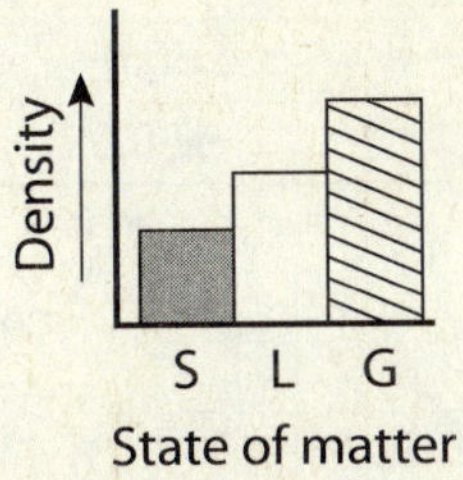

(1)

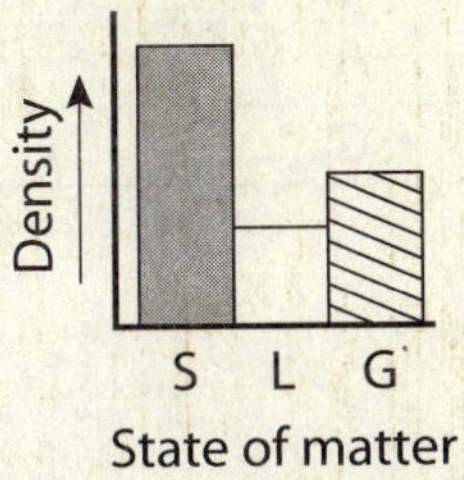

(3)

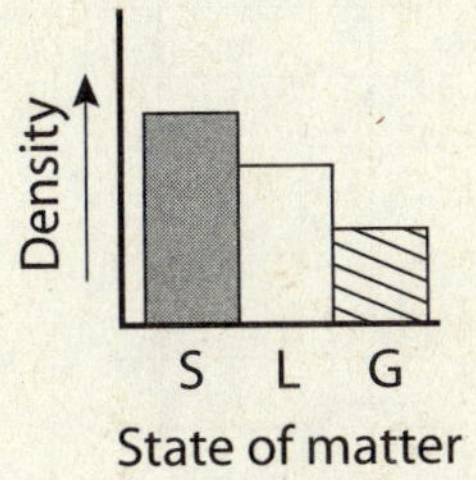

(2)

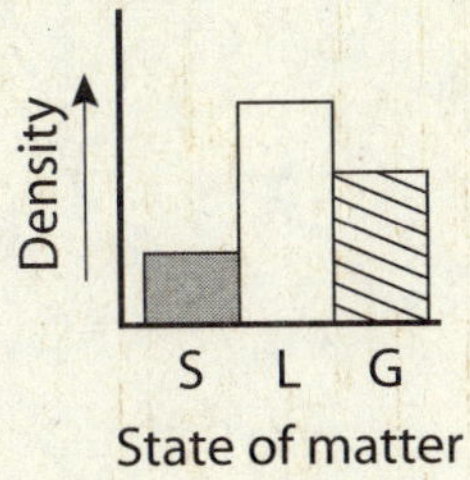

(4)

15. The mass of some pieces of granite rock is measured so their density can be calculated. The rock has water on it while its mass is being measured. How would the calculated value for density compare with the actual density?

(1) The calculated density value would be less than the actual density.

(2) The calculated density value would be greater than the actual density.

(3) The calculated density value would be the same as the actual density.

16. A prediction of next winter's weather is an example of

(1) a measurement

(2) a classification

(3) an observation

(4) an inference

17. The rising and setting of the sun are examples of

(1) noncyclic events

(2) unrelated events

(3) predictable changes

(4) random motion

18. Future changes in the environment can best be predicted from data that are

(1) highly variable and collected over short periods of time

(2) highly variable and collected over long periods of time

(3) cyclic and collected over short periods of time

(4) cyclic and collected over long periods of time

19. An interface can best be described as

(1) a zone of contact between different substances across which energy is exchanged

(2) a region in the environment with unchanging properties

(3) a process that results in changes in the environment

(4) a region beneath the surface of Earth where change is not occurring

20. Which condition exists when the rates of water flowing into and out of a lake are balanced so that the lake's depth appears to be constant?

(1) dynamic equilibrium

(2) transpiration

(3) precipitation

(4) saturation

Part B

21. A student placed a 20-cm plastic foam ball and a 5-cm piece of sandstone in a container of water. The sandstone sank and the ball floated. Propose a reason why the larger object floated and the smaller object sank. [1]

22. Three students tried to guess Joe's weight. They guessed 140 lb, 150 lb, and 160 lb. If Joe really weighed 146 lb, calculate each person's percent deviation. [5]

23. The springs and geysers in Yellowstone National Park contain enough heat energy to provide for the annual electric needs of New York City. Why is this energy not being used?

(1) Energy from hot water is not easy to convert to electricity.

(2) No electric transmission lines connect Yellowstone National Park with the rest of the United States.

(3) It has been decided to protect the natural environment of Yellowstone National Park.

(4) Using the geyser and spring water would cause a drought in the Yellowstone National Park region.

Base your answers to questions 24 through 28 on the following data table. The table shows the mass of three liquids—A, B, and C—each of which has a volume of 500 mL.

Liquid	Volume (mL)	Mass (g)
A	500	400
B	500	500
C	500	600

24. What is the density of liquid A? [1]

25. If half of liquid B is removed from its container, what will the density of the remaining liquid be in relationship to the original density? [1]

26. The accepted mass for liquid C is 600 g, but a student measures the mass as 612 g. What is the percent deviation (percent of error) of the student's measurement? [1]

27. The following graph shows the volume of liquid B as the temperature changes from 0°C to 10°C. According to the graph, at what temperature is the density of liquid B the greatest? [1]

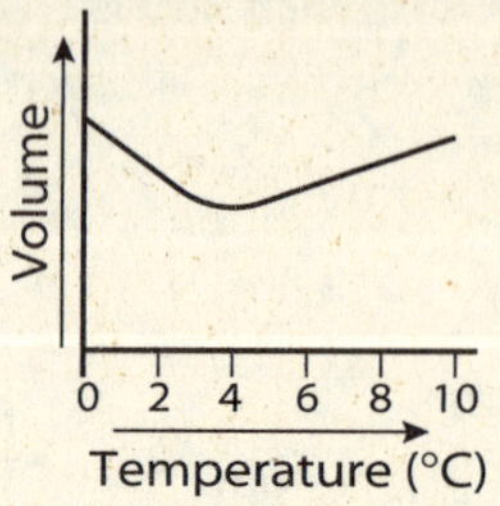

28. Liquid A is NOT water. Some of liquid A is frozen into a solid block and placed in a container of liquid A. What will probably happen to the location of the solid block in relationship to the liquid? [1]

---

29. Mount St. Helens erupted in 1980, depositing volcanic ash over much of the countryside. Why is it important for scientists to try to predict volcanic eruptions? [1]

Base your answers to questions 30 and 31 on the data table below, which shows the volume and mass of three different samples, *A, B,* and *C,* of the mineral pyrite.

	Pyrite	
Sample	**Volume** (cm3)	**Mass** (g)
A	2.5	12.5
B	6.0	30.0
C	20.0	100.0

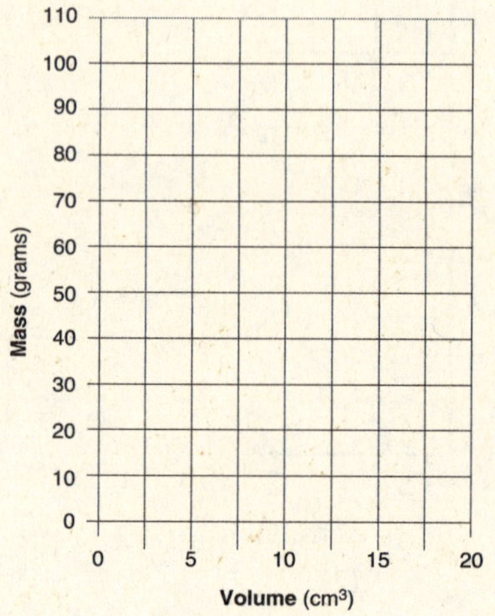

56 ______________ grams

30. On the grid provided above, plot the data (volume and mass) for the three samples of pyrite and connect the points with a line. [2]

31. State the mass of a 10.0-cm^3 sample of pyrite. [1]

Part C

The following fictional article describes an oil spill that resulted from an earthquake in Saudi Arabia. Use your knowledge of earth science and the information in the article to answer questions 32 through 36.

An earthquake in Saudi Arabia caused a break in an oil pipeline that connected oil wells to shipping facilities in the Persian Gulf. A total of 120,000 gallons of crude oil (density 0.86 g/cm^3) spilled onto the desert soil. At first it was thought that 50,000 gallons eventually flowed into a reservoir that supplies drinking water to several communities. Later it was learned that only 35,000 gallons had flowed into the reservoir.

The clean-up crews removed the crude oil from the reservoir in a little over a week. It took much longer to remove the crude oil from the sandy desert soil. After two weeks of continuous effort, only 5000 gallons of oil had been removed. This area of the desert is irrigated and used for growing crops, most of which were killed by the crude oil. It is not yet known if all the oil can be removed from the soil and if crops will grow in the region again.

32. Which physical property of the crude oil allowed it to be easily cleaned up from the water reservoir? [1]

33. Why is the earthquake in the article considered to be a natural disaster? [1]

34. Calculate the percent deviation of the first estimate of the amount of crude oil in the reservoir compared to the actual amount. Be sure to show ALL work including writing the equation, substituting values, and indicating the answer. [2]

35. What was the rate of crude oil recovery from the soil in gallons per day for the two-week period described in the article? Be sure to show ALL work including writing the equation, substituting values, and indicating the answer. [2]

36. Explain why the crude oil in the soil would be considered pollution. Be sure to use the term *concentration* in your answer. [1]

---

37. A hurricane traveled 2,600 kilometers during a 4-day period. Calculate the average rate of daily movement of a hurricane following the directions below.

a. Write the equation used to determine the rate of change

b. Substitute data into the equation [1]

c. Calculate the rate and label it with the proper units [1]"

38. The students decided to measure the speed of the stream by floating apples down a straight section of the stream. Describe the steps the students must take to determine the stream's surface rate of movement (speed) by using a stopwatch, a 10-foot rope, and several apples. Include the equation for calculating rate. [3]

TOPIC 2

Measuring Earth

VOCABULARY		
atmosphere	gradient	pauses (of atmosphere)
contour line	hydrosphere	prime meridian
coordinate system	isoline	profile
crust	latitude	topographic map
Earth's interior	lithosphere	
elevation	longitude	
equator	meridian of longitude	
field	model	

People have always wanted to know about where they live, be it Earth in general, or their part of their country. In this topic you will learn about the basic dimensions of Earth and its layered structure and composition. You will also learn about its spheres, and how to locate and model Earth positions and landscape features. People have struggled for thousands of years to find accurate and reproducible methods of locating specific positions on Earth's surface to find natural resources, decide who owns what land, and determine where crops can be grown well. The problem of how to find positions on Earth's surface has largely been solved by the latitude and longitude system—studied in this Topic—coupled with the global positioning systems (GPS), which uses measurements from Earth-orbiting satellites.

Size and Shape of Earth

Earth's shape is very close to being a perfect sphere. The actual shape is of a sphere that has a slight flattening at the polar regions and a slight bulging at the equatorial region. The human eye cannot detect a distortion from a perfect sphere, even in the most accurate globe models of Earth. A **model** is a way of representing the properties of an object or system. A model may be an object, such as a globe representing Earth, or a drawing, diagram, graph, chart, table, or mathematical formula or equation.

Evidence of Earth's spherical shape includes photographs taken from space and the way ships appear to sink as they travel past the horizon. Other evidence is the similarity of the strength of gravity at sea level all over Earth's surface. Besides being very close to a perfect sphere, Earth's surface is very smooth (has very little relief) compared to its diameter. This smoothness is evident in any photographs of Earth taken from space.

Figure 2-1 illustrates some of the basic dimensions of Earth's surface below the atmosphere. Earth's surface is the top of the liquid portion or the solid portions of Earth—described later in this topic.

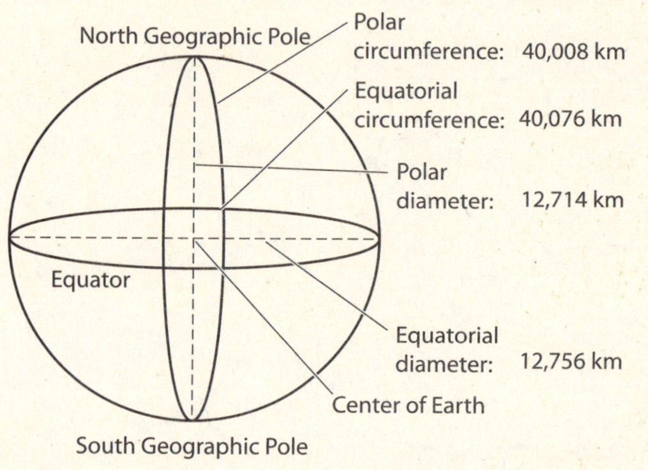

Figure 2-1. Some basic Earth dimensions: The similarities of the polar and equatorial diameters and circumferences indicate that Earth is close to being a perfect sphere. The average diameter of Earth is 12,735 kilometers.

Spheres of Earth

Earth is composed of a series of spheres held together by gravity and thus arranged from lowest to highest density moving towards Earth's center. Table 2.1 compares the spheres and Figure 2-2 is a model of this arrangement of spheres. There is no agreement as to the exact thickness of the atmosphere. The line in the model indicating Earth's surface is actually too thick to represent the hydrosphere/ocean average depth. The divisions of the atmosphere and their interfaces—the pauses—are also found with some of their properties in Selected Properties of Earth's Atmosphere in the *Earth Science Reference Tables.* The average chemical compositions of some of these spheres is shown in Average Chemical Composition of Earth's Crust, Hydrosphere, and Troposphere in the *Earth Science Reference Tables.*

Table 2–1 Comparison of the Spheres of Earth

Sphere Name	Density (in g/cm)	Phase(s) of Matter	Thickness (in km)
Atmosphere	0.0-0.0001	Gas	1000 (approximately)
Hydrosphere	1.0	Liquid	3.8 (average for oceans)
Lithosphere	2.7-3.3	Solid	100
Earth's interior	2.7-13.0	Mostly solid, some liquid	6378

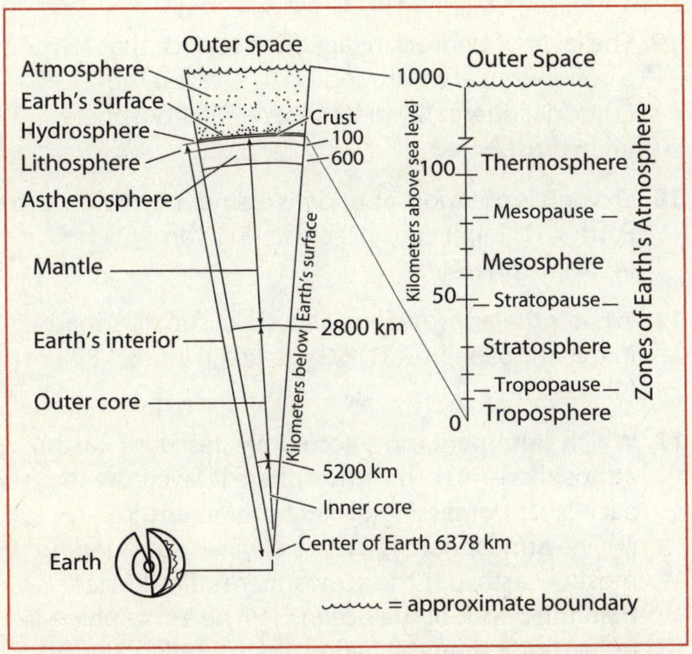

Figure 2-2. A wedge of Earth: The wedge shows the spheres of Earth drawn very close to scale.

Outer Spheres of Earth

The region that extends from the outer portion of Earth's interior to where the atmosphere blends into outer space can be considered the outer part of Earth. This portion of Earth is divided into three spheres—atmosphere, hydrosphere, and lithosphere.

ATMOSPHERE The layer of gases that surrounds Earth above the surface of liquid water and rocky material is the **atmosphere.** It is composed mostly of the gases nitrogen and oxygen, but also includes some aerosols—suspended liquids and solids. The atmosphere is the sphere farthest from Earth's center because it is the least dense, and Earth's gravity pulls on it less than on the other spheres. It extends out several hundred kilometers into space, but nearly all of its mass is confined to the few kilometers nearest Earth's surface. The atmosphere is stratified, or layered, into zones—troposphere, stratosphere, mesosphere, and thermosphere—each with its own distinct characteristics such as temperature and composition. The interfaces, or boundaries, of the layers of Earth's atmosphere are called **pauses.** The tropopause, for example, is between the troposphere and stratosphere. (See Selected Properties of Earth's Atmosphere in the *Earth Science Reference Tables.*)

HYDROSPHERE The **hydrosphere** is the layer of liquid water that lies between the atmosphere and much of the upper layer of Earth's interior. The hydrosphere consists of the oceans, which cover about 70 percent of Earth's surface, and other bodies of water such as lakes, streams, and rivers. The hydrosphere is relatively thin, averaging only 3.8 kilometers in thickness. The chemical composition of the hydrosphere is mostly liquid H_2O, thus mostly hydrogen and oxygen. Earth's surface waters also contain varying amounts of dissolved gases and dissolved and suspended solids, including life forms.

LITHOSPHERE The layer of rock that forms the solid outer shell at the top of Earth's interior is the **lithosphere.** (See Figure 2-2.) The lithosphere lies directly beneath the atmosphere or the hydrosphere. The lithosphere is approximately 100 kilometers thick and is divided into sections called lithospheric plates. The significance, movement, and interaction of these plates are a major focus of Topic 12. The upper portion of the lithosphere is called the **crust.** Where the crust is not covered by the liquid water (hydrosphere), it is usually covered by a very thin layer of broken-up rocky material. When this material contains abundant organic matter, it is called soil.

Earth's Interior

The region extending from the rocky part of Earth's surface to Earth's center is called **Earth's interior.** Topic 12 will describe the characteristics of Earth's interior in detail. Many of the features of Earth's interior are shown in the Inferred Properties of Earth's Interior in the *Earth Science Reference Tables.*

Review Questions

1. Which object best represents a true scale model of the shape of Earth? (1) a table tennis ball (2) a football (3) an egg (4) a pear

2. Which of the following is NOT evidence that supports Earth's spherical shape? (1) photographs taken from outer space (2) ships appearing to sink as they sail past the horizon (3) the changing seasons (4) the similarities in gravity at sea level all over Earth

3. In which group are the spheres of Earth listed in order of increasing density? (1) atmosphere, hydrosphere, lithosphere (2) hydrosphere, lithosphere, atmosphere (3) lithosphere, hydrosphere, atmosphere (4) lithosphere, atmosphere, hydrosphere

4. Which makes up most of Earth's surface? (1) the atmosphere (2) the lithosphere (3) the hydrosphere

5. The lower atmosphere and hydrosphere are bound to Earth by (1) magnetic fields (2) atmospheric pressure (3) the force of gravity (4) a molecular bonding at the interface

6. What do the tropopause, stratopause, and mesopause all have in common? (1) Each is a point of maximum temperature in its layer of the atmosphere. (2) Each is an interface of two layers of the atmosphere. (3) Each is a region of increasing pressure within the atmosphere. (4) Each is a zone of decreasing water vapor content within the atmosphere.

7. As depth within Earth's interior increases, the (1) density, temperature, and pressure decrease (2) density, temperature, and pressure increase (3) density and temperature decrease, but pressure increases (4) density decreases, but temperature and pressure increase

8. The following diagram shows Earth's spheres. Which spheres are zones of Earth's atmosphere? (1) lithosphere, hydrosphere, and troposphere (2) stratosphere, mesosphere, and thermosphere (3) asthenosphere, lithosphere, and hydrosphere (4) hydrosphere, troposphere, and stratosphere

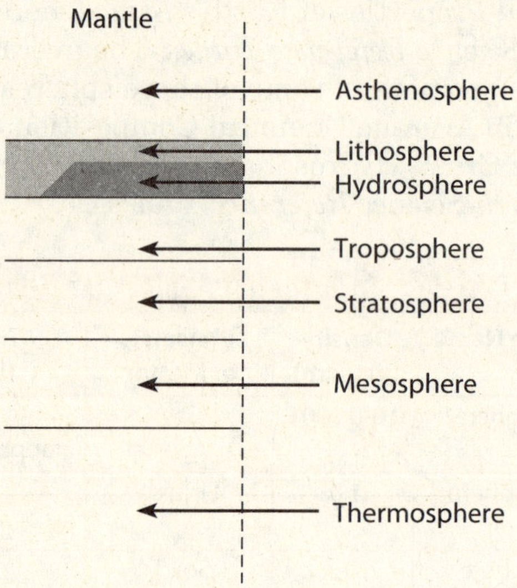

(Not drawn to scale.)

9. The layer of bedrock near Earth's surface that forms a continuous shell around Earth is called the (1) troposphere (2) stratosphere (3) lithosphere (4) hydrosphere

10. Oxygen is the most abundant element by volume in Earth's (1) inner core (2) crust (3) hydrosphere (4) troposphere

11. What is the approximate altitude of the mesopause in the atmosphere? (1) 50 km (2) 66 km (3) 82 km (4) 90 km

12. Which statement most accurately describes Earth's atmosphere? (1) The atmosphere is layered, with each layer possessing distinct characteristics. (2) The atmosphere is a mass of gases surrounding most of Earth. (3) The atmosphere's altitude is less than the depth of the ocean. (4) The atmosphere is more dense than the hydrosphere, but less dense than the lithosphere.

13. Which layer of Earth's atmosphere contains no water vapor, has an atmospheric pressure less than 10^{-4} atmosphere, and has an air temperature that increases with altitude? (1) troposphere (2) stratosphere (3) mesosphere (4) thermosphere

14. The water sphere of Earth is known as the (1) atmosphere (2) troposphere (3) lithosphere (4) hydrosphere

15. In the following diagrams, the dark zone at the surface of each wedge-shaped segment of Earth represents average ocean depth. Which segment is drawn most nearly to scale?

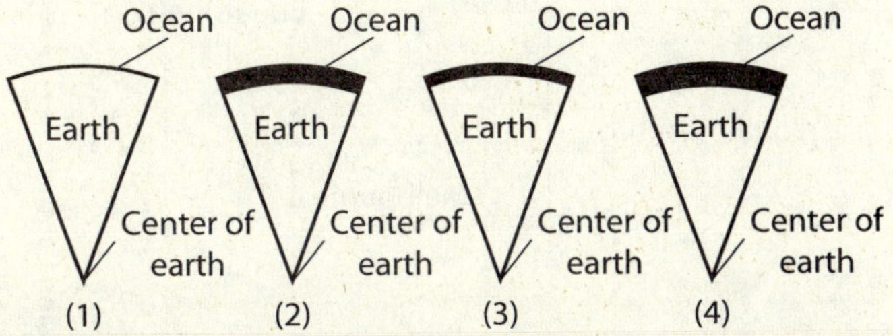

16. The following maps show Earth as viewed from above the North Pole and from above the South Pole. Which statement is best supported by these maps? (1) Most of Earth's surface is land. (2) Most of Earth's landmass is in the Southern Hemisphere. (3) The geographic North Pole is the south magnetic pole. (4) Most of Earth's surface is hydrosphere.

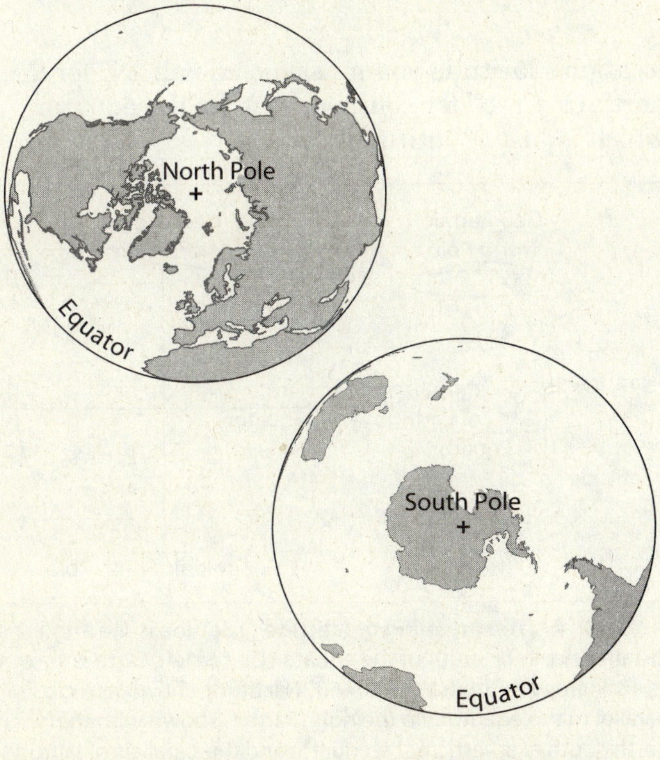

17. Which graph best represents the percentage by volume of the elements making up Earth's hydrosphere?

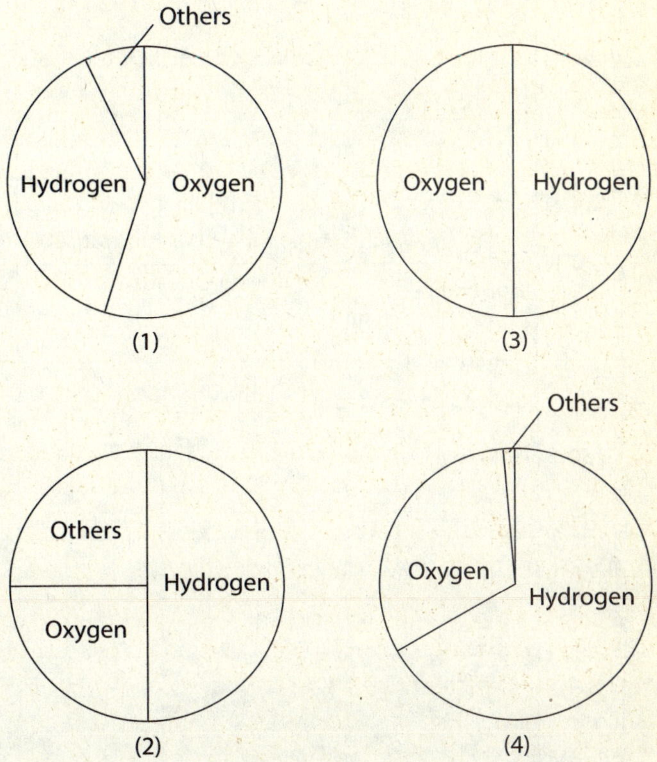

Locating Positions on Earth

To fix the location of a point on any two-dimensional surface such as the surface of Earth, two numbers, called coordinates, are needed. The system for determining the coordinates of a point is called a **coordinate system.** (See Figure 2-3 on the next page.) The latitude-longitude coordinate system is the one most commonly used to locate points on Earth's surface. Latitude and longitude are measured in angular units: degrees (°) and minutes (′) There are 60 minutes in a degree.

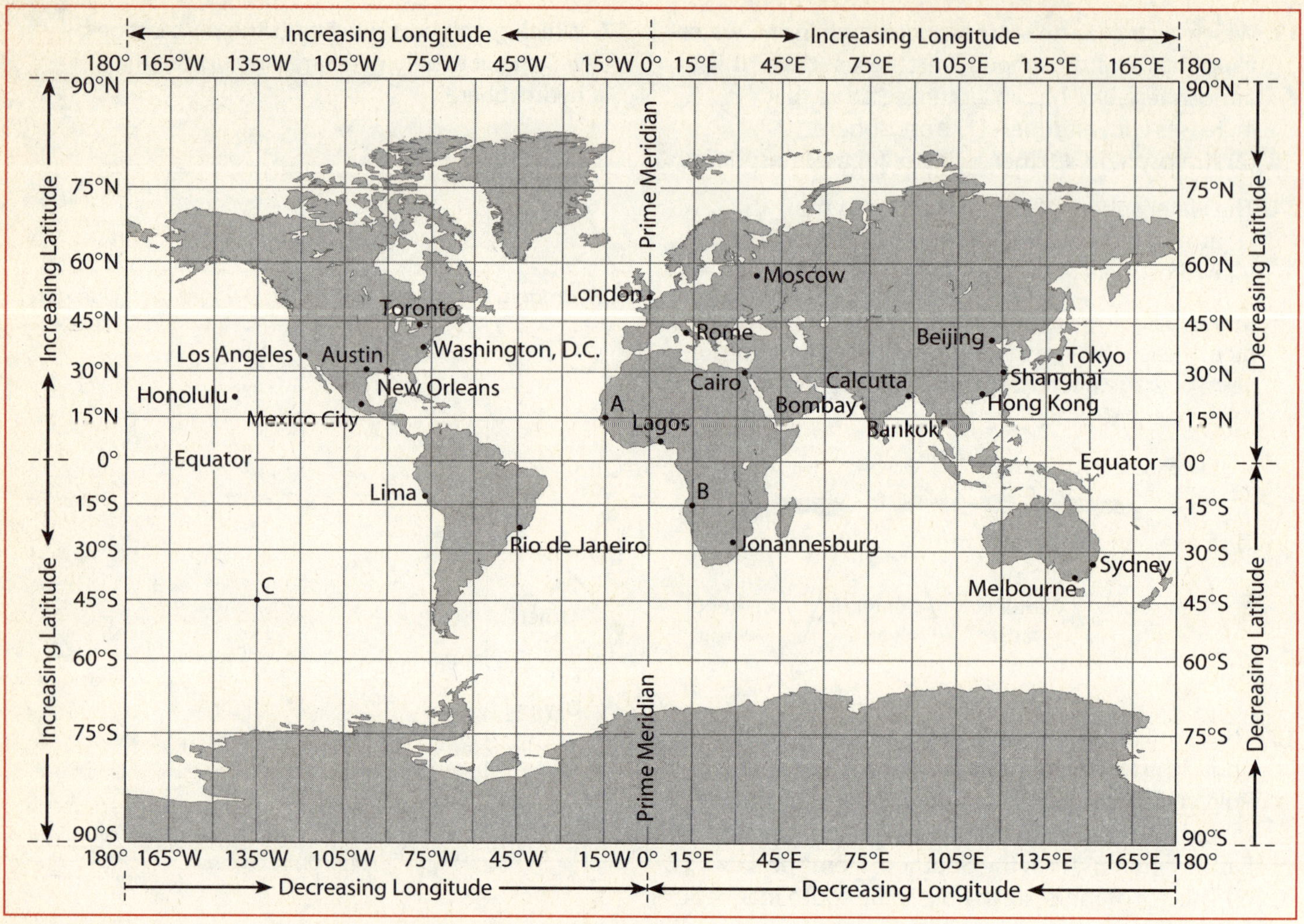

Figure 2-3. **World map with latitude-longitude coordinate system**

Latitude

The angular distance north or south of the equator is **latitude.** The **equator** marks the middle location on Earth's surface, halfway between the geographic North and South poles. If a line is drawn from any point on Earth's surface to the center of Earth, the latitude of that point is the number of degrees in the angle between that line and the plane of the equator. All points that have the same latitude lie on a circle that is parallel to the equator. These circles are called parallels of latitude or just parallels. The equator may be considered to be the parallel of latitude 0°. (See Figure 2-4.) As indicated in Figure 2-4, latitude increases north and south of the equator to a maximum of 90° at each pole.

Be sure you understand the meaning of increasing and decreasing latitudes. Increasing latitude means moving away from the equator or towards the geographic poles. Decreasing latitude means moving towards the equator or away from the geographic poles. Remember when writing a location's latitude you must include an "N" for north or an "S" for south, except for the equator, which is just 0° latitude.

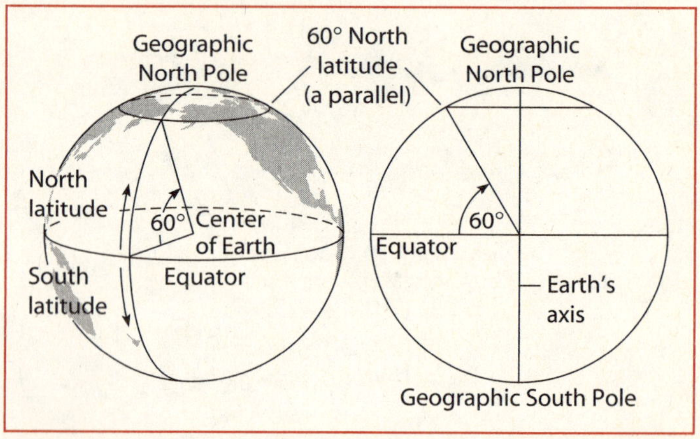

Figure 2-4. **The meaning of latitude:** Latitude is the angular distance north or south of the equator. On the left, Earth is shown as a sphere with the parallel for 60° N latitude shown as a circle parallel to the equator. On the right, Earth is shown with the north-south axis vertical; the equator and the parallels of latitude then become horizontal lines.

Measuring Latitude

As shown in Figure 2-5 the altitude of Polaris—the North Star—equals the degree of latitude in the Northern Hemisphere. The altitude of an object in the sky is its angle above the horizon. The latitude of a point on Earth's surface is its angle north or south of the equator. The star Polaris is almost directly over the geographic North Pole of Earth. The altitude of Polaris at any point in the Northern Hemisphere should be the same as the latitude of that point. That is, the altitude of Polaris should change from 0° at the equator to 90° at the geographic North Pole. In the Southern Hemisphere, latitude can be determined by measuring the altitude of certain other stars. To determine the altitude of an object in the sky, an angle-measuring instrument such as a sextant is used.

MEMORY JOGGER

The geographic North and South poles are the ends of the axis of Earth's rotation, or spinning.

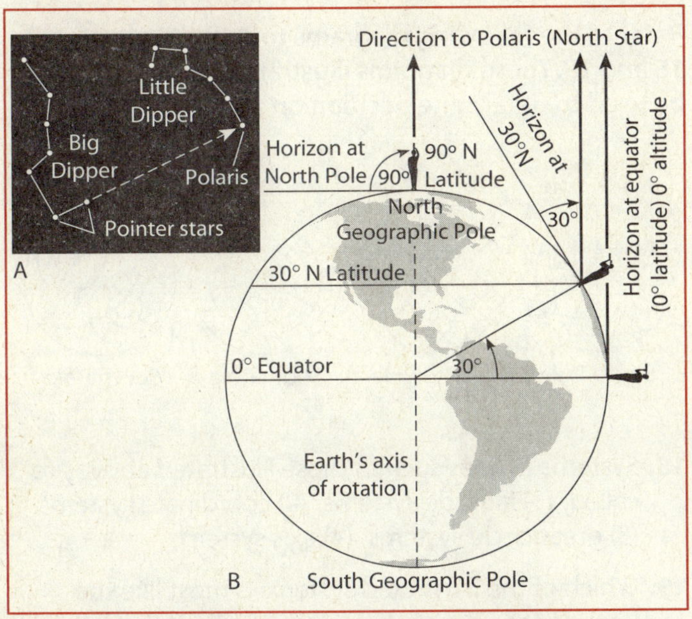

Figure 2-5. Altitude of Polaris equals latitude in the Northern Hemisphere: Diagram A shows the method used to find Polaris, or the North Star, in the night sky. Generally, face north and find the Big Dipper. The Big Dipper is a bright group of stars that is part of the constellation Ursa Major, or Big Bear. Following the pointer stars, move your eye five times the distance that separates the pointer stars. You should easily locate Polaris. The direction toward Polaris is true, or geographic north. The direction toward Polaris is the same for all locations on Earth because Polaris is trillions of miles away. Diagram B shows that the angular altitude of Polaris in the sky equals an observer's latitude in the Northern Hemisphere.

Longitude

As shown in Figure 2-6, **longitude** is an angular distance east or west of the prime meridian. A **meridian of longitude,** or meridian, is any semicircle on Earth's surface connecting the north and south geographic poles. The meridian that passes through Greenwich, England, has been designated the **prime meridian,** or the meridian of zero longitude. The longitude of any point on the prime meridian is 0°. The longitude of any other point on Earth's surface is the number of degrees between the meridian that passes through that point and the prime meridian. This angle can be measured along the equator or any parallel of latitude. Since a full circle is 360°, longitude increases east or west of the prime meridian from 0° at the prime meridian to 180°. The 180° meridian is the continuation of the prime meridian on the other side of Earth. This meridian somewhat follows the International Date Line.

In writing a location's longitude, be sure to include a west or east direction (80° W longitude; 70° E longitude) except for 0° longitude—the prime meridian—and the 180° meridian. Increasing longitude means going away from the prime meridian or towards the 180° meridian. Decreasing longitude means moving towards the prime meridian or away from the 180° meridian.

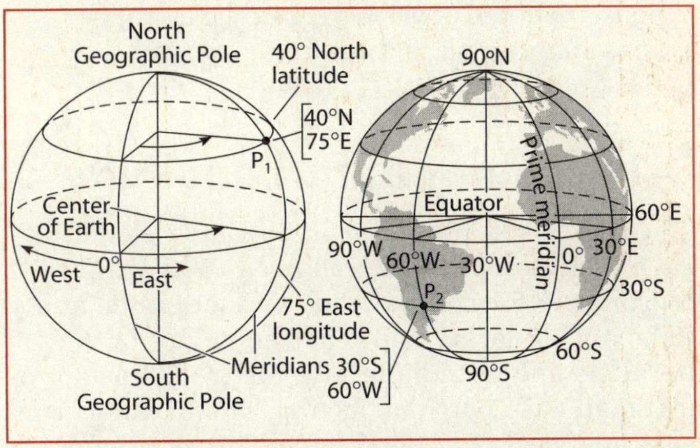

Figure 2-6. The meaning of longitude: Longitude is the angular distance east or west of the prime meridian, which is the meridian passing through Greenwich, England. Latitude and longitude together provide a system of coordinates for locating any point on Earth. The coordinates of point P_1 are 40° N latitude and 75° E longitude. Those for point P_2 are 30° S latitude and 60° W longitude.

Measuring Longitude

Local noon (12:00) at any point on Earth occurs when a line from the sun to the center of Earth cuts the meridian of that point. At that moment, the sun reaches its highest altitude of the day in the sky. Therefore, the instant of local noon can be determined by observing the sun. Since Earth rotates from west to east at the rate of one rotation per day—360° in 24 hours—it rotates 15° per hour. Therefore, the occurrence of local noon moves from east to west at the same rate of 15° per hour. Longitude can be calculated if, when local noon occurs, the observer knows what time it is at Greenwich, England—or any other location on the prime meridian. For example, if local noon occurs at 1:00 P.M. Greenwich Mean Time (GMT), one hour has passed since the sun crossed the prime meridian; the local longitude is therefore 15° W. In general, longitude can be calculated by finding the time difference in hours between local sun time and Greenwich Mean Time, and multiplying by 15°. If local time is earlier than Greenwich Mean Time, the longitude is west; if later, it is east.

DIGGING DEEPER

Greenwich Mean Time can be determined if the observer has an accurate clock that has been set to keep Greenwich Mean Time, or by means of time signals obtained from radio or satellite transmissions that are regularly broadcast.

Using Latitude and Longitude

To read or plot a location on an Earth model such as a map or globe, you need to locate the coordinates for both latitude and longitude and to be able to locate and read the values of the parallels and meridians. Use Figure 2-3 as a typical map with north at the top. The parallels run east and west (right and left). The meridians run north and south (up and down). The values for latitude are at the ends of the parallels on the left and right sides of the map. The values for longitude are at the ends of meridians at the top and bottom of the map. The accuracy of a reading or a plotting of coordinates partly depends upon the size of the map and the spacing of the meridians and parallels.

DIGGING DEEPER

Find point A on Figure 2-3, and write down the latitude and longitude as accurately as you can. Since this location is right at the intersection of a parallel and meridian, you can be quite accurate. The coordinates for point A are 15° N, 15° W, which is read as "latitude 15 degrees north and longitude 15 degrees west." Try location B. The coordinates are 15° S, 15° E. Location C is 45° S, 135° W. Rio de Janeiro is approximately 23° S, 43° W. Moscow is approximately 56° N, 37° E. Try determining the coordinates of some other locations on Figure 2-3, and check your answers in a world atlas. You can also practice placing locations on a map (such as Figure 2-3) when you know the coordinates. For example, plot 39° N, 77° W on a map. You should be at Washington, D.C. Plot 34° S, 151° E. You should be at Sydney, Australia.

Review Questions

Refer to the following diagrams to answer questions 18 and 19. These diagrams illustrate systems that can be used to determine position on a sphere.

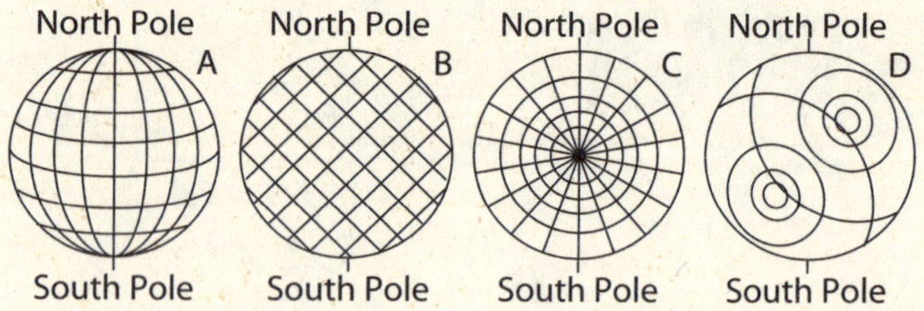

18. Systems of lines such as those illustrated above are called (1) latitude systems (2) coordinate systems (3) great circle systems (4) axis systems

19. Which of the illustrated systems is most like the latitude-longitude system used on Earth? (1) A (2) B (3) C (4) D

20. How are latitude and longitude lines drawn on a globe of Earth? (1) Latitude lines are parallel and longitude lines meet at the poles. (2) Latitude lines are parallel and longitude lines meet at the equator. (3) Longitude lines are parallel and latitude lines meet at the poles. (4) Longitude lines are parallel and latitude lines meet at the equator.

21. In the following diagram, what is the latitude of the observer?

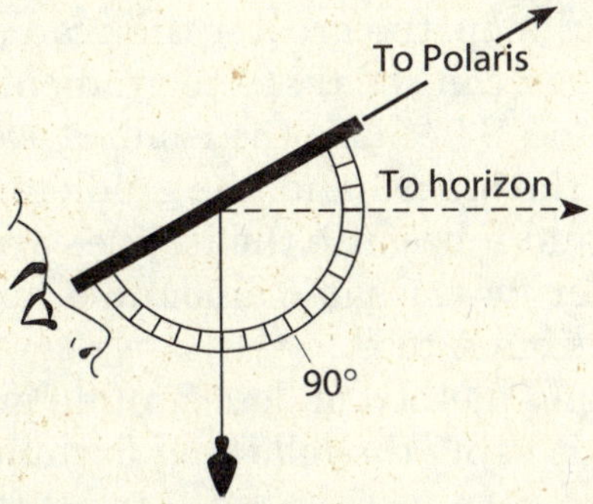

(1) 30° N (2) 60° N (3) 90° N (4) 120° N

22. What happens to the altitude of Polaris as you travel northward?

23. Polaris is used as a celestial reference point for Earth's latitude system because Polaris (1) always rises at sunset and sets at sunrise (2) is located over Earth's axis of rotation (3) can be seen from any place on Earth (4) is a very bright star

24. An airplane takes off from a location at 17°S latitude and flies to a new location 55° due north of its starting point. What latitude has the airplane reached? (1) 28° N (2) 38° N (3) 55° N (4) 72° N

25. Which diagram best shows the altitude of Polaris observed near Buffalo, New York?

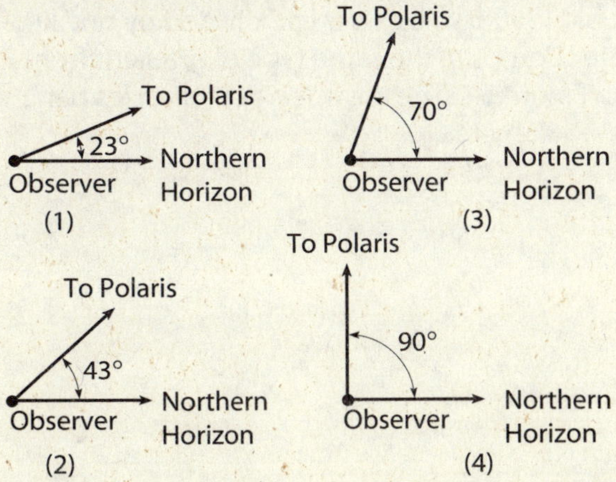

26. Which reference line passes through both the geographic North Pole and the geographic South Pole? (1) 0° latitude (2) 0° longitude (3) Tropic of Cancer ($23\frac{1}{2}$° S) (4) Tropic of Capricorn ($23\frac{1}{2}$° N)

27. A person knows the sun time on the Prime Meridian and the local sun time. What determination can be made? (1) the date (2) the altitude of Polaris (3) the longitude at which the person is located (4) the latitude at which the person is located

28. The following diagram represents a portion of Earth's latitude and longitude system. What are the approximate latitude and longitude of point A? (1) 15° S 20° W (2) 15° S 20° E (3) 15° N 20° W (4) 15° N 20° E

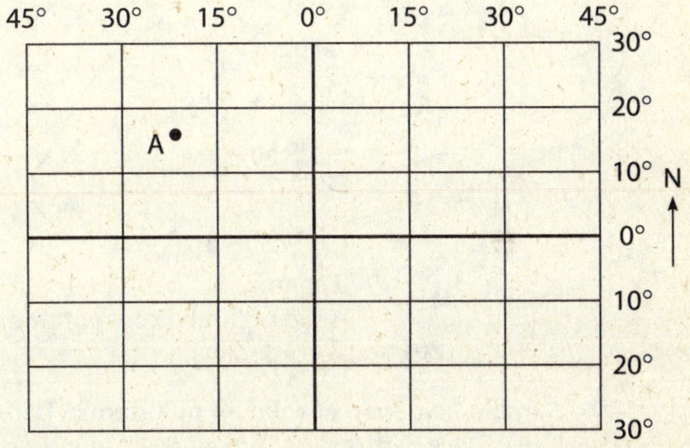

29. What is the location of Binghamton, New York? (1) 42°06′ N, 75°55′ W (2) 42°06′ S, 76°05′ W (3) 42°54′ N, 76°05′ W (4) 42°54′ N, 75°55′ W

30. Which New York landscape region includes the location 43°30′ N, 75°45′ W? (1) Adirondack Mountains (2) Erie-Ontario Lowlands (3) St. Lawrence Lowlands (4) Tug Hill Plateau

Fields

Any region of space or the environment that has some measurable value of a given quantity at every point is a **field.** Thus any aspect of matter or energy that can be measured is a type of field. Examples of field quantities are magnitudes of gravity or magnetism, temperature values, X-ray concentration, sound levels, elevation or depth, atmospheric pressure, wind speeds, acidity of the atmosphere, and relative humidity.

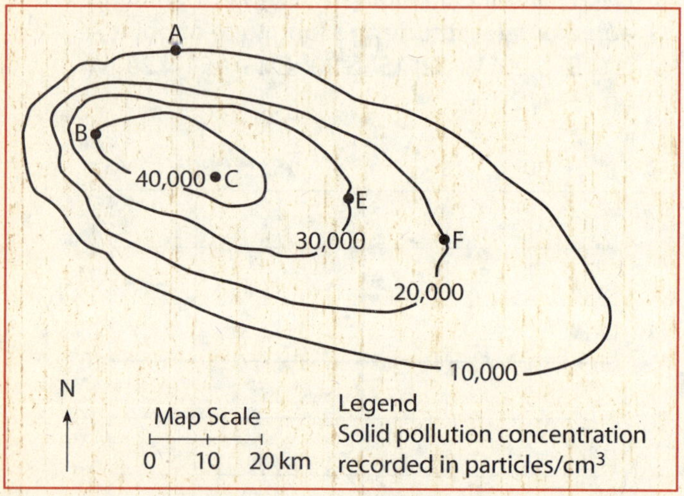

Figure 2-7. Sample field map of solid air pollutants: Data was measured just above Earth's surface of an area around a coal-fired, electric-generating plant.

Isolines

The varying values of a field are often represented on maps by the use of lines that connect points of equal field values—**isolines.** Figure 2-7 is a sample isoline map that has isolines for equal values of the amount of solid pollutants (aerosols) in the air for a portion of Earth's surface. By reading these values of air pollutants, the source of the air pollution could be inferred to be located around area B-C on the map.

Some common examples of isolines are isotherms—which connect points of equal temperature, isobars—which connect points of equal air pressure, and **contour lines**—which connect points of equal elevation.

Gradients and Changes in Fields

Since the environment is constantly changing, field characteristics change with time. This means that any model of a field, such as a weather map, shows the field for only one particular time. The rate of change from place to place within the field is called the **gradient,** or slope.

The gradient can be calculated in two ways. Gradients can be estimated by observing the closeness of the isolines. The closer together the isolines are on a field map, the greater the gradient. On Figure 2-7, you can see that the gradient is large between A and B because the isolines are relatively close. On the other hand, the gradient is small between E and F because the isolines are not as close together for the same amount of horizontal distance as from A to B.

The second method of determining gradient is to calculate it using the following formula:

$$\textbf{gradient} = \frac{\textbf{change in (amount of) field value}}{\textbf{change in distance (horizontal distance)}}$$

For example, if a weather map shows a change in temperature of 12°C between two locations that are 3 kilometers apart, the gradient between the two locations is 4°C per kilometer (12°C divided by 3 km).

DIGGING DEEPER

Compute the solid air pollution gradient between locations A and B in Figure 2-7 in number of particles per cm^3 per kilometer. For horizontal distance use the map scale. For the solid pollutant field values, read the isolines and compute the difference between the amount of solid pollutants at A and B. For example: 40,000 particles (B) minus 10,000 particles (A) equals 30,000 particles divided by 20 km equals 1500 particles per cm^3 divided by one kilometer. Compute the same type of gradient from E to F on Figure 2-7. The answer is 500 particles per $cm^3 \div$ kilometer.

Review Questions

31. The following isoline map shows the variations in the relative strength of Earth's magnetic field from 1 (strong) to 11 (weak).

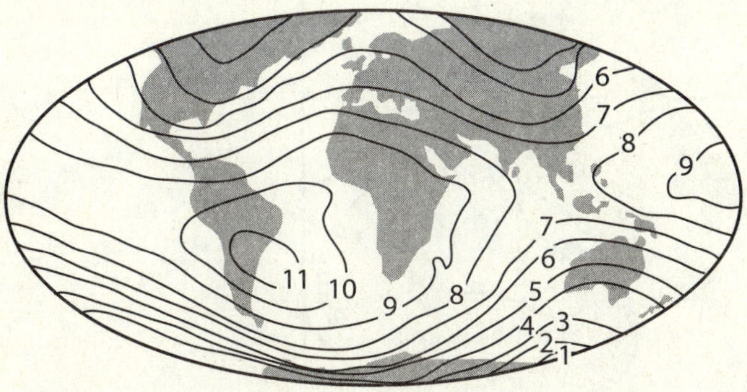

Which of Earth's plates has the weakest magnetic field strength? (1) South American plate (2) African plate (3) North American plate (4) Pacific plate

32. The following four temperature field maps represent the same region on four different occasions. All maps have been drawn to the same scale. Which map shows the greatest temperature field gradient?

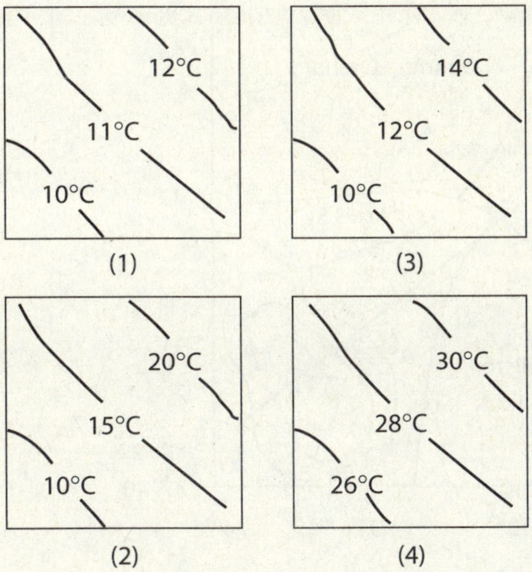

33. One thermometer that is held 2 meters above the floor shows a temperature of 30°C. A thermometer on the floor shows a temperature of 24°C. What is the temperature gradient between the two thermometers? (1) 6°C/m (2) 2°C/m (3) 3°C/m (4) 4°C/m

34. As gradient increases, what happens to the distance between isolines?

Mapping Earth's Surface

A map is a model that usually represents a portion or all of Earth's surface on a two-dimensional, or flat, surface. Maps can also model other objects or parts of Earth and the universe.

Topographic Maps

A **topographic map,** or contour map, is a commonly used model of the elevation field of the surface of Earth. The vertical distance or height above or below sea level is **elevation.** What distinguishes topographic maps from other maps is that they use contour lines to model the elevations and shapes of Earth's surface features or landforms. The contour lines are isolines that connect points of equal elevation above—and rarely, below—sea level.

There are many uses for topographic maps including: an aid in hiking or fishing, construction site selection, finding natural resources such as fresh water and trees, and planning highway locations. Topographic maps can also help you find escape routes when natural disasters are predicted.

READING CONTOUR LINES The difference in elevation between consecutive contour lines is the contour interval. The contour interval is usually listed in the key or legend. If the contour interval is not listed, subtract the difference in the value of two nearby contour lines and divide that number by the number of spaces between the known contour lines. In reading contour lines, pay attention to the following points.

- When contour lines cross a stream, they bend upward toward higher elevations. Contour lines "point" upstream. See Figure 2-8.

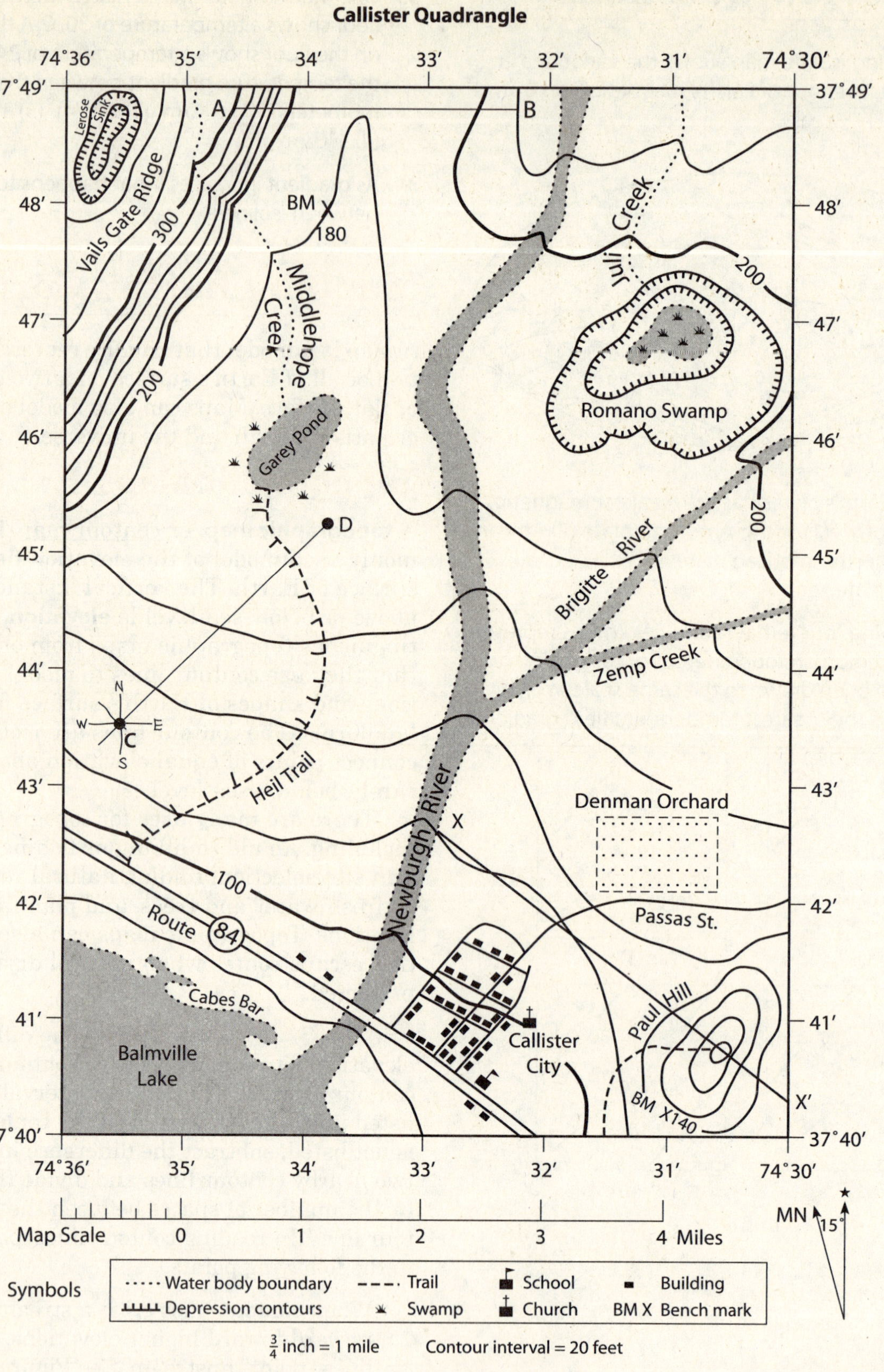

Figure 2-8. **A sample topographic (contour) map**

- When there is a series of consecutively smaller and smaller contour lines, it means that elevation is increasing toward the top of a hill or mountain. See Paul Hill on Figure 2-8.
- On most topographic maps every fourth or fifth contour line—an index contour—is thicker and bolder, and the elevation is shown on that line.
- Often maps show the location of benchmarks with the symbols BM X and an elevation value. The X marks the spot where a metal marker is in the ground labeled with an exact elevation for a location. Figure 2-8 has two benchmarks.
- Depression contour lines are marked with small lines pointing toward the center of a depression. When going uphill on a map, the first depression contour line has the same value as the last regular contour line. Each smaller and smaller depression contour line is one contour interval less than the previous line.

SYMBOLS ON TOPOGRAPHIC MAPS Various types of symbols are used to identify natural and human-made features on topographic maps. There are hundreds of graphic symbols like those shown in the key for Figure 2-8. Most government maps use colors: brown for contour lines to show landform features; red and black for human constructions such as boundaries, roads, and structures; blue for water features; green for woods or areas with trees; and purple where the map has been revised using aerial photographs.

Horizontal Distance on Maps

Contour intervals and contour lines are used to show elevations and depressions. The map scale is used to show distances along Earth's surface. A map scale is a ratio of distance between two places on a map and the actual distance on Earth's surface. Map scales can be expressed in three ways:

- verbally, such as "one inch equals a mile"
- fractionally, such as 1/500,000 or 1:500,000—which means any one unit of distance on the map equals 500,000 of the same units on Earth's surface
- graphically, as shown just below the bottom of the map in Figure 2-8 (This scale is most common.)

To measure the horizontal distance between two places, use a piece of paper with a straight edge. Place this paper on the map, and make a mark on the paper extending it onto the map to mark the beginning location. If it is just a straight-line distance you are trying to measure, then place another mark to match the ending distance on the map. Match up the marks on the piece of paper with the map scale, and read the distance between the two marks. For example, on Figure 2-8, measure the part of Route 84 from the west (left) edge of the map to the eastern (right) edge of the Newburgh River. The answer is 3 miles.

If the horizontal distance to be measured is a curved line, start by putting the straight edge of a piece of paper at the beginning location. Then draw a line on the paper extending it onto the map to mark the beginning spot. Repeat this each time you come to a significant bend on the map. Keep twisting the piece of paper and drawing lines until you come to the end of the distance to be measured. Compare the total length of the marks on the piece of paper with the map scale and measure the distance. On Figure 2-8, Heil Trail from Route 84 to Garey Pond shows the marks made to measure its distance. The length of the Heil Trail is approximately 4 miles.

Topographic Map Gradient and Profile

Gradient, or slope, on topographic maps seems unusual because you are dealing with two distances. The field value is a vertical distance, or elevation, and the other value is the horizontal distance along Earth's surface. For practice, compute the gradient—in feet per mile—of the Newburgh River in Figure 2-8 from the 180-foot contour line to the 100-foot contour line. Your result should be approximately 21 feet per mile. On any map of the elevation field, the relative amount of gradient can be estimated using the distance between the isolines. The more closely spaced the lines, the steeper the gradient. In the map example, the gradient is steep just to the east of Vails Gate Ridge, so the contour lines are close together. In the area around the Denman Orchard, the gradient is gentle, so the contour lines are far apart.

A topographic map represents three dimensions—length, width, and height, or elevation. A single contour is two-dimensional, but a series of contours shows the third dimension. A **profile** is the side view of an area's landscape. This model uses upward and downward changes of a line to show changes in elevation and slope. Drawing a profile can show the three-dimensional nature and the gradient of a field. The method for making a profile of a contour map is illustrated in Figure 2-9. The profile is drawn for the region between X and X′ in the southeast corner of the map in Figure 2-9.

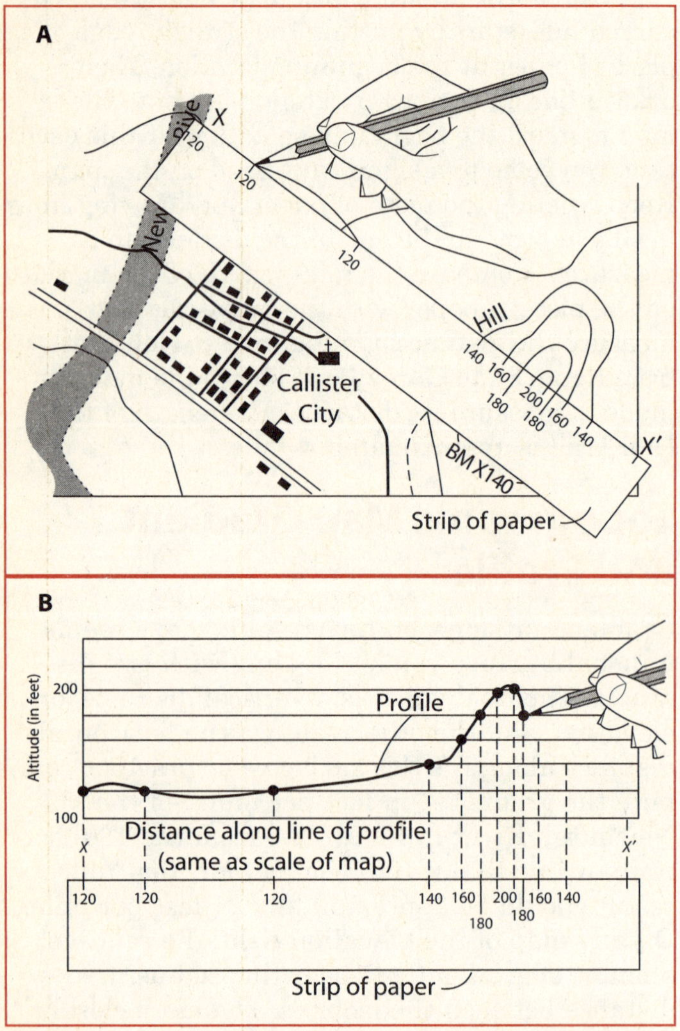

Figure 2-9. Constructing a profile along the line X-X′ on the contour map of figure 2-9: The edge of a strip of paper is placed along the line, and a mark is made wherever the paper crosses a contour line. The marks are labeled with the corresponding altitude. The marks are then projected upward to locate dots on a piece of lined paper as in drawing B. Finally the dots are connected with a smooth curved line, and the profile is complete.

Map Direction

Maps are usually constructed so that the top of the map is north, the bottom is south, the right side is east, and the left side is west. Most maps, including topographic maps, usually show directions by indicating geographic north with some type of arrow. Geographic north is the direction to Earth's geographic North Pole. The map in Figure 2-8 indicates geographic north by ★. An MN on the map indicates the direction of magnetic north. Often, instead of a north arrow, a map will have a compass rose, such as the one in Figure 2-10. To find the direction from one location to another, draw a simple compass rose on the location from which you are determining the direction. Then draw a straight line to the other location. The straight line crosses the compass rose at the correct direction. This method is illustrated in Figure 2-8 from point C to point D where the compass direction of D from C is northeast. Compass directions are usually given to the nearest of the eight cardinal directions shown on the compass rose in Figure 2-10.

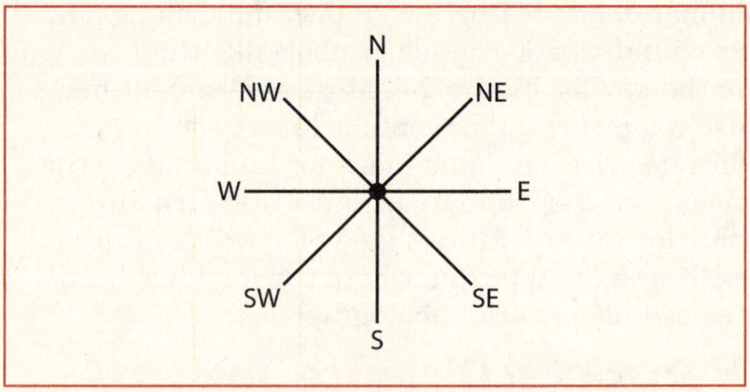

Figure 2-10. A sample compass rose: Compass roses are used to determine directions on maps and in many cases give geographic directions and not magnetic, or compass directions—which are usually different. Directions are often determined to the accuracy of the eight cardinal directions shown here.

Review Questions

Base your answers to questions 35 through 44 on the map in Figure 2-8.

35. What is the maximum altitude of Denman Orchard? (1) 140 feet (2) 150 feet (3) 159 feet (4) 161 feet

36. What is the longitude of the middle of Romano Swamp? (1) 37°47′ North (2) 37°47′ West (3) 74°31′ North (4) 74°31′ West

37. What is the distance from Passas Street at the margin of the map to where it meets Route 84? (1) 2½ miles (2) 3 miles (3) 3½ miles (4) 4 miles

38. What is the gradient of the Brigitte River from the 200-foot contour line to the contour line before the Brigitte River meets the Newburgh River? (1) 5 to 10 feet/mile (2) 15 to 20 feet/mile (3) 25 to 30 feet/mile (4) 35 to 40 feet/mile

39. What is the latitude of the church? (1) 74°32′ North (2) 74°32′ West (3) 37°41′ West (4) 37°41′ North

40. Where is the steepest slope located on the map?

41. The biggest area of gentle slopes is found in the vicinity of (1) Garey Pond (2) Denman Orchard (3) Paul Hill (4) Zemp Creek

42. Which direction is Vails Gate Ridge from Callister City? (1) north (2) northwest (3) northeast (4) west

43. Toward what direction is the Brigitte River flowing? (1) south (2) southwest (3) north (4) northeast

44. Which of the following profiles represents the shape of the landscape between A and B on the map?

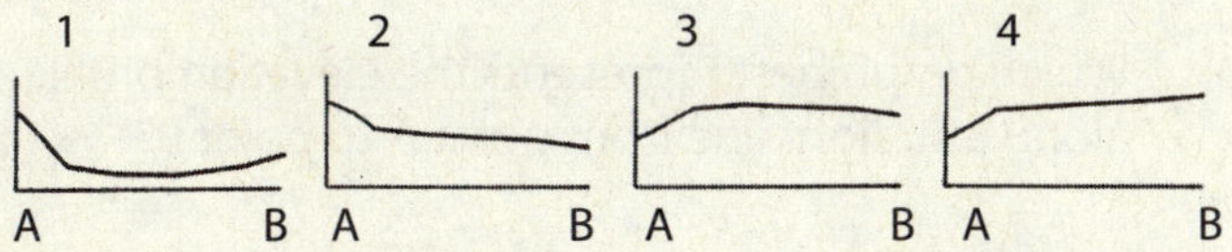

45. On the following topographic map, what is the most likely elevation of point A? (1) 1250 m (2) 1650 m (3) 1750 m (4) 1850 m

46. What is the approximate straight-line distance from Mt. Marcy in the Adirondacks to Slide Mountain in the Catskills? (1) 120 km (2) 150 km (3) 205 km (4) 235 km

Base your answers to questions 47 and 48 on the following diagram, which represents a topographic map of a hill.

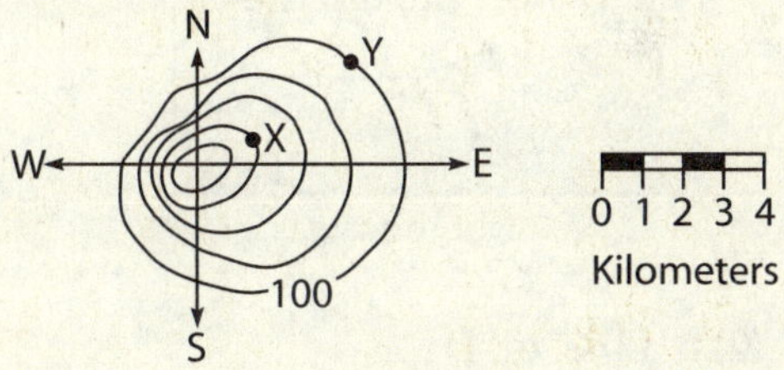

Contour interval=10 meters

47. On which side of the hill does the land have the steepest slope? (1) east (2) south (3) northeast (4) northwest

48. What is the approximate gradient of the hill between points X and Y? (1) 1 m/km (2) 10 m/km (3) 3 m/km (4) m/km

Questions for Regents Practice

Part A

1. The latitude of an observer on Earth's surface can be determined by measuring the altitude of Polaris because Earth has a

(1) nearly spherical shape

(2) nearly circular orbit around the sun

(3) variable length of day

(4) fairly constant period of revolution

2. Which line best identifies the interface of the lithosphere and the troposphere?

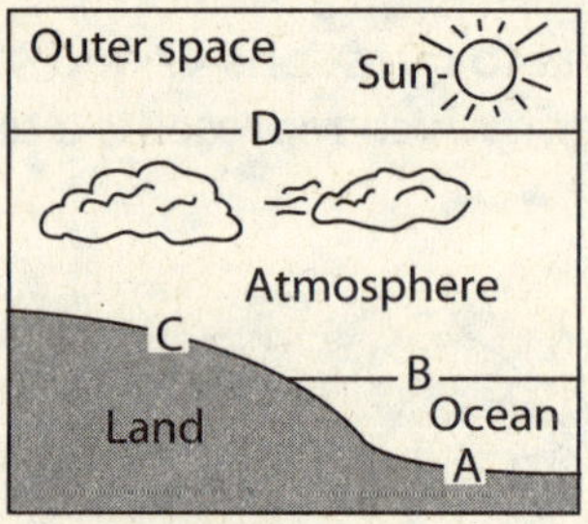

(Not to scale.)

(1) A (2) B (3) C (4) D

3. The following data table shows the altitude of Polaris as recorded by four observers at different locations on Earth.

Observer	Altitude of Polaris
A	90°
B	30°
C	30°
D	20°

Which statement is best supported by the information in the table?

(1) Observer A was at the equator.

(2) Observers A and B measured the altitude during daytime hours.

(3) Observers B and C measured the altitude at the same latitude.

(4) Observers B, C, and D were in the Southern Hemisphere.

4. Which latitude and longitude coordinates represent a location on the continent of Australia?

(1) 20° N, 135° E

(2) 20° N, 135° W

(3) 20° S, 135° E

(4) 20° S, 135° W

5. Which statement is true about an isoline on an air temperature field map?

(1) It represents an interface of high and low barometric pressures.

(2) It indicates the direction of maximum insolation.

(3) It increases in magnitude as it bends southward.

(4) It connects points of equal air temperature.

6. Which of the following items is a model?

(1) a globe

(2) a ruler

(3) a hand lens

(4) a mineral specimen

7. The following map represents an elevation field.

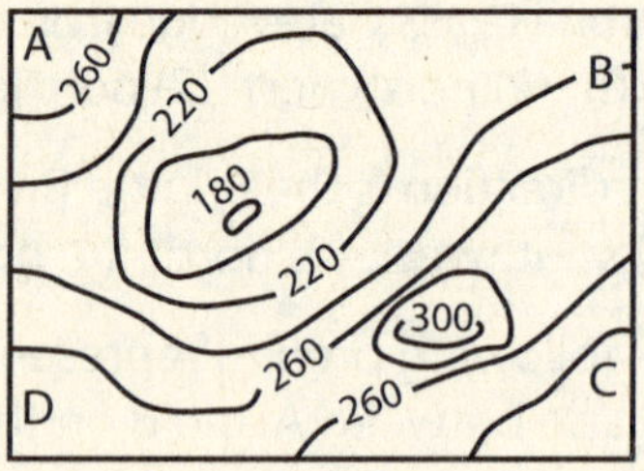

Which graph best represents the elevation profile along a straight line from point A to point B?

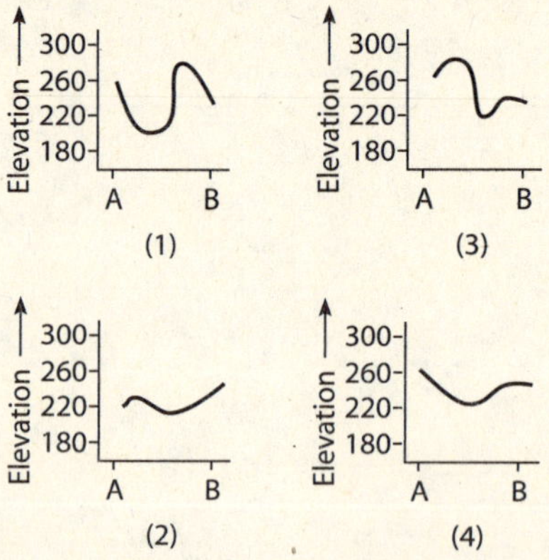

Base your answers to questions 8 through 12 on the following topographic map of an island. Points A through G represent locations on the island. Elevations are in meters.

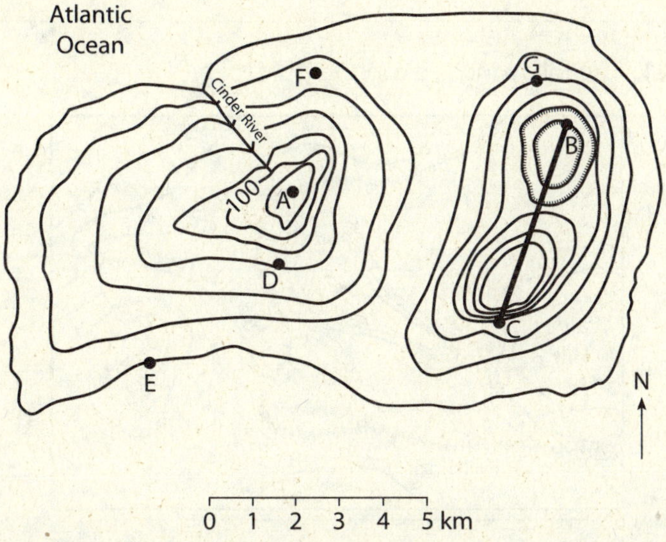

8. Which point is located on the steepest slope?

(1) F

(2) B

(3) C

(4) D

9. In which direction does the Cinder River flow?

(1) southeast

(2) southwest

(3) northeast

(4) northwest

10. What is the contour interval for this map?

(1) 10 m

(2) 15 m

(3) 20 m

(4) 25 m

11. Which two points have the same elevation?

(1) G and F

(2) B and D

(3) C and D

(4) G and C

12. Which diagram best represents the topographic profile between location C and location B?

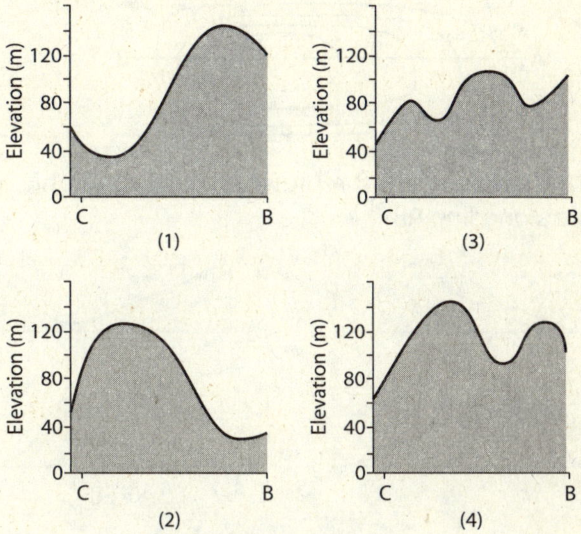

13. A movement of volcanic ash occurred at an altitude of 1.5 kilometers. In which layer of Earth's atmosphere did the ash cloud travel?

(1) troposphere

(2) stratosphere

(3) mesosphere

(4) thermosphere

14. What is the distance in degrees east or west of the prime meridian called?

(1) longitude

(2) latitude

(3) equator

(4) altitude

15. The following diagram represents contour lines on a topographic map with cross-section line AB.

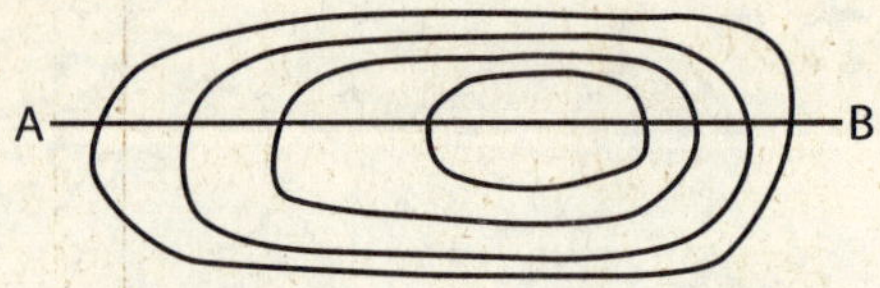

Which diagram best represents the topographic profile along line AB?

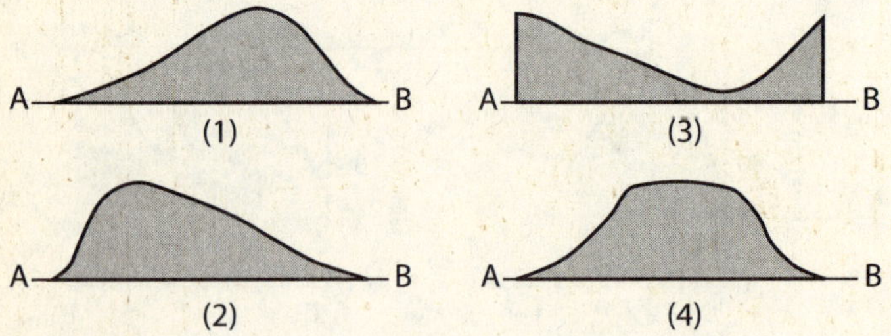

Part B

The following topographic map shows a portion of an area of inactive volcanoes in the United States. The contour interval is 20 feet. Base your answers to questions 16 through 20 on this diagram. REMEMBER: State directions or units of measure when appropriate.

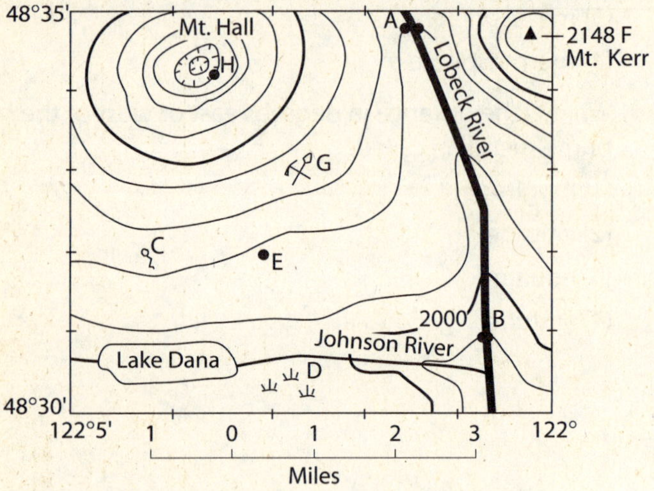

16. What is the latitude of point E to the nearest minute? [2]

17. What is the longitude of point E to the nearest minute? [2]

18. In which direction does Johnson River generally flow? [1]

19. What is the elevation of point H on the rim of the crater of Mt. Hall in feet? [1]

20. What is the distance along Lobeck River from point A to point B to the nearest quarter-mile? [1]

---

21. The following is a topographic map.

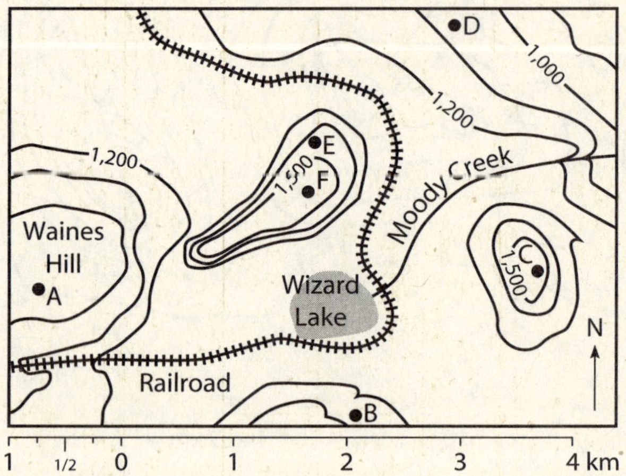

What is the approximate length in kilometers of the railroad tracks shown on the map ? [1]

22. The following diagram shows a three-dimensional model of a landscape region.

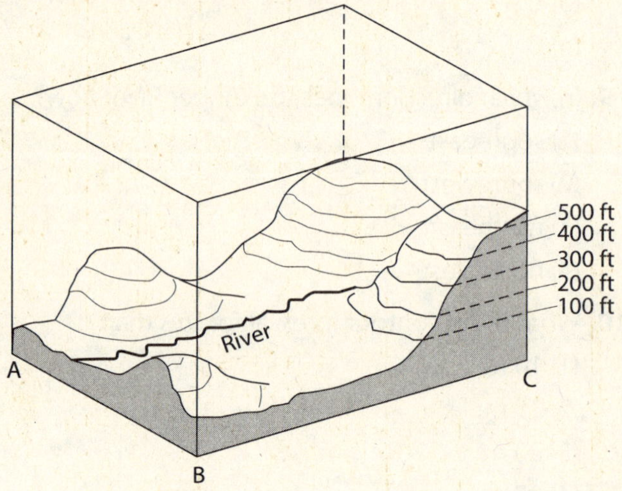

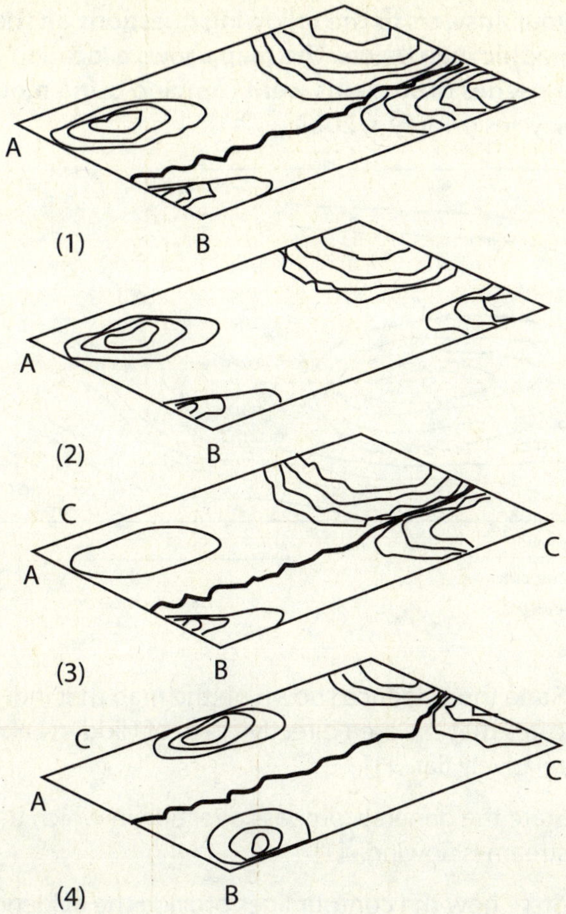

Which map view best represents the topography of this region? [1]

Base your answers to questions 23 through 26 on the following topographic map of Cottonwood, Colorado. Points A, B, X, and Y are marked for reference.

Cottonwood, Colorado

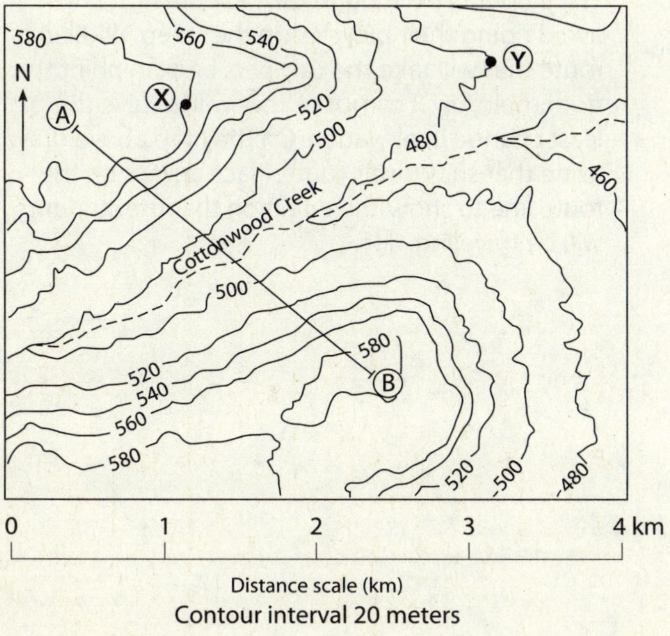

23. State the general direction in which Cottonwood Creek flows. [1]

24. State the highest possible elevation, to the nearest meter, for point B on the topographic map. [1]

25. On the following grid, draw a profile of the topography along line AB shown on the map. [2]

26. Use the following directions to calculate the gradient of the slope between points X and Y on the topographic map.

a. Write the equation for gradient. [1]

b. Substitute data from the map into the equation. [1]

c. Calculate the gradient and label it with the proper units. [1]

Part C

The planet in the following paragraph is fictional. Use the information in the paragraph and your knowledge of earth science to answer questions 27 through 32.

Glick is a planet in a solar system very similar to our own. Planet Glick is very similar to Earth except for the following differences. Glick's axis of rotation has no tilt. It rotates from east to west, approximately twice as fast as Earth. Zelda is a star that is visible directly over Glick's South Pole and Glotch is a star visible directly over Glick's North Pole. A latitude-longitude system such as that on Earth has been estabilished for Glick.

27. Why would a plastic relief replica of a continent on Glick be considered a model? [1]

28. If planet Glick has a hydrosphere, what two chemical elements is it primarily composed of? [1]

29. In relationship to Glick's atmosphere and lithosphere, what would be the relative location of its hydrosphere? [1]

30. Where would a majority of Glick's most dense substances be located in the planet? [1]

31. At a certain location on Glick, the star Zelda is observed at 37° above the horizon. From this information, what conclusion can you draw about the latitude and longitude of this location? [2]

32. If you traveled due west in the northern hemisphere of Glick at night, what changes might you observe in the altitude of the star Glotch? [1]

Base your answers to the following questions on the topographic map below. The map shows a location where a series of students went camping using mountain bicycles on July 1, 2003.

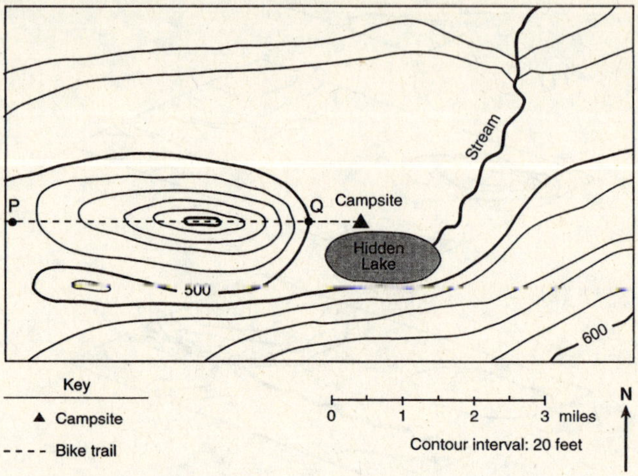

33. State the evidence shown on the map that indicates that the area directly north of Hidden Lake is relatively flat. [1]

34a. State the general compass direction in which the stream is flowing. [1]

34b. State how the contour lines provide the evidence for determining this direction. [1]

35. On July 2, 2003 the students decided to move their campsite 1 mile directly east of their original campsite that is located on the map above. On the map above place another campsite symbol to indicate the location of the July 2 campsite. [1]

36. The students decided to take a route home to avoid riding their bicycles up the steep hill. Plan a route that will take the campers back to point P from their July 2 campsite that will involve the least change in elevation. On the map above draw a line that shows this route. Place arrows on the route line to show the direction that the students will be traveling. [1]

Earth in the Universe

TOPIC 3

VOCABULARY		
asteroid	gravitation	nuclear fusion
Big Bang theory	impact crater	red shift
celestial object	impact event	revolution
comet	inertia	rotation
Doppler effect	Jovian planet	solar system
eccentricity	luminosity (of a star)	star
ellipse	meteor	terrestrial planet
focus (pl., foci)	Milky Way Galaxy	universe
galaxy	moon	

Throughout the ages, humans have observed the celestial objects in the sky and wondered about them. From Earth a **celestial object** is any object outside or above Earth's atmosphere. There are about 6000 celestial objects visible with the unaided eye from any location on Earth. Where do people fit? Are they alone? Are there other solar systems with an environment in which humans could live? In the next topic of this book, you will study the properties of Earth and its moon, while in this topic the focus is on our solar system, our galaxy, and the universe.

Origin and Age of the Universe

Over the thousands of years of human thinking, various cultures have produced a multitude of theories concerning the origin, evolution, and structure of the universe. **Universe** means everything that exists in any place—all the space, matter, and energy in existence. The majority of scientists today think that the universe is extremely vast, and that it is more than 10 billion years old—maybe as many as 17 billion years old.

Presently, the majority of scientists believe that the universe began with an event called the Big Bang and has been expanding in volume ever since. The **Big Bang theory** states that all matter and energy started out concentrated in a small area and, after a gigantic explosion, matter began to organize into subatomic particles and atoms. Most, if not all, of the earliest atoms were hydrogen and helium. Within approximately a billion years, atoms became organized into celestial bodies, and then most stars became part of gravitational groupings. As this organization of matter was occurring, the universe kept expanding in all directions, and is continuing to expand at present.

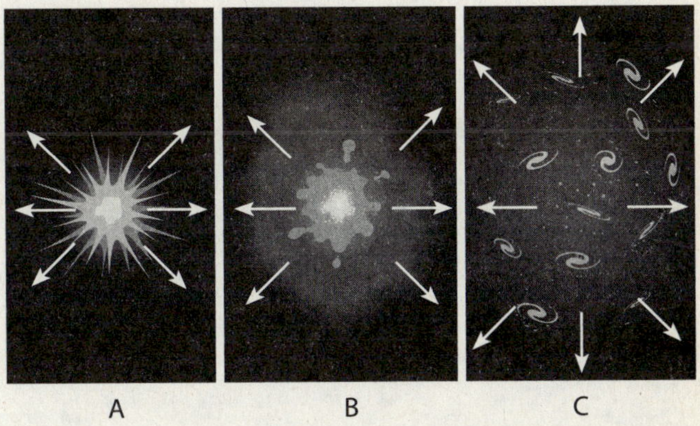

Figure 3–1. Big Bang and expanding universe: (A) Approximately 10 to 17 billion years ago, all mass and energy was concentrated in a small area that started expanding with a big explosion called the Big Bang. **(B)** The expanding, cooling universe first formed subatomic particles and finally small atoms of hydrogen and helium. **(C)** After approximately a billion years, matter clumped together forming stars and early galaxies, and the universe kept expanding.

Evidence for the Big Bang

Scientists have theorized that if a Big Bang did occur, the energy created by the explosion expanded along with the matter. Thus, there should be radiation from the Big Bang in all parts of the universe mixed with the energy given off by stars at later times. Recently, scientists have found evidence of long-wavelength background radiation—actually microwaves—that appear to be coming from all directions in the universe. This background radiation is evidence supporting the Big Bang theory.

Other evidence of the Big Bang is found in the spectrum of the radiation emitted by stars. Look over the Electromagnetic Spectrum in the *Earth Science Reference Tables.* The various types of waves that transmit energy through space are called electromagnetic energy. Each element emits energy in specific portions—wavelengths—within the electromagnetic spectrum. Because the human eye observes different wavelengths of visible light as different colors, people can distinguish specific portions of the electromagnetic spectrum. When scientists study the spectrum of electromagnetic energy coming from stars and other celestial objects, they can infer which elements are in these objects. Scientists base these inferences on the comparison of the signature wavelengths produced by elements on Earth with those coming from the other celestial objects.

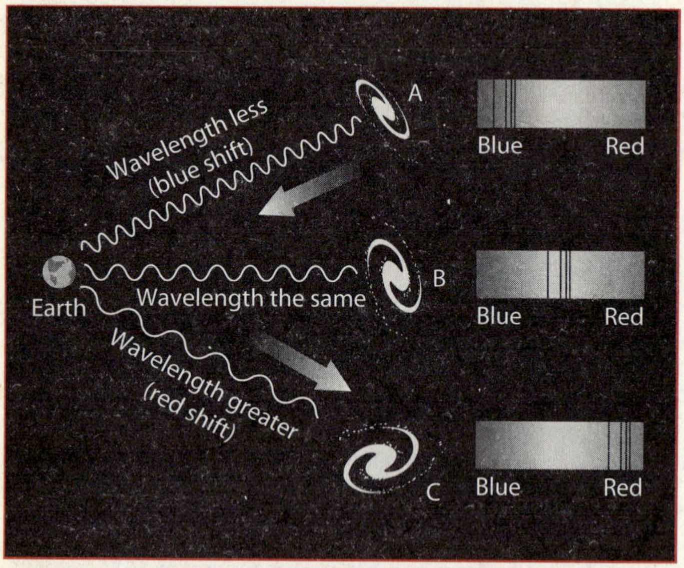

Figure 3-2. Doppler effect blue and red shifts: Galaxy B is not moving towards or away from Earth. The spectral lines of one element are the same as they would be if they were produced on Earth. Galaxy A is moving towards Earth—bunching up the light rays and shifting the spectral lines of the element to the blue end of the spectrum. Galaxy C is moving away from Earth—as most galaxies are—spreading out the light waves and shifting the spectral lines of the element to the red end—red shift.

Scientists have found that the position of the characteristic wavelengths, or colored lines, are shifted to either the shorter (blue end) or longer (red end) wavelengths. This shifting of wavelengths is called the **Doppler effect.** The relative movement between Earth and the other celestial object causes the Doppler effect. If Earth and another celestial object are coming closer together, the electromagnetic waves are bunched together, resulting in a blue shift Doppler effect. If Earth and some other celestial object are moving apart, the electromagnetic waves are spread out causing a **red shift** Doppler effect.

DIGGING DEEPER

To help you understand the Doppler effect, compare the Doppler effect of electromagnetic energy with that of sound waves. If you are standing on a street corner and a fire truck is coming towards you with the siren constantly blaring, the sound waves are bunched together as the truck gets closer to you. These shorter wavelengths are observed as a higher pitch or sound to your ears. After the truck passes, with the siren still blaring, the sound waves are stretched out (like the red shift).The sound has a lower pitch and sounds softer or muted.

The collective light from the stars in all galaxies, except for a few galaxies close to Earth, is shifted to the red end of the spectrum. The fact that almost all galaxies have a red shift indicates that the universe is expanding in all directions. This evidence further supports the Big Bang theory and the expansion from the initial explosion. In addition, the farther away a galaxy is from Earth, the greater the red shift. This fact indicates that the rate of expansion of the universe is increasing.

Review Questions

1. From your location, what is the best definition of a celestial object? (1) any object in the universe (2) any object in the universe above Earth's atmosphere (3) any object in the universe outside our solar system (4) any object in the universe outside our home galaxy

2. The age of the universe is measured in 10 to 20 (1) thousands of years (2) millions of years (3) billions of years (4) trillions of years

3. Studies of the universe generally indicate that (1) almost all galaxies appear to be moving away from Earth at tremendous velocities (2) few galaxies other than our own exist (3) all galaxies are approximately the same size (4) all galaxies are spiral in shape

4. Background radiation detected in space is believed to be evidence that (1) the universe began with an explosion (2) the universe is contracting (3) all matter in the universe is stationary (4) galaxies are evenly spaced throughout the universe

5. Which statement best describes how most galaxies generally move? (1) Galaxies move toward one another. (2) Galaxies move away from one another. (3) Galaxies move randomly. (4) Galaxies do not move.

6. The following diagram represents a standard spectrum for an element.

The spectral lines of this element are observed in light from a distant galaxy. Which diagram represents these spectral lines?

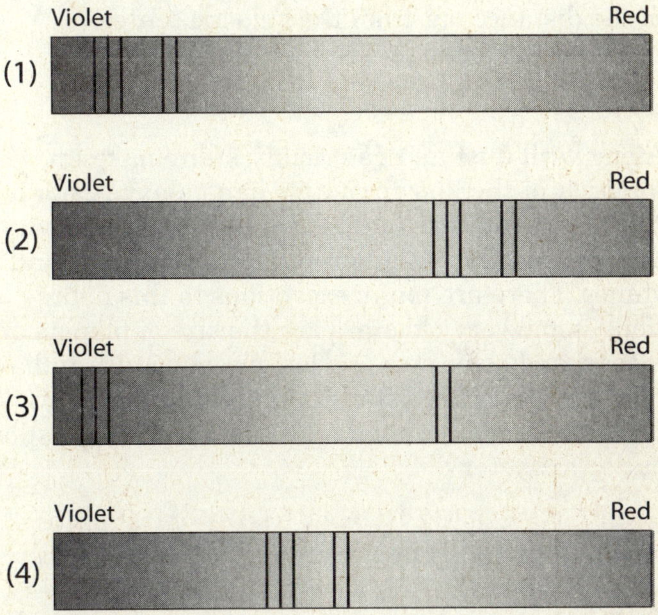

Structure of the Universe

In recent years instruments such as the Hubble space telescope have allowed scientists to infer some basic structure to the known universe.

Galaxies

The basic structural unit of matter in the universe appears to be the galaxy. A **galaxy** is a collection of billions of stars and various amounts of gas and dust

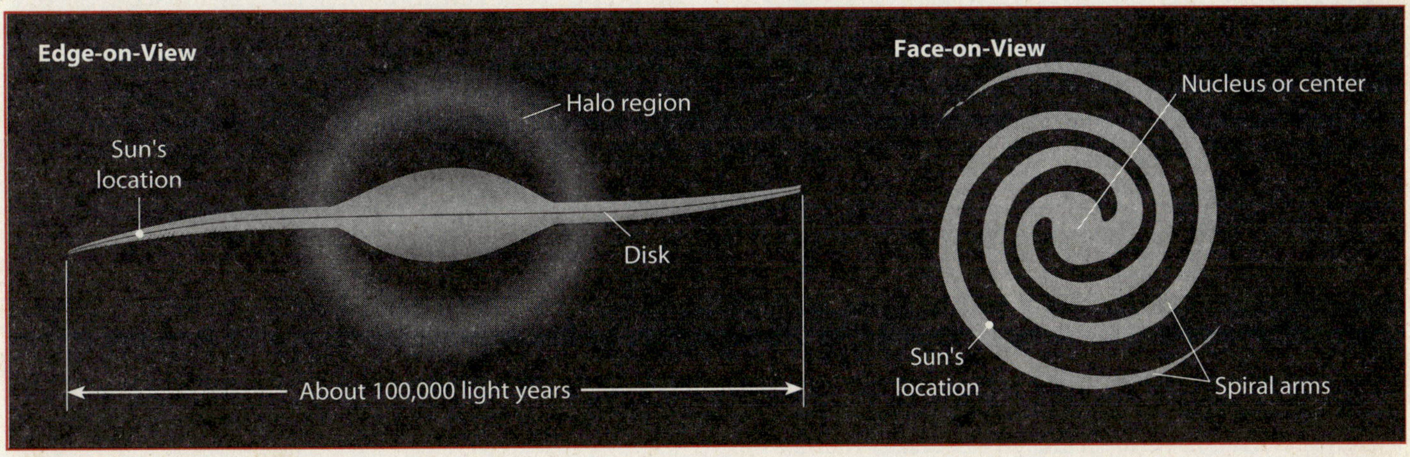

Figure 3-3. The Milky Way Galaxy: This spiral-shaped galaxy of which Earth is a part measures approximately 100,000 light years at its greatest diameter. Each light year is about 6 trillion miles or 9.5 trillion kilometers. The halo region is made up of clusters of stars that are billions of years older than our sun. The sun indicates our solar system's location near a spiral arm.

held together by gravity. An average galaxy will have over 100 billion stars, and there are over 100 billion galaxies. Galaxies have been classified into a few types largely based on shape, including elliptical (football shaped), irregular, and spiral. Our solar system is a part of a spiral-shaped galaxy called the **Milky Way Galaxy,** which has over 200 billion stars. Figure 3-3 on the previous page illustrates the spiral shape of the Milky Way Galaxy. It also shows the position of our solar system between two spiral arms about two-thirds of the distance out from the galactic center.

Stars

Along with dust and gas clouds, stars make up the majority of the known matter in a galaxy. A **star** is usually a large ball of gas held together by gravity that produces tremendous amounts of energy and shines. There are many exceptions to this definition—some very old stars are the size of planets or moons, and some stars no longer emit much radiation. The star called the sun is the dominant gravitational force associated with Earth and the rest of our solar system.

ENERGY PRODUCTION IN STARS Most of the tremendous amounts of energy produced by most stars is the result of nuclear fusion in their cores. **Nuclear fusion** is the combining of the nuclei of smaller elements to form the nuclei of larger elements with some of the mass being converted into energy. As an example, the sun converts hydrogen nuclei into helium nuclei with about 0.07 percent of the mass forming energy. There are many other types of nuclear fusion that occur in stars of different ages or sizes than the sun. Nuclear fusion can only occur in extremely high temperature and high pressure conditions like those found in star interiors. The energy of nuclear fusion in most stars is eventually radiated into space as types of electromagnetic energy.

THE LUMINOSITY AND TEMPERATURE OF STARS DIAGRAM The classification scheme of grouping stars by surface temperature compared to their luminosity is shown in the Luminosity and Temperature of Stars diagram in the *Earth Science Reference Tables.* **Luminosity** of a star measures how bright it would be in relation to the sun if all stars were the same distance from an observer. What we directly observe when we look at stars is the apparent brightness, which is dependent upon the star's absolute luminosity and its distance away from us.

An analysis of a Luminosity and Temperature of Stars diagram indicates that star properties are not random. Of the hundreds or thousands of stars plotted on the diagram, stars appear to be grouped by differences in luminosity and surface temperature as reflected by color. Just as the filament in a light bulb or the coil in an electric heater changes color with increasing temperatures, stars change color from red to blue as their surface temperatures increase.

STAR TYPES Figure 3-4 on the next page illustrates the relative size differences in some of the star groups. The diameter of the sun—a medium-sized star—is 109 times the diameter of Earth. Most stars are much bigger than Earth, but some—such as neutron stars, black holes, and some dwarfs—are smaller than Earth. These smaller stars are actually shown too big in this diagram compared to the sun. Giant stars are 10 to 100 times the size of the sun, and super giants are 100 to 1000 times the size of the sun. Some of the star types commonly identified on a luminosity and temperature of stars diagram are described here.

Main Sequence Stars About 90 percent of studied stars are located on the Luminosity and Temperature of Stars diagram in a broad band called the main sequence. Most stars spend the majority of their life span as a main sequence star. Most of these stars are average size, and as the surface temperatures of the stars increase, the luminosity increases. The luminosity increase from red to blue-white is mostly related to an increase in star size and the resulting higher temperatures. Our sun is a main sequence star of yellow color. The smallest and coolest stars of the main sequence are the very common type of star—the red dwarfs.

Giant Stars Red, orange, and yellow giant stars are a rare type of star but are commonly seen in the night sky because of their large size—10 times or more the diameter of the sun—and their high luminosity. These low-temperature stars represent a late stage in the evolution of medium to small-size main sequence stars—when they greatly expand in size.

Super Giants Super giant stars can be anywhere from 100 to 1000 times the diameter of the sun. (See Figure 3-4.) These highly luminous stars represent the late evolution of stars much more massive than the sun. Super giants usually explode in a tremendous event called a supernova. The super giant stars that are the brightest and exhibit the highest temperature are the blue super giants.

White Dwarfs Not all white dwarf stars are white, but they are all small—around the size of Earth. They are hot on the surface and low in luminosity. They represent the last luminous or shining stage of low to medium mass stars.

Black Dwarfs When a white dwarf cools and no longer emits much electromagnetic energy, it is a "dead" star—a black dwarf. Black dwarfs are probably very common in the universe because in the many billion years of the existence of white dwarf stars, many trillions of them have stopped nuclear fusion and no longer produce nuclear energy.

STAR ORIGIN AND EVOLUTION Stars, like many objects on Earth, have an origin, an evolution of features, and an ending. Figure 3-5 on the next page is a model of our current knowledge of star origin and evolution. It is believed that stars originate from clouds of gas and dust molecules. These clouds were created from the masses that evolved from the original Big Bang and/or from the mass given off by stars that have expanded, exploded, or otherwise given off mass to space. Gravity causes these gas and dust clouds to clump up, forming larger and larger balls of gas and dust molecules. When the mass of these spherical balls becomes slightly larger than the planet Jupiter, the gravitational contraction results in high enough temperatures and pressure to start nuclear fusion. Then the ball begins to shine, or radiate large amounts of electromagnetic energy, and a star is born.

The evolutionary stages of stars—after they spend most of their lives as a main sequence star—depends upon their original mass. Most

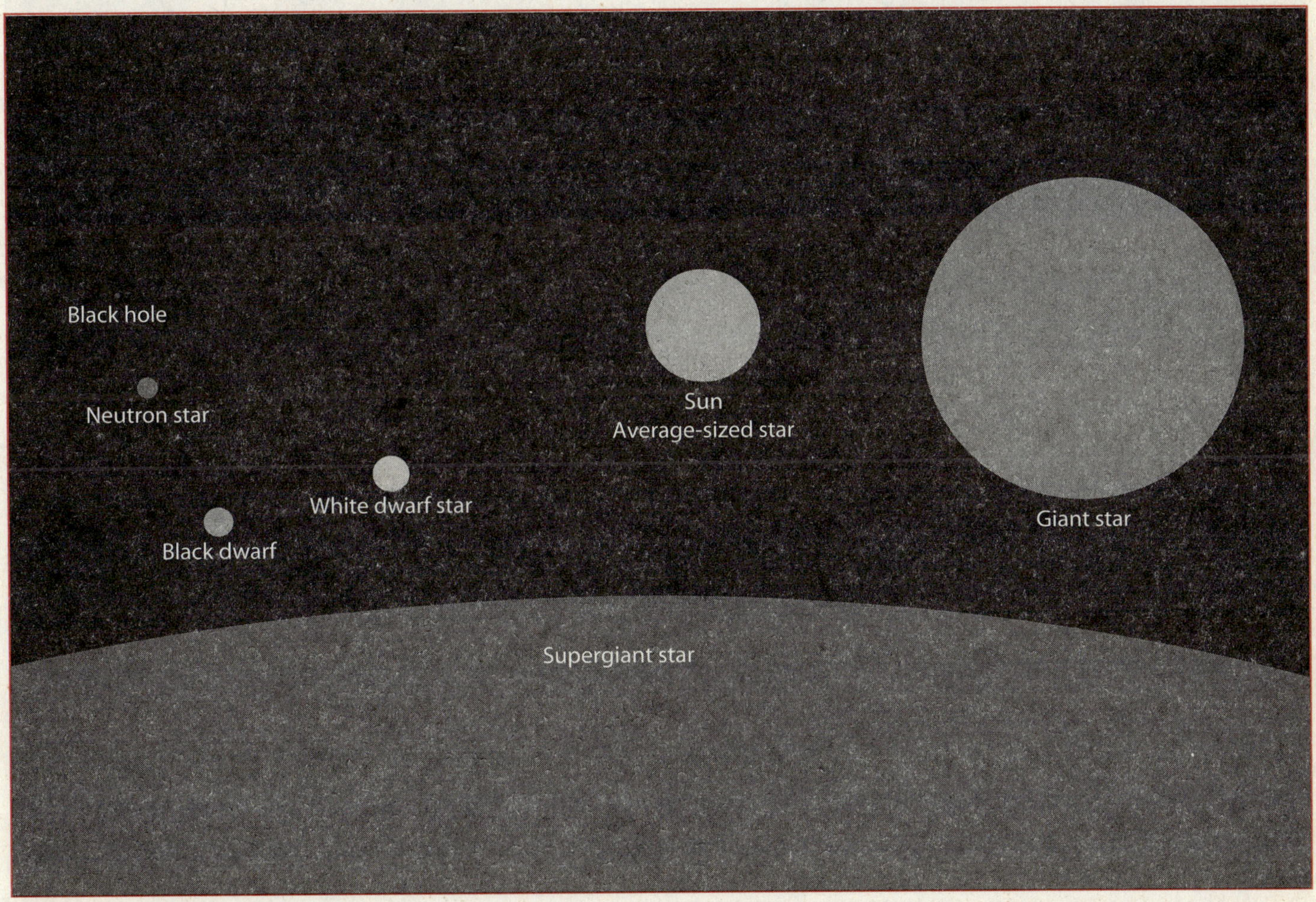

Figure 3-4. Relative sizes of stars: The smaller stars are not drawn to scale.

stars with masses similar to the sun spend billions of years as a main sequence star and eventually expand to become a red giant. As the bottom part of Figure 3.5 shows, these stars use up most of their nuclear fuel and collapse to form a white dwarf, and then slowly "die" becoming a black dwarf. This process may take many billions of years, and some stars created soon after the Big Bang may still exist.

The evolution of stars with an original mass greater than one and a half times the mass of the sun is substantially different. These stars exist for much shorter periods of time—approximately 100,000,000 years. As the upper part of Figure 3-5 indicates, these massive stars evolve into super giants after spending time as a large main sequence star. They eventually explode in a supernova event and then rapidly collapse, forming a body much smaller than a white dwarf. They are so dense that only neutrons can exist, and a neutron star is formed. When even more massive stars collapse, the density is so great that it creates an extreme gravity field that allows no visible light or any other form of energy to escape, and a black hole is formed.

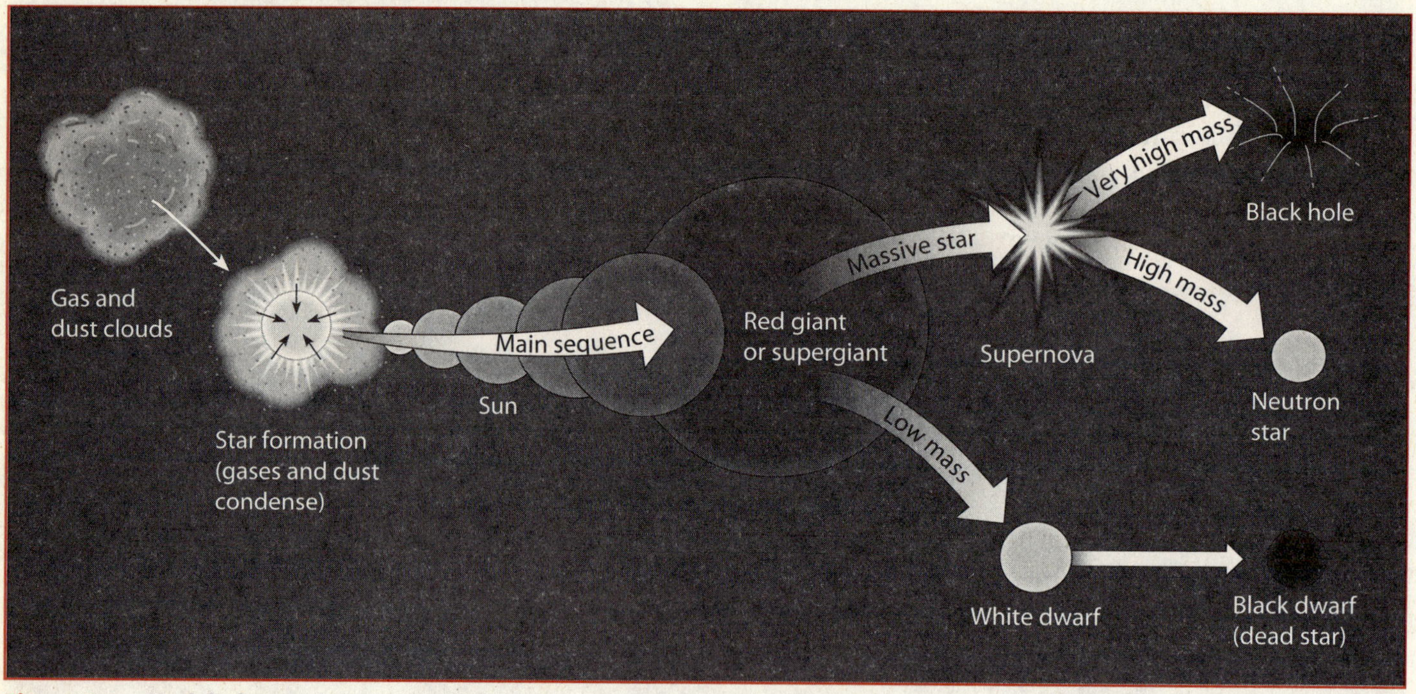

Figure 3-5. Model of star origin and evolution: The original mass of a star largely determines its evolutionary stages and how long it lasts.

Review Questions

7. Billions of stars in the same region of the universe are called (1) solar systems (2) asteroid belts (3) constellations (4) galaxies

8. The following symbols represent the Milky Way Galaxy, the solar system, the sun, and the universe.

 = Milky Way Galaxy

 = Solar System

 = Sun

 = Universe

Which arrangement of symbols is most accurate?

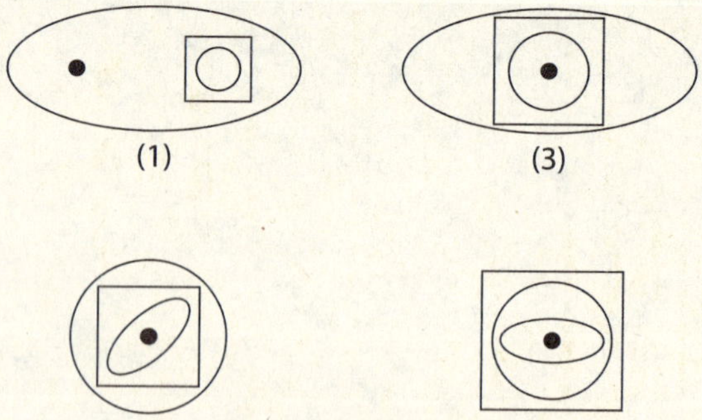

9. In which group are the parts listed in order from oldest to youngest? (1) solar system, Milky Way Galaxy, universe (2) Milky Way Galaxy, solar system, universe (3) universe, solar system, Milky Way Galaxy (4) universe, Milky Way Galaxy, solar system

10. The great system of 200 billion stars to which the sun and our solar system belong is the (1) Andromeda Galaxy (2) Large Magellanic Galaxy (3) Milky Way Galaxy (4) Orion nebula Galaxy

11. A star differs from a planet in that a star (1) has a fixed orbit (2) is self-luminous (3) revolves about the sun (4) shines by reflected light.

12. The sun's energy is most likely the result of the (1) fusion of hydrogen atoms (2) transformation of the sun's gravitational potential energy to heat energy (3) burning fossil fuels (4) radioactive decay of uranium and thorium atoms

13. Nuclear fusion can only occur in areas of (1) high temperature and low pressure (2) high temperature and high pressure (3) low temperature and low pressure (4) low temperature and high pressure

14. As star color changes from blue to red, the surface temperature of the star (1) decreases (2) increases (3) remains the same.

15. A luminosity and temperature of stars diagram classifies a star of high temperature and low luminosity as a (1) giant (2) main sequence star (3) supergiant (4) white dwarf.

Base your answers to questions 16 through 23 on the Luminosity and Temperature of Stars diagram in the *Earth Science Reference Tables.*

16. A main sequence star is 1000 times more luminous than the sun. The temperature is likely to be most nearly (1) 3000°C (2) 5000°C (3) 12,000°C (4) 25,000°C

17. A giant star has a luminosity of 300. Its color is most likely to be (1) yellow-red (2) black (3) white (4) blue-white

18. A white dwarf star has a temperature of 13,000°C. What is the probable luminosity of the star? (1) 100 (2) 10 (3) 0.1 (4) 0.01

19. A yellow star has a temperature of 4000°C and is 500,000 times as luminous as the sun. To which group does it belong? (1) giants (2) supergiants (3) white dwarfs (4) main sequence

20. The nearest star to the sun is Alpha Centauri. How does this star compare to the sun? (1) It is much hotter. (2) It has a different color. (3) It has a higher luminosity. (4) It is much smaller in diameter.

21. The sun is best described as a (1) very large star (2) medium-sized star (3) red star (4) cool star

22. The sun is brighter than any star in the group of (1) supergiants (2) giants (3) main sequence (4) white dwarfs

23. Which type of star is associated with the last stage in the evolutionary development of most stars? (1) main sequence star (2) supergiant (3) giant (4) white dwarf

Solar System

In the last few years, powerful telescopes on Earth and in space have found evidence of planets around more than 50 stars. This means that "solar system" really can refer to any star or group of two or three stars that has non-star objects orbiting it. However, in this book, the solar system is our **solar system**—the sun and all objects that orbit the sun under its gravitational influence.

Parts of the Solar System

Most of the solar system is space devoid of much mass. However, about 99 percent of the mass that does exist is contained in the sun. The sun is a medium-sized, main sequence star about 5 billion years old that is the gravitational center of our solar system. A satellite is any object that orbits or revolves around another object. The planets, asteroids, meteoroids, and comets are satellites of the sun, and the moons are satellites of planets or asteroids. Earth has one natural satellite—the moon—and more than 2,500 artificial ones.

PLANETS The nine planets are the largest objects that independently orbit the sun. Planets are generally spherical in shape. The planets will be described in more detail later in this topic. Recently, over 100 planets have been found revolving around stars other than our sun.

ASTEROIDS An **asteroid** is a solid rocky and/or metallic body that independently orbits the sun. Asteroids have an irregular shape, except for the larger ones which are spherical, and no atmosphere. A large percentage of the thousands of known asteroids are in orbits between Mars and Jupiter. Asteroids are smaller than planets (from about 100 to 1000 kilometers in diameter) and are often called minor planets.

MOONS A **moon** is a body that orbits a planet or an asteroid as those objects orbit the sun. The more than 100 known moons vary in size from larger than the smallest planets to only a few kilometers in diameter. The number of known moons for each of the planets is listed in Solar System Data in the *Earth Science Reference Tables.*

COMETS A **comet** is often compared to a snowball made from snow found along a road that has just been sanded—a dirty snowball. Comets are mainly composed of solids that easily change to gases when heated. They are largely ices of substances such as water and methane mixed with rocky or metallic solids. Most comets are only 1 to 100 kilometers in diameter. When comets get near the sun, their ices turn to gases. Some solids are released, forming spectacular tails visible in Earth's sky, sometimes for weeks at a time.

METEOROIDS Very small solid fragments that orbit the sun are called meteoroids. Most meteoroids are only the size of a dime or sand grain. When meteoroids burn up or vaporize, they leave a brief visual streak as they pass through Earth's atmosphere and are called **meteors.** If a meteoroid survives its trip through Earth's atmosphere and lands on Earth's surface, it is then called a meteorite. Some meteorites have sufficient mass to create a depression in Earth's crust called an **impact crater.**

Evolution of the Solar System

Scientists infer that our solar system started to form approximately 5 billion years ago. At first there was a gas dust cloud many times the size of the present solar system. This cloud contained remnants of stars that had exploded, giving to the cloud heavier elements to mix with the more common lighter ones. Gravitation, perhaps aided by a shock wave from an exploding star, caused the cloud to condense into one or more mass concentrations. Look at Figure 3-6 and read the caption to see how these masses formed our solar system.

Even after Earth and other planets formed, their gravitational forces pulled on the smaller clumps of matter to cause comets, asteroids, and meteoroids to collide with the planets in what is called an **impact event.** The planets, asteroids, and moons with solid surfaces bear witness to these impact events in the form of craters. Impact events have also been linked to global climate changes and mass extinctions.

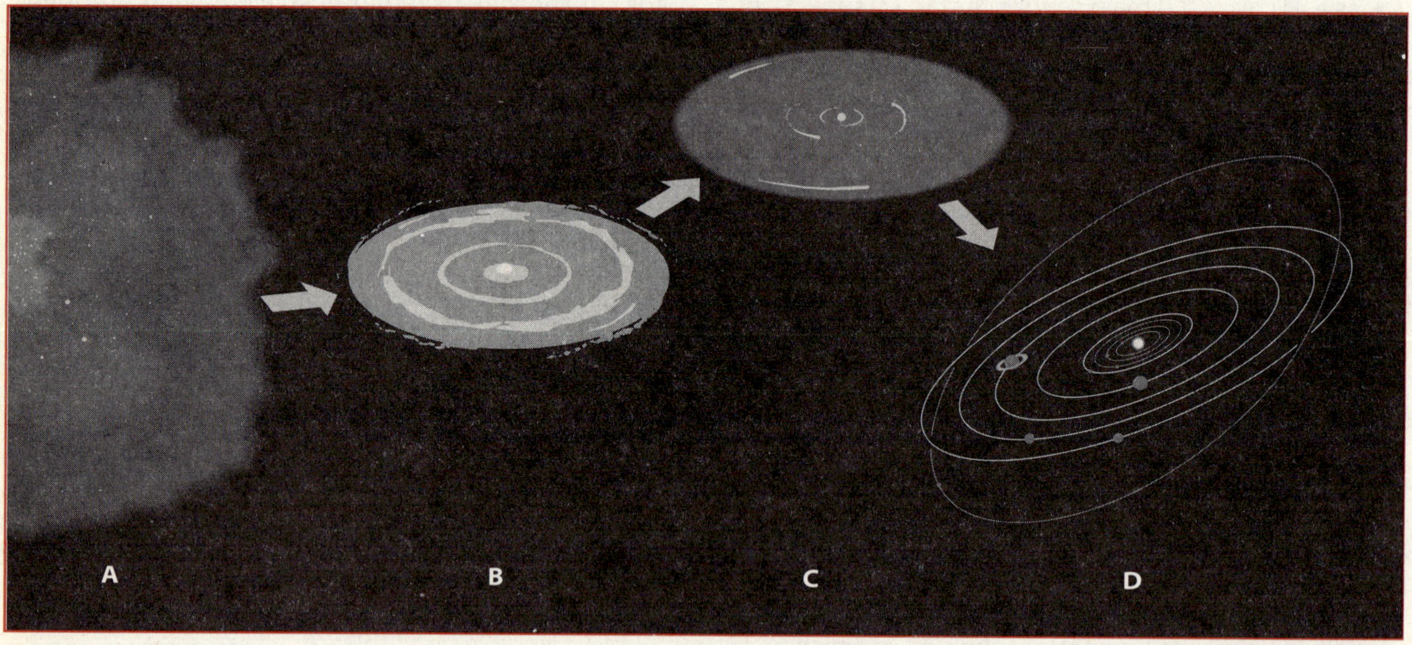

Figure 3-6. Theory of our solar system's formation: At stage A, a shock wave started a gravitational contraction of a gas-dust cloud. At stage B, most of the matter was pulled to the center to form the sun, which started nuclear fusion. Clumping of gas and dust around the sun occurred at the same time. At stage C, larger and larger clumps formed still larger masses—forming planets, moons, and asteroids. Elements having a small mass were driven out of the inner solar system leaving the terrestrial planets with only small amounts of light elements. By stage D, the solar system was similar to what it is today.

Meteorites still hit Earth, and there is concern that this type of natural disaster could destroy a large portion of the life on our planet. Presently, an intensive scientific search is being conducted to find comets, asteroids, and meteoroids that could collide with Earth and have enough mass to result in loss of life, injury, and loss of property.

The gravitational contraction of the planets and larger moons produced heat. This heat, along with the heat from radioactive decay (see Topic 5) and frictional heat from impacting meteorites, caused the newly formed planets and moons to largely melt. As a result, these bodies become layered into zones based on the density of their various elements and compounds. Topics 12 and 13 of this book have further discussions of the layering of Earth into zones, or spheres, of different density and the continuing evolution of planet Earth.

Review Questions

24. Comets are considered to belong to the solar system if they (1) glow by reflected light (2) revolve about the sun (3) have elliptical orbits (4) have uniform periods of revolution

25. Which is not included in our solar system?
(1) Polaris (2) the moon (3) meteors (4) asteroids

26. A person observes that a bright object streaks across the nighttime sky in a few seconds. What is this object most likely to be? (1) a comet (2) a meteor (3) an explosion on the sun (4) an orbiting satellite

27. A belt of asteroids is located an average distance of 503 million kilometers from the sun. Between which two planets is this belt located? (1) Mars and Jupiter (2) Mars and Earth (3) Jupiter and Saturn (4) Saturn and Uranus

28. In the last billion years when meteorites, asteroids, or comets collided with Earth, which has occurred? (1) craters have formed (2) the whole Earth's surface has melted (3) all life has been destroyed (4) the oceans have largely evaporated

29. Today it is most commonly believed that our solar system formed (1) by gravitational collapse of a gas-dust "cloud" (2) from material exploded from the sun (3) at the time of the Big Bang that formed the present universe (4) by fusion of matter between the sun and a passing star.

Planet Characteristics

A planet's distance from the sun has a major effect on its characteristics. See the Solar System Data in the *Earth Science Reference Tables* to find each planet's distance from the sun and other significant features and events. When the planets were

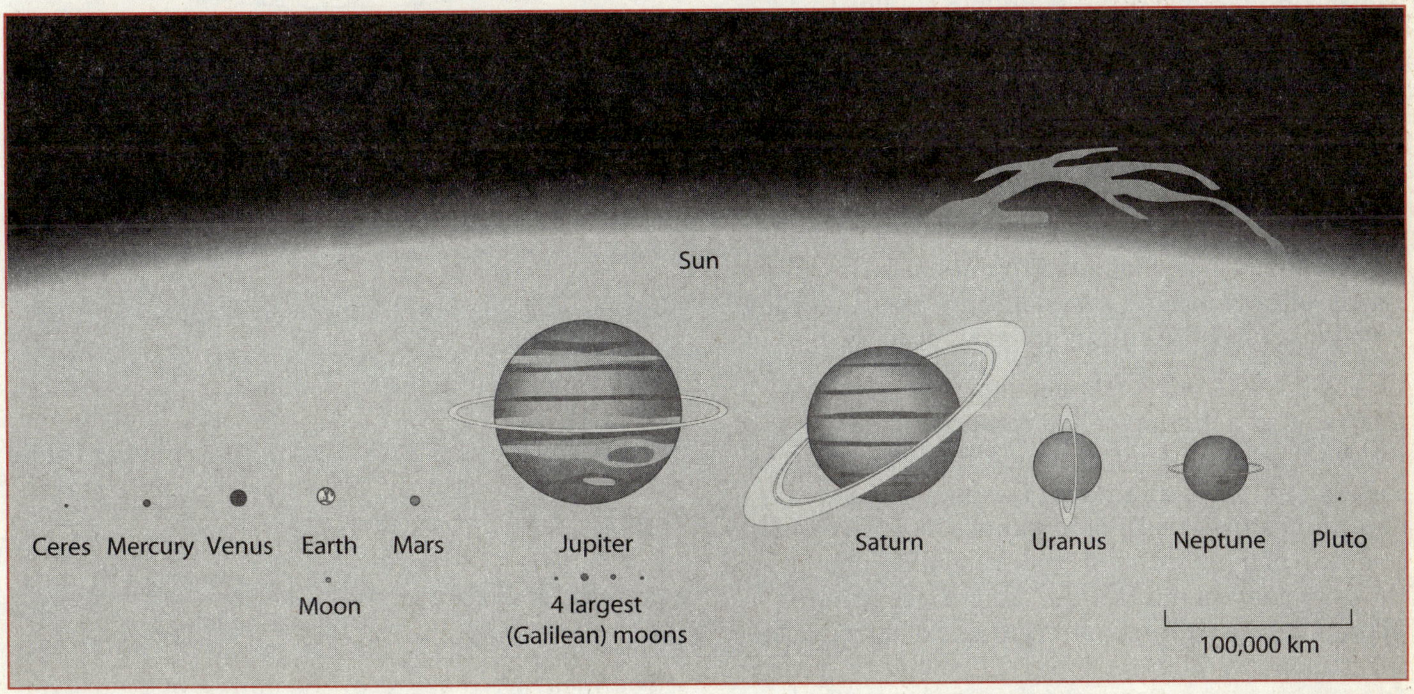

Figure 3-7. Comparison of the size (diameters) of the planets and some other solar system members: The diameter of Jupiter is about 10 times that of Earth, and the diameter of the sun is a little over 100 times that of Earth. The largest moon of Jupiter is larger than Mercury or Pluto. Ceres, with a diameter of approximately 1000 km, is the largest asteroid.

forming, the sun probably radiated more energy than it does today. The high temperatures and pressure from particles emitted by the sun drove the less dense elements and compounds away from the inner solar system. The outer parts of the solar system were not nearly so hot, and the particle pressure from the sun not nearly so great. These differences in the inner and outer solar system provide characteristics that allow the planets to be classified into the inner terrestrial planets and outer Jovian planets. Pluto is much smaller than the other planets. It has characteristics that don't fit into either classification, prompting some scientists to consider that Pluto is not a planet.

TERRESTRIAL PLANET PROPERTIES The **terrestrial planets** are close to the sun and mostly solid. They have relatively small diameters and high densities. Their rocky surfaces are dotted with impact craters. They have few or no moons and have no rings. Terrestrial planets are the planets similar to Earth and include Mercury, Venus, Earth, and Mars.

JOVIAN PLANET PROPERTIES The **Jovian planets** are far from the sun and are largely gaseous. They have relatively large diameters and low densities. They have no solid surfaces (though they may have a solid core) and thus no craters. These planets have many moons and have rings. The Jovian planets are the planets similar to Jupiter and include Jupiter, Saturn, Uranus, and Neptune.

Motions of the Planets

The planets have many different motions. They move with the solar system around the Milky Way Galaxy in periods of about 225 million years. They rotate, or spin around an imaginary axis, and they revolve around the sun in an orbit, or path.

PLANET ROTATION Planets spin on an imaginary axis in a motion called **rotation.** The period of rotation is the amount of time it takes for a planet to make one spin around its imaginary axis and determines the length of a planet's day. Seven of the nine planets rotate in the same direction as they revolve around the sun. You can tell that planets rotate through observations of surface features made at different times. The Solar System Data Table in the *Earth Science Reference Tables* lists the periods of rotation of the planets.

MEMORY JOGGER

You may recall that the length of a day on Earth is not the same as the length of a day on other planets. Since each planet has a different period of rotation, the length of a day varies from planet to planet.

PLANET REVOLUTION A planet's **revolution** is its movement around the sun in a path called an orbit. Earth's orbit (and those of the other planets) is an oval shape called an **ellipse.** Within the ellipse are two fixed points called **foci** (singular, **focus**). The sum of the distances between any point on the ellipse and the two foci is a constant; that is, the sum of those distances for one point is equal to the sum for any other point on the curve. The sun is at one of the foci of each planetary orbit in the solar system. (See Figure 3-8.) The major axis of an ellipse is the longest straight-line distance across an ellipse—the biggest diameter—and it cuts through the two foci.

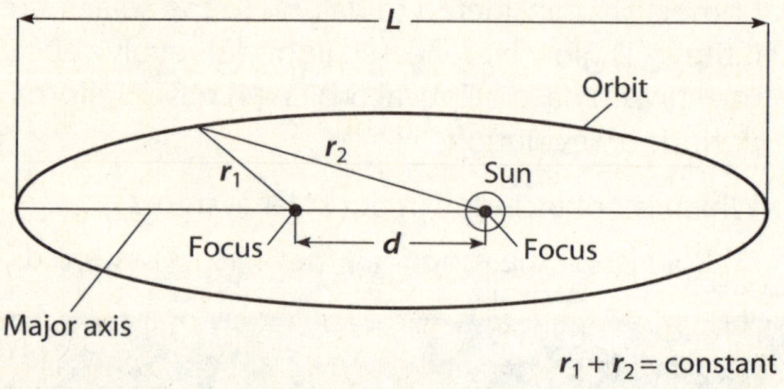

Figure 3-8. An elliptical orbit of a solar system member with the sun at one focus: Except for Mercury and Pluto, the orbits of the planets would appear to look like a circle at this scale. The eccentricity of the ellipse is 0.26 (d = 2.0 cm and L = 7.7 cm; $\frac{2}{7.7}$ = 0.26).

Eccentricity of Planet Orbits The degree of flattening or "ovalness" of an ellipse is measured by its **eccentricity.** Eccentricity can also be described as the amount of difference between an ellipse as compared to a special type of ellipse called a circle. The formula to compute eccentricity is listed below and is also found in the *Earth Science Reference Tables.*

$$\text{eccentricity of an ellipse} = \frac{\text{distance between foci}}{\text{length of major axis}} = \frac{d}{L}$$

If you measure d and L in the ellipse of Figure 3-8 and apply the formula, you should find the eccentricity to be 0.26. As the foci of an ellipse are

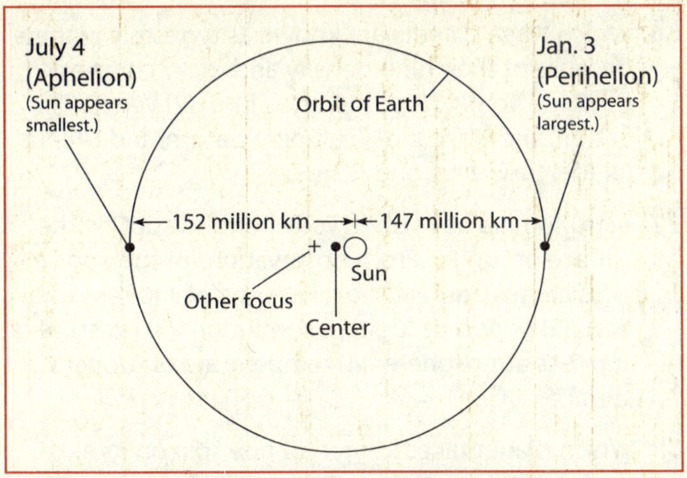

Figure 3-9. Earth's orbit drawn close to true scale: If the orbit were drawn to true scale, the two foci would be about 1 mm apart. Even at this exaggerated scale, with a greater eccentricity, the orbit appears to be a circle—thus Earth's orbit of the sun is a slightly eccentric ellipse.

brought closer together, the ellipse becomes more like a circle and the eccentricity decreases toward zero. The eccentricities of the planets are listed in the Solar System Data in the *Earth Science Reference Tables.* If you were to draw the orbits of the planets to scale they would look like circles to the human eye, except for Pluto and Mercury. As an example, Figure 3-9 shows Earth's orbit nearly drawn to scale, and it appears to be a circle.

Varying Distance of Planets From the Sun

The elliptical shape of planetary orbits causes the planets to vary in distance from the sun during a revolution. For example, Earth is 147,000,000 kilometers from the sun when closest (occurring on or about January 3). Earth is 152,000,000 kilometers from the sun when farthest away (occurring on or about July 4). The difference between these two distances is only about 5,000,000 km. It is important to note that this is NOT the factor that causes seasons on Earth.

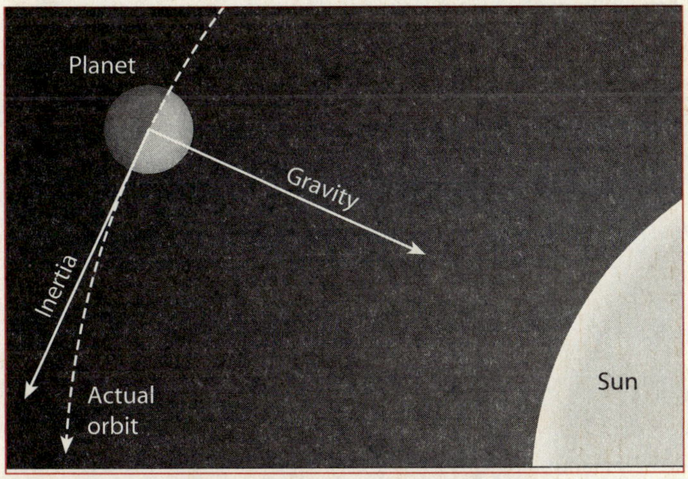

Figure 3-10. Balance between inertia and gravitation in revolution: Inertia would make a planet travel in a straight line. The mutual gravitation between the sun and a planet would pull the planet towards the sun. The actual orbit is a balance between the actions of inertia and gravity.

INERTIA, GRAVITATION, ORBITAL VELOCITY/SPEED, AND PLANET ORBITS The planets and all other bodies that orbit the sun, or any revolving body, operate under a balance between inertia and gravitation as shown in Figure 3-10. **Inertia** is the concept that an object at rest will tend to remain at rest, and that an object in motion will maintain the direction and speed of that motion unless an opposing force affects it. **Gravitation** is the attractive force that exists between any two objects in the universe. The gravitational force is proportional to the product of the masses of the objects and inversely proportional to the square of the distance between their centers. In simple terms, the greater the mass of one or both objects (such as a star and a planet), the more gravitational attraction there is between the objects. Also the closer together two objects are, the greater the gravitational attraction between them. Conversely, the farther apart two objects are, the lower the gravitational attraction between them.

The orbit of a planet around the sun, or that of any other satellite, is an example of the dynamic equilibrium, or balance, between inertia and gravitation. The shape and size of the orbit is a compromise, or balance, between the forces as illustrated in Figure 3-10.

Since each planet's orbit has some eccentricity, its distance from the sun varies during its yearly revolution. The result is that each planet's orbital speed velocity varies during its year. When a planet is closer to the sun, its orbital speed velocity is greatest, and when a planet is farthest from the sun its orbital speed velocity is slowest.

The period of revolution for a planet is the amount of time it takes the planet to make one orbit, or revolution, around the sun. This amount of time equals one year for that planet. The period of revolution of a planet is related to the planet's distance from the sun. The closer a planet is to the sun, the smaller its orbit, the shorter its period of revolution, and the faster its speed of revolution.

Review Questions

30. The Jovian planets have more gravitational pull than the terrestrial planets. Therefore, they have (1) a shorter year (2) higher average density (3) higher surface temperatures (4) more low-density gases in their atmosphere

Base your answers to questions 31 and 32 on the following graphs. The first variable mentioned is on the Y-axis and the second is on the X-axis.

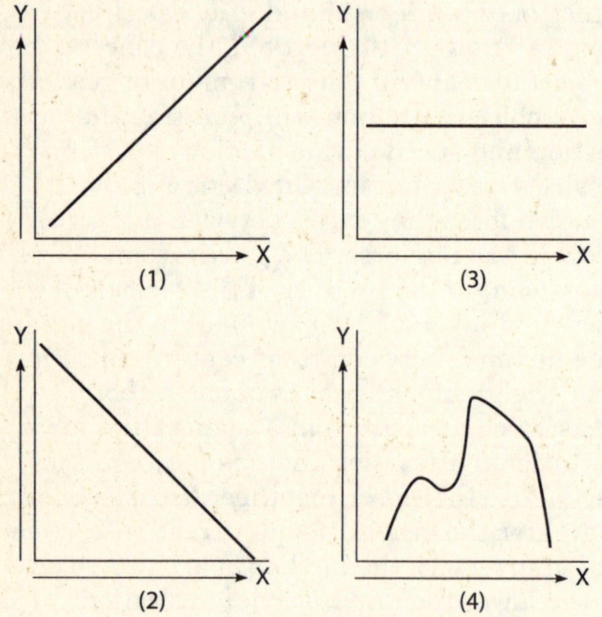

31. Which graph best illustrates the relationship between the diameter of a planet versus the distance of the planet from the sun?

32. Which graph best illustrates the relationship between the time it takes a planet to make one revolution around the sun versus the distance of the planet from the sun?

33. The planet that has the shortest day is (1) Mercury (2) Earth (3) Jupiter (4) Uranus

34. Which member of the solar system has an equatorial diameter of 3.48×10^3 kilometers? (1) the moon (2) Earth (3) the sun (4) Pluto

35. The planets known as "gas giants" or Jovians include Jupiter, Uranus, and (1) Pluto (2) Saturn (3) Mars (4) Earth

36. Which three planets are known as terrestrial planets because of their high density and rocky composition? (1) Venus, Neptune, and Pluto (2) Venus, Saturn, and Neptune (3) Jupiter, Saturn, and Uranus (4) Mercury, Mars, and Venus

37. Astronomers have observed a reddish spot on the surface of Jupiter. From observations of this spot, it is possible to estimate the (1) period of Jupiter's rotation (2) period of Jupiter's revolution (3) pressure of Jupiter's atmosphere (4) temperature of Jupiter's surface

38. Which planet takes longer for one spin on its axis than for one orbit around the sun? (1) Mercury (2) Venus (3) Earth (4) Mars

39. Which planet revolves fastest in its orbit? (1) Earth (2) Jupiter (3) Mercury (4) Pluto

40. In our solar system, the orbits of the planets are best described as (1) circular, with the planet at the center (2) circular, with the sun at the center (3) elliptical, with the planet at one of the foci (4) elliptical, with the sun at one of the foci

41. Which diagram best approximates the shape of the path of Earth as it travels around the sun?

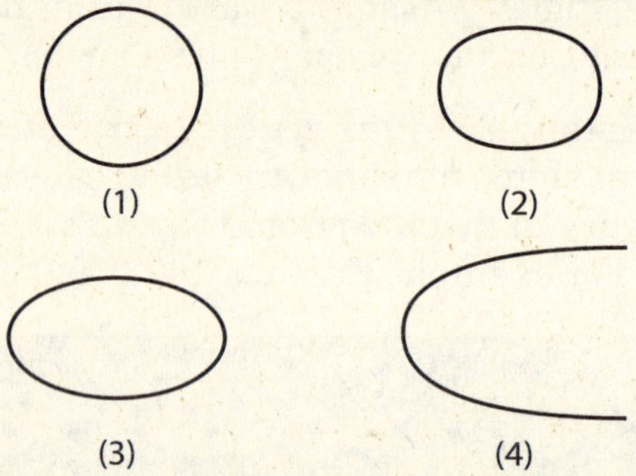

42. Based on the following diagram, what is the eccentricity of the ellipse in the diagram?

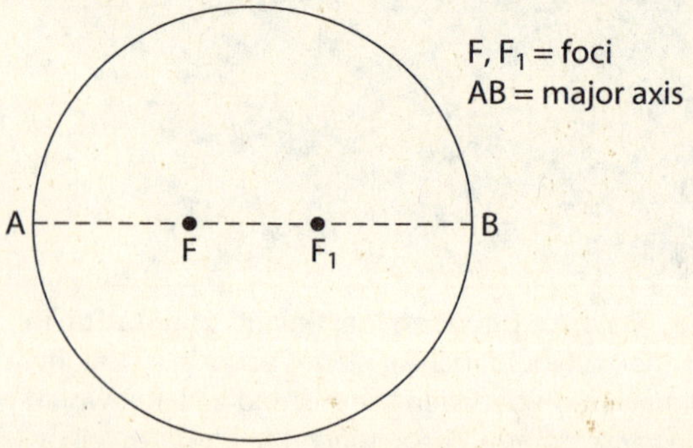

(1) 1.0 (2) 0.5 (3) 0.30 (4) 0.13

43. If the pins in the following diagram were placed closer together, the eccentricity of the ellipse being constructed would (1) decrease (2) increase (3) remain the same

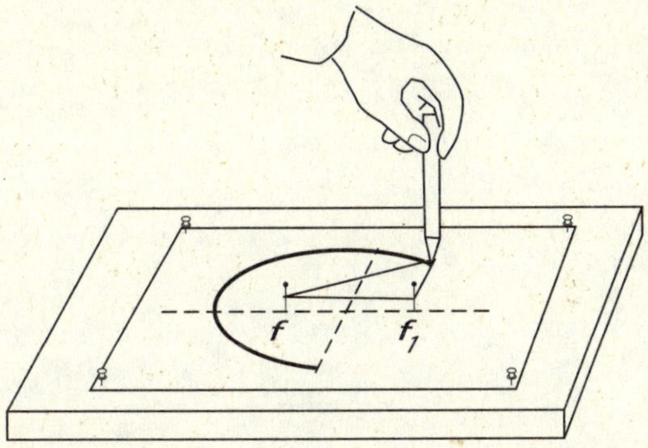

44. The following diagram represents a student's constructed laboratory drawing.

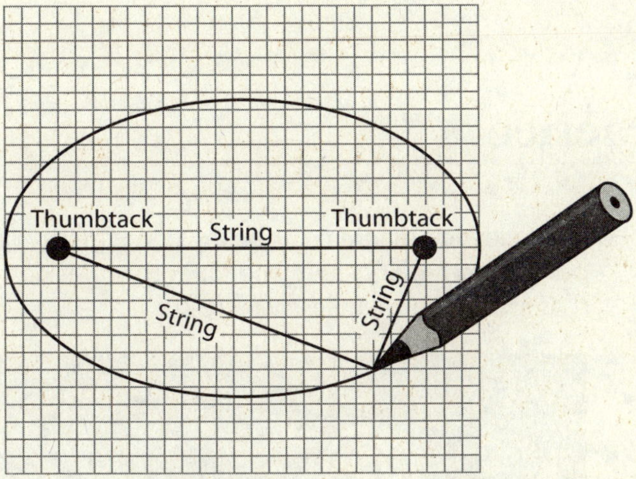

(Not drawn to scale.)

The student's drawing best represents the (1) shape of Earth's moon (2) shape of an elliptical orbit (3) path of an earthquake wave (4) path of a projectile deflected by Earth's rotation

45. The following diagram represents the construction of a model of an elliptical orbit of a planet traveling around a star.

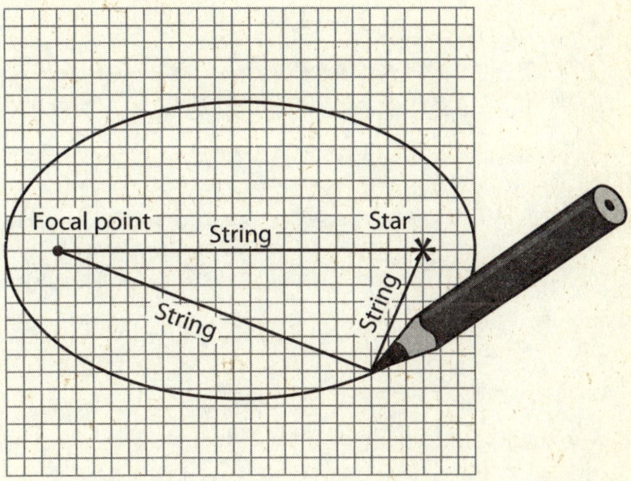

The focal point and the center of the star represent the foci of the orbit. The eccentricity of this orbit is approximately (1) 1.3 (2) 0.9 (3) 0.5 (4) 0.3

46. The following diagram shows the orbits of planets A and B in a star-planet system. The period of revolution for planet B is 40 days. The period of revolution for planet A most likely is (1) less than 40 days (2) greater than 40 days (3) 40 days

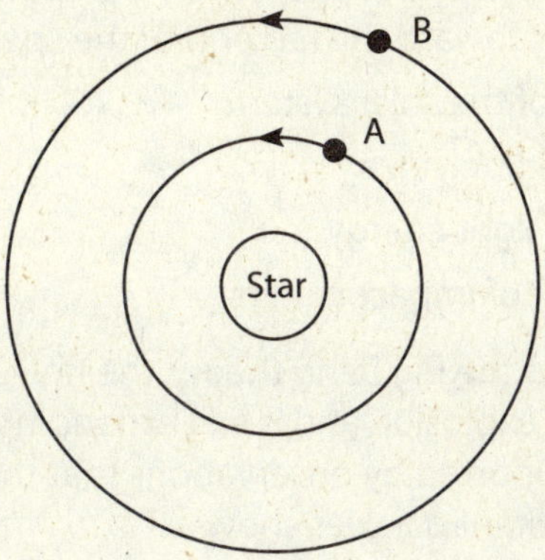

47. Which planet's orbit is most nearly circular? (1) Earth (2) Venus (3) Neptune (4) Pluto

48. As the distance between two objects in the universe increases, the gravitational attraction between these two objects (1) decreases (2) increases (3) remains the same

49. The diagram shows Earth (E) in orbit about the sun.

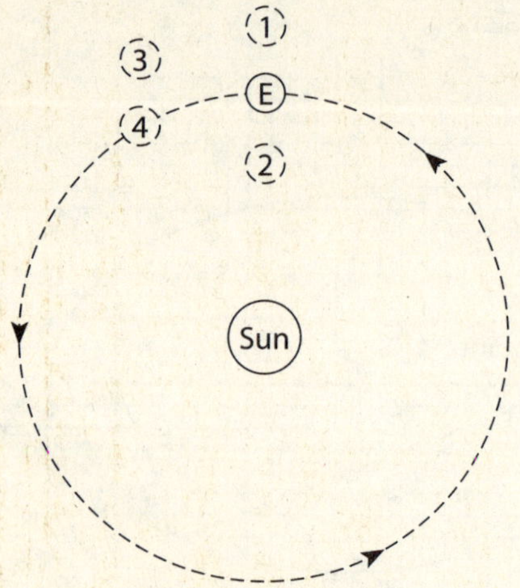

(Not drawn to scale.)

If the gravitational force between Earth and the sun were suddenly eliminated, toward which position would Earth then move? (1) 1 (2) 2 (3) 3 (4) 4

50. The force of gravity between two objects is greatest when (1) masses are small and the objects are close together (2) masses are small and the objects are far apart (3) masses are large and the objects are close together (4) masses are large and the objects are far apart.

Questions for Regents Practice

Part A

1. Which theory is best supported by a Doppler shift of spectral lines toward the red end of the spectrum?

(1) creation of the solar system

(2) expanding universe

(3) nature of light energy

(4) formation of impact craters

2. According to the Big Bang theory, the universe began as an explosion and is still expanding. This theory is supported by observations that the stellar spectra of distant galaxies show a

(1) concentration in the yellow portion of the spectrum

(2) concentration in the green portion of the spectrum

(3) shift toward the blue end of the spectrum

(4) shift toward the red end of the spectrum

3. The following diagram shows the home galaxy of our sun and solar system and four other galaxies that are trillions of miles away.

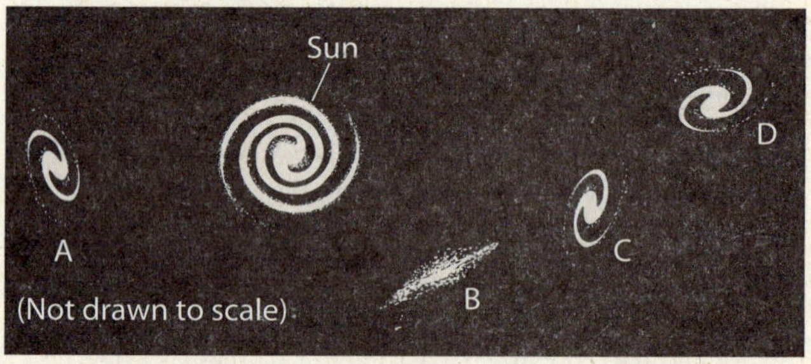

Which galaxy is most likely moving away from the sun at the greatest velocity?

(1) A (2) B (3) C (4) D

4. The following diagram represents the Milky Way Galaxy.

Which letter best represents the location of Earth's solar system?

(1) A (3) C
(2) B (4) D

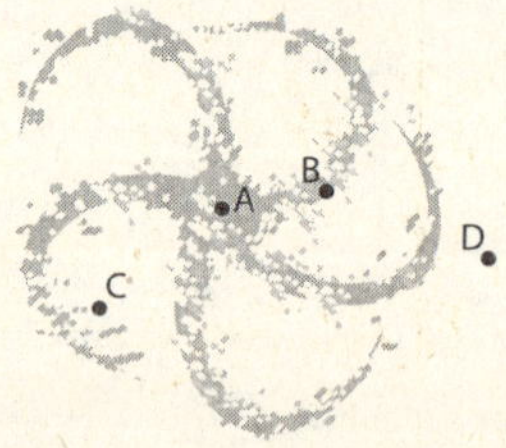

5. A star of high surface temperature and low luminosity (relative absolute magnitude) is most likely a
(1) giant star
(2) main sequence star
(3) supergiant star
(4) white dwarf star

6. Why does a star with a high amount of matter tend to have a short life?
(1) It rotates rapidly.
(2) It consumes its fuel rapidly.
(3) It has a core of heavy elements.
(4) It collapses to form a white dwarf.

7. What is the single most important factor controlling the evolution of a star?
(1) mass
(2) rotation rate
(3) element composition
(4) temperature

8. Which of the following pairs of characteristics is most important in determining the type of a planet's atmosphere?
(1) orbital speed and volume
(2) rate of rotation and diameter
(3) mass of the planet and distance from the sun
(4) density of the planet and the number of moons

9. Which planet has traveled around the sun more than once in your lifetime?
(1) Mars
(2) Uranus
(3) Neptune
(4) Pluto

10. As the distance from the sun increases, the diameter of the planets
(1) increases only
(2) decreases only
(3) both increases and decreases
(4) remains the same

11. In a scale model of the solar system, if Earth's diameter were 1 meter, the sun's diameter would be about
(1) 10 meters
(2) 100 meters
(3) 1,000 meters
(4) 10,000 meters

12. The diagram below represents Earth.

Which diagram best represents Mars, drawn to the same scale?

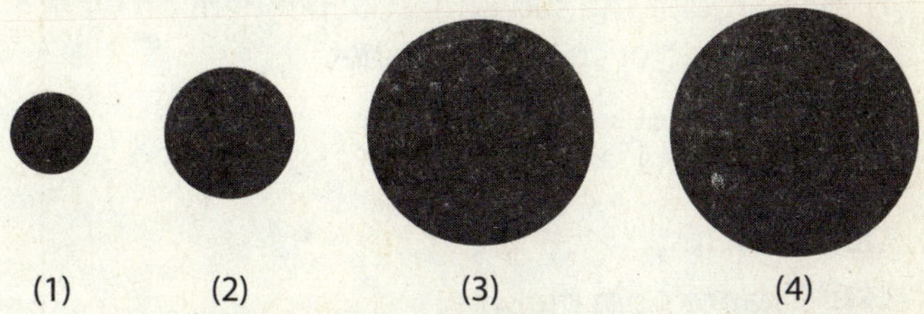

13. If the planets were all on the same side of the sun, which line best shows the correct spacing between the inner planets and their distances from the sun at a scale of 1 centimeter = 0.2 AU? (An AU or astronomical unit is Earth's average distance from the sun; the sun and planets are represented by points of identical sizes, but distances between them are drawn to scale.)

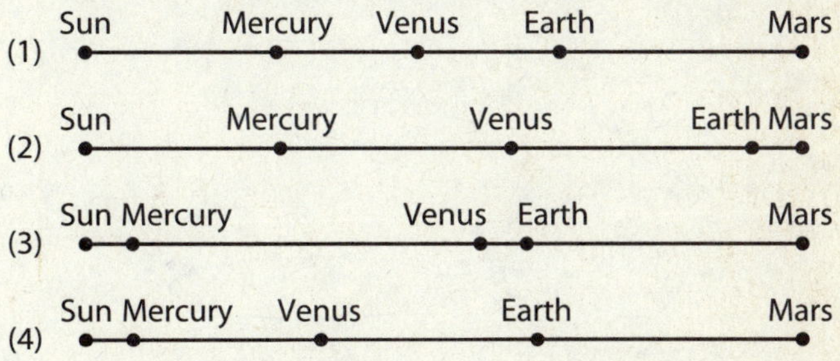

14. A planet was viewed from Earth for several hours. The following diagrams represent the appearance of the planet at four different times.

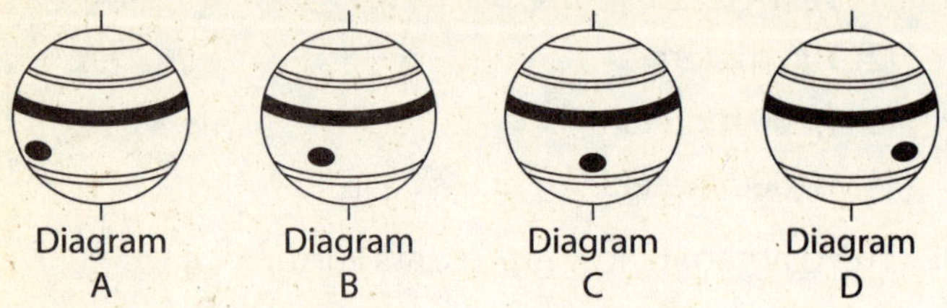

The inference that can be made based on the diagrams is that this planet is

(1) tilted on its axis

(2) changing seasons

(3) revolving

(4) rotating

15. The period of time a planet takes to make one revolution around the sun is most dependent on what other characteristic of the planet?

(1) rotation rate

(2) mass

(3) diameter

(4) distance from the sun

16. Which diagram shows a planet with the least eccentric orbit?

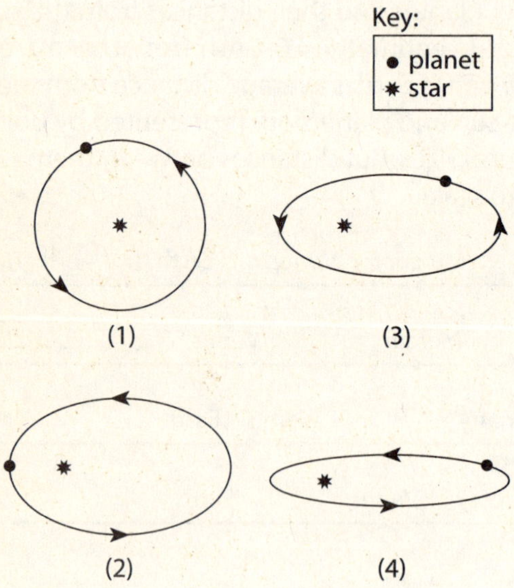

17. The following diagram shows a planet's orbit around the sun.

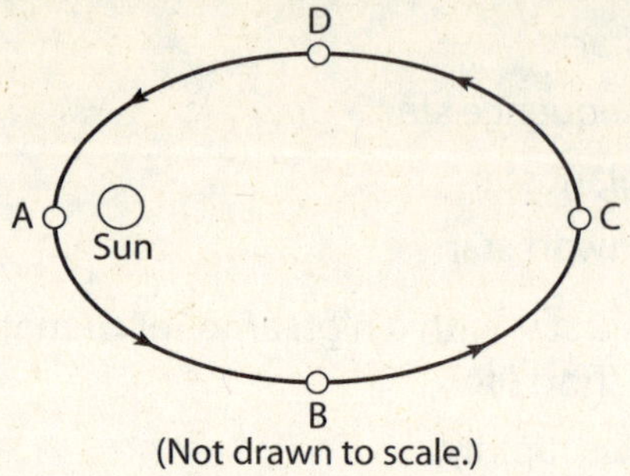

(Not drawn to scale.)

At which location is the planet's orbital speed greatest?

(1) A

(2) B

(3) C

(4) D

18. The diagram below represents a planet revolving in an elliptical orbit around the star.

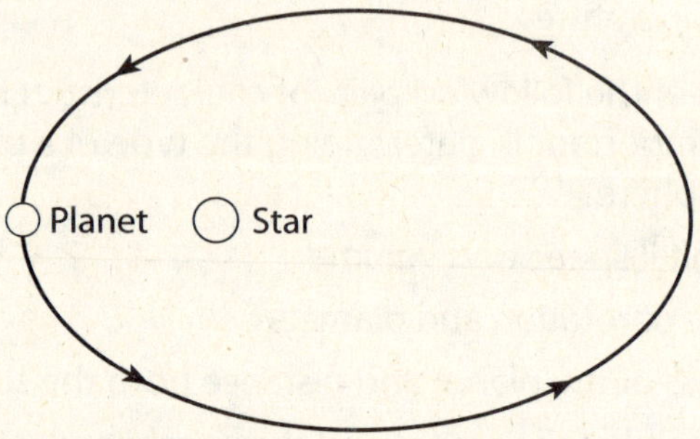

As the planet makes one complete revolution around the star, starting at the position shown, the gravitational attraction between the star and the planet will

(1) decrease, then increase

(2) increase, then decrease

(3) continually decrease

(4) remain the same

19. Planet A has a greater mean distance from the sun than planet B. On the basis of this fact, which further comparison can be correctly made between the two planets?

(1) Planet A is larger.

(2) Planet A's revolution period is longer.

(3) Planet A's speed of rotation is greater.

(4) Planet A's day is longer.

Part B

20. What is the difference between a galaxy and a constellation? [1]

21. How is a planet different from a moon? [1]

Base your answers to questions 22 through 25 on the tables below. Table 1 shows the average distance from the sun in astronomical units (AU) and the average orbital speed in kilometers per second (km/s) of the nine planets in our solar system. Table 2 lists five large asteroids and their average distances from the sun.

Table 1

Planet	Average Distance From the Sun (AU)	Average Orbital Speed (km/s)
Mercury	0.4	48.0
Venus	0.7	35.0
Earth	1.0	30.0
Mars	1.5	24.0
Jupiter	5.2	13.0
Saturn	9.6	10.0
Uranus	19.0	7.0
Neptune	30.0	5.1
Pluto	39.0	4.7

Table 2

Asteroid	Average Distance From Sun (AU)
Ceres	2.8
Pallas	2.8
Vesta	2.4
Hygiea	3.2
Juno	2.7

22. On the following grid, plot the average distance from the sun and the average orbital speed for each of the nine planets listed in Table 1. Connect the nine points with a line. [2]

Planet's Average Orbital Speed Versus Average Distance From Sun (AU)

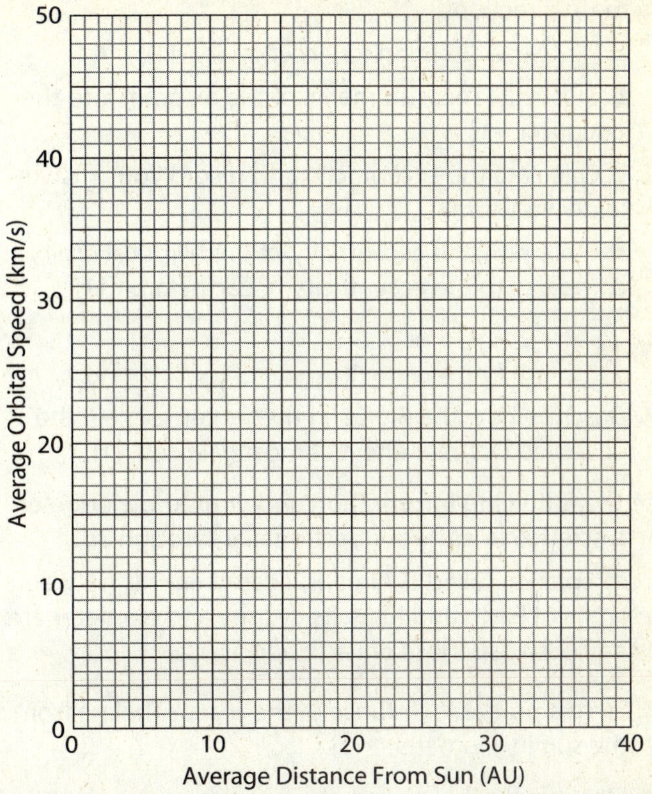

23. Describe the relationship between a planet's average distance from the sun and the planet's average orbital speed. [1]

24. The orbits of the asteroids listed in Table 2 are located between two adjacent planetary orbits. State the names of the two planets. [1]

25. Based on its distance from the sun, which of the asteroids listed in Table 2 has the greatest average orbital speed? [1]

26. The following diagram represents the elliptical orbit of a spacecraft around the sun.

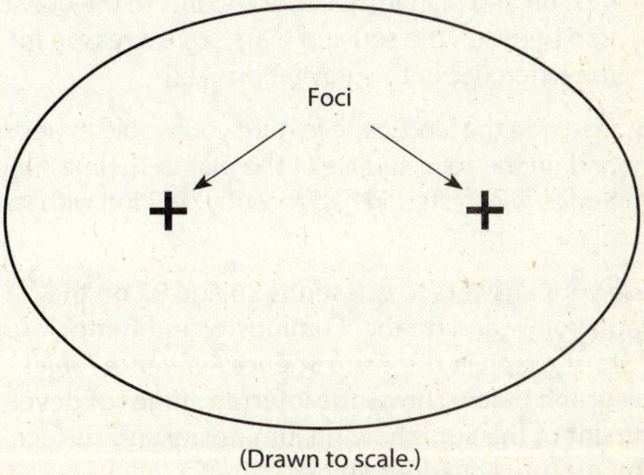

(Drawn to scale.)

Calculate the eccentricity of the spacecrafts orbit following these directions:

a. Write the equation for eccentricity. [1]

b. Subsitute measurements of the diagram into the equation. [1]

c. Calculate the eccentricity and record your answer in decimal form. [1]

d. State how the eccentricity of this elliptical orbit compares to the eccentricity of Earth's orbit. [1]

Part C

27. Describe how the Big Bang theory can explain the observed Doppler effect of most galaxies. [1]

Use the following information and your knowledge of earth science to answer questions 28 through 35.

A tenth planet, called Thelma, is located exactly between the orbits of Earth and Mars. Its density and diameter are halfway between those of Earth and Mars.

28. State a probable distance of the planet Thelma from the sun in kilometers. [1]

29. State a probable period of revolution of the planet Thelma using proper units. [1]

30. What is the most likely shape of the planet Thelma? [1]

31. In years, what is the oldest possible rock that would likely be found on planet Thelma? [1]

32. What information would you need to know to be able to determine the length of a day on the planet Thelma? [1]

33. Explain why the planet Thelma would likely be considered a terrestrial planet and NOT a Jovian planet. [1]

34. Explain how the magnitude of gravitation between the sun and planet Thelma compares to the gravitation between the sun and Mars. State a reason for the difference in the gravitations. [2]

35. Describe the landscape feature you would expect to find on the solid surface of the planet Thelma if a 5-mile-wide asteroid had recently collided with the planet. [1]

Base your answers to questions 36 and 37 on the graph below and on the "Luminosity and Temperature of Stars" graph in the *Earth Science Reference Tables.* The graph below shows the inferred stages of development of the Sun, showing luminosity and surface temperature at various stages.

Inferred Stages of Development

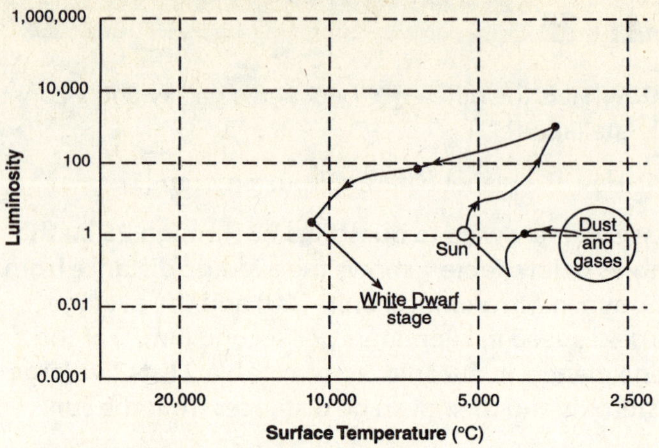

36. Describe the changes in luminosity of the Sun that will occur from its current Main Sequence stage to its final White Dwarf stage. [1]

37. Which star shown on the "Luminosity and Temperature of Stars" graph in the *Earth Science Reference Tables* is currently at the Sun's final predicted stage of development? [1]

Base your answers to questions 38 and 39 on your knowledge of Earth science and on the newspaper article shown below, written by Paul Recer and printed in the *Times Union* on October 9, 1998.

Astronomers peer closer to big bang

WASHINGTON—The faintest and most distant objects ever sighted—galaxies of stars more than 12 billion light years away—have been detected by an infrared camera on the Hubble Space Telescope.

The sighting penetrates for the first time to within about one billion light years of the very beginning of the universe, astronomers said, and shows that even at that very early time there already were galaxies with huge families of stars.

"We are seeing farther than ever before," said Rodger I. Thompson, a University of Arizona astronomer and the principal researcher in the study.

Thompson and his team focused an infrared instrument on the Hubble on a narrow patch of the sky that had been previously photographed in visible light. The instrument detected about 100 galaxies that were not seen in the visible light and 10 of these were at extreme distance.

He said the galaxies are seen as they were when the universe was only about 5 percent of its present age. Astronomers generally believe the universe began with a massive explosion, called the "big bang," that occurred about 13 billion years ago.

Since the big bang, astronomers believe that galaxies are moving rapidly away from each other, spreading out and becoming more distant.

38. The big-bang theory is widely believed by astronomers to explain the beginning of the universe. Why does the light from distant galaxies support the big-bang theory? [1]

39. Compare the age of Earth and our solar system to the age of these distant galaxies of stars. [1]

Motions of Earth, Moon, and Sun

TOPIC 4

VOCABULARY		
axis (of rotation)	Foucault pendulum	phases (of the moon)
constellation	geocentric model	tides
Coriolis effect	heliocentric model	time zone
eclipse	local time	

Throughout time, most cultures have attempted to understand the nature and the motions of celestial objects—the stars, sun, moon, planets. Many of the explanations of the nature and cause of motions of celestial objects have become part of the belief systems of past and present cultures.

In this topic you will study the nature and motion of celestial objects in relationship to Earth and our satellite, the moon. Some scientists consider Earth and its moon to be a double planet, but for the purpose of this book, the moon will be considered a satellite of Earth, and Earth a satellite of the sun.

Apparent Motions of Celestial Objects

An apparent motion is a motion that an object appears to make. Apparent motions can be real or illusions. When you observe a person spinning around, that motion is real. However, when the spinning person sees the room as spinning, that is an illusion and not a real motion. It has always been difficult for people looking into space to distinguish real motions of celestial objects from illusions. For example, the stars appear to move across the sky from east to west. However, this apparent motion is caused by Earth's rotation.

Daily Motion and Stars

Most celestial objects appear to move across the sky, rising in the eastern part of the sky and setting in the western part of the sky. The vast majority of the 6000 or so celestial objects visible with the unaided eye at any location are stars. Some of the stars near the North Star, or Polaris, appear to move in a complete circle in 24 hours. The paths of all celestial objects moving in the sky are circular, or parts of a circle called an arc. All the motion occurs at a constant rate of approximately 15° per hour or 360° in 24 hours. These movements of celestial objects over a 24-hour period are called daily motion. (See Figure 4-1.)

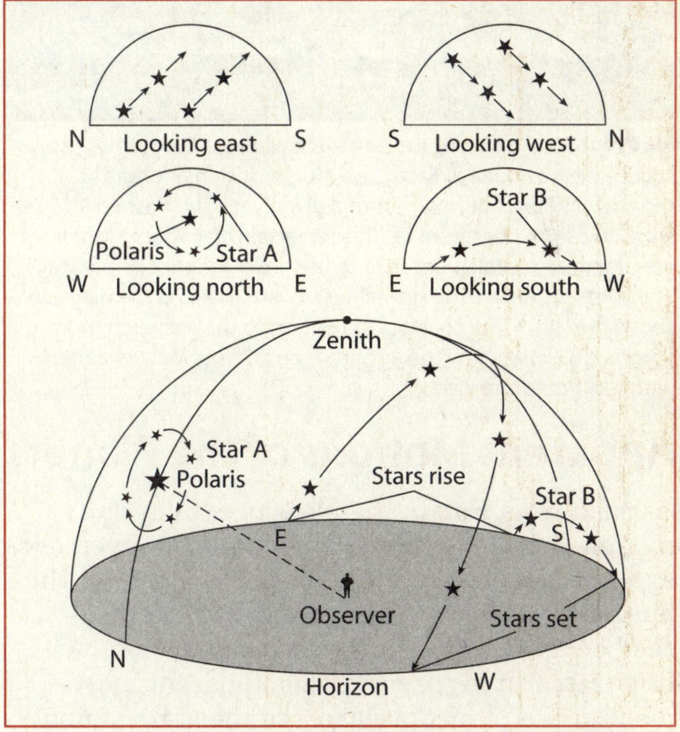

Figure 4-1. Apparent daily motion of stars in middle latitudes: To an observer in the mid-latitudes of the Northern Hemisphere—such as New York State—most stars appear to move from east to west in circular paths, or along arcs. The complete circular path can be seen for stars in the northern part of the sky around Polaris. Other stars rise over the eastern horizon and set at the western horizon in the middle latitudes.

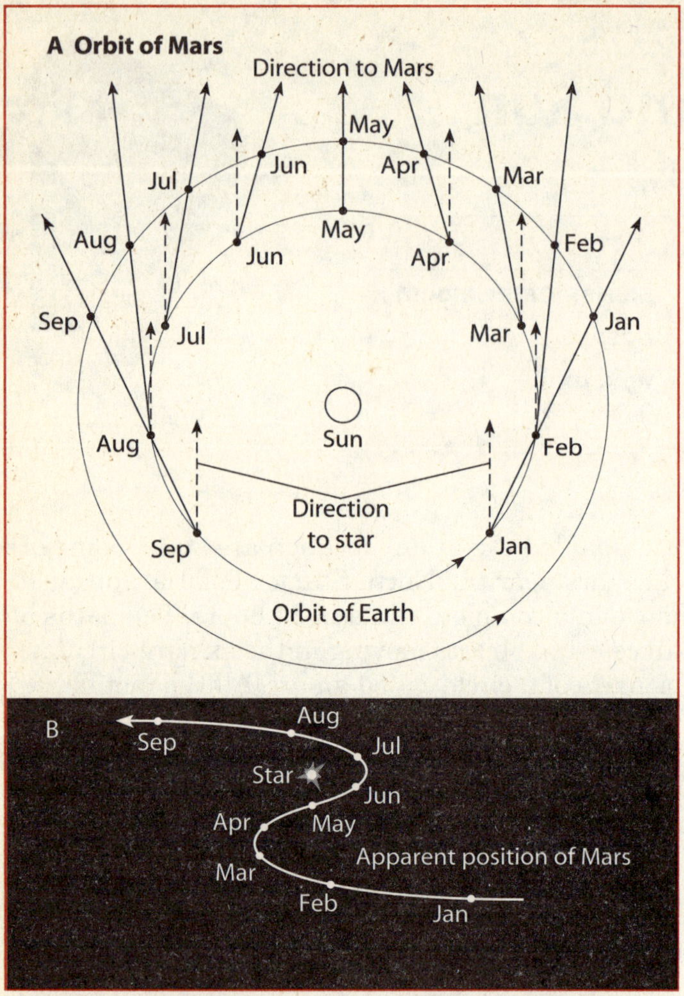

Figure 4-2. Apparent motions of a planet: Diagram A shows successive positions of Earth and Mars along their orbits at monthly intervals during part of a given year. The dashed arrows show the sight lines from Earth to a certain star. These sight lines are practically parallel because of the great distance of the star. The solid arrows are the sight lines toward Mars. As the diagram shows, the sight line changes in relation to the star each month. Diagram B shows how the apparent position of Mars changed with respect to the star.

Apparent Motions of the Planets

As seen from Earth, the planets exhibit daily motion similar to that of the stars. However, over extended periods of time (weeks to months), the planets appear to change position with respect to the background field of stars around them. As illustrated in Figure 4-2, this apparent movement of the planets relative to the stars is not uniform and often appears complicated. Each year, a similar motion would be observed, but in a different region of the sky. From a central sun perspective, the complicated motion of the planets is the result of Earth and other planets revolving around the sun in different orbits at different speeds. This causes other planets viewed from Earth to sometimes appear to make loops and move back and forth.

Apparent Motions of Earth's Moon

The moon also follows the daily east-to-west motion of the stars. However, it appears to rise about 50 minutes later each day and shifts eastward each day compared to the background field of stars.

Apparent Motions of the Sun

Like all other visible celestial objects, the sun also seems to move in the sky. The sun's apparent path, from sunrise to sunset, has the shape of an arc. As shown in Figure 4-3, the sun's path changes both its position and its length with the seasons. The greater the length of the sun's path over an area, the more hours of daylight the area has. In summer, daylight is longest, in winter shortest, and in spring and fall in-between—approximately 12 hours. Figure 4-3 also shows how the position of the sun at sunrise and sunset varies predictably with the seasons. In spring and fall, the sun rises

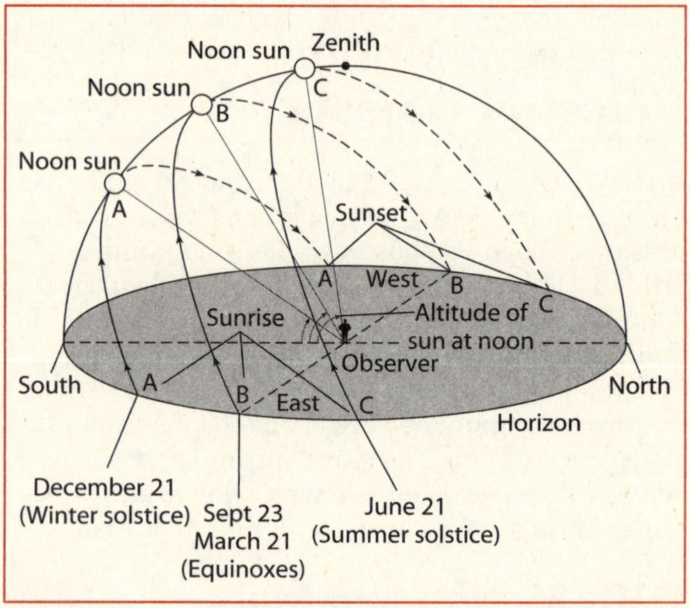

Figure 4-3. The changing apparent path of the sun in the mid-latitudes: Note that the length of the sun's apparent path is longest on June 21, shortest on December 21, and in-between on March 21 and September 23. Also, the altitude of the sun at noon is highest on June 21, shortest on December 21, and in-between on September 23 and March 21.

due, or exactly, east and sets due west. In summer, the sun rises north of east and sets north of west. In the winter, the sun rises south of east and sets south of west.

CHANGES IN THE ALTITUDE OF THE SUN AT NOON The sun always reaches its highest position in the sky at local solar noon (See page 60.) However, the altitude of the sun at noon depends on the time of year and the latitude of the observer. Only between latitudes $23\frac{1}{2}°$ N and $23\frac{1}{2}°$ S can the noon sun be directly overhead at an altitude of 90°. Thus, the noon sun is never directly overhead anywhere in the continental United States. Figure 4-4 shows how the latitude at which the sun is overhead at noon changes during a year. Topic 6 has a more detailed discussion of seasonal changes and the sun.

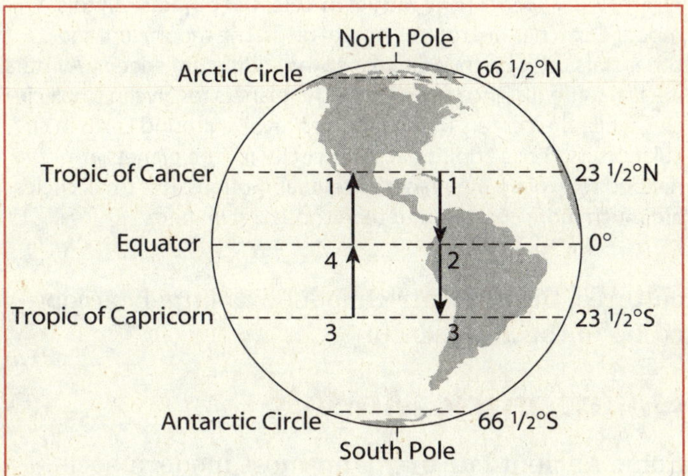

Figure 4-4. The yearly changes in latitude where the sun is directly overhead at noon—at an altitude of 90°: Point 1 is June 21, point 2 is September 23, point 3 is December 21, and point 4 is March 21.

Review Questions

1. If Earth rotated from north to south, the North Star would appear to (1) set in the south (2) set in the north (3) move in a circle in the sky (4) remain stationary

2. How would a three-hour time exposure photograph of stars in the northern sky appear if Earth did not rotate?

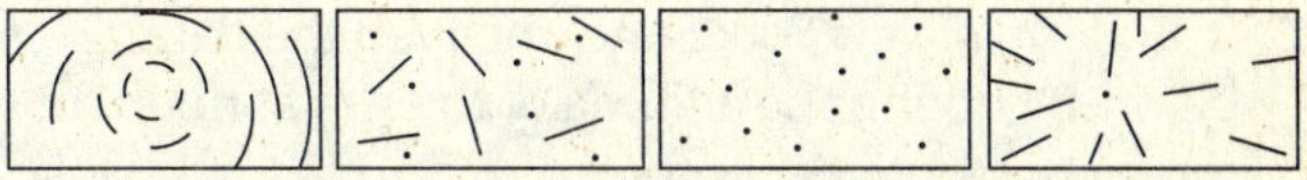

3. An observer took a time-exposure photograph of Polaris and five nearby stars. How many hours were required to show these star paths?

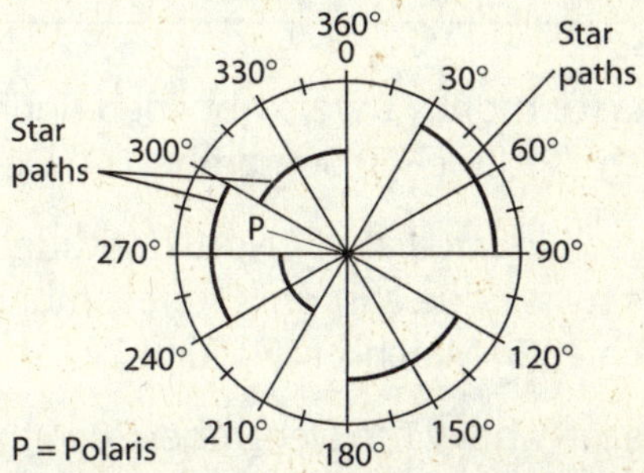

(1) 6 (2) 2 (3) 8 (4) 4

4. When observed from a location in New York State for an entire night, the North Star (Polaris) appears to (1) rise in the east and set in the west (2) rise in the west and set in the east (3) move southward along an arc-shaped path (4) remain stationary in the sky

5. Explain why most stars seem to move from east to west across the sky in New York State.

6. The star Sirius is observed in the evening sky during the month of January. At the end of 3 hours, Sirius will appear to have moved (1) 60° (2) 45° (3) 3° (4) 0°

7. On December 21, at which latitude would an observer find the sun directly overhead? (1) 0° (2) $23\frac{1}{2}°$ North (3) $23\frac{1}{2}°$ South (4) 90° South

8. The sun's apparent daily movement across the sky is caused by (1) the sun's revolution around Earth (2) Earth's revolution around the sun (3) the sun's rotation on its axis (4) Earth's rotation on its axis

9. At which location below would the sun produce an angle of 23.5° relative to an observer's head, at noon June 21st? (1) 0° (2) 23.5°N (3) 23.5°S (4) 66.5°N

10. At 40° N latitude, for how many days a year is the sun directly overhead at noon? (1) 1 (2) 2 (3) 3 (4) 0

11. In New York State, to see the sun at noon, one would look towards the (1) north (2) south (3) east (4) west

12. A student in New York State obtained the data below by noting the altitude of the sun at noon for three consecutive months.

Month	Altitude of Sun at Noon
X	27.0°
Y	23.5°
Z	25.5°

The data for month Y were obtained during (1) March (2) June (3) September (4) December

13. To a person located at 43° North latitude, the sun appears to rise due east on (1) December 22 (2) March 1 (3) March 21 (4) June 22

14. From September 21 to December 20 in the United States, the direction in which the sun sets
(1) is always directly east (2) is always directly west
(3) moves from the east towards the north
(4) moves from the west towards the south

Models That Help Explain Apparent Celestial Motions

For thousands of years, people have developed theories to explain their observations of the movements of celestial objects. An explanation of two of these theories follows.

Geocentric Models

A majority of early cultures believed what is the most obvious way to explain the daily motion of celestial objects. They assumed that Earth is stationary, and that most celestial objects revolve around it. This Earth-centered model is called the **geocentric model.** In many geocentric models, the moon and sun were thought to travel at slightly different speeds in their orbits of Earth. The geocentric model explained most of the motions of the stars, sun, and moon, but it did not explain the apparent motions of the planets. One version of the geocentric model proposed that the planets moved around in circles as the planets revolved around Earth. The geocentric models also could not explain some Earth motions and the behavior of a freely swinging pendulum. In European cultures, the geocentric model was largely accepted up to the 16th century.

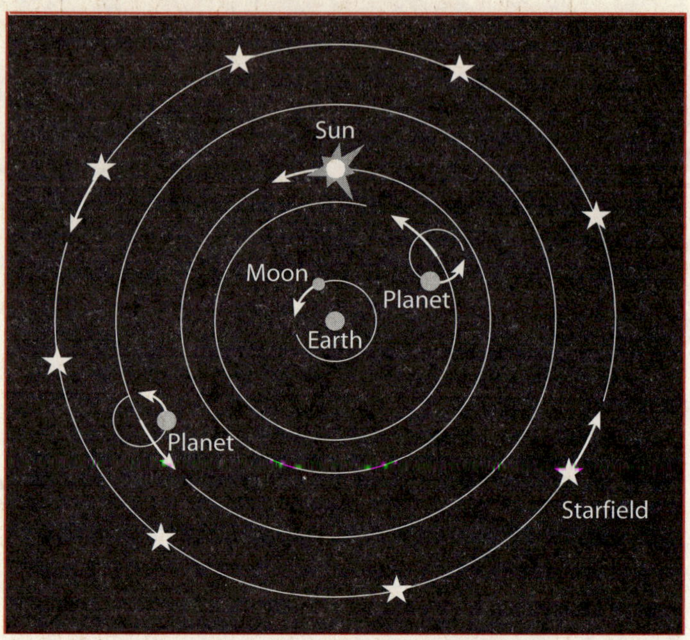

Figure 4-5. Geocentric model of celestial objects: In this model, Earth neither rotates nor revolves. The moon, sun, and other stars orbit Earth in circular paths at different speeds. All stars are the same distance from Earth. The planets revolve in small circles, while the centers of these circles revolve around Earth in circular orbits. These additional small circles for the planets are needed to explain the planet's irregular motions. The small circles are much more complicated than those shown here.

Heliocentric Models

Some ancient cultures and most modern societies use a **heliocentric model**—sun-centered model—to explain the apparent motions of celestial objects in Earth's sky. In all heliocentric models, Earth both rotates on an imaginary axis and revolves around the sun in an orbit. The moon revolves around Earth as Earth revolves around the sun. The daily motion of stars is explained by rotation of Earth. The motions of the moon are explained by the rotation of Earth, the revolution of Earth around the sun, and the revolution of the moon around Earth. The apparent motions of the sun are explained by the rotation of Earth on a tilted axis and the revolution of Earth around the sun.

Because they used circular orbits for the planets, most early heliocentric models did no better in predicting the motions of the planets than did the geocentric views. When elliptical orbits and varying orbital speeds were added to

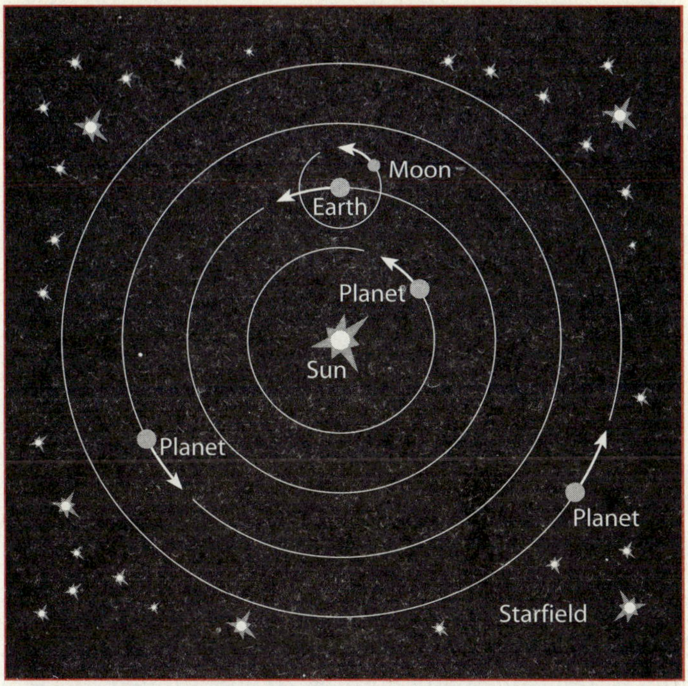

Figure 4-6: A modern heliocentric model: The sun is the center of motion. Earth and other planets rotate and revolve around the sun at different speeds in elliptical orbits with some eccentricity. The stars don't revolve around Earth and are at varying distances from the solar system.

the heliocentric model, the motions of the planets could be predicted. The heliocentric model did explain the behavior of a freely swinging pendulum.

MEMORY JOGGER

You may recall from Topic 3 that eccentricity can be described as the amount of difference between an ellipse as compared to a circle. With the exception of Pluto and Mercury, the slightly elliptical orbits of the planets, drawn to scale, look like circles to the human eye.

Review Questions

15. In the geocentric model, which motion would occur? (1) Earth would revolve around the sun. (2) Earth would rotate on its axis. (3) The moon would revolve around the sun. (4) The sun would revolve around Earth.

16. Which diagram represents a geocentric model?

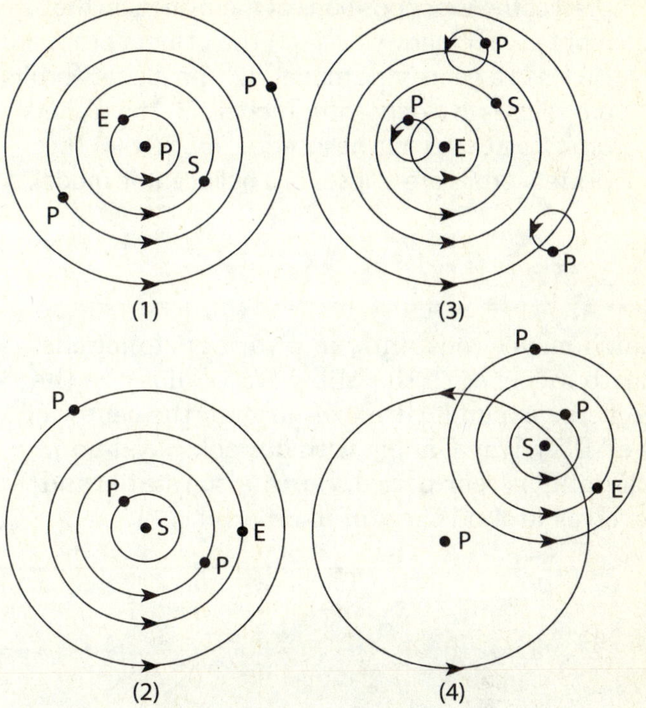

Key: E = Earth, P = Planet, S = Sun

17. The apparent rising and setting of the sun as seen from the Earth are caused by the (1) rotation of the sun (2) rotation of the Earth (3) revolution of the Earth (4) revolution of the sun

18. Which diagram best represents a portion of the heliocentric model of the solar system?

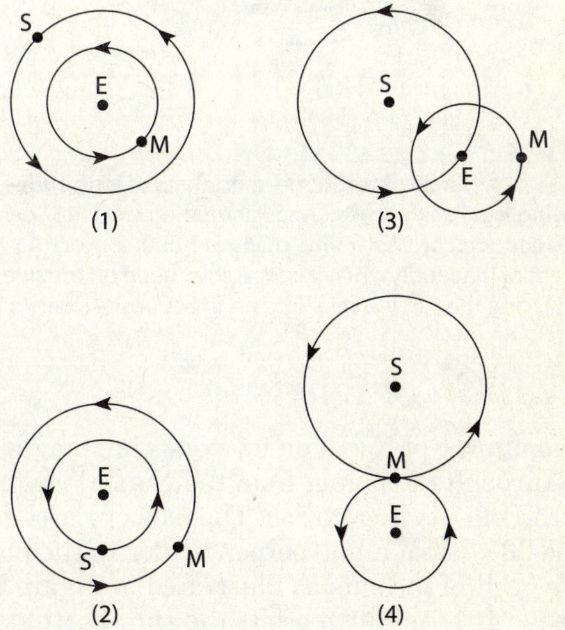

S = Sun, E = Earth, and M = Moon

19. Which planetary orbit model allows a scientist to predict the exact positions of the planets in the night sky for many years? (1) The planets' orbits are circles in a geocentric model. (2) The planets' orbits are ellipses in a geocentric model. (3) The planets' orbits are circles in a heliocentric model. (4) The planets' orbits are ellipses in a heliocentric model.

Actual Earth Motions

Earth moves constantly in a variety of motions. Earth moves with the Milky Way Galaxy as the universe expands. It moves around the center of the Milky Way Galaxy with our solar system in a 225-million-year period. Earth also rotates and revolves around the sun in an orbit.

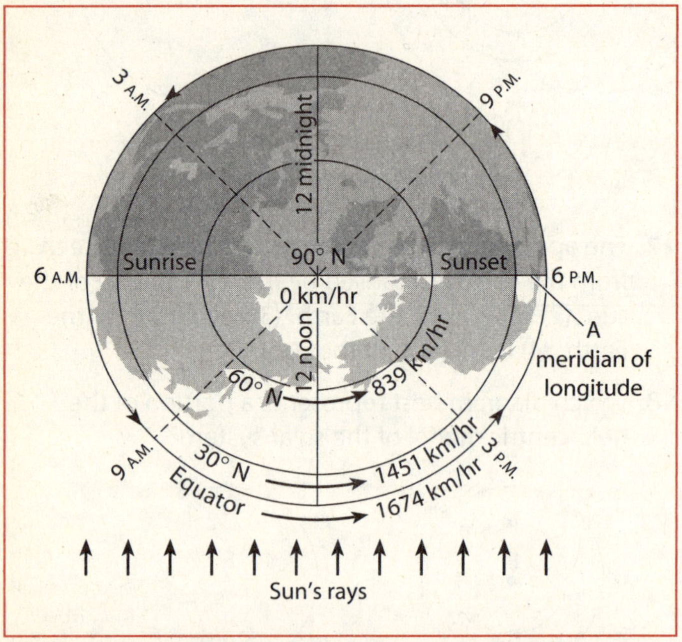

Figure 4-7. Circle of rotation—meridians of longitude—showing local solar times: Angular rotation rate is 15° per hour at any latitude. Local solar time changes 1 hour for each 15 degrees of longitude. When it is local solar noon on the side of Earth facing the sun, it is midnight on the opposite side.

Rotation of Earth

The spinning of Earth on its **axis**—the imaginary line through the planet from the North Pole to the South Pole—is its rotation. The axis of Earth is tilted $23\frac{1}{2}°$ from a line perpendicular to the plane of its orbit of the sun, as illustrated in Figure 6-16 on page 101. As Earth orbits the sun, Earth's axis remains tilted at $23\frac{1}{2}°$. The north end of the axis points towards the North Star, or Polaris, as shown in Figure 2-5 on page 19.

Earth rotates 360° from west to east in 24 hours, at an angular rate of 15° per hour. This angular rate is the same at all locations on Earth. Looking down over the North Pole, the direction of Earth's rotation is west to east, —counterclockwise—as shown in Figure 4-7.

Evidence of Earth's Rotation

Throughout much of human history, scientists have searched for evidence that Earth rotates. Earth's movement is not apparent because people don't feel very smooth motions the same way that they are aware of abrupt motions. Also, there is nothing near Earth with which to compare its motion. Only in recent history have people and satellites been able to observe the rotation of Earth.

THE FOUCAULT PENDULUM When a **Foucault pendulum** is allowed to swing freely (first observed in 1851), its path will appear to change in a predictable way, as shown in Figure 4-8. This is evidence of Earth's rotation because the pendulum—due to inertia—would continue to swing in the original path if Earth did not rotate. (See A-A′ in Figure 4-8.)

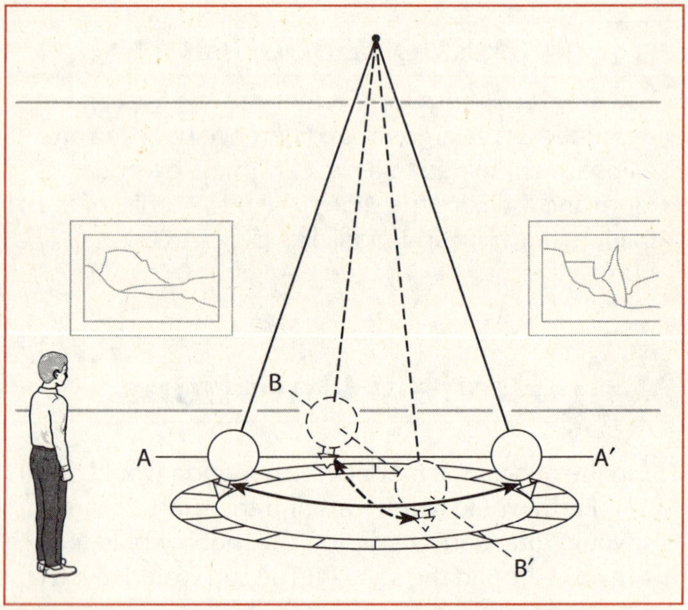

Figure 4-8: Apparent motion of a Foucault Pendulum: An observer sees a pendulum swing in the direction A-A′. Several hours later, the pendulum has changed its direction of swing to the line B-B′.

DIGGING DEEPER

The path of the Foucault pendulum only appears to change. Actually, the pendulum swings in a fixed direction in space, while Earth, carrying the observer with it, rotates under the pendulum. If you could observe the pendulum moving through a powerful telescope in space, you would see no apparent change in its path. Foucault pendulums can be found in a number of museums to recreate the experiment.

THE CORIOLIS EFFECT The tendency of all particles of matter moving at Earth's surface to be deflected, or curve away, from a straight-line path is the **Coriolis effect.** The deflection is to the right in the Northern Hemisphere and to the left in the Southern Hemisphere. This deflection occurs because Earth is rotating, and therefore Earth's surface is moving with respect to the path of the particles.

The following example can help to explain the effect. Imagine that you are at the center of a merry-go-round that is rotating counterclockwise. Your friend is near the rim of the merry-go-round. You throw a ball directly at your friend. By the time the ball reaches the rim of the merry-go-round, your friend has been carried to the left. The ball reaches the rim at a point that is now to the right of your friend. With respect to the moving merry-go-round, the ball has been deflected to the right. In a similar manner, rockets, ocean currents, large storms, and winds are deflected with respect to Earth's surface.

Since the only reasonable explanation of the Coriolis effect is that Earth rotates, the Coriolis effect is good evidence for Earth's rotation. See Figure 7-5 on page 112 for the directions of deflections caused by the Coriolis effect.

Evidence of Earth's Revolution Around the Sun

Earth revolves around the sun in a slightly eccentric elliptical orbit, or path, once a year. Just as scientists had trouble providing proof that Earth rotates, they also had difficulty finding solid evidence of Earth's revolution around the sun.

If you understand that Earth's limited change in distance from the sun during the year does not cause seasons, then the seasonal changes associated with changes of the sun's path is evidence of revolution. If Earth did not revolve around the sun, the same part of Earth would tilt toward the sun all the time. Therefore, the same part of Earth would receive the more direct rays of sunlight. The seasons would not change; each part of Earth would have the same season all the time.

Having different constellations associated with each of the four seasons is further evidence of Earth's revolution. A **constellation** is a group of stars that form a pattern and are used to help people locate celestial objects. At night you can see different constellations at different times of the year, as shown in Figure 4-9.

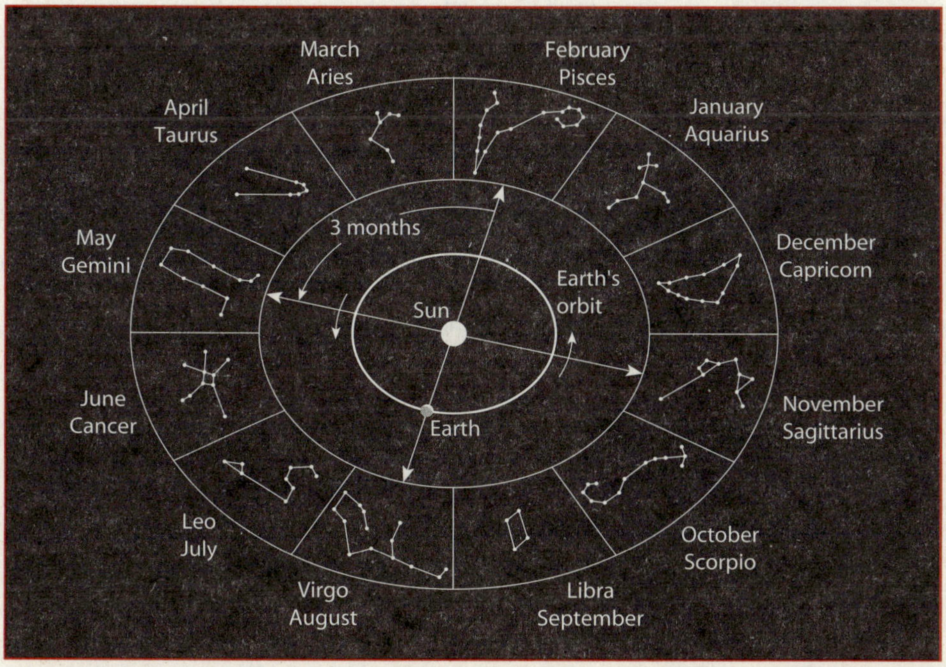

Figure 4-9. Changes in the constellations during the year: The fact that you observe different constellations in different seasons is evidence of Earth's revolution around the sun. Constellations are visible when the dark side of Earth (away from the sun) faces toward the constellation.

Earth's revolution around the sun produces other effects that can be used as evidence of revolution. During the year the apparent diameter of the sun changes in a cyclic fashion. Earth's changing distance from the sun as Earth revolves in its slightly elliptical orbit causes the size of the sun to appear to change. See Figure 3-9 on page 43.

Slight changes in the Doppler effect of stars—discussed in Topic 3—can also be used as evidence of revolution. As Earth revolves around the sun, Earth is traveling toward a given star for about half the year, and the other half of the year it is moving away from a given star. These movements toward or away from a star result in small yearly cyclic changes in blue and red shifting of starlight.

Review Questions

20. The following diagram represents part of the night sky, including the constellation Leo. The black circles represent stars. The unshaded circles represent the changing positions of one celestial object over a period of a few weeks.

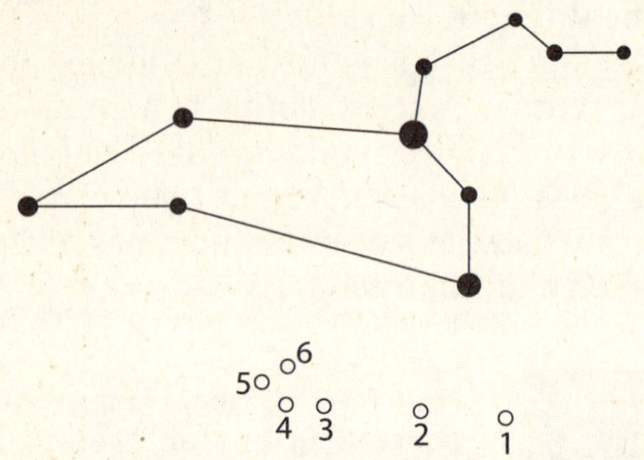

The celestial object represented by the unshaded circles most likely is (1) a galaxy (2) a planet (3) Earth's moon (4) another star

21. Why are different constellations visible in the night sky at different seasons of the year?

22. An angle of $23\frac{1}{2}°$ is formed between the axis of Earth and a line
(1) from the center of Earth to Polaris
(2) from the center of Earth to the sun
(3) perpendicular to the plane of Earth's orbit
(4) perpendicular to the plane of the moon's orbit

23. The Foucault pendulum provides evidence of Earth's (1) rotation (2) revolution (3) insolation (4) inclination

24. On Earth, surface ocean currents rotate clockwise in the Northern Hemisphere and counterclockwise in the Southern Hemisphere because of the (1) Coriolis effect (2) Seasonal effect (3) Tidal effect (4) Doppler effect

Base your answers to questions 25 and 26 on the following diagram. The diagram shows, over the course of a year, twelve constellations that are visible in the night sky to an observer in New York State. Different positions of Earth are represented by letters A through D. The arrows represent the direction of Earth's motion around the sun.

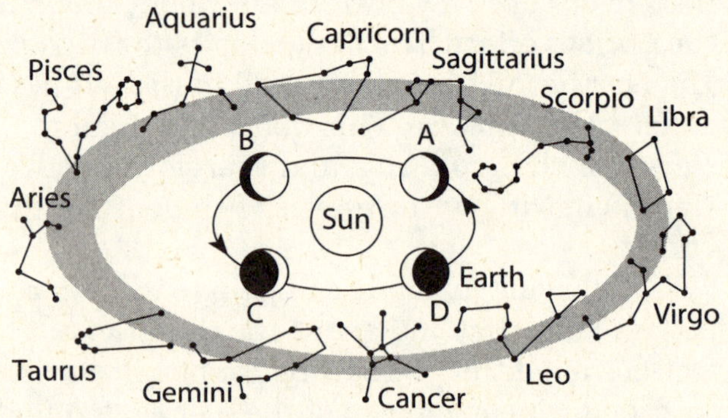

(Not drawn to scale.)

25. Which constellations are both visible at midnight to an observer in New York State when Earth is located at position D? (1) Aries and Taurus (2) Pisces and Libra (3) Leo and Virgo (4) Aquarius and Scorpio

26. The constellations observed from New York State when Earth is at position A are different from the constellations observed from New York State when Earth is located at position C because (1) Earth moves in its orbit (2) Earth is tilted on its axis (3) the lengths of day and night are different (4) the stars move around Earth as shown by star trails

27. Which statement provides the best evidence that Earth revolves around the sun? (1) The sun follows an apparent daily path, rising in the east and setting in the west. (2) A Foucault pendulum appears to shift its direction of swing in a predictable manner. (3) The stars appear to follow circular paths around the North Star (Polaris). (4) The seasons of spring, summer, fall, and winter repeat in a pattern.

Earth and Moon Motions and Time

Throughout the ages most cultures have used the actual motions of Earth and/or the moon as a frame of reference to determine time. The concept of the month is based on the length of one cycle of our moon's phases. The day is based on one rotation of Earth as measured by apparent motions of the sun. The year is based on one revolution of Earth around the sun as measured by changes in the apparent motions of the sun or other stars. Specific time within a day is more difficult to determine.

Local Time

Time based on the rotation of Earth as reflected in motions of the sun is **local time.** All places on the same north-south line, or meridian of longitude, have the same local time. However, locations even slightly east or west of the meridian have different local times, due to Earth's rotation. There are two types of time based on observations of the sun at a single location.

Figure 4-10. Local time is measured by a sundial: The position of a shadow on a sundial provides local solar time. The present reading is 12:00 noon—when the sun is highest in the sky for that day.

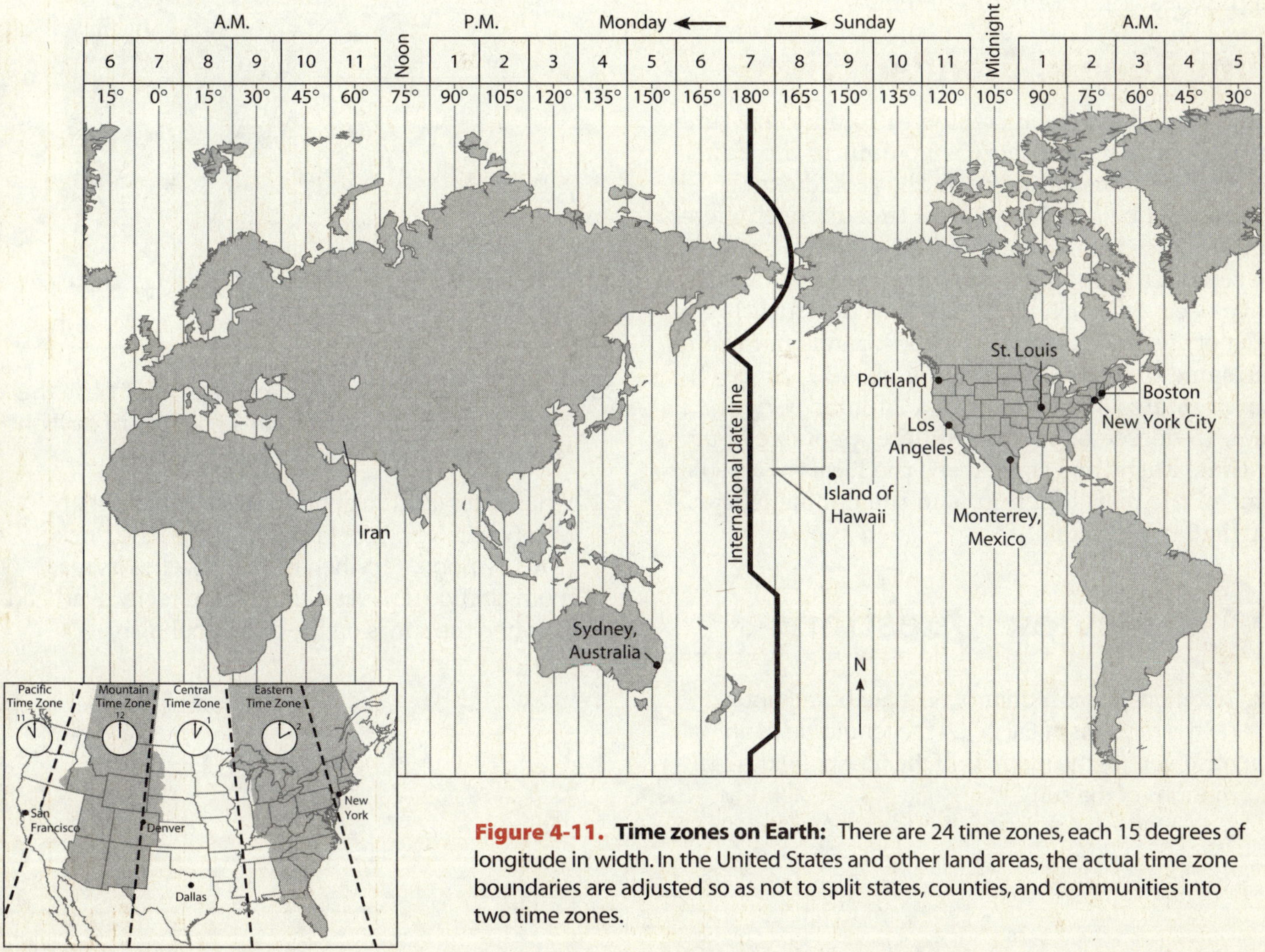

Figure 4-11. Time zones on Earth: There are 24 time zones, each 15 degrees of longitude in width. In the United States and other land areas, the actual time zone boundaries are adjusted so as not to split states, counties, and communities into two time zones.

LOCAL SOLAR TIME For any location, the sun reaches its highest point in the sky once each day at solar noon. The time it takes for Earth to rotate from solar noon to solar noon on two successive days at any fixed location on Earth is the solar day. This type of local time, based on the actual motions of the sun in the sky, is solar time, or sundial time, because a sundial is used to measure it.

If Earth moved around the sun in a circular orbit at a constant speed, the apparent solar day would have a constant length. However, Earth's speed around the sun varies in a cyclic manner in the course of the year. As a result, the length of the apparent solar day also varies during the year. Days of varying length make it very difficult to measure time using clocks and watches. Therefore, for convenience in timekeeping, a solar day of average length—the mean solar day—has been established. This mean solar day, which is the second type of local time, is mean solar time. It has been divided into exactly 24 hours. On most days of the year, the solar day is either slightly more or slightly less than 24 hours in length.

Time Zone or Standard Time

Time zones have been created to standardize time for regions. Because there is no limit to the number of meridians of longitude that could exist, there could be a different local time for each meter or kilometer where you live. To solve this problem, 24 separate **time zones** in 15-degree-wide bands have been established. All parts of Earth within each of these 15-degree longitude bands maintain the same time, as shown in Figure 4-11. Since Earth rotates 360 degrees in a 24-hour period, each 15-degree band has a difference of one hour in time. Each time zone keeps the local mean solar time of the meridian that is in the middle of that particular time zone.

Review Questions

28. Traditionally, most units of time are based upon (1) the motions of Earth (2) the longitude of an observer (3) the motions of the moon (4) the real motions of the sun

29. If Earth's rate of rotation decreased, there would be an increase in the (1) length of the seasons (2) sun's angle of insolation at noon (3) number of observable stars seen at night during the year (4) length of an Earth day

30. How long does Earth take to complete one orbit around the Sun? (1) 1 day (2) 1 month (3) 1 year (4) 1 decade

31. Units of time are based on Earth's motion relative to other celestial objects. The year is best defined as Earth's motion relative to the (1) asteroids (2) sun (3) moon (4) planets

32. According to the diagram below, the time at point C is closest to (1) 6 A.M. (2) 12 noon (3) 6 P.M. (4) 12 midnight

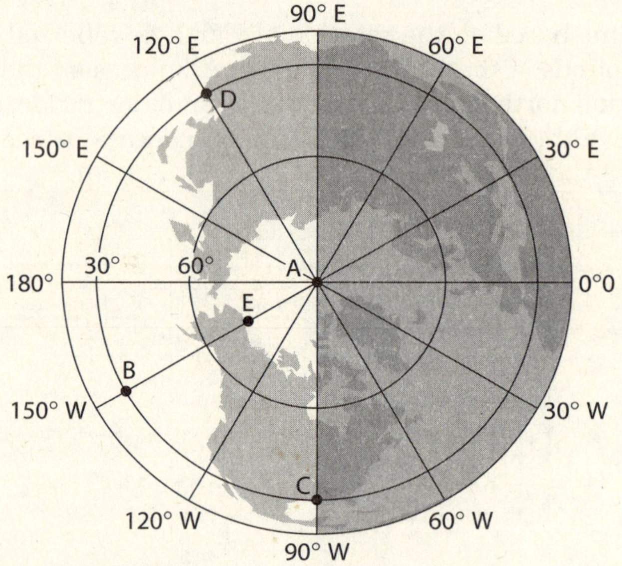

33. Cities located on the same meridian must have the same (1) altitude (2) latitude (3) length of daylight (4) local solar time

34. When does local solar noon always occur for an observer in New York State? (1) when the clock reads 12 noon (2) when the sun reaches its maximum altitude (3) when the sun is directly overhead (4) when the sun is on the prime meridian

Base your answers to questions 35 through 37 on the following diagram of Earth. Some of the latitude and longitude lines have been labeled. Points A through E represent locations on Earth's surface.

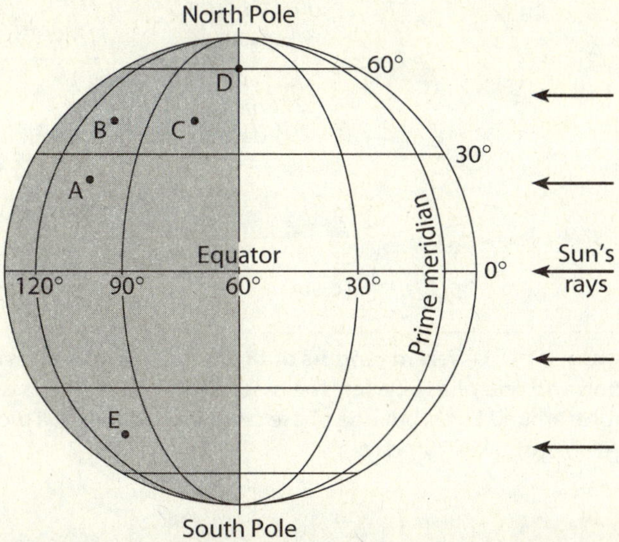

35. What do locations A, B, and E have in common? (1) They are in the same season. (2) They have the same local time. (3) They have the same prevailing wind direction. (4) They are at the same latitude.

36. The latitude and longitude of which location are closest to those of New York State? (1) A (2) B (3) C (4) D

37. What is the approximate time at location D?
(1) 6 A.M. (2) noon (3) 9 P.M. (4) midnight

38. Ship X and ship Y are sailing along the equator. The difference in local solar time between their locations is 2 hours. What is their difference in longitude?
(1) 0° (2) 15° (3) 30° (4) 45°

Base your answers to questions 39 and 40 on the following time zone map.

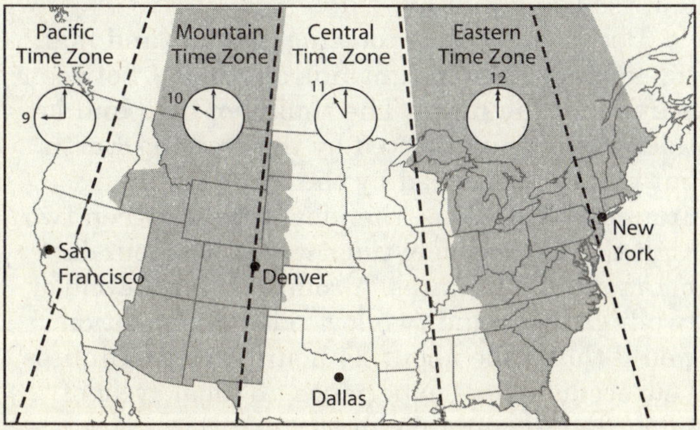

39. What is the time in San Francisco when it is 6 A.M. in Dallas? (1) 5 A.M. (2) 7 A.M. (3) 3 A.M. (4) 4 A.M.

40. The dashed boundaries between time zones are how many degrees of longitude apart? (1) 10° (2) 15° (3) $23\frac{1}{2}$° (4) 24°

Actual Motions of Earth's Moon

The revolution of the moon around Earth as Earth revolves around the sun results in many common observable events, including phases of the moon, tides, and eclipses. The moon revolves around Earth in an elliptical orbit that is tilted about 5° from Earth's orbit and that has a period of $27\frac{1}{3}$ days. Figure 4-12 provides some details of the orbital motions of Earth and its moon.

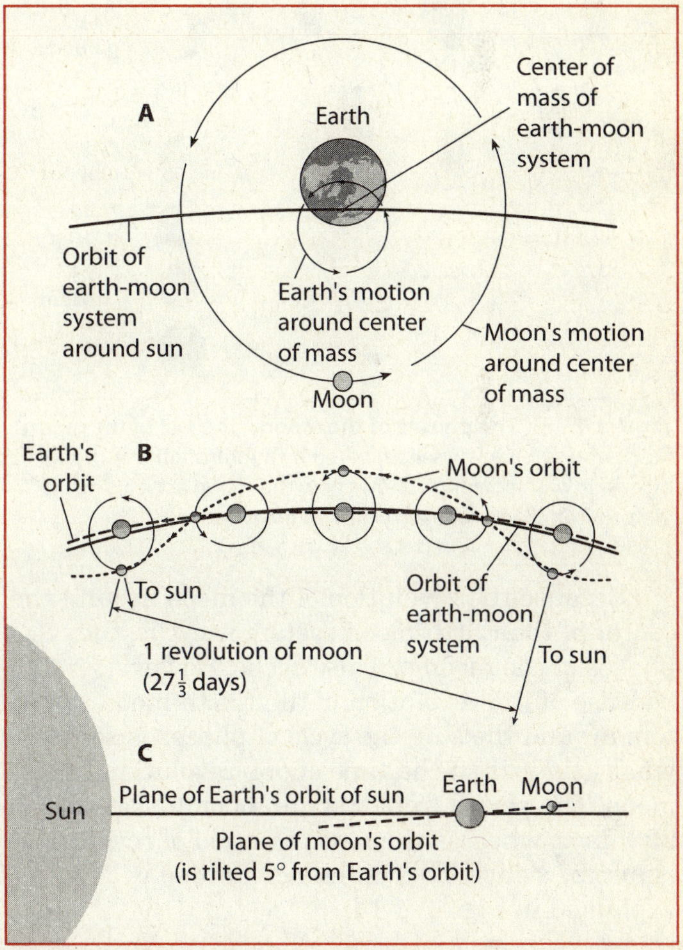

Figure 4-12. The moon's orbit: (A) The revolution of the moon around Earth is due to a balance between the inertia of the moon and the gravitational attraction of the moon to the center of the Earth-moon system. **(B)** The movements of the moon are exaggerated. **(C)** shows that the moon's orbit of Earth is tilted 5° from Earth's orbital plane of the sun.

Moon Phases

Half of the moon is always receiving light from the sun at any given time—except during lunar eclipses. Since the moon revolves around Earth, an observer on Earth sees varying amounts of this lighted half as the moon moves through its orbit. The varying amounts of the lighted moon as seen from Earth are known as the moon's **phases.** (See Figure 4-13.)

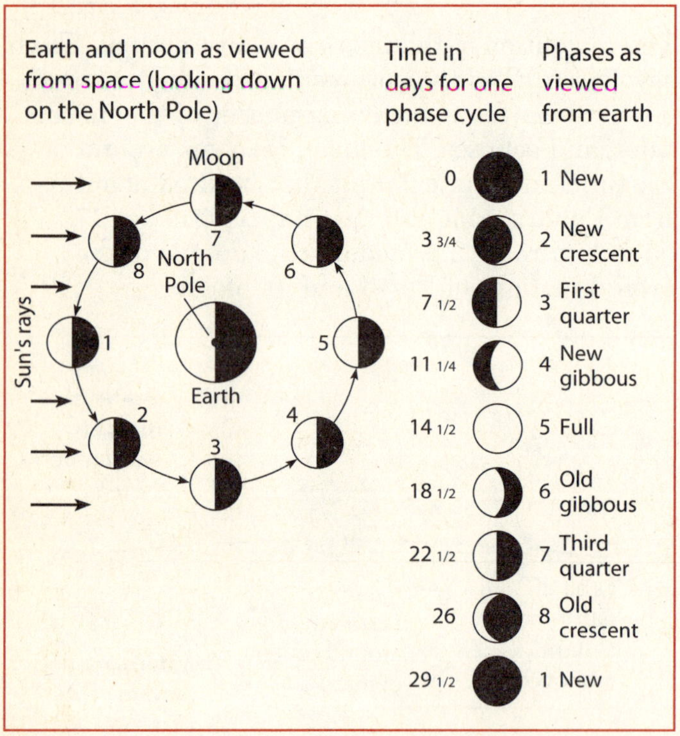

Figure 4-13. The phases of the moon: The half of the moon facing the sun is always illuminated by light from the sun. The right side of the diagram shows how the moon appears as viewed from Earth for each phase of the moon.

Because the revolution of the moon around the center of the Earth-moon system is cyclic, the phases of the moon are also cyclic. However, because of the revolution of the Earth-moon system around the sun, the cycle of phases is somewhat longer than the time of one revolution of the moon. The period from one full moon to the next is $29\frac{1}{2}$ days, whereas the moon's period of revolution is just $27\frac{1}{3}$ days. This difference in time is explained in Figure 4-14.

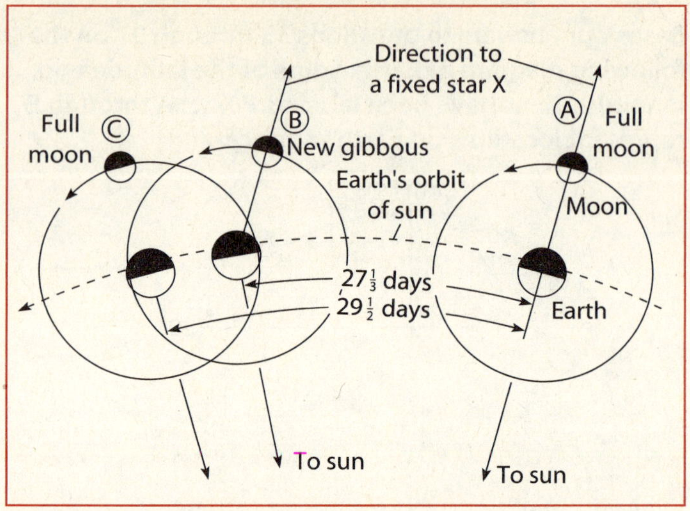

Figure 4-14. Different lengths of the moon's period of revolution and one phase cycle: The moon takes about $27\frac{1}{3}$ days to revolve around Earth, but one phase cycle, such as from full moon to full moon, takes $29\frac{1}{2}$ days.

Tides

The gravitation between the Earth, the moon, and the sun results in a cyclic rise and fall of ocean waters on Earth called **tides.** Figure 4-15 on the next page illustrates a typical tide pattern and how the varying positions of Earth, the moon, and the sun result in different levels of ocean water.

The tidal effect is caused primarily by the moon not the sun. Even though the moon is much smaller than the sun, it is about 400 times closer to Earth. Figure 4-15 illustrates that there is a bulge of water—a high tide—on Earth directly under the moon's position. There is also a high tide on the opposite side of Earth because the solid Earth is closer to the Earth-moon center of gravitation than the ocean water. This causes Earth to be pulled away from the water, leaving the high tide. At right angles to the positions of high tide, the gravitational pull of the moon is least, and the levels of ocean water are low, resulting in a low tide.

Because Earth is rotating, the low and high tides follow the straight-line alignment between Earth and the moon. The timing of high and low tides is not only affected by the rotating Earth, but is also influenced by the moon revolving around Earth. Thus, the ideal time between two high tides or two low tides would be about 12 hours and 25 minutes. A complete ideal cycle of two high tides and two low tides at a location would then take about 24 hours and 50 minutes. Few ocean shores experience an ideal cycle of tides, due to many factors.

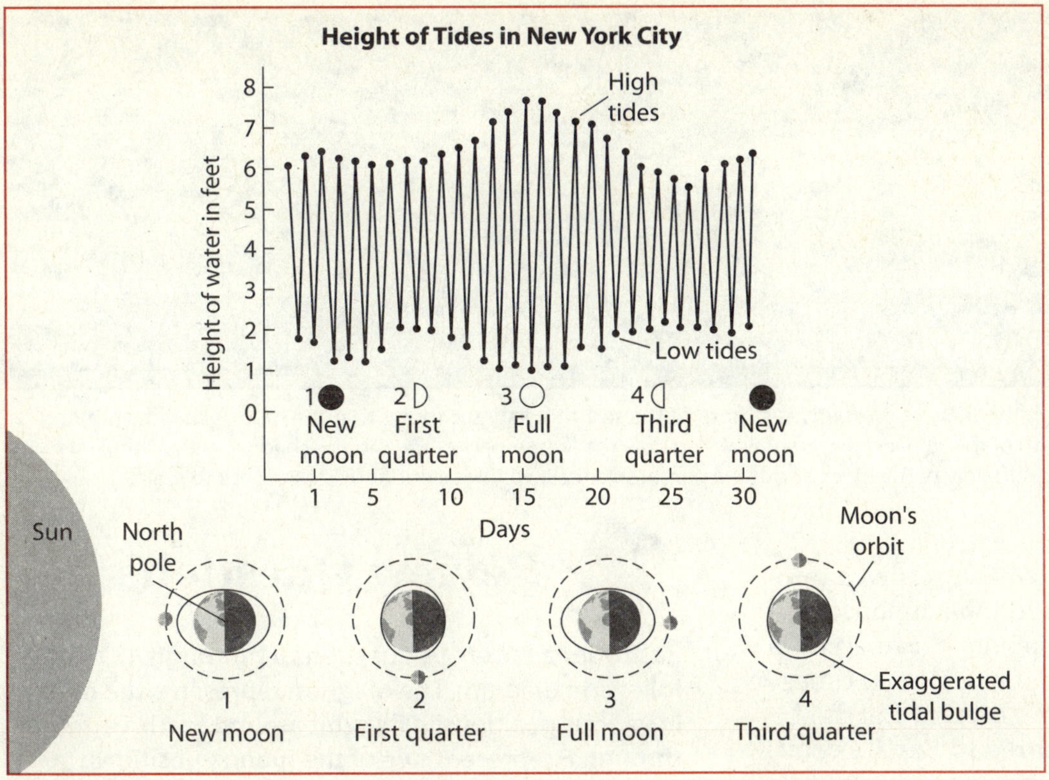

Figure 4-15. Ideal tide pattern in New York City: When the moon is in the new or full moon phase (positions 1 and 3), the gravity of the moon and sun are in a straight line. This causes higher high tides and lower low tides—a large tidal range. When the moon is in the quarter phases (positions 2 and 4), the gravitational force of the sun is at a 90° angle to the moon's force. This causes lower high tides and higher low tides—a smaller tidal range.

The effect of the sun on tides is largely that of reducing or enhancing their height. When Earth, the moon, and the sun are in a straight line, the sun enhances the tidal effect, and high tides are higher and low tides are lower. This large tidal range happens during new and full moon phases. When the moon is in one of the quarter phases, the sun is pulling at right angles to the moon. As a result, high tides are then lower, and low tides are higher, producing a smaller tidal range.

Eclipses

Earth and the moon, like all other opaque objects that don't produce their own light, cast long shadows into space. When a celestial object partly or completely comes into the shadow of another celestial object, there is an event called an **eclipse.** (See Figure 4-16.)

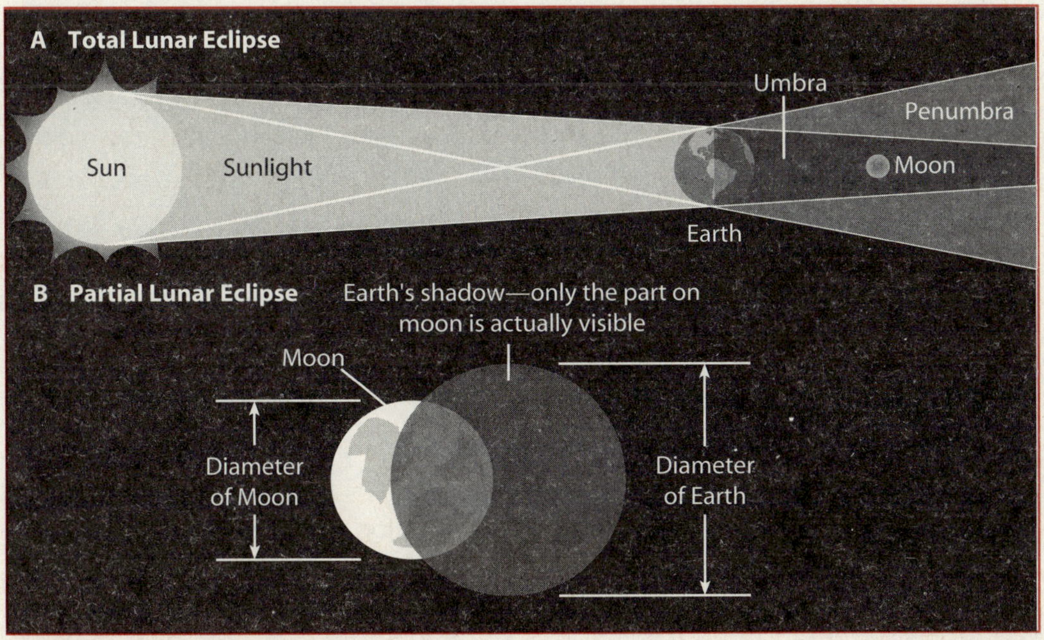

Figure 4-16. Lunar eclipse: Diagram A shows the condition of a total eclipse when the moon is completely covered by Earth's shadow (umbra). Diagram B shows a partial lunar eclipse as viewed from Earth when the moon is only partly in Earth's shadow. (Distance and size are not drawn to scale.)

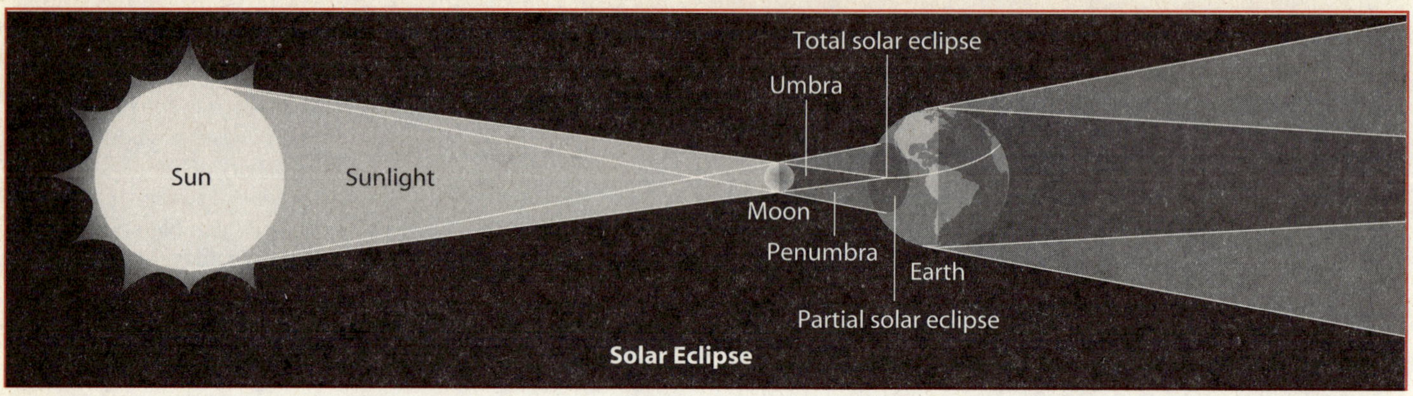

Figure 4-17. Solar eclipses: When the new moon revolves in front of the sun, the diagram shows that the moon's shadow (umbra) just barely reaches Earth and the path of the total eclipse is small. When only the lighter part of the moon's shadow (penumbra) hits Earth, people in this shadow see the sun only partly blocked out—a partial solar eclipse. (Distances and sizes are not to scale.)

LUNAR ECLIPSE When the moon revolves into the shadow of Earth, at the full moon phase, a lunar eclipse, or eclipse of the moon, can occur. A lunar eclipse doesn't happen during most times of full moon, because the moon's orbit of Earth is tilted about 5 degrees compared to Earth's orbit. If Earth's shadow (umbra) covers only part of the moon, there is a partial eclipse. The curved shadow of the Earth on the moon during a partial lunar eclipse is evidence for the Earth's spherical shape.

When there is a total lunar eclipse, the moon is completely covered by Earth's shadow and all people on the dark half of Earth can view the eclipse for up to about 100 minutes. The complicated cycle of lunar eclipses results in an average of two total lunar eclipses a year.

SOLAR ECLIPSE Under rare ideal conditions at the new moon phase, the moon can just barely block out the sun, casting a shadow on Earth that causes a solar eclipse. As seen in Figure 4-17, the moon's shadow can just barely reach Earth, making solar eclipses a rare event. A total eclipse of the sun can only be observed at any one location on Earth for up to $7\frac{1}{2}$ minutes once every 200 years or so.

Review Questions

Base your answers to questions 41 through 47 on the following diagram. The diagram represents the moon in various positions in its orbit around Earth. Letters A through E represent five of the moon's positions.

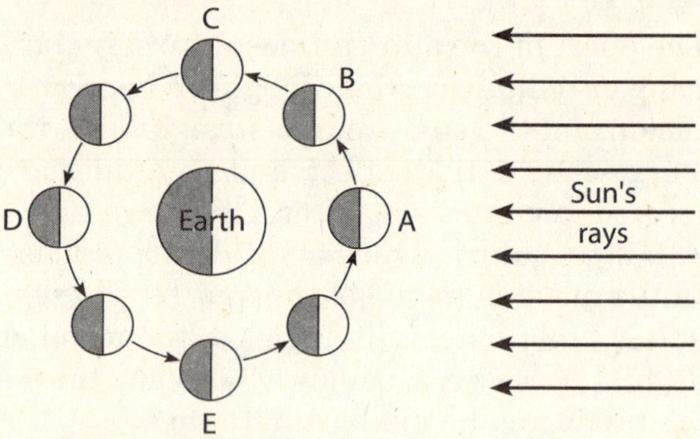

(Not drawn to scale.)

41. Which diagram best represents the appearance of the moon to an observer on Earth when the moon is at position B?

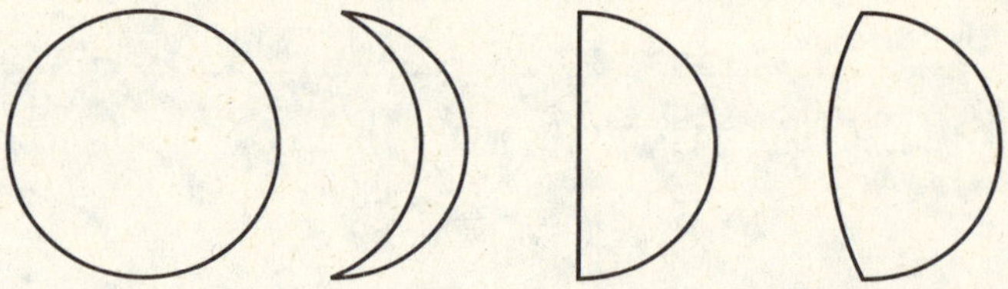

(1) (2) (3) (4)

42. The moon would not be visible from the Earth when the moon is at position (1) A (2) E (3) C (4) D

43. Why would an observer on Earth see a complete cycle of phases of the moon in approximately one month? (1) The moon rotates on its axis. (2) The moon revolves around Earth. (3) Earth rotates on its axis. (4) Earth revolves around the sun.

44. If the distance of the moon from Earth were to increase, the length of time the moon would take to complete one revolution around Earth would (1) decrease (2) increase (3) remain the same

45. How does the moon appear to an observer in New York State when the moon is located at position E?

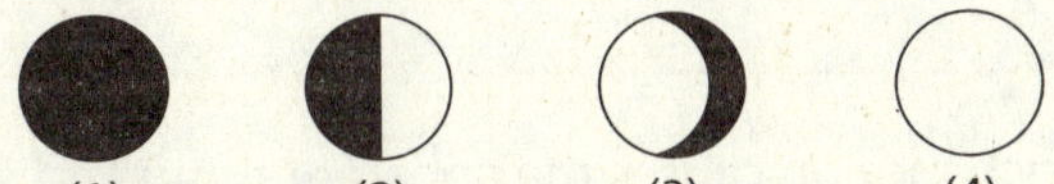

46. The shadow of the moon may fall on Earth at position (1) A (2) B (3) C (4) D

47. At which positions of the moon do the smallest high tides occur on Earth? (1) C and E (2) A and D (3) B and C (4) A and B

---

48. If a full moon occurs on November 1, the next full moon phase will occur about (1) November 22 (2) November 30 (3) December 6 (4) December 13

49. In the open ocean, a bulge of water two to three feet high that follows the moon's movement around Earth is a (1) tide (2) tsunami (3) wind wave (4) surface ocean current

50. The following diagram represents a north polar view of Earth in relation to the moon and the sun.

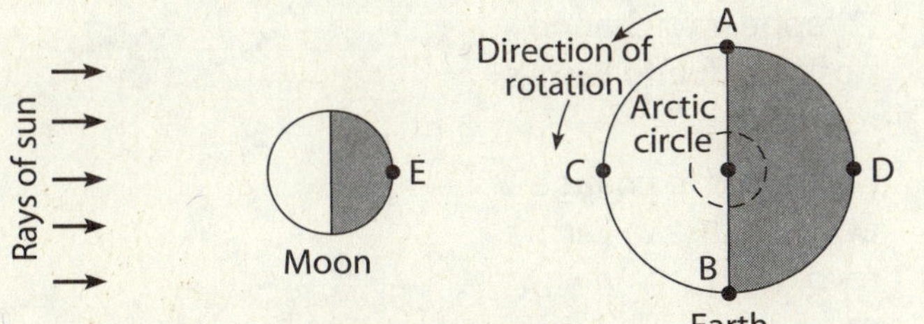

(Not drawn to scale.)

The tides that occur when the sun, moon, and Earth are in the relative positions indicated by the diagram are best described as (1) the highest high tides and the lowest low tides (2) the highest high tides and the highest low tides (3) the lowest high tides and the lowest low tides (4) the lowest high tides and the highest low tides

51. A high tide occurs at Boston on a certain day at 3 P.M. The next high tide may be expected to occur at (1) 9:26 P.M. the same day (2) 3:26 A.M. the following day (3) 9:52 A.M. the following day (4) 3:52 P.M. the following day

52. During which event does our moon receive the least amount of sunlight? (1) a Jupiter eclipse (2) an Earth eclipse (3) a solar eclipse (4) a lunar eclipse

53. During which phase of the moon do solar eclipses occur? (1) new moon (2) first quarter moon (3) last quarter moon (4) full moon

The following diagram represents the landscape on the moon with Earth shown in the background. Use this diagram to answer questions 54 through 57.

54. As viewed from the Earth, what is the phase of the moon? (1) new (2) full (3) quarter (4) gibbous

55. With respect to the diagram, which statement is true? (1) No eclipse could occur. (2) A lunar eclipse might occur. (3) A partial solar eclipse might occur. (4) A total solar eclipse might occur.

56. In the week following the period shown in the diagram, the amount of the moon's surface that is illuminated by the sun will (1) increase (2) decrease (3) remain the same (4) increase, then decrease

57. What is the most probable cause of the lunar features indicated by A?

Questions for Regents Practice

Part A

Base your answers to questions 1 through 4 on the following diagram. The diagram represents a plastic hemisphere upon which lines have been drawn to show the apparent paths of the sun on four days at a location in New York State. Two of the days are December 21 and June 21. The protractor is placed over the north-south line.

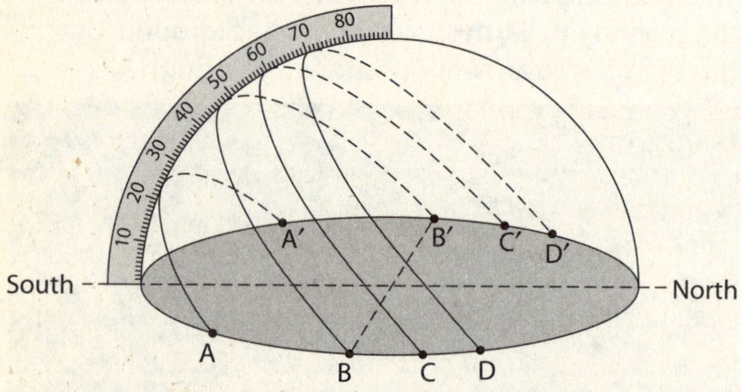

1. On which two dates could the sun have followed path C-C′?

(1) October 22 and March 28

(2) September 9 and January 7

(3) January 27 and August 21

(4) May 7 and August 1

2. Which path was recorded on a day that had 12 hours of daylight and 12 hours of darkness?

(1) A-A′

(2) B-B′

(3) C-C′

(4) D-D′

3. Which would be the approximate length of daylight for the observer, when the sun travels along the entire length of path A-A′?

(1) 9 hours

(2) 12 hours

(3) 15 hours

(4) 18 hours

4. Which observation about the sun's apparent path on June 21 is best supported by the diagram?

(1) The sun appears to move across the sky at a rate of 1° per hour.

(2) The sun's total daytime path is shortest on this date.

(3) Sunrise occurs north of east.

(4) Sunset occurs south of west.

5. Which motion causes the apparent rising and setting of the moon each day as seen from a location in New York State?

(1) Earth revolving around the sun

(2) the moon revolving around Earth

(3) Earth rotating on its axis

(4) the moon rotating on its axis

6. As Earth revolves in orbit from its January position to its July position, the angle between its axis and orbital plane will

(1) decrease

(2) increase

(3) remain the same

7. The following diagram represents a Foucault pendulum in a building in New York State.

Points A and A′ are fixed points on the floor. As the pendulum swings for six hours, it will

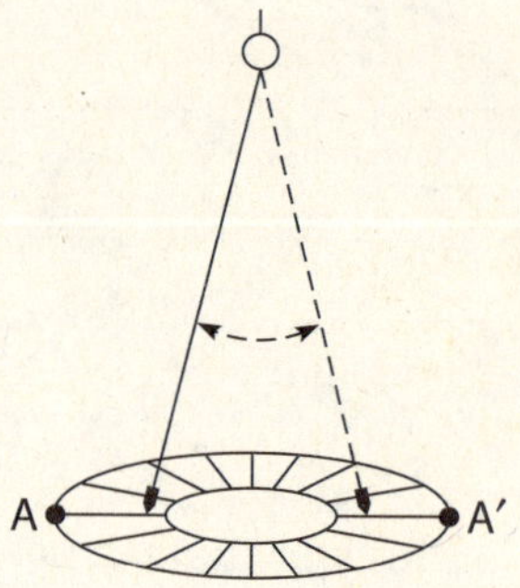

(1) appear to change position due to Earth's rotation

(2) appear to change position due to Earth's revolution

(3) continue to swing between A and A′ due to inertia

(4) continue to swing between A and A′ due to air pressure

Base your answers to questions 8 through 12, on the following diagram.

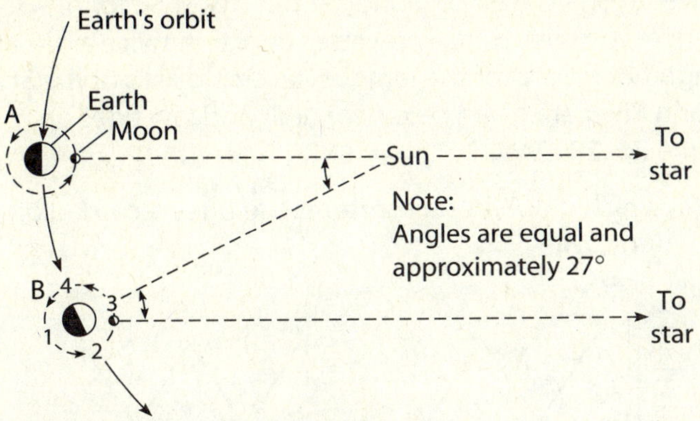

8. Earth will travel from A to B in

(1) 1 day

(2) 2.5 days

(3) $27\frac{1}{3}$ days

(4) $29\frac{1}{2}$ days

9. In respect to Earth at position B, in which phase is the moon?

(1) full

(2) new

(3) gibbous

(4) crescent

10. When Earth is at position B, how long must the moon travel beyond where it is shown to be in the same phase as it was in position A?

(1) 1 day

(2) 2 days

(3) 6 hours

(4) 7 days

11. What is the main reason why the moon must travel a greater distance to be in the same phase at position B as it was at position A?

(1) The sun's rays are essentially parallel when they reach Earth.

(2) The moon's periods of revolution and rotation are the same.

(3) Earth is constantly revolving around the sun.

(4) Earth's axis is inclined to the plane of its orbit.

12. At position B, the highest tides on Earth will occur when the moon is at points

(1) 1 and 2

(2) 1 and 3

(3) 2 and 4

(4) 3 and 4

---

13. The Coriolis effect provides evidence that Earth

(1) has a magnetic field

(2) has an elliptical orbit

(3) revolves around the sun

(4) rotates on its axis

14. A projectile is launched from a point near the North Pole toward the equator. Which diagram best represents the apparent path of the projectile, if it were viewed from Earth?

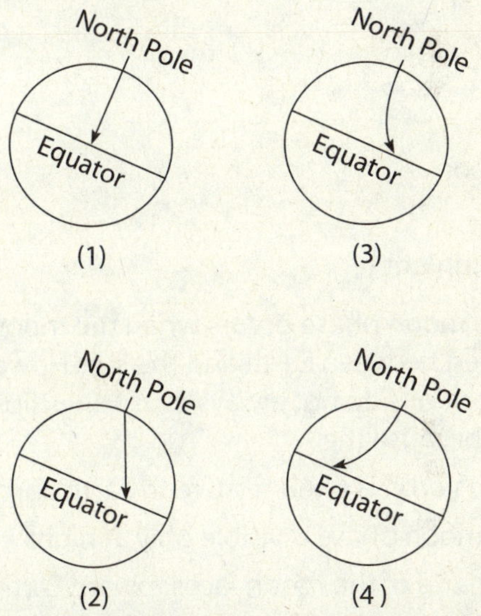

15. Some stars that can be seen in New York State on a summer night cannot be seen on a winter night. This fact is a result of the

(1) rotation of Earth on its axis

(2) rotation of the stars around Polaris

(3) revolution of Polaris around Earth

(4) revolution of Earth around the sun

16. To an observer on Earth, the planet Venus does not appear at one fixed position among the stars because Venus

(1) rotates on its axis

(2) revolves around the sun

(3) shows an apparent motion around Earth

(4) shows a complete cycle of phases

17. The diagram below shows the rotation of Earth as it would appear from a satellite over the North Pole. The time at point X is closest to

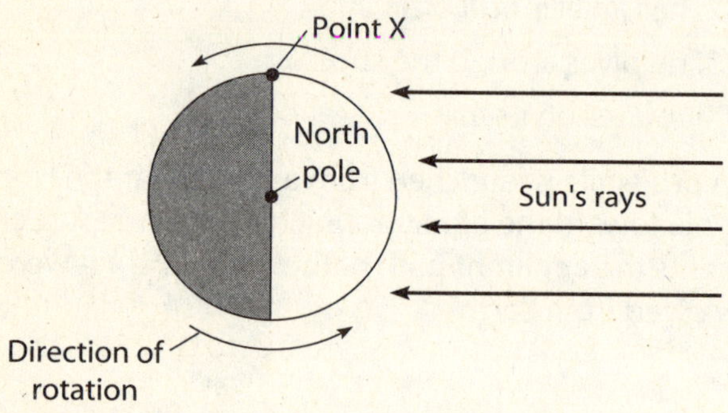

(1) 6 A.M.

(2) 12 noon

(3) 6 P.M.

(4) 12 midnight

18. The new moon phase occurs when the moon is positioned between Earth and the sun. However, these positions do not always cause an eclipse of the sun because the

(1) moon's orbit is tilted relative to Earth's orbit

(2) new moon phase is visible only at night

(3) night side of the moon faces toward Earth

(4) apparent diameter of the moon is greatest during the new moon phase

Part B

Base your answers to questions 19 through 23 on the following table and information.

Tidal Record for Reversing Falls, St. John River				
Date	**Time of First High Tide**	**Time of First Low Tide**	**Time of Second High Tide**	**Time of Second Low Tide**
June 26	2:25 A.M.	8:45 A.M.	2:55 P.M.	9:05 P.M.
June 27	3:15 A.M.	9:35 A.M.	3:45 P.M.	9:55 P.M.
June 28	4:05 A.M.	10:25 A.M.	4:35 P.M.	10:45 P.M.

The Bay of Fundy, located on the east coast of Canada, has the highest ocean tides in the world. The St. John River enters the Bay of Fundy at the city of St. John, where the river actually reverses direction twice a day at high tides. Data for the famous Reversing Falls of the St. John River are given below for high and low tides on June 26–28, 1994.

19. Which graph best represents the tides recorded on June 28?

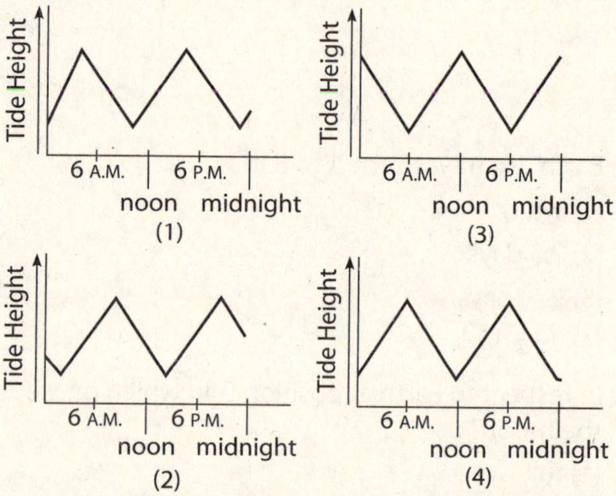

20. Compared to the first high tide on June 26, how much later in the day did the first high tide occur on June 27? [1]

21. Tides in the Bay of Fundy are best described as

(1) predictable and noncyclic

(2) predictable and cyclic

(3) unpredictable and noncyclic

(4) unpredictable and cyclic

22. Complete the following sentence: The moon has a greater effect on Earth's ocean tides than the sun has because the

23. On the United States time zone map provided below, indicate the standard time in *each* time zone when it is 9 a.m. in the Central Time Zone. The dashed lines represent the standard time meridians for each time zone. Be sure to indicate the time for all *three* zones.

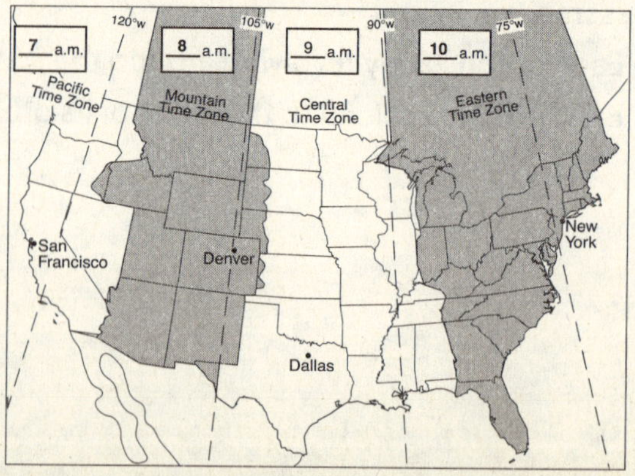

24. Which model of the sun, Earth (E), and moon (M) best represents a position that would cause the highest ocean tides in the Bay of Fundy?

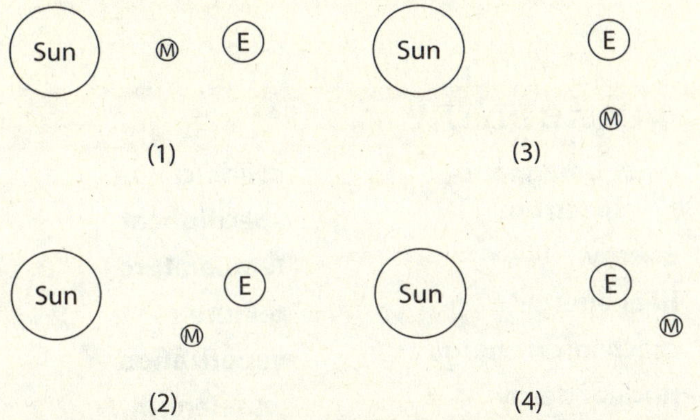

25. Draw and fully label a diagram showing a total eclipse of the sun. Include labels for the shadows, all solar system members shown, and phase of the moon. [4]

Use this information to answer questions 26 through 29. The following diagram represents the apparent path of the sun for an observer in New York State on the dates indicated. The diagram also shows the angle of Polaris above the horizon.

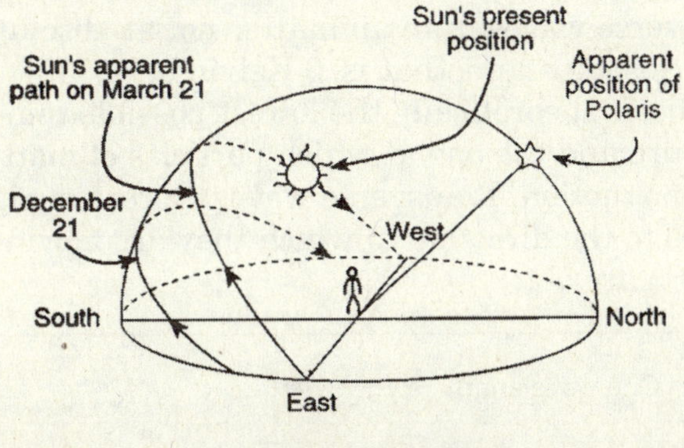

(Not drawn to scale)

26. On the diagram, draw the apparent path of the sun on May 21. [1]

27. On the diagram, mark and label the position of sunrise on May 21. [1]

28. State the latitude of the location represented by the diagram to the nearest degree. Include the latitude direction in your answer. [1]

29. At approximately what hour of day would the sun be at the position shown in the diagram? [1]

Part C

Using the following statement, answer questions 30 through 33.

Suppose that Earth does not have and has never had a moon.

30. Why would Earth still have ocean water tides? [1]

31. How would Earth's ocean tides be different than they are with the presence of the moon? [1]

32. Would people on Earth experience lunar and/or solar eclipses? Explain why or why not. [2]

33. Infer how the lack of a moon would have affected time keeping throughout the history of humans. What specific time period would have been affected? [1]

Using the following statement, answer questions 34 and 35.

Through accurate measurements of time and studies of growth patterns of fossil life forms, it has been determined that Earth's rate of rotation is slowing down.

34. What unit of time would be most altered by this change in Earth's rotation? [1]

35. List two completely different types of apparent motions observed on Earth today that would not exist if the Earth were to stop rotating. [2]

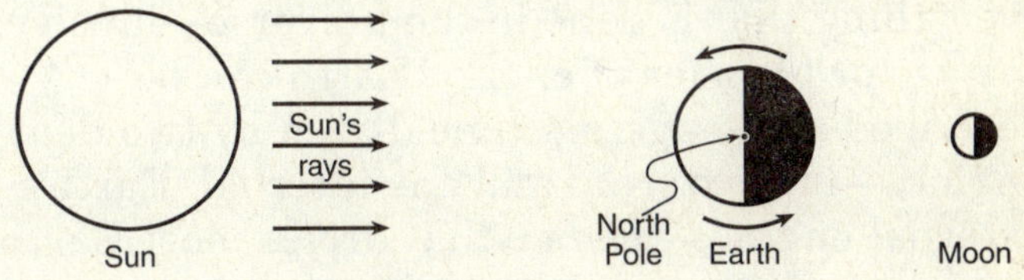

(Not drawn to scale)

36. The diagram provided below represents the Sun and Earth as viewed from space on a certain date.

(a) Using a symbol for the Moon of approximately this size (◯), draw the position of the Moon on the diagram provided *above* at the time when the full-Moon phase is observed from Earth. [1]

(b) Draw an arrow on the diagram provided *above* that shows the Earth motion that causes surface ocean currents and surface winds to curve (Coriolis effect). [1]

37. The photo below shows the mile wide Barringer impact crater located in Diablo Canyon, Arizona. Describe the event that produced this crater. [1]

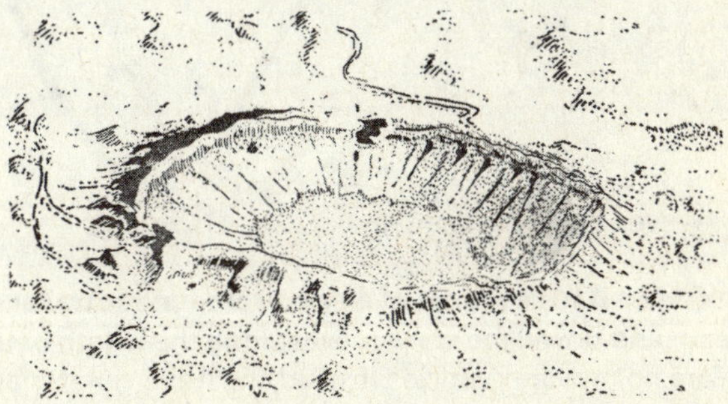

TOPIC 5

Energy in Earth Processes

VOCABULARY		
calorie	electromagnetic spectrum	solidification
condensation	energy	specific heat
conduction	heat energy	temperature
convection	mechanical energy	texture
crystallization	nuclear decay	vaporization
electromagnetic energy	radiation	wavelength

Energy, like mass, time, or distance, is one of the basic attributes of the universe. **Energy** is the ability to do work. Everything you do—in fact, everything that is done in the universe—involves the use or transfer of energy. Earth can be described as a huge machine driven by two heat engines—one external and one internal. Earth's external engine—the one that drives most Earth surface processes—is powered mainly by energy from the sun. Heat from the interior of Earth and the resulting mechanical energy powers Earth's internal heat engine.

Electromagnetic Energy

Visible light is the type of electromagnetic energy you are probably most familiar with. Visible light waves are the parts of the electromagnetic energy that is radiated by the sun and observed with the human eye. **Electromagnetic energy** is a type of energy that is radiated, or given off, in the form of transverse waves from all matter not at absolute zero. Absolute zero, that is, 0 Kelvin or –273° Celsius, is theoretically the lowest possible temperature and the one at which particles of matter have no motion. Transverse waves vibrate at right angles to the direction in which they are moving. (See Figure 5-1.)

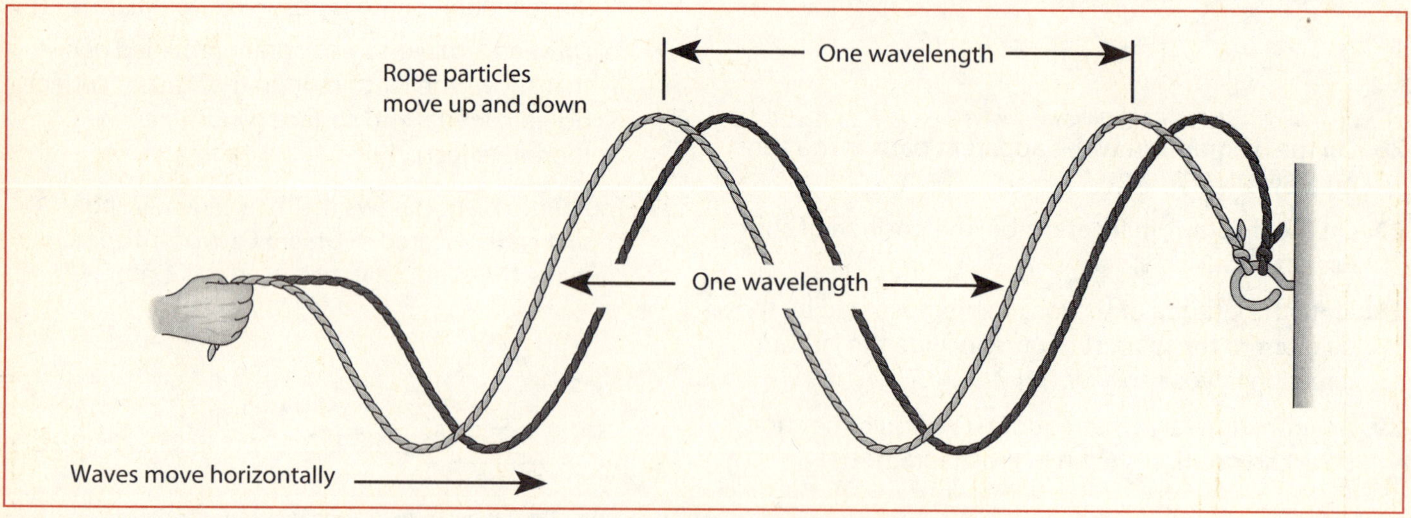

Figure 5-1. A rope model of an electromagnetic transverse wave: As particles of the rope move up and down, the wave moves along the length of the rope. In an electromagnetic wave, there are no moving particles. Instead, the wave vibrates at right angles to the direction in which the wave is moving.

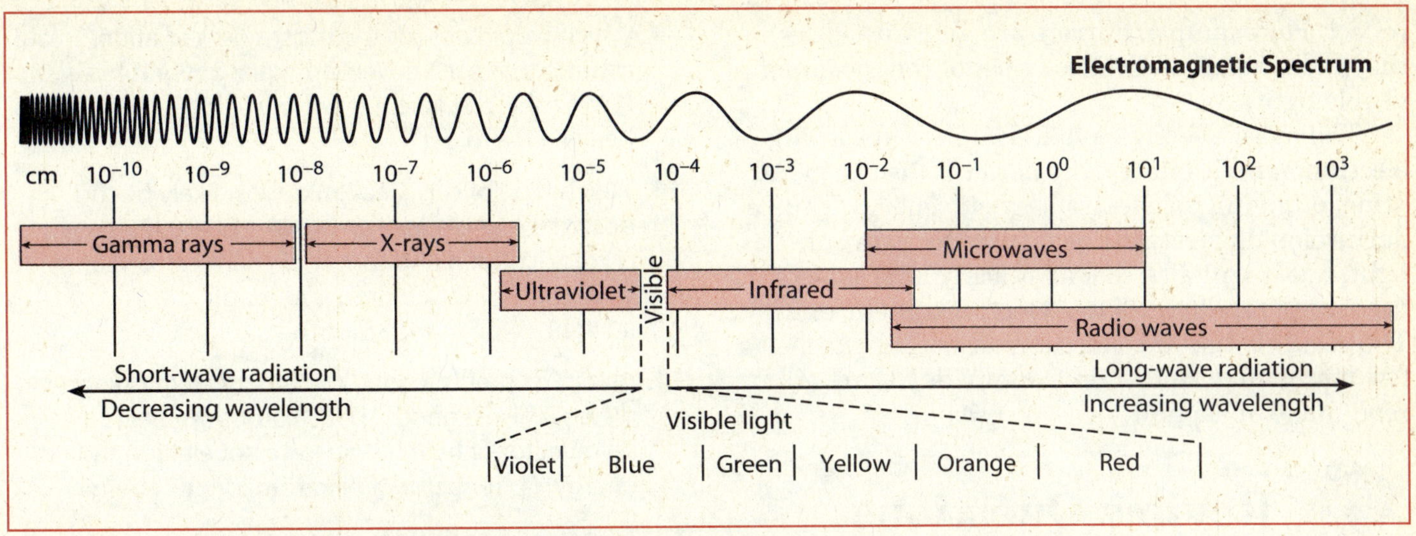

Figure 5-2. An electromagnetic spectrum: Wavelengths are indicated on a logarithmic scale, in which each mark indicates a wavelength 10 times as long as the preceding one. There is no zero point on such a scale. Also see a similar chart in the *Earth Science Reference Tables*.

Characteristics and Types of Electromagnetic Energy

Different types of electromagnetic energy are distinguished by their different wavelengths. The **wavelength** is the distance from one crest of a wave to the next crest. A wavelength may also be described as the distance between any two corresponding points on successive cycles, as shown in Figure 5-1. Figure 5-2 shows the various types of electromagnetic energy in order of increasing wavelength. Any model that shows the types of electromagnetic energy is known as an **electromagnetic spectrum.**

Visible light is the only part of the electromagnetic spectrum that can be seen by the human eye. Infrared energy is often felt due to its heating effects. Instruments must be used to observe most forms of electromagnetic energy, including ultraviolet energy, which tans or burns the skin. Electromagnetic energy can be separated into two groups by comparing its wavelengths to those of visible light. Electromagnetic energy with a wavelength longer than visible light is called long-wave electromagnetic energy. Electromagnetic energy with a wavelength shorter than visible light is called a short wave. Some scientists include visible light as a type of short-wave radiation.

Interactions Between Electromagnetic Energy and an Environment

When electromagnetic energy comes in contact with a material, the waves interact with the material. The waves may be

- refracted, or bent, in their passage through materials of varying density so that the direction of the waves is changed
- reflected, or bounced off the material
- scattered, or refracted and/or reflected in various directions
- transmitted, or passed through the material
- absorbed, or taken into the material

SURFACE PROPERTIES AND ABSORPTION

Characteristics of a surface determine the amount of electromagnetic energy that can be absorbed. For example, the darker the color of the surface of a material, the more visible light it will absorb. This is the reason that a dark-colored asphalt road feels hotter to the bare foot on a sunny day than a light-colored concrete road. Color is the way the human eye distinguishes the various wavelengths of visible light.

The **texture**—roughness or smoothness—of a surface also affects the amount of electromagnetic energy absorbed. The rougher a surface is, the more energy it will absorb and the less it will reflect. For example, mirrors are shiny because they are smooth and reflect most of the incoming visible light.

The more effective a material is at absorbing electromagnetic energy, the better it also is at radiating, or giving off, electromagnetic energy. Thus a dark-colored object will heat up quickly in sunlight, but it will also cool off quickly after sunset because it rapidly radiates electromagnetic energy. The result is that a concrete road will be warmer a few hours after sunset on a sunny day than will a road made of asphalt.

Review Questions

1. Which is the major source of energy for most of Earth's processes? (1) radioactive decay within the Earth's interior (2) convection currents in the Earth's mantle (3) radiation received from the sun (4) earthquakes along fault zones

2. All objects warmer than 0 Kelvin (absolute zero) must be (1) radiating electromagnetic energy (2) condensing to form a gas (3) warmer than 0°C (4) expanding in size

3. The various types of electromagnetic energy are distinguished from one another by their (1) temperature (2) wavelengths (3) height of waves (4) speed of travel in space

4. What is an approximate wavelength of infrared radiation? (1) 10^{-10} cm (2) 10^{-8} cm (3) 10^{-3} cm (4) 10^{-1} cm

5. In the visible spectrum, which color has the longest wavelength? (1) red (2) green (3) orange (4) violet

6. Which statement about electromagnetic energy is correct? (1) Violet light has a longer wavelength than red light. (2) X-rays have a longer wavelength than infrared waves. (3) Radio waves have a shorter wavelength than ultraviolet rays. (4) Gamma rays have a shorter wavelength than visible light.

7. As electromagnetic energy from a heat source interacts with its environment, it is being (1) reflected only (2) refracted only (3) absorbed only (4) reflected, refracted, and absorbed

8. When electromagnetic energy travels from air into water, the waves are bent due to the density differences between the air and water. What is this bending called?

9. Which would absorb the most solar radiation, if you assume that each covers an equal geographic area? (1) a freshwater lake (2) a snow field (3) a sandy beach (4) a forest

10. Which type of surface would most likely be the best reflector of electromagnetic energy? (1) dark-colored and rough (2) dark-colored and smooth (3) light-colored and rough (4) light-colored and smooth

11. An object that is a good radiator of electromagnetic waves is also a good (1) insulator from heat (2) reflector of heat (3) absorber of electromagnetic energy (4) refractor of electromagnetic energy

12. A homeowner decides to install carpet in a room that receives the most sunlight. A carpet with which characteristics will absorb the most radiation from the sun? (1) smooth texture and light color (2) smooth texture and dark color (3) rough texture and light color (4) rough texture and dark color

Transfer of Energy

Energy moves away from a region of high concentration, or source, to a region of low concentration, or sink. Energy transfer from a region of higher temperature to a region of lower temperature is called **heat energy**. Heat transfers thermal energy, the energy of the motions of atoms and molecules.

Achieving Dynamic Equilibrium

Heat will continue to move from the source to the sink until their energies are equal, establishing a dynamic equilibrium. At dynamic equilibrium a region loses and gains equal amounts of energy. If the dynamic equilibrium is between all forms of energy, then the temperature of the region or system will remain constant.

DIGGING DEEPER

The portion of space that separates regions of different properties is called an interface, or boundary. Energy is transferred across interfaces, such as the metal bar shown in Figure 5-3, or the region of space between the sun and Earth.

Methods of Energy Transfer

Heat is transferred from an area of high concentration to an area of low concentration by one of three methods—conduction, convection, or radiation.

CONDUCTION The transfer of heat energy from atom to atom or molecule to molecule when vibrating atoms or molecules collide is **conduction.** Conduction is most effective in solids—especially metals—because the atoms or molecules are closer together than in gases and liquids. However, some heat conduction can take place in liquids and gases. In Figure 5-3 conduction is the main method of heat transfer from the source to the sink through the metal bar.

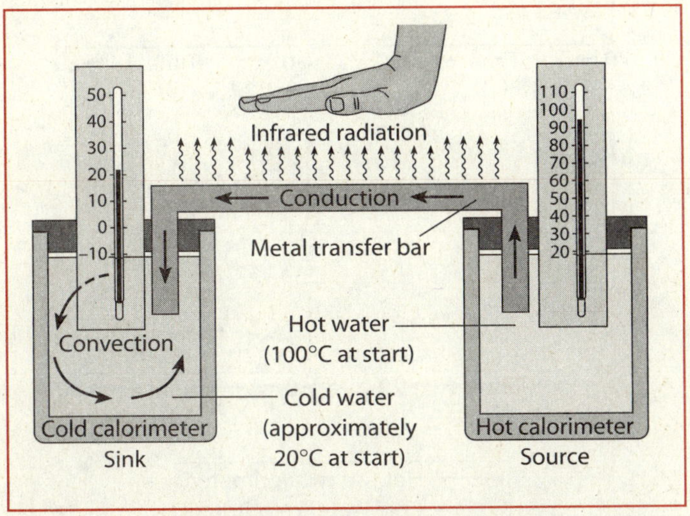

Figure 5-3. An experiment in heat energy transfer: In this experiment, the amounts of heat transferred from the hot calorimeter (the source) to the cold calorimeter (the sink) and the methods of that transfer are investigated. A calorimeter is an insulated container used in energy experiments.

CONVECTION The transfer of heat by movement in fluids—gases and liquids—caused by differences in density within the fluids is **convection.** Warmer portions of the fluid usually have lower density and tend to rise above the cooler portions. The reason for the rise of less dense fluids is that gravity pulls less on objects that are lower in density and pulls more on objects of higher density. Higher-density portions of a fluid are pulled down and displace less dense objects, pushing them upward. The result is a convection current—a circulatory motion—that transfers heat energy from one place to another. (See Figures 5-3 and 5-4.) Convection currents transfer heat throughout Earth's atmosphere, hydrosphere, and most likely below the lithosphere.

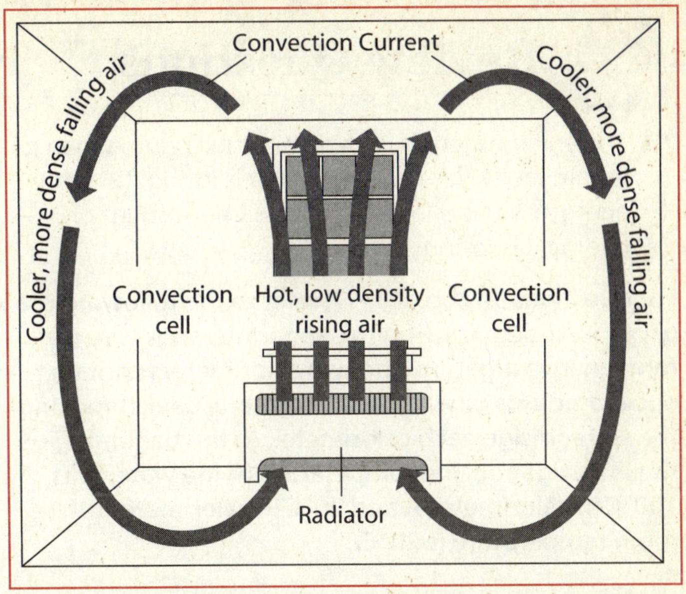

Figure 5-4. Heating a room by convection: The air around a radiator is heated by conduction and by infrared radiation emitted by the radiator. Greenhouse gases (carbon dioxide and water vapor) in the room absorb the radiation. The heat is then carried around the room by convection currents.

RADIATION The method by which heat is transferred via electromagnetic waves is **radiation.** No medium is needed to transfer the transverse waves that carry electromagnetic energy; this energy can radiate from its source across empty space. Radiation can also occur in liquids, solids, and gases—such as sunlight being transmitted through the atmosphere (gas) or through window glass (solid). X-ray electromagnetic radiation can pass through all but the densest materials. Radiation is also the method by which the sun's electromagnetic energy moves through space to other objects in the solar system. The higher an object's temperature, the more electromagnetic energy it gives off. In Figure 5-3, the higher the temperature of the metal bar, the more infrared energy it will radiate to the hand. (See Figure 5-2 and the Electromagnetic Spectrum in the *Earth Science Reference Tables*.)

Review Questions

13. The environment is in dynamic equilibrium when it is gaining (1) less energy than it is losing (2) more energy than it is losing (3) the same amount of energy it is losing

For questions 14 through 17, refer to the following diagram. A student is using the apparatus shown to perform an investigation. The two calorimeters contain equal amounts of water, and the metal bar is touching the water inside each calorimeter. At the beginning of the investigation, the temperature of the water was 100°C in calorimeter A and 20°C in calorimeter B. The room temperature is 20°C.

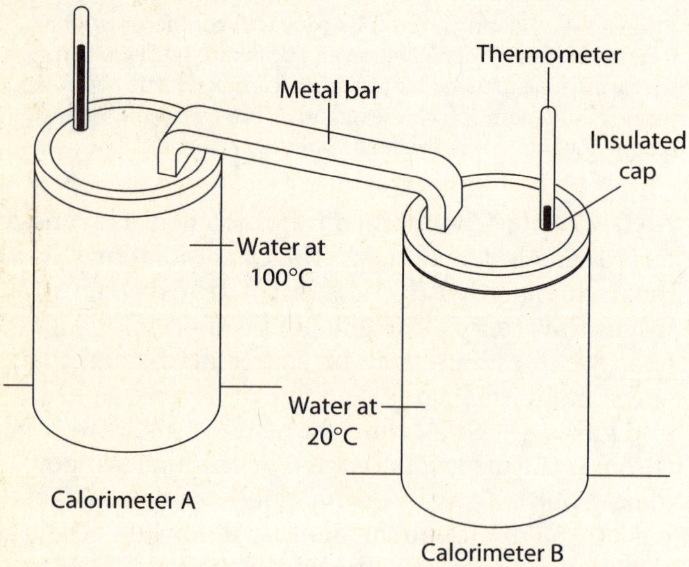

14. If this were a closed system, what would be the temperature when the system reaches equilibrium? (1) 100°C (2) 75°C (3) 60°C (4) 40°C

15. Which conclusion should the student make after performing this investigation? (1) The energy gained by the cold water equaled the energy lost by the hot water. (2) The energy gained by the cold water was less than the energy lost by the hot water. (3) The change in temperature of the cold-water thermometer equaled the change in temperature of the hot-water thermometer. (4) Energy was transferred between the two calorimeters primarily by radiation.

16. Which procedure would best increase the amount of heat energy that is actually gained by calorimeter B? (1) increasing the length of the metal bar (2) increasing the thickness of the metal bar (3) circulating air over the metal bar (4) placing insulation around the metal bar

17. Which graph best represents the probable relationship between the temperatures of the two calorimeters and the time for this heat transfer investigation?

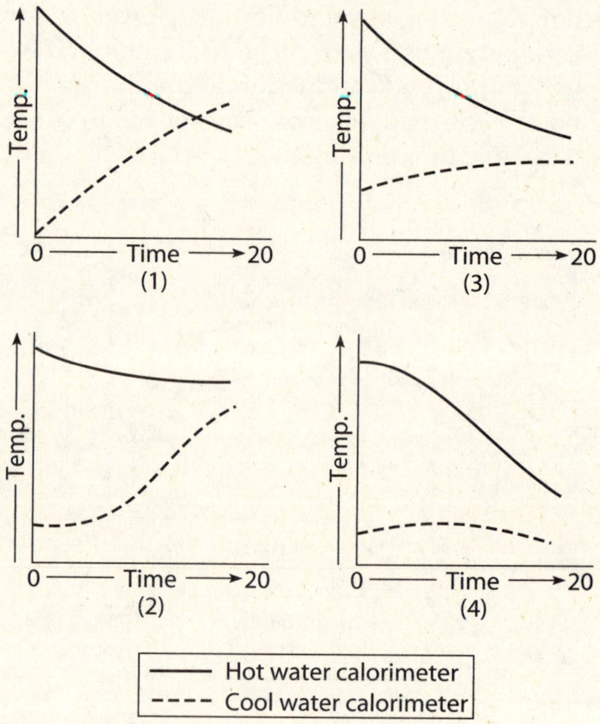

18. Differences in density cause energy to be transferred by which method? (1) absorption (2) conduction (3) convection (4) radiation

19. Water is being heated in a beaker as shown.

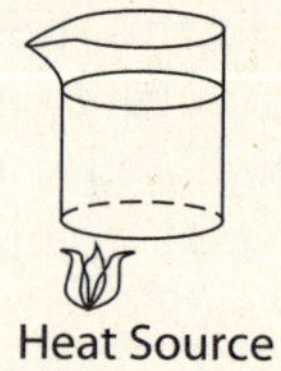

Which of the following drawings shows the most probable movement of water in the beaker due to the heating?

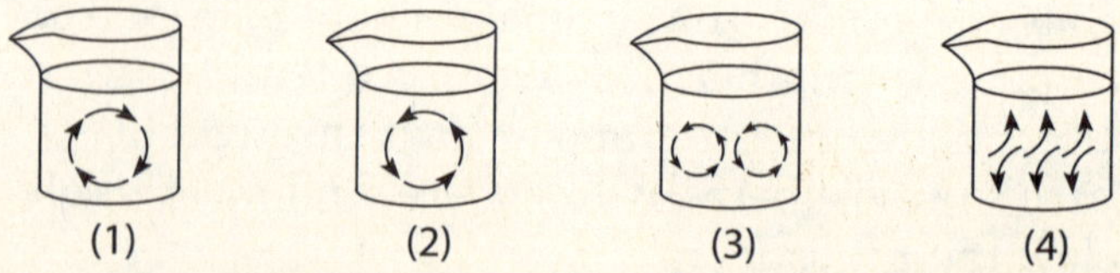

20. At which temperature will an object radiate the greatest amount of electromagnetic energy? (1) 0° F (2) 5°C (3) 10°F (4) 230 K

21. What method of energy transfer requires no medium for transfer? (1) conduction (2) convection (3) refraction (4) radiation

22. By which process does starlight travel through space? (1) absorption (2) conduction (3) vaporization (4) radiation

23. By which process do light rays pass through window glass? (1) conduction (2) convection (3) radiation (4) compression

24. Which diagram correctly indicates why convection currents form in water when water is heated?

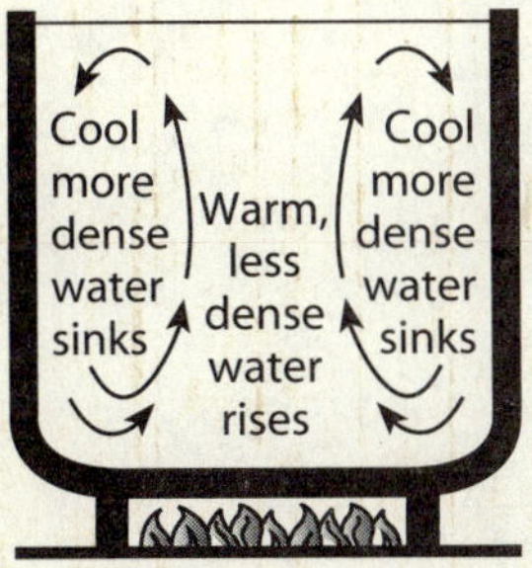

(1)

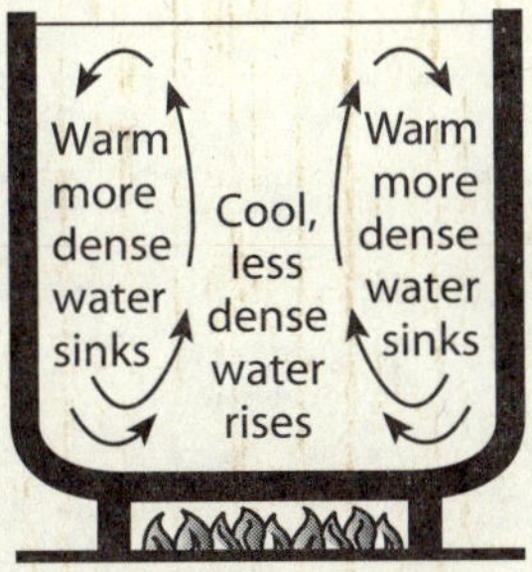

(3)

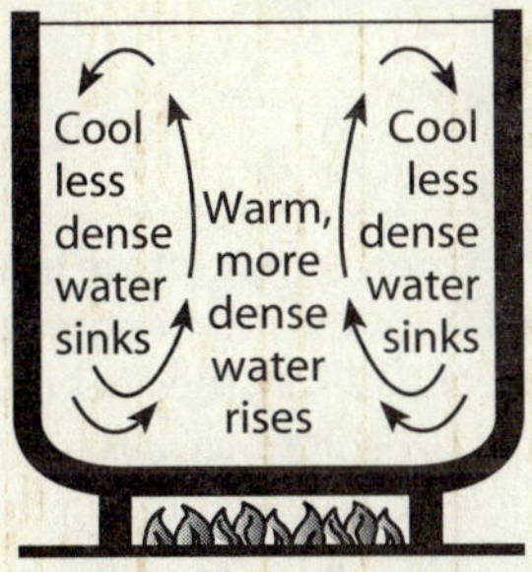

(2)

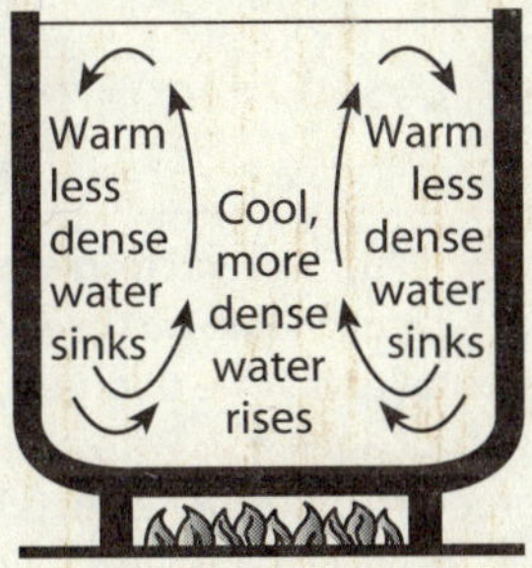

(4)

Transformation of Energy

A transformation of energy is the changing of one type of energy into another type of energy. Some of these transformations are discussed in the following paragraphs.

Heat Production

Transformations of energy often occur when there is friction. For example, as a glacier flows, some of its kinetic energy is transformed into heat energy at the interface of the glacier and its valley walls. This transformation occurs because of friction between the moving glacier and the rock of the valley walls. Transformation of energy also occurs when wind blows over the ocean, creating waves and surface ocean currents. At the interface of the atmosphere and hydrosphere, heat is formed.

Transformations of Mechanical Energy

All the energy of an object or system not related to the individual motions of atoms and molecules is **mechanical energy**. It can also be defined as the total of the potential and kinetic energy of an object or system. An object in motion has a kind of energy called kinetic energy. The faster something moves and the more mass it has, the greater its kinetic energy. Objects or systems can also have a kind of energy called potential energy—energy related to position or phase. It may be thought of as "stored" energy. The higher an object is above the center of Earth, the greater its potential to fall—the greater its potential energy. The more mass that is involved, the more potential energy that is present.

Either kinetic or potential energy can be transformed into the other. For example, water at the top of a waterfall has potential energy because of its position relative to Earth's center. As the water falls to a lower level, some of its potential energy is transformed into kinetic energy, resulting in an increase in speed.

Transformation of Electromagnetic Wavelength

One example of wavelength transformation is when electromagnetic energy is absorbed by an object and reradiated at a longer wavelength. This occurs because energy moves from regions of greater to lower concentrations, and the region of greater concentration usually has a higher temperature and emits shorter wavelengths. This type of transformation is very common at Earth's surface. There the relatively short-wavelength ultraviolet and visible radiations from the sun are absorbed and reradiated as longer wavelength infrared radiations. Since Earth's surface is much lower in temperature than the sun, it emits longer wavelengths than the sun's surface.

Temperature and Heat

The temperature of an object or region is directly related to the amount of heat, or thermal energy, in the object or region.

Temperature

Temperature is a measure of the average kinetic energy of the particles of a body of matter—and is NOT a type of energy. According to the theory of matter, the particles of every material are in continuous, random motion, and therefore have kinetic energy. At any moment, some of the particles have more kinetic energy than others. The greater the average kinetic energy of the particles of matter, the higher the temperature.

Your senses respond to temperature by sensations of hot and cold. However, to accurately measure temperature you need to use a thermometer. A thermometer indicates temperature on a scale marked in degrees. Three different temperature scales are in use in the United States. Relationships among the three temperature scales—Fahrenheit, Celsius, and Kelvin—are given in the *Earth Science Reference Tables*. Figure 5-5 illustrates the method of converting among the three temperature scales.

Heat and Thermal Energy

As explained earlier, thermal energy is the energy of the motion of atoms and molecules. When the thermal energy of one object is greater—has a higher temperature—than another, some of the thermal energy will be transferred from the hotter body to the colder one. The type of energy that is transferred from hotter to colder objects is heat energy. It is measured in a unit called a **calorie**—the quantity of heat needed to raise the temperature of one gram of liquid water by one degree Celsius.

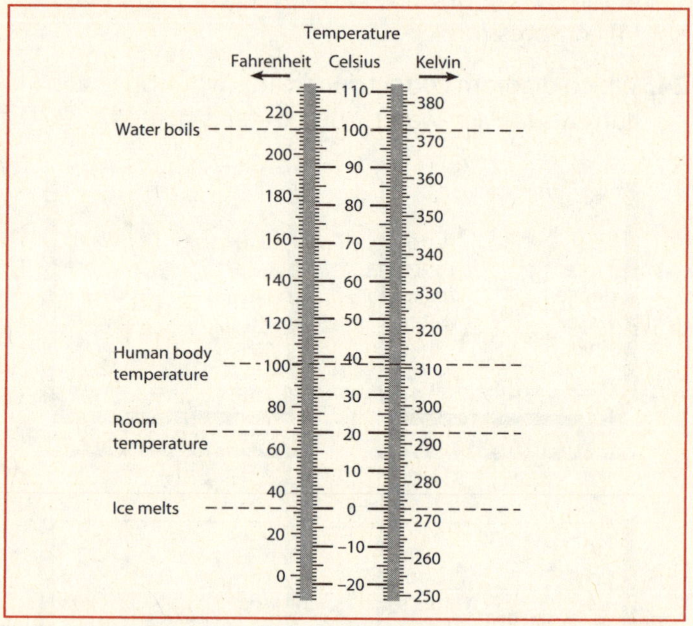

Figure 5-5. Converting among the three temperature scales: Note that each line on the thermometer scales is two degrees for Fahrenheit and one degree for Celsius and Kelvin. The lines are so close that a reading within one degree of Fahrenheit and 0.5 degree of Celsius or Kelvin is acceptable. To convert among the scales, hold a straight edge on the value you are converting from. Then read the value of the other temperature scale on the line. For example, room temperature is 68°F, which converts to 20°C and 293 K. Ice melts at 32°F, 0°C, and 273 K. Water boils at 212°F, 100°C, and 373 K.

Specific Heat

It takes one calorie to raise the temperature of one gram of liquid water one degree Celsius. It takes only about 0.2 calorie (one fifth as much) to raise the temperature of one gram of a typical rock (basalt) one degree Celsius. The quantity of heat needed to raise the temperature of one gram of any substance one degree Celsius is called the **specific heat** of that substance. In simpler terms, specific heat is the resistance a material presents to heating up or cooling off.

Liquid water has the highest specific heat of naturally occurring substances, which is why large bodies of water have a major moderating effect on weather and climate. All other naturally occurring substances have a specific heat less than that of liquid water. Therefore, gaining or losing the same amount of heat causes equal masses of these other substances to heat up or cool off, respectively, faster than water. See Specific Heats of Some Common Materials in the *Earth Science Reference Tables*.

Review Questions

25. How do the wavelengths of electromagnetic energy absorbed by materials on Earth compare to the wavelengths radiated by materials on Earth?
(1) The reradiated wavelengths are shorter.
(2) The reradiated wavelengths are longer.
(3) The absorbed and reradiated wavelengths are the same.

26. A boulder falls freely from point A to point B, as shown in the following diagram.

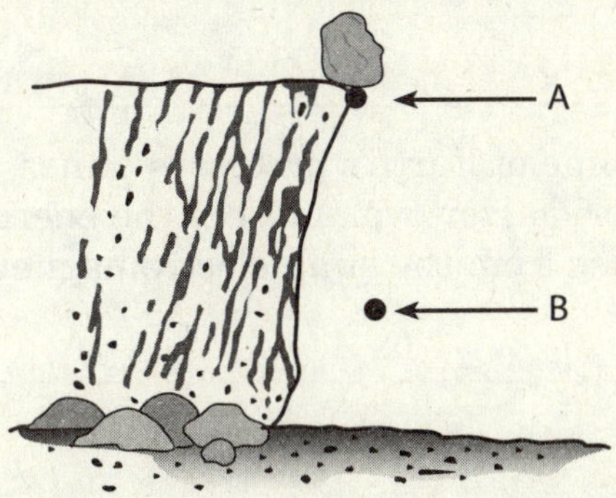

Which statement best explains the relationship between the height of the boulder and its potential and kinetic energy as it falls? (1) Its potential energy decreases, and its kinetic energy increases. (2) Its potential energy increases, and its kinetic energy decreases. (3) Its potential energy and kinetic energy both decrease. (4) Its potential energy and kinetic energy both increase.

27. The temperature of an object is determined by the
(1) average kinetic energy of its molecules
(2) average potential energy of its molecules
(3) total kinetic energy of the object
(4) total potential energy of the object

28. The temperature of a stovetop is 65°C. What is the equivalent Fahrenheit temperature? (1) 126°F (2) 132°F (3) 144°F (4) 149°F

29. Equal masses of lead, granite, basalt, and water at 5°C are exposed to equal quantities of heat energy. Which would be the first to show a temperature rise of 10°C? (1) lead (2) granite (3) basalt (4) water

Heat Energy and Changes of State

Matter may exist in the solid, liquid, or gaseous states, or phases. An increase or decrease in energy and temperature of matter can cause the matter to go from one state to another.

Types of Change of State

The following are some changes of state.

- Melting is the changing of a solid to a liquid.
- **Solidification**, or freezing, is the changing of a liquid to a solid. If solidification results in a solid with an ordered pattern of atoms, the process is **crystallization**.
- Evaporation, or **vaporization** is the changing of a liquid to a gas, or vapor.
- **Condensation** is the changing of a gas, or vapor, to a liquid.
- Sublimation is the changing of a gas directly to a solid, or from a solid directly to a gas—without going through a liquid state.

The changes in state of water are shown in Figure 5-6 on the next page. When water changes state from a solid to a liquid to a gas, it absorbs heat, molecular movement speeds up, and the tightly bound water molecules become less and less tightly bound. When water changes state from a gas to a liquid to a solid, water releases stored heat, molecular movements slow down, and molecules become more tightly bound.

Stored Heat and Changes of State

When a material is in one of the three states, its temperature rises as heat is added to it. If, however, the material is in the process of changing state,

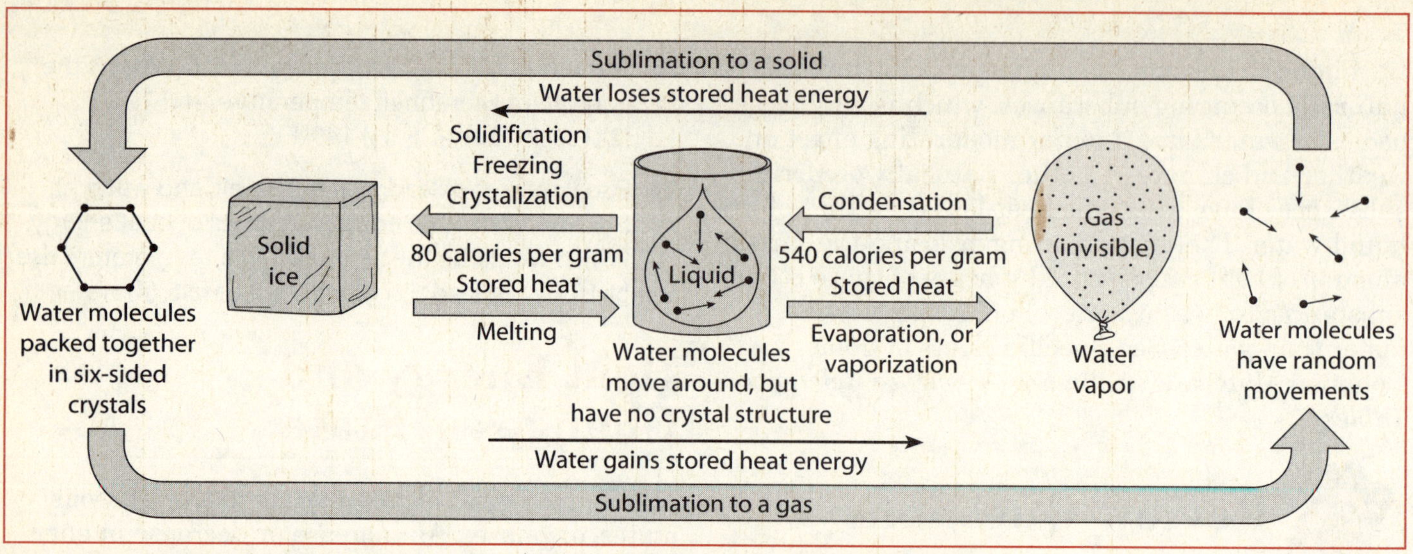

Figure 5-6. Changes in state of water: When water changes phase from a solid to a liquid to a gas, it absorbs heat and changes tightly bound water molecules of a solid to less tightly bound molecules of liquid water and gaseous water. To go from a gas to a liquid to a solid, water releases stored heat and the molecular movements slow down.

its temperature remains the same as it is heated. During the change of state, the added heat energy is not increasing the kinetic energy of the atoms or molecules, and therefore, the temperature does not change. The added heat energy is being converted to a kind of potential energy, or stored heat.

In order for a change of state from a solid to a liquid, or from a liquid to a gas to occur, the substance must gain heat. When the change of state is from a gas to a liquid, or from a liquid to a solid, the substance must lose stored heat. The amount of stored heat, gained or lost, varies with the particular substance and type of change of state.

Stored Heat and Changes of State of Water

Figure 5-7 shows how the temperature of a fixed amount of water changes as it is heated at a constant rate from ice—solid water—at –100°C to gaseous water—water vapor—at 200°C. The temperature is plotted against time in minutes (upper scale of the graph) and the corresponding amount of added heat (lower scale of the graph). You can see that the temperature remains constant at 0°C for almost 2 minutes while the water is changing from the solid state to the liquid state. There is another constant-temperature interval, of more than 10 minutes, at 100°C while the water is changing from the liquid state to the gaseous phase. The reason for these intervals of constant temperature is that the heat being added at those times is being changed to stored heat (potential energy). To find the amount of energy gained or released by water as it changes state, refer to Figure 5-6 or to Properties of Water in the *Earth Science Reference Tables*.

Earth's Energy Supply

The energy for some of Earth's processes comes from Earth's interior. However, most of the energy Earth needs comes from the sun's electromagnetic radiation.

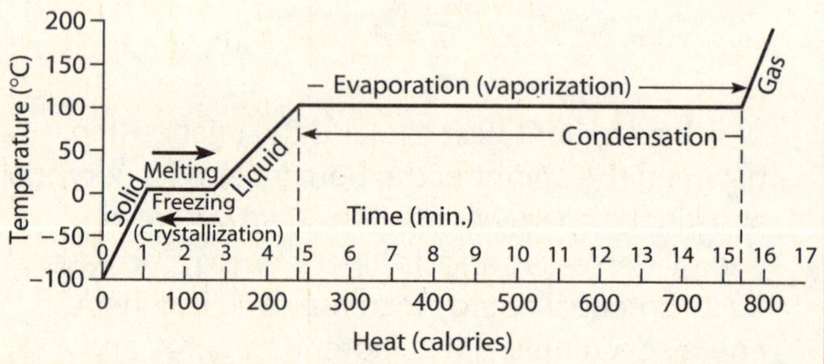

Figure 5-7. Heating or cooling of water: When read from left to right, the graph shows the temperature change of one gram of water as heat is added at a constant rate—50 calories per minute. When read from right to left, the graph shows the temperature change if 50 calories per gram is released from the water. The flat portion of the graph shows data as water absorbs or releases stored heat associated with a change of state. The steeply sloped portions of the graph show data as water is increasing or decreasing in temperature, but not experiencing a change of state.

Solar Energy

The sun radiates, or gives off, and Earth receives a wide range of electromagnetic energy of various wavelengths. This solar electromagnetic spectrum includes X-rays, ultraviolet rays, visible light, and infrared rays. Of all the types of electromagnetic radiation from the sun, the one with the greatest intensity is visible light. (See Figure 6-1 on page 85.)

Solar energy is produced by nuclear fusion—a process in which the nuclei (centers) of atoms are combined to form larger atomic nuclei—releasing great amounts of energy in the process. Nuclear fusion can only occur under the conditions of very high temperature and pressure, for example, in the interior of a star. Figure 5-8 illustrates the type of fusion believed to be most common in the sun. A small fraction of the energy produced by nuclear fusion is radiated as electromagnetic energy and received by Earth—thus "fueling" most Earth processes.

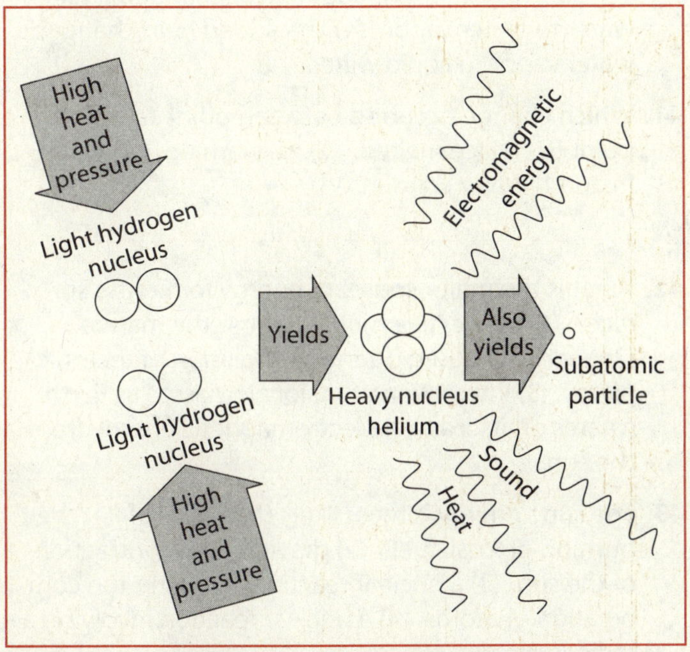

Figure 5-8. A simple model of nuclear fusion in our sun: The high heat and pressure of our sun combine the nuclei of hydrogen to produce helium, energy, and subatomic particles.

Earth's Own Energy

Heat energy from Earth's interior is converted into mechanical energy for mountain building, volcanic eruptions, plate movements, and other internal movements. Some of this heat may remain from when Earth formed, and some may be produced as the result of materials being pulled by gravity towards Earth's center. Probably most of Earth's interior energy comes from the nuclear decay of radioactive materials within Earth. **Nuclear decay,** also called nuclear fission or radioactive decay, is the process by which unstable or radioactive atomic nuclei of elements, such as uranium and radium, split to form lighter elements. In the process, large amounts of energy are released. Nuclear decay is the process employed in the atomic reactors used to produce a portion of the nation's electric supply. Figure 5-9 illustrates the process of nuclear decay.

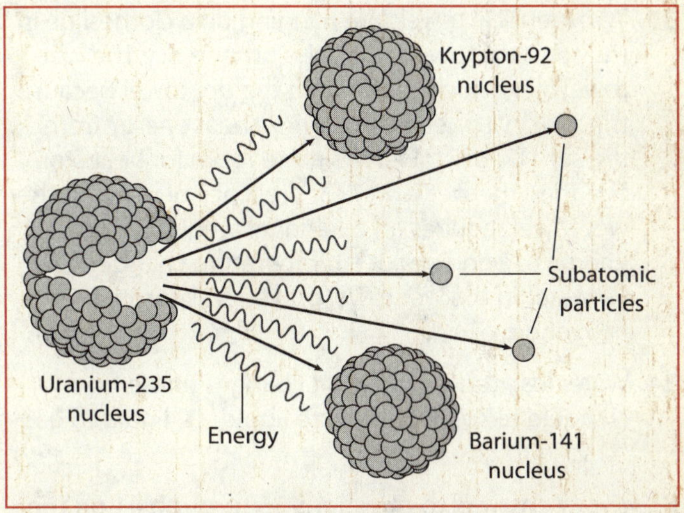

Figure 5-9. The process of nuclear decay (of uranium-235): The nuclei of radioactive elements are unstable, and they can split to form elements with smaller nuclei, subatomic particles, and large amounts of energy.

Review Questions

30. At which temperature will ice melt under normal conditions? (1) 0 K (2) 32 K (3) 212 K (4) 273 K

31. The change from the gas, or vapor phase, to the liquid phase is called (1) evaporation (2) condensation (3) precipitation (4) transpiration

32. Which diagram correctly shows the processes that change the phases of matter?

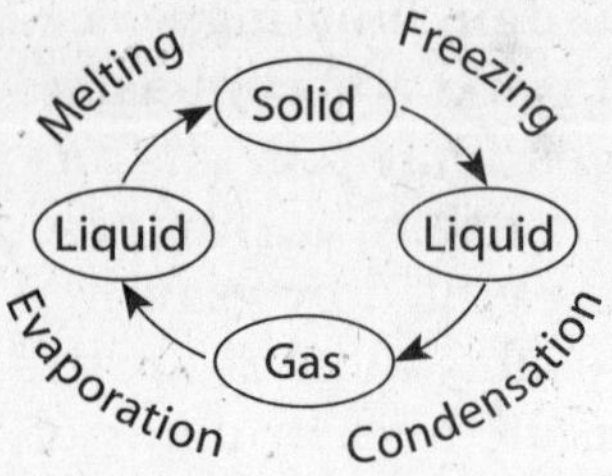

(1)

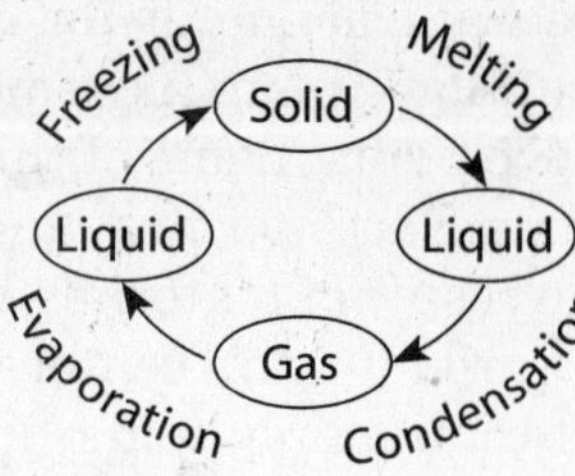

(3)

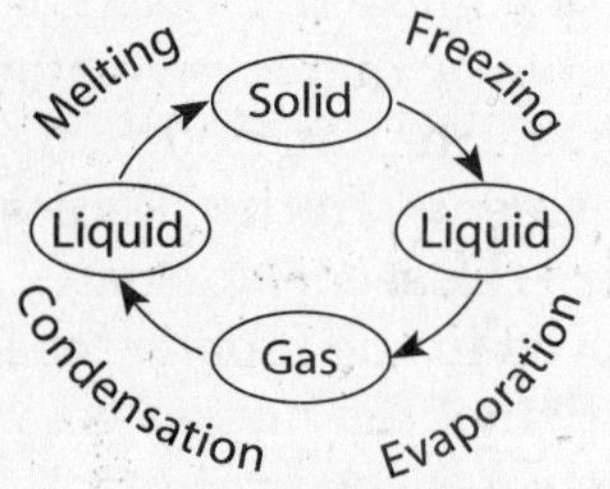

(2)

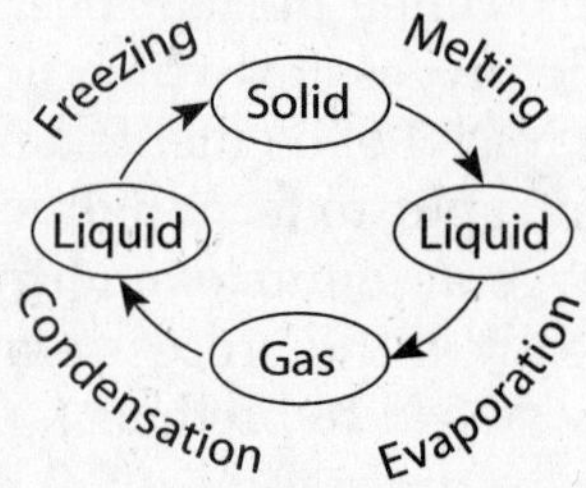

(4)

33. Two identical towels are hanging on a clothesline in the sun. One towel is wet; the other is dry. The wet towel feels much cooler than the dry towel because the (1) dry towel receives more heat energy from the sun (2) dry towel has more room for heat storage than the wet towel (3) presence of water in the wet towel requires an additional amount of heat energy to bring about a temperature change (4) water in the wet towel prevents absorption of heat energy

34. Water loses energy when it changes phase from (1) liquid to solid (2) solid to liquid (3) liquid to gas (4) solid to gas

Base your answers to questions 35 through 41 on the following graph. The graph shows the temperatures recorded when a sample of water was heated from −100°C to 200°C. The water received the same amount of heat every minute.

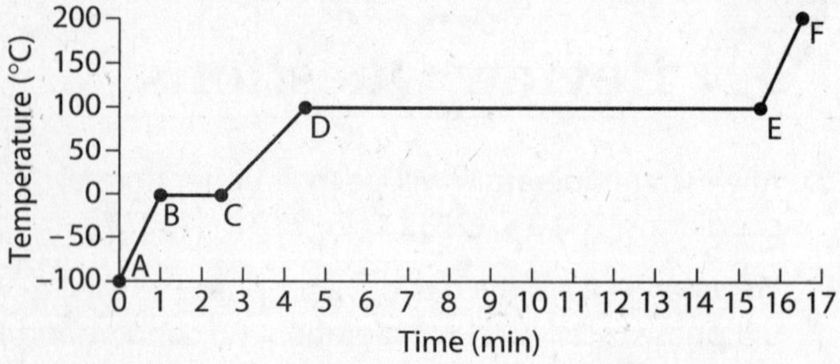

35. For the time on the graph represented by the line from point B to point C, the water was (1) freezing (2) melting (3) condensing (4) boiling

36. At which point in time would most of the water be in the liquid phase? (1) 1 minute (2) 14 minutes (3) 16 minutes (4) 4 minutes

37. What is the rate of temperature change between points C and D? (1) 10°C/min (2) 25°C/min (3)50°C/min (4)150° C/min

38. The greatest amount of energy was absorbed by the water between points (1) A and B (2) B and C (3) C and D (4) D and E

39. During which time interval was the rate of temperature change the greatest? (1) A to B (2) B to C (3) C to D (4) D to E

40. What is the most probable explanation for the constant temperature between points D and E on the graph? (1) The added heat was radiated as fast as it was absorbed. (2) The added heat was lost to the surroundings. (3) The added heat changed liquid water to water vapor. (4) The added heat changed water vapor to liquid water.

41. Which change occurred between point A and point B? (1) Ice melted. (2) Ice warmed. (3) Water froze. (4) Water condensed.

---

42. What is the major source of energy for Earth's surface? (1) convective motion within the mantle (2) energy released during the collision of tectonic plates (3) interaction of major air masses as Earth rotates on its axis (4) electromagnetic energy from the sun

43. The sun's energy is most likely the result of (1) the burning of fossil fuels (2) gravitational contraction of the sun (3) a nuclear reaction involving the combination of atoms (4) a nuclear reaction involving a radioactive decay

44. A source of energy for the high temperatures found deep within Earth is (1) tidal friction (2) incoming solar radiation (3) nuclear decay of radioactive materials (4) meteorite bombardment of Earth

Questions for Regents Practice

Part A

1. A concrete play area is resurfaced with dark-colored asphalt. Compared with the amount of heat energy that was absorbed by the old concrete surface, the amount of heat energy absorbed by the dark-colored asphalt surface will most probably be

1) less

(2) greater

(3) sometimes the same and sometimes less

(4) sometimes greater and sometimes less

2. The following diagram represents the path of visible light as it travels from air to water to air through a glass container of water.

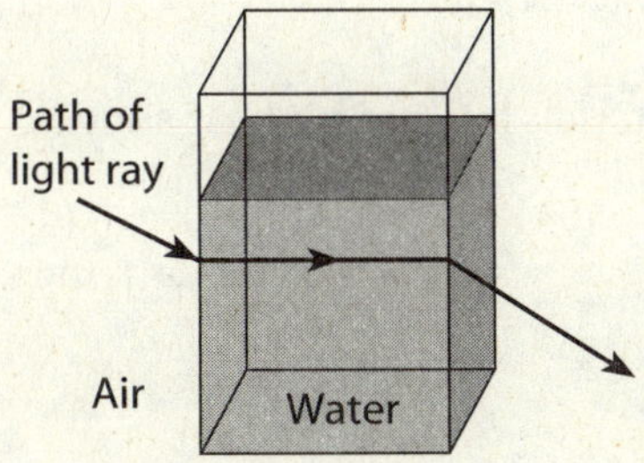

The light does not travel in a straight line because of

(1) convection

(2) scattering

(3) absorption

(4) refraction

3. An ice cube is placed in a glass of water at room temperature. Which heat exchange occurs between the ice and the water within the first minute?

(1) The ice cube gains heat and the water loses heat.

(2) The ice cube loses heat and the water gains heat

(3) Both the ice cube and the water gain heat.

(4) Both the ice cube and the water lose heat.

4. The following map shows four locations in a temperature field. The temperature of each location is given in degrees Celsius. Heat energy will normally flow from

(1) A to B

(2) A to C

(3) B to D

(4) D to C

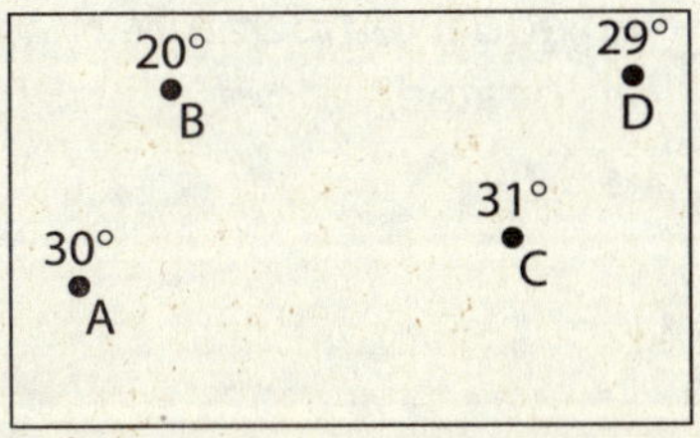

5. A barefoot student steps on a hot concrete surface. Most of the heat transferred to the student's skin by this contact is by the process of

(1) convection

(2) conduction

(3) vaporization

(4) radiation

6. During a volcanic eruption, a rock is thrust upward into the air from point A to point B as shown in the following diagram.

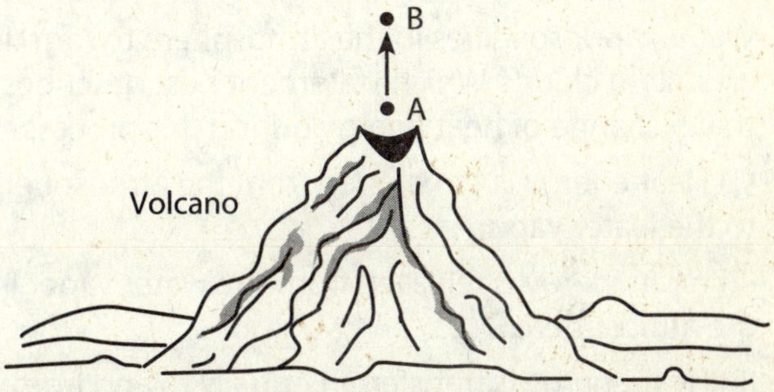

Which graph best represents the relationship between the height of the rock and its potential energy (PE) as it rises?

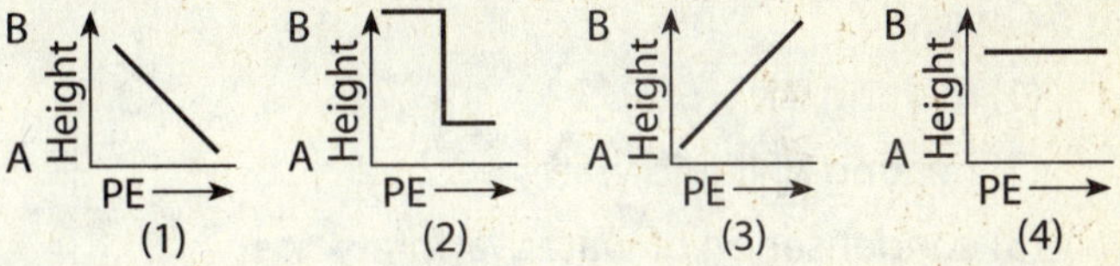

7. The transfer of heat from the sun to Earth takes place by

(1) conduction

(2) convection

(3) radiation

(4) evaporation

8. A camper takes a 100-gram piece of basalt rock from a campfire and places it in a cup holding 250 milliliters of water. The temperature of the rock is 300°C and the temperature of the water is 20°C. Air temperature is 20°C. In the process of heating the water with the basalt, the temperature of the basalt decreased more than the temperature of the water increased. This difference most likely occurred because

(1) water has a higher specific heat than basalt

(2) water has a higher density than basalt

(3) the basalt had a greater mass than the water did

(4) the basalt had a higher starting temperature than the water did

9. Water vapor solidifies in the atmosphere to form ice crystals in clouds. Which statement best describes the exchange of heat energy during this process?

(1) Heat energy is transferred from the atmosphere to the water vapor.

(2) Heat energy is released from the water vapor into the atmosphere.

(3) Heat energy is transferred equally to and from the water vapor.

(4) No heat energy is exchanged between the atmosphere and the water vapor.

10. Which process results in a release of stored heat energy?

(1) melting ice

(2) heating of liquid water

(3) condensation of water vapor

(4) evaporation of water

11. Which process requires the most absorption of energy by water?

(1) melting 1 gram of ice

(2) condensing 1 gram of water vapor

(3) vaporizing 1 gram of liquid water

(4) freezing 1 gram of liquid water

12. The sun's energy is often converted into potential energy by

(1) evaporation of water from the oceans

(2) formation of fog in a valley

(3) freezing of water droplets on a highway

(4) precipitation of rain from a thunderstorm

Base your answers to questions 13 through 15 on the following graph. The graph shows the amount of heat energy (calories) needed to raise the temperature of 1-gram samples of four different materials.

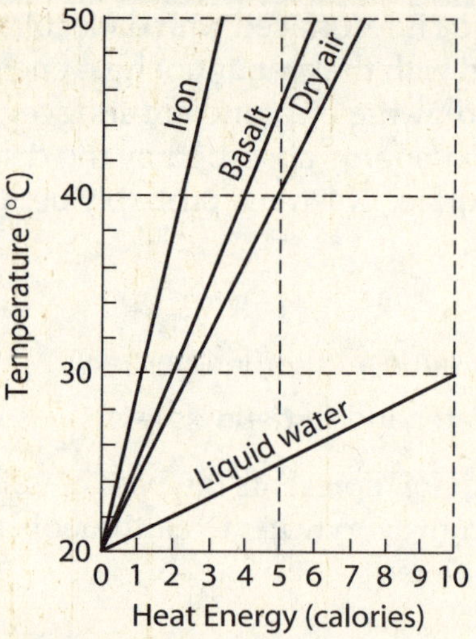

13. Which of these materials has the highest specific heat?

(1) liquid water

(2) dry air

(3) basalt

(4) iron

14. If all four materials were heated to 100°C and then allowed to cool, which material would show the most rapid drop in temperature?

(1) basalt

(2) iron

(3) dry air

(4) liquid water

15. Which statement is best supported by the graph?

(1) The same amount of heat energy is required to raise the temperature of each material by 10°C.

(2) The temperature of a material with a high specific heat is raised faster than that of a material with a low specific heat.

(3) Three of the four materials have the same specific heat.

(4) The amount of heat energy needed to produce an equal temperature change varies with the materials heated.

Part B

Base your answers to questions 16 through 20 on the following diagram and data tables. The diagram represents a light source located equal distance from two air-filled metal cans. One can is shiny, and the other is black. Both cans were heated for 10 minutes. The lamp was removed, and both cans were allowed to cool for 10 minutes. Temperature readings of the cans' interiors were recorded each minute for the 20 minutes of heating and cooling. The data tables show these temperatures. A graph is provided to allow you to plot the given data and to assist you in answering the questions.

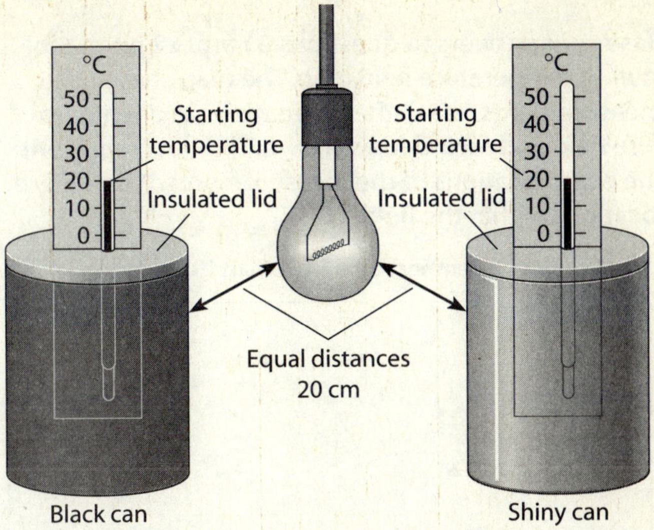

Lamp On											
Time (min)	0	1	2	3	4	5	6	7	8	9	10
Temperature of Shiny Can (°C)	20	20	21	22	23	24	24.5	25	25.5	26	26.5
Temperature of Black Can (°C)	20	21	23	25	27	28.5	29.5	30.5	31	31.5	32
Lamp Off											
Time (min)	11	12	13	14	15	16	17	18	19	20	
Temperature of Shiny Can (°C)	26.5	26	25.5	25	24.5	24	24	23.5	23.5	23	
Temperature of Black Can (°C)	31.5	31	30.5	29.5	28.5	27.5	26.5	25.5	25	24.5	

16. On the graph at the right, plot the data given in the tables for each can. [4]

17. At which time during the investigation was the temperature in the black can 31°C and the temperature in the shiny can 25.5°C? [1]

18. At what rate did the temperature of the black can change during the first 10 minutes of the investigation? [1]

19. By which process was most of the heat energy transferred from the light source to the cans during the first 10 minutes of the investigation? [1]

20. Why did the black can become warmer than the shiny can during the heating period? [1]

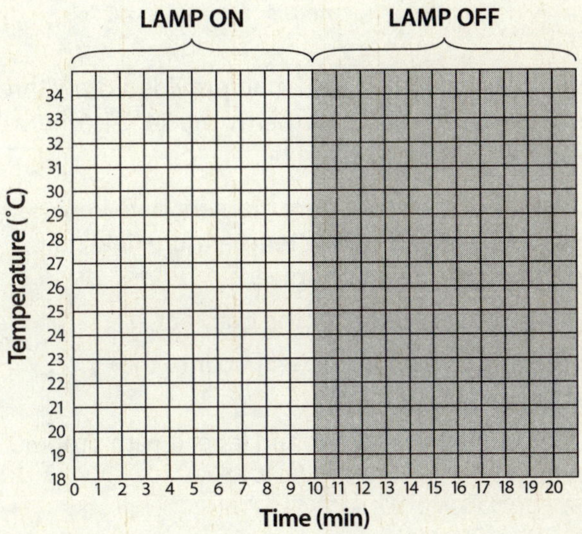

Base your answers to questions 21 and 22 on the following temperature field map. The map shows 25 measurements (in °C) that were made in a temperature field and recorded as shown. The dots represent the exact locations of the measurements. A and B are locations within the field.

Temperature Field Map (°C)

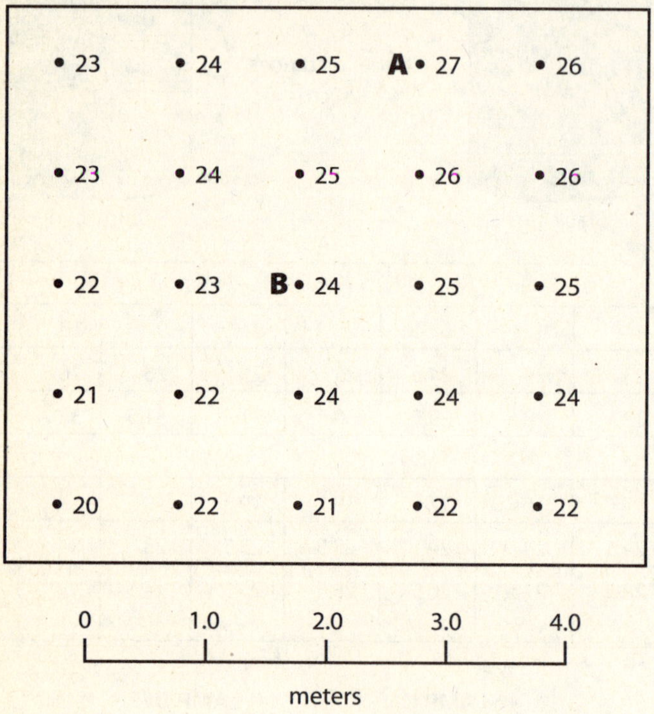

21. On the temperature field map provided, draw three isotherms—the 23°C isotherm, the 24°C isotherm, and the 25°C isotherm. [2]

22. Calculate the temperature gradient between locations A and B on the temperature field map, following the directions given.

a. Write the equation for the gradient. [1]

b. Substitute data from the map into the equation. [1]

c. Calculate the gradient, and label it with proper units. [1]

Part C

The following paragraph provides information about geothermal energy as an alternate energy source. Use data from the paragraph and your knowledge of earth science to answer questions 23 through 26.

One alternate source of energy humans can substitute for declining supplies of fossil fuels is geothermal energy. Geothermal energy is the heat energy usually stored in rocks and water beneath Earth's solid surface. It is generally believed that hot rocks are available at any location if you drill deep enough, but drilling more than few thousand feet is usually too expensive and not economically practical. Where hot rocks with hot gaseous water (steam) and/or hot liquid water are found in large quantities, within a few thousand feet of the Earth's surface, the geothermal energy may be used in place of fossil fuels. The hot gaseous or liquid water is either used directly to heat homes and industries, or it is used to generate electricity in a facility similar to ones that use fossil fuels to heat water to produce electricity.

23. Explain what nuclear, or atomic, process most likely results in geothermal energy. Also state why this nuclear energy will last longer than the supply of fossil fuels. [2]

24. Explain how the nuclear process that results in geothermal energy is different from the process that produces energy within the sun. [1]

25. What evidence in the *Earth Science Reference Tables* suggests the existence of geothermal energy? [1]

26. There are two geothermal deposits of equal mass and depth. One deposit is all steam and the other is all hot liquid water. Which deposit would most likely produce the greater amount of geothermal energy? Why? [1]

27. About 5 percent of the energy used by people in the United States comes from hydroelectric facilities. These facilities use the power of falling water from streams and lakes to turn mechanical devices to generate electricity. Using the information above and your knowledge of earth science, explain how energy produced by hydroelectric facilities can be considered the result of both solar energy and gravitational energy. [2]

Insolation and the Seasons

TOPIC 6

VOCABULARY		
angle of incidence	**greenhouse gases**	**ozone**
deforestation	**heat budget**	**sunspot**
El Niño	**ice ages**	**transpiration**
global warming	**insolation**	

Without energy from the sun, conditions on Earth would be drastically different. Energy from the sun drives global wind patterns, ocean currents, and the water cycle. Life processes of plants and animals also depend almost entirely on energy from the sun. Without solar energy, Earth would be a frozen wasteland, as all water and all the gases of Earth's atmosphere would exist only in the solid state. There would be no life and no rock cycle as it is now known. Very little weathering and erosion would occur. Both Earth and its moon would be totally dark.

Solar Radiation and Insolation

As discussed in Topic 5, a body of matter that is not at absolute zero emits electromagnetic energy. This energy is radiated over a portion of the electromagnetic spectrum, with more of the energy radiated at some wavelengths than at others. The rate at which energy is radiated is called the intensity of radiation. In general, the higher the temperature of a body of matter, the shorter the wavelength at which the maximum intensity of radiation occurs. At the sun's temperature, the maximum intensity of radiation occurs in the range of visible wavelengths of electromagnetic energy. (See Figure 6-1.)

Insolation (**IN**coming **SOL**ar radi**ATION**) is the portion of the sun's output of electromagnetic radiation that is received by Earth at the outermost part of our atmosphere. The intensity of insolation is the relative strength of the sun's radiation that reaches a specific area of Earth in a specific amount of time. Figure 6-1 shows how the intensity of insolation varies with wavelength. The maximum intensity of insolation occurs in the range of wavelengths of visible light. However, approximately 48 percent of the total energy received at the outermost part of our atmosphere is infrared, a type of long-wave radiation.

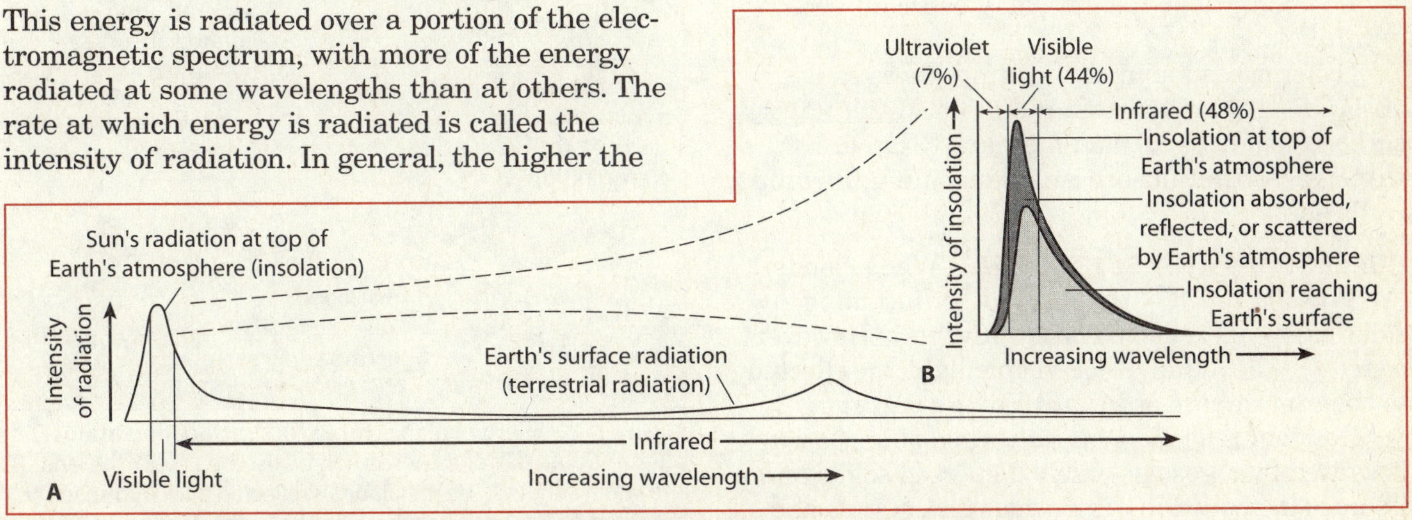

Figure 6-1. Intensity of insolation received by Earth and energy radiated from Earth into space: **(A)** The sun, being much hotter than Earth (sun 5800°C; Earth 15°C) emits shorter wavelengths of radiation than Earth does. Earth's surface emits only long wavelengths—infrared radiation. **(B)** The total amount of energy received by Earth's surface equals approximately 50 percent of the insolation that reaches the outermost part of Earth's atmosphere.

MEMORY JOGGER

You may recall from Topic 5 that a wavelength in an electromagnetic wave, as in other waves, is the distance between two successive peaks.

Effects of Earth's Atmosphere on Insolation

The insolation reaching Earth's surface is different from the insolation entering Earth's upper atmosphere, as shown in Figure 6-1. Since the atmosphere is mostly transparent to visible light, it transmits most of the visible light from the sun. However, insolation may be absorbed, reflected, or scattered before reaching Earth's surface. The atmosphere affects insolation in several ways.

ABSORPTION OF ULTRAVIOLET AND INFRARED Most incoming ultraviolet radiation and other short wave radiation are absorbed by Earth's atmosphere. Nearly all ultraviolet radiation is absorbed in Earth's upper atmosphere by **ozone** (O_3), a form of oxygen gas. In recent years, the amount of ozone in the upper atmosphere has been reduced by the chlorine and fluorine that have escaped into the atmosphere—mostly due to human activities. The result is that more ultraviolet radiation has been reaching Earth's surface. Ultraviolet radiation can be lethal to many forms of life and is a direct cause of skin cancer in humans. Regions in which the amount of ozone has been greatly reduced—popularly called "ozone holes"—occur near Earth's poles, especially the South Pole.

Other gases found in the atmosphere can absorb long-wave infrared radiation. Water vapor, carbon dioxide (CO_2), and methane (CH_4) are three gases that absorb much incoming infrared radiation.

REFLECTION AND SCATTERING When clouds are present, much of the incident, or incoming, solar energy is reflected back into space. However, some of it, including some visible light, is reflected to the atmosphere or toward Earth's surface.

Random reflection, or scattering, of insolation is caused by aerosols—finely dispersed solids and liquids suspended in air. Besides ice crystals and water droplets, the dispersed materials include dust, volcanic ash, and various other air pollutants. As the concentration of aerosols increases, the scattering of insolation also increases, thus reducing the amount of insolation that reaches Earth's surface.

DIGGING DEEPER

On a typical day, it is estimated that approximately 25 percent of incident solar energy is reflected back into space by the cloud cover. This is the cause of the bright clouds seen in images of Earth taken from space.

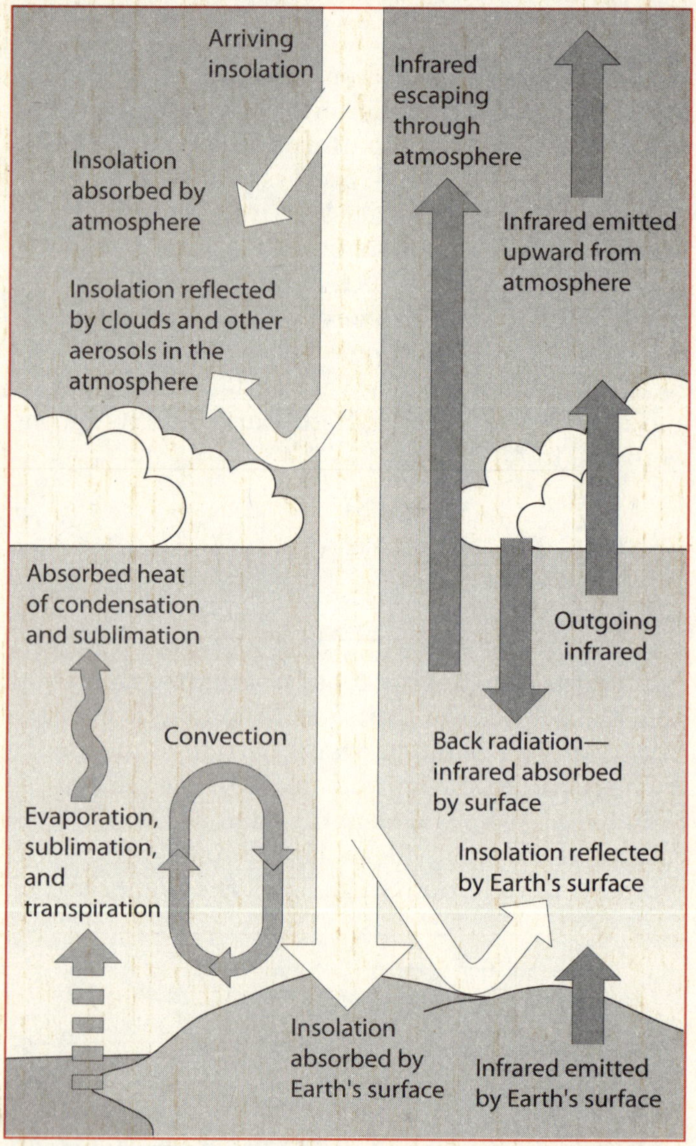

Figure 6-2. Energy balance for Earth's surface and atmosphere: About half of the insolation is absorbed by Earth's land and water surfaces. The remainder is reflected back into space or absorbed by Earth's atmosphere. Earth's surface and atmosphere also absorb energy from reradiation which occurs as some of the infrared radiation emitted by Earth's surface is radiated back by greenhouse gases in the atmosphere.

Balance of Energy from Insolation and Earth's Surface Radiation

About half of the insolation that strikes Earth's upper atmosphere reaches Earth's land and water surfaces, as shown in Figure 6-2. Some of the insolation is reflected and some is absorbed. Insolation that is absorbed by Earth's surface is converted to heat and tends to raise the temperature of the surface. However, Earth's surface also radiates electromagnetic energy, thus tending to lower its temperature. Over time, the amount of energy absorbed from insolation is generally equal to Earth's surface radiation, and thus temperature and heat are balanced.

Factors Affecting Absorption and Reflection of Insolation

When insolation reaches Earth's surface, many factors affect whether it is absorbed or reflected. These factors include the angle at which the insolation reaches Earth's surface, the characteristics of the surface, and how the energy of the insolation interacts with Earth's surface materials and living things.

ANGLE OF INCIDENCE One important factor affecting the absorption of insolation is the angle at which the insolation strikes Earth's surface—called the **angle of incidence.** The altitude of the sun—which varies with time of day, latitude, and season—determines the angle of incidence. The higher the sun is in the sky, the higher the angle of incidence, and the more insolation is absorbed. Generally, a lower angle of incidence means that more insolation is reflected and less is absorbed.

SURFACE CHARACTERISTICS Texture and color also affect the absorption of insolation by Earth's surfaces. When a surface has a rough or uneven texture, more insolation is absorbed and less is reflected. Surfaces with darker colors tend to absorb more insolation than they reflect. Surfaces with lighter colors tend to reflect more insolation than they absorb. For example, ice and snow reflect almost all of the insolation that strikes them. This is one reason the polar regions stay cool even during summer's six months of continuous sunshine.

CHANGE OF STATE AND TRANSPIRATION When energy from insolation reaches Earth's surfaces, it interacts with the materials and living things found there. For example, energy from insolation can cause water to change state—melting of ice or snow into liquid water or evaporating liquid water to form water vapor. Energy from insolation also fuels plant growth and increases **transpiration**—a process by which plants release water vapor into the atmosphere as part of their life functions. When energy from insolation is involved in change of state and transpiration, it is not available to raise the temperature of Earth's surface. Instead, it is transformed into potential energy.

Land and Water Heating

With equal amounts of insolation, an area of Earth's surface made up of liquid water will heat up and cool off more slowly than an equal area of land. These rates in heating and cooling differ for the following reasons:

- Water has a much higher specific heat than the rocks and soil that make up land.
- Water is highly transparent to insolation. Some insolation penetrates up to 300 feet in clear water but usually to less than a foot in rocks and soil. Rocks and soil heat rapidly because the energy of insolation is concentrated in a thin zone.
- Water can flow freely, so convection can occur. Convection currents in water can distribute absorbed energy through a large volume.
- When insolation strikes water surfaces, much of its energy is converted to stored heat as some water evaporates. Thus, less energy is available to increase the water temperature.

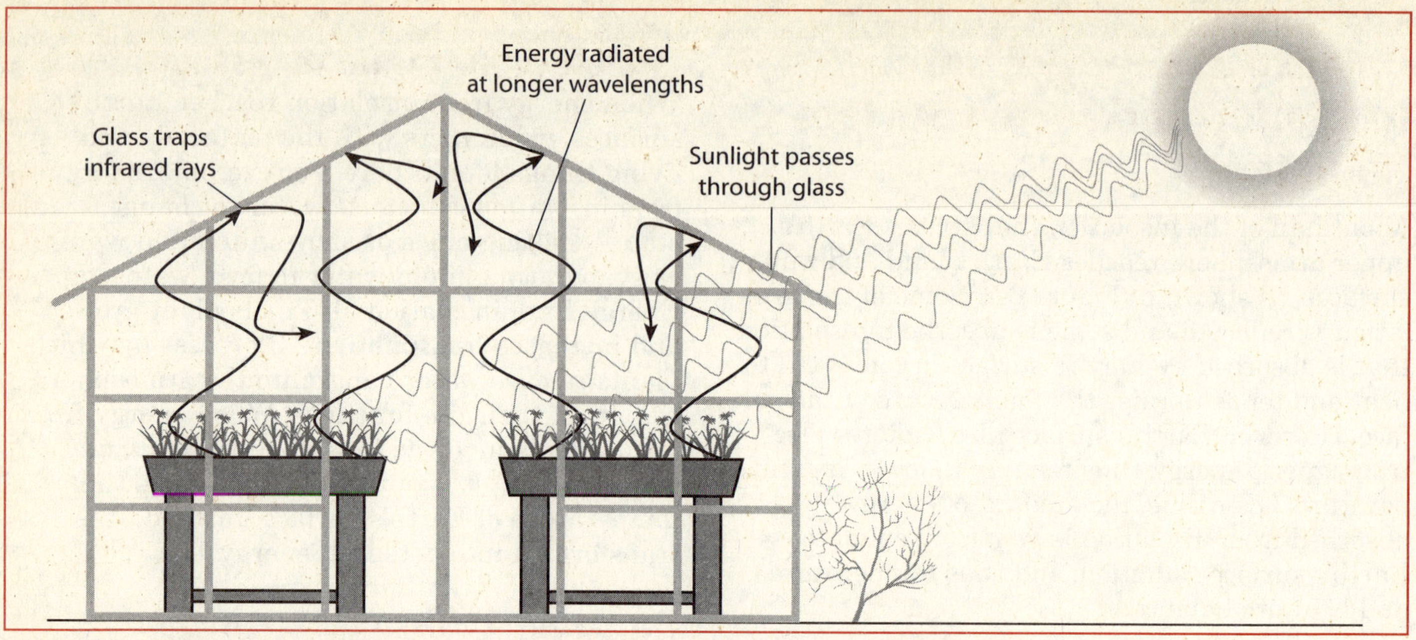

Figure 6-3. The greenhouse effect: Insolation is transmitted through the glass of a greenhouse, just as it is transmitted through Earth's atmosphere. Objects inside a greenhouse absorb energy from the insolation and are heated. The heated objects then give off long wave infrared radiation, which cannot pass outward through the glass. The infrared radiation is trapped by the glass, and is partly absorbed by the air and by objects inside the greenhouse—raising their temperatures. In the same way, radiation emitted from Earth's surface is trapped in the atmosphere by greenhouse gases—increasing Earth's surface temperatures.

The Greenhouse Effect

Although Earth's atmosphere allows much insolation to pass through to Earth's surface, most of the radiation from Earth's surface does not escape into outer space. Instead, it is absorbed by the atmosphere. This occurs because Earth's surface radiation consists mostly of long infrared radiation, which can be absorbed by carbon dioxide, water vapor, methane, and other gases in the atmosphere. These gases are called **greenhouse gases.** This absorption warms the atmosphere and causes it to act as a heat "blanket." This blanket reduces energy loss to outer space and makes Earth's surface about 59°F (33°C) warmer than it would otherwise be. Without any greenhouse gases, the average temperature of Earth would be about 0°F (–18°C).

The process by which the atmosphere transmits short-wave radiation from insolation and absorbs long-wave reradiation is called the greenhouse effect. (See Figure 6-3.) In recent years there has been concern that the greenhouse effect may be causing **global warming**—an increase in average Earth temperatures. The burning of wood and fossil fuels releases additional carbon dioxide and other greenhouse gases into the atmosphere. This increase in carbon dioxide and other greenhouse gases is thought to intensify the greenhouse effect—thus raising Earth's surface temperatures. Global warming could result in environmental problems such as rising sea levels caused by melting glaciers and sea ice, and the shifting of Earth's climatic zones. (See Figure 6-10.)

Review Questions

1. Electromagnetic energy that reaches Earth from the sun is called (1) insolation (2) conduction (3) specific heat (4) terrestrial radiation

2. The graph on the next page represents the relationship between the intensity and wavelength of the sun's electromagnetic radiation.

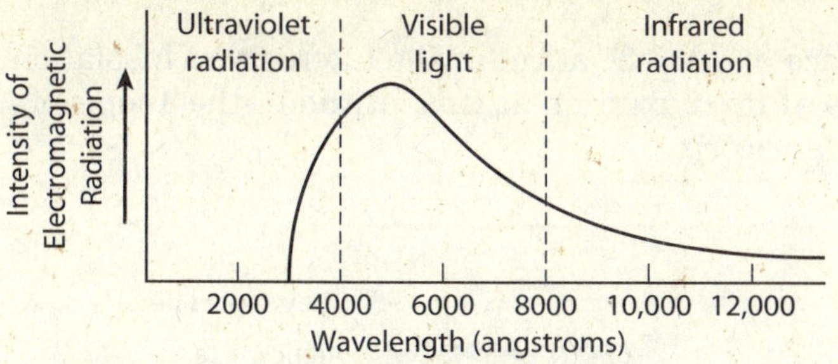

Which statement is best supported by the graph? (1) The infrared radiation given off by the sun occurs at a wavelength of 2000 angstroms. (2) The maximum intensity of radiation given off by the sun occurs in the visible region. (3) The infrared radiation given off by the sun has a shorter wavelength than ultraviolet radiation. (4) The electromagnetic energy given off by the sun consists of a single wavelength.

3. Water vapor and carbon dioxide in Earth's atmosphere are good absorbers of (1) visible radiation (2) infrared radiation (3) ultraviolet radiation (4) X-rays

4. If dust particles are added to the atmosphere, the amount of insolation reaching the ground will probably (1) decrease (2) increase (3) stay the same (4) increase, then decrease

5. Ozone is important to life on Earth because ozone (1) cools refrigerators and air conditioners (2) absorbs energy that is radiated by Earth (3) absorbs harmful ultraviolet radiation (4) destroys excess atmospheric carbon dioxide

6. Which form of electromagnetic energy is radiated from Earth's surface with the greatest intensity? (1) X-rays (2) infrared rays (3) ultraviolet rays (4) visible light rays

7. In land areas of equal size located at the same latitude, the most solar radiation would probably be reflected by a (1) snow field (2) sandy desert (3) grassy field (4) forest

8. Insolation is changed into potential energy by (1) evaporation and transportation of water from Earth's surface (2) formation of clouds and fog (3) freezing of lake and ocean water (4) rain and snow forming in clouds

9. How do the rates of warming and cooling of land surfaces compare to the rates of warming and cooling of ocean surfaces? (1) Land surfaces warm more quickly and cool more slowly. (2) Land surfaces warm more slowly and cool more quickly. (3) Land surfaces warm more quickly and cool more quickly. (4) Land surfaces warm more slowly and cool more slowly.

10. Which model best represents how a greenhouse remains warm as a result of insolation from the sun?

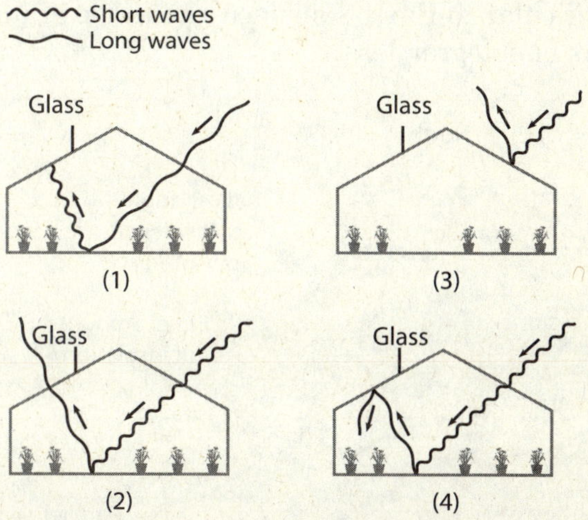

Variation of Insolation

Insolation varies from place to place on Earth's surface. It also varies in any given place over the course of a year. Insolation can vary in two general ways—by intensity (strength) and by duration (length of time).

Variation in Intensity of Insolation

The rate at which solar energy is received by a given area per unit time is the intensity of insolation. Intensity of insolation can be measured in calories per square meter per second. The intensity of insolation received at a particular area of Earth's surface varies for several reasons because of the angle of incidence. When insolation strikes a surface at an angle of 90°, or perpendicular, it has maximum intensity, because the insolation is concentrated in the smallest possible area. As the angle of incidence decreases from 90° toward 0°, the same amount of insolation is spread out over greater and greater areas, as shown in Figure 6-4. Therefore, at smaller angles of incidence, the intensity of insolation decreases the heating effect.

EFFECT OF EARTH'S SHAPE The sun's energy reaches Earth as a bundle of parallel rays. If Earth were flat and positioned perpendicular to these rays, the intensity of insolation would be the same everywhere on Earth's surface. However, Earth is shaped like a sphere. As a result, at any given time, there is just one place where insolation arrives perpendicular to Earth's surface. At all other places on Earth the angle of incidence is less than 90°. Generally, for each degree of latitude North or South of the place where insolation is perpendicular to Earth's surface, the angle of incidence is one degree less.

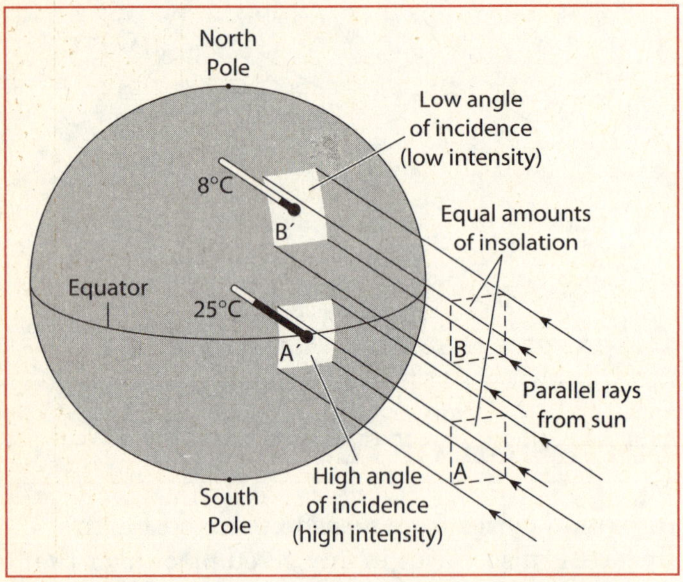

Figure 6-4. Earth's shape and the intensity of insolation: Because of Earth's spherical shape, the parallel rays from the sun strike different parts of Earth's surface at different angles. Near Earth's poles where the angle of incidence is smaller, the same amount of insolation spreads out over a larger area. As a result, the intensity of insolation is much less than that in regions with a higher angle of incidence of the sun's rays.

EFFECT OF LATITUDE The intensity of insolation is greatest at the equator at each equinox, and decreases with increasing latitude (north or south). As shown in Fig. 6-5, insolation is perpendicular to Earth's surface at the equator at the time of each equinox. The spring equinox is March 21, and the fall equinox is September 23.

On the first day of summer in the Northern Hemisphere (summer solstice), insolation is perpendicular to Earth's surface at $23\frac{1}{2}°$ N latitude. The intensity of insolation is at its maximum at this latitude—the Tropic of Cancer. On the first day of winter in the Northern Hemisphere (winter solstice), insolation is perpendicular to Earth's surface at $23\frac{1}{2}°$ S latitude. The intensity of insolation is at its maximum at this latitude—the Tropic of Capricorn.

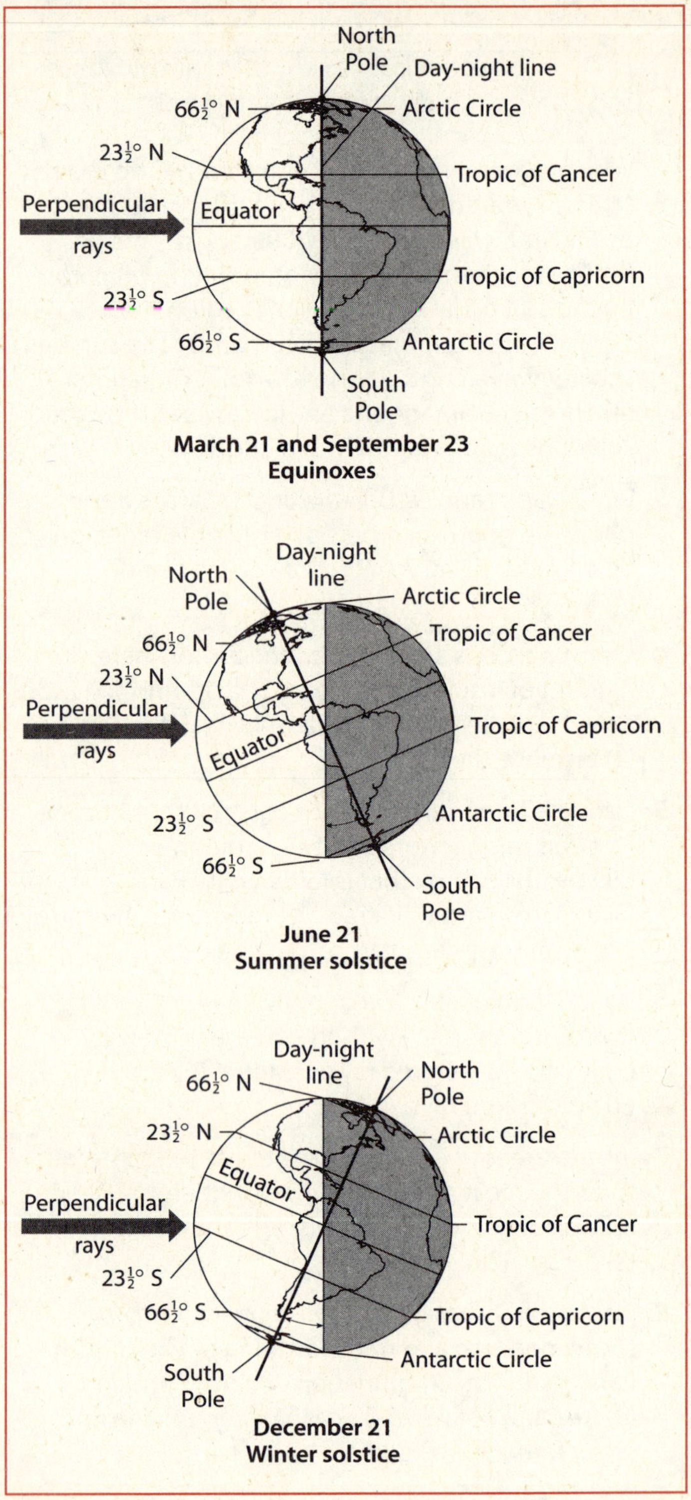

Figure 6-5. Seasonal changes in insolation at different latitudes: At any given time, the intensity of insolation is at a maximum at the latitude where the angle of incidence is 90°. The latitude of maximum intensity of insolation shifts with the seasons between $23\frac{1}{2}°$ N on June 21 to $23\frac{1}{2}°$ S on December 21.

EFFECT OF SEASONAL CHANGES As Earth travels in its yearly orbit around the sun, the angle of incidence for insolation at any given latitude varies with the seasons. As shown in Figure 6-6, the lowest angle of incidence at 42° N latitude occurs at the winter solstice and the highest angle at the summer solstice. As shown in Table 6-1 on this page, the angle of incidence for insolation decreases by one degree for each degree of latitude away from the point of perpendicular insolation.

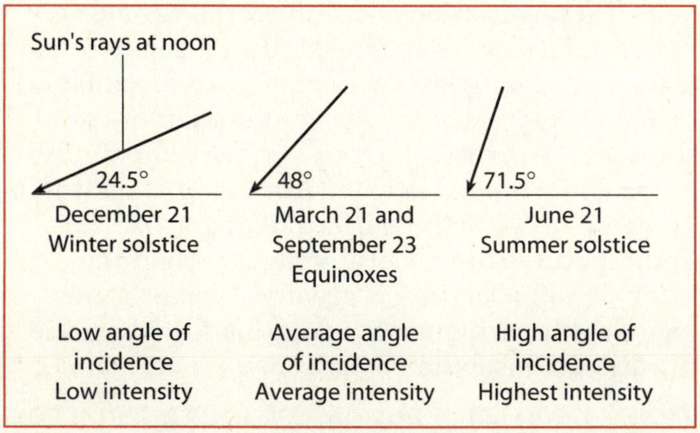

Figure 6-6. Maximum angles of incidence at 42° N latitude in different seasons: Values were measured at solar noon in New York State on the first day of each of the four seasons.

EFFECT OF TIME OF DAY On any given day, the altitude of the sun varies from zero at sunrise, to a maximum at apparent solar noon, and back to zero again at sunset. In a similar way, the angle of incidence and the intensity of insolation also vary during the day as shown in Figure 6-7. The maximum angle and intensity occur at solar noon.

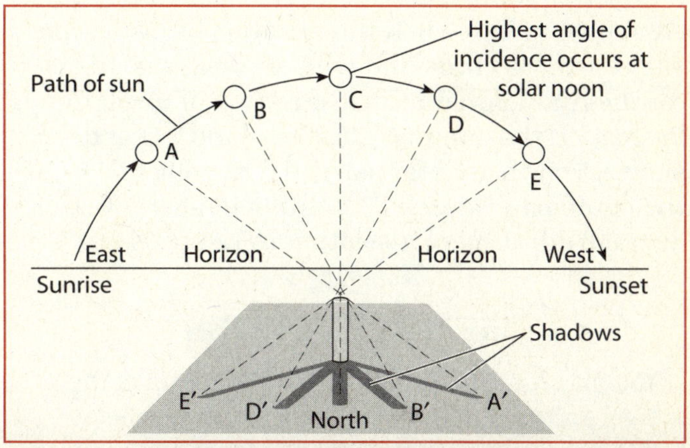

Figure 6-7. Changes in length and direction of shadows during a day: The shadows shown here correspond to those occurring on an equinox day in the mid-latitudes of the Northern Hemisphere. The shadow of a vertical post indicates how the angle of incidence varies throughout the day. A higher angle of incidence produces a shorter shadow and a greater intensity of insolation.

Table 6-1. Variation in Insolation by Latitude and Season

Latitude	Summer Solstice June 21		Equinoxes: March 21, September 23		Winter Solstice December 21	
	Angle of Incidence at 12:00 Noon	Duration of Insolation	Angle of incidence at 12:00 Noon	Duration of Insolation	Angle of Incidence at 12:00 Noon	Duration of Insolation
90° N	$23\frac{1}{2}$°	24 Hours	0°	12 Hours	—	0 Hours
80° N	$33\frac{1}{2}$°	24	10°	12	—	0
70° N	$43\frac{1}{2}$°	24	20°	12	—	0
$66\frac{1}{2}$° N	47°	24	$23\frac{1}{2}$°	12	0°	0
60° N	$53\frac{1}{2}$°	$18\frac{1}{2}$	30°	12	$6\frac{1}{2}$°	$5\frac{1}{2}$
50° N	$63\frac{1}{2}$°	$16\frac{1}{4}$	40°	12	$16\frac{1}{2}$°	$7\frac{3}{4}$
40° N	$73\frac{1}{2}$°	15	50°	12	$26\frac{1}{2}$°	9
30° N	$83\frac{1}{2}$°	14	60°	12	$36\frac{1}{2}$°	10
$23\frac{1}{2}$° N	90°	$13\frac{1}{2}$	$66\frac{1}{2}$°	12	43°	$10\frac{1}{2}$
20° N	$86\frac{1}{2}$°	$13\frac{1}{4}$	70°	12	$46\frac{1}{2}$°	$10\frac{3}{4}$
10° N	$76\frac{1}{2}$°	$12\frac{1}{2}$	80°	12	$56\frac{1}{2}$°	$11\frac{1}{2}$
0°	$66\frac{1}{2}$°	12	90°	12	$66\frac{1}{2}$°	12
10° S	$56\frac{1}{2}$°	$11\frac{1}{2}$	80°	12	$76\frac{1}{2}$°	$12\frac{1}{2}$
20° S	$46\frac{1}{2}$°	$10\frac{3}{4}$	70°	12	$86\frac{1}{2}$°	$13\frac{1}{4}$
$23\frac{1}{2}$° S	43°	$10\frac{1}{2}$	$66\frac{1}{2}$°	12	90°	$13\frac{1}{2}$
30° S	$36\frac{1}{2}$°	10	60°	12	$83\frac{1}{2}$°	14
40° S	$26\frac{1}{2}$°	9	50°	12	$73\frac{1}{2}$°	15
50° S	$16\frac{1}{2}$°	$7\frac{3}{4}$	40°	12	$63\frac{1}{2}$°	$16\frac{1}{4}$
60° S	$6\frac{1}{2}$°	$5\frac{1}{2}$	30°	12	$53\frac{1}{2}$°	$18\frac{1}{2}$
$66\frac{1}{2}$° S	0°	0	$23\frac{1}{2}$°	12	47°	24
70° S	—	0	20°	12	$43\frac{1}{2}$°	24
80° S	—	0	10°	12	$33\frac{1}{2}$°	24
90° S	—	0	0°	12	$23\frac{1}{2}$°	24

Variation in Duration of Insolation

The length of time that insolation is received each day—or the time between sunrise and sunset—is called the duration of insolation. The duration of insolation at a given location corresponds to the number of hours that the sun is visible each day. As shown in Table 6-1, the duration of insolation varies with latitude and with the seasons. In the continental United States, duration of insolation is longest on the summer solstice (June 21) and shortest on the winter solstice (December 21). On the equinoxes (March 21 and September 23), the average duration of insolation is 12 hours.

> **MEMORY JOGGER**
>
> You may recall from Topic 4 that the length of the sun's apparent path through the sky varies during the year. Day length, or duration of insolation, is proportional to the length of the sun's path. There is one hour of insolation for each 15° of path.

EFFECTS OF LATITUDE AND SEASON A review of Table 6-1 will reveal how duration of insolation varies with latitude and season. At the time of the summer solstice, northern areas of the contiguous United States have more hours of daylight than do southern areas. The reverse happens at the time of the winter solstice when southern areas of the contiguous United States have more hours of daylight than the northern areas. Only during the equinoxes is the duration of insolation the same everywhere—12 hours.

Relationship of Surface Temperatures to Insolation

The temperature of Earth's surface at a given location varies throughout the day. Similarly, the average daily temperature varies throughout the year. These variations depend on the balance between energy being gained from insolation and the energy being lost by Earth's surface radiation. When energy is being gained at a greater rate than it is being lost, Earth's surface temperature rises. When energy is being lost faster than it is being gained, Earth's surface temperature falls. Temperatures are generally higher when the intensity of insolation is greater. Temperatures also tend to be higher at a location for which the duration of insolation is longer.

TIMES OF YEARLY MAXIMUM AND MINIMUM TEMPERATURES Each year, the times of maximum and minimum temperatures in a given place occur somewhat later than the times of maximum and minimum insolation. For example, in mid-latitude areas of the contiguous United States (49° N to 25° N) the maximum intensity of insolation and the maximum duration of insolation occur on June 21. However, the period of highest daily temperatures occurs later—toward the end of July or early in August. (See Figure 6-8.)

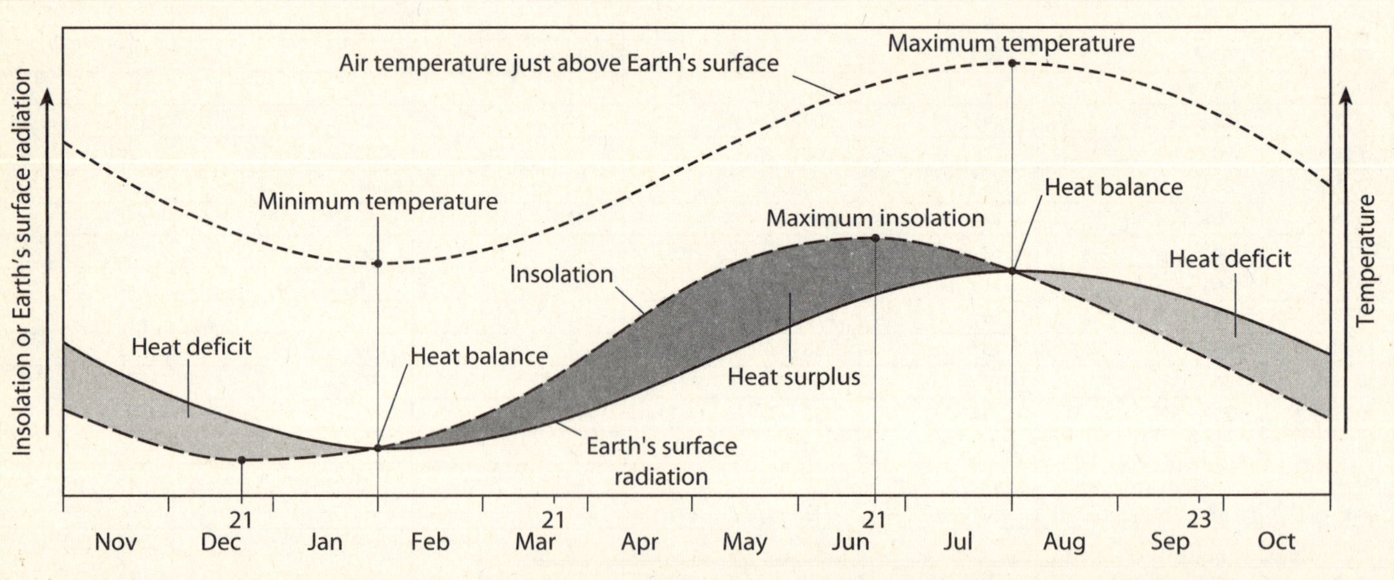

Figure 6-8. **Average daily temperatures, insolation, and Earth's surface radiation for mid–latitudes of the Northern Hemisphere during a year**

The reason for this delay relates to the changing balance between energy gain from insolation and energy loss from Earth's surface radiation. During spring and early summer in the contiguous United States, the intensity and duration of insolation increase. Each day the surface receives more energy of insolation than it loses by radiation and the average temperature goes up slightly. After June 21, a little less energy is received, but the amount of energy received is still more than the amount lost. Therefore, the temperature continues to go up until the rate of incoming energy finally drops below the rate of energy loss by Earth's surface radiation. Similarly, mid-latitude areas of the contiguous United States are not at their coldest when the duration of insolation and angle of incidence are at their minimum on December 21. These areas continue to lose more energy from surface radiation than they gain by insolation until late January or early February—when the coldest temperatures occur.

TIMES OF DAILY MAXIMUM AND MINIMUM TEMPERATURES The hottest part of an average day is some time in mid-afternoon, not at solar noon when the intensity of insolation is greatest. (See Figure 6-9.) Each day, although insolation is greatest at solar noon, Earth's surface continues to gain more energy from the sun than it loses from surface radiation until mid-afternoon. In a similar way, the coolest temperatures on an average day usually occur just slightly after sunrise. Earth's surface continues to lose heat throughout the night until after insolation begins at sunrise.

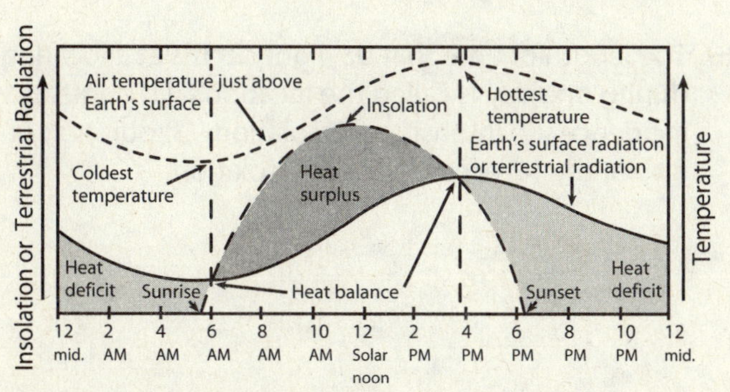

Figure 6-9. Typical daily variations in insolation, surface radiation, and near-surface air temperature

Review Questions

11. In which diagram does the incoming solar radiation reaching Earth's surface have the greatest intensity?

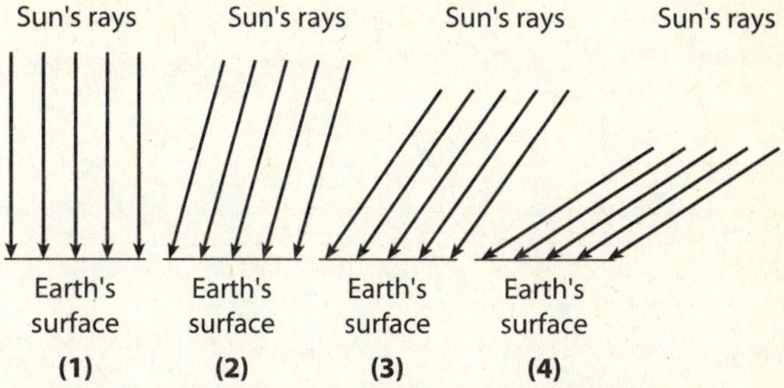

12. The following diagram represents a portion of Earth's surface that is receiving insolation. Positions A, B, C, and D are located on Earth's surface.

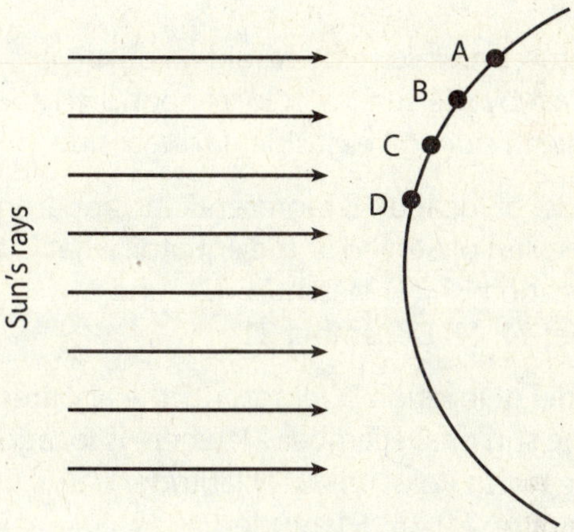

At which position would the intensity of insolation be greatest? (1) A (2) B (3) C (4) D

13. Over a period of one year, which location would probably have the greatest average intensity of insolation per unit area? Assume that there is equal atmospheric transparency at each location.
(1) Tropic of Cancer, $23\frac{1}{2}°$ N (2) New York City, 41° N (3) the Arctic Circle, $66\frac{1}{2}°$ N (4) the North Pole, 90° N

14. The most logical conclusion that can be made from the relationship between the altitude of the sun throughout the day and the amount of insolation is that, as the sun's altitude (1) increases, the insolation increases (2) increases, the insolation decreases (3) decreases, the insolation increases (4) decreases, the insolation remains the same

Base your answers to questions 15 through 18 on the following diagrams. The diagrams represent models of the apparent path of the sun across the sky for observers at four different locations, A through D, on Earth's surface.

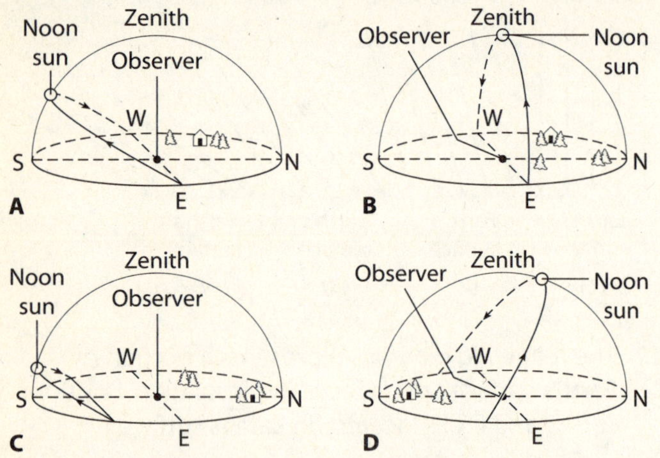

15. At location A, on which side would an observer's shadow cast by the sun fall at local noon? (1) south side (2) north side (3) east side (4) west side

16. If the model of location B represents the apparent path of the sun observed at the equator, what is the date at location B? (1) March 21 (2) June 21 (3) October 21 (4) December 21

17. If the model of location D represents the apparent path of the sun on December 21, where is location D? (1) the North Pole (2) 45° N latitude (3) the equator (4) 45° S latitude

18. At location B three months later, how would the altitude of the noon sun compare to its present altitude? (1) The altitude would be less than shown. (2) The altitude would be greater than shown. (3) The altitude would be the same as shown.

---

19. At which time of day would an observer's shadow cast by the sun be shortest? (1) 6:00 A.M. (2) 12:00 noon (3) 3:00 P.M. (4) 6:00 P.M.

20. At the time of the fall equinox, the number of hours of daylight in New York City is generally about (1) 9 (2) 12 (3) 15 (4) 18

Base your answers to questions 21 through 25 on the following diagram. The diagram represents Earth at a specific time in its orbit. The dashed lines indicate radiation from the sun. Points A through H are locations on Earth's surface.

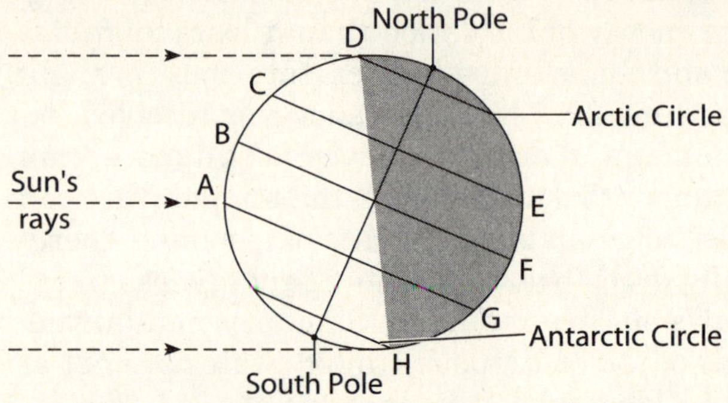

21. Which line represents the equator? (1) AB (2) BF (3) CE (4) DH

22. What is the season in the Northern Hemisphere when Earth is in the position shown in the diagram? (1) spring (2) summer (3) fall (4) winter

23. When the sun is in the position shown in the diagram, how many hours of daylight would occur at the North Pole during one complete rotation? (1) 0 (2) 8 (3) 12 (4) 24

24. In which direction would a person located at position H have to look to see the sun at the time shown in the diagram? (1) north (2) east (3) south (4) west

25. Six months after the date indicated by the diagram, which point would receive the sun's vertical rays at noon? (1) A (2) B (3) C (4) D

---

26. The length of time that daylight occurs at a location during one day is called the location's (1) angle of incidence (2) intensity of insolation (3) duration of insolation (4) eccentricity of insolation

27. Which graph best represents the duration of insolation during the year at Earth's equator?

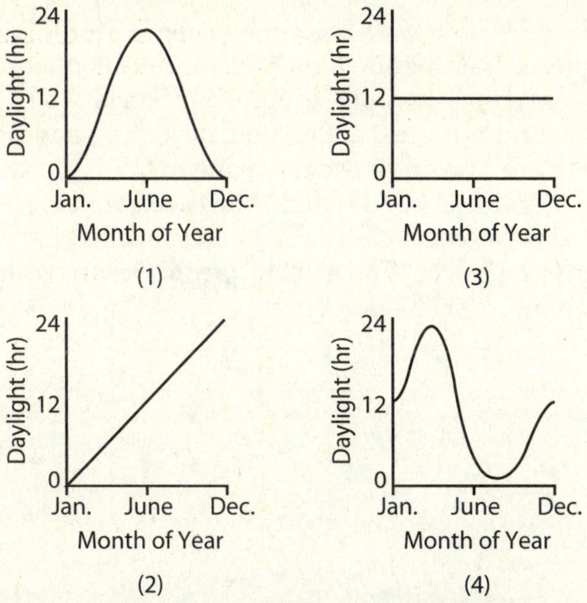

28. What is the primary reason New York State is warmer in May than in February? (1) Earth is traveling faster in its orbit in February. (2) The altitude of the noon sun is greater in February. (3) The insolation in New York is greater in May. (4) Earth is closer to the sun in May.

29. In New York State, the maximum total daily insolation occurs during June. Which statement best explains why the maximum annual temperature is usually observed about a month later, in July? (1) Earth is closer to the sun in June than it is in July. (2) Earth is farther from the sun in June than it is in July. (3) New York State loses far more energy than it receives from the sun during most of July. (4) New York State receives more energy from the sun than it loses during most of July.

30. What is the usual cause of the drop in temperature that occurs between sunset and sunrise at most New York State locations? (1) strong winds (2) ground radiation (3) cloud formation (4) heavy precipitation

31. The following graph illustrates the relationship between insolation and Earth's surface radiation during a 24-hour period in New York State on March 21. At what time did the maximum air temperature probably occur?

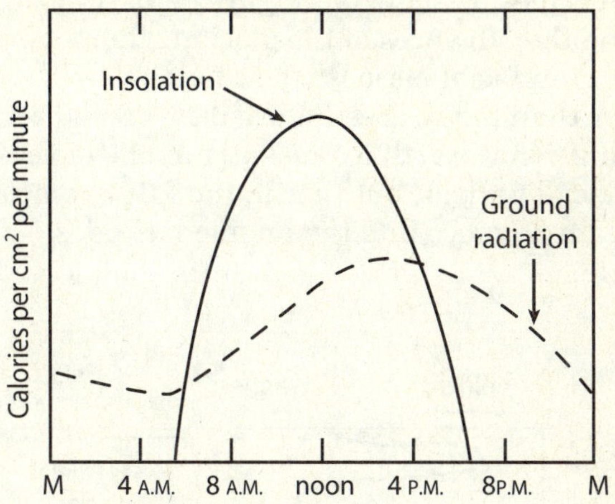

(1) 6 A.M. (2) 12 noon (3) 4 P.M. (4) 6 P.M.

Heat Budget and Climate Change

An object's **heat budget** is the result of the balance between the total amount of energy it receives and the total energy it emits or loses. Heat budget can be measured as the average temperature of an object. Earth's heat budget is the result of a radiation balance between the radiation from the sun and Earth's interior compared to the radiation Earth gives off to space.

Examples of Climate Change

Long-term changes in Earth's heat budget caused by Earth heating up or cooling off can result in climatic changes. The sections that follow describe some of the types of changes in Earth's heat budget and climate and some possible reasons for these changes.

ICE AGES AND LONG, WARM PERIODS ON EARTH Earth's heat budget has shifted in the past for long periods of time. Some of these shifts in heat balance produced warm periods when Earth probably had few or no glaciers. Other shifts in heat balance produced periods called **ice ages** when glaciers advanced into the middle latitudes.

According to Geologic History of New York State in the *Earth Science Reference Tables,* the last ice age—the Pleistocene—lasted from about 1.6 million to approximately 10 thousand years ago. During the Pleistocene, glaciers advanced four or more times as shown in Figure 6-10. At times between the advances of the glaciers, there were warm interglacial periods when Earth was warmer than today. (See Figure 6-10.) Some scientists think that Earth is presently in one of these interglacial periods, but that in the future glaciers will again advance into the middle latitudes.

EL NIÑO AND LA NIÑA EVENTS Every 2 to 10 years, the normally cold waters of the eastern Pacific Ocean off western South America are replaced with a vast area of warmer waters. This warming event, known as **El Niño,** causes major climatic repercussions around the world. As shown in Figure 6-11, flooding, droughts, and heat waves are known to occur during an El Niño. Recently it has been noted that there are also periods of exceptionally cold water—called La Niña events—in the eastern Pacific Ocean, which also affect worldwide climate.

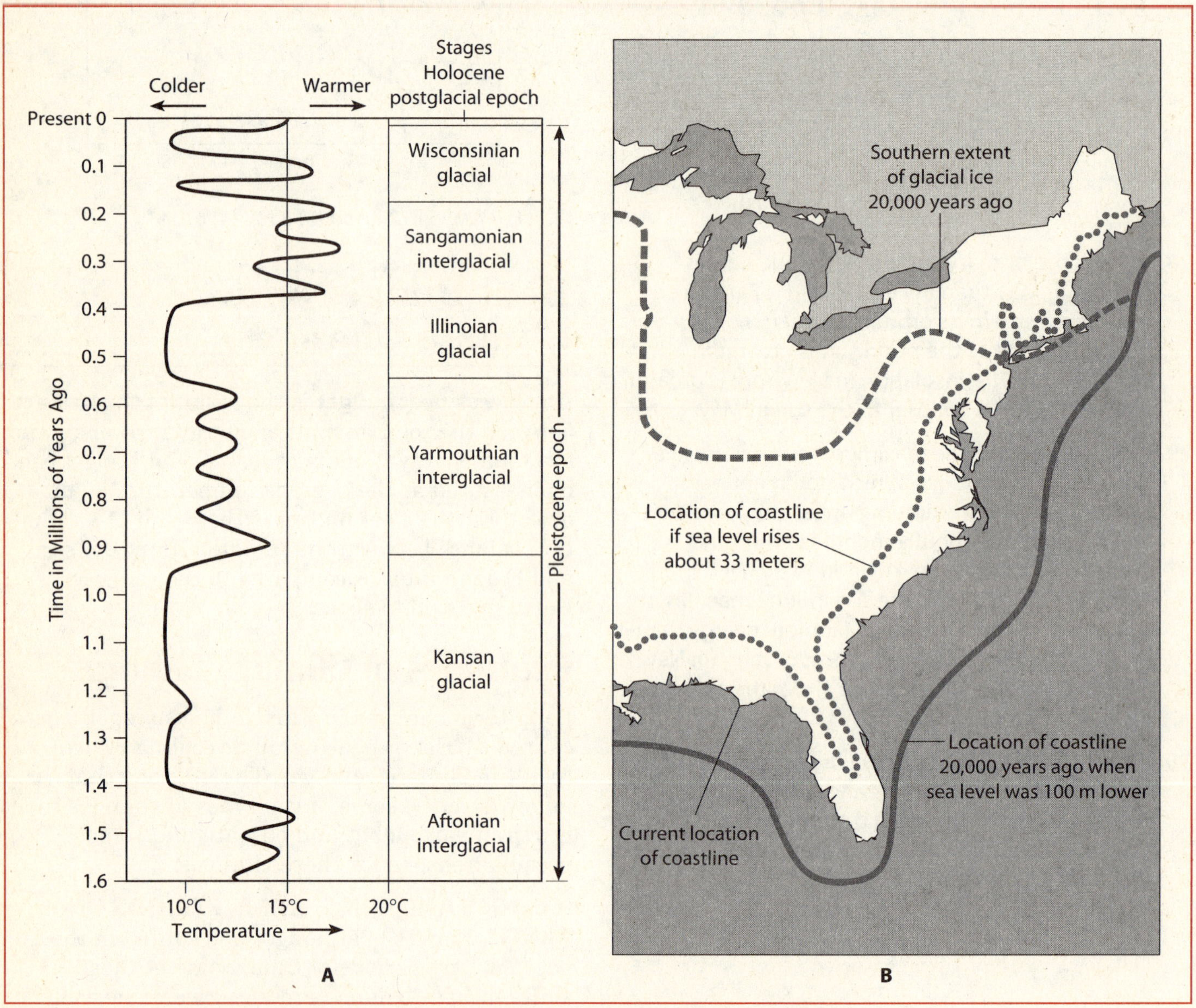

Figure 6-10. Changes in temperature, sea level, and position of glaciers associated with the Pleistocene Ice Age and present times: Diagram A shows estimated temperature variations from today's average Earth temperature of 15°C. Diagram B shows the maximum extent of glaciation about 20,000 years ago and the extended coastline of the area that is now the eastern United States. Sea levels 20,000 years ago were 100 meters lower, due to the large amounts of water stored as ice. Diagram B also shows how sea levels would rise if global warming continues—melting today's glaciers and ice caps.

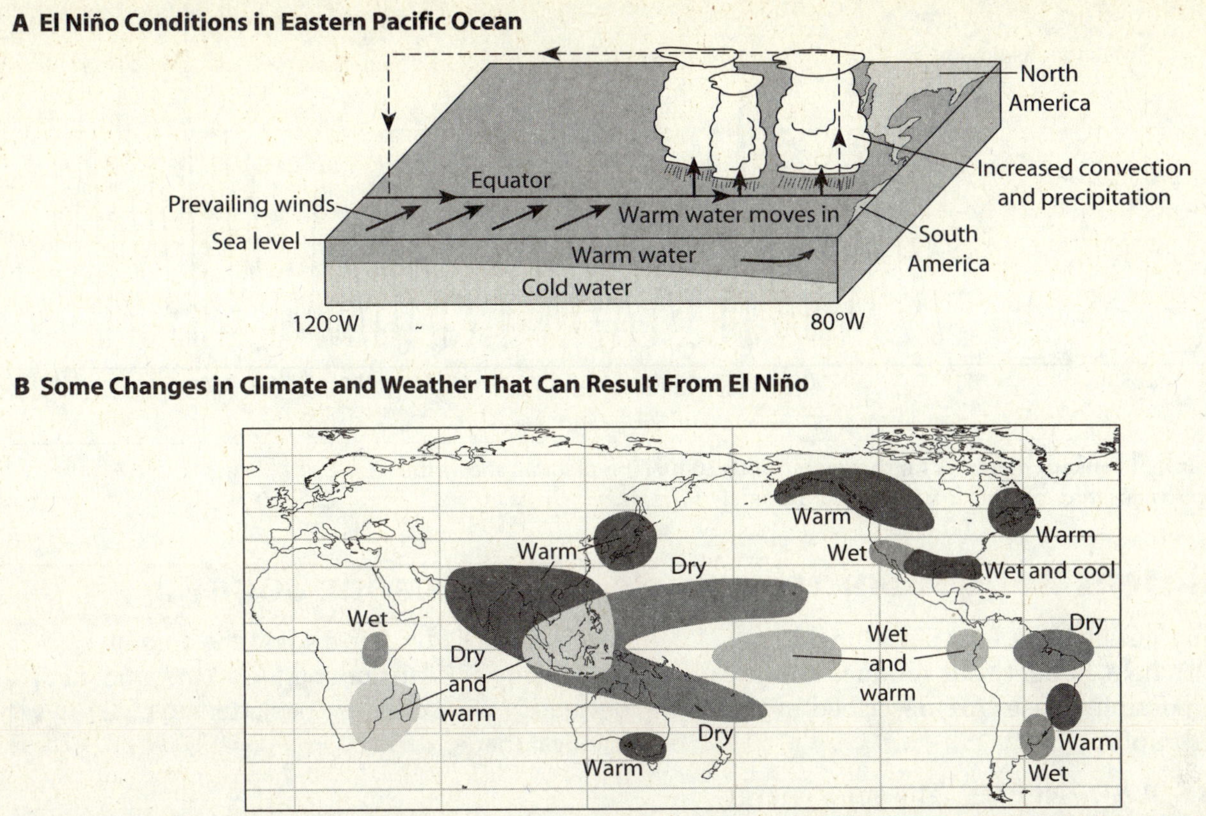

Figure 6-11. El Niño conditions: Diagram A shows how every 2 to 10 years, changes in prevailing winds push a huge mass of warm ocean water toward the west coast of South America—replacing the normally cold waters of the eastern Pacific Ocean. The influx of warm water results in a temperature rise and an increase in air convection currents and precipitation in a region with a normally cool and dry climate. Map B shows some of the worldwide conditions linked to El Niño events, such as major flooding in southern California and droughts in southern Africa.

GLOBAL WARMING

From the early 1980s to the present time, there has been a trend of rising temperatures known as global warming. In 1998, worldwide temperatures were the warmest ever recorded. Figure 6-12 shows the changes in Earth's average surface temperature over the last 120 years. It remains to be seen whether worldwide temperatures will continue to rise, or whether they just reflect a temperature fluctuation similar to the one that occurred during the 1930s and 1940s.

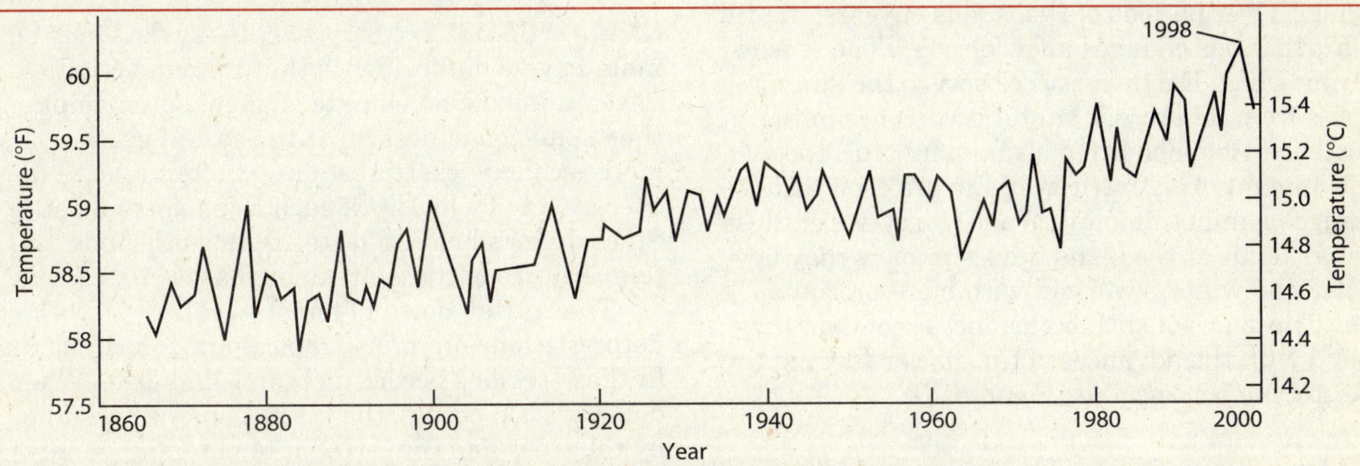

Figure 6-12. Worldwide average temperatures from 1865 to 1999: The increasing temperatures of the 1980s and 1990s have convinced many scientists that global warming is occurring and may worsen as human activities continue to add greenhouse gases to the atmosphere.

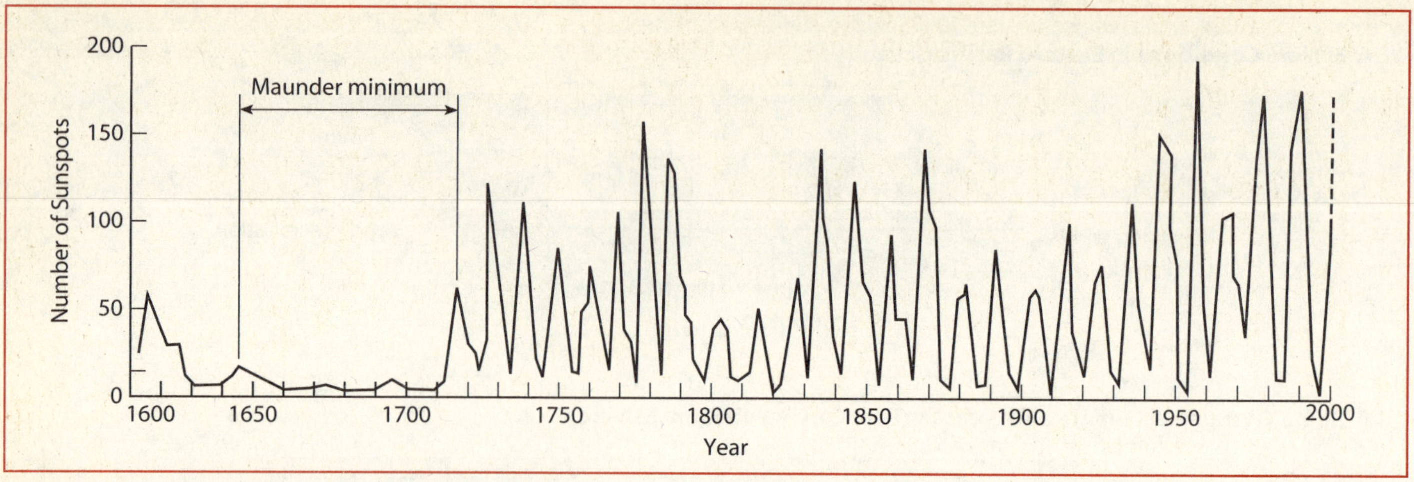

Figure 6-13. Graph of sunspot numbers for the years 1600–2000: This diagram shows the number of sunspots varies through a solar cycle of approximately 11 years.

Causes of Heat Budget Shifts

There are many possible explanations for the shifts in Earth's heat budget that produce climate change. Some possible causes are described in the following paragraphs.

CHANGES IN SOLAR ENERGY The sun's energy output and related sunspot activity follow an 11-year cycle, as shown in Figure 6-13. A **sunspot** is a darker region of the sun's visible surface. When there are large numbers of sunspots, the sun emits from 0.1 to 1.0 percent more electromagnetic energy. As a result, Earth receives more insolation at these times. Note that the high temperatures of the 1990s, shown in Figure 6-12, show a correlation with an increase in the number of sunspots as indicated in Figure 6-13.

CHANGES IN EARTH'S ORBIT AND AXIS TILT Changes in Earth's axis and orbit over many thousands of years may be a major influence in climate change. Over periods of thousands of years, the tilt of Earth's axis changes a few degrees. The season during which Earth passes closest to the sun varies from its present winter occurrence in the Northern Hemisphere to a summer occurrence. The amount of eccentricity of Earth's orbit also changes a minor amount. When the cycles of these factors occur at the same time, summers may be cooler and winters warmer with more precipitation. The times of such occurrences seem to correspond with the advances of the glaciers during the Pleistocene Ice Age. (See Figure 6-10.)

MEMORY JOGGER

You may recall from Topic 4 that the amount of eccentricity of Earth's orbit refers to the amount of deviation from a circle, or the flattening of its elliptical shape.

VOLCANIC ERUPTIONS AND CLIMATE CHANGES In the early 1990s, Earth's atmosphere underwent cooling that was probably related to the eruption of the Pinatubo volcano in the Philippines. (See Figure 6-12.) When there are major volcanic eruptions, aerosols such as volcanic ash and sulfur compounds are propelled into the stratosphere and stay there for months or years. The aerosols make Earth's atmosphere less transparent to insolation and reflect a greater than normal amount of insolation back into space. As a result, less energy arrives at Earth's surface, and cooler temperatures occur.

HUMAN CAUSES Many scientists who study climate have found evidence that human activities have an impact on climate change. For example, the expansion of deserts at the cost of grasslands—called desertification—is often related to overgrazing by livestock such as sheep and cattle. Since deserts heat up faster than grasslands, the temperature of these areas increases.

The cutting down of forests—called **deforestation**—in many tropical areas has resulted in these regions becoming hotter and drier. When deforestation occurs, the insolation once absorbed by the trees and converted into potential energy instead goes into heating Earth's surface. Additionally, without the trees, there is less transpiration of water vapor to increase the humidity of the atmosphere—resulting in less precipitation.

As discussed previously, human activities that produce greenhouse gases may be a major reason for the present global warming. For example, the increase in atmospheric carbon dioxide in recent years is largely due to the burning of fossil fuels. Another human factor is the building of cities—urbanization—which results in clearing land of plants and trees to construct buildings. See figure 14-10 on page 277. Cities also increase carbon dioxide because of an increased use of fossil fuels.

Review Questions

32. Between the years 1850 and 1900, records indicate that Earth's mean surface temperature showed little variation. This would support the inference that (1) Earth was in heat balance (2) another ice age was approaching (3) Earth was gaining more energy than it was losing (4) the sun was emitting more energy

33. Air temperature is regulated partly by the percentage of carbon dioxide (CO_2) in the air. Which of the following statements is correct? (1) The CO_2 content of the air during the glacial period was very high because cold water absorbs more CO_2 than warm water. (2) During periods of great volcanic activity, the CO_2 content of the air decreases because the clouds reduce plant growth. (3) The CO_2 content of the air is increasing due to the burning of fossil fuels. (4) The CO_2 content of the air is increasing due to the deposition of limestone from groundwater.

34. Some scientists predict that an increase in atmospheric carbon dioxide would cause a worldwide increase in temperature. Which of the following could result if a worldwide temperature increase occurred? (1) Continental drift will increase. (2) Isotherms will shift toward the equator. (3) Additional landmasses will form. (4) Ice caps at Earth's poles will melt.

35. A heating of the normally cold ocean waters of the eastern Pacific Ocean off the coast of South America is called (1) the greenhouse effect (2) El Niño (3) ozone effect (4) global warming

36. Which event would most likely cause a new ice age in North America? (1) a decrease in the energy produced by the sun (2) a decrease in the light reflected by Earth's surface (3) an increase of carbon dioxide in Earth's atmosphere (4) an El Niño event

37. When were large parts of North America covered by ice sheets? (1) only once, early in Earth's geologic history (2) only once, in the recent geologic past (3) once early in Earth's geologic history and once in the recent geologic past (4) many times during Earth's geologic history

38. Another ice age would probably result in a change in (1) sea level (2) moon phases (3) the speed at which Earth moves in its orbit (4) the time between high tides

Seasons

New York State and other mid-latitude areas of Earth's Northern Hemisphere experience four seasons—winter, spring, summer, and fall. These seasons are distinguished by differences in temperature, moisture, and vegetation. Some land regions near Earth's equator have little seasonal change. In these areas, the intensity of insolation remains relatively high, and the duration of insolation is about 12 hours per day throughout the year.

Direct Causes of the Seasons

Seasonal changes in temperature, moisture, and other weather conditions result from the cyclic variations in the angle of incidence and the intensity and duration of insolation that occur during the year. (See Figure 6-14 on the next page.) Generally the seasons follow the north-south shift in the direct rays of insolation. This shift, caused by the tilt of Earth's axis, occurs between $23\frac{1}{2}°$ N latitude and $23\frac{1}{2}°$ S latitude as discussed previously. (See Figure 6-5 and Table 6-1 on pages 90 and 91.)

Climate zones, and their associated wind patterns and surface ocean currents, also shift in a north-south pattern. However, this shift lags about a month behind the shift of direct rays of insolation. Figure 6-15 on the next page shows the effects of the shifting rays of the sun on the seasonal positions of the wind and moisture belts on Earth.

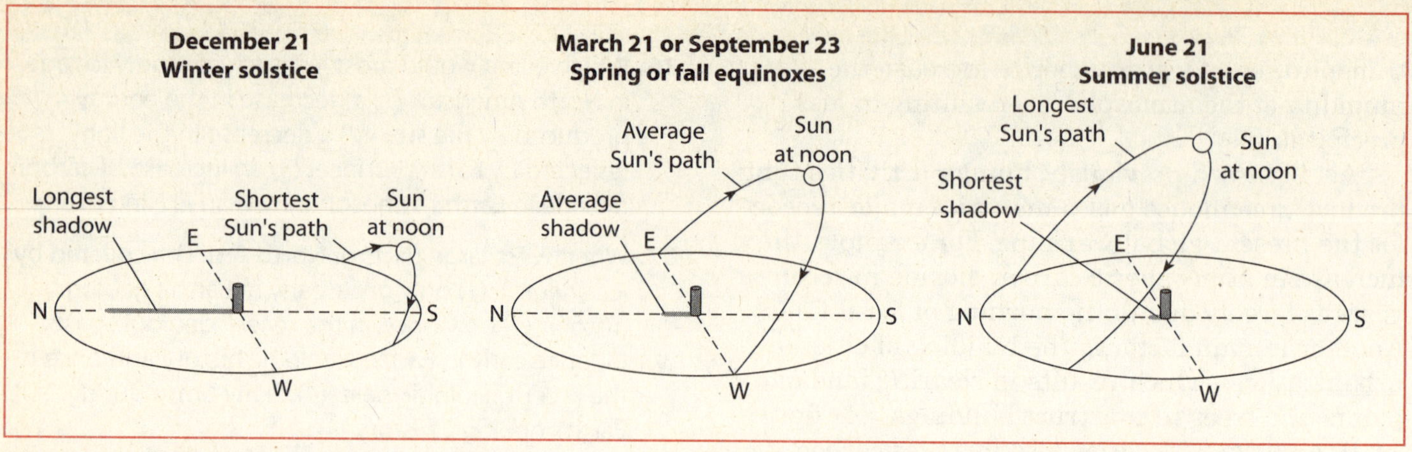

Figure 6-14. The direct causes of seasons: Shown for the mid-latitudes of the northern hemisphere include changes in angle of incidence or height of the sun in the sky and the changes in duration of the sun's daily motion path. Note how the length of the shadow of the sun at noon changes with the seasons.

Astronomical Causes of the Seasons

The variations in insolation that directly cause the seasons are themselves the result of factors that may be called the astronomical, or indirect, causes of the seasons. These indirect causes of seasons include the tilt and parallelism of Earth's axis and Earth's revolution around the sun.

TILT OF EARTH'S AXIS As shown in Figure 6-16, Earth's rotational axis is tilted at an angle of $23\frac{1}{2}°$ with respect to a line perpendicular to the plane of its orbit of the sun. This tilt, coupled with the other reasons described here, means that perpendicular insolation from the sun shifts between $23\frac{1}{2}°$ N and $23\frac{1}{2}°$ S latitudes. If the tilt of Earth's axis were greater, perpendicular insolation would reach farther north and south, resulting in generally warmer summers and colder winters for much of Earth. If the tilt of Earth's axis were less than $23\frac{1}{2}°$, then perpendicular insolation would not shift as much. As a result, seasonal effects would be less pronounced—there would be cooler summers and warmer winters.

PARALLELISM OF EARTH'S AXIS Regardless of the position of Earth in its orbit, Earth's axis always points in the same direction in space. The north end of Earth's axis—the North Pole—always points to the present-day North Star—Polaris. Figure 6-16 shows that as Earth orbits the sun, the position of its axis at any given time is always parallel to its position at any other time—this condition is called parallelism.

REVOLUTION OF EARTH AROUND THE SUN As Earth revolves around the sun, the direction of Earth's axis with respect to the sun varies because of its tilt and parallelism. For example, on June 21, the North Pole is inclined toward the sun at an angle of $23\frac{1}{2}°$ from the perpendicular. On December 21, the North Pole is inclined away from the sun at this angle. On March 21 and September 23, the axis is still inclined $23\frac{1}{2}°$ from the perpendicular, but neither toward nor away from the sun. This cycle of variations causes variations in angle of incidence and duration of insolation throughout the year and results in seasonal changes. (See Figures 6-14 and 6-16.)

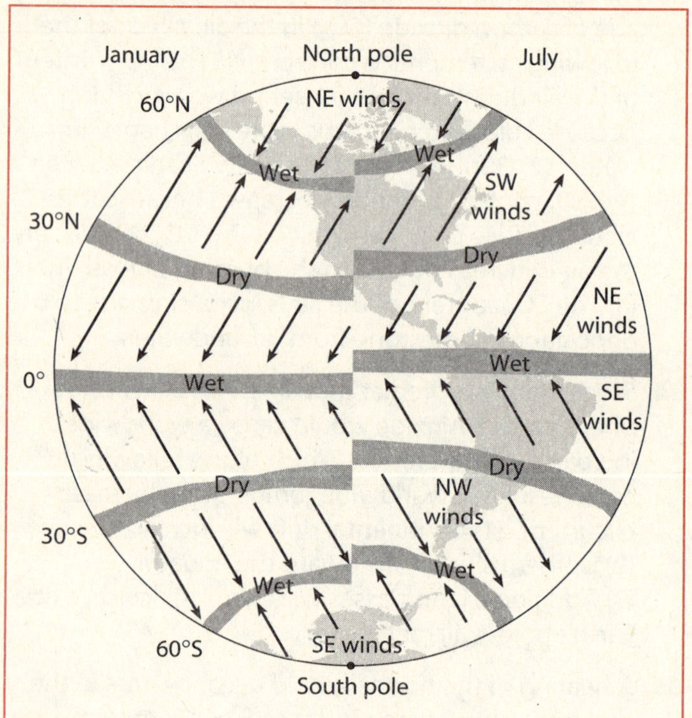

Figure 6-15. Seasonal shifting of Earth's wind and moisture belts: By comparing the latitudes of the wind and moisture belts in January and July, you can see how the wind and moisture belts shift north and south during the year following the direct rays of insolation. Also, look at the diagram Planetary Wind and Moisture Belts in the Troposphere in the *Earth Science Reference Tables* that represents conditions near the time of the equinoxes.

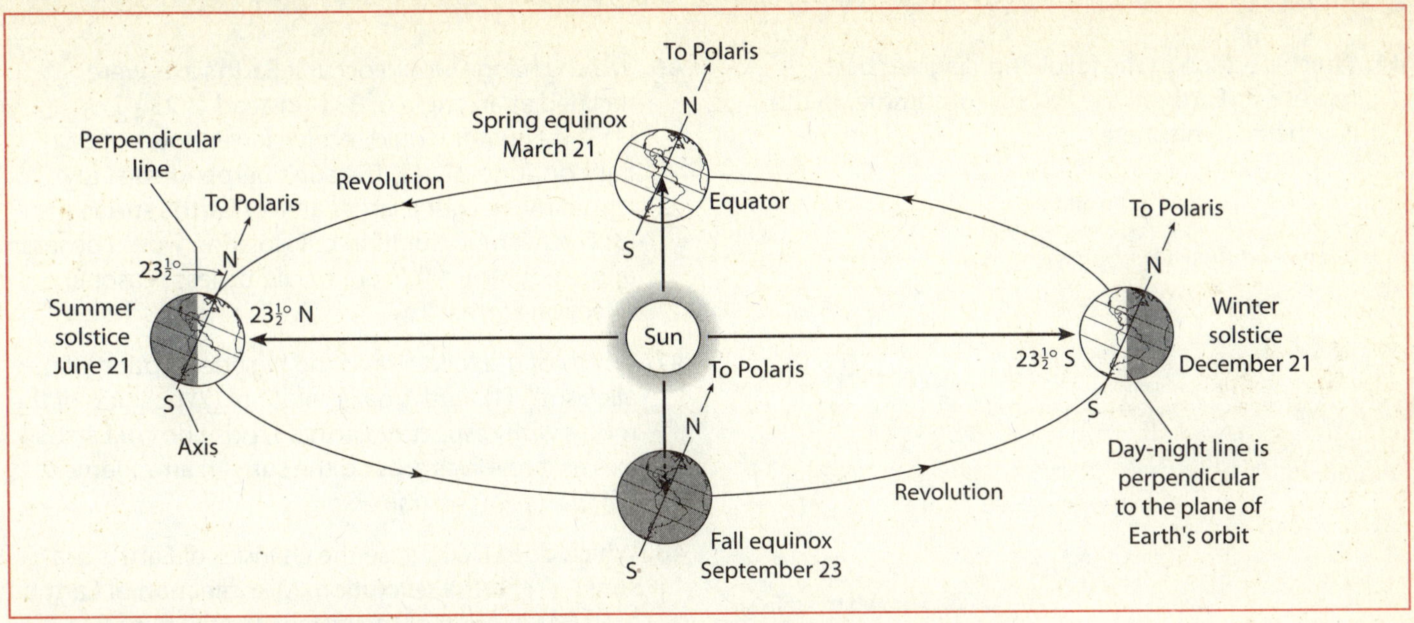

Figure 6-16. Indirect causes of the seasons: The astronomical, or indirect, causes for the seasons described in the text, result in the changes in angle, intensity, and duration of insolation.

SMALL SEASONAL EFFECT OF EARTH'S ELLIPTICAL ORBIT Because of Earth's slightly elliptical orbit, there is a variation in distance between Earth and the sun as Earth revolves around the sun. (See Figure 3-9 on page 43.) The small changes in distance between Earth and the sun during the year are too small to have a significant effect on the seasons. For example, winter in the Northern Hemisphere occurs at a time when Earth is actually nearest the sun. Earth as a whole receives about seven percent more energy when it is closest to the sun than it does when it is farthest away. This seven percent difference would only be significant if Earth's axis was not tilted at an angle of $23\frac{1}{2}°$.

Review Questions

39. The factor that contributes most to the seasonal temperature changes during a year in New York State is the changing (1) speed at which Earth travels in its orbit around the sun (2) angle at which sun's rays strike Earth's surface (3) the distance between Earth and the sun (4) energy given off by the sun

40. On which two dates could all locations on Earth have equal hours of day and night? (1) September 23 and December 21 (2) December 21 and March 21 (3) March 21 and June 21 (4) March 21 and September 23

41. At which latitude does Earth receive the greatest intensity of insolation on June 21? (1) 0° (2) $23\frac{1}{2}°$ S (3) $23\frac{1}{2}°$ N (4) 90° N

Base your answers to questions 42 and 43 on the following diagram, which shows Earth as viewed from space. The shaded side represents Earth's nighttime side.

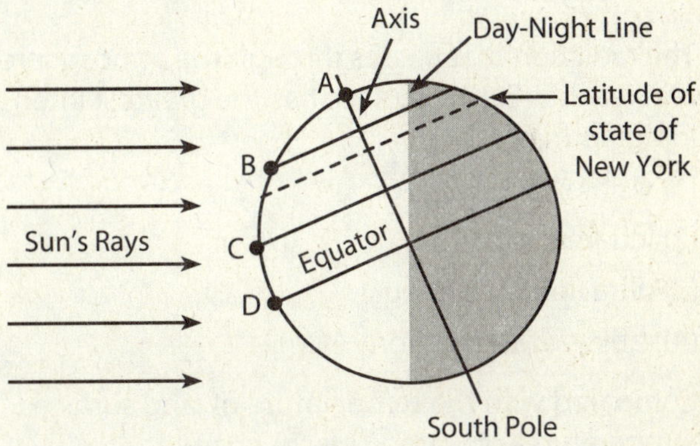

42. Which point on Earth's surface is receiving the greatest intensity of insolation? (1) A (2) B (3) C (4) D

43. The total number of hours of daylight received by New York State on the days represented in the diagram is closest to (1) 9 hr (2) 12 hr (3) 15 hr (4) 20 hr

44. Which position in the following diagram best represents Earth on the first day of summer in the Northern Hemisphere?

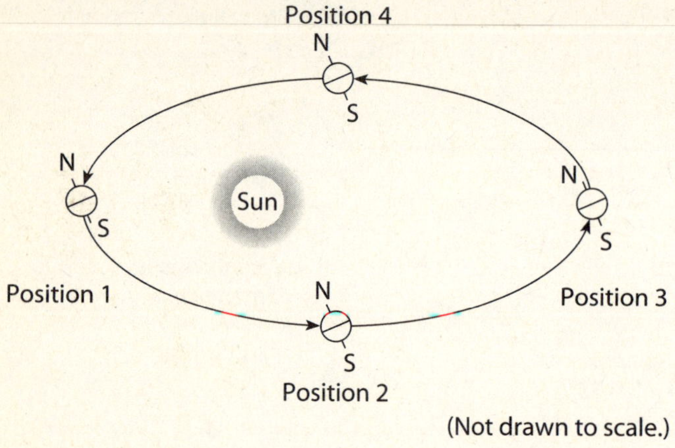

(1) 1 (2) 2 (3) 3 (4) 4

45. At which latitude would the duration of insolation be greatest on December 21? (1) $23\frac{1}{2}°$ S (2) 0° (3) 10° N (4) $23\frac{1}{2}°$ N

46. Which change would occur if Earth's axis were inclined at an angle of $33\frac{1}{2}°$ instead of $23\frac{1}{2}°$? (1) The equator would receive fewer hours of daylight on June 21. (2) The sun's perpendicular rays would move over a larger area of Earth's surface. (3) The average duration of insolation would be less at the equator. (4) There would be less seasonal effect on Earth.

47. Which of the following is the best description of parallelism? (1) Earth's daily rotation (2) position of the moon with respect to Earth (3) positions of Earth's axis as it revolves around the sun (4) arrangement of galaxies in the universe

48. Which does NOT cause the changes of Earth's seasons? (1) Earth's revolution (2) inclination of Earth's axis (3) variation of the distance to the sun (4) parallelism of Earth's axis

Questions for Regents Practice

Part A

1. The radiation that passes through the atmosphere and reaches Earth's surface has the greatest intensity in the form of

(1) visible light radiation

(2) infrared radiation

(3) ultraviolet radiation

(4) radio-wave radiation

2. Compared with the temperature of land surfaces, temperatures of water surfaces change

(1) faster because water has a higher specific heat
(2) faster because water has a lower specific heat
(3) slower because water has a higher specific heat
(4) slower because water has a lower specific heat

3. Most of the energy radiated by Earth's surface at night is in the form of

(1) infrared rays

(2) ultraviolet rays

(3) visible light rays

(4) X-rays

4. Under which conditions will the greatest amount of cooling by Earth's surface radiation occur?

(1) a clear night with low humidity

(2) a clear night with high humidity

(3) a cloudy night with low humidity

(4) a cloudy night with high humidity

5. The sun's rays have a greater effect on land temperature than they have on water temperature. This difference can be explained partly by which of the following?

(1) Some energy from the sun is used to evaporate ocean water.

(2) Less energy is reflected by the water's surface than by the land's surface.

(3) Land materials have a higher specific heat than water has.

(4) Land materials require more heat energy than water does to raise their temperature 1°C.

6. Adding more carbon dioxide to the atmosphere increases the amount of

(1) radiant energy reflected by Earth

(2) radiation from the sun absorbed by the oceans

(3) radiation from Earth absorbed by the atmosphere

(4) ultraviolet rays striking Earth

7. Which graph best illustrates the relationship between the intensity of insolation and the angle of incidence?

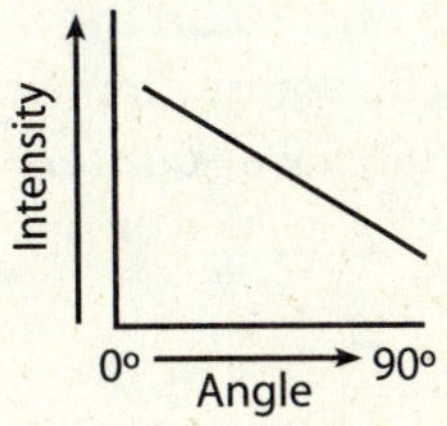

(1)

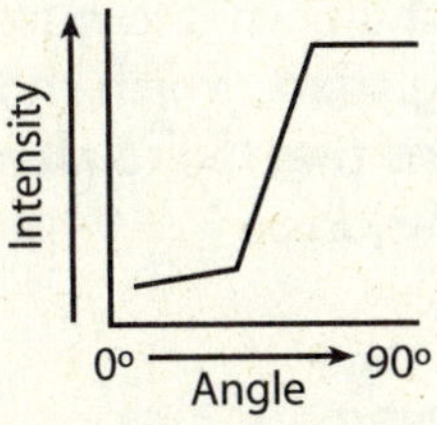

(3)

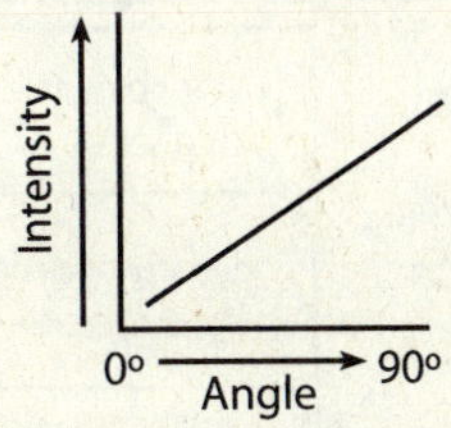

(2)

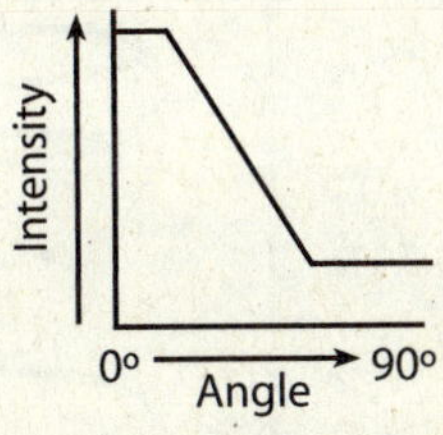

(4)

8. The seasonal temperature changes in the climate of New York State are most influenced by the changing

(1) CO_2 content of the air

(2) angle of incidence at which the sun's rays strike Earth's surface

(3) distance from the sun to Earth

4) speed at which Earth revolves around the sun

9. The activity shown in the following diagram was used to test the effect of the angle of incidence on temperature. A student placed four thermometers—A, B, C and D—on a large globe. The bulb of each thermometer was placed against a black plastic square directly on a line representing latitude. The thermometers were then exposed to direct sunlight for 10 minutes at the angles shown.

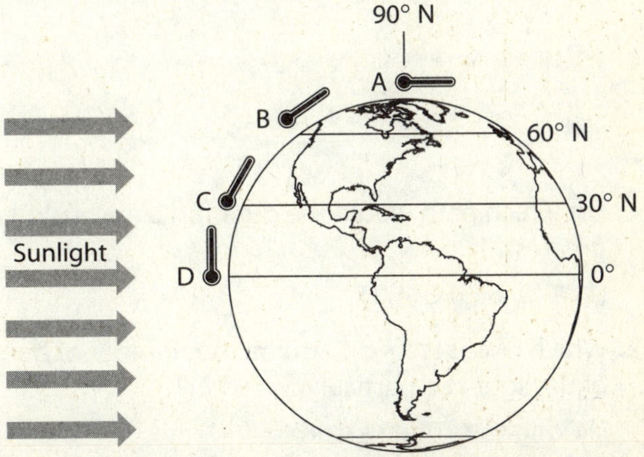

(Not drawn to scale.)

Which thermometer will show the greatest increase in temperature?

(1) A (2) B (3) C (4) D

10. In New York State at 3 P.M. on September 21, the vertical pole shown on the following diagram casts a shadow. Which line best approximates the position of that shadow?

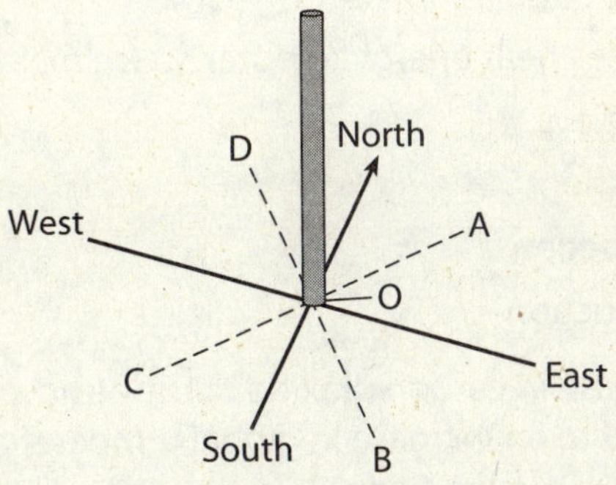

(1) OA (2) OB (3) OC (4) OD

11. The following diagram represents a model of the sun's apparent path across the sky in New York State for selected dates.

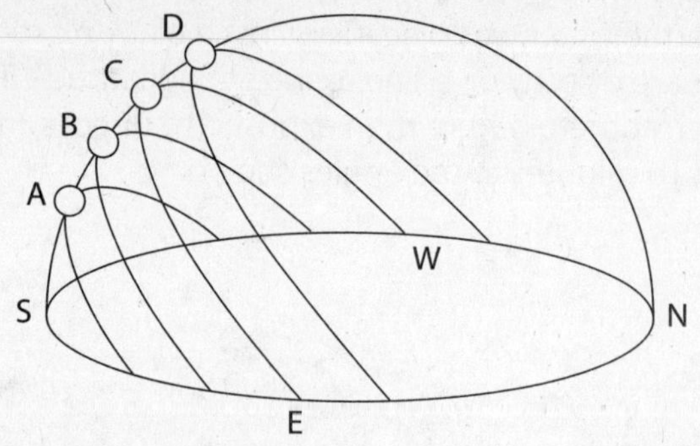

For which path would the duration of insolation be greatest?

(1) A (2) B (3) C (4) D

12. Which two factors determine the number of hours of daylight at a particular location?

(1) longitude and season

(2) longitude and Earth's average diameter

(3) latitude and season

(4) latitude and Earth's average diameter

13. When do maximum surface temperatures usually occur in the Northern Hemisphere?

(1) early June to mid-June

(2) mid-July to early August

(3) late August to mid-September

(4) mid-September to early October

14. Earth loses heat energy to outer space mainly by

(1) radiation

(2) reflection

(3) convection

(4) conduction

15. An increase in carbon dioxide in the atmosphere may lead to an increase in global temperatures because carbon dioxide

(1) absorbs infrared radiation, thus preventing it from escaping Earth's atmosphere

(2) blocks harmful ultraviolet radiation from reaching Earth's surface and being absorbed

(3) interferes with the formation of clouds that block sunlight and cause cooling

(4) enters the atmosphere at a high temperature from combustion and respiration

16. It is known that less ozone in Earth's stratosphere can result in

(1) ice ages

(2) an increase in carbon dioxide

(3) an increase in skin cancer

(4) a decrease in rock weathering caused by ultraviolet rays

17. Which angle of the sun above the horizon produces the greatest intensity of sunlight per unit area?

(1) 25° (2) 40° (3) 60° (4) 70°

Base your answers to questions 18 and 19 on the following chart, which shows the changing climatic conditions that led to alternating glacial and interglacial periods.

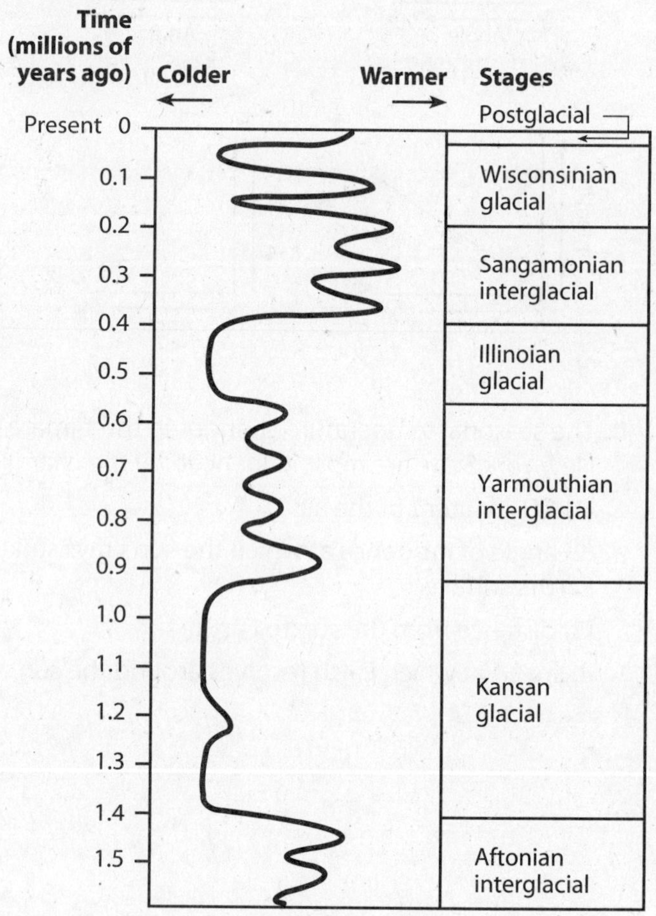

18. The interglacial stages were most likely caused by

(1) a drop in worldwide sea levels

(2) an increase in average worldwide temperature

(3) the movement of Earth's plates

(4) a large increase in the amount of snowfall

19. The chart represents climatic conditions that occurred mostly during which geological time period?

(1) Triassic Period

(2) Ordovician Period

(3) Quaternary Period

(4) Cretaceous Period

Part B

Base your answers to questions 20 and 21 on the following diagram. It represents a field showing the amount of insolation received at Earth's surface on a clear day in the morning—in calories per square centimeter per minute.

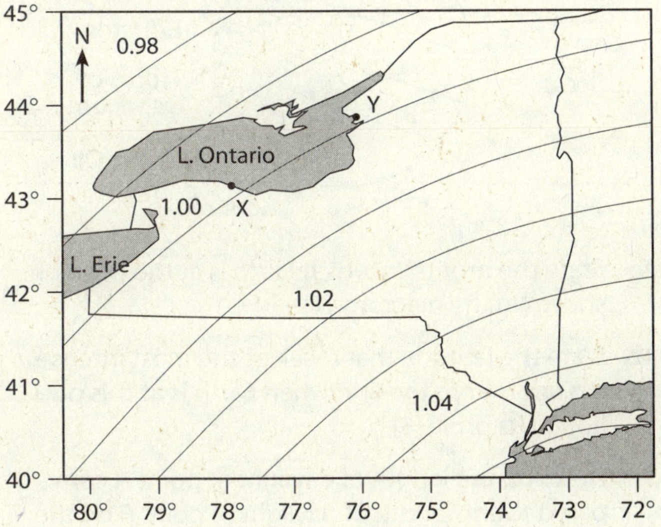

20. If the insolation value at the outer edge of Earth's atmosphere over New York State equals 1.85 calories per square centimeter per minute, why are the values lower at the surface in New York State? [1]

21. What change in atmospheric conditions within the dashed circle could cause the different pattern of isolines as indicated in that area in the following diagram? [1]

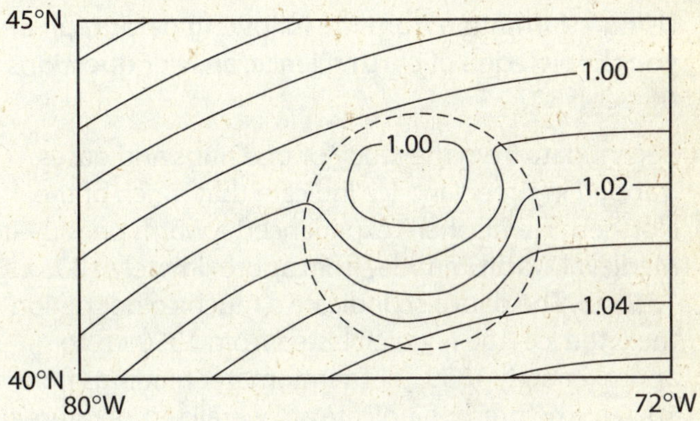

Base your answers to questions 22 and 23 on the table below, which provides information about several gases and their relationship to the greenhouse effect.

22. State one reason that a scientist would recommend restricting the emission of chlorofluorocarbon CFC-12 into the atmosphere. [1]

23. State which of the following actions would contribute more to the greenhouse effect and give a reason for your answer. [2]

(1) releasing 1 kilogram of methane directly into the atmosphere, OR

(2) burning 1 kilogram of methane, resulting in the release of about 3 kilograms of carbon dioxide into the atmosphere

Name of Gas	Symbol	Concentration in Troposphere (parts per billion)	Relative Greenhouse Effect per Kilogram of Gas (in equal concentrations)	Decay Time (years)
Carbon dioxide	CO_2	353,000	1	120
Methane	CH_4	1700	70	10
Ozone	O_3	10-50	1800	0.1
Chlorofluorocarbon	CFC-11	0.28	4000	6.5
Chlorofluorocarbon	CFC-12	0.48	6000	120

24. Why is reflectivity of insolation at high latitudes greater in winter than in summer? [1]

25. In New York State, ski trails on a slope that faces north usually retain their snow later in the spring than those on a slope that faces south. Why does this occur? [1]

26. Why are equatorial areas equal in size to Earth's polar regions heated much more intensely by the sun? [1]

27. How does the way that noontime shadows are cast in New York State change with the approach of the summer solstice? [1]

28. In late summer 1991, Mt. Pinatubo, a volcano in the Philippines, exploded and sent thousands of tons of volcanic dust into the atmosphere. According to scientists, what effects did this event have on Earth's average temperatures in the months that followed? [1]

Part C

Using information from the following paragraph and your knowledge of earth science, answer questions 29 through 33.

Recent data from the study of tree rings and gases trapped in Greenland ice indicate that much of the Northern Hemisphere experienced a warm period—the Medieval Warm Period—from approximately 1000 A.D. to 1350 A.D. The data also indicate a much colder period—the Little Ice Age—which lasted from 1350 A.D. to approximately 1850 A.D. Thermometer readings from 1850 to the present indicate an overall temperature rise, often called global warming.

29. What human activities could have caused the overall change in temperature after 1850 and how could these activities cause the temperature change? [2]

30. Propose two methods of slowing down or reversing the changes in temperature that have occurred since 1850 and explain how these methods might accomplish the goal. [2]

31. At the end of the Medieval Warm Period, Vikings abandoned their settlements in North America and Greenland. What are two possible natural climate-related causes and their effects that might explain this action by the Vikings? [2]

32. Some scientists propose that the Little Ice Age was in part the result of a few exceptionally cold winters increasing the ice and snow coverage in the Northern Hemisphere. How is it possible that more ice and snow would make it colder for such an extended number of years? [1]

33. Some scientists have suggested that an increase in solid aerosols—produced by burning wood and coal—deposited on Earth's surface from the atmosphere, at least in part, resulted in the end of the Little Ice Age. Explain how the deposited aerosols might produce the end of the Little Ice Age. [1]

Base your answers to questions 34 through 36 on the diagram below, which represents Earth at a specific position in its orbit as viewed from space. The shaded area represents nighttime. Points *A* and *B* are locations on Earth's surface.

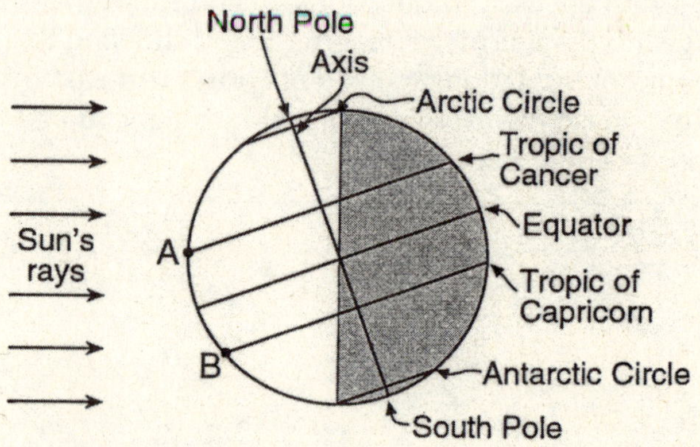

34a. State the month in which Earth is at the position shown in the diagram. [1]

34b. State the latitude that receives the most intense radiation from the Sun when Earth is at this position in its orbit. [1]

35. Describe the length of daylight at point *A* compared to the length of daylight at point *B* on the day represented by the diagram. [1]

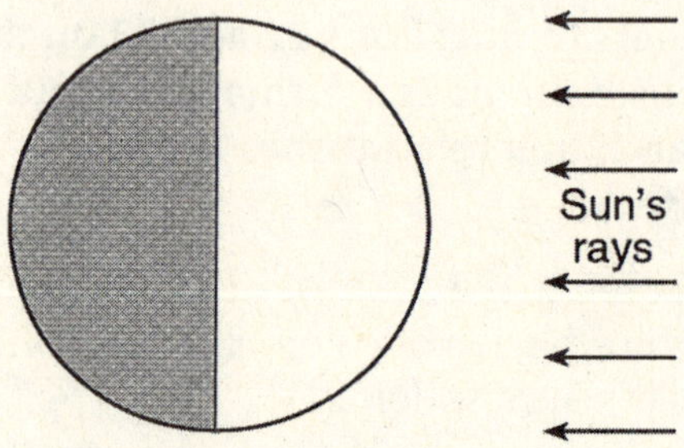

Earth's position in its orbit 6 months later

36. The model of Earth provided *above* represents Earth in its orbit *6 months later.* On the model shown *in your answer booklet*

- draw the position of Earth's axis and label the axis [1]
- label the North Pole [1]
- draw the position of Earth's Equator and label the Equator [1]

Weather

TOPIC 7

VOCABULARY		
air mass	**cyclonic storm**	**probability**
air pressure gradient	**dew point**	**psychrometer**
anemometer	**front**	**radar**
atmospheric, barometric, or air pressure	**humidity**	**relative humidity**
atmospheric transparency	**isobar**	**stationary front**
barometer	**jet stream**	**station model**
cloud cover	**monsoon**	**troposphere**
cold front	**occluded front**	**visibility**
cyclone	**planetary wind belt**	**warm front**
	polar front	**water vapor**
	precipitation	**weather variables**

Are you going to have a picnic or tennis match today? Will the airplane be able to take off for Florida before the approaching ice storm arrives? Will there be a snow day and no school? So much of what you do and how you feel depends on the weather. These conditions affect such things as purchases, your health, how crops grow, the costs of heating and cooling, and the clothing you wear.

Weather is the state or condition of the variables of the atmosphere at any given location for a short period of time. **Weather variables** include temperature, air pressure, wind, moisture conditions, cloud cover, precipitation, and storms. Most of the weather changes occur in the **troposphere,** the part of the atmosphere immediately above Earth's surface.

Variations in insolation (discussed in Topic 6) cause heat energy to be unevenly distributed in the atmosphere. This heat energy tends to move toward a condition of more uniform distribution. That movement of heat energy results in the constant changes in the atmosphere that are a major cause of weather.

Atmospheric Temperature

Most of the world uses the Celsius scale to measure temperatures, but in the United States people still use the Fahrenheit scale. However, people who work in medical fields and other areas of science often use either the Celsius or Kelvin scales. You should be able to convert between the Fahrenheit, Celsius, and Kelvin scales using the Temperature section of the *Earth Science Reference Tables*.

Air temperature is usually measured using a liquid-filled glass tube called a thermometer. As the temperature of the liquid in the thermometer rises, the liquid expands and moves up the scale. Often temperature is modeled on maps and charts by the use of isolines called isotherms.

MEMORY JOGGER

You may recall from Topic 5 that temperature is the measure of the average kinetic energy of particles of a body of matter. Therefore, the greater the average kinetic energy of the particles in the air, the higher the atmospheric temperature.

You may also recall from Topic 6 that insolation (INcoming SOLar radiATION) is the portion of the sun's radiation that reaches the top of Earth's atmosphere.

Heating of the Atmosphere

The sun is the original source of almost all the heat in the atmosphere. Generally, the more insolation at a location, the warmer Earth's surface and the atmosphere will be. The atmosphere acquires much of its heat directly by contact with Earth's surfaces, but it also gains energy in other ways. The many ways in which the atmosphere is heated include the following:

- Conduction moves heat from Earth into the atmosphere as air touches Earth's surface.
- Direct absorption of insolation from the sun by gases and aerosols moves heat into the atmosphere. (An aerosol is any liquid or solid particle, such as solid smoke particles or liquid cloud droplets, suspended or floating within a gas.)
- Absorption of long-wave infrared radiations from Earth's surface move heat into the atmosphere. Much of the infrared absorption by the atmosphere is due to the presence of water vapor, methane, and carbon dioxide; the larger the amounts of these greenhouse gases, the more heat is absorbed by the atmosphere.
- Condensation (change of water vapor to liquid water) and sublimation (change of water vapor directly to ice) release large amounts of stored heat, directly heating the atmosphere. Condensation and sublimation release this energy when clouds, fog, dew, and frost form.
- The Coriolis effect, which results from the rotation of Earth and wind—causes friction where the atmosphere and Earth's surface meet. This friction produces heat, some of which is absorbed by the atmosphere.

CONVECTIONAL TRANSFER OF HEAT IN THE ATMOSPHERE Heat energy is transferred within the atmosphere by convection. Differences in air density cause differences in air pressure. The air pressure differences in turn cause air to move in circular patterns—convection currents, or cells. These air convectional movements transfer heat energy within the atmosphere, especially the troposphere. Winds are the parts of convection currents that are parallel to Earth's surface.

HEATING AND COOLING OF AIR BY EXPANSION AND COMPRESSION When a gas expands, its temperature decreases; when a gas is compressed—or contracted—its temperature increases. Thus when air rises in the atmosphere, it expands and its temperature decreases; similarly, when it descends, it is compressed and its temperature increases. Under average conditions air temperature in the troposphere decreases with increasing altitude because as air rises there is less air above it and less air pressure and thus the air expands and cools. (See Selected Properties of Earth's Atmosphere in the *Earth Science Reference Tables*.)

Review Questions

1. Where does the greatest amount of daily weather change take place? (1) tropopause (2) stratopause (3) troposphere (4) stratosphere

2. The highest air temperature ever recorded in Albany, New York, is 104°F, which occurred on July 4, 1911. This temperature is equal to (1) 35°C (2) 40°C (3) 45°C (4) 50°C

3. Which source provides the most energy for atmospheric weather changes? (1) radiation from the sun (2) radioactivity from Earth's interior (3) heat stored in ocean water (4) heat stored in polar ice caps

4. The transfer of heat energy within the troposphere occurs primarily by (1) insolation (2) conduction (3) radiation (4) convection

5. As a mass of air rises in the troposphere its temperature will usually (1) decrease due to expansion (2) decrease due to compression (3) increase due to expansion (4) increase due to compression

6. Which graph best represents what most likely happens to the temperature of Earth's atmosphere as the amount of carbon dioxide in the atmosphere increases over a period of many years?

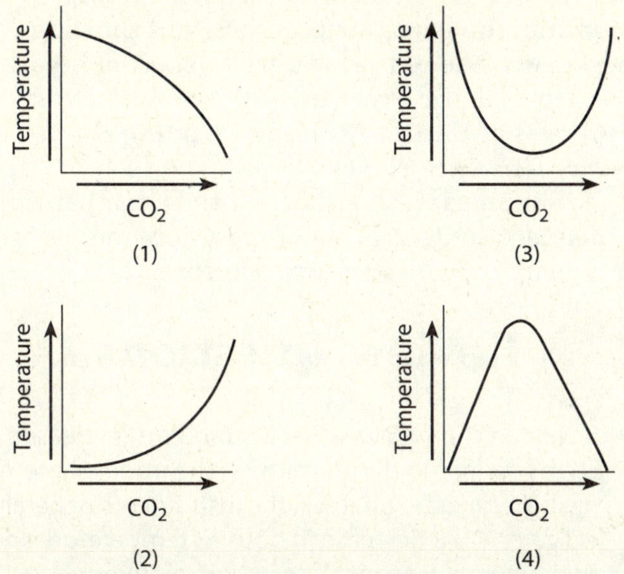

7. The diagram below shows the direction of movement of air over a mountain. As the air moves down the leeward side of the mountain, the air will (1) warm due to compression (2) warm due to expansion (3) cool due to compression (4) cool due to expansion

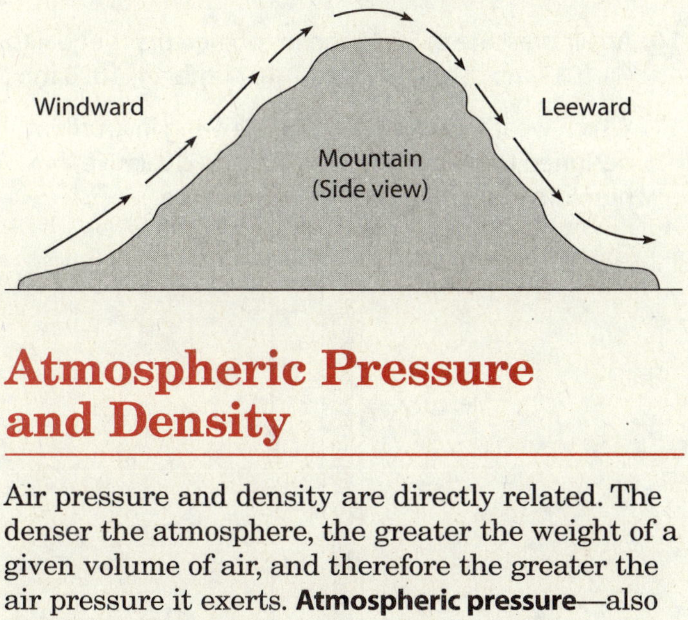

Atmospheric Pressure and Density

Air pressure and density are directly related. The denser the atmosphere, the greater the weight of a given volume of air, and therefore the greater the air pressure it exerts. **Atmospheric pressure**—also called **barometric pressure** and **air pressure**—is the pressure due to the weight of the overlying atmosphere pushing down on any given area.

Measurement of and Changes in Air Pressure

Instruments called **barometers** are used to measure air pressure. Two types of barometers are used—the mercury barometer and the aneroid barometer. The mercury barometer is the standard reference instrument, but it is difficult to transport, and there is the danger of mercury vapor. Standard air pressure (one atmosphere) at sea level is 14.7 pounds per square inch, 29.92 inches of mercury, or 1013.2 millibars. Barometric pressure in millibars can be converted to inches of mercury using the Pressure chart in the *Earth Science Reference Tables*. (See Figure 7-1.) The barometric trend—shown on weather maps—is how the magnitude of the air pressure has changed over the past three hours. Air pressure is often shown on weather maps by the use of isolines called **isobars,** as shown in Figure 7-4 on page 112.

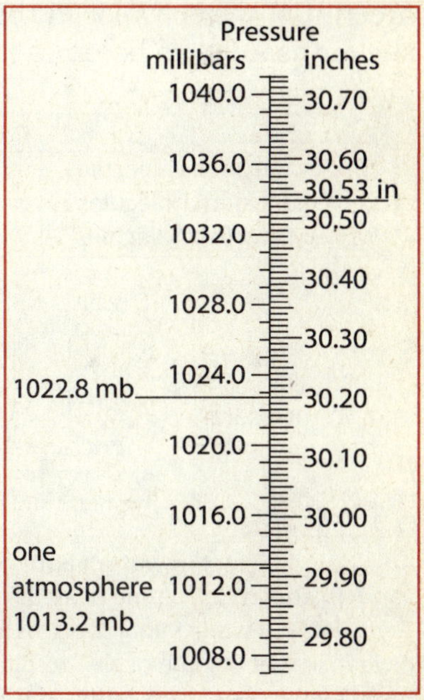

Figure 7-1. Method of converting from millibars to inches of mercury in air pressure readings: This is a portion of the chart from the *Earth Science Reference Tables.* The millibar part of the table has divisions for each whole millibar. You should be able to estimate to the nearest tenth of a millibar within three tenths. The inch part of the scale's big divisions is for each tenth of an inch, and smaller divisions are for hundredths of an inch. Because of the smallness of these divisions, a measurement to the nearest hundredth of an inch is a reasonable answer. Some conversions are 1013.2 millibars (mb) = 29.92 inches, 1022.8 mb = 30.20 inches, 30.53 inches = 1034.0 mb, 29.81 inches = 1009.5 mb.

EFFECT OF TEMPERATURE ON AIR PRESSURE

Changes in the temperature of the air cause changes in air pressure. As the temperature of air increases, the air expands and its density and pressure decrease. Decreasing temperature has the reverse effect.

MEMORY JOGGER

You may recall that in a mercury barometer, a tube with a closed top is placed in a container of mercury. When a column of air from above pushes down on the mercury in the container, the mercury rises in the tube. The air pressure is read in inches of mercury. In an aneroid barometer, a metal can that is a partial vacuum expands and contracts with changes in air pressure, and a series of mechanical parts display a reading on a dial.

EFFECT OF WATER VAPOR ON AIR PRESSURE

The greater the amount of water vapor in the air (called moisture content, absolute humidity, and humidity), the lower the air density and pressure. Each water molecule in the atmosphere replaces another molecule of air, usually oxygen or nitrogen. Since a water molecule weighs less than either an oxygen or a nitrogen molecule, the greater the amount of water vapor in the air, the less dense the air as a whole becomes (See Figure 7-2).

Type of Gas	Relative Molecular Weight
N = N_2, nitrogen molecules	14
O = O_2, oxygen molecules	16
W = H_2O, water molecules	10

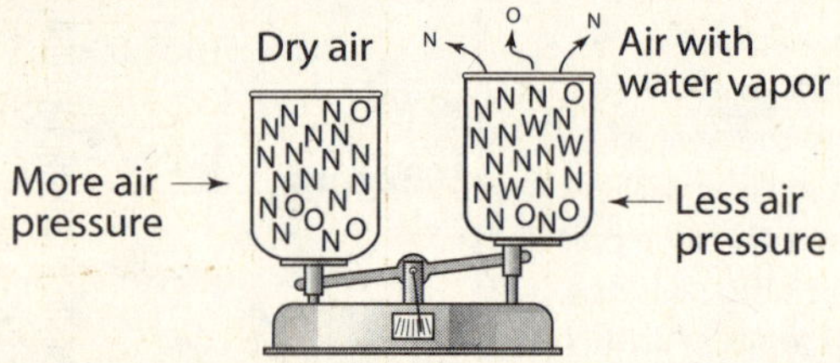

Figure 7-2. Effect of water vapor on air pressure: When water vapor is "added" to volume of air, it is not really added. A molecule of water vapor replaces a molecule of nitrogen or oxygen in that volume of air—resulting in the same number of molecules as there were before the water vapor was added. Since each molecule of water has less weight and mass than nitrogen and oxygen, the more water vapor in a given volume of air, the less the air's weight, density, and pressure. The effect shown in the diagram is greatly exaggerated.

EFFECT OF ALTITUDE ON ATMOSPHERIC PRESSURE As altitude or elevation increases, atmospheric density and pressure decrease. As you go up higher in the atmosphere, the less gas there is above you. If you have ever climbed a high mountain, the lower air pressure, and thus less oxygen, was the reason you were "short of breath." (See Atmospheric Pressure with Selected Properties of Earth's Atmosphere in the *Earth Science Reference Tables*.)

To summarize, as either altitude, temperature, or moisture content in the atmosphere increases, air density and air pressure decrease.

Review Questions

8. A balloon carrying weather instruments is released at Earth's surface and rises through the troposphere. As the balloon rises, what will the instruments generally indicate? (1) a decrease in both air temperature and air pressure (2) an increase in both air temperature and air pressure (3) an increase in air temperature and a decrease in air pressure (4) a decrease in air temperature and an increase in air pressure

9. An air pressure of 1023 millibars is equal to how many inches of mercury? (1) 30.10 (2) 30.15 (3) 30.19 (4) 30.21

10. An air pressure of 30.15 inches of mercury is equal to (1) 1017 mb (2) 1019 mb (3) 1021 mb (4) 1023 mb

11. Which weather variable is the following instrument designed to measure? (1) visibility (2) relative humidity (3) dew point (4) air pressure

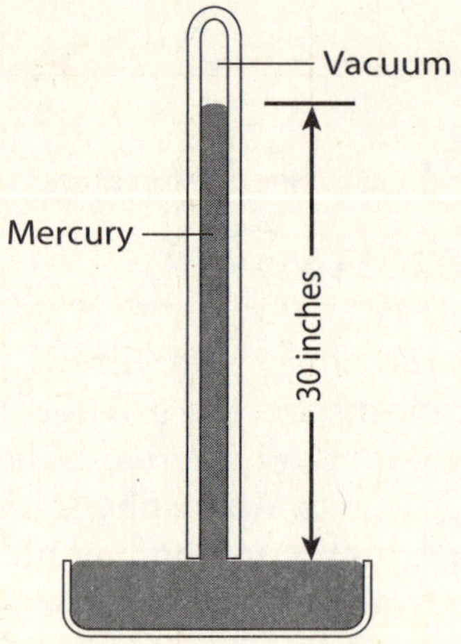

12. A barometer indicates a pressure of 30 inches of mercury at sea level; at $3\frac{1}{2}$ miles above sea level, it indicates a pressure of 15 inches. What is the best conclusion you can draw from these data? (1) humidity affects the pressure (2) temperature affects the pressure (3) of the total mass of air, about 99 percent is within $3\frac{1}{2}$ miles of sea level (4) of the total mass of air, about 50 percent is less than $3\frac{1}{2}$ miles above sea level

13. At sea level, as the temperature of the atmosphere decreases, the air pressure usually (1) decreases (2) increases (3) remains the same

14. Which graph best represents the relationship between air temperature and air density in the atmosphere?

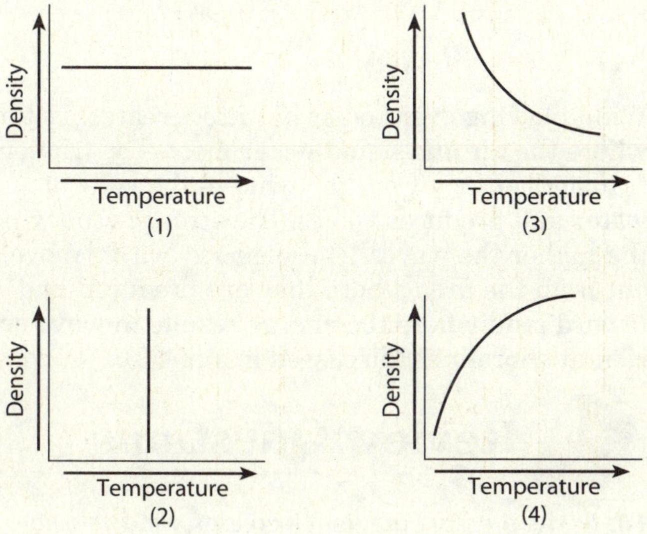

15. If the air humidity increases and the temperature remains the same, the air pressure will (1) decrease (2) remain the same (3) increase (4) first increase, then decrease

16. What is the approximate elevation above sea level at which atmospheric pressure is 10^{-1} atmosphere? (1) 7.0 km (2) 17 km (3) 27 km (4) 37 km

17. As a parcel of air increases in altitude, its pressure will (1) decrease (2) increase (3) remain the same

Wind

Horizontal movement of air parallel to Earth's surface is called wind. Wind is a type of field called vector because it requires a magnitude and direction to totally describe it.

Wind Speed

Winds are caused by differences in air pressure, which are often caused by changing atmospheric conditions of temperature and water vapor content. Figure 7-3 illustrates the cause of wind using local land and sea breezes as an example. The difference in air pressure for a specific distance is called the **air pressure gradient.** The closer together the isobars on a weather map are, the greater (steeper) the pressure gradient. The greater the pressure gradient, the faster the wind speed. (See Figure 7-4.) An instrument called an **anemometer** is used to measure wind speed. Wind speed is measured in miles per hour and knots—nautical miles per hour. One knot equals 1.15 miles per hour (mi/hr); thus, 5 knots equals about 5.75 mi/hr.

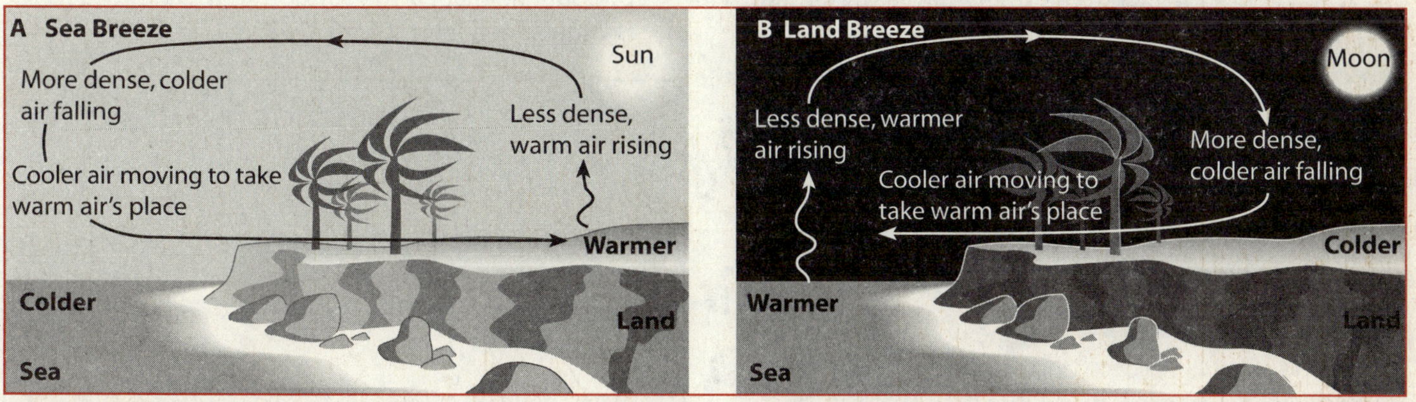

Figure 7-3. Cause of wind illustrated by land and sea breezes: In diagram A, a sea breeze develops because the land heats up faster than the water during the day—resulting in warm, less dense, low-pressure air that rises. Over the ocean cooler, denser, higher-pressure air forms. The cooler, denser, higher-pressure air from the sea replaces the rising warmer air over land—resulting in a sea breeze. In diagram B, at night, reverse conditions result in a land breeze. Note that land and sea breezes can be part of a convection cell.

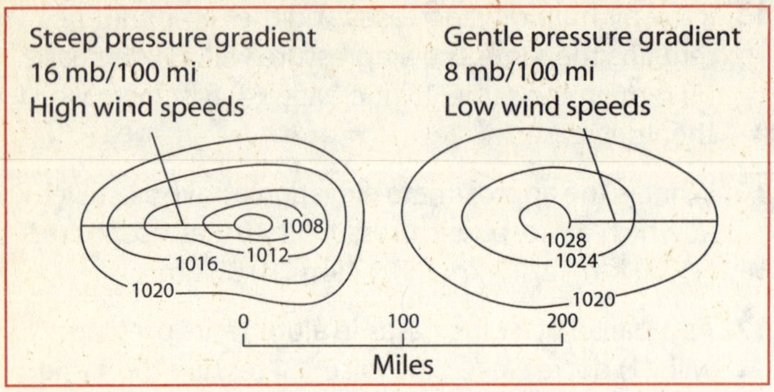

Figure 7-4. Pressure gradient as indicated by closeness of isobars on a weather map: The closer together the isobars, the greater the air pressure gradient. The greater the pressure gradient, the greater the wind speed.

MEMORY JOGGER

You may recall that an anemometer usually has four cups placed on the ends of two intersecting rods. As the wind blows, the cups catch the wind and cause the rods to spin. The faster the wind is blowing, the greater the spinning of the anemometer. The spinning rods cause a shaft—to which they are attached—to spin. The shaft is attached to an indicator that shows the wind speed.

Wind Direction

Air (wind) moves from areas of high pressure to areas of low pressure. However, the Coriolis effect (caused by Earth's rotation) modifies that pattern of movement, deflecting winds to the right in the Northern Hemisphere and to the left in the Southern Hemisphere. Figure 7-5 illustrates how wind directions are modified by Earth's rotation.

A wind is named for the direction from which it comes; for example, a wind blowing from the south toward the north is a south wind. A wind blowing towards the southeast is a northwest wind—the direction the wind is coming from. An instrument called a wind vane, or weather vane, is often used to determine wind direction.

MEMORY JOGGER

You may recall seeing a wind vane or a picture of one. The arrowhead of a wind vane points into the wind—that is, it points in the direction that the wind is blowing from.

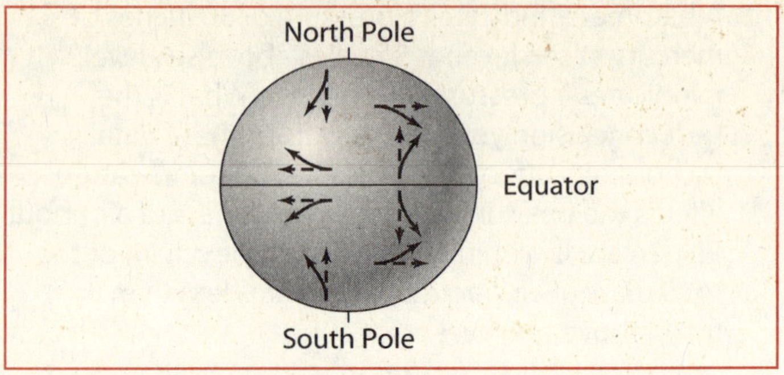

Figure 7-5. Deflection of winds by the Coriolis effect: The dashed arrow in each case shows the direction the wind would blow if there were no Coriolis effect. The solid arrow shows the actual path of the wind. If you face toward the direction the wind is blowing toward, it is always deflected to the right in the Northern Hemisphere and to the left in the Southern Hemisphere.

Formation of Waves on Surface Water

Wind blowing over bodies of water creates friction where the air and liquid water meet. The friction transfers energy from the wind to the body of water and produces waves. The stronger the wind, the higher the waves. These energy waves move out from the area where they are produced and toward land where the energy results in wave erosion at shorelines (discussed in Topic 9).

Review Questions

18. As the pressure gradient increases, wind velocity (1) decreases (2) increases (3) remains the same

19. The following map represents a portion of an air-pressure field at Earth's surface. At which position is wind speed lowest? (1) A (2) B (3) C (4) D

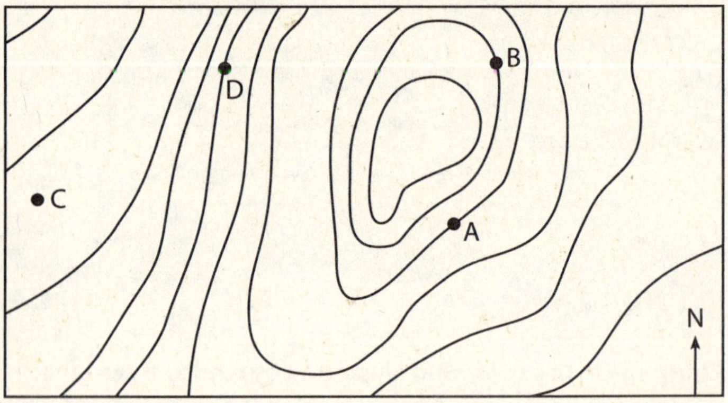

20. Which cross-section best shows the normal movement of the air over Oswego, New York, on a very hot summer afternoon?

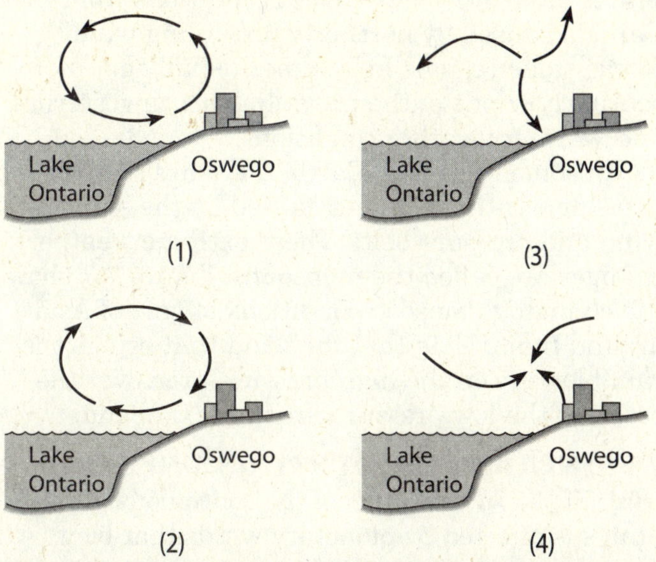

21. The weather element determined by an anemometer is (1) pressure (2) humidity (3) visibility (4) wind velocity

22. Winds blow from regions of (1) high temperature to regions of low air temperature (2) high air pressure to regions of low air pressure (3) low temperature to regions of high temperature (4) low air pressure to regions of high air pressure

23. In the Northern Hemisphere, a wind blowing from the south will be deflected towards the (1) northwest (2) northeast (3) southwest (4) southeast

24. Which is NOT true about wind? (1) Wind direction is named for the direction toward which the wind blows. (2) Wind moves from regions of higher pressure to lower pressure. (3) The steeper the gradient, the greater the wind speed. (4) Wind is horizontal movement of air.

General Circulation of the Air in the Troposphere

Differences in air temperature and pressure in various portions of the troposphere cause air to be in constant motion. This circulation of air produces convection cells throughout the troposphere.

Convection Cells

The unequal distribution of insolation on Earth results in unequal heating and differences in air pressure. Cooler air, being denser, sinks toward Earth under the influence of gravity, causing the less dense, warmer air to rise. The result is a series of convection cells around Earth at various latitudes, as shown in Figure 7-6. As indicated by the solid arrows, there are upward currents in the vicinity of 0° latitude (the equator) and 60° North and South latitudes. Downward currents exist near 30° and 90° North and South latitudes. Regions where air comes together to form vertical currents are regions of convergence. Regions where air spreads out from the vertical currents are regions of divergence.

As part of these tropospheric convection cells, there are bands of easterly moving air at the top of the troposphere, called **jet streams.** The winds of the jet stream can blow 200 miles an hour or more. Commercial airplanes flying with the jet stream can save a half hour of flight time traveling from the west to east coasts of the contiguous United States, or lose an hour going in the other direction, if the jet stream is not avoided. The jet streams have an important influence on the formation and direction of weather pattern movement in mid-latitude regions such as the United States. The jet streams help steer the air masses and low pressure centers. The locations of the jet streams are shown on Figure 7-6 and on the Planetary Wind and Moisture Belts in the Troposphere in the *Earth Science Reference Tables*.

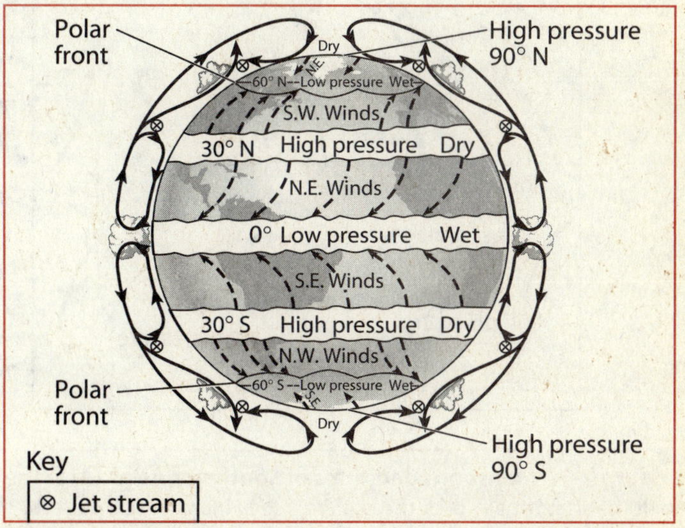

Figure 7-6. Planetary wind, pressure, and moisture belts in the troposphere: The drawing shows the locations of the belts near the time of an equinox. The locations shift somewhat with the changing latitude of the sun's vertical rays.

Planetary Wind and Pressure Belts

At Earth's surface, winds blow horizontally away from regions of divergence and high air pressure, and toward regions of convergence and low air pressure. Figure 7-6 shows the pressure belts of Earth with generally consistent high or low air pressure. Because of the Coriolis effect, winds moving away from high air pressure are deflected to the right in the Northern Hemisphere and to the left in the Southern Hemisphere, as indicated by the dashed arrows in Figure 7-5 on page 112. The result is a series of **planetary wind belts** within which the winds move generally in a specific direction much of the time—thus they are prevailing winds.

SEASONAL SHIFTING OF THE WIND AND PRESSURE BELTS Because the vertical rays of the sun shift north and south with the seasons, the pressure belts and the resulting wind belts generally follow the vertical rays of the sun. This results in many parts of Earth being under the influence of different wind and/or pressure belts in different seasons. Difference in air pressure and prevailing winds can often mean quite different weather conditions. As an example, in the winter the northern portions of the contiguous United States often get cold weather conditions from Canada caused by northerly prevailing winds. These same regions in summer usually get more southerly winds of the prevailing southwesterlies and warmer weather conditions.

In some regions of Earth, there are regular and extreme weather changes caused by the shifting wind and pressure belts. These extreme weather changes are called the **monsoons.** Figure 7-7 shows the changing monsoon conditions of part of Asia around India. Note that the wind that brings the rain blows from the high pressure area over the ocean to the low pressure area over the land.

WEATHER MOVEMENT IN THE CONTIGUOUS UNITED STATES Much of the contiguous United States is affected by planetary winds that blow from the southwest to the northeast—called the prevailing southwesterly winds. Therefore, weather changes in the United States move generally from a southwesterly direction to a northeasterly direction.

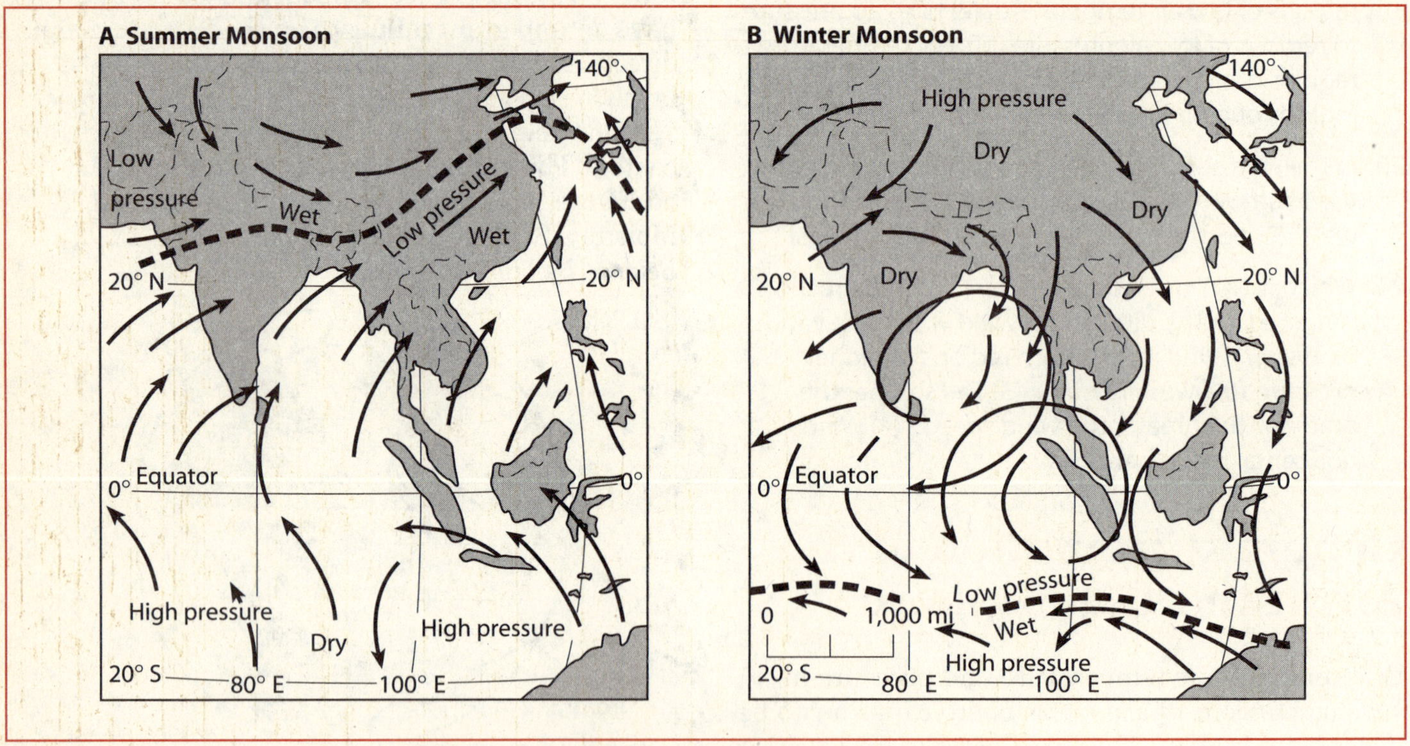

Figure 7-7. Monsoon conditions of Southeast Asia: (A) In the summer, the wet, low pressure belt that is normally over the equator (see Figure 7-6 on page 113) moves northward and brings high amounts of rain to much of Southeast Asia, including India. **(B)** In the winter, the high pressure area is over the land, and this brings dry air from Central Asia over Southeast Asia towards the low pressure belt.

FORMATION OF SURFACE OCEAN CURRENTS

Surface ocean currents are caused by wind blowing over the oceans and transferring energy to the water. The direction of these currents is affected by the direction of the planetary winds. Compare the map Surface Ocean Currents with the diagram Planetary Winds and Pressure Belts in the *Earth Science Reference Tables* as well as with Figure 7-8. The directions of the surface ocean currents are also affected by blocking by landmasses and the rotation of Earth through the Coriolis effect. Note that the maps of the surface ocean currents show that they are often part of a circular pattern that spins clockwise in the Northern Hemisphere and counterclockwise in the Southern Hemisphere. This difference in spinning direction is caused by the Coriolis effect's deflection to the right in the Northern Hemisphere and to the left in the Southern Hemisphere.

Since the surface ocean currents are a direct result of the transfer of energy from the movement of the prevailing winds, they can seasonally shift their positions. Some of the surface ocean currents shift position north and south as the winds of the prevailing wind belt do—following the motions of the vertical rays of the sun.

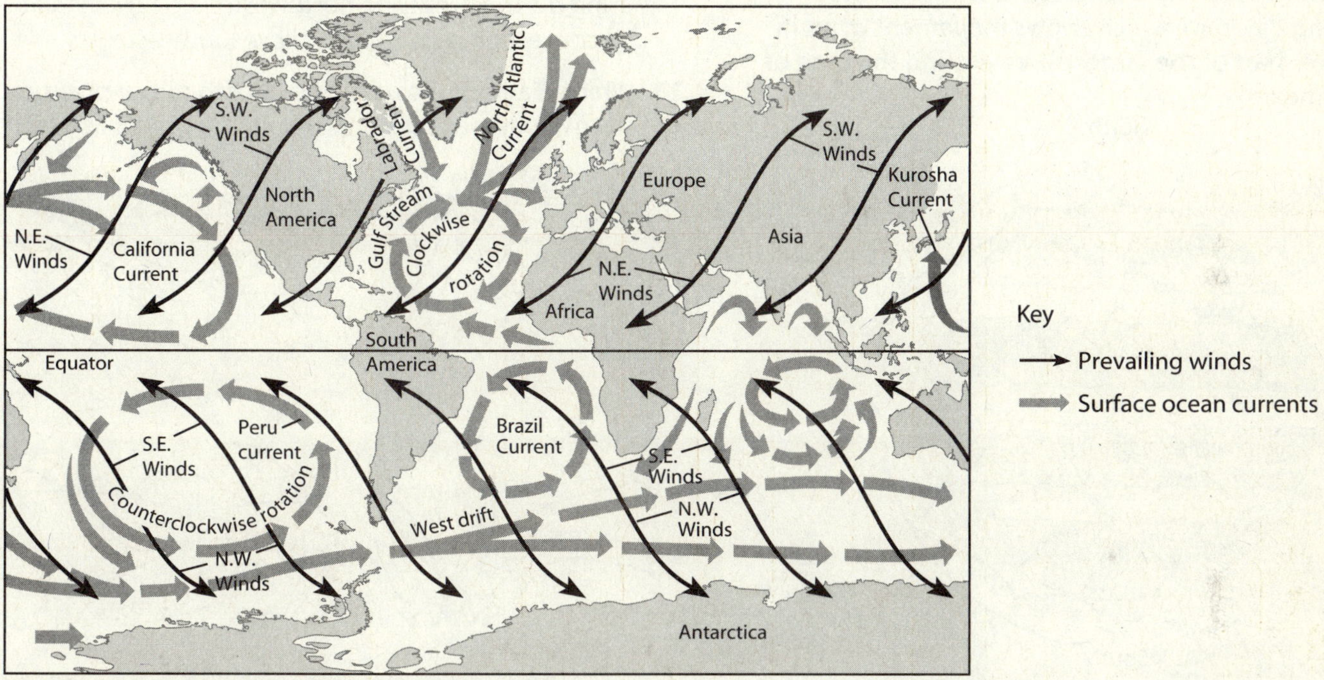

Figure 7-8. Cause of surface ocean currents: When you observe a comparison of the directions of Earth's prevailing winds and the location of the surface ocean currents, it is obvious that they are related. Prevailing winds transmit kinetic energy and "drive" the surface ocean currents. The Coriolis effect twists the direction of the currents to the right in the Northern Hemisphere creating a clockwise rotation of the currents there. In the Southern Hemisphere, the Coriolis effect shifts currents to the left (west), creating a counterclockwise rotation of the surface ocean currents. Note that without the blocking effect of the continents, surface ocean currents could circle the world like the west wind drift north of Antarctica.

Review Questions

25. Planetary winds do not blow directly north or south because of (1) the Coriolis effect (2) gravitational force (3) magnetic force (4) Earth's revolution

26. On a certain day in Syracuse, New York, there is a high-pressure center directly north and a low-pressure center directly south. Because of the effect of Earth's rotation, the air in Syracuse will (1) blow directly to the south (2) blow directly to the north (3) be deflected toward the southwest (4) be deflected toward the northeast

To answer questions 27 through 31, refer to the following diagram, which shows movement of air in the lower part of the atmosphere around the time of an equinox.

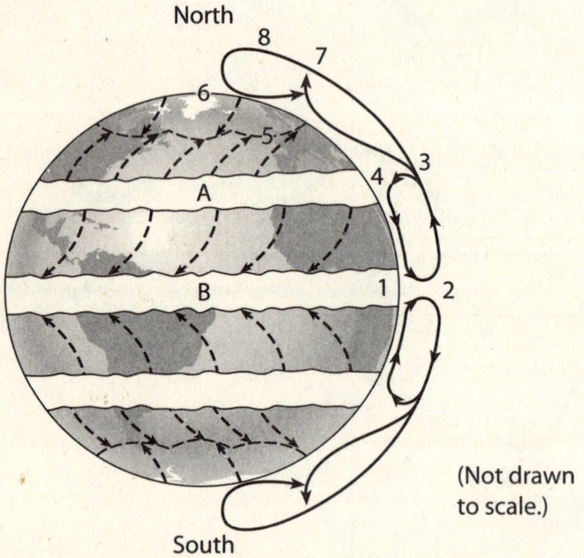

27. The movement of air from 1 to 2 to 3 to 4 to 1 would be called a (1) high-pressure cell (2) contiguous cell (3) convection cell (4) Coriolis cell

28. What is the basic underlying reason for the movement of air shown? (1) differences in gravity (2) differences in air density (3) differences in pressure gradient (4) differences in magnetism

29. What would be the most likely cause of the movement of air at position 1? (1) low moisture content and low temperature (2) high moisture content and high temperature (3) low moisture content and high temperature (4) high moisture content and low temperature

30. The air moves at the surface of Earth, from position A to position B, because (1) positions A and B have low pressure (2) positions A and B have high pressure (3) position A has low pressure, and position B has high pressure (4) position A has high pressure, and position B has low pressure

31. Condensation would most likely occur at position (1) 3 (2) 4 (3) 6 (4) 7

---

32. Pilots flying from the west coast of the United States to New York may sometimes shorten their flight time by using a high-speed tailwind. They are most likely using the (1) prevailing northwesterlies (2) jet stream (3) convection cells (4) prevailing southeasterlies

33. Which diagram shows the usual path of low-pressure storm centers as they pass across the United States?

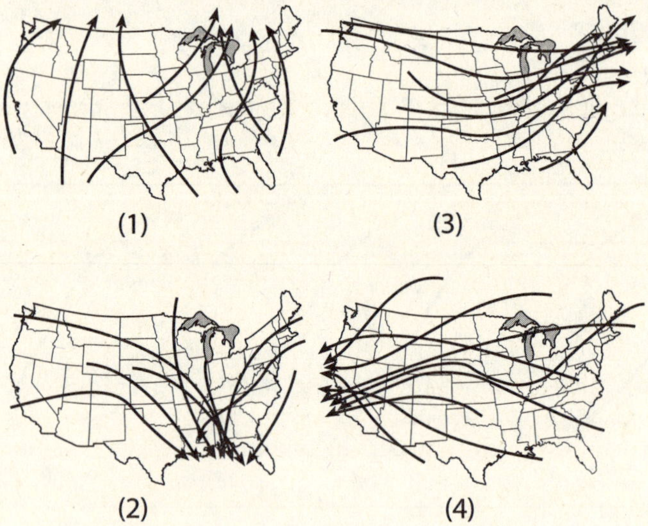

34. What is the primary source of energy for ocean currents and waves? (1) the moon (2) the atmosphere (3) the continents (4) the sun

35. What is the most direct cause of major surface ocean currents? (1) prevailing winds (2) gravity (3) tides (4) salinity differences

Atmospheric Moisture

Atmospheric moisture—water—exists in three states of matter: liquid, solid, and gas. Gaseous water in the atmosphere is called **water vapor.** The oceans, which cover about 70 percent of Earth's surface, are the source of most atmospheric moisture.

Moisture in the form of water vapor enters the atmosphere by evaporation, sublimation to a gas, and transpiration. Evaporation is the process by which a liquid changes to a gas; transpiration is the process by which plants release water vapor as part of their life functions. Collectively, evaporation and transpiration are called evapotranspiration. Sublimation to a gas is the process by which ice changes directly to water vapor without forming any liquid water.

Energy of Evaporation and Transpiration of Water

Large amounts of energy (approximately 540 calories/gram) are required to change liquid water into water vapor during the processes of evaporation and transpiration. Almost all of this energy comes from insolation. The solar energy is converted into a type of potential or stored energy in the water vapor. When evaporation or transpiration occurs, the more energetic water molecules leave the liquid and form water vapor. Since the molecules leaving the liquid are the more energetic ones, the average kinetic energy of the molecules remaining in the liquid decreases. As a result, the temperature of an evaporating liquid is somewhat lower than its surroundings. You have noticed this effect of evaporation when you step out of the shower and you feel chilled.

Process of Evaporation

Figure 7-9 helps to explain what happens when water—or any other liquid—evaporates. Diagram A shows a closed container of air into which some water has just been added. We assume there are no water vapor molecules in the air at this moment. However, as soon as water is placed in the container, some of its more energetic molecules begin to escape into the air above and mingle with the other gas molecules that are present. Diagram B shows the situation a short time later. There are water molecules in the air. We see two more molecules in the process of leaving the water. However,

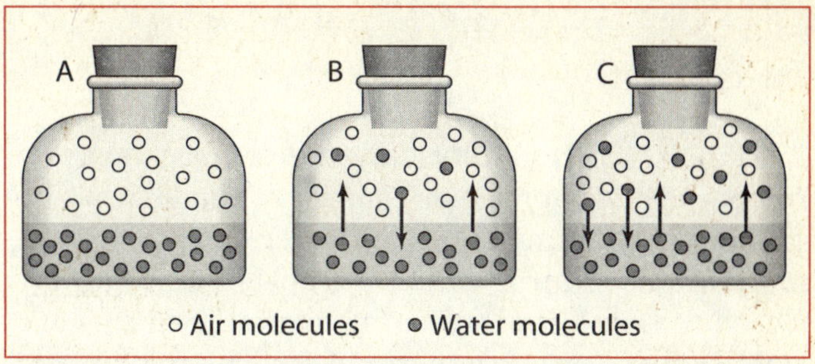

Figure 7-9. The saturation of the air with water vapor and evaporating liquid water: In diagram A, no evaporation has occurred. In diagram B, evaporation is proceeding faster than condensation. In diagram C, a dynamic equilibrium between evaporation and condensation has been reached, and the air is saturated with water vapor.

one molecule in the air is about to enter the water, so that there will be five molecules in the air. Diagram C shows the situation some time later. There are now eight water molecules in the air. Again, two more water molecules are about to leave the water, but now two molecules in the air are about to return to the water. The net result will be eight molecules in the air—the same number as before. We have a condition of dynamic equilibrium in which the rate at which water is entering the gaseous state (evaporation) equals the rate at which water vapor is returning to the liquid state (condensation).

Factors Affecting Evaporation Rates of Water

The net evaporation rate of water at a location is determined by

- amount of energy available: the more energy available, and thus the higher the temperature, the faster the evaporation of available water
- surface area of the water: the more spread out the water, the greater the air-water interface and the faster the evaporation
- degree of saturation of the air with water vapor: the higher the water vapor content of the air, the closer the air is to being saturated with water vapor and the slower the net evaporation rate
- wind speed: wind often moves air at the water-air interface that is high in water vapor and replaces it with air that is less saturated and thus speeds up evaporation rates. The greater the speed of the wind, the more air replacement occurs.

Humidity, Temperature, and Dew Point

Like moisture, **humidity** is a general term that refers to the water vapor content of the atmosphere. The amount (mass) of water vapor in each unit volume of air is called the absolute humidity. Absolute humidity is often measured in grams of water vapor per cubic meter of air. The maximum absolute humidity, or moisture capacity, increases rapidly with an increase in air temperature, as shown in the graph in Figure 7-10. This means that the total amount of water vapor the air can hold—its capacity—increases with an increase in temperature, or in other words, hotter air can hold more water vapor than colder air.

The ratio of the amount of water vapor in the air (absolute humidity) to the maximum amount it can hold (its moisture capacity) is **relative humidity.** It is usually expressed as a percent. In simpler terms, relative humidity is the amount of water vapor in the air compared to the amount of water vapor the air can hold at a specific temperature. The closer the absolute humidity is to the maximum absolute humidity (capacity) the higher the relative humidity. Note that relative humidity is a percentage of saturation, or capacity, NOT how much water vapor there is in a volume of air.

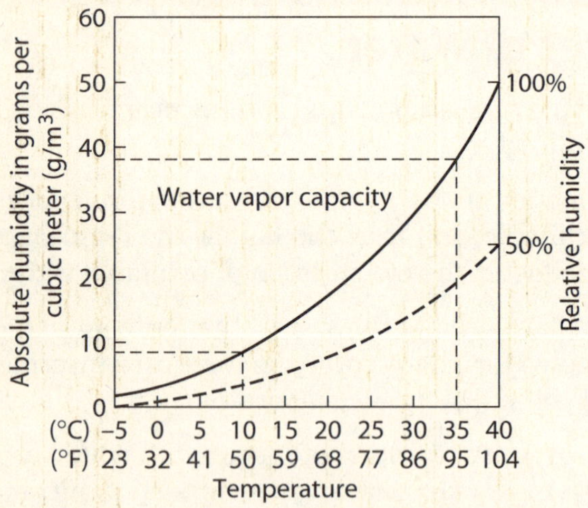

Figure 7-10. Relationship between air's capacity for water vapor and temperature: Note that air with a temperature of 35°C can hold about four times more water vapor than air at 10°C can. Note that when the air at 10°C has 50 percent relative humidity, it has only about 1/4 as much water vapor as air at 35°C with 50 percent relative humidity.

RELATIVE HUMIDITY AND TEMPERATURE

At any given time and place, the air has a certain amount of water vapor (humidity), with a corresponding absolute humidity. If the temperature of the air changes but the amount of water vapor remains the same, the relative humidity will change. For example, if the temperature increases, the relative humidity will decrease. This occurs because the capacity increases with the increase in temperature, while the absolute humidity remains the same. On the other hand, if the temperature decreases while the absolute humidity remains the same, the relative humidity will increase.

If the temperature of the air remains constant, but more water vapor is added to the air by evaporation or transpiration, both the absolute humidity and the relative humidity will increase. The relative humidity increases because the air becomes closer to its capacity, or saturation point, when more water vapor is added.

DEW POINT If the temperature of the air decreases while the absolute humidity remains the same, the temperature will eventually reach a point at which the absolute humidity equals capacity. At this temperature, the relative humidity will be 100 percent. This temperature is called the **dew point**—the temperature at which air is filled with water vapor. If the air temperature drops below the dew point, water vapor in the air will condense to liquid water or sublimate to a solid.

The dew point depends on the absolute humidity and not on the relative humidity. As the amount of water vapor in the air increases, the dew point also rises because the more water vapor in the air, the closer the air is to its saturation point—the dew point.

39. The primary source of most of the moisture in Earth's atmosphere is (1) soil-moisture storage (2) rivers and lakes (3) melting glaciers (4) oceans

40. As the exposed area of a moist object decreases, the rate of evaporation of the liquid from that object (1) decreases (2) increases (3) remains the same

41. As the amount of light energy striking a moist object increases, the rate of evaporation of the liquid from that object (1) decreases (2) increases (3) remains the same

42. As the amount of water vapor in a given volume of air increases, the rate of evaporation from a moist object (1) decreases (2) increases (3) remains the same

43. There will most likely be an increase in the rate at which water evaporates from a lake if there is (1) a decrease in the temperature of the air (2) a decrease in the altitude of the sun (3) an increase in the moisture content of the air (4) an increase in the wind velocity

44. In the following diagram, at which location would the amount of water vapor in the air most likely be greater? (1) A (2) B (3) C (4) D

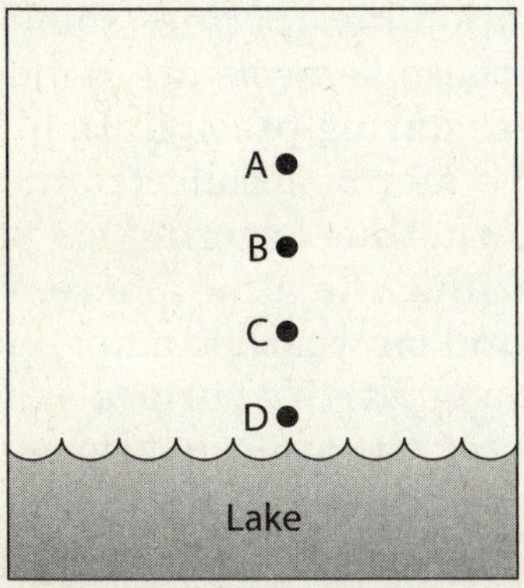

45. In the closed aquarium shown in the following diagram, the amount of water evaporating is equal to the amount of water vapor condensing.

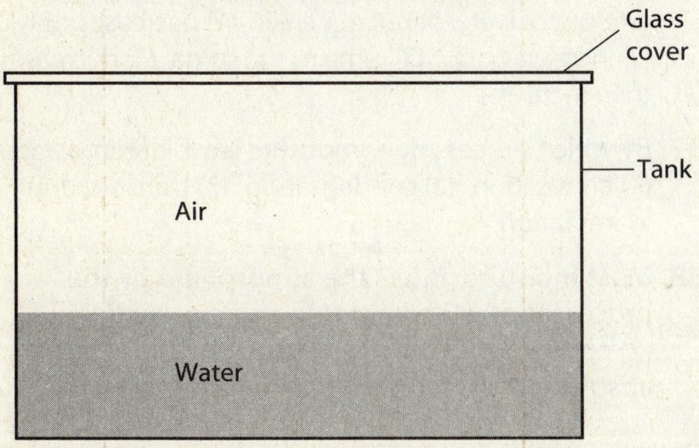

Which statement best explains why these amounts are equal? (1) The glass sides of the aquarium are warmer than the water. (2) The air in the aquarium is 50 percent saturated. (3) The relative humidity outside the aquarium is 100 percent. (4) The air in the aquarium is saturated.

46. Which graph best represents the relationship between the moisture-holding capacity of the atmosphere and atmospheric temperature?

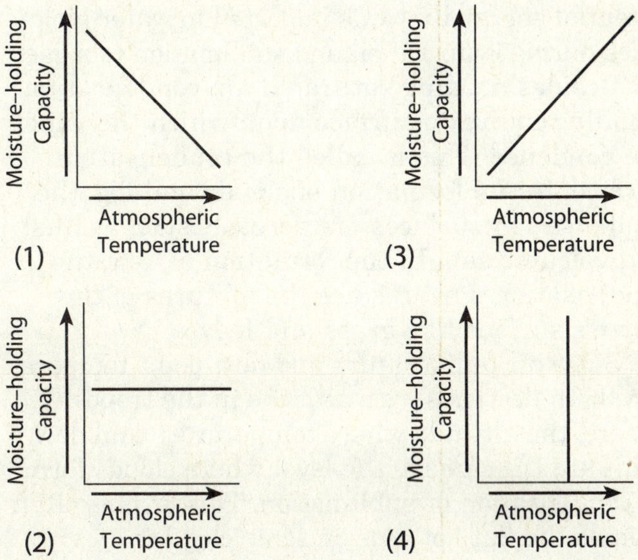

47. To say that the relative humidity on a given day is 70 percent means that the air (1) is composed of 70 percent water vapor (2) holds 70 percent of its water vapor capacity (3) contains 70 parts of water to 100 parts of dry air (4) contains the same amount of water that it would contain at 70°F

48. If the amount of water vapor in the air increases, then the dew point of the air will (1) decrease (2) increase (3) remain the same

49. During which part of the day is the relative humidity usually lowest? (1) morning (2) midafternoon (3) evening (4) late night

50. As the air temperature rises, the relative humidity (1) increases (2) decreases (3) remains the same

51. Air with a temperature of 60°F and a relative humidity of 51 percent was warmed to a temperature of 70°F, but the relative humidity remained at 51 percent. How did the moisture content change? (1) The moisture content decreased. (2) The moisture content increased. (3) The moisture content remained the same.

MEASURING RELATIVE HUMIDITY It is difficult to directly measure absolute humidity or determine the air's capacity to hold water vapor. Relative humidity is therefore measured by indirect methods. An instrument used to measure relative humidity is called a **psychrometer.** One indirect method of measuring relative humidity uses an instrument called a sling psychrometer, as shown in Figure 7-11. This instrument contains an ordinary thermometer called a dry-bulb thermometer, and another thermometer with a wick around its bulb, called the wet-bulb thermometer. When the wick is moistened and the thermometers are whirled in the air, the temperature of the wet bulb drops because of the cooling effect of the evaporation of the water. The amount of cooling depends on the rate of evaporation, and is therefore related to the relative humidity.

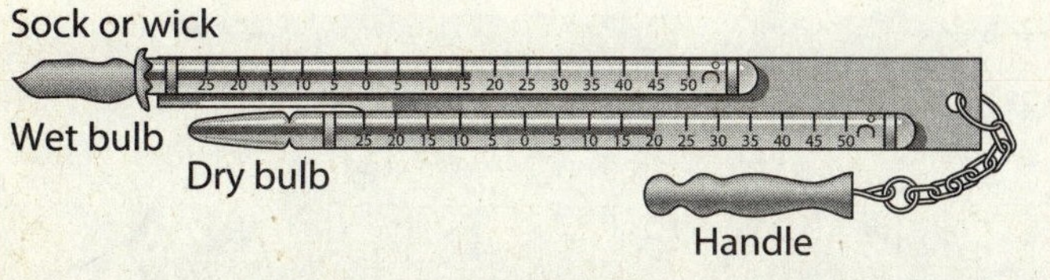

Figure 7-11. Sling psychrometer: The difference between the two temperature readings is used to find the relative humidity and dew point. Note that the dry-bulb reading at 20°C is the same as a regular temperature reading at 20°C.

To determine percent relative humidity, use the Relative Humidity (%) tables in the *Earth Science Reference Tables* and the following procedure.

- Locate the dry-bulb reading on the left-hand side of the Relative Humidity (%) chart in the *Earth Science Reference Tables.*
- Subtract the wet-bulb reading from the dry-bulb reading.
- Locate the difference between the wet-bulb and dry-bulb readings across the top of the chart.
- Follow the horizontal row for the dry-bulb reading to the right until it meets the vertical column running down from the top. This number is the relative humidity.

See Figure 7-12 for an example for determining relative humidity.

Dry-bulb Temperature (°C)	Difference Between Wet-Bulb and Dry-Bulb Temperatures (°C)									
	1	2	3	4	5	6	7	8	9	10
−20	28									
−18	40									
−16	48	0								
−14	55	11								
−12	61	23								
−10	66	33	0							
−8	71	41	13							
−6	73	48	20	0						
−4	77	54	32	11						
−2	79	58	37	20	1					
0	81	63	45	28	11					
2	83	67	51	36	20	6				
4 --→	→85	→70	→56	→42	→27	→14	= 14% Relative humidity			
6	86	72	59	46	35	22	10	0		
8	87	74	62	51	39	28	17	6		
10	88	76	65	54	43	33	24	13	4	

Figure 7-12. Determining relative humidity: Suppose the dry-bulb reading is 4°C and the wet-bulb reading is −2°C. The difference is 6°C. Reading across from 4°C and down from 6°C, you find a relative humidity of 14%.

DETERMINING DEW POINT To determine the dew point, you need the dry-bulb and wet-bulb readings from a sling psychrometer and the Dewpoint Temperatures chart in the *Earth Science Reference Tables*. Use the following procedure which is very similar to the procedure used to determine the relative humidity:

- Locate the dry-bulb reading on the left-hand side of the chart.
- Subtract the wet-bulb reading from the dry-bulb reading.
- Locate the difference between the wet-bulb and dry-bulb readings across the top of the chart.
- Follow the horizontal row for the dry-bulb reading to the right until it meets the vertical column running down from the difference between the wet-bulb and dry-bulb readings. This number is the dew point temperature.

For example, suppose the dry-bulb reading is 8°C, and the wet-bulb reading is 6°C. The difference is 2°C. Reading across from 8°C and down from 2°C, you find a dew point temperature of 3°C.

Precipitation

The falling of liquid or solid water from clouds toward the surface of Earth is **precipitation.** For precipitation to occur, the ice crystals or water droplets in clouds must come together to become big enough so that they will fall under the influence of gravity. The forms in which precipitation occur are rain, drizzle, snow, sleet, freezing rain, and hail. The conditions that result in the formations of these different types of precipitation are shown in Figure 7-13.

A rain gauge, a type of precipitation gauge, is used to measure liquid precipitation. It is measured in depth—usually inches or centimeters. Snow depth is often measured with a ruler, or it is gently warmed and the depth of liquid is measured and reported as precipitation. A rain shower is brief, rapidly forming, rapidly ending, and often heavy rain associated with thunderstorm-type clouds. A snow shower is a brief, heavy snowfall.

Cloud Formation

If the temperature of the air cools (often by expanding) below the dew point, the water vapor will usually condense to a liquid or sublimate to a solid, changing to microscopic liquid water droplets or ice crystals, respectively. At Earth's surface, condensation produces liquid dew, and sublimation produces the solid frost. If the temperature in the atmosphere is above 0°C, condensation produces water droplets, which appear as a cloud when grouped. Some clouds are composed of ice crystals, which form when water vapor sublimates to a solid. A cloud is therefore a collection of liquid water

Figure 7-13. Forms of precipitation

Rain Falling liquid droplets larger than 0.2 mm in diameter; may be melted snow

Drizzle Falling liquid droplets from 0.2 to 0.5 mm in diameter

Snow Falling ice crystals formed by combining cloud ice crystals

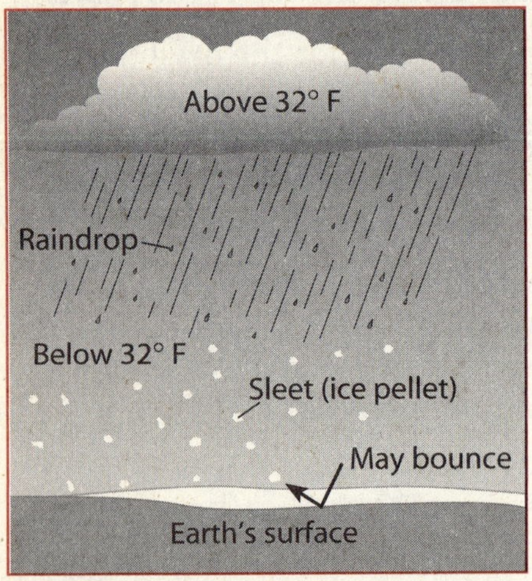

Sleet Solid pellets of ice that form by freezing of rain drops as they fall

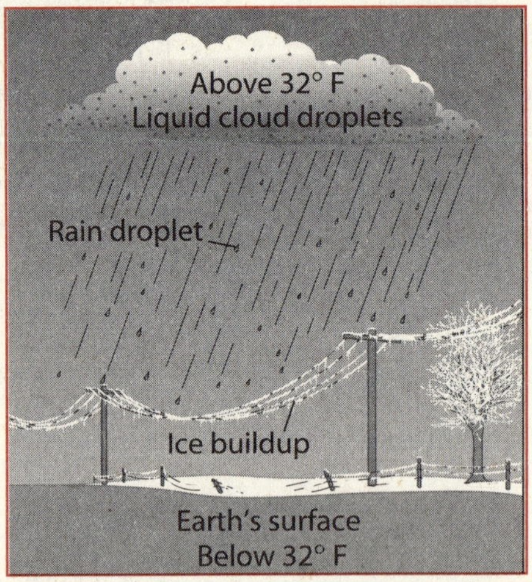

Freezing rain Rain or drizzle that freezes on contact with features of Earth's surface

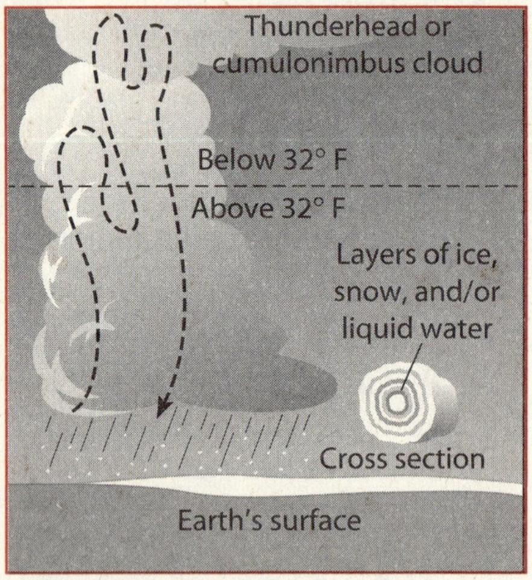

Hail Layers of ice, snow, and water formed by many up-and-down movements in a thunderstorm cloud; may be grapefruit-size

droplets and/or ice crystals suspended in the atmosphere and dense enough to be visible. If there is only a low density of water droplets or ice crystals, a condition called haze may result. **Cloud cover**—found on some weather maps—is the fraction or percent of the total sky at a location that is covered by clouds. If a cloud is on, or just above, Earth's surface, it is called fog. The condensation or sublimation that forms clouds and fog releases tremendous amounts of energy into the troposphere. This is the potential energy that was imparted to water molecules during evaporation and sublimation to a gas.

Besides needing saturated air, condensation usually requires a surface upon which the vapor can condense. This is called the condensation surface. In the formation of clouds and fog, the condensation surfaces are aerosols such as dust and volcanic ash. In the formation of dew, the condensation surfaces are the features of the Earth's surface like grass and leaves.

Since air temperature and dew point temperature both decrease with altitude in the troposphere, the altitude where temperature and dew point are the same is the level where clouds form by condensation or sublimation. This level is often seen as the flat bottoms on some clouds.

Calculating Cloud Base Altitude

Since temperature and dewpoint temperature both decrease with elevation or altitude in the troposphere, the altitude where temperature and dewpoint are the same can be calculated to obtain the altitude of the base of the clouds. Refer to the Generalized Graph for Determining Cloud Base Altitude in Appendix 3 on page 304. On this graph the dashed lines represent the decrease in dewpoint with increasing altitude (about 2C°/km). The solid diagonal lines represent the decrease in temperature with increasing altitude (about 10C°/km) above the Earth's surface. If you know the temperature and dewpoint at the Earth's surface, you can determine the altitude of condensation (or sublimation) and therefore the altitude of the bottom of the clouds above a location.

Suppose the surface temperature is 20°C and the surface dewpoint is 12°C at a location. Follow these steps to find the level of the base of the clouds above the location.

1. Find the 20°C mark and the 12°C mark along the bottom edge of the graph on page 303.
2. From the 20°C mark, follow the solid diagonal line upward and to the left.
3. From the 12°C mark, follow the dashed line upward.
4. The place where the two lines intersect represents the altitude at which clouds will form. In this case the base of the clouds is 1.0 km above the Earth's surface.

Atmospheric Transparency and Precipitation

The more pollutants added to the atmosphere by the activities of people and nature, the more aerosols are present. The more aerosols in the air, the less transparent the atmosphere is to insolation. How transparent the atmosphere is to insolation is called **atmospheric transparency.** When the atmosphere is less transparent, it means that more energy from the sun is absorbed or reflected by the atmosphere, resulting in less insolation reaching Earth's surface.

When the atmosphere has an aerosol content so high that distant images are blurred, and a cloudless sky does not appear blue, the condition is called haze. A haze or fog, usually brownish, which is highly polluted is smog. Condensation in cloud formation incorporates some of the aerosols, and these aerosols are removed from the atmosphere during precipitation. The falling liquid or solid water also collects other aerosols on the way down, thus lowering air pollution levels and cleaning the atmosphere. **Visibility**—sometimes found on weather maps—is how far you can see along Earth's surface expressed in miles. The poorer the atmospheric transparency, the lower the visibility.

Review Questions

36. As altitude within the troposphere increases, the amount of water vapor generally (1) decreases, only (2) increases, only (3) remains the same (4) decreases, then increases

37. By which process does moisture leave green plants? (1) convection (2) condensation (3) transpiration (4) radiation

38. Most moisture enters the atmosphere by the processes of (1) convection and conduction (2) condensation and radiation (3) reflection and absorption (4) transpiration and evaporation

MEASURING RELATIVE HUMIDITY It is difficult to directly measure absolute humidity or determine the air's capacity to hold water vapor. Relative humidity is therefore measured by indirect methods. An instrument used to measure relative humidity is called a **psychrometer.** One indirect method of measuring relative humidity uses an instrument called a sling psychrometer, as shown in Figure 7-11. This instrument contains an ordinary thermometer called a dry-bulb thermometer, and another thermometer with a wick around its bulb, called the wet-bulb thermometer. When the wick is moistened and the thermometers are whirled in the air, the temperature of the wet bulb drops because of the cooling effect of the evaporation of the water. The amount of cooling depends on the rate of evaporation, and is therefore related to the relative humidity.

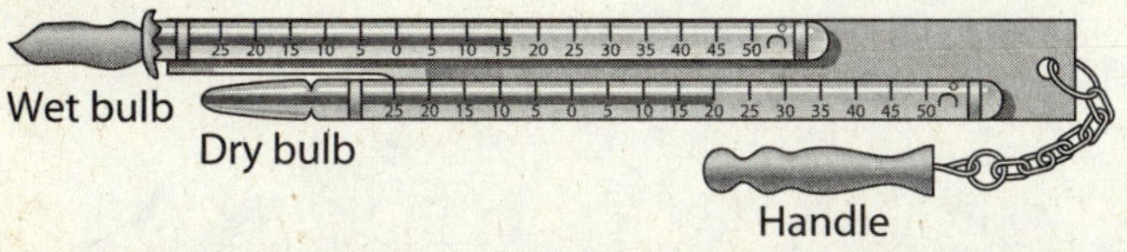

Figure 7-11. Sling psychrometer: The difference between the two temperature readings is used to find the relative humidity and dew point. Note that the dry-bulb reading at 20°C is the same as a regular temperature reading at 20°C.

To determine percent relative humidity, use the Relative Humidity (%) tables in the *Earth Science Reference Tables* and the following procedure.

- Locate the dry-bulb reading on the left-hand side of the Relative Humidity (%) chart in the *Earth Science Reference Tables.*
- Subtract the wet-bulb reading from the dry-bulb reading.
- Locate the difference between the wet-bulb and dry-bulb readings across the top of the chart.
- Follow the horizontal row for the dry-bulb reading to the right until it meets the vertical column running down from the top. This number is the relative humidity.

See Figure 7-12 for an example for determining relative humidity.

Dry-bulb Temperature (°C)	Difference Between Wet-Bulb and Dry-Bulb Temperatures (°C)									
	1	2	3	4	5	6	7	8	9	10
–20	28									
–18	40									
–16	48	0								
–14	55	11								
–12	61	23								
–10	66	33	0							
–8	71	41	13							
–6	73	48	20	0						
–4	77	54	32	11						
–2	79	58	37	20	1					
0	81	63	45	28	11					
2	83	67	51	36	20	6				
4 ‑ ‑ →	→ 85	→ 70	→ 56	→ 42	→ 27	→ 14	= 14% Relative humidity			
6	86	72	59	46	35	22	10	0		
8	87	74	62	51	39	28	17	6		
10	88	76	65	54	43	33	24	13	4	

Figure 7-12. Determining relative humidity: Suppose the dry-bulb reading is 4°C and the wet-bulb reading is –2°C. The difference is 6°C. Reading across from 4°C and down from 6°C, you find a relative humidity of 14%.

DETERMINING DEW POINT To determine the dew point, you need the dry-bulb and wet-bulb readings from a sling psychrometer and the Dewpoint Temperatures chart in the *Earth Science Reference Tables*. Use the following procedure which is very similar to the procedure used to determine the relative humidity:

- Locate the dry-bulb reading on the left-hand side of the chart.
- Subtract the wet-bulb reading from the dry-bulb reading.
- Locate the difference between the wet-bulb and dry-bulb readings across the top of the chart.
- Follow the horizontal row for the dry-bulb reading to the right until it meets the vertical column running down from the difference between the wet-bulb and dry-bulb readings. This number is the dew point temperature.

For example, suppose the dry-bulb reading is 8°C, and the wet-bulb reading is 6°C. The difference is 2°C. Reading across from 8°C and down from 2°C, you find a dew point temperature of 3°C.

Precipitation

The falling of liquid or solid water from clouds toward the surface of Earth is **precipitation.** For precipitation to occur, the ice crystals or water droplets in clouds must come together to become big enough so that they will fall under the influence of gravity. The forms in which precipitation occur are rain, drizzle, snow, sleet, freezing rain, and hail. The conditions that result in the formations of these different types of precipitation are shown in Figure 7-13.

A rain gauge, a type of precipitation gauge, is used to measure liquid precipitation. It is measured in depth—usually inches or centimeters. Snow depth is often measured with a ruler, or it is gently warmed and the depth of liquid is measured and reported as precipitation. A rain shower is brief, rapidly forming, rapidly ending, and often heavy rain associated with thunderstorm-type clouds. A snow shower is a brief, heavy snowfall.

Cloud Formation

If the temperature of the air cools (often by expanding) below the dew point, the water vapor will usually condense to a liquid or sublimate to a solid, changing to microscopic liquid water droplets or ice crystals, respectively. At Earth's surface, condensation produces liquid dew, and sublimation produces the solid frost. If the temperature in the atmosphere is above 0°C, condensation produces water droplets, which appear as a cloud when grouped. Some clouds are composed of ice crystals, which form when water vapor sublimates to a solid. A cloud is therefore a collection of liquid water

Figure 7-13. Forms of precipitation

Rain Falling liquid droplets larger than 0.2 mm in diameter; may be melted snow

Drizzle Falling liquid droplets from 0.2 to 0.5 mm in diameter

Snow Falling ice crystals formed by combining cloud ice crystals

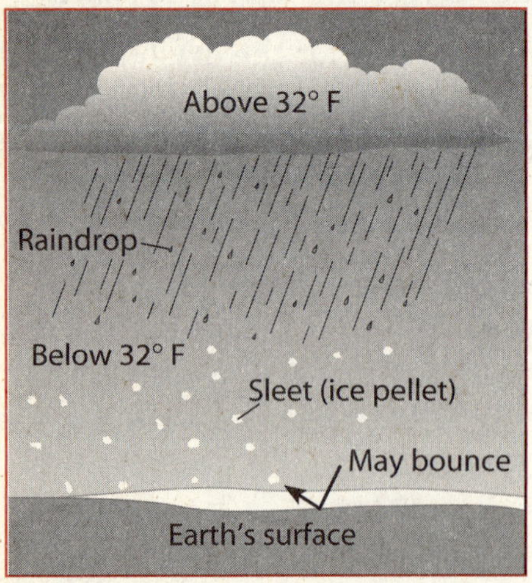

Sleet Solid pellets of ice that form by freezing of rain drops as they fall

Freezing rain Rain or drizzle that freezes on contact with features of Earth's surface

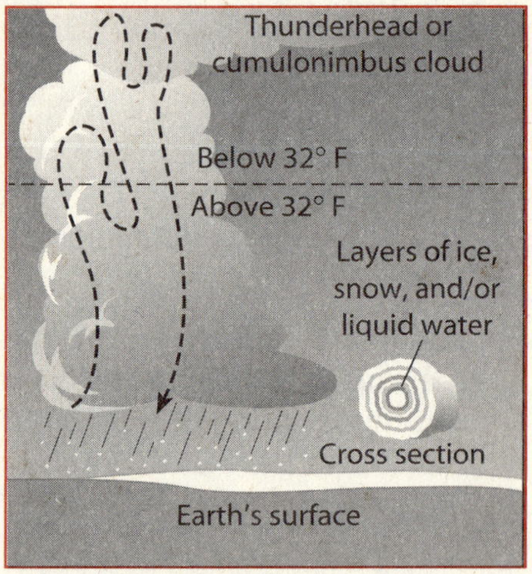

Hail Layers of ice, snow, and water formed by many up-and-down movements in a thunderstorm cloud; may be grapefruit-size

droplets and/or ice crystals suspended in the atmosphere and dense enough to be visible. If there is only a low density of water droplets or ice crystals, a condition called haze may result. **Cloud cover**—found on some weather maps—is the fraction or percent of the total sky at a location that is covered by clouds. If a cloud is on, or just above, Earth's surface, it is called fog. The condensation or sublimation that forms clouds and fog releases tremendous amounts of energy into the troposphere. This is the potential energy that was imparted to water molecules during evaporation and sublimation to a gas.

Besides needing saturated air, condensation usually requires a surface upon which the vapor can condense. This is called the condensation surface. In the formation of clouds and fog, the condensation surfaces are aerosols such as dust and volcanic ash. In the formation of dew, the condensation surfaces are the features of the Earth's surface like grass and leaves.

Since air temperature and dew point temperature both decrease with altitude in the troposphere, the altitude where temperature and dew point are the same is the level where clouds form by condensation or sublimation. This level is often seen as the flat bottoms on some clouds.

Calculating Cloud Base Altitude

Since temperature and dewpoint temperature both decrease with elevation or altitude in the troposphere, the altitude where temperature and dewpoint are the same can be calculated to obtain the altitude of the base of the clouds. Refer to the Generalized Graph for Determining Cloud Base Altitude in Appendix 3 on page 304. On this graph the dashed lines represent the decrease in dewpoint with increasing altitude (about 2C°/km). The solid diagonal lines represent the decrease in temperature with increasing altitude (about 10C°/km) above the Earth's surface. If you know the temperature and dewpoint at the Earth's surface, you can determine the altitude of condensation (or sublimation) and therefore the altitude of the bottom of the clouds above a location.

Suppose the surface temperature is 20°C and the surface dewpoint is 12°C at a location. Follow these steps to find the level of the base of the clouds above the location.

1. Find the 20°C mark and the 12°C mark along the bottom edge of the graph on page 303.
2. From the 20°C mark, follow the solid diagonal line upward and to the left.
3. From the 12°C mark, follow the dashed line upward.
4. The place where the two lines intersect represents the altitude at which clouds will form. In this case the base of the clouds is 1.0 km above the Earth's surface.

Atmospheric Transparency and Precipitation

The more pollutants added to the atmosphere by the activities of people and nature, the more aerosols are present. The more aerosols in the air, the less transparent the atmosphere is to insolation. How transparent the atmosphere is to insolation is called **atmospheric transparency.** When the atmosphere is less transparent, it means that more energy from the sun is absorbed or reflected by the atmosphere, resulting in less insolation reaching Earth's surface.

When the atmosphere has an aerosol content so high that distant images are blurred, and a cloudless sky does not appear blue, the condition is called haze. A haze or fog, usually brownish, which is highly polluted is smog. Condensation in cloud formation incorporates some of the aerosols, and these aerosols are removed from the atmosphere during precipitation. The falling liquid or solid water also collects other aerosols on the way down, thus lowering air pollution levels and cleaning the atmosphere. **Visibility**—sometimes found on weather maps—is how far you can see along Earth's surface expressed in miles. The poorer the atmospheric transparency, the lower the visibility.

Review Questions

36. As altitude within the troposphere increases, the amount of water vapor generally (1) decreases, only (2) increases, only (3) remains the same (4) decreases, then increases

37. By which process does moisture leave green plants? (1) convection (2) condensation (3) transpiration (4) radiation

38. Most moisture enters the atmosphere by the processes of (1) convection and conduction (2) condensation and radiation (3) reflection and absorption (4) transpiration and evaporation

39. The primary source of most of the moisture in Earth's atmosphere is (1) soil-moisture storage (2) rivers and lakes (3) melting glaciers (4) oceans

40. As the exposed area of a moist object decreases, the rate of evaporation of the liquid from that object (1) decreases (2) increases (3) remains the same

41. As the amount of light energy striking a moist object increases, the rate of evaporation of the liquid from that object (1) decreases (2) increases (3) remains the same

42. As the amount of water vapor in a given volume of air increases, the rate of evaporation from a moist object (1) decreases (2) increases (3) remains the same

43. There will most likely be an increase in the rate at which water evaporates from a lake if there is (1) a decrease in the temperature of the air (2) a decrease in the altitude of the sun (3) an increase in the moisture content of the air (4) an increase in the wind velocity

44. In the following diagram, at which location would the amount of water vapor in the air most likely be greater? (1) A (2) B (3) C (4) D

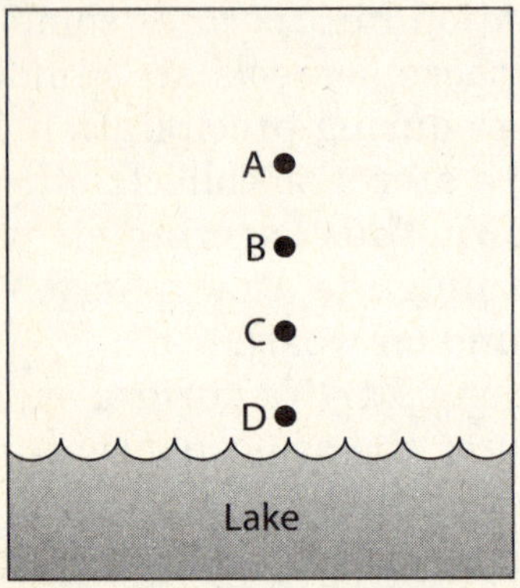

45. In the closed aquarium shown in the following diagram, the amount of water evaporating is equal to the amount of water vapor condensing.

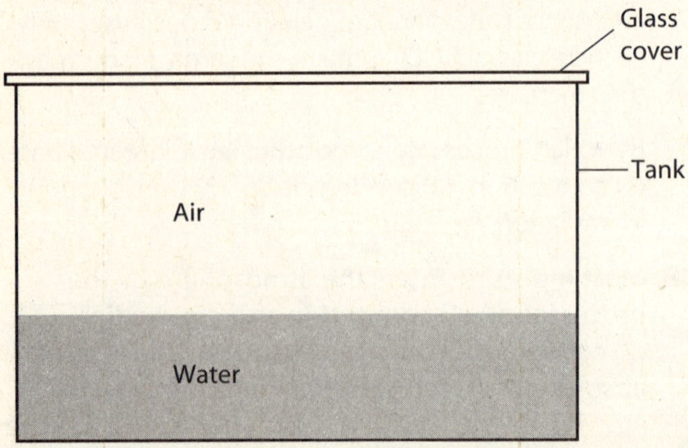

Which statement best explains why these amounts are equal? (1) The glass sides of the aquarium are warmer than the water. (2) The air in the aquarium is 50 percent saturated. (3) The relative humidity outside the aquarium is 100 percent. (4) The air in the aquarium is saturated.

46. Which graph best represents the relationship between the moisture-holding capacity of the atmosphere and atmospheric temperature?

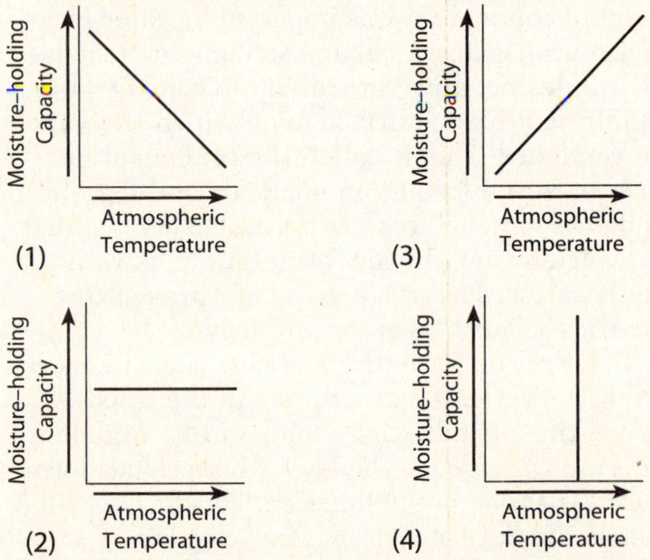

47. To say that the relative humidity on a given day is 70 percent means that the air (1) is composed of 70 percent water vapor (2) holds 70 percent of its water vapor capacity (3) contains 70 parts of water to 100 parts of dry air (4) contains the same amount of water that it would contain at 70°F

48. If the amount of water vapor in the air increases, then the dew point of the air will (1) decrease (2) increase (3) remain the same

49. During which part of the day is the relative humidity usually lowest? (1) morning (2) midafternoon (3) evening (4) late night

50. As the air temperature rises, the relative humidity (1) increases (2) decreases (3) remains the same

51. Air with a temperature of 60°F and a relative humidity of 51 percent was warmed to a temperature of 70°F, but the relative humidity remained at 51 percent. How did the moisture content change? (1) The moisture content decreased. (2) The moisture content increased. (3) The moisture content remained the same.

52. Which statement best explains why the wet-bulb thermometer of a sling psychrometer usually shows a lower temperature than the dry-bulb thermometer? (1) Water evaporates from the wet-bulb thermometer. (2) Water vapor condenses on the wet-bulb thermometer. (3) The air around the wet-bulb prevents absorption of heat. (4) The air around the dry bulb prevents absorption of heat.

53. When the dry-bulb temperature is 20°C, and the wet-bulb temperature is 16°C, the relative humidity is (1) 42 percent (2) 62 percent (3) 66 percent (4) 69 percent

54. The dry-bulb temperature is 20°C. The wet-bulb temperature is 17°C. What is the dew point? (1) 12°C (2) 13°C (3) 14°C (4) 15°C

Base your answers to questions 55 and 56 on the weather instrument shown in the following diagram.

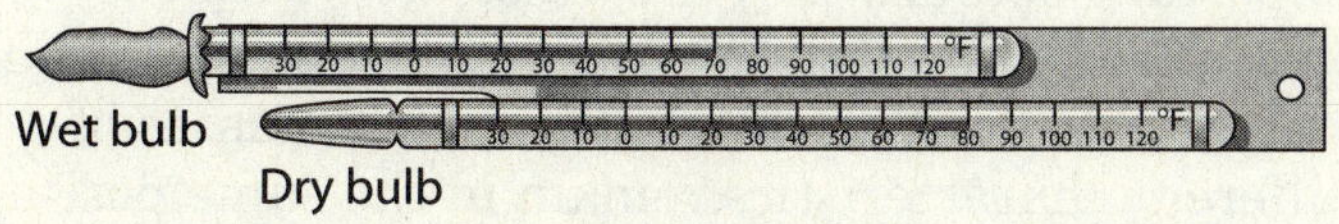

55. What are the equivalent Celsius temperature readings for the Fahrenheit readings shown? (1) wet 21°C, dry 27°C (2) wet 26°C, dry 37°C (3) wet 70°C, dry 80°C (4) wet 158°C, dry 176°C

56. Which weather variables are most easily determined by using this weather instrument and the *Earth Science Reference Tables?* (1) air temperature and wind speed (2) visibility and wind direction (3) relative humidity and dew point (4) air pressure and cloud type

---

57. Which process most directly results in cloud formation? (1) condensation (2) transpiration (3) precipitation (4) radiation

58. At which temperature could water vapor in the atmosphere change directly into solid ice crystals? (1) 20°F (2) 40°F (3) 10°C (4) 100°C

59. Which statement best explains how atmospheric dust particles influence the water cycle? (1) Dust particles are the main source of dissolved salts in the sea. (2) Dust particles increase the capacity of the atmosphere to hold water vapor. (3) Dust particles increase the amount of evaporation that takes place. (4) Dust particles provide surfaces on which water vapor can condense.

60. Atmospheric transparency is most likely to increase after (1) volcanic eruption (2) forest fires (3) industrial activity (4) precipitation

Air Masses and Fronts

Much of the weather of the contiguous United States is the result of the invasion of air masses and their interactions at their boundaries. An **air mass** is a large body of air in the troposphere with similar characteristics of pressure, moisture, and temperature.

Characteristics of Air Masses

An air mass forms when a large mass of air remains stationary over a part of Earth's surface for a period of time and thus acquires some of the characteristics of that surface. The geographic regions in which air masses are formed are called source regions. The source regions for the air masses that affect weather in the United States are shown in Figure 7-14.

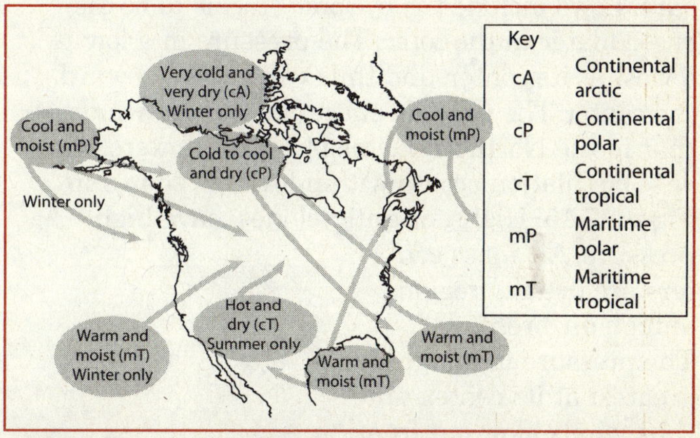

Figure 7-14. Source regions and tracks of air masses that affect weather of the continental United States

If the source region is at a high latitude, the air mass will have a low temperature; if it is at a low latitude, the air mass will have a high temperature. If the source region is land, the air mass will be dry; if it is water, the air mass will be moist. These air mass symbols are found in the *Earth Science Reference Tables*. These symbols and the air mass types that affect the weather of the contiguous United States are shown on Figure 7-14.

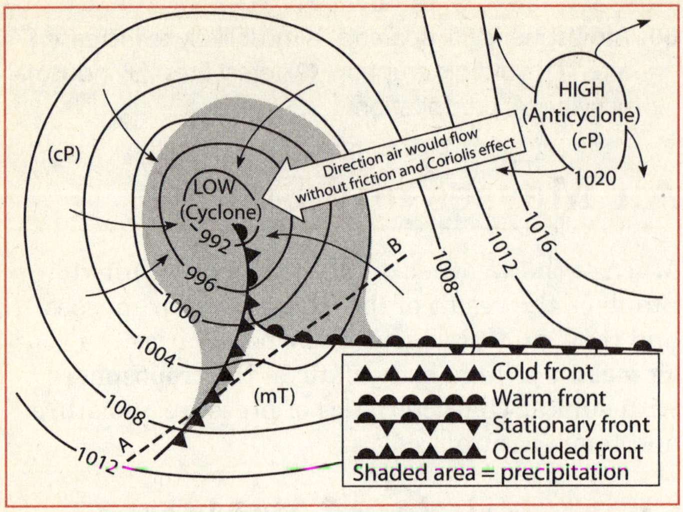

Figure 7-15. Circulation of winds and interaction of high-and low-pressure areas in the Northern Hemisphere: Air circulates clockwise in a high and counterclockwise in a low. Winds blow out from a high and blow into a low.

Lows and Highs

The troposphere is divided into two portions according to air pressure and direction of circulation of winds. Lows, also called **cyclones,** have low pressure. Lows include the cyclonic storms to be discussed later in the topic. The pressure in a low is lowest at its center, and thus winds blow toward the center. The Coriolis effect deflects the winds, so that in the Northern Hemisphere a counterclockwise circulation occurs within lows, as shown in Figure 7-15. Highs, or anticyclones, have high pressure. Air mass centers are usually regions of high air pressure. The pressure in a high is greatest at its center, and thus winds blow out from the center. The general circulation around highs is clockwise in the Northern Hemisphere. Lows are usually associated with stormy weather with high cloud cover, much precipitation, and high winds. Highs are usually associated with relatively clear skies and little or no precipitation.

Fronts

Where two air masses of different characteristics meet, an interface, or boundary, called a **front** develops. (See Figure 7-16.) Most of the storms and severe weather of the mid-latitude United States, except for hurricanes, are usually formed in association with fronts.

TYPES OF FRONTS A **cold front** is the boundary of an advancing cold air mass and a warmer air mass, where the underlying cold air pushes forward like a wedge. A **warm front** is the boundary of an advancing warm air mass and a retreating wedge of a cooler air mass. Because the cooler air is heavier, the warm air mass is forced to rise as it advances. An **occluded front** is the boundary of opposing wedges of cold air masses formed when a cold front overtakes a warm front, lifting the warm air mass off the ground. Occluded fronts are important because they are associated with the formation of mid-latitude cyclones (lows), as shown in Figure 7-15. When two adjacent air masses of different characteristics remain in the same positions, they form a **stationary front.** The **polar front** (see Figure 7-6) is an ever-changing boundary in the middle latitudes between the colder air masses toward the poles and the warmer air masses toward the equator.

At fronts between air masses of different temperatures, the warmer air, being less dense, is forced to rise. This results in the unstable

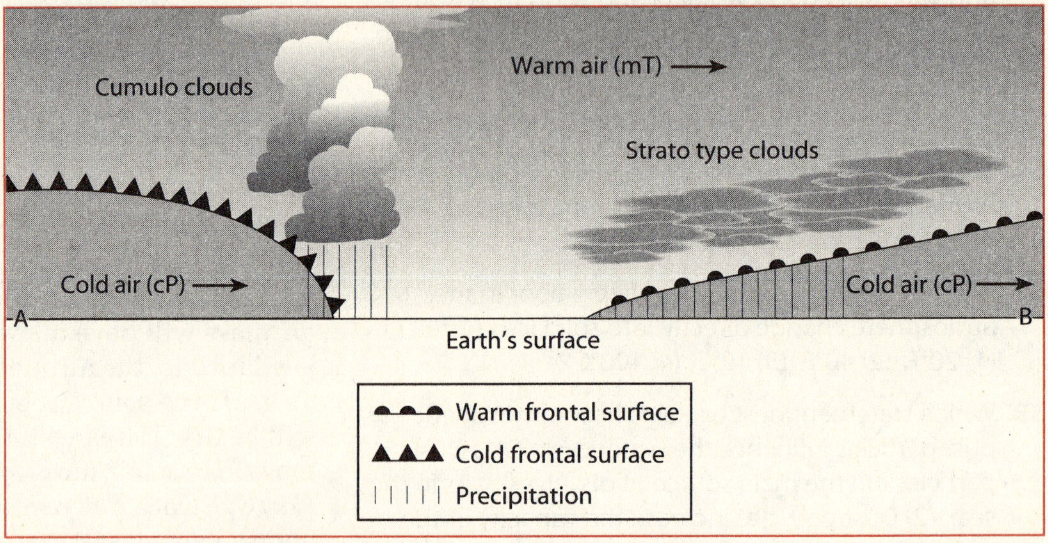

Figure 7-16. Cross-sectional view of the (A- - - -B) line in Figure 7-15: The conditions of the cold front at left are: steep slope, cumulus-type, or puffy thunderstorm type clouds, and precipitation just before and after the point at which the front meets Earth's surface. The conditions of the warm front at right are gentle frontal surface, stratus-type or layered clouds, and a broad band of precipitation preceding the point at which the front meets Earth's surface.

conditions that are characteristic of fronts and that produce much of the precipitation and stormy weather of the contiguous United States.

FRONTS AND WEATHER MAPS On weather maps, half circles and/or triangles are used on the lines representing fronts to indicate the type of front. See Figure 7-15 or the Front Symbol section of Weather Map Information in the *Earth Science Reference Tables* for specific front symbols. The triangles and half circles point in the direction the fronts and associated air masses are moving.

TRACKS OF AIR MASSES AND FRONTS In the United States, the tracks (paths) and rate of movement of air masses and fronts can be predicted on the basis of past observations. Thus, the succession of weather changes that accompanies such movements can be reasonably forecasted. The arrows in Figures 7-14 and in Figure 8-10 indicate the direction of some of these tracks. Most of the tracks follow a southwest to northeast or westerly to easterly route in the contiguous United States because of our location in the prevailing southwest winds belt.

Review Questions

61. In order for a large mass of air to acquire uniform characteristics, it must (1) stagnate over a large land or water surface (2) descend from the upper troposphere (3) move rapidly with the prevailing westerlies (4) move in the general planetary circulation

62. During the summer a warm, moist air mass moved over Texas. This air mass probably originated over (1) northern Canada (2) the Pacific Ocean (3) southern Arizona (4) the Gulf of Mexico

63. On a weather map, an airmass that is very warm and dry would be labeled (1) mP (2) mT (3) cP (4) cT

64. The movement of air in a low-pressure cyclonic system in the Northern Hemisphere is best described as (1) toward the center and clockwise (2) toward the center and counterclockwise (3) away from the center and clockwise (4) away from the center and counterclockwise

65. Why do clouds usually form at the leading edge of a cold air mass? (1) Cold air contains more water vapor than warm air does. (2) Cold air contains more dust particles than warm air does. (3) Cold air flows over warm air, causing warm air to descend and cool. (4) Cool air flows under warm air, causing the warm air to rise and cool.

66. The original characteristics of an air mass are determined by the (1) surface over which it is formed (2) pressure of the air mass (3) insolation it receives (4) rotation of Earth

Base your answers to questions 67 and 68 on the following diagram, which shows the frontal boundary between mT and cP air masses.

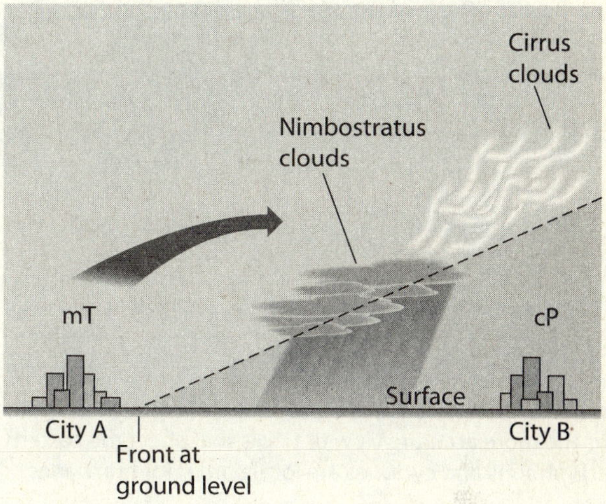

67. If the front at ground level is moving toward city B, which type of weather front is shown? (1) cold front (2) warm front (3) occluded front (4) stationary front

68. Why do cloud and precipitation usually occur along the frontal surface? (1) The warm air rises, expands, and cools. (2) The warm air sinks, expands, and warms. (3) The cool air rises, compresses, and cools. (4) The cool air sinks, compresses, and warms.

69. A storm system centered over Elmira, New York, will most often track toward (1) Albany (2) Jamestown (3) Rochester (4) New York City

Storms and Severe Weather

A storm is violent or severe disturbance of the atmosphere that usually creates dangerous, destructive, or unpleasant conditions on Earth's surface. Most storms are associated with high winds, heavy precipitation, fronts, and the low air pressure of cyclonic conditions. Types of storms include mid-latitude cyclones, hurricanes, thunderstorms, tornadoes, and blizzards. People need to be aware that storms can cause loss of life, injuries, and loss of property.

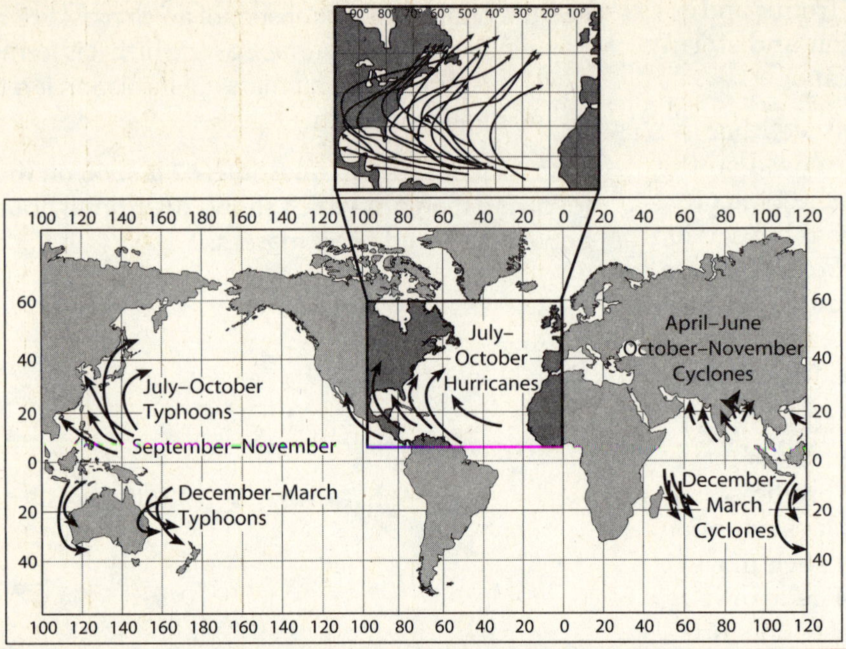

Figure 7-17. Hurricanes around the world and their more common paths or tracks: The insert shows the more accurate view of tracks that affect the eastern United States, including New York State. Typhoons and cyclones are local names for hurricanes.

Mid-latitude or Cyclonic Storms

Movements of air masses along the polar fronts of the middle latitudes can result in the formation of low-pressure storm systems called mid-latitude cyclones, wave cyclones, or just **cyclonic storms.** These storms bring much of the precipitation of the middle latitudes, including the contiguous United States. These storms form as a result of a variety of conditions associated with polar fronts that eventually result in an occluded front—the lifting of a warm air mass as it develops into a low-pressure cyclone. The rotation is the result of the Coriolis effect caused by Earth's rotation. Mid-latitude cyclones are often associated with many thunderstorms and associated hail and tornadoes.

Hurricanes

Usually in late summer or early fall, low pressure centers form over tropical ocean waters. These low pressure centers gain their energy from the condensation of millions of gallons of water vapor that acquired its energy from solar evaporation of ocean water. When a tropical low pressure center gains enough energy and a large enough pressure gradient to sustain winds of 74 miles an hour or more, it is called a hurricane. Hurricanes have different names in different parts of the world as shown in Figure 7-17. They are all characterized by rotating bands of precipitating clouds, a cloudless eye in the center that has the lowest air pressure, and high winds often over 150 miles an hour. (See Figure 7-18 on the next page.) When a hurricane passes over land or cold ocean water, it loses its energy source of evaporating water and slowly dissipates. Moving over land hurricanes lose energy due to friction with ground.

Hurricanes are a threat to people on ships from high winds and ocean waves, but the most significant danger is when a hurricane strikes land. The biggest danger in the United States is often in areas along the southern and eastern coasts as shown in Figures 7-17 and 7-18. Hurricanes sometimes bring high winds that can blow down all but the strongest buildings and trees, and in doing so, cause injury and death from falling or flying debris. Hurricanes also bring a danger to humans due to flooding. When a hurricane travels over land, the heavy rains can overwhelm streams and the ground's ability for infiltration and runoff. Flooding along coastlines can also be caused by a storm surge, which is a wall of ocean water that invades low coastal areas due to the low pressure and high winds of the hurricane. If you are near the ocean coast, when a hurricane is predicted, the best tactic is usually to evacuate using the emergency plans and routes developed by government agencies.

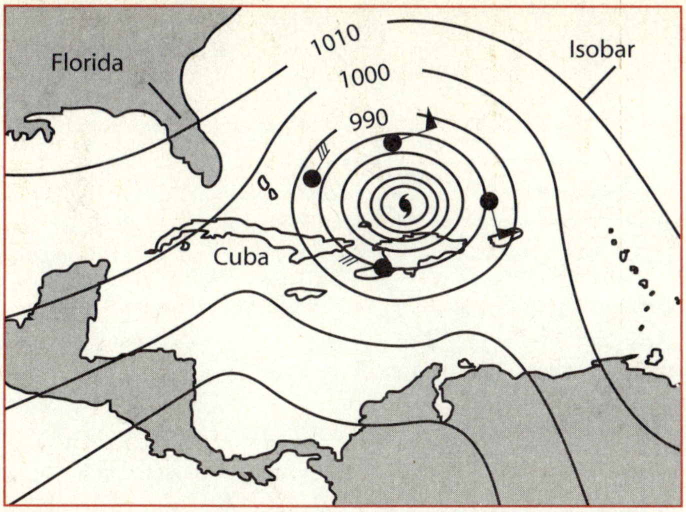

Figure 7-18. A simple weather map of a hurricane in the Atlantic Ocean: Note the location of the hurricane's eye.

DIGGING DEEPER

Hurricanes are the largest of storms—typically over 350 miles in diameter. They are also the longest lasting storms—usually a week or more.

Thunderstorms

At any one time over 2000 thunderstorms are happening on Earth. Thunderstorms are heavy rainstorms accompanied by thunder and lightning from thunderstorm-type clouds. Only thunderstorm clouds have enough up-and-down air movement to produce the type of precipitation called hail. Thunderstorms often form from the uplifting of warm air associated with fronts, but they also form as part of hurricanes and within air masses when local heating causes much convective rising of warm, moist air. Figure 7-19 is a model for the formation of a thunderstorm.

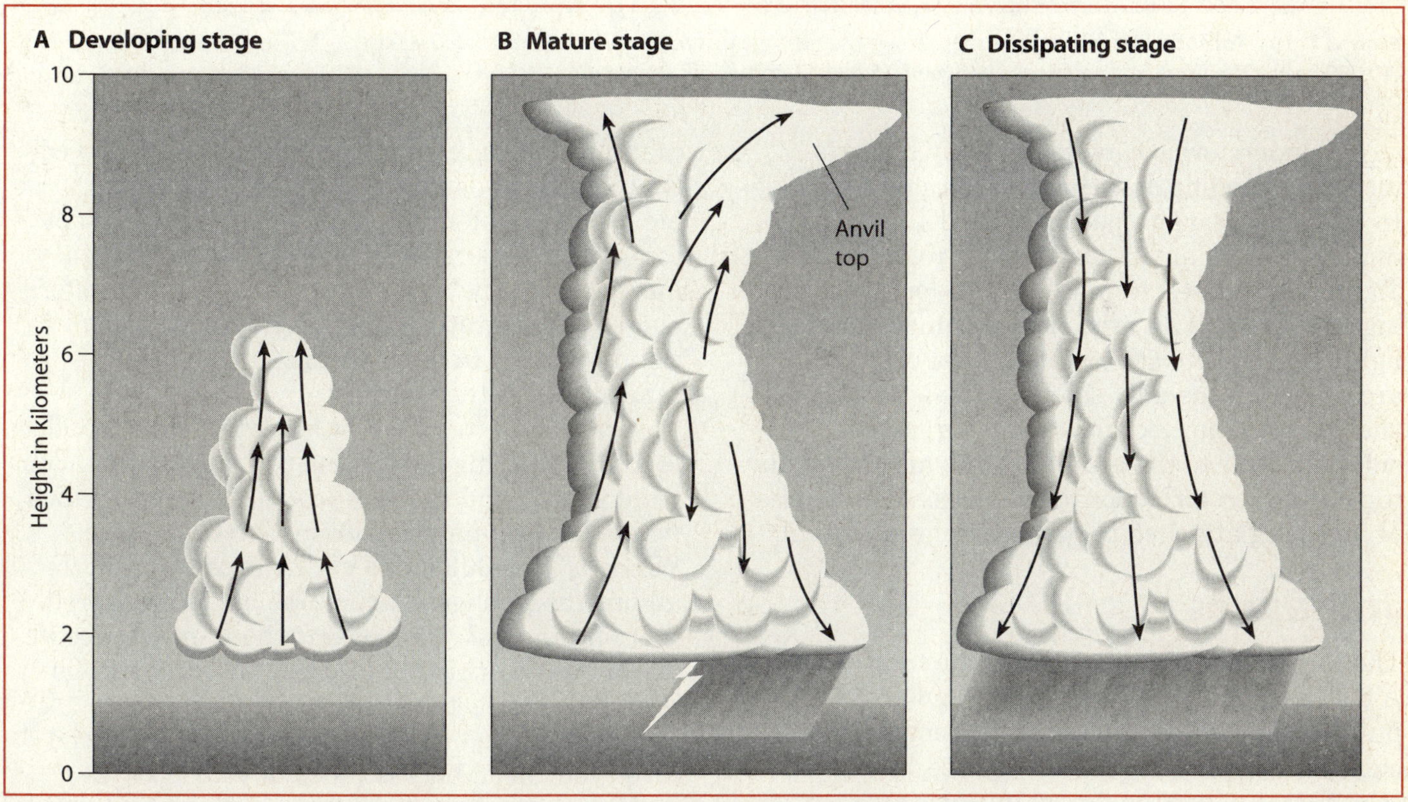

Figure 7-19. Formation of a thunderstorm cloud: This cloud, often called a thunderhead or cumulonimbus, and associated storm often follow a series of stages. In stage A, the towering puffy cloud indicates rising air. There is occasional lightning and usually little rain falls. In stage B, most likely hail, heavy rain, frequent lightning, strong winds, and tornadoes occur. The storm occasionally has a black or dark green appearance. It lasts an average of 10 to 20 minutes, but may last longer. In stage C, rainfall decreases in intensity. Some thunderstorms produce a burst of strong winds during this stage; lightning remains a danger. The cloud eventually changes to water vapor—usually after a few hours.

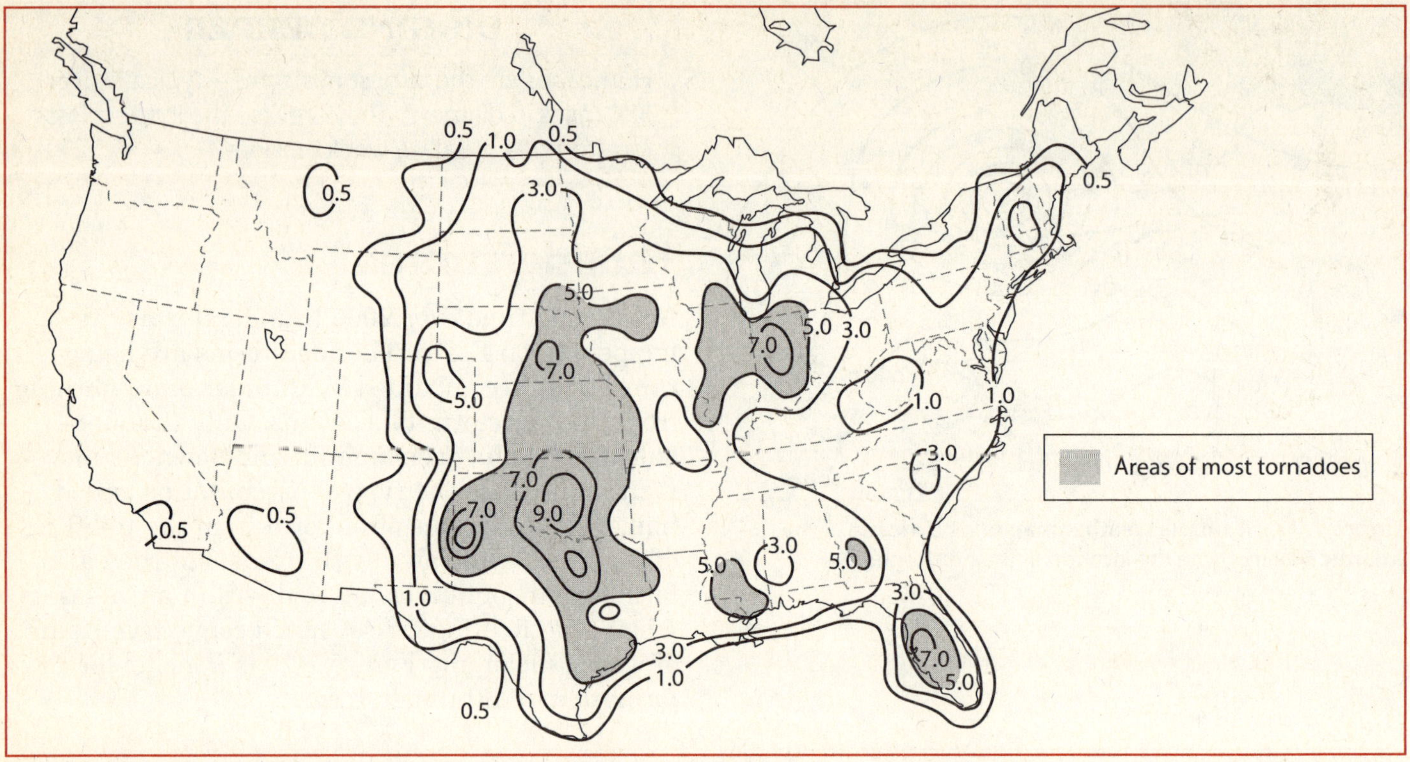

Figure 7-20. Tornado incidence: The map shows the average tornado incidence per 26,000 sq km (10,000 sq mi). New York State averages between 0.5 and 1.0 tornadoes per year.

Thunderstorms can create many hazards including flooding, hail, high winds, and lightning. In many years, more people get killed from lightning than from any other weather-related cause. To protect yourself from lightning, don't be the highest object in an area or stand under a tree, as lightning tends to strike the highest object in an area. Cars are usually safe because a car's metal shell directs the electrical charge around the outside of the car to the ground. If you are inside during a thunderstorm, don't touch electrical devices or anything attached to the building's plumbing.

Tornadoes

The most violent of all storms is the tornado. A tornado is a rapidly rotating, extremely low-pressure funnel that hangs down from thunderstorm clouds towards Earth's surface. Some scientists say that the funnel must touch Earth's surface to be a true tornado. Tornadoes are usually quite narrow—about 100 feet or so in diameter—but they may exceed a half-mile. Most tornadoes only last a few minutes, but some can last over an hour. Radar has calculated the wind speed in a tornado at 318 miles an hour—the record for measured surface wind speed—but it is believed that wind speeds over 350 miles an hour are possible. The actual formation of tornadoes is still not completely understood, but we do know that they form at fronts between air masses with big differences in temperature from thunderstorm-type clouds, usually in later afternoon when the atmosphere is the warmest.

The greatest danger to humans from tornadoes is from the flying debris caused by the tremendous wind speeds and the very low air pressure, which "vacuums" up materials from Earth's surface. Winds of 200–300 miles per hour can tear down all but the strongest structures and trees, shred them into pieces, and convert them into projectiles. At the first warning or sighting of a tornado, go into the strongest available structure, in the strongest room, at the lowest possible level, and cover your head and as much of your body as possible with something like a coat or blanket. These steps can provide the best chance of avoiding injury or death during a tornado. The locations in the United States where tornadoes are most common are shown in Figure 7-20.

Blizzards

A blizzard is a storm with winds of 35 miles an hour or greater associated with considerable amounts of falling and/or blowing snow. Blizzards are usually associated with the frontal conditions of mid-latitude cyclones. Blizzards are an environmental hazard because the drifting snows can shut down all types of transportation, disrupt telephone and electrical service, and prevent people from obtaining needed services and products such as food and fuel. People caught out in a blizzard without proper shelter can suffer many conditions including frostbite, hypothermia, and freezing to death. To prepare for a blizzard, people should have food and other supplies and stay indoors.

Emergency Preparedness for Storms

With accurate weather predictions and the knowledge of how to react to the different types of storms, we can greatly reduce the chances of loss of life, personal injury, and loss of property. This emergency preparedness often includes prior planning by individuals, families, communities, and government agencies to design evacuation routes out of the area or to safer areas within a building. The preparedness also includes having emergency supplies such as food, fresh water, lighting, heat, and backup electric power available before a severe storm approaches.

Review Questions

70. Most storms are associated with high winds, heavy precipitation, and (1) high air pressure (2) slowly rising air pressure (3) low air pressure or cyclonic conditions (4) rapidly rising air pressure

71. The characteristic of a blizzard that makes it an environmental hazard is (1) heavy rainfall (2) drifting snow (3) freezing rain (4) flying debris

72. Along the coast, the most dangerous aspect of hurricanes is the (1) lightning and hail (2) flooding by high waves (3) large central eye (4) suction effect of the funnel

Base your answers to questions 73 through 75 on the data table below. The data table shows a classification system for hurricanes. A storm surge is a dome of water 65 to 80 kilometers wide that sweeps ashore at the coast near the point where the storm center (eye) reaches land.

73. Which characteristic must a tropical storm have to be classified as a hurricane on the Safir-Simpson scale? (1) enough strength to cause catastrophic damage (2) a storm surge of at least 2.0 m (3) central air pressure over 980 mb (4) a wind speed of at least 119 km/hr

74. A hurricane with a central air pressure recorded at 28.70 inches has an expected storm surge of (1) 1.3 m (2) 2.0 m (3) 3.3 m (4) 4.0 m

75. The difference between the wind speed of a category-1 hurricane and the wind speed of a category-5 hurricane is primarily caused by the differences in (1) types of clouds (2) amounts of precipitation (3) air-pressure gradients (4) air-temperature gradients

Safir-Simpson Hurricane Scale

Hurricane Category	Central Air Pressure (mb)	Wind Speed (km/hr)	Expected Storm Surge Height (m)	Expected Damage
1	Over 979	119-153	1.2-1.5	Minimal
2	965-979	154-177	1.6-2.4	Moderate
3	945-964	178-209	2.5-3.6	Extensive
4	920-944	210-250	3.7-5.4	Extreme
5	Below 920	Over 250	Over 5.4	Catastrophic

76. The funnel of a tornado can "vacuum" up materials from Earth because of its extremely (1) high pressure (2) high temperature (3) low pressure (4) low wind speed

77. The most dangerous place to be during a thunderstorm is (1) in a car (2) under a high tree (3) in a ditch (4) in your bed

Weather Prediction and Probability

Accurate forecasts are important in giving people time to prepare for severe weather. Given accurate forecasts, people can shore up buildings, move to higher ground, move to safer parts of a building, go to stronger buildings, evacuate, or take other measures—all of which can potentially save them from death, injury, and loss of property.

Computers and Probability Prediction

Many weather predictions, or forecasts, are based on the **probability**—chance—of occurrence of types of weather based on relationships between weather variables. Computer models that use data collected over many years provide these predictions today. For example, suppose that during a twenty-day period the air pressure drops on ten days, and the air temperature increases on eight of those ten days. What is the probability that air temperature will increase if air pressure is decreasing? Based on the data above, there is an 8 in 10 chance, or 80 percent probability, of increasing temperature. In computing probability of weather events, the Weather Service bases its predictions on large amounts of data collected over many years.

Computer modeling has aided people's ability to work out the probabilities of a vast number of possible relationships between variables. Weather forecasting has become more accurate in recent years because of the ability of computers to rapidly compare present conditions around the world with data from several previous years. After rapidly calculating the probabilities, timely forecasts are produced. The following are some of the simple relationships between weather variables:

- Air pressure is related to temperature changes. As temperature increases pressure decreases, and vice versa.
- The chance of precipitation increases as the air temperature gets closer to the dew point. The smaller the difference between the air temperature and the dew point, the nearer the air is to saturation, condensation, and precipitation.
- The greater the air pressure gradient in an area, the faster the wind speeds.
- Rapidly falling air pressure at a location indicates that a storm system or low pressure center is approaching and that the chances of precipitation are very high in the very near future.

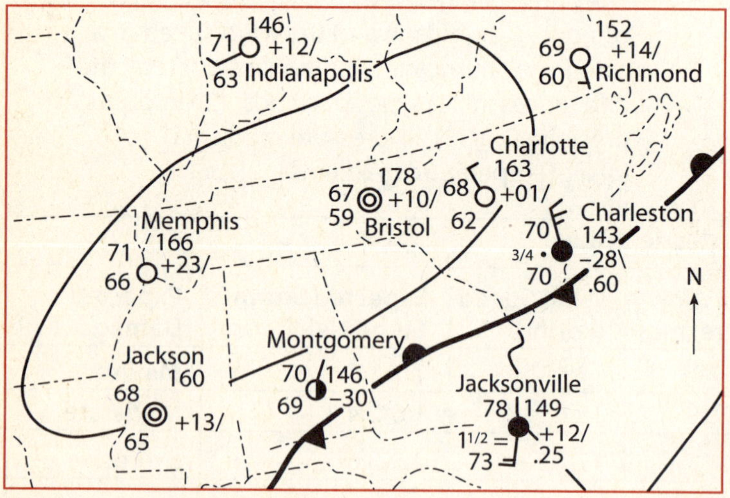

Figure 7-21. Weather map:

Weather Maps and Station Models

For over a hundred years, weather data has been collected at various locations and sent by electronic means to a central location where it is plotted to produce weather maps. These maps provide a synoptic weather maps, which is a summary of weather variables for a short period of time. This information becomes the basis for future forecasts. The National Weather Service has specific sites that collect weather information from different areas. Weather maps often show this information plotted in a circle around these sites. The circle indicates where the collection site—usually a city—is located. Often the weather data is represented by symbols to save space. The symbols and the circle make up a **station model,** which indicates a site's weather variables for a specific time of day. For a sample of a station model see Figure 7-22.

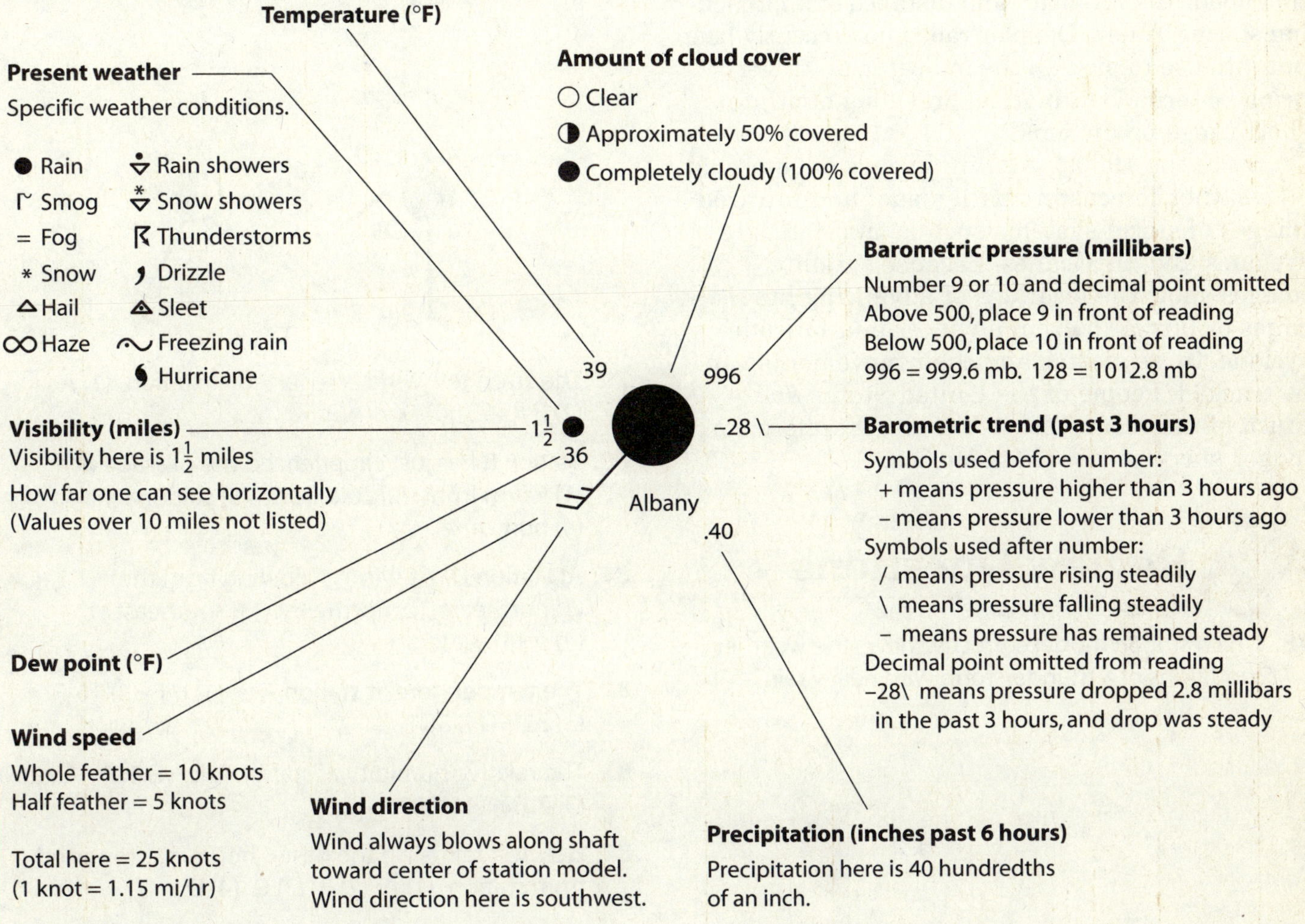

Symbol													
Wind speed (mph)	Calm	1–4	5–8	9–14	15–20	21–25	26–31	32–37	38–43	44–49	50–54	55–60	61–66

Figure 7-22. Sample station model for weather map: Also see the charts of Weather Map Information in the *Earth Science Reference Tables.*

Other Aids in Weather Prediction

Weather scientists have many tools available to aid them in preparing their forecasts. Atmospheric cross-sections—such as in Figure 7-16 on page 124—help give a three-dimensional view of weather conditions. **Radar**—which stands for **RA**dio **D**etection **A**nd **R**anging instrument—reflects radio electromagnetic energy off the aerosols of precipitation and clouds. This gives forecasters information about the intensity and distance of approaching storms. A new Doppler radar has recently been put into use to pick up the indicator, or "fingerprint" of tornado formation, providing accurate short-range predictions.

Since the 1960s, satellite images have provided weather forecasters with visual and infrared images of storm systems such as mid-latitude cyclones and hurricanes. Because satellite images show the locations of advancing hurricanes or blizzards from a winter mid-latitude cyclone, these storms and their movements can be tracked. People of the United States and much of the rest of the world are, therefore, no longer surprised by these storms.

Review Questions

78. Which station model correctly shows the weather conditions of a thunderstorm with heavy rain?

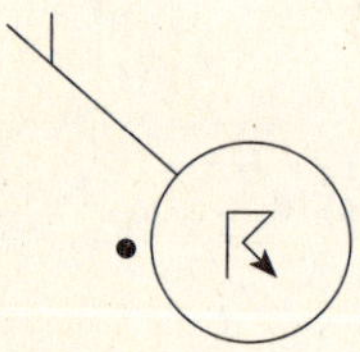

(1)

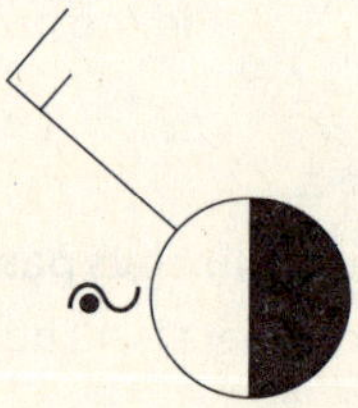

(3)

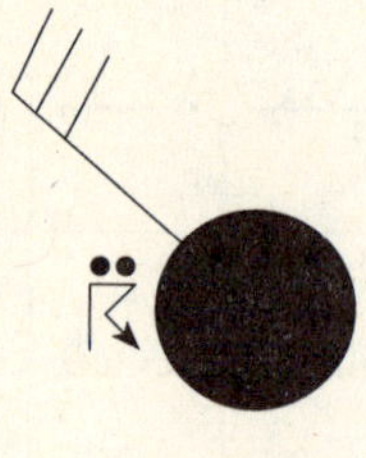

(2)

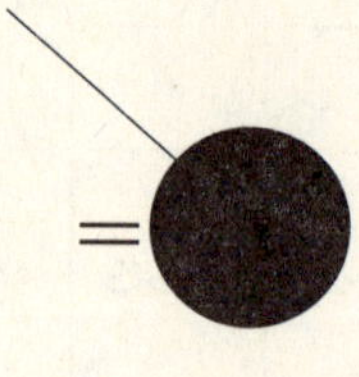

(4)

To answer questions 79 through 87 use the following diagram, which shows four station models, at the same instant in time, identified by letters A–D. The stations are located on a west-east line at fifty-mile intervals in the United States.

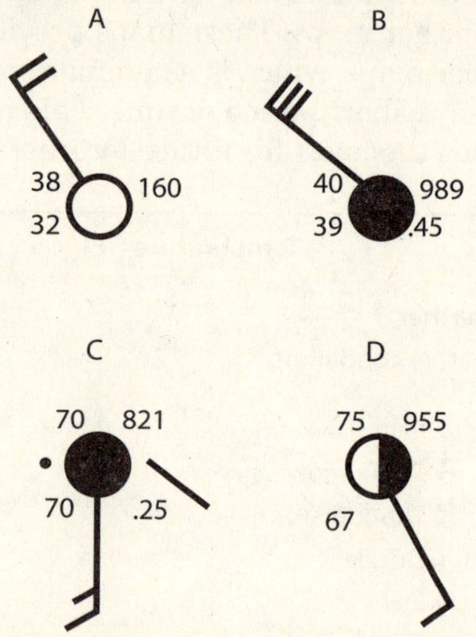

79. The strongest wind is reported at station (1) A (2) B (3) C (4) D

80. Station B has just experienced the passage of a (1) warm front (2) cold front (3) stationary front (4) hurricane

81. At station D the wind is blowing from the (1) northeast (2) northwest (3) southeast (4) southwest

82. The temperature at station A is (1) 6°F (2) 16°F (3) 32°F (4) 38°F

83. The relative humidity at station C is (1) 10% (2) 25% (3) 70% (4) 100%

84. Heaviest rainfall in the past 6 hours was reported from station (1) A (2) B (3) C (4) D

85. Clear skies are reported from station (1) A (2) B (3) C (4) D

86. What is the wind speed at station B in mi/hr? (1) 3 mi/hr (2) 23 mi/hr (3) 34.5mi/hr (4) 69 mi/hr

87. What is the wind speed at Station A in knots? (1) 5 knots (2) 10 knots (3) 15 knots (4) 20 knots

Questions for Regents Practice

Part A

1. As the temperature of a given volume of saturated air decreases the condensation

(1) decreases

(2) increases

(3) remains the same

2. By which process does water vapor leave the atmosphere and form dew?

(1) condensation

(2) transpiration

(3) convection

(4) precipitation

3. Clouds and fog differ principally in

(1) amount of water droplets

(2) density

(3) amount of dust

(4) height above the ground

4. Why does fog disappear after sunrise?

(1) air pressure rises

(2) fog evaporates

(3) air pressure falls

(4) fog condenses

5. Which is a form of precipitation?

(1) frost

(2) snow

(3) dew

(4) fog

6. Rain passing through layers of air below 32°F may form

(1) dew

(2) frost

(3) sleet

(4) snow

7. Which graph best represents the relationship between water droplet size and the chance of precipitation?

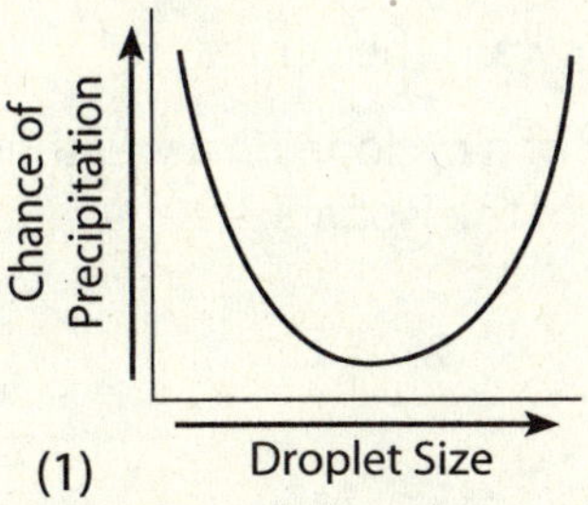

(1)

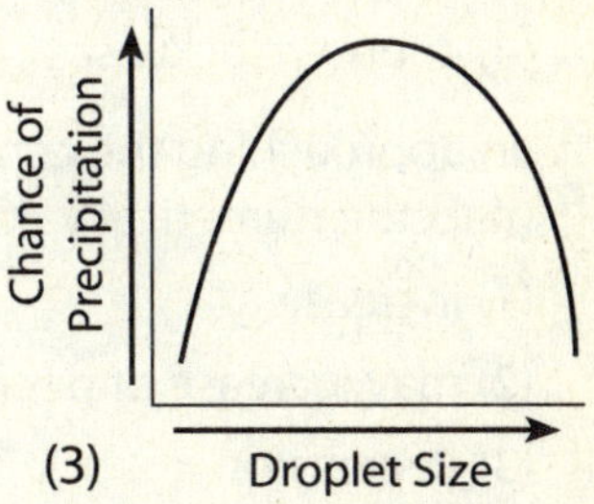

(3)

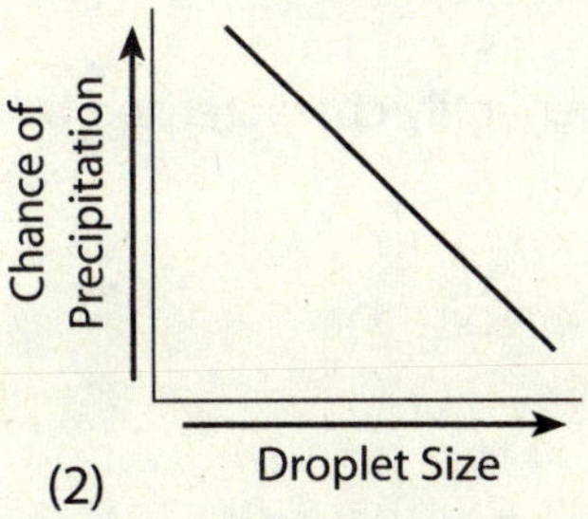

(2)

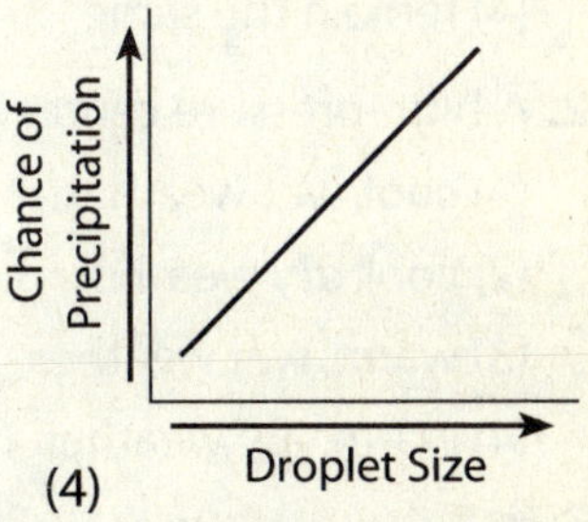

(4)

8. Which process is most likely to remove pollutants from the air?

(1) precipitation

(2) evaporation

(3) transpiration

(4) runoff

9. Which graph best shows the relationship between atmospheric transparency and the concentration of pollution particles in the air?

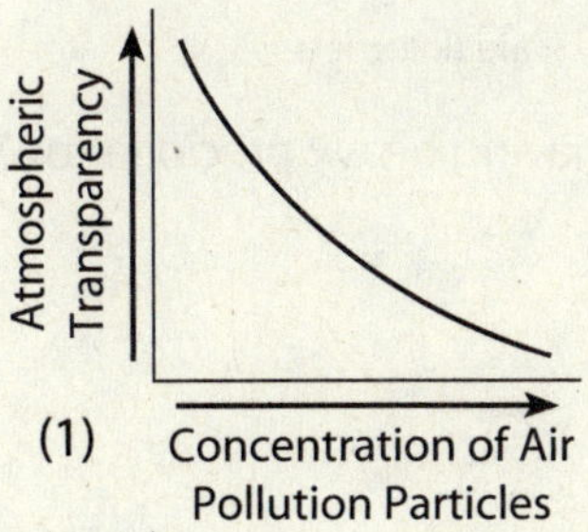

(1)

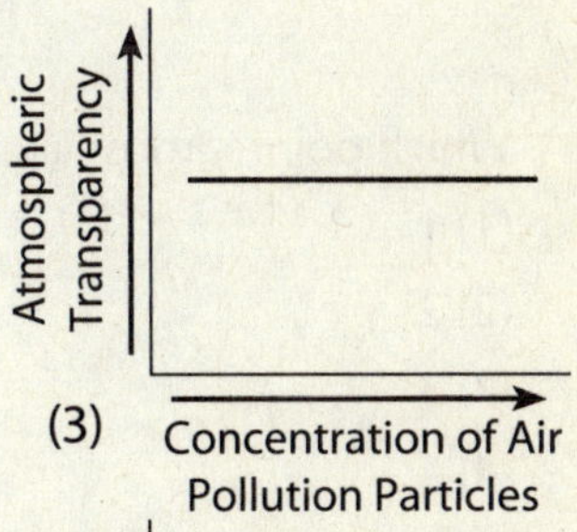

(3)

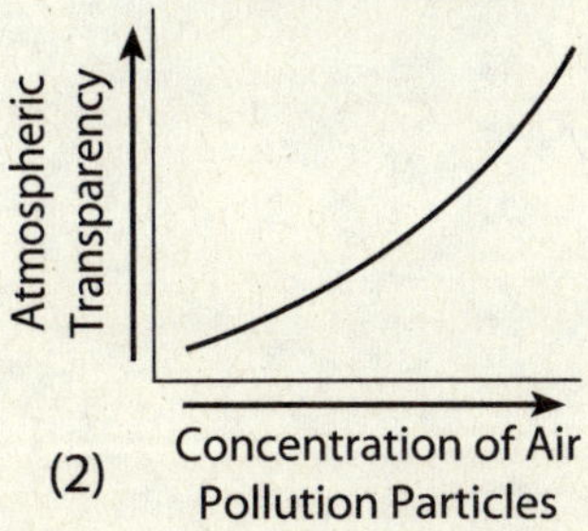

(2)

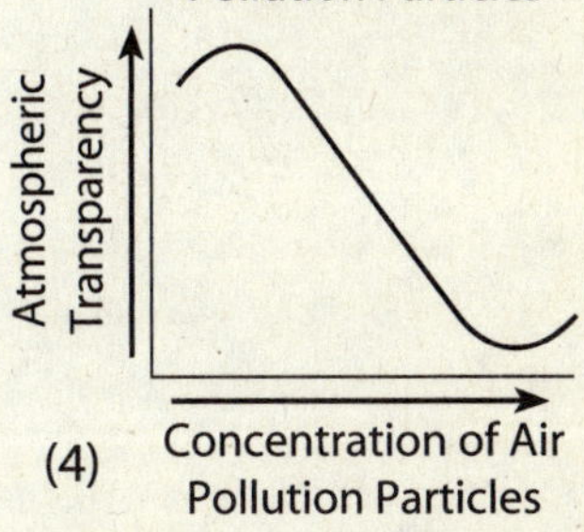

(4)

10. Which best describes the movement of air in a high-pressure air mass (anticyclone) in the Northern Hemisphere?

(1) clockwise and away from the center

(2) clockwise and toward the center

(3) counterclockwise and away from the center

(4) counterclockwise and toward the center

11. In approaching the center of a cyclone on a weather map, the numerical values of the isobars

(1) increase

(2) may increase or decrease

(3) decrease

(4) remain the same

12. A high-pressure center is generally characterized by

(1) cool, wet weather

(2) cool, dry weather

(3) warm, wet weather

(4) warm, dry weather

13. The following diagram shows four points on a map with their positions relative to a low-pressure weather system.

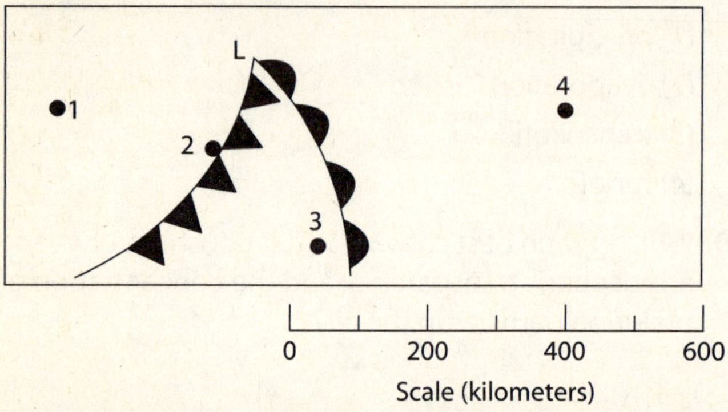

Which point is most likely having heavy precipitation?

(1) 1

(2) 2

(3) 3

(4) 4

14. A large number of thunderstorms occur in the southeastern United States. Which type of air mass is most likely the main source of the moisture that produces these thunderstorm?

(1) cP

(2) mP

(3) mT

(4) cT

15. Which statement about a major hurricane is an inference?

(1) The wind speed is measured at 200 km/hr.

(2) The central air pressure is recorded at 946.0 mb.

(3) A rain gauge records three inches of rain in less than one hour.

(4) Damage from the storm is expected to be extensive.

16. The highest wind velocities are usually encountered in

(1) monsoons

(2) hurricanes

(3) blizzards

(4) tornadoes

17. On a July afternoon in New York State, the barometric pressure is 29.85 inches and falling. This reading most likely indicates

(1) an approaching storm

(2) rapidly clearing skies

(3) continuing fair weather

(4) gradually improving conditions

18. The chief reason for plotting a surface weather map is to

(1) locate major wind belts

(2) identify cold fronts

(3) forecast weather conditions

(4) record climatic data

19. Tornadoes are usually associated with

(1) cold fronts

(2) tropical fronts

(3) stationary fronts

(4) warm fronts

Part B

To answer questions 20 through 22 refer to the following graph, which shows the hourly surface air temperature, dew point, and relative humidity for a 24-hour period during the month of May at Washington, DC.

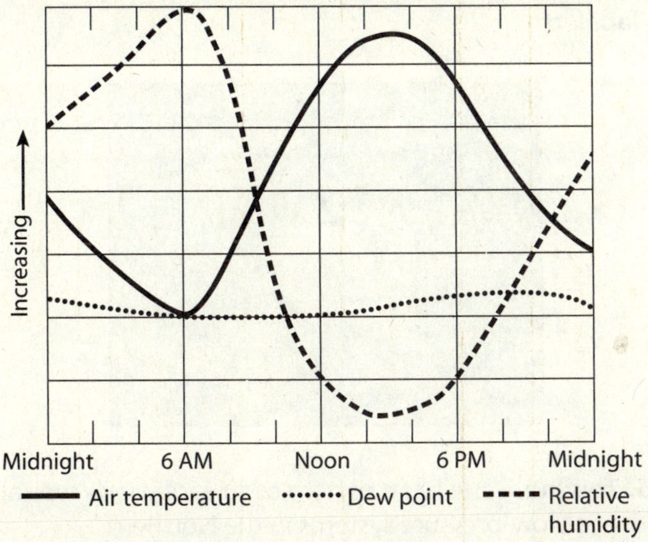

20. The greatest net change in air temperature occurred during what period? [1]

21. What does the graph indicate about the relationship between air temperature and relative humidity? [1]

22. At approximately what time is condensation most likely to occur? [1]

Base your answer to questions 23 through 26 on the following diagram, which represents a vertical cross section of a frontal system moving from west to east in an area of New York State. Temperatures for seven weather stations, A through G, are shown on the diagram.

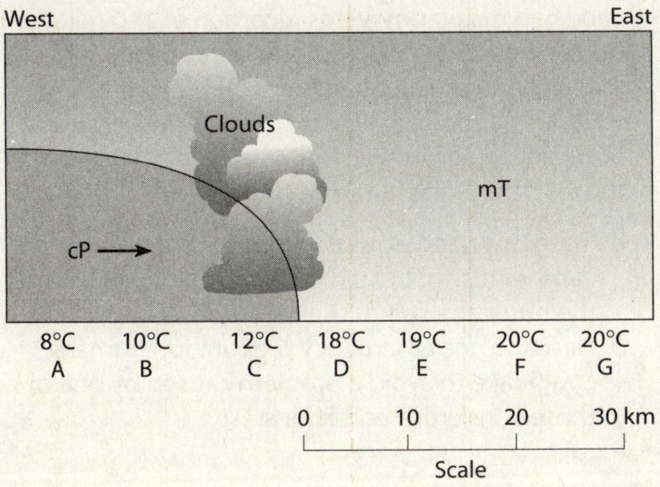

23. How will the cloud cover change at Station E as the front approaches and how will it change after the front passes? [2]

24. What barometric pressure change will occur as the front passes station E? [1]

25. Will the temperature increase or decrease at station E after the front passes? [1]

26. Describe how the motion of the air masses causes clouds to form along the front. [1]

Base your answers to questions 27 and 28 on the meteorological conditions shown in the following table and partial station model, as reported by the weather bureau in the city of Oswego, New York.

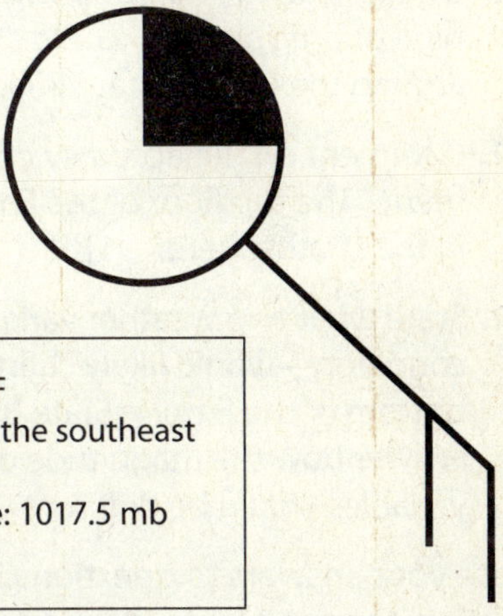

Air temperature: 65°F
Wind direction: from the southeast
Windspeed: 20 knots
Barometeric pressure: 1017.5 mb
Dewpoint: 53°F

27. Using the meteorological conditions given, copy and complete the partial station model by recording the air temperature, dew point, and barometric pressure in the proper format. [2]

28. State the sky conditions, or amount of cloud cover, over Oswego as shown by the station model. [1]

Part C

Use the information in the following paragraph and your knowledge of earth science to answer questions 29 through 32.

A person on the 102nd floor (with an elevation of 1250 feet) of the Empire State Building in New York City observes that it is snowing. At the same time, another person on the 50th floor observes that it is raining. Also at the same time, a third person at street level notices that the sky is completely covered by clouds, but there is no precipitation.

29. Explain the differences regarding precipitation that the three people observed at the three levels of the Empire State Building. [3]

30. During a gentle rain, at a temperature of 25°F, what type of precipitation would strike the sidewalk around the Empire State Building? [1]

31. Comment on the accuracy of the following statement. "The top floor of the Empire State Building is in the stratosphere." [1]

32. State what two weather variables—besides moisture condition—would likely change from the top to the bottom of the Empire State Building. Include in your answer how the magnitude of these two weather variables would be different. [4]

Base your answers to questions 33 and 34 on the cross section provided in your answer booklet, which represents a house at an ocean shoreline at night. Smoke from the chimney is blowing out to sea.

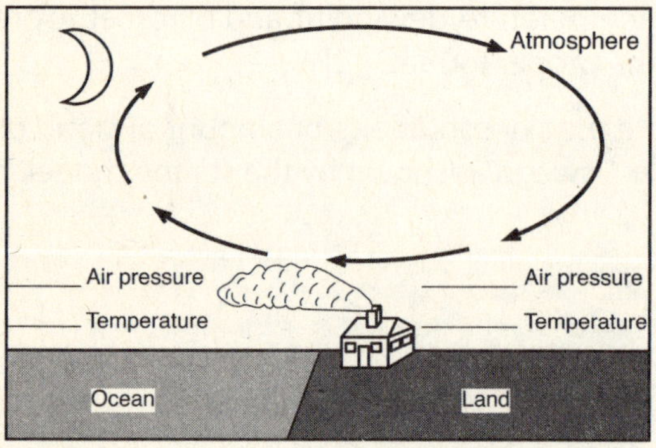

33. Label the *two* lines provided on the cross section *above* to show where air pressure is relatively "high" and where it is relatively "low." [1]

34. Assume that the wind blowing out to sea on this night is caused by local air-temperature conditions. Label the *two* lines provided on the cross section *above* to show where Earth's surface air temperature is relatively "warm" and where it is relatively "cool." [1]

Base your answers to questions 35 through 39 on the weather satellite photograph of a portion of the United States and Mexico provided below. The photograph shows the clouds of a major hurricane approaching the eastern coastline of Texas and Mexico. The calm center of the hurricane, the eye, is labeled.

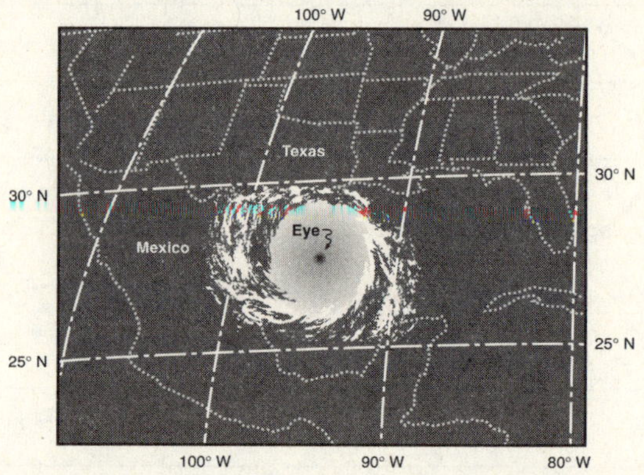

35. This hurricane has a pattern of surface winds typical of all low-pressure systems in the Northern Hemisphere. On the satellite photograph provided, draw *three* arrows on the clouds to show the direction of the surface wind movement outside the eye of the hurricane. [1]

36. Cloud droplets form around small particles in the atmosphere. Describe how the hurricane clouds formed from water vapor. Include the terms "dew-point" and either "condensation" or "condense" in your answer. [1]

37. State the latitude and longitude of the hurricane's eye. The compass directions must be included in the answer. [1]

38. At the location shown in the photograph, the hurricane had maximum winds recorded at 110 miles per hour. Within a 24-hour period, the hurricane moved 150 miles inland and had maximum winds of only 65 miles per hour. State why the wind velocity of a hurricane usually decreases when the hurricane moves over a land surface. [1]

39. *a* State *two* dangerous conditions, other than hurricane winds, that could cause human fatalities as the hurricane strikes the coast. [2]

b Describe *one* emergency preparation humans could take to avoid a problem caused by one of these dangerous conditions. [1]

Water and Climate

TOPIC 8

VOCABULARY		
capillarity	**porosity**	**unsorted**
climate	**prevailing winds**	**urbanization**
ground water	**runoff**	**water cycle**
hydrologic cycle	**seep**	**water retention**
infiltrate	**sorted**	**water table**
permeability	**stream discharge**	

While weather is the study of the short-term events of our atmosphere, climate is the study of the conditions of the atmosphere over long periods of time—tens, hundreds, thousands of years.

Climate greatly affects the kind of natural vegetation, the type of landscape features, and the crops and animals that are native to a region. Water is one of Earth's most important natural resources—one that people need to sustain their lives, help grow crops, and provide transportation. Water also has an effect on weather and climate and erodes Earth's surface. Since liquid water covers about 70 percent of Earth's surface, it creates environments for a large percentage of the life forms on our planet in streams, ponds, lakes, seas, and oceans.

The Water Cycle

Earth has been recycling its water supply for approximately 4 billion years—ever since the outgassing of water vapor (from Earth's solid surface) and its cooling to form liquid water. It is believed that most of Earth's water supply was created when Earth formed and that large amounts have not been gained or lost since. Figure 8-1 illustrates the locations of Earth's water. This figure shows that the oceans contain most of the water, while the majority of the water people can drink—fresh water—is in glacier ice and floating on ocean tops in cold regions.

One model used to illustrate the movement and the phase changes of water at and near Earth's surface is the **water cycle**—also called the **hydrologic cycle.** (See Figure 8-2 on the next page.) The water cycle is "fueled" by solar energy—insolation—which changes liquid or solid water to water vapor, and by gravity pulling water down in the atmosphere, hydrosphere, and upper parts of the lithosphere. The ultimate source of most water on land is the evaporation of the oceans' water. The moisture gets to the land from the oceans by way of the atmosphere. When precipitation falls on the land, four things can happen to it.

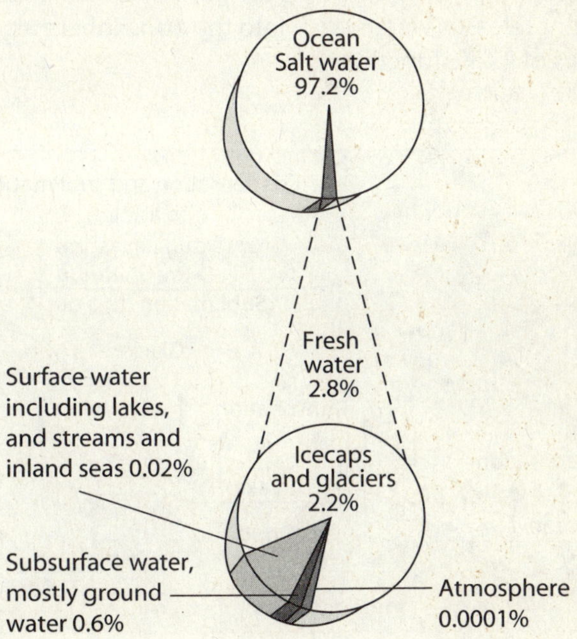

Figure 8-1. Graph of distribution of Earth's water by volume: Most of Earth's water is located in the ocean and is too salty for most human uses. Most fresh water is in the ice of glaciers or frozen on top of the oceans. Ground water is the biggest readily available supply of the natural resource fresh water.

- Precipitation can be stored or retained on the land surface as ice or snow—**water retention.** Water can also be retained on the leaves of trees and other plants.
- Precipitation can **infiltrate,** or **seep** (sink into) the upper parts of Earth's lithosphere under the influence of gravity. All the water beneath Earth's surface is called subsurface water.
- Precipitation can flow over Earth's surface as **runoff.** About one third of the precipitation falling on land is returned to the water cycle through runoff.
- The combination of evaporation and transpiration discussed in Topic 7 is evapotranspiration. It usually includes the sublimation of ice and snow directly into water vapor. Approximately two thirds of the precipitation that falls on land is evaporated or transpired back into the atmosphere.

MEMORY JOGGER

You may recall from Topic 5 that evaporation is the change in phase from liquid to a gas, such as liquid water into water vapor. You may also recall from Topic 6 that transpiration is the process by which a plant releases water vapor into the atmosphere as part of its life function.

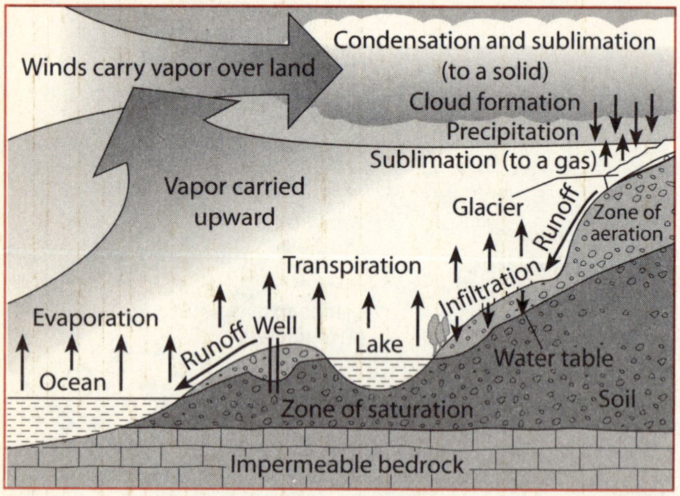

Figure 8-2. The water cycle—a model of the movement and phase changes of water at and near Earth's surface: The water table is an interface of changing position between the zone of saturation, where the pores of Earth's loose materials are filled with water (ground water), and the zone of aeration, where the pores are only partly filled with water (capillary water). Withdrawal of water from wells causes the water table to be lower.

Factors Affecting Infiltration

Most infiltration occurs in the loose material (including soil) that directly covers much of Earth's solid surface or is beneath a thin vegetation zone. Some infiltration can also occur into rock exposed at or near Earth's solid surface. The amount of water that can infiltrate when precipitation occurs depends on several variables.

SLOPE OF THE LAND The steeper the slope (gradient) of Earth's solid surface, the less the infiltration or seepage.

DEGREE OF SATURATION The more saturated the loose materials at Earth's surface, the less the infiltration. Figure 8-2 shows that Earth's surface is divided into two zones: the zone of saturation, where the pores or openings between solid particles are filled with water, and the zone of aeration, where the the pores are partly filled with air and partly filled with water (capillary water). Water infiltrates down under the influence of gravity until it meets the interface between the zone of saturation and the zone of aeration. This interface is called the **water table.** The depth of the water table below the surface varies with the amount of infiltration. The subsurface water below the water table is called **ground water.** It is ground water—through wells—that supplies a large portion of human water requirements. Withdrawing water from wells faster than it can be replenished lowers the water table.

POROSITY The percentage of open space (pores and cracks) in a material compared to its total volume is its **porosity.** Generally, the greater the porosity of the loose materials at Earth's surface and the rocks, the greater the amount of infiltration that can occur. The shape, packing, and sorting of the particles composing a material determine its porosity. (See Figure 8-3 on the next page.)

Shape Well-rounded particles have greater porosity than angular or plate-shaped particles because round shapes do not fit together as tightly as the other shapes do.

Packing The more closely packed the particles, the lower the porosity. The constant passage of people, animals, or vehicles in the same place or path will pack the soil, lowering the rate of infiltration.

Sorting If all the particles in a material are about the same size, they are said to be **sorted;** if the particles are of mixed sizes, they are said to

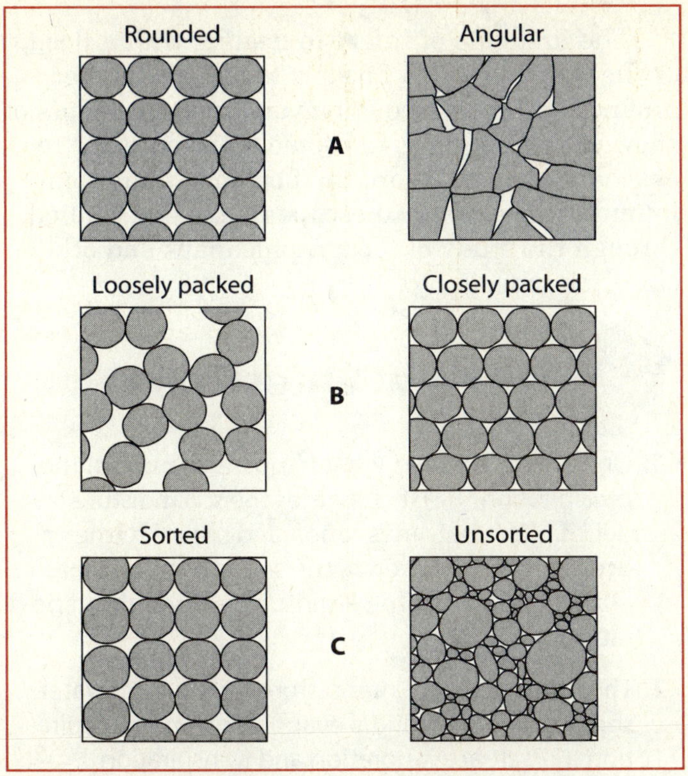

Figure 8-3. Effect of particle or sediment shape, packing, and sorting on the porosity of loose materials on Earth's surface: White regions are areas of porosity. **(A)** Well-rounded particles have more porosity than particles with angular shapes. **(B)** Loosely packed particles have more porosity than closely packed particles. **(C)** Sorted particles have more porosity than unsorted particles.

be **unsorted.** The more unsorted the particles, the lower the porosity because the small particles can fit into the spaces between the larger particles. It should be noted that particle size by itself does not affect porosity. For example, a material with large particles may have about the same porosity as one with smaller particles if the shape, packing, and sorting of both particle sizes are about equal. If there were two rooms of equal size, one filled with marbles and the other filled with basketballs, the total volume of pore space would be about the same for both rooms.

PERMEABILITY The ability of a material to allow fluids such as water to pass through it is **permeability.** The permeability rate is the speed at which fluids will flow through a material. A material can be porous and yet impermeable—not permeable. It is often how well connected the pores or openings in rock and loose materials are that will determine the degree of permeability, not just the volume of openings. Impermeability may be due to tight packing or cementing of particles, which seals off the pores from one another so that water cannot enter them. In the winter the cementing is often due to ice. Ground water can deposit some of its dissolved minerals, adding cement to loose materials and rocks, and thus reducing porosity and permeability. In loose particles, the larger the particle size, the faster the permeability rate and infiltration because the size of the pores increases.

CAPILLARITY During infiltration, some water is stopped from moving downward by the attractive force between water molecules and the surrounding Earth materials. This attractive force is **capillarity.** The water that is stored in small openings in the zone of aeration is capillary water. Capillarity also causes water to move up from the water table, against gravity, toward Earth's surface in the zone of aeration. This upward movement is capillary migration (capillary action). Capillary migration is extremely important in supplying water to plants when the soil is not saturated with water, which is the usual situation. When the particle size of loose particles become smaller, the capillarity and capillary migration become greater. (See Figure 8-4.) When the size of the particles increases the size of pores increases, and the effects of capillarity are reduced.

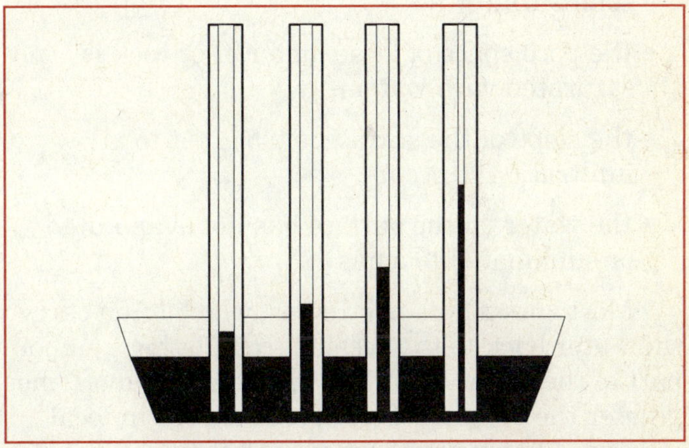

Figure 8-4. Capillary action, or migration, as a result of pore size: Capillarity causes water to move up against gravity in these glass tubes in the same way it moves up in Earth's loose materials and rocks by capillary action. The smaller the grain or sediment sizes are, the smaller the pores. The smaller the pores are, the higher the capillary action.

VEGETATION The amount and type of vegetation in an area influences the amount of water that infiltrates the ground. Grasses, trees, and other plant types intercept falling precipitation, reducing its velocity and temporarily storing some of the water above Earth's surface. The reduced speed of the precipitation gives the ground more time to absorb the water. Plant litter—dead and partly decayed vegetation—serves the same function. Some of the water stored on vegetation and plant litter falls to the ground after precipitation has stopped, and this can increase infiltration and reduce runoff. Ground without vegetation usually has high runoff and low infiltration rates.

LAND USE How land is used by people can greatly influence the amount of water seeping into the ground. Roads, parking lots, and buildings cover otherwise permeable ground and create impermeable surfaces. These impermeable surfaces often channel runoff directly to streams or drainage pipes, reducing the chances of water seeping into the ground. Farming, cutting down trees, and the grazing of domesticated animals often either reduce the amount of plant life in an area or compact the soil, making it less permeable.

Factors Affecting Runoff and Stream Discharge

Surface runoff can occur when

- the rate of precipitation exceeds the permeability rate (or infiltration rate) of Earth's solid surface
- the pore space of loose material or rock is saturated with water
- the slope of the surface is too great to allow infiltration to occur
- the water on the surface has not evaporated or sublimated to a gas

Most runoff gets to streams, which often carry the water back to the oceans—completing a major part of the water cycle. The greater the runoff, the greater the amount of stream discharge in local streams. **Stream discharge** is the volume of water flowing past a certain spot in a stream in a specific amount of time, and is expressed in units such as cubic meters/second or liters/minute.

FLOODING Flooding occurs when a stream overflows its normal channel. Flooding also happens when precipitation exceeds the ability of the ground to infiltrate the water, and evapotranspiration is not rapid enough, and when water is not able to move to the streams quickly enough. Flooding can also be the result of a hurricane's storm surge, coastal storms, the rising of sea level, sinking land, or tides moving water onto the land.

The first rule of safety in dealing with a flood, or the possibility of a flood, is to move to higher ground. Accurate predictions and measurements of snow or rain might give people sufficient time to evacuate an area before the flooding. Many communities have planned escape routes determined through the study of topographic maps and other elevation models.

Review Questions

1. On Earth as a whole, what happens to most of the precipitation? (1) It recharges the soil moisture deficit. (2) It becomes runoff and moves to the oceans. (3) It is stored in the soil as capillary water. (4) It is returned to the atmosphere through evaporation and transpiration.

2. The primary source of moisture for the local water supplies is (1) potential evaporation and transpiration (2) actual evaporation and transpiration (3) ground water storage (4) precipitation

3. Which event is a direct result of evaporation and transpiration? (1) The atmosphere warms. (2) Cloud cover decreases. (3) Moisture enters the atmosphere. (4) Moisture leaves the atmosphere.

4. What is the largest source of the moisture for the atmosphere?

5. The following diagram is a cross-sectional view of rain falling on a farm field and then moving to the water table.

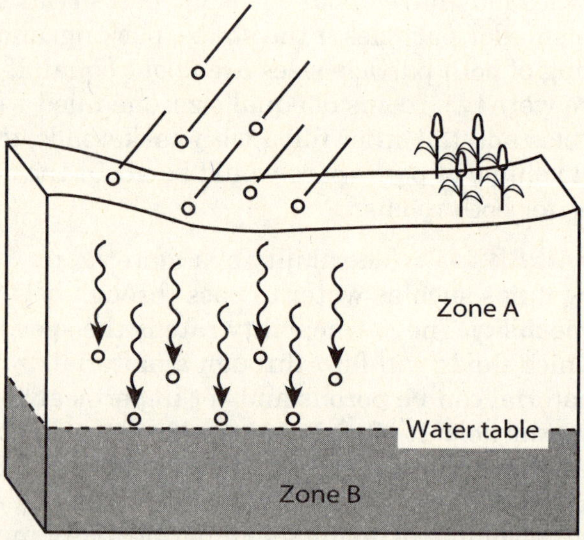

Which word best describes the movement of the rainwater through zone A? (1) runoff (2) saturation (3) infiltration (4) precipitation

6. Which graph best illustrates the relationship between the slope of the land and the amount of surface runoff?

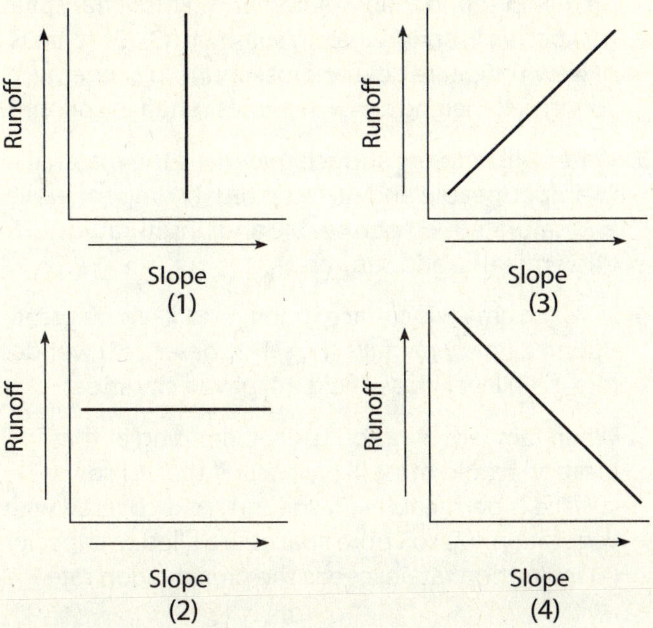

Base your answers to questions 7 through 9 on the following diagrams. The diagrams show two soil cross sections from adjacent fields in New York State. Both soils are the same except that human activities have removed the vegetation from the surface of field B. Each field has been receiving rain for several hours.

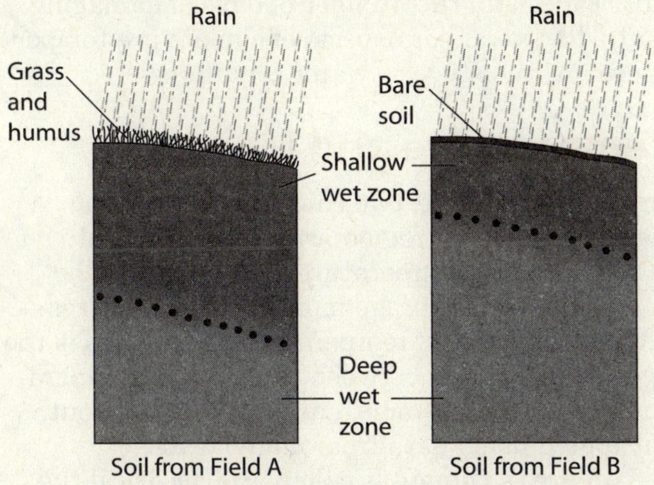

7. The soils in field B would have a higher rate of permeability if the soils (1) had lower porosity (2) had steeper surface slope (3) were composed of larger rock particles (4) were compacted by machinery traveling over the field

8. How would the amount of water that would infiltrate be affected by replanting vegetation in field B?

9. If the skies remain clear for the week following this rain, the water in the wet zone in field B will (1) mostly become surface runoff (2) partially evaporate into the air (3) all remain as stored water along the surface (4) mostly transpire into the ground

---

10. A rock with a high porosity will probably (1) be resistant to weathering (2) be composed of large grains (3) have a large percentage of space between particles (4) have a small percentage of rounded particles

11. Soil with the greatest porosity has particles that are (1) poorly sorted and densely packed (2) poorly sorted and loosely packed (3) well-sorted and densely packed (4) well-sorted and loosely packed

12. Which graph best represents the relationship between soil permeability rate and infiltration when all other conditions are the same?

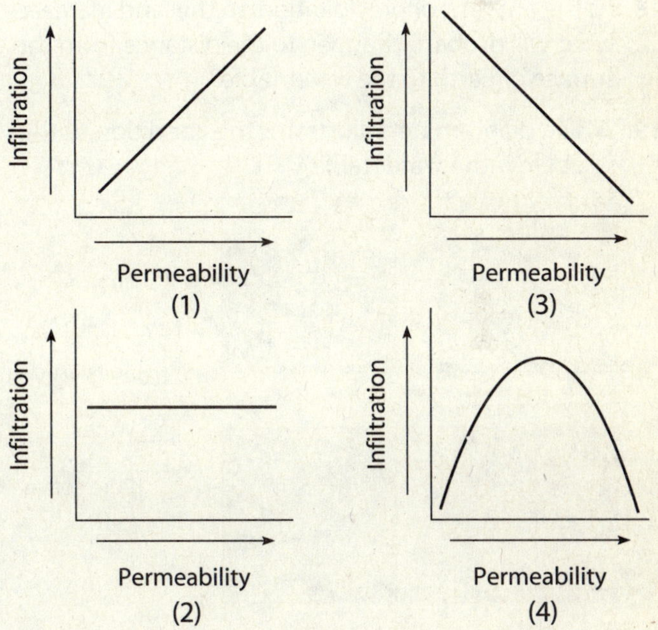

13. Which Earth material covering the surface of a landfill would permit the least rainwater to infiltrate the surface? (1) silt (2) clay (3) sand (4) pebbles

14. As the temperature of soil decreases from 10°C to – 5°C, what will most likely happen to the infiltration rate of water through it?

15. In which tube will capillary action be the greatest?

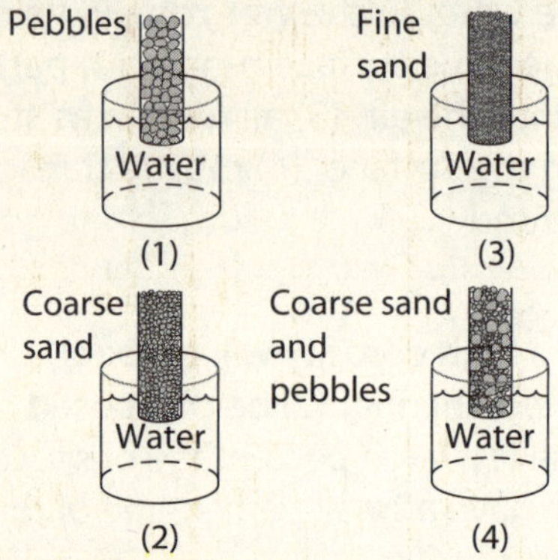

16. Water moves upward through the soil because of (1) capillary action (2) permeability (3) porosity (4) infiltration

17. The rate of flow of water through rock is determined by the (1) types of minerals that make up the rock (2) total amount of space between mineral grains (3) number of connected pore spaces and their sizes (4) position of the rock with respect to the water table

18. As the amount of precipitation on the land increases, what will probably happen to the distance from the surface of Earth to the water table?

19. Which diagram best illustrates the condition of the soil below the water table?

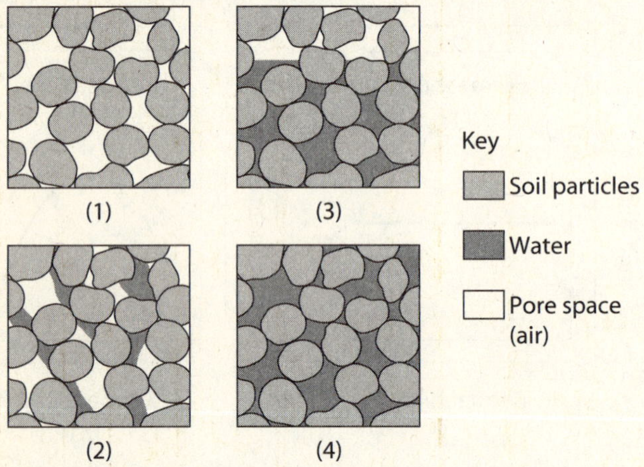

20. Surface runoff of precipitation occurs when (1) porosity is exceeded by permeability (2) the infiltration rate is greater than the precipitation rate (3) the precipitation rate is greater than the infiltration rate (4) the evaporation rate is increased

21. Surface runoff will generally be greatest when the (1) rainfall is light and the ground is permeable (2) infiltration rate is greater than the rainfall rate (3) slope of the land is too great to permit infiltration (4) ground is permeable and unsaturated

22. Why do most streams in New York State have a greater stream discharge runoff in spring than in summer? (1) Potential evaporation and transpiration is greater in spring than in summer. (2) More transpiration occurs in spring than in summer. (3) Most areas of New York State get their maximum solar energy in spring. (4) Melting snow increases runoff in spring.

23. Water will infiltrate surface material if the material is (1) impermeable and unsaturated (2) impermeable and saturated (3) permeable and unsaturated (4) permeable and saturated

24. In which area will surface runoff most likely be greater during a heavy rainfall? (1) sandy desert (2) wooded forest (3) level grassy field (4) paved city street

25. When rain falls on a soil surface, flooding at that location would most likely occur if the (1) soil surface is permeable (2) soil surface is covered with vegetation (3) soil pore spaces are filled to capacity (4) infiltration rate exceeds the precipitation rate

Climate

The overall view of a region's weather conditions over a long period of time—tens to thousands of years—is its **climate.** Climate is not just average weather; it includes extremes like the number of hurricanes and the number of days of damaging hail. The two major aspects of climate are temperature and moisture or water conditions.

Temperature and Moisture

In terms of climate, two characteristics of the temperatures of a region are most important: (1) the average temperature over the year, and (2) the range of average monthly temperatures—often called annual temperature range. This is the average difference between the average temperature of the hottest month and the coldest month, calculated using data from many years.

An area's climate is called arid, or dry, if the average total precipitation during the year is less than the average yearly potential evapotranspiration. Potential evapotranspiration is the amount of water vapor that would evapotranspirate from an area if the water were available. Potential evapotranspiration is largely determined by an area's amount of energy available for evapotranspiration—thus temperature. An area's average temperature is often determined by the solar energy or insolation it receives. An analysis of Figures 8-5 and 8-6 will show the direct relationship between insolation received at Earth's surface and potential evapotranspiration.

An area's climate is humid, or wet, if the average precipitation for the year is greater than the average potential evapotranspiration. Thus, whether a climate is said to be arid or humid depends not on the amount of precipitation, but on the difference between the amount of moisture available from precipitation and the potential need for water as determined largely by temperature. As illustrated in Figure 8-7, a region can have very little precipitation and still have a humid climate. Regions where the potential evapotranspiration is much greater than what an area receives as precipitation are called deserts. If an area has somewhat equal amounts of precipitation and potential evapotranspiration, the term semi-arid or sub-humid can be used to describe the climate.

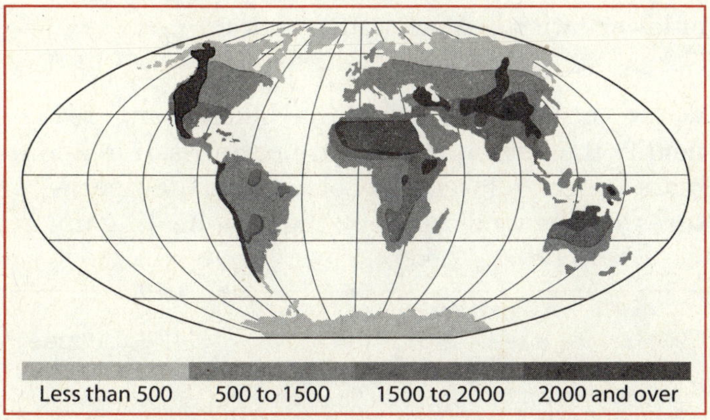

Figure 8-5. World map of insolation at Earth's surface in thousands of calories/cm²/year: Note how yearly insolation generally increases with decreasing latitude. Due to many factors, however, especially differences in cloud coverage, it is not a perfect correlation. (Darkest areas have very high elevation.)

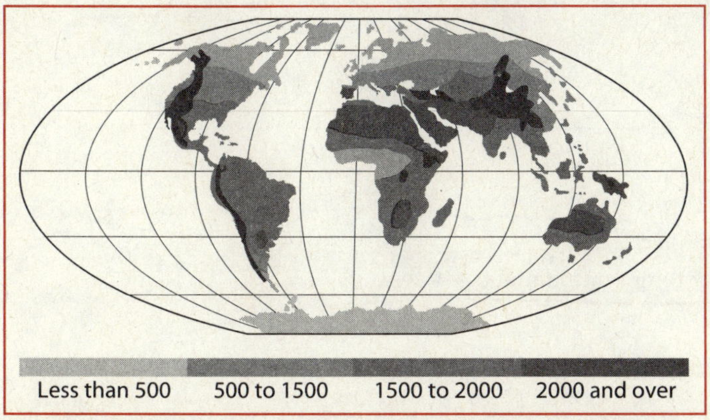

Figure 8-6. World map of potential for evaporation and transpiration in centimeters per year: Note the similarities to the yearly insolation map above. It is not a perfect correlation, because local heat values are affected by many other factors, such as surface ocean currents and tropospheric convection currents. (Darkest areas have very high elevation.)

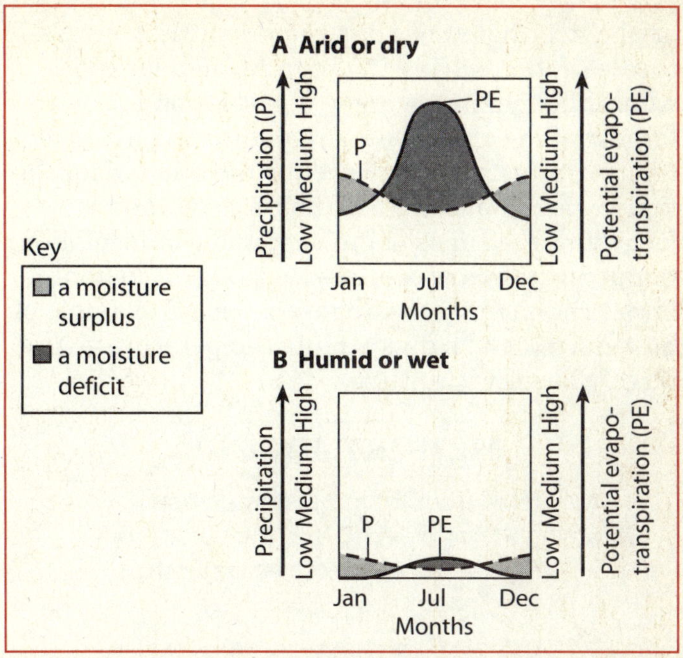

Figure 8-7. Examples of an arid and a humid climate: Graph A shows the relationship between precipitation (P) and potential evapotranspiration (PE) for a region of moderate amounts of rainfall but high potential evapotranspiration. Since the yearly totals of P are less than the yearly PE values for the year—there is a yearly moisture deficit—the region has an arid climate. Graph B represents a climate with low yearly amounts of P, but even smaller yearly amounts of PE. Since there is more P than PE for the year—there is a yearly moisture surplus—the climate is humid, or wet.

Factors Affecting Climate

The factors that determine climate include

- latitude
- planetary wind and pressure belts
- oceans and other large bodies of water
- ocean currents
- mountains
- elevation
- amount of cloud cover
- vegetation

Descriptions of these factors that determine the type of climate of an area follow. Often, many factors interact to determine an area's type of climate.

LATITUDE AND TEMPERATURE Latitude is a major factor in determining climates because of its influence on both temperature and moisture conditions. Temperature characteristics vary with latitude because of the relationships between angle, intensity, and duration of insolation and latitude discussed in Topic 6. At low latitudes, where the maximum angle of insolation is always high, average temperatures are high throughout the year. Because the duration of insolation is fairly constant at about 12 hours per day, there is little temperature variation during the year. At high latitudes, where the maximum angle of insolation is never large and in some months remains zero, average temperatures are low. Since the duration of insolation varies from zero to 24 hours a day, and is longest at the times of greatest angle of insolation, temperatures vary over a wide range from winter lows to summer highs. Figure 8-8 is an example of how Earth can be divided into temperature zones largely based on latitude.

MEMORY JOGGER

You may recall from Topic 6 that insolation (**IN**coming **SOL**ar radi**ATION**) is the part of the sun's radiation that is received by Earth.

LATITUDE AND MOISTURE Moisture conditions vary with latitude because of the location of the planetary wind and pressure belts illustrated on Figure 7-6 on page 113. Where there is low pressure at Earth's surface, such as near the equator and in mid-latitudes, the air rises. The cooling of the rising air results in many clouds, much precipitation, and humid climates. When air falls toward Earth's surface, it creates high pressure, warming of the air, and often arid or desert climates. These conditions result in the belts of arid climates around 30° latitude and around the geographic poles.

The map Planetary Wind and Moisture Belts in the Troposphere in the Earth Science Reference Tables shows the locations of these dry belts and the wet belts.

LATITUDINAL CLIMATIC PATTERNS The combination of the temperature and the moisture effects of latitude results in a basic distribution of climate types around the world—latitudinal climate patterns—shown in Figure 8-9. Note that with increasing distance from the equator (increasing latitude) the annual temperature range increases. This is largely because higher latitudes have a greater amount of change in duration of insolation during the year than areas of lower latitude nearer the equator.

LARGE BODIES OF WATER Large bodies of water—oceans, seas, and large lakes—serve to modify the latitudinal climate patterns. If a land mass is near a large body of water, its temperatures will be moderated by the slow heating up

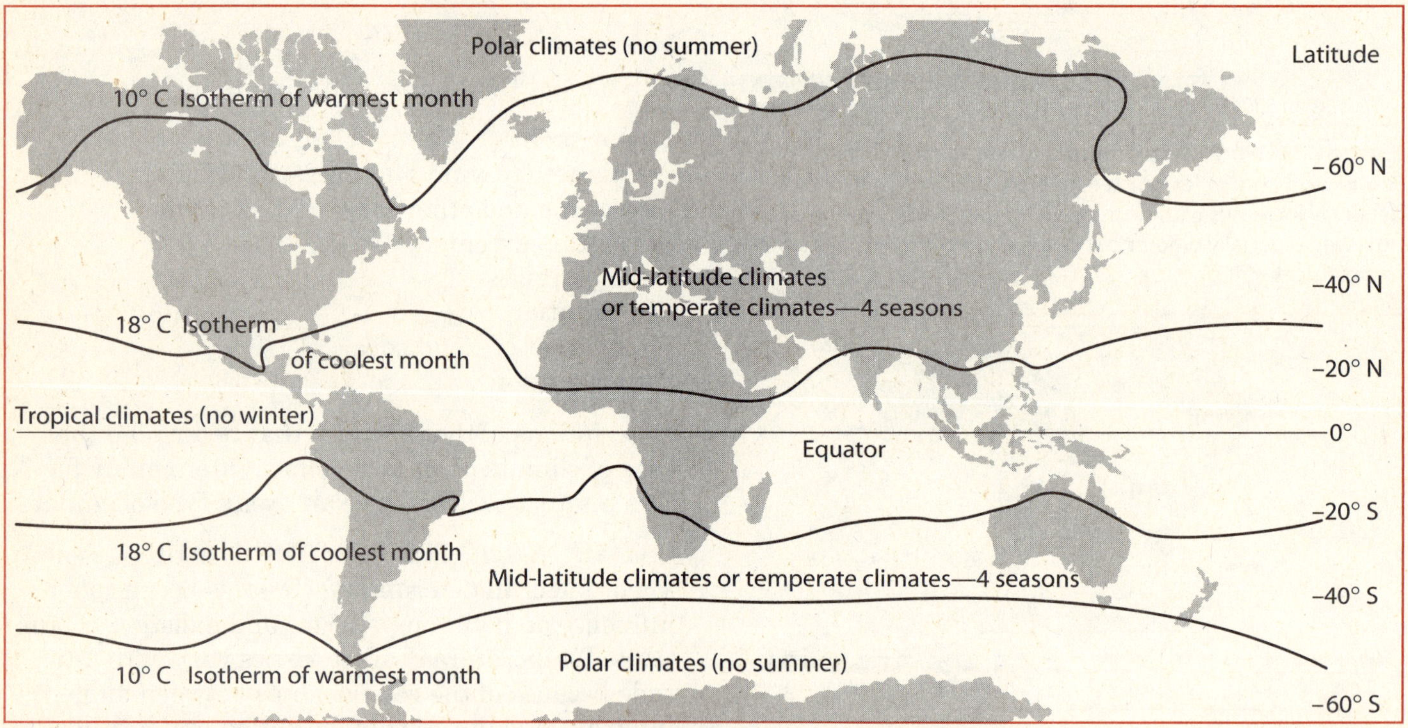

Figure 8-8. Earth's latitudinal climate zones based on temperature: The lines on the map are isotherms—isolines of equal temperature. This is a very basic climate classification based entirely on temperature. In the tropical climate zone, the average monthly temperatures at sea level never drop below 18°C. In the polar climate zones, the average monthly temperatures never rise above 10°C. Effects of altitude have been omitted.

and cooling off of the water body. This is especially true if the prevailing winds blow toward the landmass much of the year.

An area with temperatures moderated by proximity to a body of water is said to have a marine climate. Marine climates have cooler summers and warmer winters than inland areas, and thus a small annual range of temperatures compared to inland areas at the same latitude. Inland areas away from the effects of large bodies of water have cooler winters and warmer summers and a wider annual range of temperatures. Such areas are said to have a continental climate. The larger a landmass is, and the farther it is from the effects of large bodies of water, the wider the annual temperature range. This is why central Eurasia has a more pronounced continental climate than North America. Figure 8-9 illustrates the differences in the temperatures of regions with continental versus marine climates.

PREVAILING WINDS Movements of air over Earth's surface that blow in the same direction most of the time are **prevailing winds.** Figure 7-6 on page 113 and Planetary Wind and Moisture Belts in the Troposphere in the *Earth Science Reference Tables* show the latitudinal locations of the prevailing winds—planetary winds—on Earth. These belts of prevailing winds are caused by the air pressure differences (from unequal heating of Earth by the sun) and the effects of Earth's rotation (the Coriolis effect.)

Most of the contiguous United States is located in the prevailing southwesterly wind belt for much of the year, which means that most weather conditions move from southwest to northeast—more generally west to east. The west coast of the contiguous United States has much more of a marine climate than the east coast because the west coast's prevailing winds are from the ocean. Since the east coast has its prevailing winds blow from the center of the North American continent, it has continental climates almost to the shore of the Atlantic Ocean. Long Island, New York, has some marine moderation of climate compared to the rest of the state, as do some western parts of New York State downwind of Lake Erie and Lake Ontario.

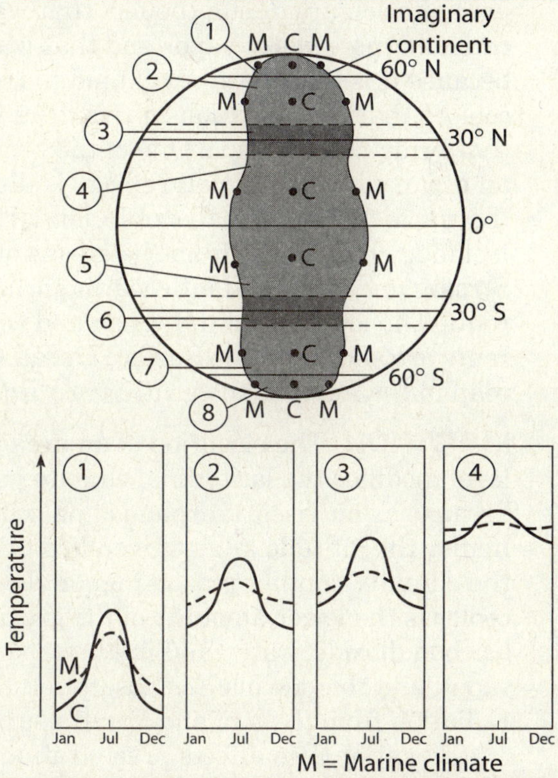

Zones	Climate types
1,8	Polar Cold, humid to arid, with large temperature variation
2,7	Temperate or mid latitude Moderate temperatures, humid, with moderate temperature variation
3,6	Tropical desert Hot, arid
4,5	Tropical rainforest Hot, humid, with little temperature variation

Figure 8-9. Latitudinal climatic zones and continental and marine climates based on temperature and moisture conditions of an imaginary continent on a planet similar to Earth: Compare the zones here to the more simplified pattern in Figure 8-8. Note that as the latitude increases, the annual range of temperatures increases. Also note that marine climates—with ocean influences—have a much narrower annual temperature range than continental regions at similar latitudes. Notice the opposite temperature patterns in northern and southern hemispheres, which are the result of opposite seasons in the two hemispheres.

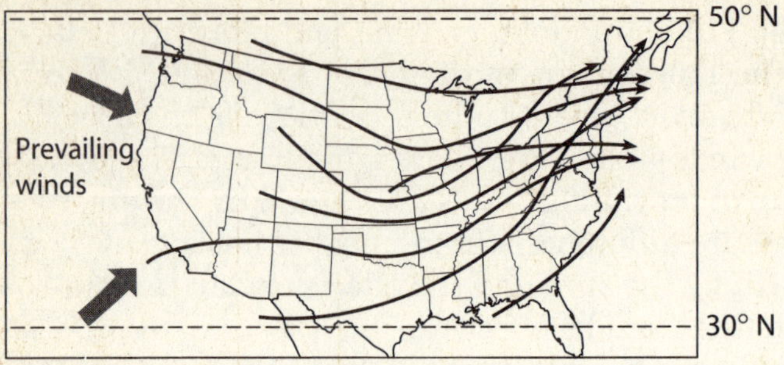

Figure 8-10. Tracks or paths of air masses and low-pressure storms across the contiguous United States: The weather systems are pushed across the United States by the prevailing southwest winds. These air masses and storms tend to follow similar tracks. Certain tracks are associated with specific parts of the year or seasonal shifting winds and pressure belts.

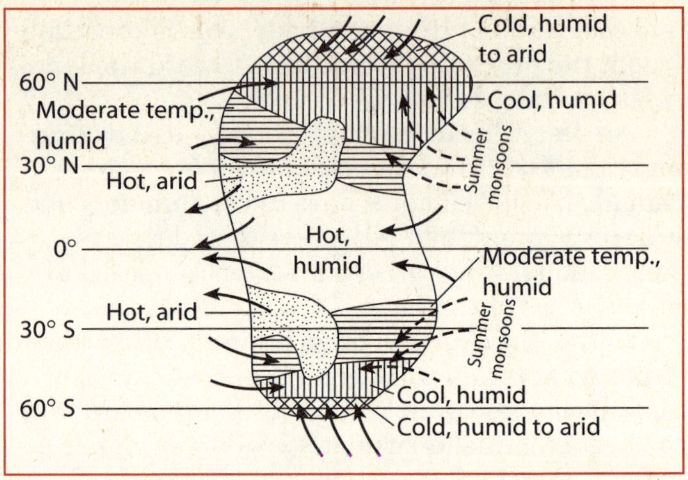

Figure 8-11. Climate zone of an imaginary continent modified from Figure 8-9 by addition of prevailing winds and their shifting latitudes during the year: Seasonal winds and their associated weather are often called monsoons. The summer winds shown on the diagram would bring humid weather with much precipitation into areas that would be much drier without the monsoons.

The prevailing southwesterly winds also have an important large climatic effect by moving the air masses and frontal low pressure storms across the contiguous United States generally from west to east. The paths of these air masses and storm movements are often called "tracks." Much of the precipitation in the mid-latitudes is the result of these tracks and associated weather. Western New York State is noted for the lake effect snows caused by winds and storms blowing over the Great Lakes, picking up moisture, and then dropping the moisture as deep snows.

Monsoons are the weather changes caused by the seasonal shifts in the direction of prevailing winds that are also discussed in Topic 7. Though monsoons are usually associated with the wet summers and dry winters of southeast Asia—especially India—other areas experience a monsoon effect. Portions of southern Europe, northern Australia, and even the southeastern United States experience shifting prevailing winds and changing seasonal precipitation associated with the north and south seasonal movements of the vertical rays of the sun.

Besides the prevailing southwesterly wind belt, other wind belts also cause similar climatic effects in their associated latitudinal zones. The effects of the prevailing winds and their shiftings compared to the simple latitudinal climatic pattern of Figure 8-9 result in the more realistic climate zones illustrated in Figure 8-11.

SURFACE OCEAN CURRENTS Coastal climates are often modified by surface ocean currents. Currents flowing away from the equator carry warm water to higher latitudes, while currents flowing toward the equator carry cool water to lower latitudes. A cool ocean current will cause a coastal area to have cooler temperatures and less precipitation. One reason there is less precipitation is that cool water results in cool air, which cannot hold much water vapor. A warm ocean current will cause a coastal area's climate to be warmer and have more precipitation. The greater amounts of precipitation are due to the fact that warmer air can hold more water vapor and that warmer air—because it is less dense—will tend to rise and then cool, so that condensation can produce the clouds that bring precipitation. One of the best examples of an ocean current's climatic effects is illustrated in the warm, humid climates of Iceland, England, and Ireland, which result from the effects of the Gulf Stream and North Atlantic warm surface ocean currents. The locations of the warm and cool ocean currents are shown on the Surface Ocean Currents map in the *Earth Science Reference Tables*.

ELEVATION The elevation of an area above sea level modifies the latitudinal climate pattern because as air rises, it expands and cools. Thus, the higher the altitude at any given latitude, the cooler the climate. Another reason higher elevations are cooler is the lower amounts of the greenhouse gases (carbon dioxide, water, and methane) at higher altitudes, and this means less absorption of long-wave radiation from the sun and Earth's surface.

Elevation also affects precipitation. As the elevation increases, the temperature and water vapor capacity decrease; the air thus approaches the dew point, condensation forms clouds, and precipitation often occurs. Therefore, areas at higher altitudes generally have more precipitation than lower areas. In New York State, the Catskills and the

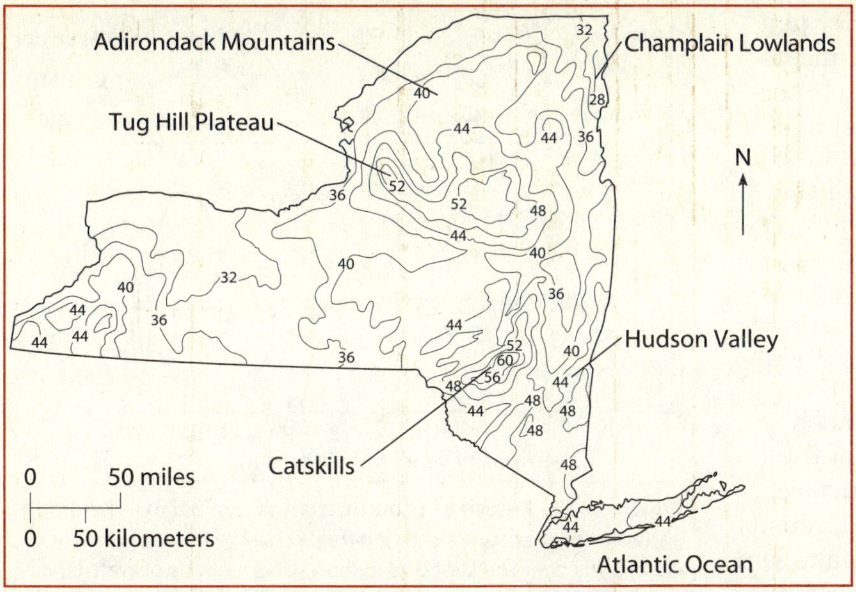

Figure 8-12. Isolines of average annual precipitation in New York State: Even though southeastern New York is on the ocean, this region doesn't have the highest amount of precipitation, because the prevailing southwest winds blow towards the ocean, not away from it. The prevailing winds cause air masses to deposit the most precipitation in the three highest regions of the state—the Catskills, Tug Hill Plateau, and the Adirondack Mountains. Areas downwind from these high elevation regions—such as the Hudson Valley and Champlain Lowlands—receive lower amounts of precipitation.

Adirondack Mountains are generally cooler and have more precipitation than most surrounding areas. (See Figure 8-12 and Landscape Regions of New York State and Their Characteristics on the inside back cover.)

MOUNTAINS Mountains that intersect prevailing winds, such as those associated with the planetary wind belts, can modify the latitudinal climate pattern. Figure 8-13 shows a cross section of a mountain against which the prevailing winds blow from the left. This side is the windward side. As the wind strikes the windward side of the mountain, the air is forced to rise. As it does so, it expands and cools. When rising air cools to the dew point, condensation releases stored heat energy, which lowers the rate of cooling with increasing elevation. At elevations above the place the dew point is reached, condensation results in clouds and then precipitation on the windward side. On the opposite side of the mountain—the leeward side—the air begins to descend. As the air descends, it is compressed and warmed. The warming of the air raises the temperature above the dew point, and condensation and precipitation stop. As a result, the leeward side is warmer than the windward side at any given altitude. The leeward side also has much less precipitation because the air has lost much of its moisture, and its water vapor capacity rises as its temperature increases. A good example of this mountain effect is the fact that the Champlain Lowlands get the least amount of precipitation in New York State. The Champlain Lowlands are on the leeward side of the Adirondack Mountains.

Another way that mountains modify climate is by acting as barriers to moving air masses, preventing cold air or warm air from crossing the mountain to the other side. As a result, opposite sides of a mountain can have different temperature patterns. In southeastern New York, the Hudson Highlands often keep warmer air south and colder air north. Therefore, sometimes during winter when it rains in New York City, there is either snow or no precipitation in the Hudson Valley north of the Hudson Highlands.

VEGETATION An area's natural vegetation amounts and types are largely determined by the climate. Thus, an arid area has desert vegetation and a hot, humid climate has tropical rainforest vegetation. On the other hand, the vegetation of an area can help determine the climate of an area. When rainforests are cut down and converted into farms or grazing land, the climate often becomes hotter and less humid. Part of the reason for this is that there is more runoff and less transpiration to add humidity. Also, without all the trees to absorb solar energy and convert it into potential food energy, the solar energy heats the land instead, which then heats the atmosphere it touches. Have

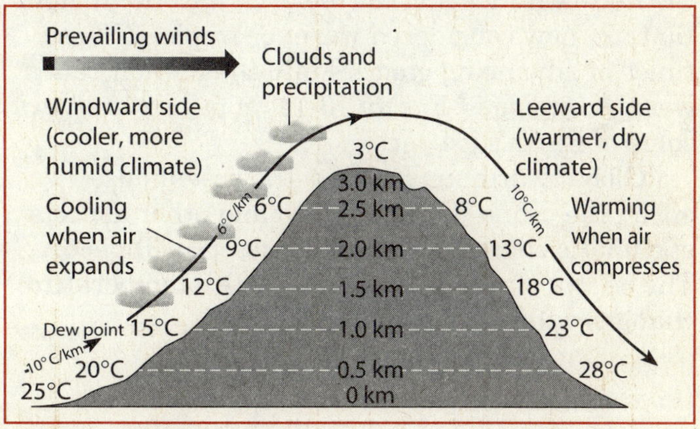

Figure 8-13. Climate effect when mountains intersect prevailing moist winds: The windward side of mountains has much precipitation while the leeward is dry.

you ever noticed how much cooler it is under just one shade tree than it is when you are out in the open? Cutting down or deforestation of whole sections of forest would just multiply the effect of moving out from under a tree. Building cities—**urbanization**—also causes a decrease in forests and grasslands. See pages 276–277 and Figure 14-10 for a description of some of the changes in climate and weather caused by urbanization.

CLOUD COVER If an area has a high percentage of cloudy days, this will have a major climatic effect. Without clouds, the atmosphere is much more transparent to insolation, and more of the sun's energy can reach and heat Earth's surface. The high angle of incidence and thus high intensity of the sun's rays near the equator would make you think that areas at or near the equator would be the hottest on Earth, but this is not true. Thunderstorm clouds often cover the area around the equator, and these clouds absorb a large percentage of the insolation or reflect it back into space. On the other hand, the desert regions that exist around or at 30 degrees of latitude seldom have clouds. These dry latitudes without much cloud cover are often hotter than regions closer to the equator. See the dry and wet latitude belts on the Planetary Wind and Moisture Belts in the Troposphere in the *Earth Science Reference Tables.*

Climatic Change

A study of Figure 8-14 will indicate that Earth has gone through many times of overall increasing or decreasing temperatures. Some of the reasons for these warming and cooling times on Earth were discussed in Topic 6. During the times of cooling, glaciers have advanced from the polar regions into the mid-latitudes and maybe even into the regions that are now considered warm or tropical. These times of advancing glaciers into the mid-latitudes were the ice ages. Figure 13-11 on page 259 shows some of Earth's ice ages.

Global warming, El Niño, the greenhouse effect, the ozone "hole" problem and other aspects of climatic change are also discussed in Topic 6. The effects of volcanic ash on Earth's temperature changes will be discussed in Topic 12.

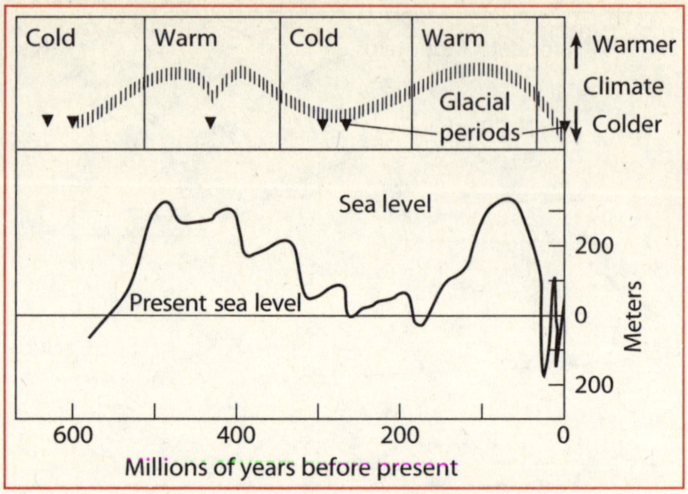

Figure 8-14. Graph of climatic changes on Earth: The graph shows a comparison of the generalized sea levels and climatic conditions over the past 600 million years. Periods of warmer climate generally correspond with times of higher sea levels and periods of colder climates, with ice ages, generally correspond to the times of lower sea levels.

Review Questions

26. Which would cause the potential evaporation and transpiration to increase in a given month? (1) below-normal precipitation (2) a major flood (3) a month-long cold spell (4) higher amount of solar energy

27. Although New York City is at approximately the same latitude as Omaha, Nebraska, New York City's winter months are warmer and summer months are cooler. Which statement best explains why this is so? (1) The sun's rays shine more directly on New York City in the winter. (2) Nebraska is nearer to the Rocky Mountains. (3) The water around New York City has a moderating effect on the temperature (4) The prevailing westerlies have a greater effect on Omaha than on New York City.

28. Which would cause the potential evaporation and transpiration to decrease in a given month? (1) below-normal precipitation (2) drilling of a large well (3) a month-long cold spell (4) a high rainfall

29. Which generally has the greatest effect in determining the climate of an area? (1) degrees of longitude (2) extent of vegetation (3) distance from equator (4) month of the year

30. Two locations, one in northern Canada and one in the southwestern United States, receive the same amount of precipitation each year. The location in Canada is classified as a humid climate. Why would the location in the United States be classified as an arid climate?

31. The planetary wind and moisture belts indicate that large amounts of rainfall occur at Earth's equator because air at Earth's surface is (1) converging and rising (2) converging and sinking (3) diverging and rising (4) diverging and sinking

32. Which climate conditions are typical of regions near the North Pole and the South Pole? (1) low temperature and low precipitation (2) low temperature and high precipitation (3) high temperature and low precipitation (4) high temperature and high precipitation

33. Describe the temperature differences between a coastal city and an interior city, at the same latitude, during summer and winter.

34. As the degrees of latitude from the equator increases, what generally happens to the yearly average temperature?

35. On one of the Hawaiian Islands, the annual rainfall is 200 inches per year on one side of the island and less than 20 inches per year on the opposite side of the same island. This difference is most likely caused by (1) jetstreams (2) hurricanes or typhoons (3) monsoons (4) prevailing winds and mountains

36. Which two climate factors are most directly responsible for the amount of snowfall normally received in Buffalo, New York? (1) ocean currents and storm tracks (2) mountain barriers and average temperatures (3) elevation and potential evaporation and transpiration (4) prevailing wind direction and proximity to a large body of water

37. Which planetary wind pattern is present in many areas of little rainfall? (1) Air sinks and winds converge. (2) Air rises and winds converge. (3) Air sinks and winds diverge. (4) Air rises and winds diverge.

38. Which area of New York State would probably have the lowest annual temperature range? (1) Long Island (2) the Catskills (3) the Adirondack peaks (4) the Mohawk Valley

39. Bodies of water have a moderating effect on climate primarily because (1) Water gains heat more rapidly than land does. (2) Water surfaces are flatter than land surfaces. (3) Water temperatures are always lower than land temperatures. (4) Water temperatures change more slowly than land temperatures do.

40. Which is a characteristic of water that helps the oceans to moderate the climates of Earth? (1) Water is a fluid with a high specific heat. (2) Water can exist as a high-density solid. (3) Water can dissolve and transport minerals. (4) Water can flow into loose sediments to deposit mineral cements.

41. If Lake Michigan were to vanish, the winters in Chicago would probably (1) remain the same in temperature and humidity (2) remain the same in temperature but become drier (3) become colder and drier (4) become warmer and drier

42. Which locality has the greatest annual range of temperature? (1) Seattle, Washington (2) Bismarck, North Dakota (3) New York City (4) Miami, Florida

43. Compared to an inland location of the same elevation and latitude, a coastal location is likely to have (1) warmer summers and cooler winters (2) warmer summers and warmer winters (3) cooler summers and cooler winters (4) cooler summers and warmer winters

44. A low-pressure storm center located over New York State will most likely move toward the (1) southeast (2) southwest (3) east (4) northwest

45. Arrows on the following map represent ocean currents.

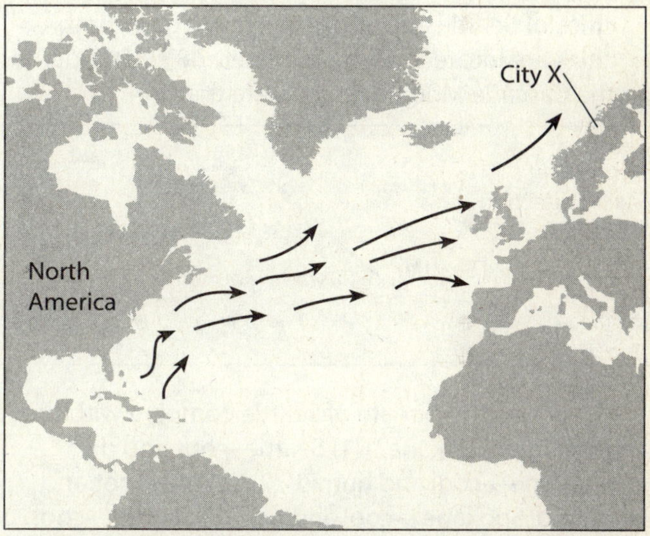

These ocean currents affect the climate pattern of city X by (1) decreasing the average annual cloud cover (2) decreasing the average annual evaporation and transpiration (3) increasing the average annual temperature (4) increasing the average annual air pressure

46. Two coastal cities have the same latitude and elevation, but are located near different oceans. Which statement best explains why the two cities have different climates? (1) They have different longitudes. (2) They are near different ocean currents. (3) They have different angles of insolation. (4) They have different numbers of daylight hours.

47. As a parcel of air rises, its temperature will (1) decrease due to expansion (2) decrease due to compression (3) increase due to expansion (4) increase due to compression

48. Which statement best explains why a cloud is forming in the following diagram?

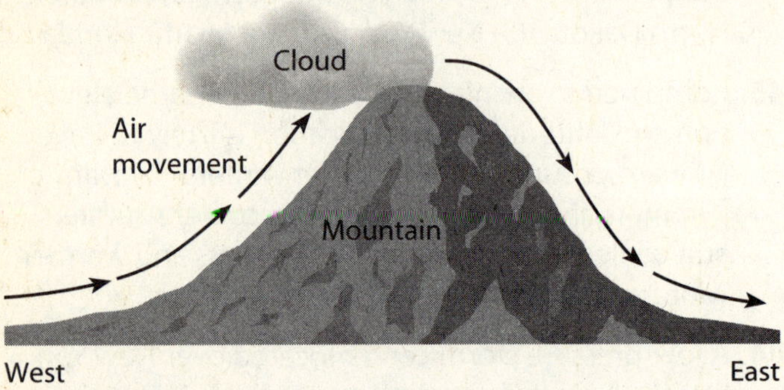

(1) Water vapor is condensing. (2) Moisture is evaporating. (3) Cold air rises and compresses. (4) Warm air sinks and expands.

49. The following diagram shows the locations of the cities of Seattle and Spokane, Washington. Both cities are located at approximately 48°N latitude, and the Cascade Mountains separate them.

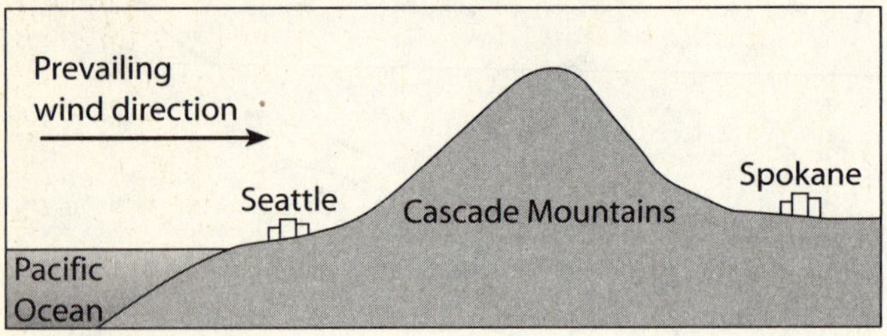

How does the climate of Seattle compare with the climate of Spokane? (1) Seattle—hot and dry; Spokane—cool and humid (2) Seattle—hot and humid; Spokane—cool and dry (3) Seattle—cool and humid; Spokane—warm and dry (4) Seattle—cool and dry; Spokane—warm and humid

Refer to the following diagram to answer questions 50 through 53.

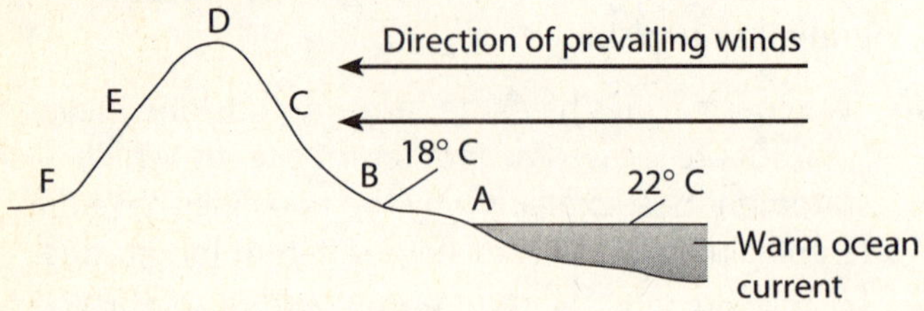

50. Which location would have the coolest average temperature?

51. Which location would have the most condensation in its atmosphere during the year? (1) B (2) C (3) E (4) F

52. Which location would most likely be driest.

53. How is the potential evaporation and transpiration of area A-B affected by the ocean current?

Base your answers to questions 54 through 57 on your knowledge of earth science and on the following diagram. The diagram represents an imaginary continent on Earth surrounded by water. The arrows indicate the direction of the prevailing winds. Two large mountain regions are also indicated. Points A, B, E, and H are located at sea level; C, D, and F are in the foothills of the mountains; G is high in the mountains.

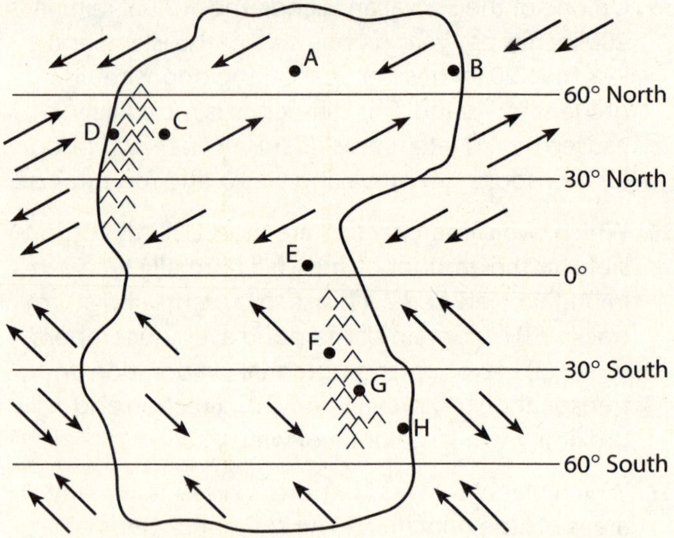

54. Which physical characteristic would cause location G to have a colder yearly climate than any other location? (1) the nearness of location G to a large ocean (2) the location of G with respect to the prevailing winds (3) the elevation of location G above sea level (4) the distance of location G from the equator

55. Which location probably has the greatest annual rainfall? (1) A (2) F (3) C (4) D

56. Which location probably has the widest range in temperature during the year? (1) A (2) B (3) H (4) D

57. Which location will probably record its highest potential evaporation and transpiration values for the year during January? (1) A (2) F (3) C (4) D

Questions for Regents Practice

Part A

1. The following flowchart shows one process by which moisture enters the atmosphere.

The last step of this process is known as

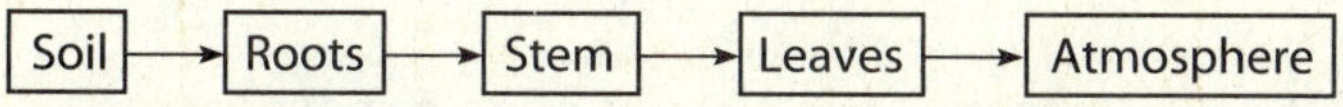

(1) condensation

(2) convection

(3) radiation

(4) transpiration

2. Replanting forests usually results in an increase in

(1) floods

(2) ground water

(3) run-off

(4) soil erosion

3. Which graph best represents the relationship between porosity and particle size for soil samples of uniform size, shape, and packing?

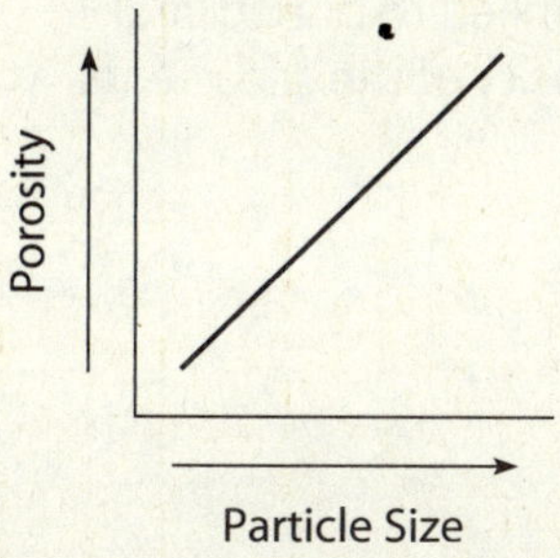

(1)

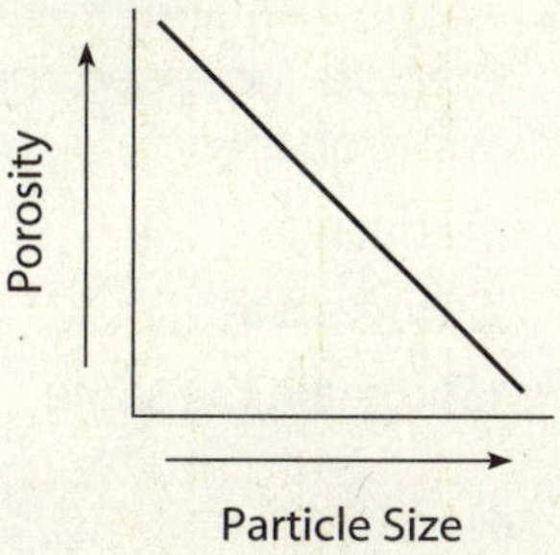

(3)

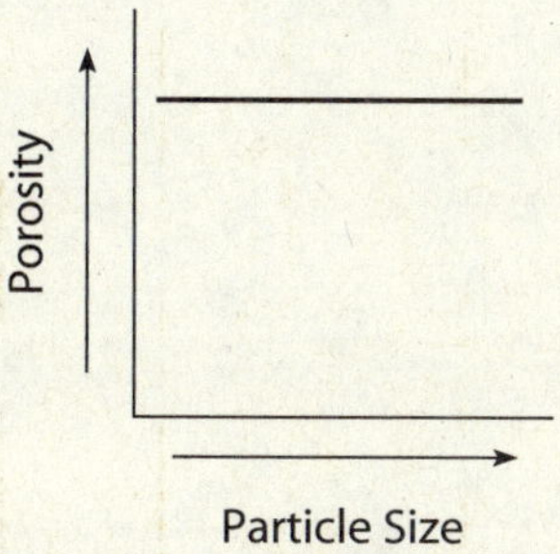

(2)

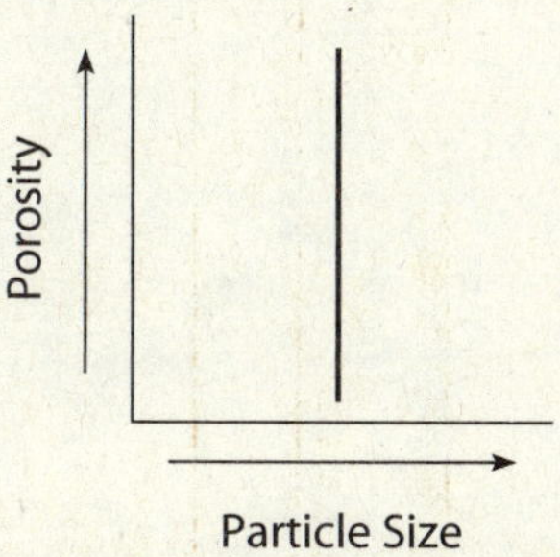

(4)

4. The following diagram represents a cross-sectional view of a tunnel cut through a mountain. The area where the mountain is located receives heavy rainfall. If the shale layers are impermeable, at which point would the most water seep through the roof of the tunnel?

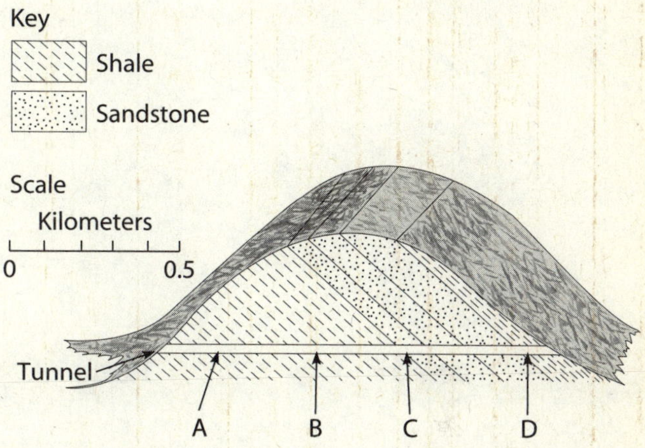

(1) A (2) B (3) C (4) D

5. Which statement best explains why Plattsburgh has a low average annual precipitation compared to most of New York State?

(1) Ocean currents bring warm, moist air to Plattsburgh.

(2) A planetary low-pressure belt is located over Plattsburgh.

(3) High latitudes cause a warm, dry climate near Plattsburgh.

(4) Prevailing winds travel over the Adirondack Mountains before they reach Plattsburgh.

6. Under which condition would a climate be classified as humid?

(1) When the soil moisture is being used all year long.

(2) When the annual precipitation equals the annual potenital evapotranspiration .

(3) When the annual precipitation is much greater than the annual evapotranspiration .

(4) When the annual precipitation is much less than the annual potential evapotranspiration.

Base your answers to questions 7 through 11 on the following diagram. The diagram represents two identical barrels, each filled to the same level, one with BBs and the other with marbles.

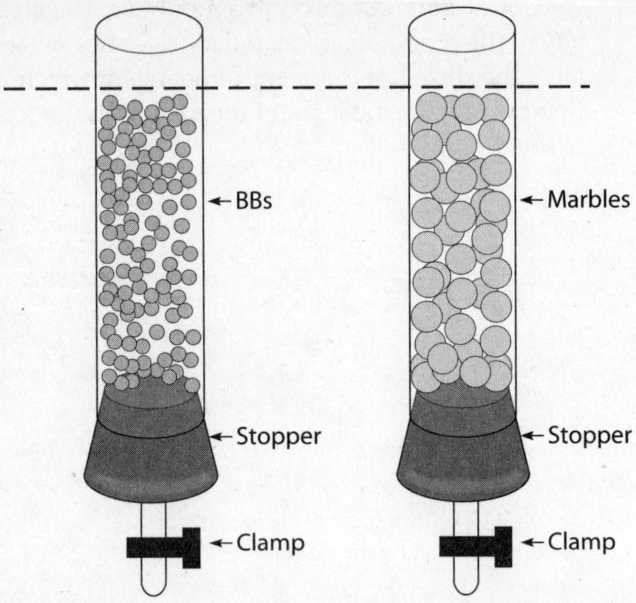

7. If water were added to each barrel to the height of the dashed line and then the clamps opened, which graph best illustrates how fast the water would run through each barrel?

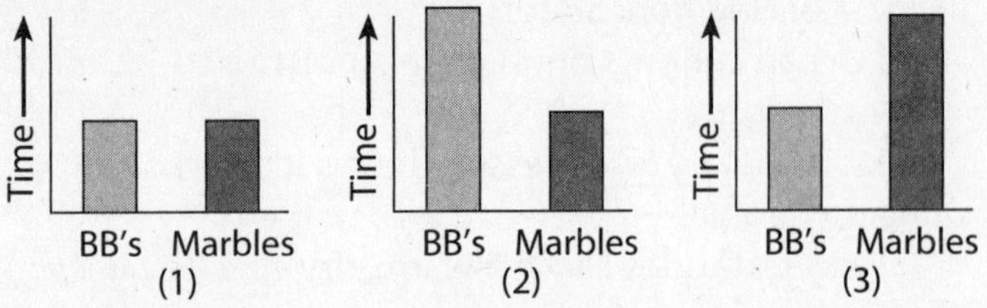

8. After the water has been allowed to pass freely through the barrels, which graph best illustrates the amount of water retained by each barrel?

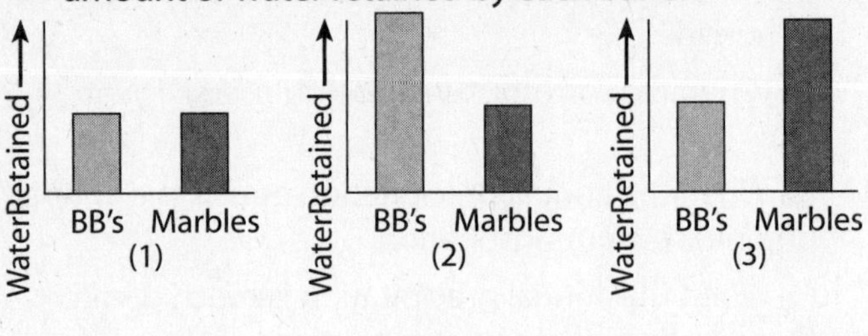

9. If the barrels and their contents were dried of all water and then arranged in such a way as to show capillary action, which graph best illustrates the height to which water would rise in each barrel?

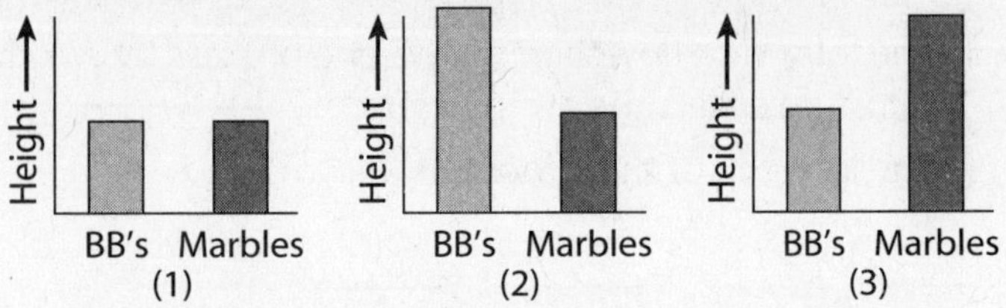

10. How does the total amount of pore space in the barrel with the BBs compare with the total amount of pore space in the barrel with the marbles?

(1) It is much more.

(2) It is much less.

(3) It is approximately the same.

11. If an identical third barrel were filled with a mixture of BBs and marbles to the same level as the other two barrels, how would the size of the pore spaces in this barrel compare with the size of the pore spaces in the barrel of marbles?

(1) It would be less.

(2) It would be greater.

(3) It would be the same.

---

12. As the heat and pressure applied to buried sediments is increased, the porosity of the sediments will probably

(1) decrease

(2) increase

(3) remain the same

Base your answers to questions 13 through 15 on the following map. The map shows the climate classifications of some countries in eastern Africa.

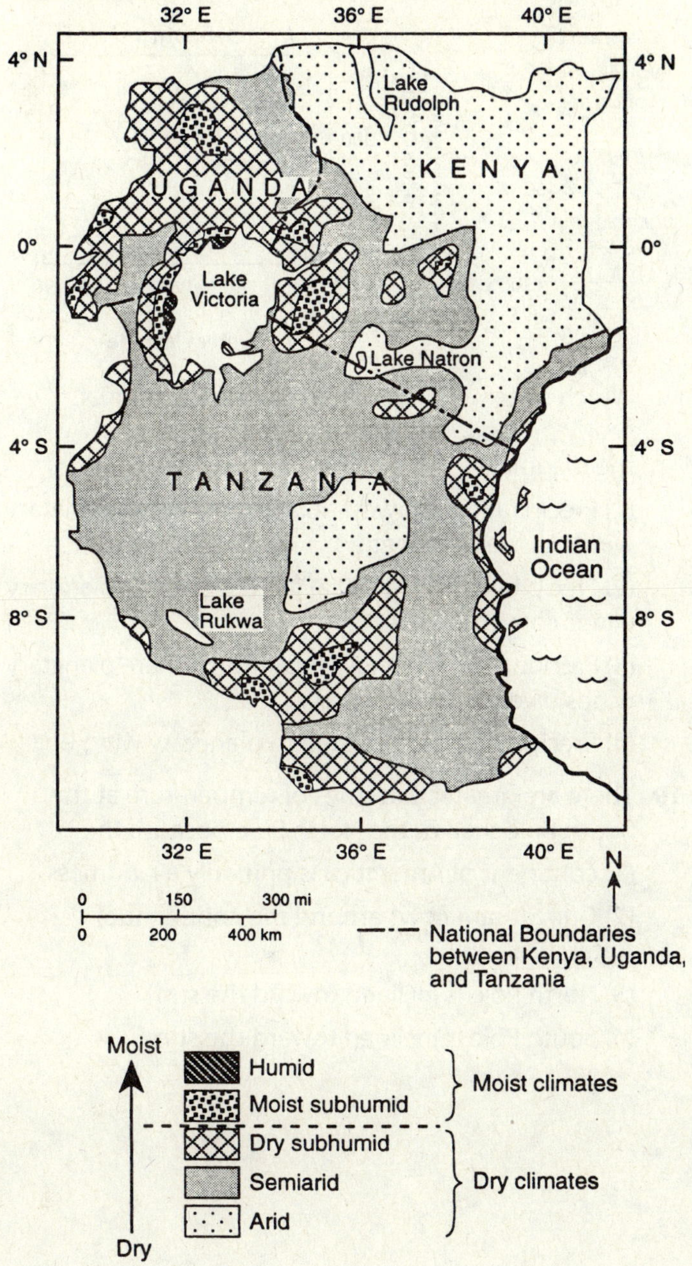

13. Which inference best explains the cause of the climate of the northeastern section of Kenya?

(1) The air reaching this section comes from a dry landmass region.

(2) Warm ocean currents bring moisture to this coastal section.

(3) Mountains force air to rise over this section.

(4) Cloud cover reflects most insolation in this section.

14. Which body of water appears to produce the greatest increase in the climate humidity of the adjacent land area?

(1) Indian Ocean

(2) Lake Rudolph

(3) Lake Rukwa

(4) Lake Victoria

15. Which pattern represents the general surface planetary winds during March in this region?

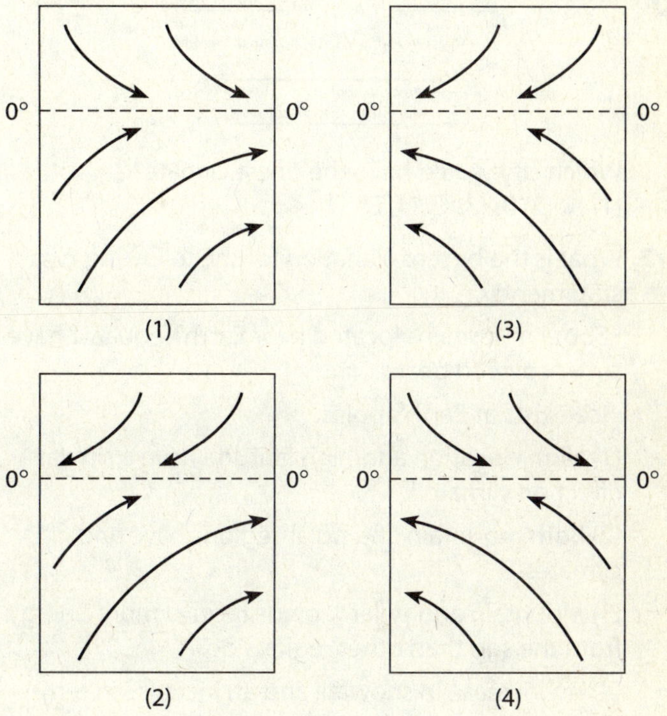

16. The following graphs show some climate data for four cities: A, B, C, and D.

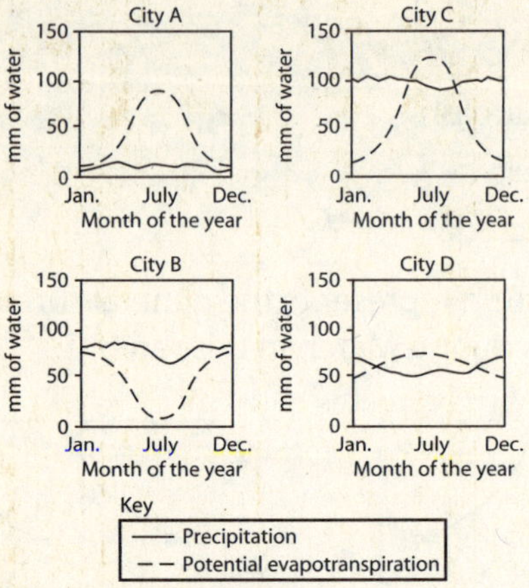

Which city would have the driest climate?

(1) A (2) B (3) C (4) D

17. What is the best explanation for the following two statements?

• Some mountains located near Earth's equator have snow-covered peaks.

• Ice exists at Earth's poles.

(1) High elevation and high latitude have a similar effect on climate.

(2) Both mountain and polar regions have arid climates.

(3) Mountain and polar regions receive more energy from the sun than other regions do.

(4) An increase in snowfall and an increase in temperature have a similar effect on climate.

18. Diagram I shows the planetary wind belts of Earth. Diagram II is a graph of the average yearly precipitation for locations between 90° N latitude and 90° S latitude.

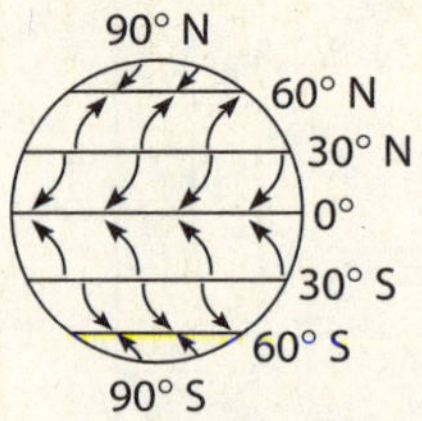

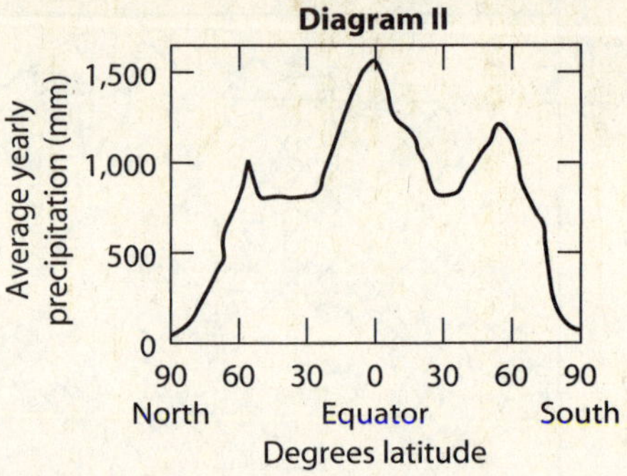

Which statement about Earth's average annual yearly precipitation is best supported by the diagrams?

(1) Precipitation is lowest at latitudes where planetary winds converge (meet).

(2) Precipitation is highest at latitudes where planetary winds converge (meet).

(3) Precipitation is highest at latitudes where planetary winds diverge (move apart).

(4) Precipitation is unrelated to planetary wind belts.

19. There are greater extremes of temperature at the South Pole than at the North Pole because the

(1) continent of Antarctica is primarily a landmass

(2) Gulf Stream flows around the continent of Antarctica

(3) North Pole is inclined toward the sun

(4) South Pole is inclined toward the sun

Part B

Base your answers to questions 20 through 22 on the data table below. The table shows the elevation and average annual precipitation at ten weather stations, *A* through *J*, located along a highway that passes over a mountain.

Data Table

Weather Station	Elevation (m)	Average Annual Precipitation (cm)
A	1,350	20
B	1,400	24
C	1,500	50
D	1,740	90
E	2,200	170
F	1,500	140
G	800	122
H	420	60
I	300	40
J	0	65

Symbol Chart

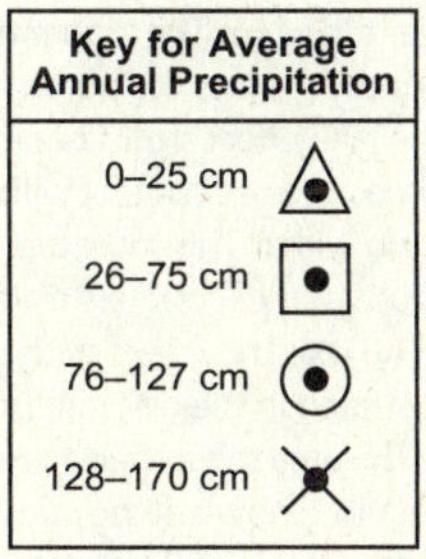

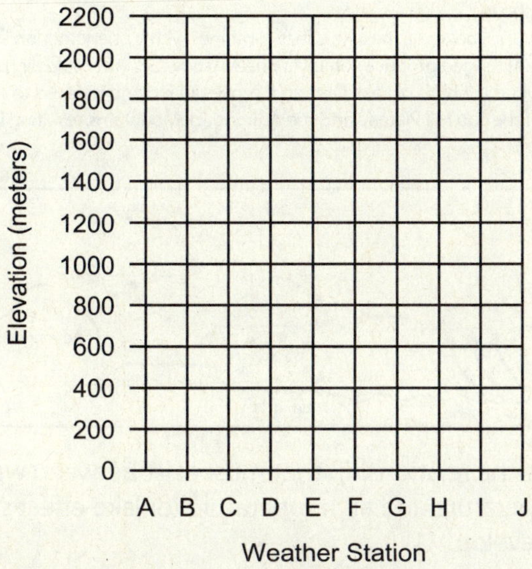

20. On the grid above, graph the data shown on the data table by following the directions below.

a Mark the grid with a point showing the elevation of *each* weather station. [1]

b Surround *each* point with the proper symbol from the symbol chart to show the amount of average annual precipitation for the weather station. [1]

21. State the relationship between the elevation of weather stations *A* through *E* and the average annual precipitation at these weather stations. [1]

22. Although stations *C* and *F* are at the same elevation, they have very different amounts of average annual precipitation. Explain how the prevailing wind direction might cause this difference. [1]

Base your answers to questions 23 through 34 on the map below. The map represents an imaginary continent and an island on Earth, which are surrounded by oceans. Locations E and H are on a coast a sea level, locations B, D, F, I, and J are about 500 meters in elevation, locations A and C are at the base of mountains at about 500 meters elevation, and Location G is at the top of a mountain at an elevation of 3000 meters. The arrows represent the direction of the prevailing winds. Some of these arrows have been omitted. Three mountain ranges with elevations over 1,000 meters are indicated.

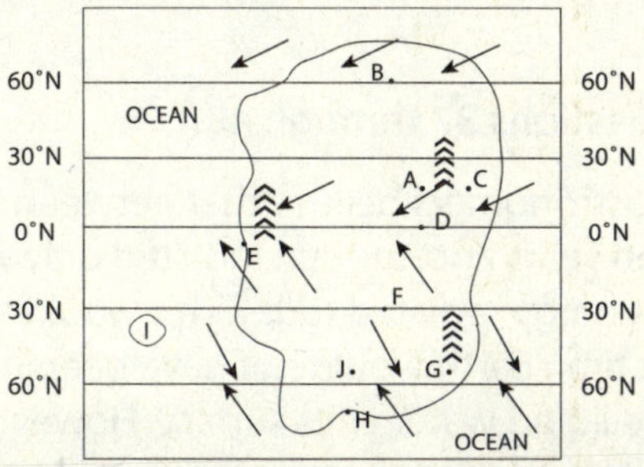

23. What is the main reason why location *B* is colder in January than in July? [1]

24. Which location, with an elevation below 600 meters, most likely has the lowest average yearly temperature? [1]

25. Which non-polar location has the most arid (dry) climate due to falling air from the upper troposphere? [1]

26. On the blank graph below draw a line that best represents the relationship between time of year and the amount of insolation at location *F*. [1]

27. Why would location *F* have cooler winters and warmer summers than location *I*? [1]

28. On the map draw three arrows to indicate the prevailing wind direction between 35°N and 50°N. [1]

29. Draw a big **W** in the ocean just off the west coast of the imaginary continent where there is most likely the location of a cold water ocean current just off the coast. [1]

30. How would a warm water ocean current just off of location *C* affect the amount of precipitation at this location? [1]

31. Which location on the map has the combination of highest yearly temperatures and high yearly amounts of precipitation? [1]

32. Why is location *A* much more arid (drier) than location *C*? [1]

33. What physical aspect of location *G* would cause it to be much colder than location *J*? [1]

34. Which location has higher than normal amounts of precipitation for its latitude because moist air is forced to rise, expand, cool, producing large volumes of clouds? [1]

35. State two ways in which the amount of ground water may be increased. [2]

36. Explain how the slope and the porosity of the surface of the ground determine how much rainfall will seep into the ground. [1]

Part C

Answer questions 37 through 40.

37. Many astronomers believe that between 3 billion and 4 billion years ago, the sun radiated only 70 percent of the energy radiated today. This would suggest that at that time Earth would not have been hot enough to have liquid water at its surface. However, geologists cite evidence of sedimentary rocks and fossils that formed in liquid water during this time. Suggest a way in which Earth's surface could have been warm enough to have liquid water and life forms with much less energy coming from the sun. [1]

38. In a region of flat landscape, one family is obtaining drinking water from a well drilled 150 feet below the surface and 50 feet below the water table—all in sediment. Another family, in the same landscape area 2 miles away, has the same type of well drilled to the same depth, but this well produced almost no water. Describe a sediment condition that could account for the differences in water production from the two wells. [1]

39. In the last few years drought in the southwestern United States has been associated with sinking land and development of cracks in Earth's surface. Some of the mile-long cracks are over 12 feet deep and over 20 feet in width. Soil scientists have suggested that the cracks and sinking land are due to the lowering of the water table and changes in the sediments between the new and old water-table levels. Describe how sediments above and below the water table have changed and how the changes could account for the sinking land and the formation of surface cracks. [2]

40. You are at an ocean beach where it hasn't rained in weeks. You observe that for miles along the beach the sand is saturated with water about 5 feet inland and about 2 feet higher than any recent ocean levels. Describe a common condition that would account for the water-saturated sand above ocean levels. (Don't consider splashing of waves.) [1]

Base your answers to questions 41 through 46 on the magazine article and diagram below.

Lake-Effect Snow

During the cold months of the year, the words "lake effect" are very much a part of the weather picture in many locations in New York State. Snow created by the lake effect may represent more than half the season's snowfall in some areas.

In order for heavy lake-effect snow to develop, the temperature of the water at the surface of the lake must be higher than the temperature of the air flowing over the water. The higher the water temperature and the lower the air temperature, the greater the potential for lake-effect snow.

A lake-effect storm begins when air flowing accross the lake is warmed as it comes in close contact with the water. The warmed air rises and takes moisture along with it. This moisture, which is water vapor from the lake, is turned into clouds as it encounters much colder air above. When the clouds reach the shore of the lake, they deposit their snow on nearby land. A typical lake-effect storm is illustrated in the diagram below.

The area most likely to receive snow from a lake is called a "snowbelt." Lake Ontario's snowbelt includes the counties along the eastern and southeastern ends of the lake. Beacause the lake runs lengthwise from west to east, the prevailing westerly winds are able to gather the maximum amount of moisture as they flow across the entire length of the lake. There can be lake-effect snowfall anywhere around the lake, but the heaviest and most frequent snowfalls occur near the eastern shore.

In parts of the snowbelt, the lake effect combines with a phenomenon known as orographic lifting to produce some very heavy snowfalls. After cold air has streamed over the length of Lake Ontario, it moves inland and is forced to climb the slopes of the Tug Hill Plateau and the Adirondack Mountains, resulting in very heavy snowfall.

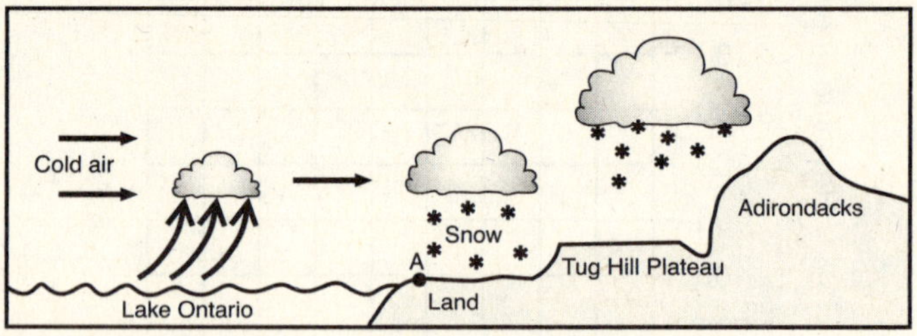

41. State the relationship that must exist between water temperature and air temperature for lake-effect snow to develop. [1]

42. State why locations east and southeast of Lake Ontario are more likely to receive lake-effect snow than are locations west of the lake. [1]

43. State where on the diagram above the annual precipitation would most likely be the lowest. [1]

44. State why very heavy snowfall occurs in the Tug Hill Plateau region. [1]

45. State where it would be the safest location to plant crops that would be killed by large extremes of temperature. [1]

46. State two reasons why the land around location *A* is cooler in the summers compared to areas fifty miles directly east of location *A*. [1]

Weathering and Erosion

TOPIC 9

VOCABULARY

abrasion	glacial parallel scratches	stream
breaking wave	glacier	stream abrasion
chemical weathering	mass movement	stream channel shape
delta	meander	tributary
erosion	physical weathering	U-shaped valley
finger lake	sandbar	V-shaped valley
flood plain	sandblasting	watershed
glacial groove	sediment	weathering

Weather conditions, such as temperature changes and precipitation, play an important role in the rock cycle. Weather conditions, the actions of living things, and other factors can make Earth's surface rocks break down into smaller particles. Moving water or wind can then carry the rock particles away as **sediments**. Eventually, sediments are deposited in environments such as the ocean floor, where they become part of sedimentary rocks.

Weathering

The chemical and physical breakdown of rocks at or near Earth's surface is called **weathering.** Weathering occurs when rocks are exposed to air, water, and the actions of living things. The origins of most of Earth's landscape features and soils can be traced to the weathering of exposed rocks. Weathering processes also affect structures such as buildings and roads.

Types of Weathering

Like other types of matter, rocks can undergo physical or chemical changes. Weathering processes are classified into two types—those that cause chemical changes in rock and those that cause physical changes.

CHEMICAL WEATHERING The breakdown of rock through a change in mineral or chemical composition is called **chemical weathering.** One form of chemical weathering is oxidation—such as when iron combines with oxygen to make iron oxide, or rust.

MEMORY JOGGER

You may recall that oxidation is the chemical combining of oxygen from the air with other elements. You have probably seen rust (iron oxide) on old nails and other iron objects. The iron in rocks and soil can also oxidize, combining with oxygen to form red, brown, and yellow pigments.

Another form of chemical weathering is the effect of water on minerals. Water is often called the universal solvent because, over time, water can dissolve most rock materials. Acids from decaying organic matter mix with groundwater and aid in dissolving rocks. Water can also combine with carbon dioxide from air to form carbonic acid. Carbonic acid in surface water and ground water easily dissolves some rocks—especially limestone and marble.

PHYSICAL WEATHERING The breakdown of rock into smaller pieces without chemical change is called **physical weathering.** Physical weathering occurs when frost action, plant root growth, or abrupt temperature changes cause rocks to crack apart.

Frost action breaks up rocks in a way similar to the formation of road potholes, as shown in Figure 9-1. When water freezes, its volume expands nearly ten percent. At temperatures below 0°C, water trapped in cracks freezes and expands, enlarging the cracks. After many cycles of freezing and thawing, the rock crumbles.

Physical weathering also occurs when rock particles grind against rock in a process called **abrasion.** Abrasion occurs as sediments moved by ice, gravity, running water, or air come into contact with other rocks. The characteristic rounded shape of rocks from river beds and beaches is caused by abrasion.

Factors Affecting the Rate and Type of Weathering

Many factors determine the rate and type of weathering that will occur in a given location on Earth's surface. These factors include variations in the location, composition, and size of rocks, as well as in local weather conditions.

EXPOSURE Both the rate and type of weathering are dependent on the exposure of rocks to air, water, and the actions of living things. Generally, the closer a rock is to Earth's surface, the faster it will weather.

PARTICLE SIZE Weathering results in the formation of rock particles of different sizes. As shown in Figure 9-2 on the next page, when the rock particles are smaller, the total surface area per unit volume exposed to weathering is greater. Therefore, the rate of weathering is faster. To understand the effect of increased surface area, consider what happens when you add sugar to a drink. Granulated sugar dissolves more quickly than an equal mass of sugar cubes because the granulated sugar has a greater surface area exposed to the liquid around it.

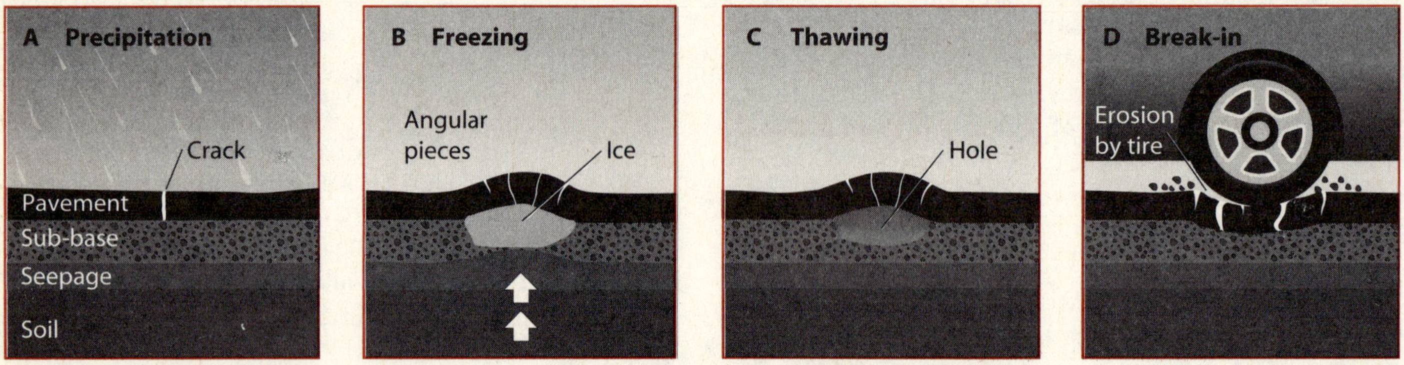

Figure 9-1. Formation of a road pothole by frost action: (A) In the winter, water can seep into cracks in pavement. **(B)** When the water freezes, it expands and pushes upward because the resistance is least in that direction. The pavement starts to crack into sharp, angular pieces. **(C)** When the water thaws, it seeps into the ground and leaves a hole. **(D)** The pavement breaks down further due to the weight of passing cars. The more cycles of freezing and thawing, the more damaging the frost action.

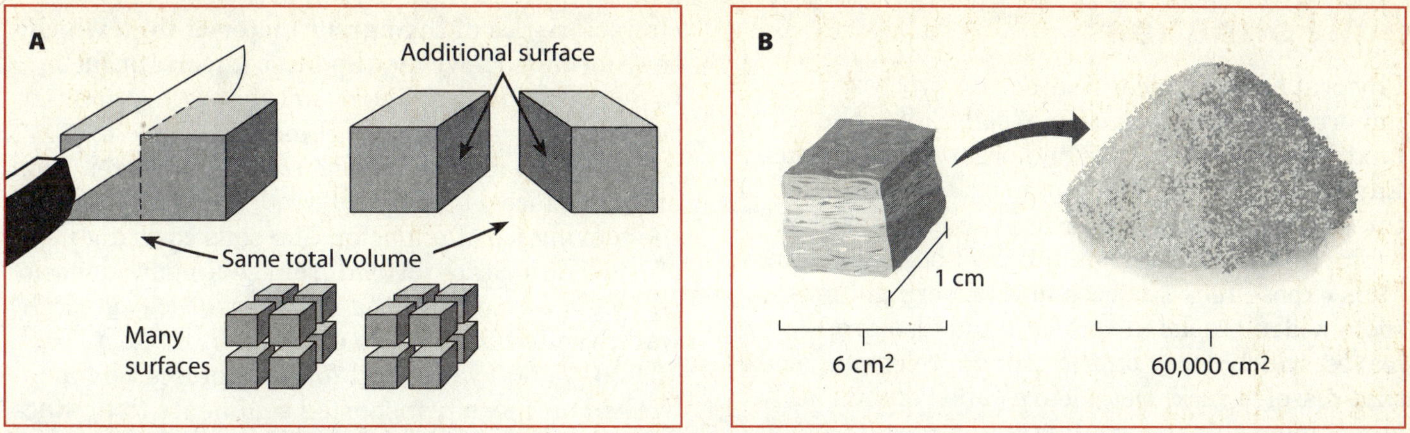

Figure 9-2. Explaining why surface area increases as particle size decreases: In diagram A, a single division of the block at the left exposes two new surface areas. If the two blocks at the bottom are divided again and again into smaller and smaller pieces, the total surface area will increase rapidly, although the total mass of material will remain the same. In diagram B, when a rock is broken up, the surface area of the material which forms it increases.

MINERAL COMPOSITION A rock's mineral composition affects the rate of weathering because different minerals have different physical and chemical properties. Some minerals, such as calcite in limestone and marble, weather rapidly because they dissolve in slightly acidic water. In contrast, the mineral quartz does not react chemically with most substances in the environment. Thus, quartz is largely unaffected by chemical weathering and is commonly found as sand in many environments.

CLIMATE As shown in Figure 9-3, the rate and type of weathering in a given location are greatly influenced by climate. Chemical weathering is most pronounced in warm, moist climates. Generally, the higher the average temperature and humidity, the more rapid the chemical weathering. In cold climates, frost action is the most common form of weathering. Frost action is especially intense in moist climates with temperature variations that lead to repeated cycles of freezing and thawing.

MEMORY JOGGER

You may recall from Topic 8 that the factors influencing climate include latitude, nearness to large bodies of water, ocean currents, and elevation.

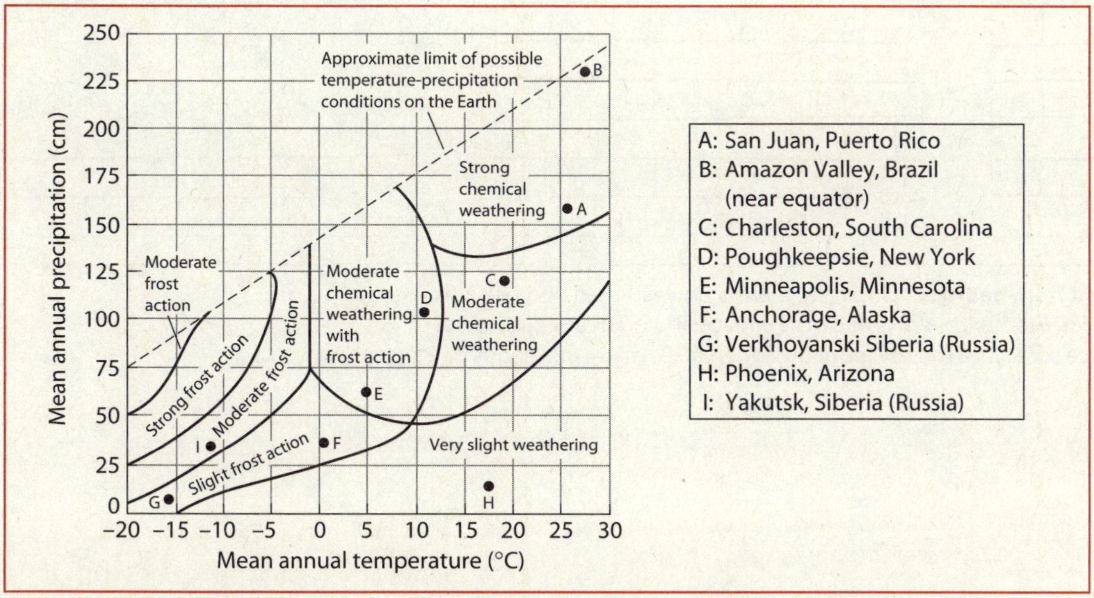

Figure 9-3. Dominant type of weathering for various climatic conditions: Notice that for an area with a mean annual temperature of 15°C and a mean annual precipitation of 100 cm, moderate chemical weathering is the dominant type. The upper left region of the graph represents extreme conditions that almost never occur on Earth.

Soil Formation

Physical and chemical weathering processes are important in the formation of soil. Soil is the mixture of rock particles and organic matter on Earth's surface that supports rooted plants. Living things play a significant role in soil formation. For example, the activities of ants and burrowing animals expose rock surfaces to weathering. Plant roots widen cracks in rocks. Plants also drop leaves, which decay on the ground, releasing acids that dissolve rock. Dead leaves and remains of other living things form humus—a part of soil that serves as a source of plant nutrients.

Soils can form in place from a parent material, such as bedrock. Over time, soils develop horizontal layers with a distinctive profile, as shown in Figure 9-4. Soils formed in place have some characteristics of the parent material from which they formed. However, the most important factor in determining the type of soil that forms in a particular place is the area's climate.

Soils can also be transported into an area from another place—usually by wind, moving water, or the moving ice of a glacier. The soils covering most of New York State formed from sediments deposited by glaciers during the most recent ice age, or by water melted from these glaciers. In the last 10,000–15,000 years, soil-forming processes have worked on these transported sediments to produce the present-day soils. Most of these soils have a soil profile similar to the immature soil shown in Figure 9-4. New York does have some true transported soils deposited by the flooding of rivers and streams.

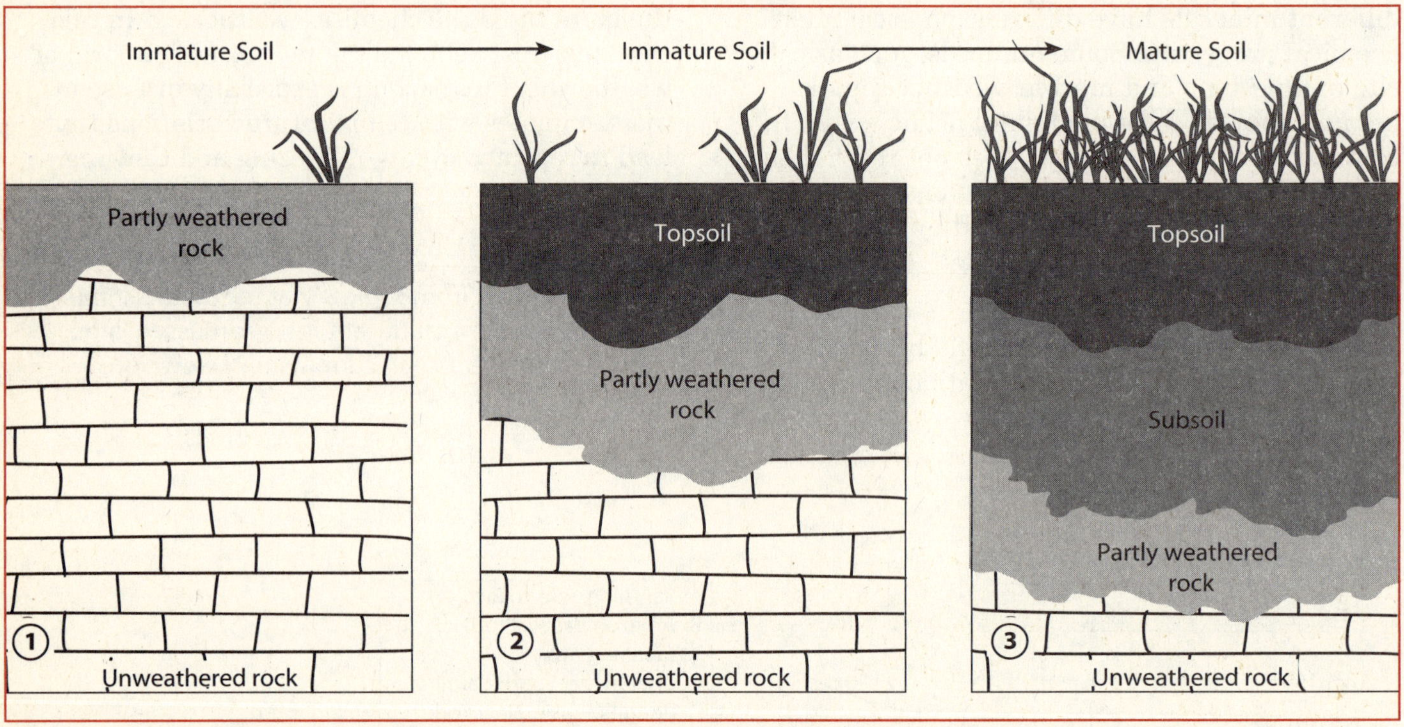

Figure 9-4. Soil development from bedrock: Originally, there is unweathered bedrock at the surface. Soil development, by way of weathering, starts at Earth's surface and gradually progresses downward over time to produce soil layers. The soil formation often takes many thousands of years.

Review Questions

1. What is the main difference between chemical and physical weathering? (1) Chemical weathering alters the composition of minerals, and physical weathering does not. (2) Chemical weathering increases the surface area of minerals, and physical weathering does not. (3) Physical weathering alters the composition of minerals, and chemical weathering does not. (4) Physical weathering increases the surface area of minerals, and chemical weathering does not.

2. Which is the best example of physical weathering? (1) the cracking of rock caused by the freezing and thawing of water (2) the transportation of sediment in a stream (3) the reaction of limestone with acid rainwater (4) the formation of a sandbar along the side of a stream

3. Water is a major agent of chemical weathering because water (1) cools the surroundings when it evaporates (2) dissolves many of the minerals that make up rocks (3) has a density of about one gram per cubic centimeter (4) has the highest specific heat of all common earth materials

4. In which climate would the chemical weathering of limestone occur most rapidly? (1) cold and dry (2) cold and humid (3) warm and dry (4) warm and humid

5. Which type of climate has the greatest amount of rock weathering caused by frost action? (1) a wet climate in which temperatures remain below freezing (2) a wet climate in which temperatures alternate between below freezing and above freezing (3) a dry climate in which temperatures remain below freezing (4) a dry climate in which temperatures alternate between below freezing and above freezing

6. A variety of soil types are found in New York State primarily because areas of the state differ in their (1) amounts of insolation (2) distances from the ocean (3) underlying bedrock and sediments (4) amounts of human activities

7. Describe the specific temperature and moisture characteristics of a climate in which frost action is dominant in the winter season.

8. The chemical composition of a soil formed in a certain area from the bedrock beneath it is determined by the (1) method by which the soil was transported to the area (2) slope of the land and the particle size of the soil (3) length of time since the last crustal movement in the area occurred (4) minerals in the bedrock beneath the soil and the climate of the area

9. What is the general relationship between the depth of a soil formed from underlying bedrock and the length of the weathering period responsible for its formation?

10. New York State soils are most likely composed of rock particles that have been (1) weathered from the underlying bedrock (2) weathered from the bedrock of northern Pennsylvania (3) transported by glaciers or water (4) transported by wind

Erosion

Weathering of rock produces particles that are transported as sediments through the process of **erosion.** Over time, erosion shapes and lowers Earth's surface. Sediments displaced from their source are evidence of erosion. Such sediments can be seen in muddy streams, on beaches, in rock fields near glaciers, along riverbanks, and in soils.

Transporting Systems of Erosion

Erosion involves a transporting system with several components:

- an agent of erosion, such as a stream, glacier, wave, current, wind, or human activity
- the sediments being moved
- a driving force

The driving force for most types of erosion comes from gravity. Rocks and sediments at higher elevations have gravitational potential energy, part of which changes into kinetic energy when gravity pulls them downward. The process is ongoing because heat energy input from Earth's interior raises landmasses upward.

Energy from the sun plays an important indirect role in erosion. The sun's energy drives the water cycle, producing precipitation that results in running water and glaciers. Insolation also "fuels" wind patterns that create ocean currents and waves.

GRAVITY EROSION As explained previously, gravity is the driving force behind erosion. It causes sediments to move downslope by way of running water, glaciers, and underwater currents. However, gravity can also act largely on its own by pulling rocks and sediments downhill in **mass movements.**

Mass movement—sometimes called direct gravity erosion—involves two major opposing forces. One force is the constant downward pull of gravity. The other force is friction, which attempts to keep objects in place. Rocks and other loose materials on Earth's surface slide downward when the forces holding them in place become weaker than the downward pull of gravity. For example, when a heavy rainstorm saturates the ground, frictional forces are greatly lowered. Under these conditions, mass movements often occur. Other conditions leading to mass movements include earthquakes, wave erosion, side cutting by streams, and road-building activities.

As shown in Figure 9-5, great variation exists among the types and speeds of mass movements from the slow downward movement of soil (soil creep) to the rapid fall of rock from a cliff. The rock piles found at the base of many steep slopes result from frost action and rock fall mass movements over time.

RUNNING WATER EROSION AND STREAMS

Running water is the most common agent of erosion on Earth's solid surface. As explained in the last topic, some liquid water from precipitation becomes runoff. Running water erosion often begins with the splashing of raindrops, which may move silt and clay a few centimeters. During rainstorms, thin sheets of water often flow over the land, also causing running water erosion. These sheets of water often become confined to a channel of a temporary stream valley called a rill, or gully.

When running water erosion is confined to a channel, a **stream** exists. In this book, water flowing through a channel—whether a large river or a narrow creek—is referred to as a stream. A smaller stream that flows into a larger stream is called a **tributary.** If water flows through a particular channel for the major part of the year, the stream is considered permanent.

Streams carry sediments in different ways. Dissolved minerals are carried in solution. Solid sediments of small size—including clay-sized sediments—are carried suspended in the water. Larger solid sediments are usually carried by rolling, sliding, or bouncing along the stream bottom, resulting in rounding of sediments by **stream abrasion.** The way in which a particular type of solid sediment moves varies with how fast the stream flows.

Process	Soil Creep	Debris Flow	Mud Flow (debris flow)	Rock Fall, or Debris Fall
Illustration	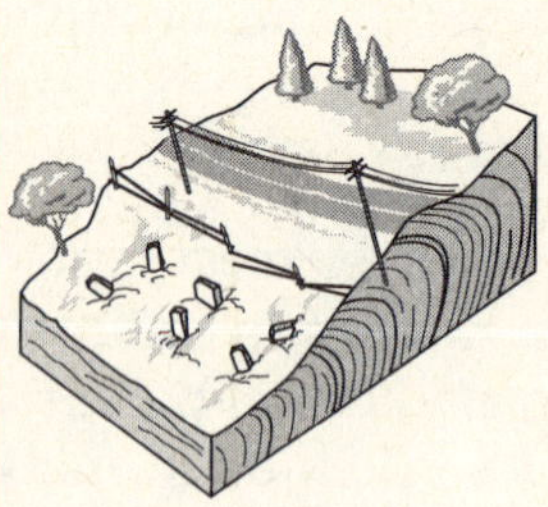	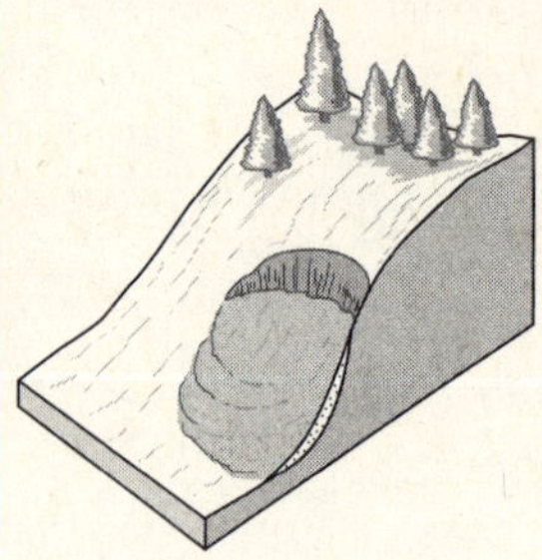		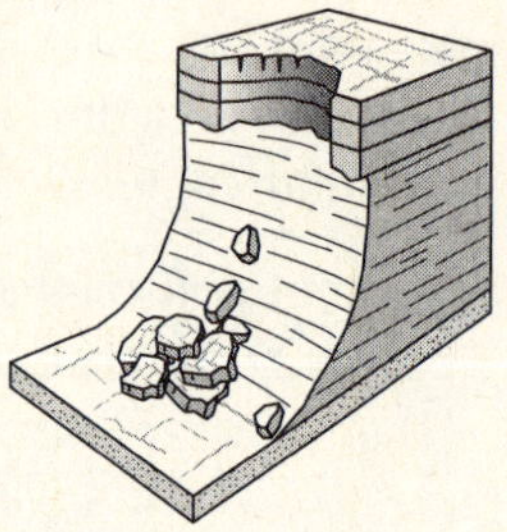
Description	Gradual downhill movement of soil	Rapid downslope plastic flow of a mass of debris	Downward flow of fine particles (mud) and large amounts of water	Rapid falling of pieces of rock from a cliff or steep slope
Velocity	Less than 1 cm/year	1 mm/day to 1 km/hr	1 to 5 km/hr	Greater than 4 km/hr
Slowest		**Increasing Velocities**		**Fastest**

Figure 9-5. Types of mass movement: Gravity causes many types of mass movements, from gradual soil creep to rapid rock falls.

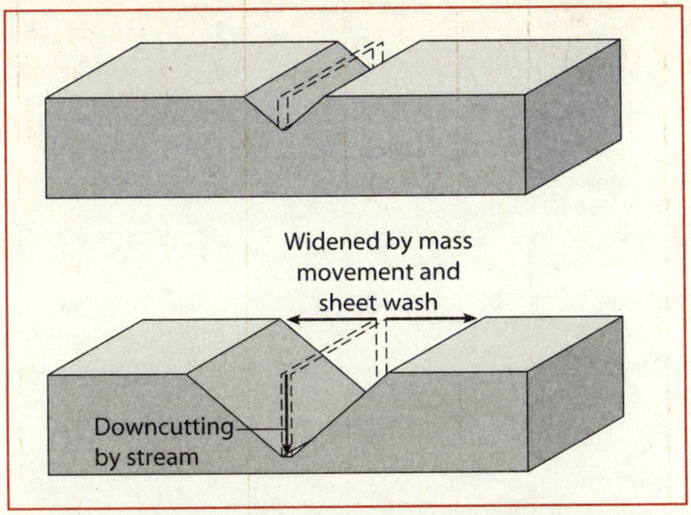

Figure 9-6. Formation of a stream's V-shaped valley: The action of stream downcutting alone would produce a very narrow vertical channel. However, mass movements and run-off erode the sides of a stream valley, creating the characteristic V-shape.

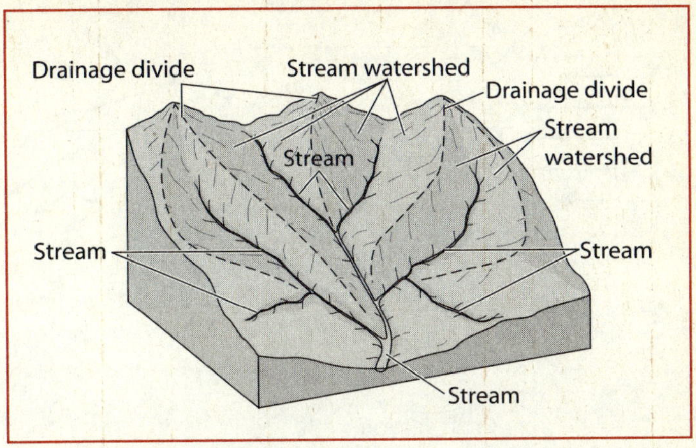

Figure 9-7. Stream watersheds or drainage basins: The portion of Earth's surface drained by one stream is a watershed.

Over time, streams carve deeper channels. As shown in Figure 9-6, a **V-shaped valley** forms. The V-shape results from the combined actions of stream downcutting, runoff, and mass movement, which causes the sides of the valley to "cave in." The area of land drained by any one stream is called its **watershed,** or drainage area. An example is shown in Figure 9-7. As streams vary in size, watersheds do also. A watershed can be as small as the size of this book or can cover a large portion of a continent—such as the area drained by the Mississippi River and all its tributaries. A region of higher land that separates one watershed from another is called a divide. In the United States, the Continental Divide follows the crest of the Rocky Mountains. Watersheds on the eastern side of the continental divide drain toward the Atlantic Ocean, while watersheds on the western side drain toward the Pacific Ocean.

Stream Velocity Three factors are most important in determining the average velocity of a stream. One factor is the gradient, or slope of the stream. Another factor is the discharge, or volume of water in the stream. Generally, as either the slope or the discharge of a stream increases, the velocity increases. The third factor involved is the **stream channel shape**—the shape of the bed of rock or loose materials that confine the stream.

Figure 9-8 shows how stream channel shape affects stream velocity. Notice that a wide, flat stream channel has a large surface in contact with moving water—and therefore, a large amount of

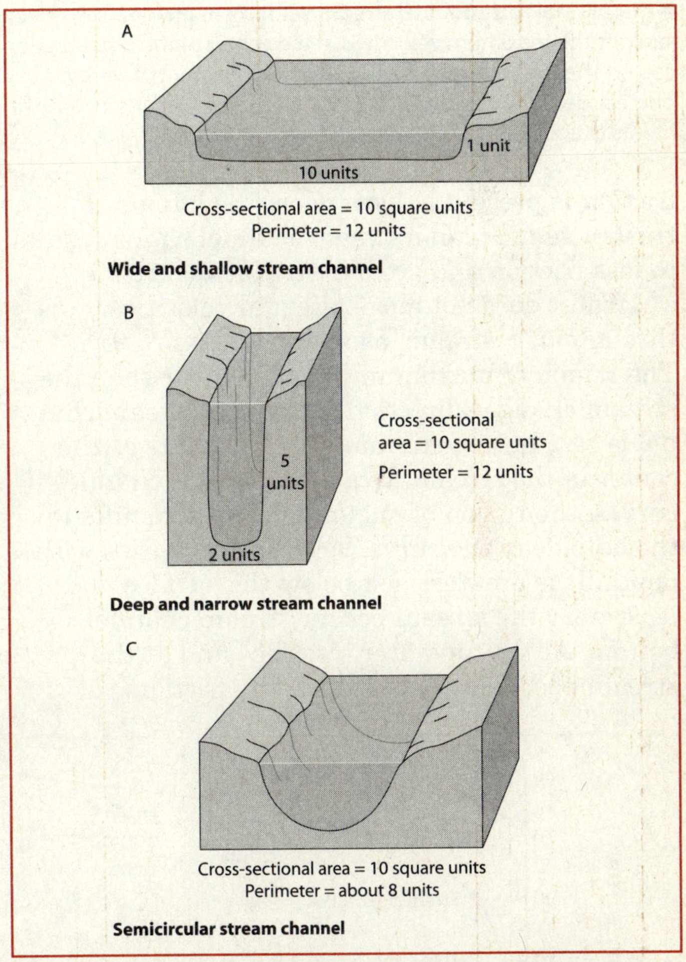

Figure 9-8. Effect of channel shape on stream velocity: Although the cross-sectional area of the three streams is the same, the semicircular shape in diagram C has less water in contact with the material of the stream channel. If all other factors for the stream system are the same, less contact means less friction and higher stream velocity.

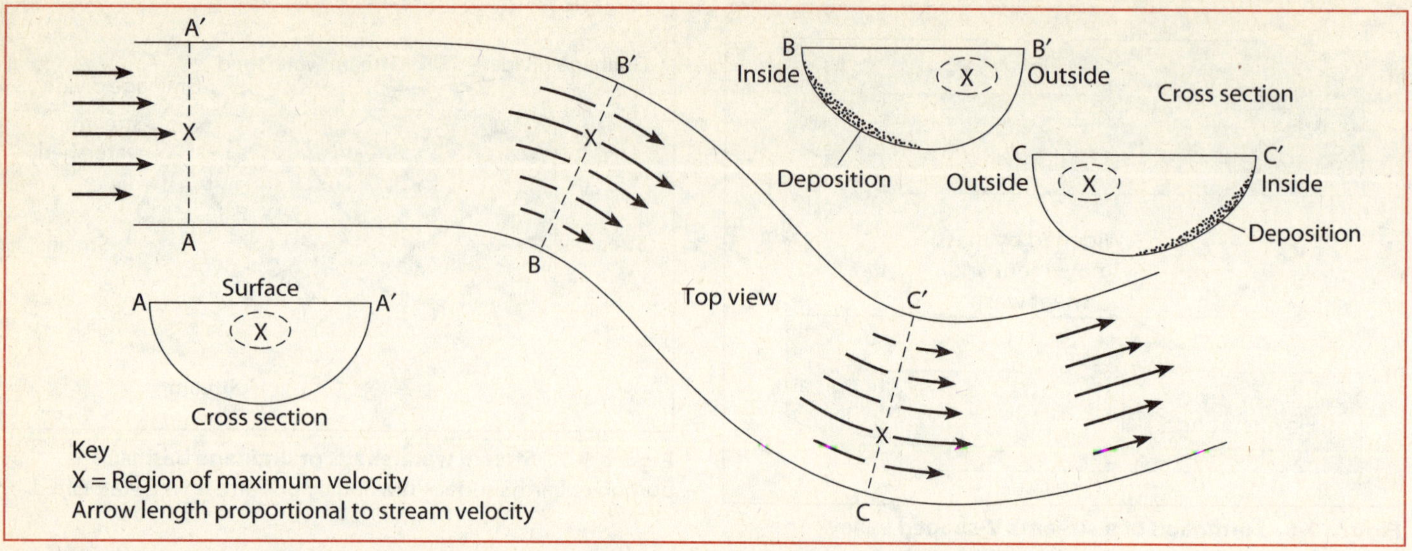

Figure 9-9. Variations in stream velocity in an idealized stream: (A) Where the stream course is straight the maximum velocity is at the center of the cross section. **(B)** and **(C)** Where the stream course changes direction, the location of maximum velocity moves toward the outside of the curve, or meander. Low velocity at the inside of curves results in deposition of the larger and denser sediments there. (Also see Figure 10-6 on page 180.)

friction is present to slow down the stream. In contrast, a semicircular stream channel shape results in less friction and greater stream velocity.

Water does not move at equal velocities throughout a stream, as shown in Figure 9-9. The region of maximum velocity shifts when the stream changes direction. When the stream channel is straight, maximum velocity occurs at the center of the stream. When the stream channel curves, the region of maximum velocity shifts to the outside of the curve. Velocity also varies with depth. It is greatest just below the surface and least near the stream bed, or stream channel bottom. At the air-water interface and at the stream bed, velocity is reduced by friction.

Generally, the solid sediments being transported by a stream move much more slowly than the stream itself. The greater the velocity of a stream, the larger the sediment particles it can carry, and the greater the velocity, the more total sediments it can carry. (See Figure 9-10.) Also see Relationship of Transported Particle Size to Water Velocity in the *Earth Science Reference Tables.*

A stream with greater velocity can carry larger sediment particles, as shown in Figure 9-10. For example, a stream moving faster than 800 cm per second can transport large boulders. In addition, streams with greater velocities can also move a greater total amount of sediments.

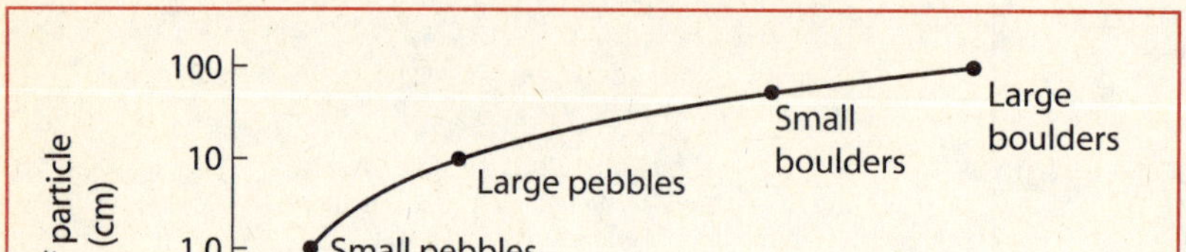

Figure 9-10. Relationship between stream velocity and particle size

Evolution of a Stream Streams slowly change their characteristics over time in response to changes in elevation, gradient, volume of water, and available sediments. Figure 9-11 shows a typical New York stream in various stages of its evolution. In its uppermost part, this stream starts at a **finger lake.** Finger lakes are bodies of water in U-shaped valleys carved by glaciers and often partly dammed at one end by a pile of glacially deposited sediments.

In the early stages of the stream, most of its energy is used to downcut a narrow V-shaped valley. There usually is a steep slope and many abrupt changes in elevation resulting in waterfalls and rapids. Farther downstream, more water from tributaries and from ground water increases the size of the stream.

As the stream grows larger, it begins to shift its course in a series of bends or turns called **meanders**. During flood stages, some meanders may form small lakes when deposits of sediments close off water flow from the new stream course. In shifting from side to side, the stream carves a wider valley. During times of flooding, the stream may flow out onto this wider valley and deposit sediments—forming a **flood plain.**

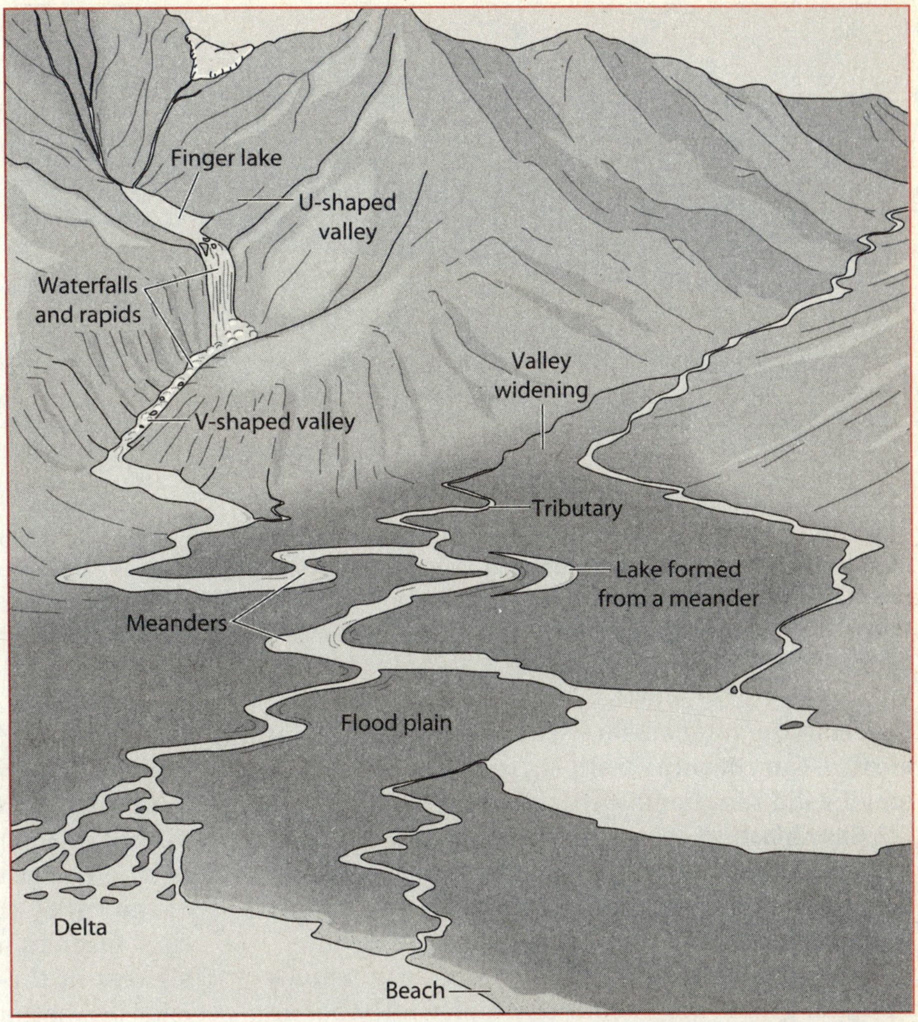

Figure 9-11. Features of an evolving stream: A stream's characteristics vary along its course from its source to its end, or mouth. Where the stream flows across easily eroded sediment, its channel bends from side to side in a series of meanders.

DIGGING DEEPER

Farmers often locate on flood plains because they have fertile soil for growing crops. The flatness of flood plains also makes them convenient for constructing buildings. However, the periodic flooding that occurs on flood plains can damage the crops and buildings.

Near the end or mouth of a stream, the discharge of water is often great, but there is little slope to the stream channel. At this stage the stream cuts a very wide valley and an equally wide flood plain. Deposition of sediments along the banks of a stream commonly accompanies frequent flooding—in some cases forming sediment mounds, called levees. Sediments are also deposited on the flood plain. At the very end of a stream, there is often a fan-shaped **delta.** A delta forms from sediments deposited over time from the ever-shifting channel at the stream's mouth.

Wind Erosion Another agent of erosion is wind. When loose sediments the size of sand or smaller are present, nearby solid surfaces are susceptible to erosion by wind. Arid regions, such as deserts, and lake or ocean coastlines are the most common places where loose sediments are available for wind erosion. There are two main aspects to wind erosion—deflation and sandblasting, or abrasion. Examples of each are illustrated in Figure 9-12 on the next page. Deflation takes place in areas with small, loose sediments exposed to the atmosphere. Winds readily blow away the loose sediments,

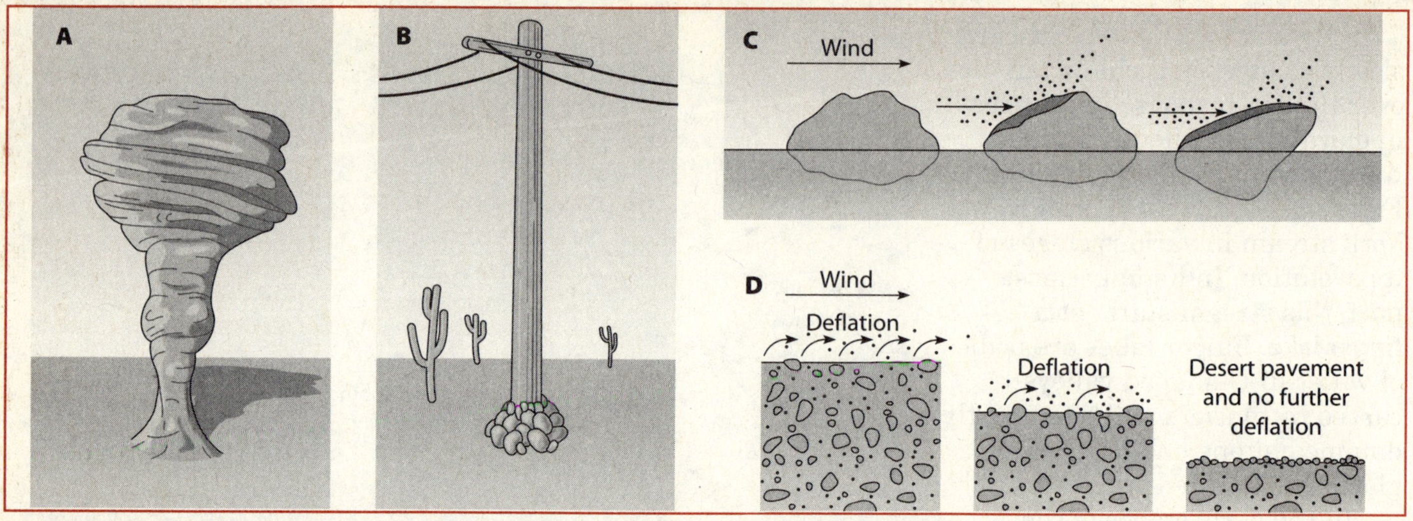

Figure 9-12. Features of wind erosion: (A) Abrasion by wind blown sand often sandblasts rocks—especially at lower levels—leaving a "mushroom rock." **(B)** The base of a utility pole has to be protected from sand blasting so it will not be "sawed" off. **(C)** Prevailing winds can abrade rocks into sediments with flat sides. **(D)** Wind erosion can lower an area's landscape by removing sand and smaller sediments.

lowering the land surface. The lowering of the land surface can continue until there are no longer any small solid sediments exposed to the winds.

Sandblasting occurs when winds blow sand or silt grains against rocks and other objects. The pelting by the grains can erode, or abrade, rocks in a manner similar to the process used to clean brick and stone buildings. Since wind can usually lift sand grains only about one meter into the air, the lower portions of exposed rocks become more eroded than the upper portions, as shown in diagram A in Figure 9-12. When sand grains collide, sandblasting also results in abrasion, and the frosted, or pitted, appearance of the sand grains themselves.

GLACIAL EROSION A **glacier** is a naturally formed mass of ice and snow that moves downhill on land under the influence of gravity. Mountain glaciers are found in high mountain valleys around the world. Ice-sheet, or continental, glaciers cover huge landmass regions. The area covered by a glacier has an ever-changing boundary determined by the accumulation of snow and ice and the wasting away of the glacier by melting and sublimation. If more snow and ice accumulates than wastes away, the glacier will advance. If the opposite occurs, the glacier will retreat. It should be noted that ice within a glacier is always moving in a particular direction, even if the glacier retreats as its overall boundaries shrink.

Glacial movement is by plastic flow. The glacial ice acts like a fluid, and many of its motions are like those of a stream. As a result, glaciers are often called "rivers of ice." Figure 9-13C on the next page shows that a valley glacier moves fastest in the middle and slowest at the sides. This is true because there is more friction at the sides of the valley than in the middle. In the cross-section, glaciers usually flow fastest near the top and slowest at the bottom because the greatest friction is at the bottom.

MEMORY JOGGER

You may remember from Topic 5 that sublimation is a phase change in which water vapor can change directly to ice. Through sublimation, glacial ice can waste away by changing directly from a solid to a gas without going through the liquid state.

Types and Features of Glacial Erosion When a glacier moves over the land, loose rocks and other materials beneath it freeze into the ice and are dragged along. The rocks and loose sediments frozen to the bottom of the glacier act like gougers and sandpaper on rock surfaces beneath them. These gouging and sanding actions produce **glacial grooves** and **glacial parallel scratches,** along with more loose sediments. The direction of exposed glacial grooves and scratches on bedrock shows the direction of former glacial movement.

Many of the sediments carried by a glacier are carried on or near its surface. These sediments are generally rock fragments broken off from valley walls by frost action and deposited onto the glacier

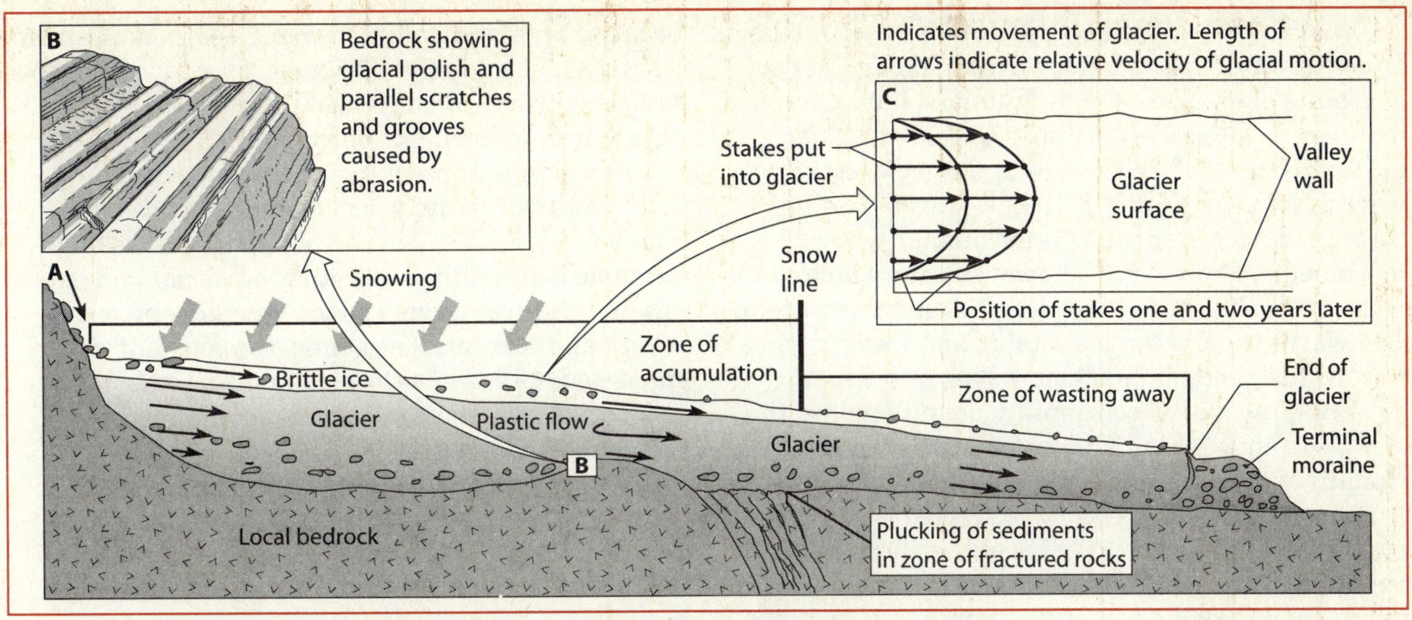

Figure 9-13. Movement of a glacier: (A) A glacier constantly flows forward. As a glacier moves forward, it transports sediments and abrades and plucks the bedrock beneath. The length of the arrows on the glacier shows comparative velocity of glacial motion. The greatest velocity is near the top of the glacier, and the slowest velocity is near the bottom. **(B)** Abrasion forms polished, scratched, and grooved bedrock. **(C)** As seen from above, glacial movement is fastest near the middle of the glacier and slowest at the interface with valley walls.

by mass movements. Often the rock fragments are very large—up to the size of a small house. Glacial sediments can be carried hundreds of miles. When deposited, they may differ greatly from the bedrock in their new location. Such contrasting sediments, called erratics, are found in much of the northern United States and most of New York State.

The wide, thick ice of a glacier erodes its confining valley walls as much as it does the rock beneath it. The result is a characteristic glacial **U-shaped valley.** The Finger Lakes of western New York occupy former V-shaped stream valleys that were altered to a U-shape by ice sheets during the last ice age. New York State also has other erosional features associated with continental and mountain glaciers from the last ice age, as shown in Figure 9-14.

DIGGING DEEPER

Found today in Greenland and Antarctica, continental glaciers covered much of North America during past ice ages. Continental glaciers are thick enough to cover everything except for the highest mountains. They flow outward from their centers like pancake batter poured on a griddle—along a downhill slope created by the thickness of the glacier itself and not by the underlying landscape.

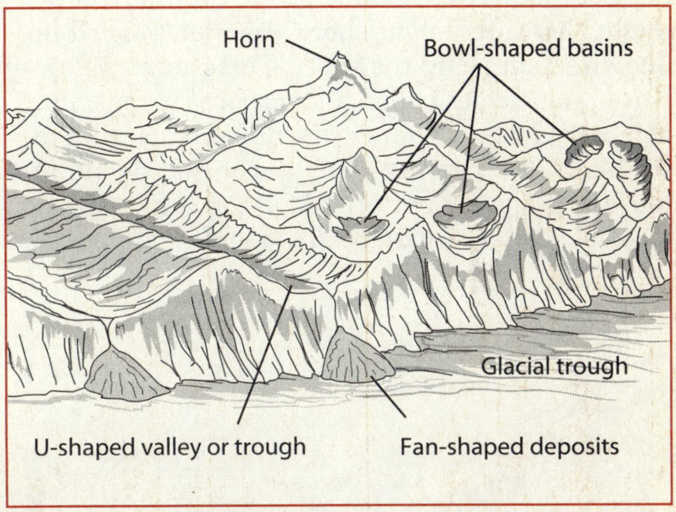

Figure 9-14. Some landscape features of glacial erosion: The sharp and angular features shown here are characteristic of mountain glaciers. The fan-shaped deposits (land deltas) are features of stream deposition. These mountain features can be seen in higher regions of New York State, such as the Adirondack Mountains, the Catskills, and the Hudson Highlands.

WAVE AND CURRENT EROSION Where lake or ocean waters meet land, waves and currents act as agents of erosion—creating unique landscape features. As described in Topic 7, winds transfer energy to the surface of lakes and oceans, creating waves. Notice in Figure 9-15 how waves passing through deep water cause brief circular water movements. However, when waves enter shallow water near the shore, they drag against the bottom. The waves then become unstable and water rushes toward the shore as **breaking waves,** or surf.

Breaking waves continually pound against the shoreline, unleashing a great amount of energy. Usually, waves arrive at an angle to the shore. Arriving waves often become refracted or bent, concentrating energy on parts of the shore that extend out into the water. Over time, these parts of the shore erode away. If the coastline is straight, the wave energy is more evenly spread out, and the waves pound the entire shoreline. If the shore has no depositional features to protect it, such as a beach or barrier island, wave action may carve a sea cliff with the aid of mass movement.

Because waves usually strike the shore at an angle, the waters near shore are often pushed in one direction along the shore. This causes a flow of water, sometimes called a longshore current, that can carry large amounts of sediments parallel to the shore. Over time, sediments on shore tend to move in a zig-zag pattern along the same direction as the longshore current. This movement occurs because breaking waves, bearing sediments, strike the beach at an angle. However, as water from the breaking waves washes back into the lake or ocean, it flows straight downhill.

In the zone of breaking waves, deposited sediments can create a series of mounds called a **sandbar.** Figure 9-16 on the next page shows an example of a sandbar and other erosional and depositional features formed at the shoreline by ocean waves and currents. Over time, the constant movement and abrasion of the sediments at a shoreline causes the sediments to become more rounded, smoother, and smaller.

Sediment Features and Erosional Agents

Each agent of erosion produces distinctive characteristics in the sediment it transports, as illustrated in Figure 9-17 on the next page. A solid sediment in wind, beach, or stream erosion will become smooth, rounded, and smaller. The longer the time the sediment is eroded, the smaller, smoother, and rounder it becomes. Wind-blown sediments are often more frosted (pitted) than stream sediments. Glacial sediments are only partly rounded (sub-rounded), and often have scratches of various sizes and directions on them. Sediments produced by physical weathering or moved directly by gravity are often very angular in shape.

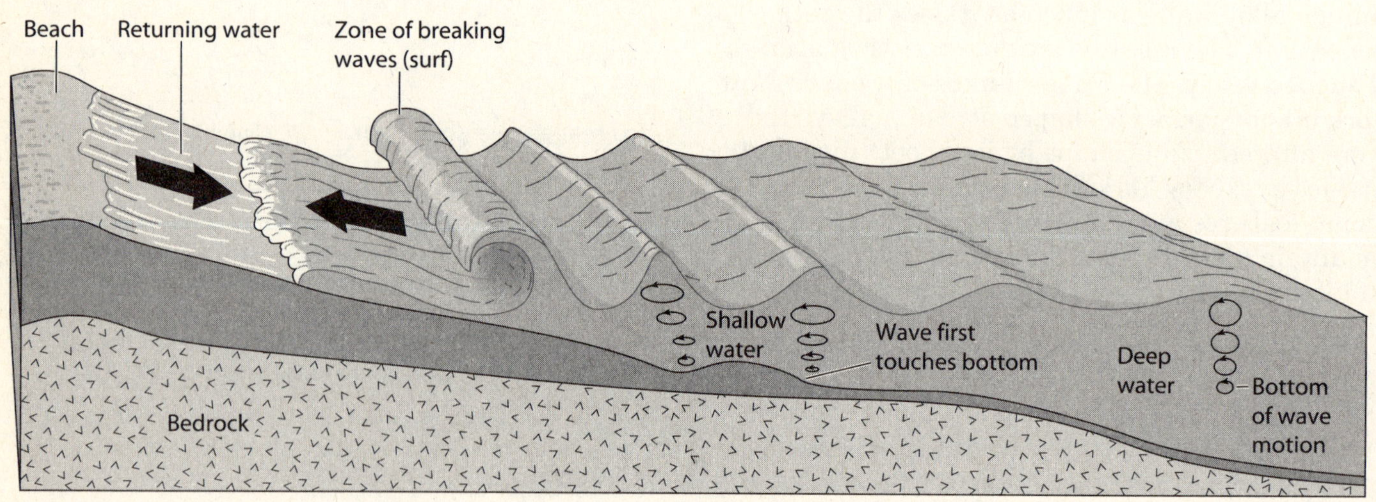

Figure 9-15. Motion of waves in deep and shallow water: When ocean waves move in deep water, there is no forward movement of the water—only circular movements that diminish rapidly with depth. When a wave comes near shore, the shallower water depth causes the wave to drag on the bottom. This dragging causes the water in the waves to actually fall forward as the waves bunch together, rise up, and break against the shore.

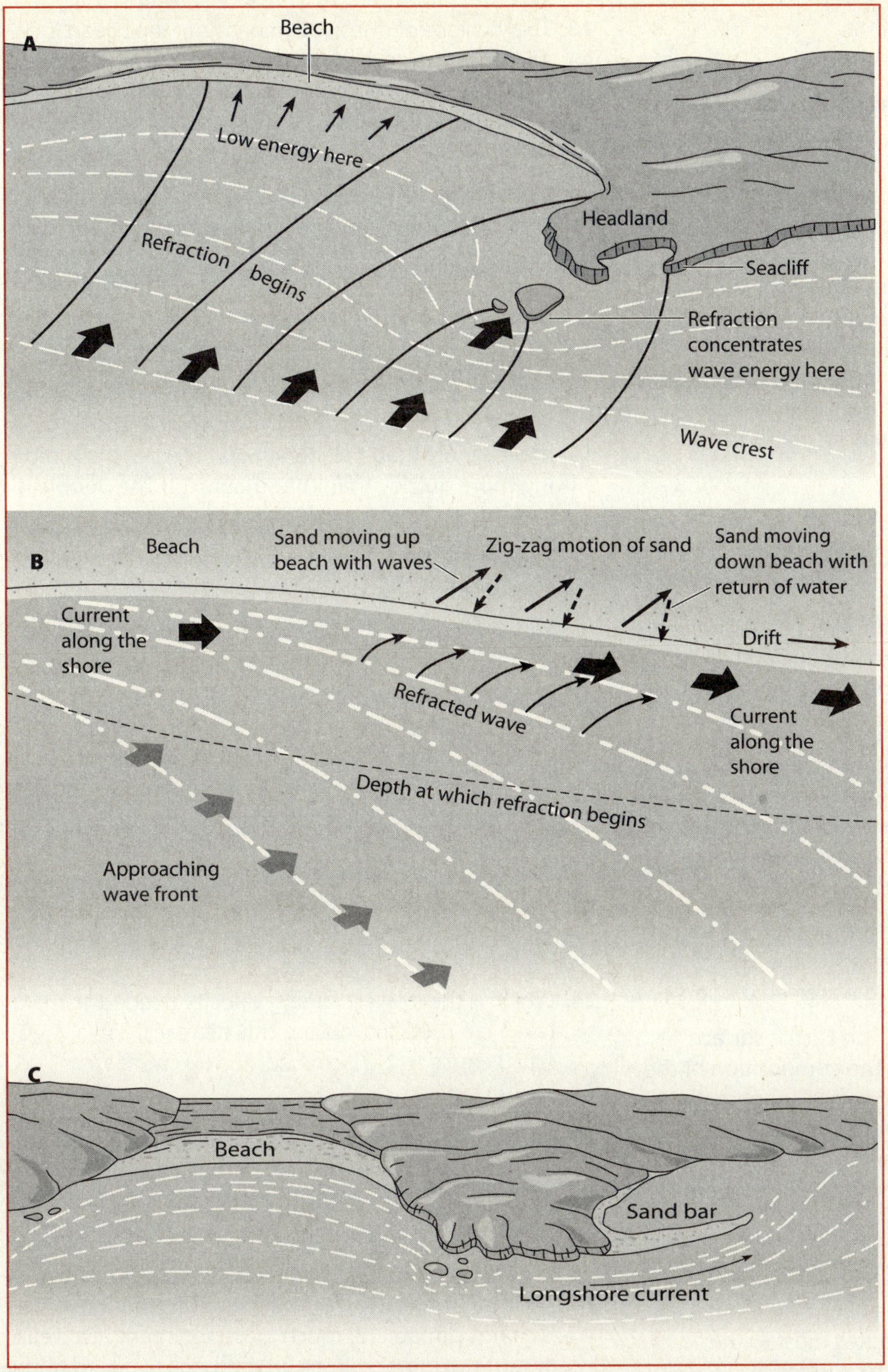

Figure 9-16. Erosional and depositional features of a shoreline: (A) Most waves strike a coastline at an angle and when the bottom of the wave drags, they are refracted, or bent. This refraction concentrates the energy of erosion on the areas of the coast that extend into the water. **(B)** The bent waves result in sand moving in a zig-zag pattern as they strike the coast. The angled water movement also results in a longshore current flowing along the shore in one direction **(C)** When the current slows, sediments are deposited, creating such depositional features as beaches and sandbars.

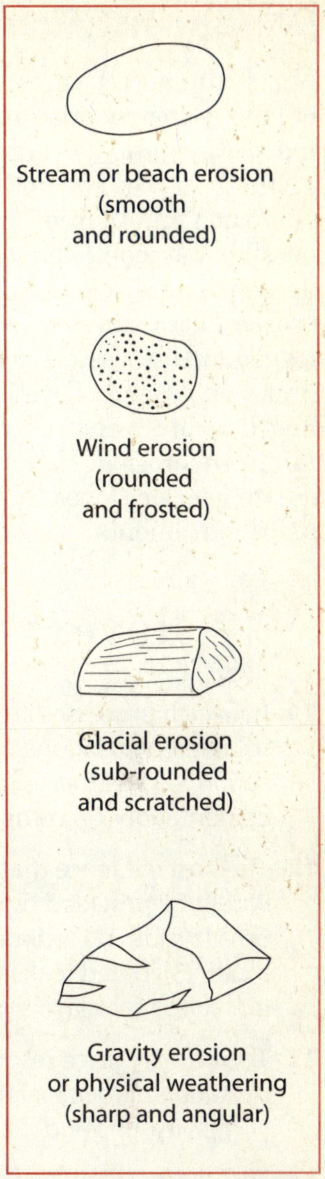

Figure 9-17. Sediment features reflect the erosional or weathering agents that produced or transported them.

People and Erosion

Many human activities greatly increase the erosion of land. Construction projects, road-building, mining, deforestation, overgrazing, and poor farming methods expose soil and loose materials to agents of erosion and thereby increase erosion rates. It is possible for people to make decisions to restore past damage and discontinue activities that increase erosion. For example, people can choose to replant logged forests, restore coastal vegetation, and set limits on grazing. Farming techniques such as no-till farming and contour plowing can be used to minimize erosion. Highways can be designed to create gentler slopes that are less likely to produce mass movements.

Review Questions

11. By which processes are rocks broken up and moved to different locations? (1) evaporation and condensation (2) weathering and erosion (3) burial and cementation (4) compaction and transportation

12. The best evidence that erosion has taken place would be provided by (1) deep soil formed in place on a hillside (2) sediment observed at the bottom of a cliff (3) folded rock layers observed on a mountain (4) faulted rock layers observed on a plateau

13. The primary force responsible for most of the transportation of rock material on the surface of Earth is (1) gravity (2) wind (3) running water (4) glaciers

14. The composition of sediments on Earth's surface usually is quite different from the composition of the underlying bedrock. This observation suggests that most (1) bedrock is formed from sediments (2) bedrock is reduced by weathering (3) sediments are formed from local bedrock (4) sediments are transported

15. Describe the relationship between mass movement and gravity.

16. Which agent probably contributes most to the general wearing down of Earth's surface? (1) wind (2) glaciers (3) running water (4) ocean waves and currents

17. The area drained by a river and its tributaries is called its (1) watershed (2) mouth (3) source (4) divide

18. The increase of dissolved materials in the ocean is primarily the result of the (1) abrasion of the ocean floor (2) transporting of material by rivers (3) melting of continental glaciers (4) deposition of materials by ground water

19. The velocity of a stream at a particular location is controlled mainly by the (1) elevation of the stream at the location (2) distance of the location from the source (3) slope of the stream channel at the location (4) amount of sediments carried at the location

20. A pebble is being transported in a stream by rolling. How does the velocity of the pebble compare to the velocity of the stream? (1) The pebble is moving more slowly than the stream. (2) The pebble is moving more quickly than the stream. (3) The pebble is moving at the same velocity as the stream.

21. The velocity of a stream depends on the stream's (1) slope, only (2) discharge, only (3) slope and discharge, only (4) slope, discharge, and stream channel shape

22. Which statement best describes a stream with a steep gradient? (1) It flows slowly, producing a V-shaped valley. (2) It flows slowly, producing a U-shaped valley. (3) It flows rapidly, producing a V-shaped valley. (4) It flows rapidly, producing a U-shaped valley.

23. Which is the largest sediment type that could be carried by a steam flowing at a velocity of 75 centimeters per second?

24. Which is the slowest stream velocity needed to maintain 1 cm particles moving downstream? (1) 50 cm/sec (2) 75 cm/sec (3) 100 cm/sec (4) 125 cm/sec

25. Much of the erosion by wind is caused by (1) pushing sediments over cliffs (2) abrasion by blowing sand (3) deposition of silt and clay in dunes (4) scraping and gouging deep depressions in Earth's surface.

26. Wind erosion is most active in regions that have (1) a hot climate (2) much loose sediment (3) weak or non-resistant bedrock (4) bedrock exposed to the atmosphere.

27. Which agent of erosion formed the long U-shaped valleys now occupied by the Finger Lakes in central New York State? (1) running water (2) ocean currents (3) wind (4) glacial ice

28. The mushroom rock illustrated in Figure 9-12A shows sign of sandblasting. Explain why the rock has been eroded more at the base than at the top.

29. Ocean waves do most of their erosion (1) at the bottom of deep ocean water (2) at and near the zone of breaking waves (3) at the highest parts of beaches (4) at the bottom of sand bars where sand is most dense

30. What is the most important reason that sediments at beaches are usually rounded and smooth? (1) The water dissolves the edges of the sediments. (2) The high heat at beaches partly melts the sediments. (3) The moving water causes sediments to abrade against each other. (4) The coastal storms move rounded sediments toward the beach from deep ocean areas.

Questions for Regents Practice

Part A

1. In addition to its effects on living organisms, acid rain may cause changes in the landscape by
   (1) decreasing chemical weathering due to an increase in destruction of vegetation
   (2) decreasing physical weathering due to a decrease in frost action
   (3) increasing the breakdown of rock material due to an increase in chemical weathering
   (4) increasing physical weathering of rock material due to an increase in the circulation of ground water

2. Chemical weathering will occur most rapidly when rocks are exposed to the
   (1) hydrosphere and lithosphere
   (2) mesosphere and thermosphere
   (3) hydrosphere and atmosphere
   (4) lithosphere and atmosphere

3. Which factor has the most influence on the development of soil?
   (1) climate
   (2) longitude
   (3) amount of rounded sediment
   (4) age of the bedrock

4. The following diagram represents a geologic cross section.

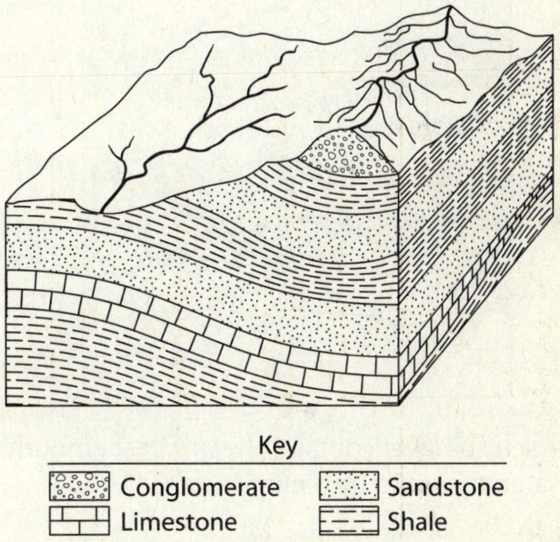

Which statement best explains why the conglomerate appears to be the most weathered bedrock?
   (1) The conglomerate particles are large.
   (2) The conglomerate is composed of quartz cobbles and pebbles.
   (3) The conglomerate underlies a steeper-sloped surface.
   (4) The conglomerate has been exposed to weathering the longest.

5. Which change would cause the topsoil in New York State to increase in thickness?

(1) an increase in slope
(2) an increase in biologic activity
(3) a decrease in rainfall
(4) a decrease in air temperature

6. The following diagram shows soil layers formed in an area of granite bedrock. Four different soil layers—A, B, C, and D—are shown.

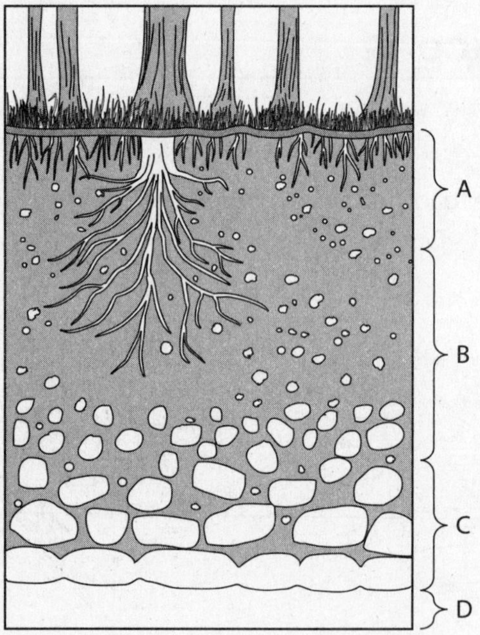

Which soil layer contains the greatest amount of material formed by biological activity?

(1) A (3) C
(2) B (4) D

7. Which is the best evidence that erosion has occurred?

(1) a soil rich in lime on top of a limestone bedrock
(2) a layer of basalt found on the floor of the ocean
(3) sediments found in a sandbar of a river
(4) a large number of fossils embedded in limestone

8. Which erosional force acts alone to produce avalanches and landslides?

(1) gravity
(2) winds
(3) running water
(4) sea waves

9. How are dissolved particles of sediment carried in a river?

(1) by bouncing and rolling
(2) by precipitation
(3) in solution
(4) in suspension

10. Which material would most easily be carried in suspension by a slow-moving stream?

(1) clay (3) sand
(2) silt (4) gravel

11. In summer, a small stream has a depth of 3 meters and a velocity of 0.5 meter per second. In spring, the same stream has a depth of 5 meters. The velocity of the stream in spring is more likely closest to

(1) 0.1 m/sec
(2) 0.2 m/sec
(3) 0.5 m/sec
(4) 0.8 m/sec

12. Two streams, A and B, carry the same volume of water, but stream A has a greater velocity. The most likely cause of this greater velocity would be that stream A

(1) has more tributaries
(2) has a wider streambed
(3) flows down a steeper slope
(4) travels over less resistant bedrock

13. Which graph best represents the relationship between the maximum particle size that can be carried by a stream and the velocity of the stream?

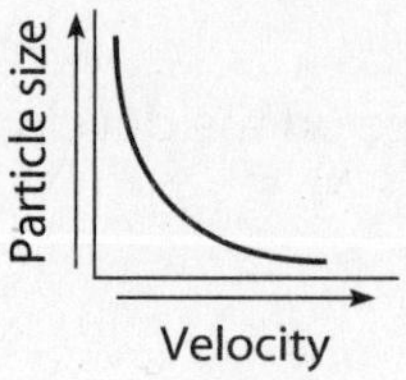

(1)

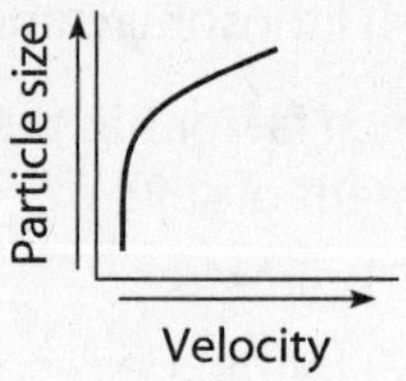

(3)

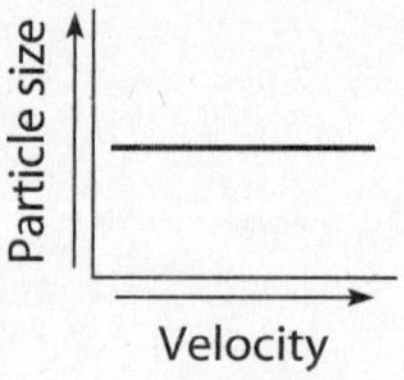

(2)

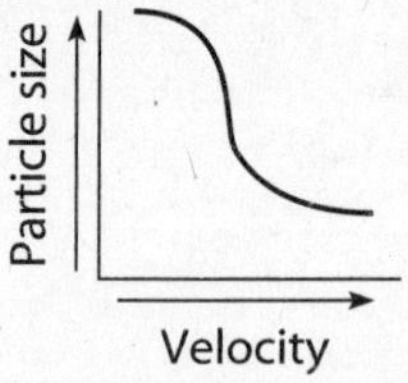

(4)

14. Which evidence best supports the inference that the movement of the ice sheet was generally from north to south over New York State during the Pleistocene glaciation?

(1) pieces of anorthositic rock from the Adirondacks found near Albany

(2) scratches aligned east to west on bedrock near Utica

(3) rocks of Devonian age found near Elmira

(4) the direction of flow of the Niagara River at Niagara Falls

15. The following diagram is a map view of a stream flowing through an area of loose sediments. Arrows show the location of the strongest current.

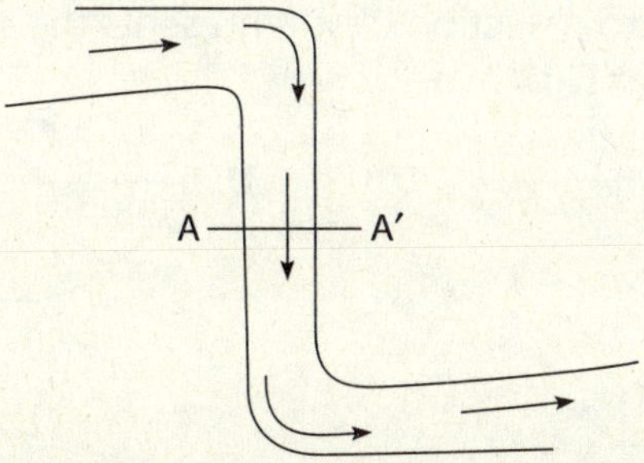

Which stream profile represents the cross section from **A** to **A′**?

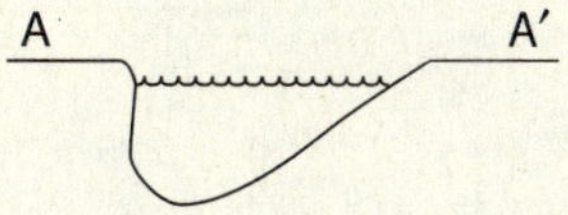

(1)

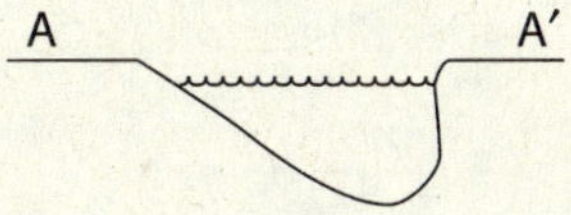

(3)

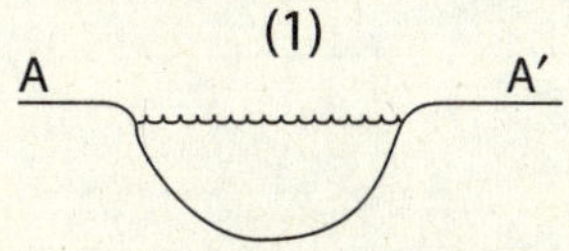

(2)

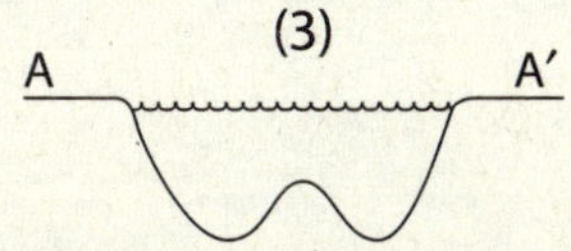

(4)

17. In the following diagrams, which rock fragment shows the least evidence of erosion?

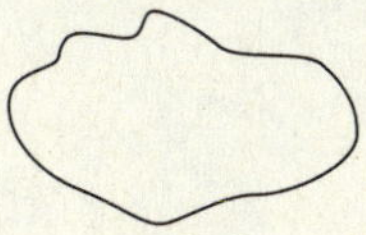

(1)

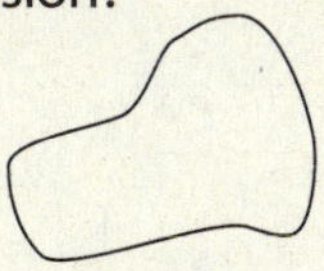

(3)

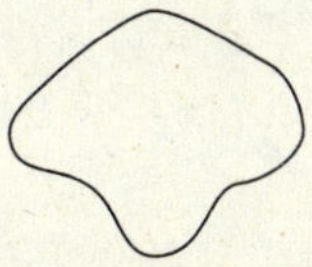

(2)

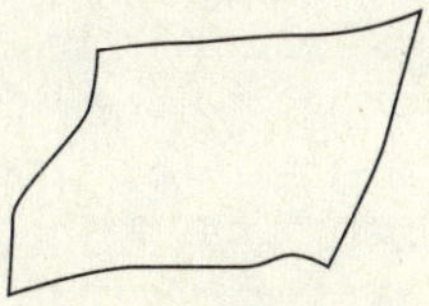

(4)

16. In the following diagram of a straight-flowing stream, the lengths of the arrows represent differences in relative stream velocity on the stream's surface.

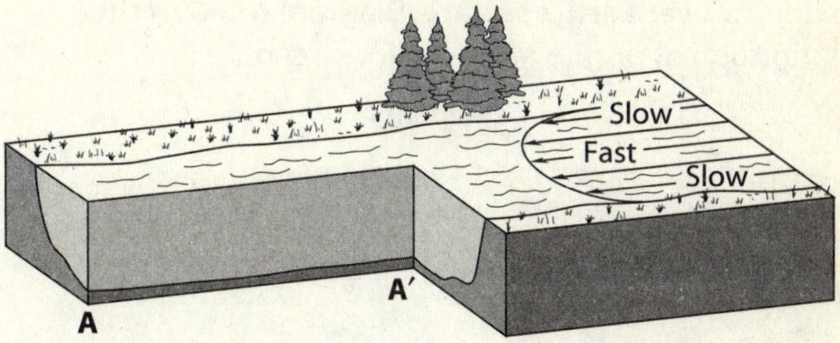

Which diagram best represents the relative stream velocity from the surface to the bottom of the stream for the cross section from **A** to **A′**?

(1)

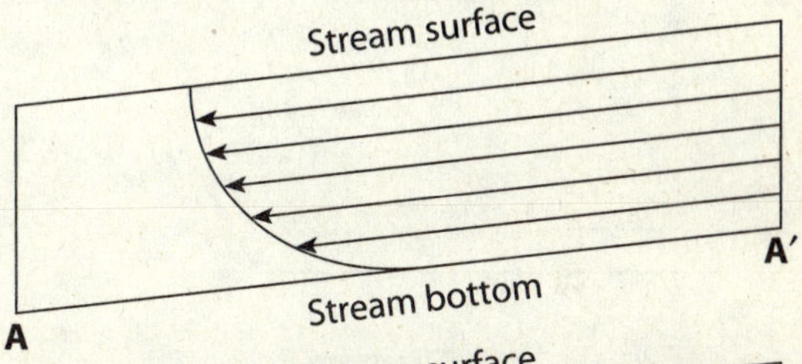

(2)

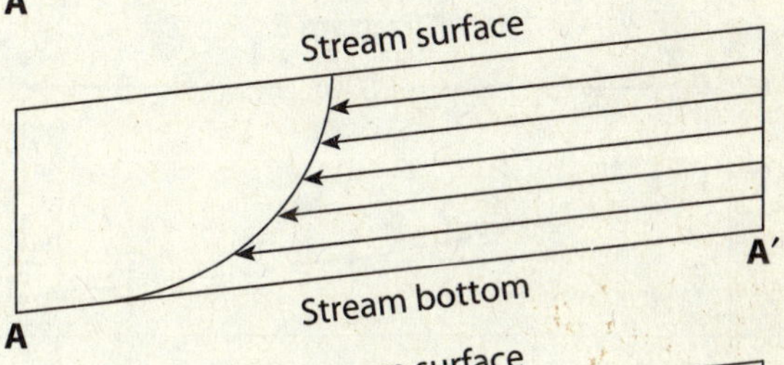

(3)

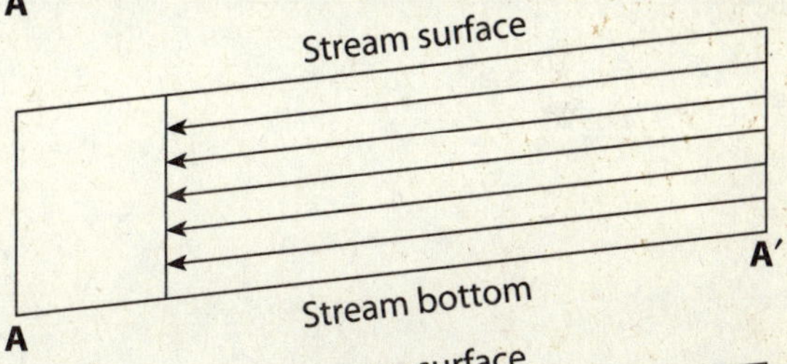

(4)

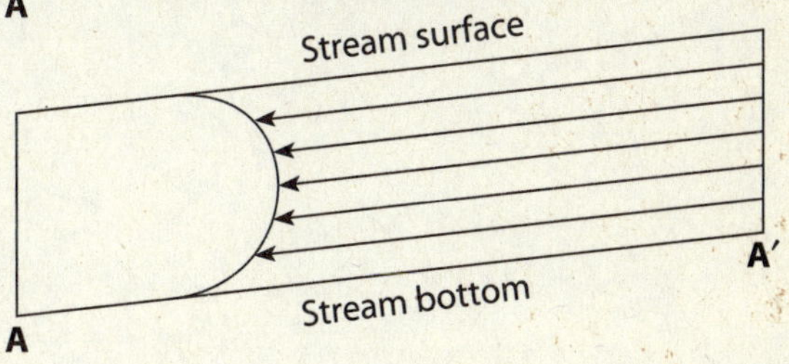

Part B

Base your answers to questions 18 through 21 on the following diagrams. Diagram A shows the paths of two streams over Earth's surface. Diagram B shows the longitudinal profile of the major stream.

Diagram A

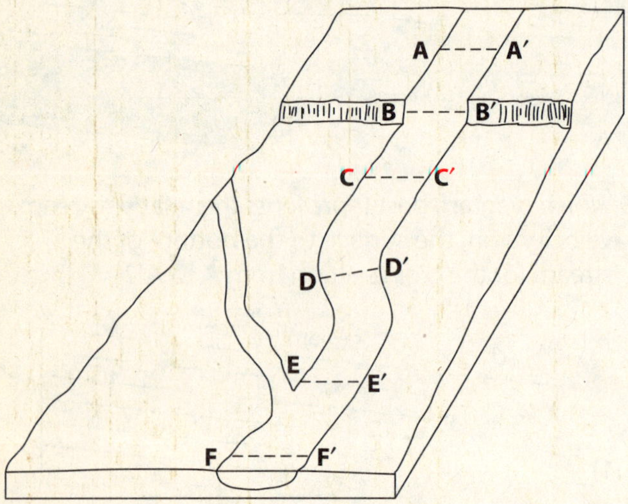

Diagram B

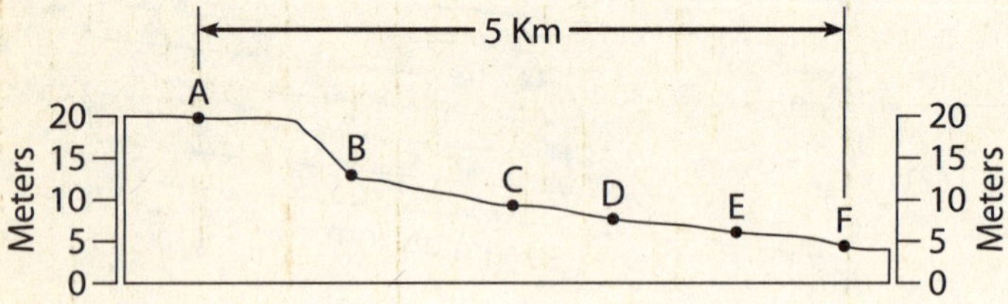

18. The following diagram shows the cross section of the stream at **C-C′**. At which position in the stream channel would the velocity of the water be greatest?

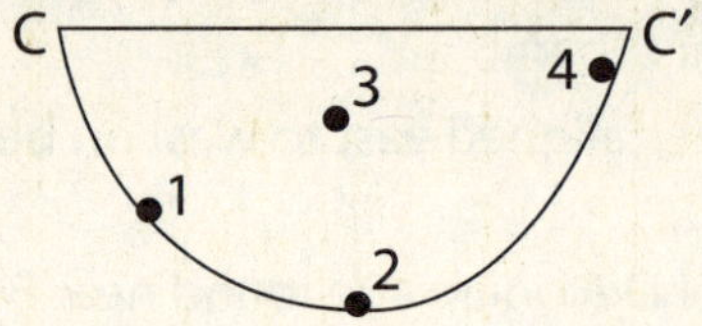

19. What is the approximate average gradient of this stream between points **A** and **F**? [1]

20. The greatest volume of water would most likely be moving past which location? [1]

21. Which cross-section best represents the shape of the stream at **D-D′**?

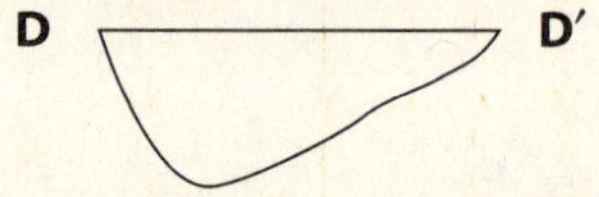

(1)

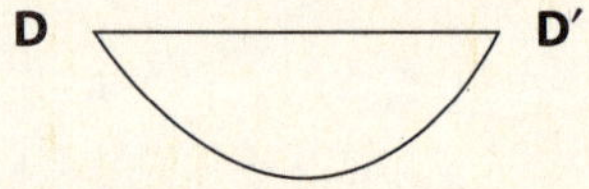

(3)

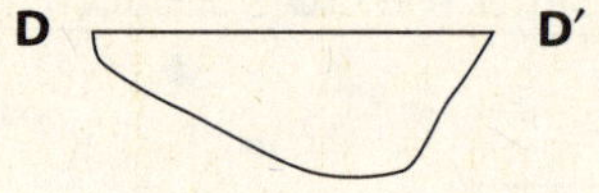

(2)

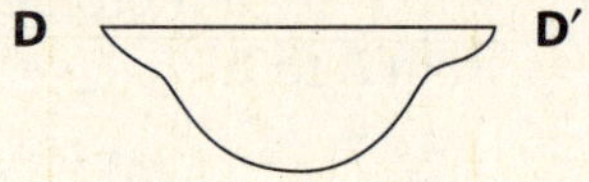

(4)

Base your answers to questions 22 through 27 on the following data table. Samples of three different rock materials, A, B, and C, were placed in different containers of water and shaken vigorously for 20 minutes. At 5-minute intervals, the contents of each container were strained through a sieve screening. The mass of the material remaining in the sieve was measured and recorded as shown in the data table.

Mass of Material Remaining in Sieve

Shaking Time (minutes)	**Rock Material A** (grams)	**Rock Material B** (grams)	**Rock Material C** (grams)
0	25.0	25.0	25.0
5	24.5	20.0	17.5
10	24.0	18.5	12.5
15	23.5	17.0	7.5
20	23.5	12.5	5.0

Following the directions in questions 22 through 24, use the information in the data table to construct a line graph on a grid such as the one shown.

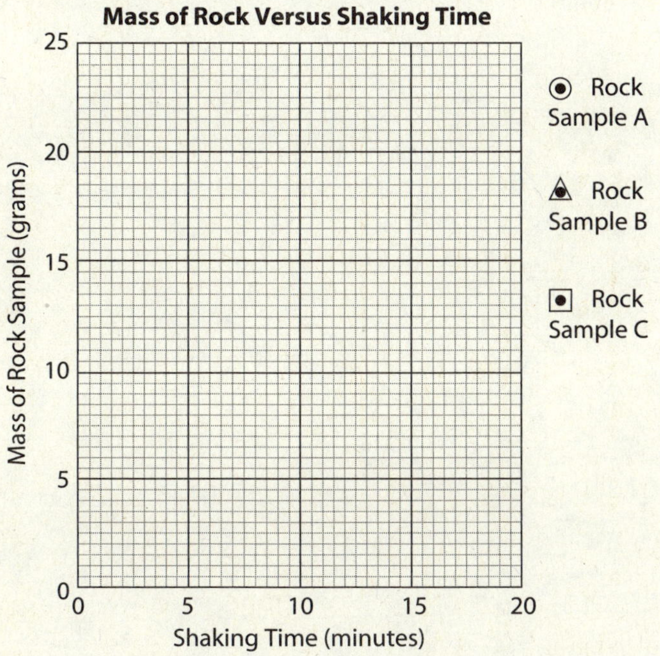

22. Plot the data for rock sample A for the 20 minutes of the investigation. Surround each point with a small circle and connect the points. [1]

Example: 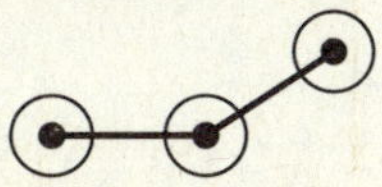

23. Plot the data for rock sample B for the 20 minutes of the investigation. Surround each point with a small triangle and connect the points. [1]

Example: 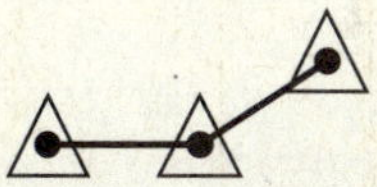

24. Plot the data for rock sample C for the 20 minutes of the investigation. Surround each point with a small square and connect the points. [1]

Example: 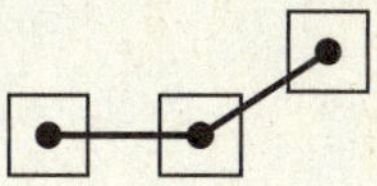

25. State the most likely reason for the differences in the weathering rate of the three rock materials. [1]

26. Describe the most likely appearance of the corners and edges of rock material C at the end of the 20 minutes. [1]

27. Calculate the rate of change in the mass of rock material C at the end of the 20 minutes. [1]

Part C

The following paragraph provides information about sediment. Use data from the paragraph and your knowledge of earth science to answer questions 28 through 36.

A cobble-sized sediment is naturally broken off the steep sides of a valley in a mountainous region. The sediment falls onto a glacier that occupies much of the valley. Over several years, the sediment "falls through" the glacier and is then dragged along the glacier's bottom. Next, the sediment is deposited in a pile of debris at the end of the glacier. A few years later, the sediment becomes part of a sand dune many miles from the glacier. The sediment is then transported to a stream from which, after a few months, it is deposited in a delta out in the ocean. Finally, the sediment becomes part of a child's sand castle on a beach tens of miles from the delta.

28. What natural cause broke the sediment off the side of the valley? [1]

29. What is the most likely shape of the sediment just after it fell onto the glacier? Explain the reason for your answer. [2]

30. Explain how the sediment can be considered a "tool" or aid to erosion while it is in the glacier and describe a landscape feature the sediment could help create. [2]

31. Tell how the sediment would most likely end up in a sand dune due to natural processes. [1]

32. Provide three reasons to explain why the stream would transport the sediment at varying speeds. [3]

33. Explain how the sediment could have naturally moved from the delta to the beach. [2]

34. What evidence in the paragraph suggests that a human was an agent of erosion for the sediment? [1]

35. Based on your reading of the above paragraph, list five natural agents of erosion that could have been responsible for moving the sediment. [2]

36. Describe three likely changes in the appearance of sediment as it was transported from the valley wall to the beach, and indicate the cause or causes of the changes. [3]

Deposition

TOPIC 10

VOCABULARY		
barrier island	**kettle lake**	**sand dune**
deposition	**moraine**	**sorted sediments**
drumlin	**outwash plain**	**unsorted sediments**

The process by which sediments are released, settled from, or dropped from an erosional system is **deposition,** or sedimentation. Deposition includes the releasing of solid sediments and the process of precipitation—the releasing of dissolved minerals, hardness, or salts from a water solution to form chemical sedimentary rocks.

The deposition of sediments has many effects on people. For example, you might play on a beach or marvel at the formations in caves (which are the product of precipitation of rocks). People mine sand and gravel from wave, glacier, and stream deposits. They also mine chemical sedimentary rocks, such as rock salt, rock gypsum, and dolostone. Most people spend a large part of their lives on sediment depositional features such as beaches, flood plains, deltas, marine coastal plains, and glacial deposits, or on the sedimentary rocks formed from deposition of sediments.

Deposition is part of a series of processes that start with the formation and movement of sediments by weathering and erosion as explained in Topic 9. In deposition the sediments are placed in locations where they may form beds, or layers, of sedimentary rocks. (See Topic 11.) Most final deposition occurs in large water bodies (such as lakes and oceans) because running water is the most important erosional system. Before sediments reach the large water bodies, many of them are temporarily deposited in different environments by running water, wind, glaciers, or other such means. Many of these depositional environments are shown in Figure 10-1.

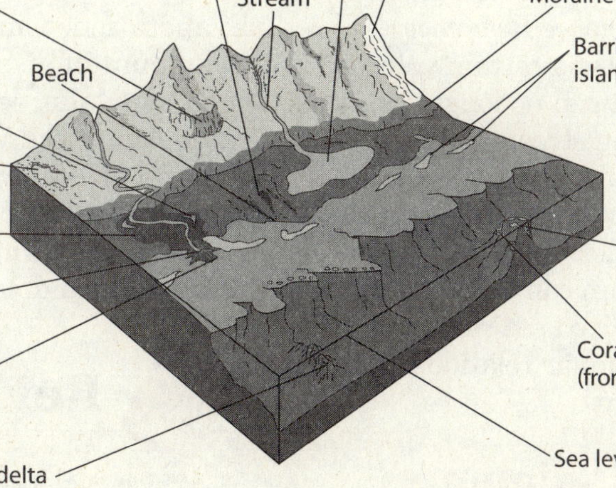

Figure 10-1. Types of depositional environments: The delta is formed from streams and the moraines from glaciers. Beaches, sandbars, and barrier islands are formed by ocean waves and long-shore ocean currents. Sand dunes are formed by wind. Some sloped deposits are formed by mass movement (rock falls). Underwater deltas are formed by sediment-laden density currents.

MEMORY JOGGER

You may recall the types of solid sediments—such as sand, silt, and clay—from Figure 9-10 or from the Scheme for Sedimentary Rock Identification in the *Earth Science Reference Tables*.

Factors Causing Deposition

Deposition usually occurs when the velocity, or speed, of the stream, wind, or other erosional system decreases or just stops moving. Think of a piece of paper that has been picked up by a gust of wind. The piece of paper may blow all over, but when the wind slows down or stops blowing the paper will settle or be deposited. The time it takes for sediments to be deposited—the rate of deposition—varies. The faster the rate of deposition, the less time deposition takes; and the slower the rate of deposition, the more time deposition takes. Two types of factors—the velocity of the erosional system and characteristics of the sediments themselves—affect the rate of deposition.

Velocity of an Erosional System and Rate of Deposition

The faster a stream flows, the larger size sediments it can carry. Refer to Figure 9-10 on page 164 and the Relationship of Transported Particle Size to Water Velocity in the *Earth Science Reference Tables*. If the stream flows below a certain velocity, it will deposit the sediments it can no longer carry. As an example, if a fast moving stream flowing at 350 cm/sec (78.3 miles per hour) slows down to 50 cm/sec (11.2 miles per hour) it will deposit all the pebbles, cobbles, and boulders it may be transporting. At 50 cm/sec the stream can continue to transport sand, silt, and clay-sized sediments of average density and shape. Similar relationships exist for other erosional systems.

Characteristics of Sediments and Rate of Deposition

Many aspects of the sediments themselves affect how fast they will be deposited in air or water environments. These factors include size, shape, and density, and saturation of dissolved minerals.

SIZE All other factors being equal, the larger sediments settle out first when wind or running water slows down. This occurs because the larger sediment particles are heavier and therefore sink faster. Very small particles, such as clay-sized particles (less than 0.0004 cm in diameter), may remain suspended in water almost indefinitely. Clay suspended in water is the "mud" of muddy or cloudy water.

SHAPE The shape of a particle helps determine how fast it will be deposited from wind or running water. All other factors being equal, the more rounded a sediment, the faster it will settle out. The more flattened it is, the greater its resistance—caused by friction—to deposition.

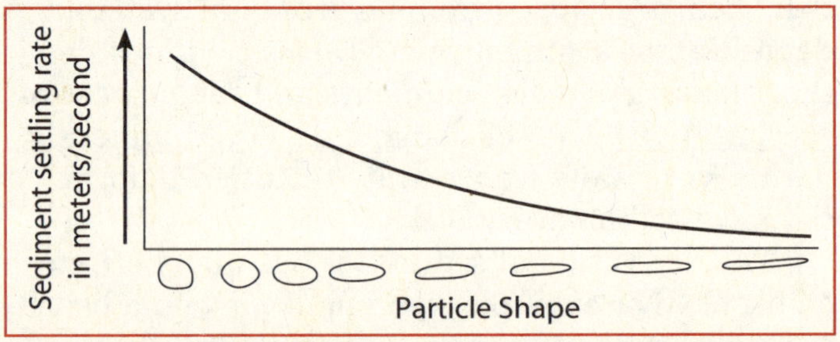

Figure 10-2. Settling rates of sediments of various shapes: As sediments of equal volume and density become less rounded and more flattened, it takes more time for them to settle, so the settling rate decreases.

DENSITY All other factors being the same, the higher the density of a sediment, the faster it will settle out of air or water. If two particles have the same size and shape, the denser one will be heavier.

SATURATION OF DISSOLVED MINERALS Evaporation, temperature changes, or increases in amount of dissolved minerals in a water body such as a lake, sea, or ocean may result in a saturated condition. When this happens, the dissolved mineral or minerals will settle or precipitate out of the dissolved condition and crystallize. As a result, rocks composed of one mineral, such as rock salt and dolostone, may form.

Review Questions

1. Which evidence best supports the idea that a vast inland sea once covered the Great Plains area of the United States? (1) considerable erosion has occurred there (2) extensive igneous intrusions are presently exposed there (3) extensive sedimentary rock layers have been formed there (4) numerous earthquakes occur there

2. A stream is carrying sediment particles ranging from 0.0004 to 25.6 centimeters. When the stream's velocity decreases from 300 to 100 centimeters per second, the stream will most probably deposit (1) silt and clay (2) sand and silt (3) pebbles and sand (4) cobbles and pebbles

3. When a river enters the ocean, sediment is deposited. Describe the change in kinetic energy of the river that leads up to this event.

4. The rate at which particles are deposited by a stream is least affected by the (1) size and shape of the particles (2) velocity of the stream (3) stream's elevation above sea level (4) density of the particles

Sorting of Sediments and Deposition

During deposition, sediments of similar size, shape, or density get separated or sorted by types. At any one time the majority of sediments being moved are similar in density and shape, and thus sediment size is the most common type of sorting performed by most agents of deposition. If a deposit or layer of sediment has particles that are similar in size (or density, or shape), they are considered **sorted sediments.** The greater the similarity of size, the more sorted the sediments are said to be. If sediments are very mixed in size (or density, or shape), they are considered **unsorted sediments.**

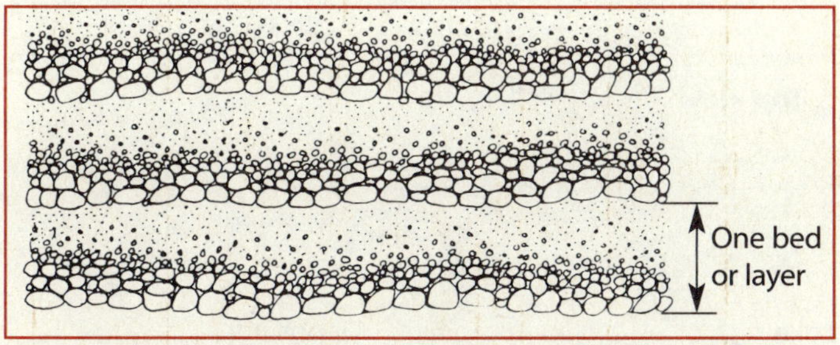

Figure 10-3. Layers of sediment with graded bedding: In each bed or layer, sediment sizes decrease from bottom to top. Graded bedding results from rapid deposition in an erosional system, such as by a flooding stream. Each bed is the result of a single flood by a stream.

When a mixture of sediment sizes in water settles out rapidly, a horizontal bed, or layer, develops with the sediment size decreasing from the bottom to the top. Such an arrangement in a sediment layer is called "graded bedding." (See Figure 10-3.) Graded bedding is most often associated with sediment-laden density currents. These currents are most common on the sloped ocean bottoms off the coasts of continents, and on a lake where flooding streams rapidly decrease in velocity when they enter.

When the velocity of a wind or water erosional system gradually decreases, such as when a stream flows into the ocean at a delta, the larger, denser, and more rounded sediments settle out first. This results in layers with horizontal sorting, in which the sediment size, roundness, and density generally decrease in the direction toward which the erosional system was moving. (See Figure 10-4.) Horizontal sorting is a major reason why most sediment deposits are sorted.

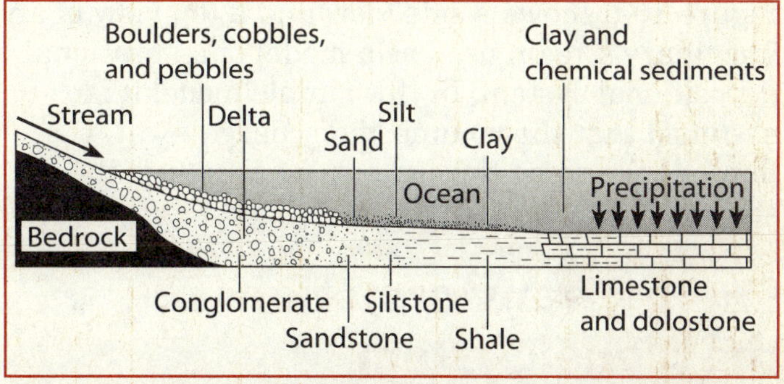

Figure 10-4. Horizontal sorting where a stream enters the ocean or lake forming a delta: The larger sediments settle first (nearer the shore). Sediments become smaller as distance from the shore increases. Precipitation nearer the shore may provide the cement for the formation of the types of sedimentary rocks indicated.

Unsorted Glacial and Mass Movement Deposits

In a solid erosional system such as a glacier, sediments of all sizes, shapes, and densities are deposited together when the glacier melts. This deposition results in the unsorted and unlayered characteristic of direct glacial deposits that cover much of New York State. (See Figure 10-5.)

Figure 10-5. Unsorted glacial deposits: In direct glacial deposition, there is no fluid medium in which sediments can become sorted. Thus, direct glacial deposits are characterized by a random distribution of sediment sizes (unsorted) and no bedding, or layering.

When there is mass movement, such as a rock fall, landslide, or avalanche, the sediments are usually dumped together in a random deposit that is unsorted and unlayered, similar to glacial deposits. Remember the jumbled masses of ice, rock, snow, and trees that you have observed in avalanches on television. You may have observed another example—a pile of sediments at the base of a cliff along a highway.

A Model of an Erosional-Depositional System

Figure 10-6 shows a side view and a top view of an imaginary stream used as a model of an erosional-depositional system. In this simple model it is assumed that the volume of discharge—water flow—is the same throughout the length of the stream.

Energy Transformations in the Model System

At the source, or beginning, of the stream the system has a maximum of potential energy. (See A in Figure 10-6.) As the stream flows toward its mouth, or end, potential energy is continuously being transformed into kinetic energy. (See D in Figure 10-6.) The kinetic energy is at the same time being lost to the environment in the form of heat produced by friction. Where the slope of the stream is steep, the transformation of energy occurs most rapidly, the stream has its greatest velocity, and the system has its greatest kinetic energy. Where the slope is small, the rate of energy transformation decreases, the stream slows down, and the kinetic energy of the system decreases. At the mouth of the stream, the velocity drops to zero, and the system has zero kinetic energy. Since the system has less potential energy because of its lower elevation, there has been a net loss of energy between the source and the mouth.

EROSION AND DEPOSITION IN RELATION TO ENERGY CHANGES Wherever the kinetic energy of the system is large, erosion is the dominant process. Where the kinetic energy is small, deposition is the dominant process. Thus, erosion occurs in regions of steep slope or high discharge, and deposition occurs in regions of gentle slope or low discharge. Deposition is particularly rapid at the mouth of the stream, where the kinetic energy becomes zero. Stream velocity is faster at the outside of curves, or meanders, and slower at the inside. Therefore, erosion usually occurs at the outside of meanders, and deposition occurs at the inside.

If one or more tributaries were to enter a stream, this would increase the mass or discharge of water, thus increasing the potential energy of the stream. The velocity of the stream would increase, causing erosion to increase and deposition of solid sediments to decrease.

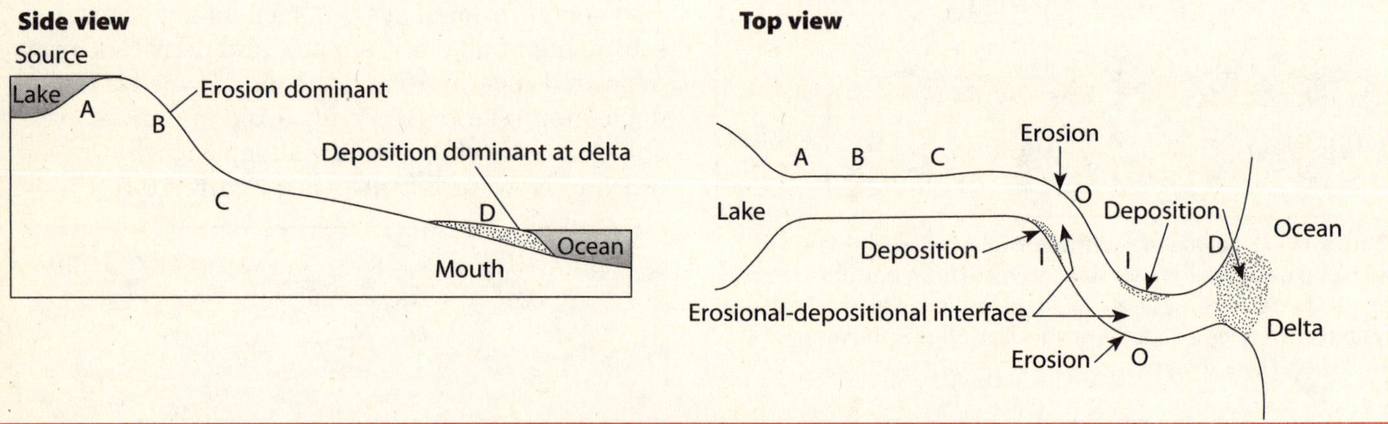

Figure 10-6. Model of a stream representing an erosional-depositional system: O indicates the outside of a curve or meander and **I** indicates the inside. Some deposition may occur at **C,** where the slope and velocity decrease. Erosional-depositional interfaces will then exist between **B** and **C,** between **C** and **D,** and between each pair of **O**'s and **I**'s. Also see Figure 9-11.

EROSIONAL-DEPOSITIONAL INTERFACES

Since there are regions of erosion and regions of deposition along the length of the stream, an interface between an erosional and a depositional state can often be located in the system. Interfaces between erosion and deposition exist at the meanders in the model stream, and between the source and the mouth of the stream. They may also be found where changes in slope occur, as between points B and C in Figure 10-6 on the previous page.

DYNAMIC EQUILIBRIUM OF AN EROSIONAL SYSTEM Since all sediments picked up by the stream during erosion must eventually be deposited, the system is in a state of dynamic equilibrium. Although erosion and deposition are occurring continuously, the rate of erosion equals the rate of deposition by the system as a whole. If a flood occurs, a stream will erode—pick up and transport—more sediment, but it will also deposit an equally increased volume of sediment, thus establishing a new balance of equilibrium.

Review Questions

5. Small spheres that are identical in shape and size are composed of one of four different kinds of substances: A, B, C, or D. The spheres are mixed together and poured into a clear plastic tube filled with water. Which property of the spheres caused them to settle in the tube as shown in the diagram below? (1) their size (2) their shape (3) their density (4) their hardness

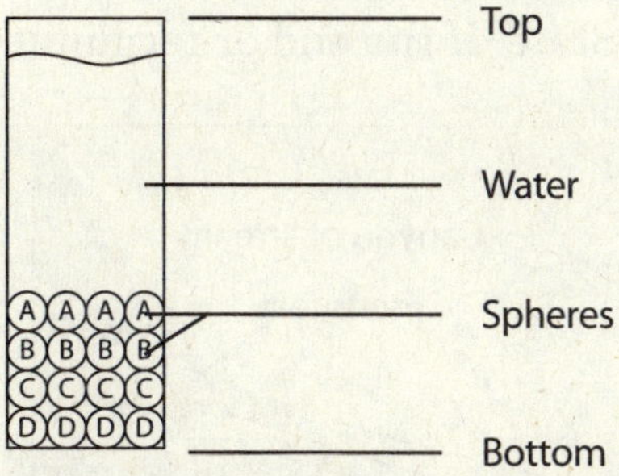

6. Much of the sediment that covers New York State was deposited by glaciers. Describe a depositional characteristic that these sediments should have in common.

7. The following diagram represents a cross section of sedimentary deposits. Where would this type of deposition most likely occur? (1) at the base of the shifting sand dune (2) at the rapids in a stream (3) beneath a large glacier (4) in a lake fed by a stream that often floods

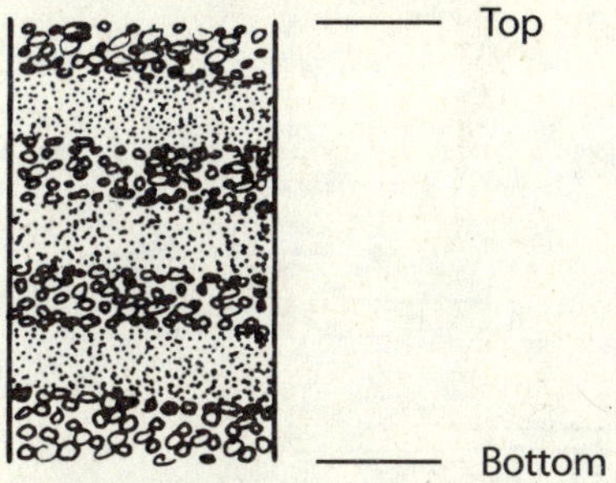

Base your answers to questions 8 and 9 on the following diagram. The diagram shows points A, B, C, and D on a meandering stream.

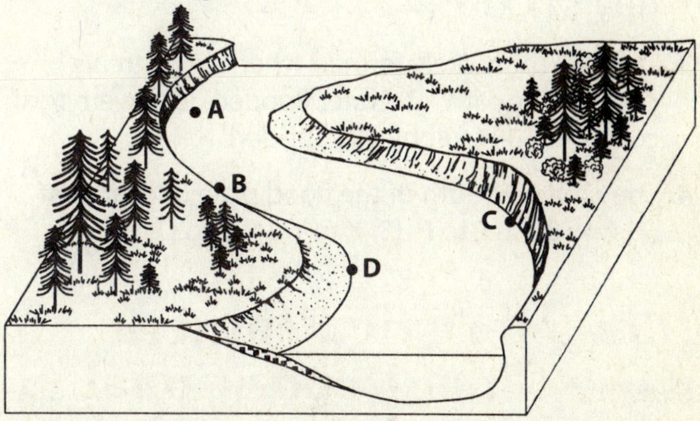

8. Which material is most likely to be transported in suspension during periods of slower stream velocity? (1) gravel (2) sand (3) silt (4) clay

9. At which point is the amount of deposition more than the amount of erosion? (1) A (2) B (3) C (4) D

10. More deposition than erosion will take place in a streambed when the (1) density of the rock particles carried by the stream decreases (2) slope of the stream increases (3) discharge of the stream increases (4) velocity of the stream decreases

11. Which is the most probable description of the energy of a particle in an erosional-depositional system? (1) Particles gain kinetic energy during erosion and lose kinetic energy during deposition. (2) Particles lose kinetic energy during erosion and lose kinetic energy during deposition. (3) Particles gain potential energy during erosion and gain potential energy during deposition. (4) Particles lose potential energy during erosion and gain potential energy during deposition.

Base your answers to questions 12 through 14 on the *Earth Science Reference Tables* and the diagram below. The diagram represents the landscape features associated with a meandering stream. An arrow shows the direction the stream is flowing.

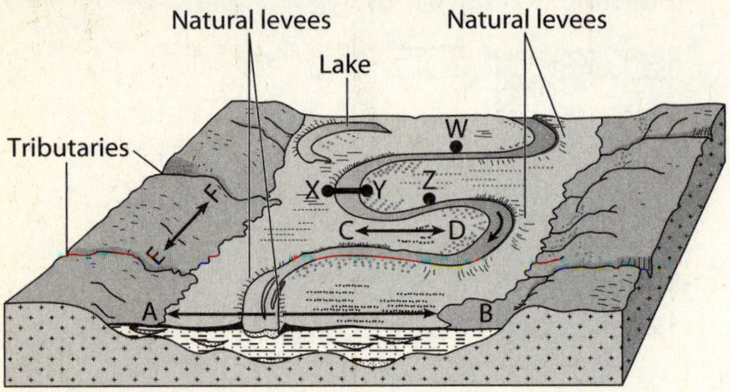

12. At which location is stream erosion greatest? (1) W (2) X (3) Y (4) Z

13. The lake most likely formed when a (1) stream changed its path (2) crater flooded (3) cavern roof collapsed (4) fault block subsided

14. The greatest width of the flood plain is located at (1) A to B (2) E to F (3) X to Y (4) C to D

Characteristic Features of the Chief Depositional Agents

Deposition occurs mainly by streams, glaciers, water waves and currents, wind, and mass movements. Each of these depositional agents has its own characteristic features.

Deposition by Streams

Streams make their deposits in different locations, many of which are illustrated in Figure 9-11 on page 165. In the stream course itself sediments are deposited on the inside of the meanders where stream velocity is slow. When a stream floods and overflows its banks, some of the sediment forms a mound at the edge of the river called a levee, but most of the sediment is spread over a relatively flat region to the sides of a stream forming a <u>flood plain</u>. Often the high velocity of a flooding stream will cut a straighter path through a meander. Depositions at the side of a stream will then separate the former meander from the main stream—forming a curved lake. The sediment that makes it to a lake or ocean will be deposited. This deposition at the end of a stream, with characteristic horizontal sorting as shown in Figure 10-4 on page 179, is a <u>delta</u>. If there are strong ocean or lake currents, a delta may not form because the velocity of the water is not reduced enough at this location. Figure 10-7 shows the formation of a delta in arid regions.

Deposition by Glaciers

At the end of a glacier where there is a balance between melting and forward movement, the sediments it carries are just dropped in unsorted sheets or piles called a **moraine.** Figure 10-8 illustrates this unsorted, unlayered nature of direct glacial deposits. If the moraine is a thin sheet deposited from the bottom of the glacier, it is called a ground moraine. Ground moraines cover much of the northern United States including much of New York State. If the end or terminus of

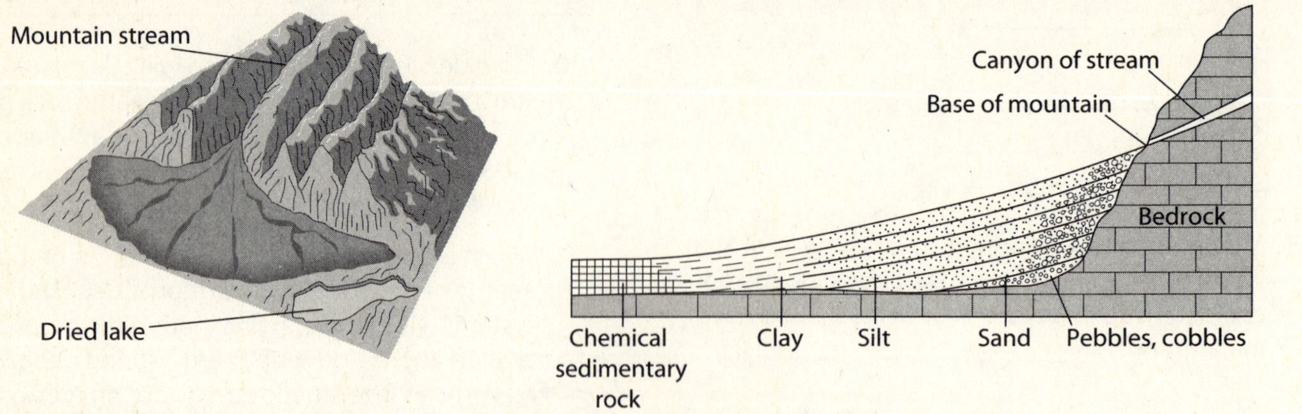

Figure 10-7. A land delta: In arid to semi-arid climates, streams flow off mountains—usually from flash floods—onto more level land and deposit most, if not all, of their sediment load. This fan-shaped deposition forms a land delta. Note the horizontal sorting of the sediment which is very similar to a regular delta, such as that in Figure 10-4.

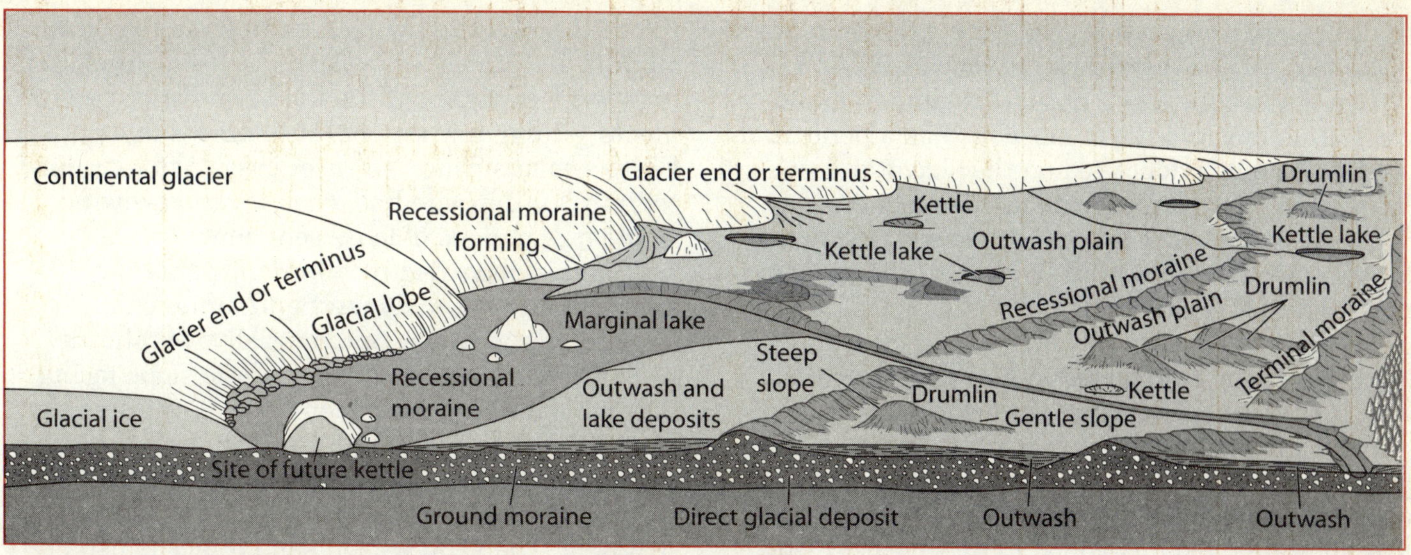

Figure 10-8. Depositional features associated with glaciers: When glaciers melt, some of the sediments are directly deposited into a single layer or small hills. These direct glacial deposits are unlayered and unsorted. They include the ground moraine, terminal moraine, recessional moraines, and drumlins shown here. Water running out of a glacier results in sorted, layered deposits found in outwash plains.

a glacier stays in one location for some time, the glacier builds up a pile of sediments called a terminal moraine. Parts of central Long Island, New York are the terminal moraine of the last stage of the last ice age. Moraines can also be formed at the sides of a glacier. If a glacier mounds up the ground moraine into a streamlined oval shape—something similar to the shape of the end of an inverted spoon—that feature is a **drumlin.** The drumlins indicate the direction a glacier came from by being oriented with direction of former glacial movement. Drumlins have steeper slopes pointing to the direction the glaciers came from. (See Figure 10-8.) Sometimes as a glacier melts back, it leaves blocks of ice in the terminal or ground moraine. When these blocks of ice melt, they leave behind circular depressions called kettle holes or kettles. If these kettles intersect the water tables, they will become filled with water and are then called **kettle lakes**.

Wherever glaciers melt, running water will carry sediments from the glaciers to produce layered and sorted sediment deposits when the water slows down or stops. One such feature is an **outwash plain**—a broad delta-like feature. Much of southern Long Island, New York, is an outwash plain formed by running water when the last sheets of the ice age left this region.

Deposition on Coastlines by Water Waves and Currents

When wind-generated ocean or lake waves slow down as they drag bottom approaching the shore, they tend to move sediment towards shore. This wave movement towards the shore often builds up a strip of sediment at the coastline called a beach.

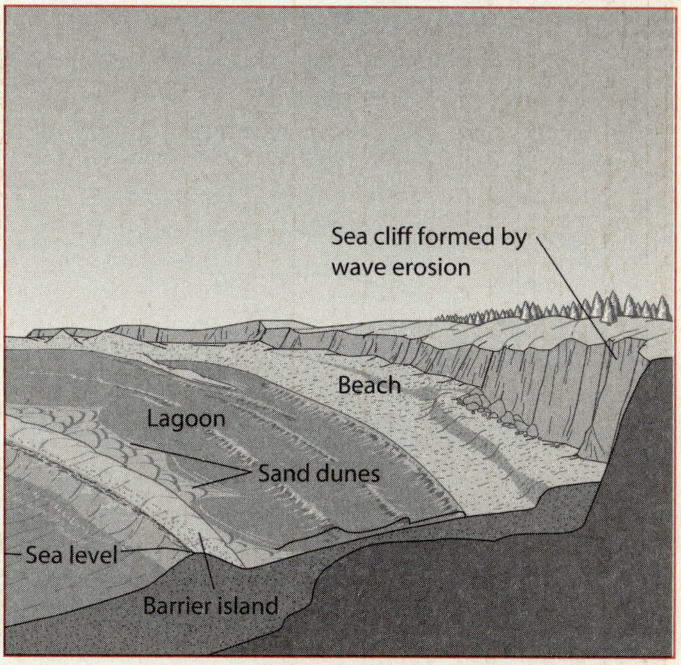

Figure 10-9. Beaches and barrier islands: Two features formed by deposition of sediments by waves and longshore currents are the beach and barrier islands. Barrier islands often protect the shore from direct pounding by storm waves.

Beaches are composed of whatever sediments are available, but most beaches are composed of sand. When a shoreline has a natural or human-made projection out into the ocean—such as a peninsula or pier—sand is deposited on the side of the projection facing an oncoming longshore current. This deposition occurs because the water slows down when it reaches the projection, and the sand settles out. The side of the projection facing away from the longshore current usually gets eroded because it is not protected by the sand deposits. When the waves are more powerful, such as during storms, the beach tends to erode more and the sand is carried back into the ocean. There it is usually deposited as underwater bars parallel to the shoreline. Sand is added to these features by the longshore currents. These longshore currents transport sediment—sand—and when the velocity of the water slows, deposition creates various types of depositional features such as sandbars. (See Figure 9-16.) If this sandbar rises above average sea level, winds will help to pile up the sediments. Then vegetation can stabilize this offshore sediment pile, creating a **barrier island**. Barrier islands are common along the east and southeast coast of the United States. (See Figure 10-9.)

Deposition by Wind

When wind slows down or stops, the sediments it is carrying are dropped. Because air has such a low density compared to liquid water, it usually transports and deposits only sand and smaller sized sediments. The sediments smaller than sand (silt and clay) are commonly called dust. Much of the dust picked up by wind or given to the air in volcanic eruptions—as volcanic ash—is deposited over large expanses of the land and water environments, adding to the general sediment layer of Earth's solid surface. In rarer cases a fine-grained sediment is deposited as a soil layer downwind from large regions of loose sediment.

Sand is deposited by wind in layers or in mounds called **sand dunes**. Sand dunes can assume many shapes but generally—as shown in Figure 10-10—they have a gentle slope facing into the wind, and a steeper slope on the side the wind is blowing towards. Sand dunes migrate as a body downwind—in the direction the wind is blowing to—creating layers of sloping sorted sediment. The sand within sand dunes is usually very rounded and frosted in appearance due to numerous collisions of sand grains during erosion and weathering.

Deposition by Mass Movement

When an avalanche or other type of mass movement hits the ground and stops moving, the sediments it was transporting are deposited as a landform feature of Earth's surface. These features are usually composed of unsorted and unlayered sediments. Depending on the history of the sediments, many will not be very rounded in shape but will often have sharp sides. This is especially true if the sediments have been recently produced by frost action. The most recognizable depositional feature of mass movement is a pile of sediments that is the result of falling sediments often found at the base of cliffs. (See Figure 9-5 on page 162.)

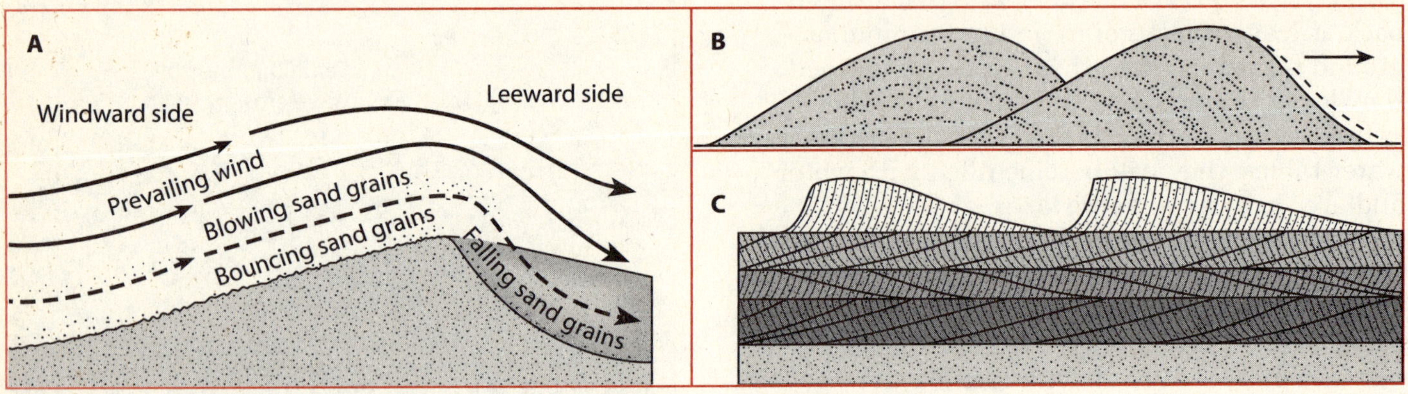

Figure 10-10. Features and movement of sand dunes: When there is loose sand, wind will bounce it along the surface and carry some a few feet into the air. When the wind slows down, it may deposit the sand in a mound called a sand dune. **A** Shows that sand dunes have a gentle slope facing into the wind—windward side—and have a steep slope on the opposite side—leeward side. **B** Shows that dunes migrate up to a few feet a day. **C** If the winds change direction, beds of sand will cross one another at various angles producing a feature called crossbedding. Most sand in dunes is well sorted in sloping layers, and the sand grains are well rounded and have a frosted appearance.

Review Questions

15. Deltas form where (1) a ground moraine is deposited (2) stream velocity is reduced (3) small streams empty into larger streams (4) extensive ground water action has occurred

Base your answers to questions 16 through 19 on the following diagram. The diagram represents a glacier moving out of a mountain valley. Water from the melting glacier is flowing into a lake. Letters A through F identify points within the erosional-depositional system.

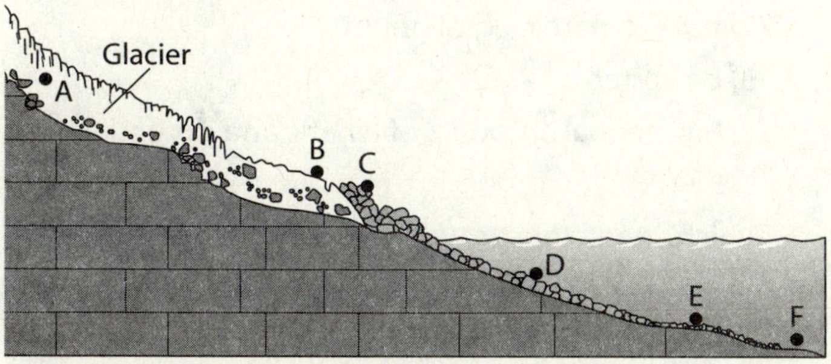

16. Deposits of unsorted sediments would probably be found at location (1) E (2) F (3) C (4) D

17. An interface between erosion and deposition by the ice is most likely located between points (1) A and B (2) B and C (3) C and D (4) D and F

18. The clay-sized sediment particles carried by water are most probably being deposited at point (1) F (2) B (3) C (4) D

19. Which graph best represents the speed of a sediment particle as it moves from point D to point F?

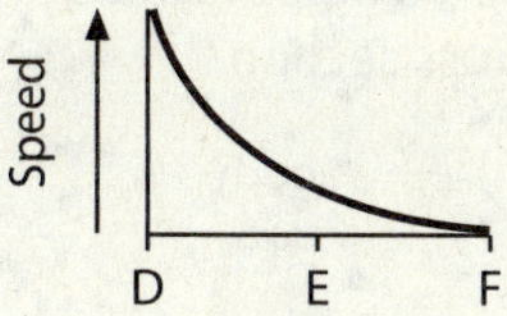

(1)

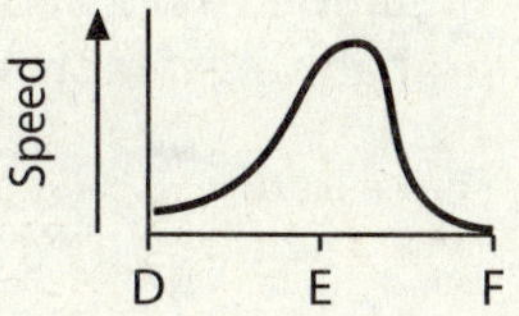

(3)

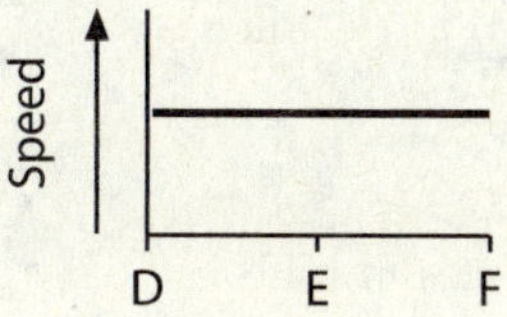

(2)

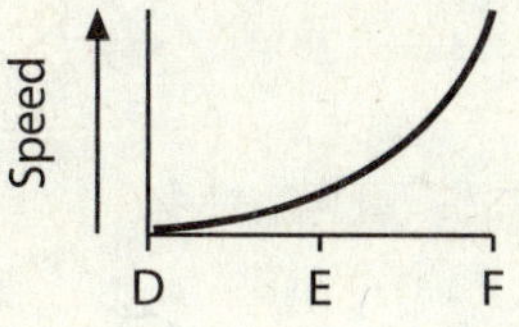

(4)

Base your answers to questions 20 through 25 on the following diagrams. Diagram I shows melting ice lobes of a continental glacier during the Pleistocene Epoch. Diagram II represents the landscape features of the same region at present, after the retreat of the continental ice sheet. Letters A through G indicate surface features in this region.

Diagram I

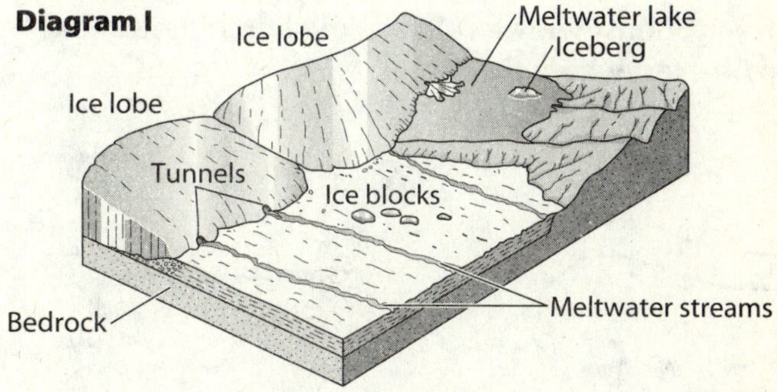

Diagram II

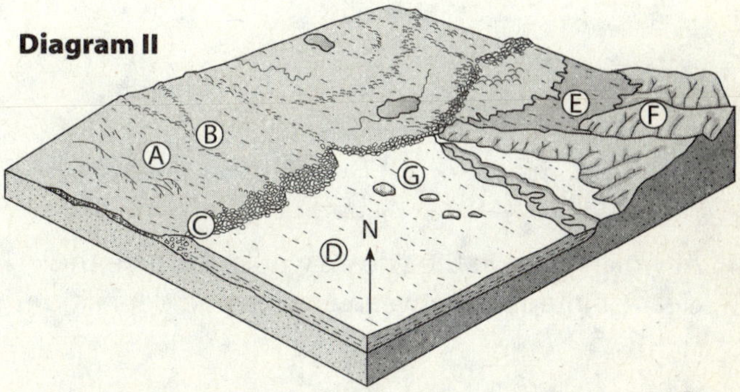

20. Which features in Diagram II are composed of sediment directly deposited by the glacial ice? (1) A and C (2) B and D (3) C and E (4) E and F

21. Kettle lakes are located nearest to (1) A (2) B (3) F (4) G

22. Drumlins that formed under a lobe of ice are located at (1) A (2) B (3) C (4) D

23. An outwash plain of sorted layered sediment is located at (1) A (2) B (3) C (4) D

24. A moraine deposit formed at a former interface of a glacier and no glacier is located at location (1) A (2) C (3) F (4) G

25. Which arrow best represents the directions of movement that formed the deposits shown in Diagram II?

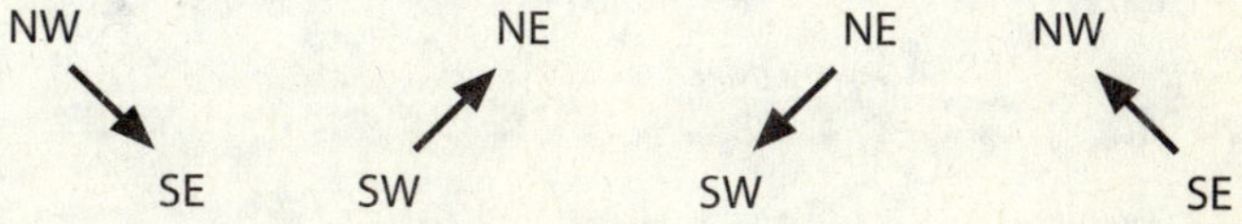

(1) (2) (3) (4)

Questions for Regents Practice

Part A

To answer questions 1 through 9 refer to the following diagram, which shows the top and side views of a stream. Consider the volume of water in the stream to be everywhere the same.

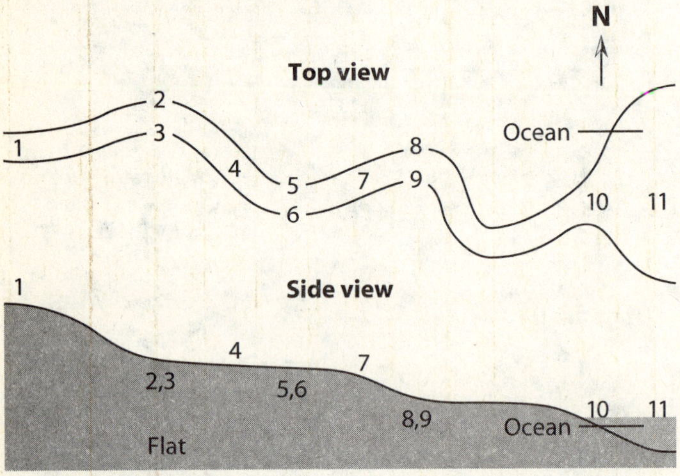

1. At which place in the stream would there be the clearest interface between erosion and deposition?

(1) 1 (3) 3-4
(2) 2-3 (4) 7

2. What happens to the total amount of energy of this stream as it flows from 1 to 10?

(1) It steadily increases.
(2) It steadily decreases.
(3) It increases and decreases.
(4) It remains the same.

3. Between what two points would the most potential energy be converted to kinetic energy?

(1) 1-2 (3) 5-8
(2) 2-5 (4) 8-10

4. At which location would the most deposition occur because of loss of energy?

(1) 4 (3) 9
(2) 6 (4) 10

5. At which location would the stream be doing the most eroding of its side and banks?

(1) 1 (3) 5
(2) 3 (4) 8

6. If the stream is eroding as much as it depositing between 8 and 10, the stream at this location can be said to

(1) be an interface
(2) have dynamic equilibrium
(3) be sorted
(4) have equal amounts of kinetic and potential energy

7. Which characteristic would usually decrease the most between locations 10 and 11?

(1) the amount of salt solution
(2) the size of the sediments
(3) the density of the water
(4) the depth of the water

8. As the water of the stream flows from 1 to 10, the total potential energy of the stream will

(1) increase
(2) decrease
(3) remain the same

9. An observer looks downstream from a location just above location 8-9 and draws a cross section of the streambed at location 8-9. Which diagram would probably best represent this cross section?

(1)
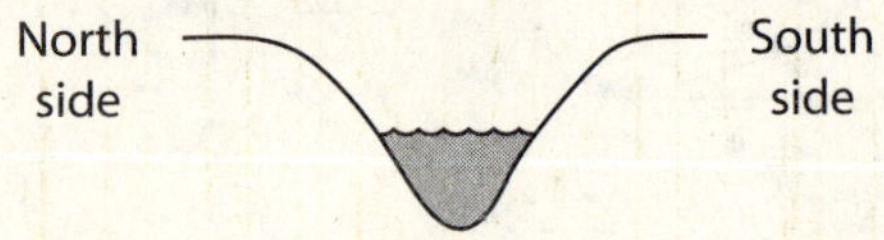

(2)
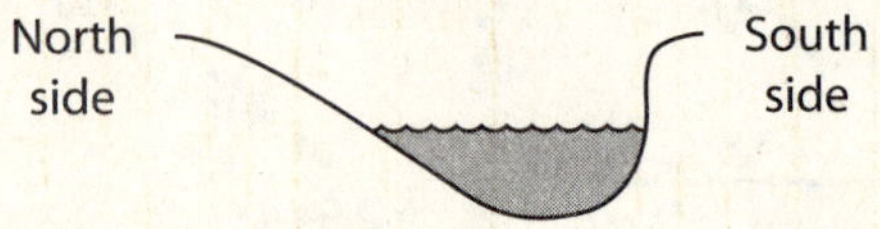

(3)
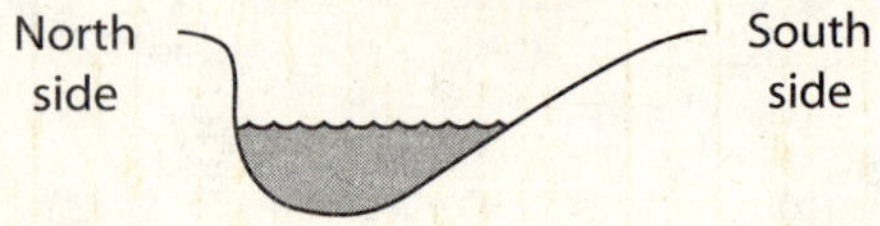

(4)

10. The following graph shows the relationship between particle shape and settling rate.

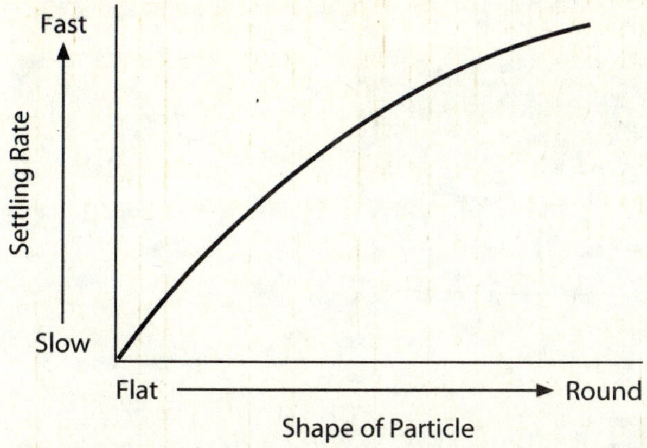

Which statement best describes the relationship shown?

(1) flatter particles settle more slowly than rounder particles

(2) flatter particles settle at the same speed

(3) all particles settle at the same speed

(4) particle shape does not affect settling rate

11. A mixture of sand, pebbles, clay, and silt, of uniform shape and density, is dropped from a boat into a calm lake. Which material most likely would reach the bottom of the lake first?

(1) sand
(2) pebbles
(3) clay
(4) silt

12. A low hill is composed of unsorted sediments that have mixed grain sizes. This hill was probably deposited by

(1) a glacier

(2) the wind

(3) running water

(4) wave action

Part B

Match each of the deposition or erosional features in Column 1 with the phrase in Column 2 that best applies to that feature. Briefly describe HOW each feature forms and WHERE it forms. A CHOICE MAY BE USED MORE THAN ONCE AND SOME CHOICES MAY NOT BE USED.

Column 1	Column 2
13. outwash plain [3]	(a) deposited by glaciers
14. sand dune [3]	(b) deposited mostly by water from melting glaciers
15. offshore bar [3]	(c) formed by glacial erosion
16. moraine [3]	(d) not related to glaciation

Match each feature in Column 1 with the erosional-depositional agent in Column 2 that is chiefly responsible for producing that feature. Briefly describe how each land feature is produced. A CHOICE MAY BE USED MORE THAN ONCE AND SOME CHOICES MAY NOT BE USED.

Column 1	Column 2
17. drumlin [2]	(a) glaciers
18. barrier island [2]	(b) gravity
19. sand dune [2]	(c) streams
20. flood plain [2]	(d) waves and shore currents
	(e) wind

Base your answers to questions 21 through 26 on the diagram below, which shows ocean waves approaching a shoreline. A short wall of rocks perpendicular to the shoreline and an offshore structure have been recently constructed along the beach. Letters A, B, C, D, and E represent locations in the area.

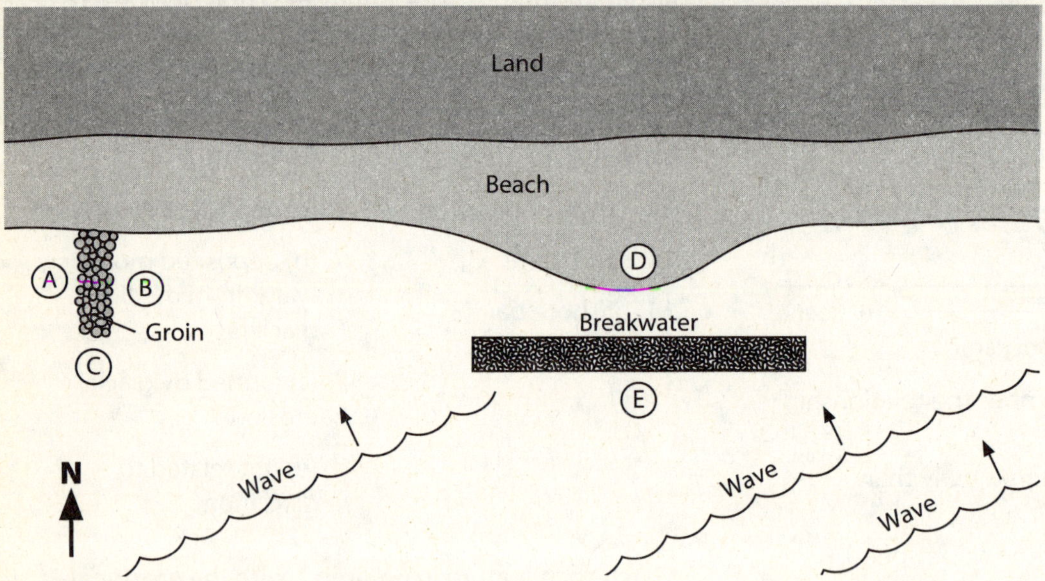

21. What is the most common cause of the approaching waves?

(1) underwater earthquakes

(2) variations in ocean-water density

(3) the gravitational effect of the moon

(4) winds at the ocean surface

22. At which location will the beach first begin to widen due to sand deposition?

(1) A (3) C

(2) B (4) E

23. The size of the bulge in the beach at position D will

(1) decrease

(2) increase

(3) remain the same

24. How was this beach directly formed?

(1) uplifting from an earthquake

(2) sinking of land during a landslide

(3) deposition of sediments by ocean waves and currents

(4) erosion of bedrock by ocean waves and currents

25. If sand dunes were located on the beach, they would have been formed by

(1) deposition by strong winter ocean waves

(2) erosion by mild summer ocean waves

(3) deposition by wind

(4) erosion by surf and ocean shore currents

26. Which statement best describes the ocean current that is modifying this coastline?

(1) the current is flowing northward at a right angle to the shoreline

(2) the current is flowing southward at a right angle away from the shoreline

(3) the current is flowing eastward parallel to the shoreline

(4) the current is flowing westward parallel to the shoreline

---

27. If the prevailing winds are from the east, a sand dune will migrate toward the

(1) east, and have a gentle slope on the east side

(2) east, and have a gentle slope on the west side

(3) west, and have a gentle slope on the east side

(4) west, and have a gentle slope on the west side

Part C

28. A person finds gold nuggets near a meander, or curve, of a stream in California. Explain where in the meander area the gold nuggets were most likely found. Why is gold (density 19.3 g/cm^3) deposited at this location and not minerals such as halite (density 2.2 g/cm^3) and gypsum (density 2.3 g/cm^3)? [2]

29. You are investigating a dry stream bed at the point it once entered a lake. As you walk out into the lake bed from the mouth of the stream, you notice that the sediments below your feet become smaller. You also notice that as you dig vertically in the same location, the uncovered sediments gradually increase in size. Using your knowledge of erosion and deposition, state a hypothesis to explain both observations. [2]

30. Your friend offers the following description of sediment she observed in a V-shaped canyon on a camping trip in Arizona. "The pebbles were all piled together evenly and they were very uniform in size. They were rounded, and smooth, with almost no markings of any kind. I think that they must have been deposited long ago by a glacier." Do you agree with her inference? Cite four clear pieces of evidence to either support or refute her conclusion. [5]

Base your answers to questions 31 through 35 on the notes below written by a student during field trips to three different locations in New York State.

NOTES

Location A

Good view from this hilltop; chilly and windy. We rested to catch our breath, then collected samples. Rocks are visible everywhere. There are boulders, cobbles, and pebbles of many sizes and shapes mixed together. These surface rock fragments are composed of metamorphic rock sitting on the limestone bedrock. The teacher showed us parallel scratches in the bedrock. I saw almost no soil.

Location B

It is rocky and the streambank is steep. Where we are standing, we can see a waterfall and rapids. It is cool by the water. From the streambed we collected pebbles and cobbles—some red, some white, others a mixture of many colors. The streambed is full of rocks of all sizes. The teacher warned us to be careful of the strong stream current.

Location C

It is cool in the shade, and the rock cliff above us still has some ice on it from winter. The rocks we are sitting on have sharp edges. Rock fragments at the bottom of the cliff are the same color as the cliff. Our teacher warned us to watch out for falling rocks.

31. *a* State the agent of erosion that deposited most of the sediment found at location *A*. [1]

31. *b* State *one* observation recorded by the student that supports this conclusion. [1]

32. Some samples of sediment collected from the streambed at location *B* are shown below.

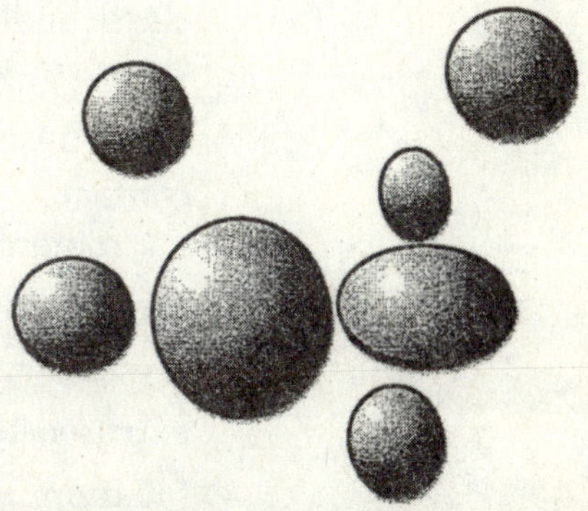

Explain why these samples are smooth and have rounded shapes. [1]

33. Explain how ice in cracks on the cliff at location *C* may have helped cause weathering of the bedrock on the face of the cliff. [1]

34. What agent of erosion resulted in the deposition of the rocks with the sharp edges at location C?

35. What is one feature of wind deposited sediments that would differ from rocks (sediments) deposited at locations A and B?

TOPIC 11

Earth Materials—Minerals, Rocks, and Mineral Resources

VOCABULARY		
bioclastic sedimentary rocks	fossil	mineral resources
chemical sedimentary rocks	fracture	organic
clastic sedimentary rock	hardness	precipitation (of minerals)
cleavage	igneous rock	regional metamorphism
contact metamorphism	inorganic	rock cycle
crystal shape	intrusive igneous rock	sedimentary rocks
crystal structure	luster	streak
extrusive igneous rock	magma	texture
foliation	metamorphic rocks	
	metamorphism	
	mineral	
	mineral crystal	

Rocks and minerals are the source of much of the material and energy that people want or need. If you make a list of about one hundred objects you used or wanted to use today, most likely 95 to 100 percent of them come from rocks and minerals. Earth materials—minerals, rocks, and mineral resources—are of value to people in many ways. Earth materials fuel our industrial society as extracted fossil fuels. They provide the raw materials for the building of homes and other construction projects. Rocks and minerals make up Earth's solid surface—the lithosphere—that you live on. When Earth's solid surface is weathered and eroded, the end results are the landscape features that people live, work, and play on.

Minerals

Minerals have characteristic physical and chemical properties. Some of these properties are color, streak, luster, hardness, density, cleavage and crystal structure.

What a Mineral Is

A **mineral** is a naturally occurring, inorganic, crystalline solid having a definite chemical composition. A mineral is considered to be naturally occurring because it is formed in nature and not made by people. It is **inorganic** because it has not been made by or composed of life forms. Thus fossil fuels or a pearl from an oyster are NOT minerals.

A mineral is crystalline because its atoms have a specific arrangement. This arrangement of atoms is called **crystal structure.** Each mineral has its own distinctive crystal structure that can lead to very accurate identifications through the use of X-ray. Figure 11-1 on the next page illustrates the crystal structure of a mineral.

All minerals are solids that are composed of one or more chemical elements. The chemical composition of a mineral describes the types and ratios of elements that make up the mineral. Some minerals contain only one element and others are compounds of two or more elements. You can find the characteristics of some minerals in the Properties of Common Minerals in the *Earth Science Reference Tables.*

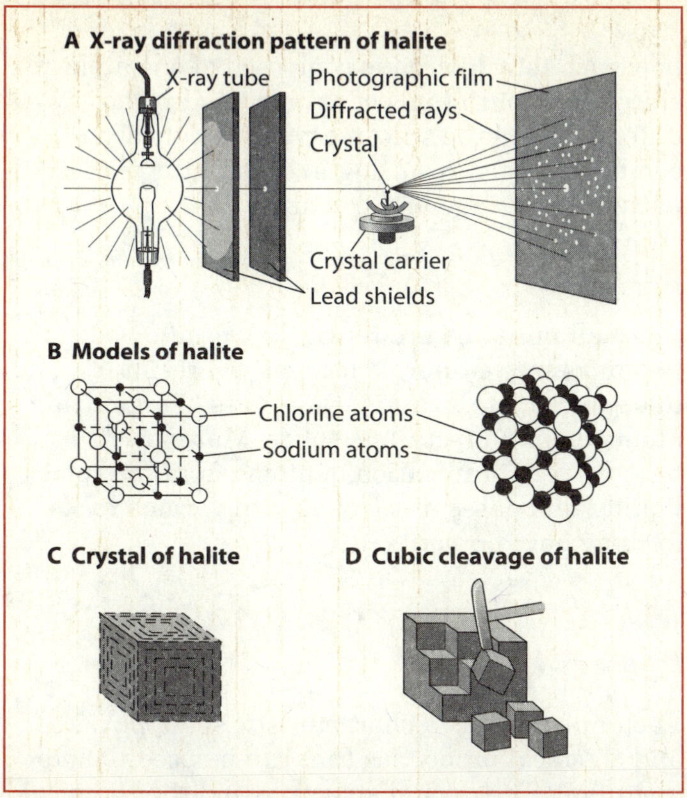

Figure 11-1. Crystal structure and properties of the mineral halite: Halite is the mineral with the formula NaCl (sodium chloride). It is the one mineral of rock salt and common table salt. In diagram A, a small crystal of halite is exposed to X-rays, which produces a characteristic and unique pattern on a photographic plate. Diagram B shows two interpretations of the pattern on the photographic plate.

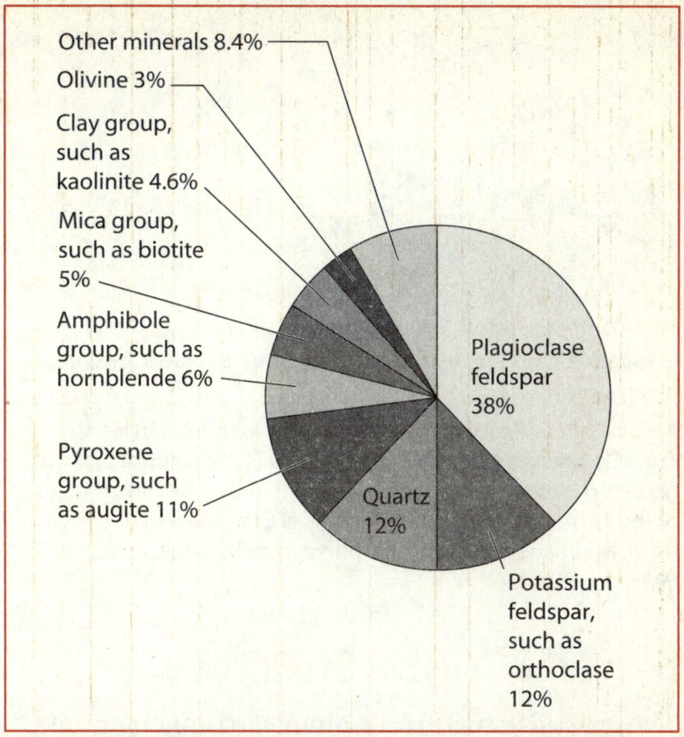

Figure 11-2. Pie graph of the most common minerals of Earth's crust: 90 percent of Earth's crust by weight is composed of eight minerals or groups of minerals—all silicates. These common minerals are called rock-forming minerals.

Relation of Minerals to Rocks

All minerals are rocks, but not all rocks are minerals nor are they all composed of minerals. A rock is any naturally formed solid that is part of Earth or any other celestial body. Though a large percentage of rocks are composed of minerals, many rocks are composed of organic or glassy materials that are not minerals. Glasses are not minerals because their atoms are not arranged in a specific pattern. The majority of rocks are made of two or more minerals—multiple-mineral rocks. Some rocks are composed of only one mineral—single-mineral rocks.

A review of the three schemes for rock identification—sedimentary, metamorphic, and igneous—in the *Earth Science Reference Tables* and Figure 11-2 indicates that only a small number of minerals are commonly found in rocks. These 20 to 30 very common minerals, found in rocks, are called the rock-forming minerals. Many of these rock-forming minerals are listed in the Properties of Common Minerals in the *Earth Science Reference Tables.*

Element Composition of Earth's Crust

The chemical element composition of Earth's crust is shown in Figure 11-3 on the next page.

- The graph indicates that over 99 percent of Earth's crust and its minerals are, by volume and mass, composed of only 8 of the 90 naturally occurring elements found on Earth.
- Silicon is the second most abundant element by mass, but the element potassium is number two in crustal abundance by volume because of its lower density-higher volume.

MEMORY JOGGER

You may recall from Topic 1 that density is the ratio of the mass of an object to its volume. Therefore, an element that has a lower density than another element has a smaller mass per volume than an element with higher density.

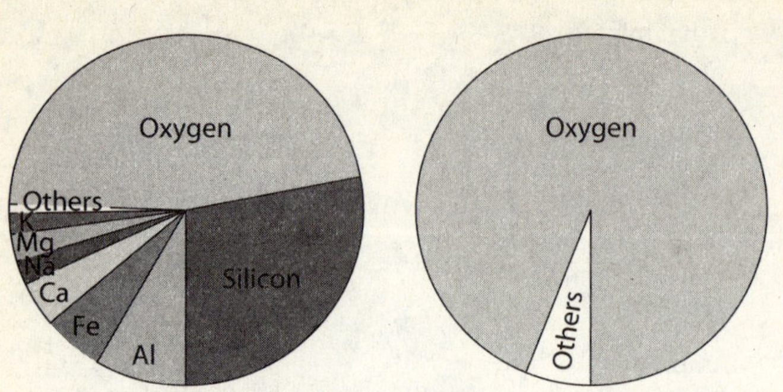

Figure 11-3. Percentages of the chief elements in Earth's crust by mass and by volume: Volume is the amount of space occupied by the atoms of each element in the solid substances of the crust. Also see Average Chemical Composition of Earth's Crust, Hydrosphere, and Troposphere in the *Earth Science Reference Tables.*

Mineral Crystal Structure

The crystal structure, or atomic arrangement of the atoms, that comprise minerals is responsible for many of their chemical and physical properties, such as crystal form, breaking pattern, and hardness. Most rock-forming minerals are silicates. Silicate minerals have a structure that results from various arrangements of a tetrahedron-shaped (4-sided) unit of oxygen and silicon called the silicon-oxygen tetrahedron. Figure 11-4 shows how each tetrahedron is composed of one atom of silicon and four atoms of oxygen. It also shows different ways the silicon-oxygen tetrahedron can be arranged resulting in different breaking patterns (cleavage and fracture).

Mineral Formation

Since all minerals are rocks, they form by one of two processes. Minerals form as the result of inorganic crystallization—a process of organizing atoms to form crystalline solids. Minerals also form by recrystallization of atoms from the solids, liquids, and gases associated with various rock-forming environments.

Mineral Properties and Identification

Each mineral has a characteristic set of physical and chemical properties that can be used to help identify it. The crystal structure and the chemical composition of minerals largely determine these properties. Some properties, such as color, are often caused by impurities. The mineral corundum, when pure, is colorless. However, with slight chemical impurities, corundum becomes the blue sapphire or red ruby.

Silicate structure	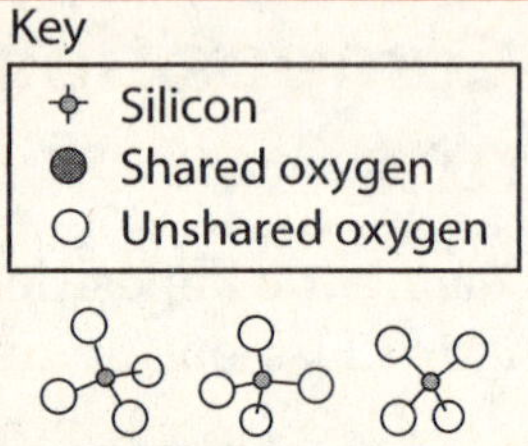	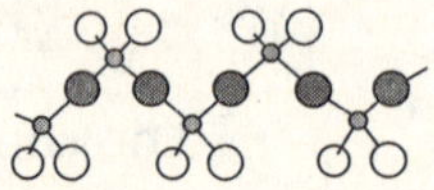	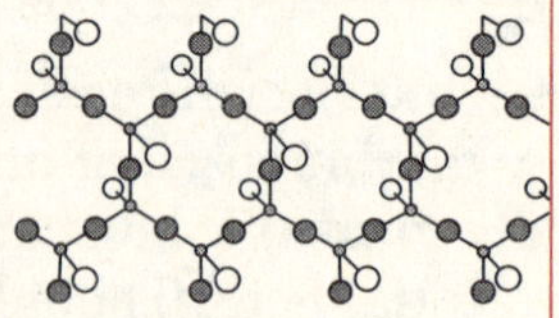	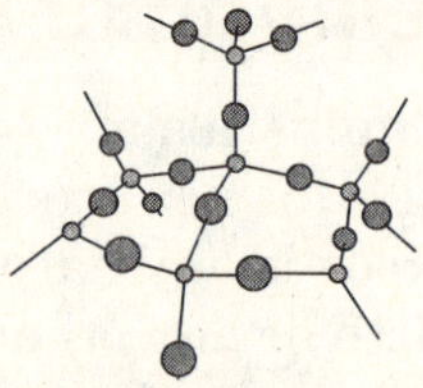
Silicate structure	Single tetrahedron	Chains of tetrahedra	Sheets of tetrahedra	A network of tetrahedra in three dimensions
Mineral example	Olivine—peridot is gem variety	Pyroxene, such as augite	Mica, such as biotite	Quartz
Cleavage or fracture type	Curved or conchoidal fracture	Two directions of cleavage into blocky or splinter shapes and fracture	One direction of cleavage into sheets and fracture	Curved or conchoidal fracture
Drawing	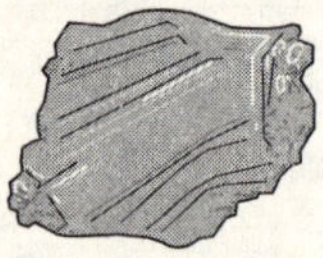	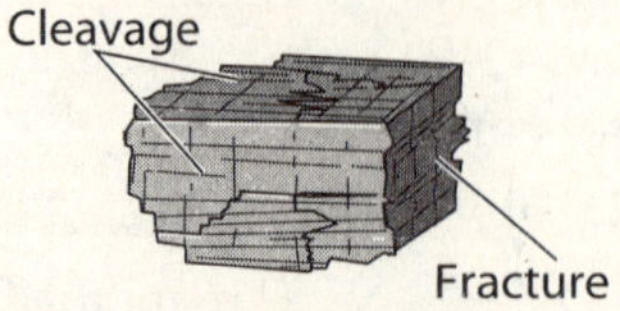	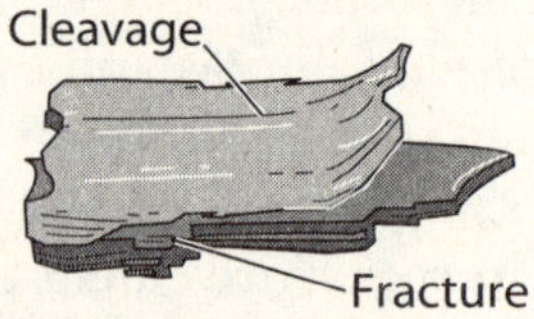	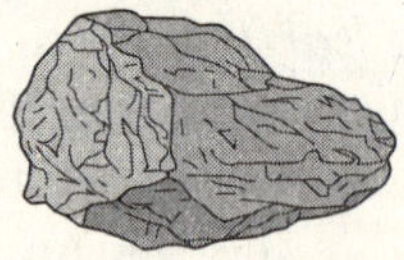

Figure 11-4. Various arrangements of the silicon-oxygen tetrahedron in silicate minerals: The tetrahedra combine with themselves and other elements in different atomic structures. The different combinations affect the physical properties of the minerals—including cleavage and fracture patterns shown in the illustration.

The most accurate method for identifying minerals is by the use of X-ray diffraction instruments and other machines not available to most individuals. Therefore, simple tests and mineral identification charts are relied on. An example of a mineral identification key, or chart, is found in Properties of Common Minerals in the *Earth Science Reference Tables*.

COLOR The color of a mineral is one of its most obvious properties. However, in most cases color is not useful because many minerals have the same color. In addition, the color of many minerals varies due to impurities, and many minerals are clear or colorless when pure. In a few cases however, such as in the yellow of sulfur, the gray of graphite (pencil lead) and galena, or the brassy yellow of pyrite (fool's gold), the mineral's color is usually consistent.

STREAK The color of finely crushed residue or powder of a mineral is its **streak**. When you write on a chalkboard, you observe the streak of the rock chalk. The streak of a mineral is usually quite consistent; thus streak color is much more useful than mineral color. For example, the iron ore mineral, hematite, can be various shades of silver-gray to red in color, but the streak is a consistent red.

LUSTER The shine from an unweathered mineral's surface, or the way a mineral looks in reflected light, is **luster**. There are two broad groups of luster—metallic and nonmetallic. Minerals with a metallic luster, such as pyrite and galena, shine like the surface of a clean stainless steel pot. Most minerals have a nonmetallic luster. There are many types of nonmetallic luster, such as the glassy luster of black hornblende and clear quartz "rhinestone," or the pearly luster of muscovite mica.

HARDNESS The resistance a mineral offers to being scratched is its **hardness**—the scratchability of a mineral, not how easily the mineral breaks. Diamond is the hardest mineral, but drop an unmounted diamond on a tile floor and it will likely shatter. On the other hand, if the very soft mineral graphite is dropped, only a small amount will chip off, or cleave.

Figure 11-5 shows the Mohs hardness scale and some other common materials that are often used to determine hardness. Mohs hardness scale is arranged from the softest #1 (talc) to the hardest #10 (diamond). A quick way to determine

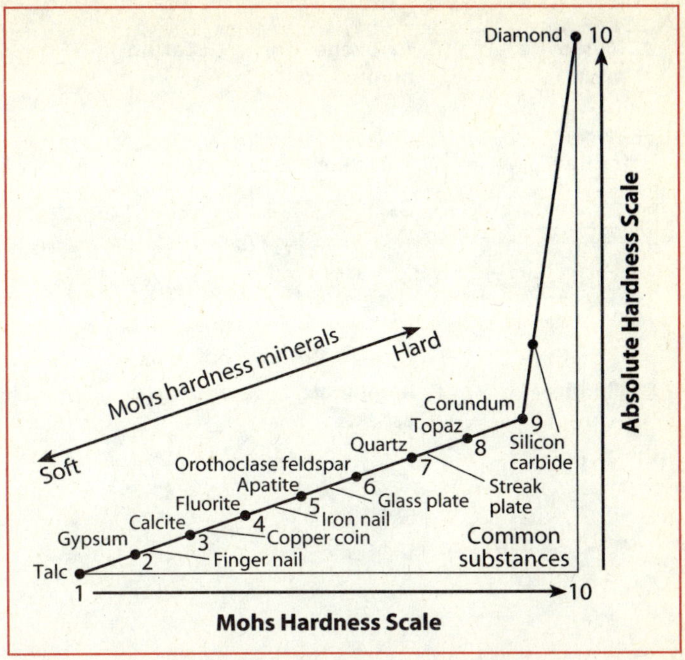

Figure 11-5. Mohs and absolute hardness scale: The differences in the hardnesses on the Mohs scale vary, as shown by the comparison to an absolute scale of hardness. Note that on the absolute scale, the difference of hardness between diamond and corundum—ruby or sapphire—is more than all the way from talc to corundum.

relative hardness is to use a piece of window glass. If a mineral scratches the glass, the mineral is hard, and if it doesn't, it is soft.

DENSITY Each mineral has a specific density or a small range of densities—for those minerals that vary in mineral composition. Often in mineral studies, density is stated as specific gravity, a value without units. Specific gravity is the density of a mineral compared to the density of water. Specific gravity is a good test to distinguish gemstones, because it doesn't harm the samples like hardness or cleavage tests do. In mining and refining processes, differences in the densities of various minerals allow them to be separated. A common example is the panning of high density gold. For a detailed description of density see Topic 1.

CLEAVAGE The tendency of a mineral to break along the zones of weakness and form smooth to semi-smooth parallel sides, or surfaces, is called **cleavage.** Cleavage surfaces can often be distinguished from sides without cleavage by having a shinier or more brilliant luster (smooth surfaces reflect better). If a mineral lacks preferred zones of weakness in the crystal structure, then it will demonstrate uneven breaking surfaces called **fracture**. For example, some types of fracture are irregular (earthy), fibrous (splintery), and curved

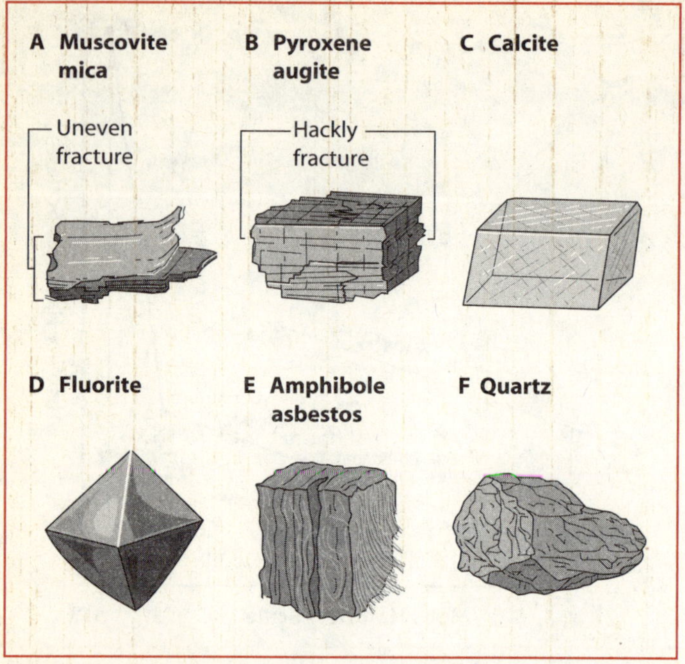

Figure 11-6. Types of cleavage and fracture: **(A)** shows one direction of cleavage and some uneven fracture. **(B)** shows two directions of cleavage and a hackly (bumpy) fracture. **(C)** shows three directions of cleavage. **(D)** shows four directions of cleavage. **(E)** shows fibrous fracture. **(F)** shows curved fracture.

(conchodial). The curved surfaces in a type of quartz called flint make this material very useful in making knives and arrowheads. Often a mineral will have both cleavage and fracture on different sides, such as the silicates hornblende and the feldspars. (See Figures 11-4 and 11-6)

CRYSTAL STRUCTURE The outward geometric shape of a mineral, the crystal form, or **crystal shape**, reflects the crystal structure—orderly arrangement of the atoms in the mineral. It is only when individual mineral grains have the room to freely grow that this crystal shape, with its smooth sides or faces, can take shape. This is the reason most mineral samples found in nature don't illustrate the crystal forms; the use of crystal form in mineral identification is limited. Another problem is that even though the internal crystal structure of minerals is unique, the outward crystal shape, such as the cubic shape of halite, galena, and fluorite, isn't unique. Also, any mineral can have many different crystal shapes.

OTHER MINERAL PROPERTIES Besides physical properties, some chemical properties of minerals are also used for identification. One of these chemical properties is the reaction of a mineral with acid. When a small amount of dilute hydrochloric acid is placed on a mineral or rock containing calcite (CaCO3), the mineral or rock will bubble (effervesce)—giving off carbon dioxide. The mineral dolomite can be distinguished from calcite, because dolomite will bubble in acid only after the mineral is powdered.

Many other chemical and physical properties are used to identify minerals. Many of the properties only apply to a few minerals and will often be the key to a mineral's identification. For example, some minerals such as thin pieces of muscovite and biotite micas are flexible. This means that they can be bent and will snap back to their original shape. Other properties used for identification are found in Properties of Common Minerals in the *Earth Science Reference Tables*.

DIGGING DEEPER

A few minerals have very fine lines—called striations—on cleavage surfaces and on the faces of their crystal form. These striations can be used to distinguish the mineral plagioclase feldspar from the potassium feldspars such as orthoclase. Striations are an outward expression of the internal arrangement of the atoms within a mineral.

Review Questions

1. A mineral CANNOT be (1) organic (2) crystalline (3) a solid (4) formed in nature

2. Only a small number of Earth's minerals are commonly found in rocks. This fact indicates that most (1) minerals weather before they can be identified (2) minerals have properties that are difficult to identify (3) rocks have a number of minerals in common (4) exposed surface rocks are mostly igneous

3. Which rock is usually composed of several different minerals? (1) rock gypsum (2) limestone (3) quartzite (4) gneiss

4. The data table shows the composition of six common rock-forming minerals.

Mineral	Composition
Mica	$KAl_3Si_3O_10$
Olivine	$(FeMg)_2SiO_4$
Orthoclase	$KAlSi_3O_8$
Plagioclase	$NaAlSi_3O_8$
Pyroxene	$CaMgSi_2O_6$
Quartz	SiO_2

The data table provides evidence that (1) the same elements are found in all minerals (2) a few elements are found in many minerals (3) all elements are found in only a few minerals (4) all elements are found in all minerals

5. What are the four most abundant elements, by volume, in Earth's crust? (1) oxygen, potassium, sodium, and calcium (2) hydrogen, oxygen, nitrogen, and potassium (3) aluminum, iron, silicon, and magnesium (4) aluminum, calcium, hydrogen, and iron

6. Diamonds and graphite are both minerals that are composed of the element carbon. Diamond has a hardness of 10, while graphite has a hardness of 1. Based on your knowledge of earth science, what is the most probable cause of this difference in hardness?

7. Minerals are composed of (1) one or more rocks (2) only one rock (3) one or more chemical elements (4) only one metal

8. The cubic shape of a mineral crystal is most likely the result of that crystal's (1) hardness (2) density distribution (3) internal arrangement of atoms (4) intensity of radioactive decay

9. The following diagrams represent four different mineral samples.

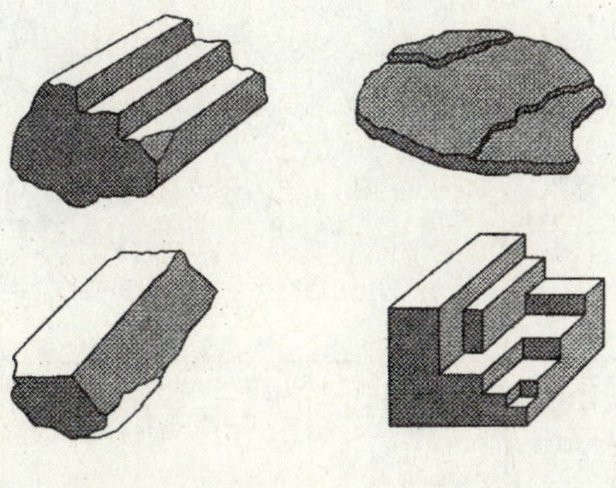

Which mineral property is best represented by the samples?
(1) density (2) cleavage (3) hardness (4) streak

10. Minerals are identified on the basis of (1) the method by which they were formed (2) the type of rock in which they are found (3) the size of their crystals (4) their physical and chemical properties

11. A six-sided mineral crystal with a pyramid end is a very hard mineral called (1) hornblende (2) orthoclase feldspar (3) quartz (4) biotite mica

12. The relative hardness of a mineral can best be tested by (1) scratching the mineral across a glass plate (2) squeezing the mineral with calibrated pliers (3) determining the density of the mineral (4) breaking the mineral with a hammer

13. What property would a mineral have if it appears like a new quarter in reflected light? (1) a metallic luster (2) metallic element composition (3) magnetic (4) a high density

14. What information about a mineral is needed to determine its density? (1) shape and volume (2) shape and mass (3) volume and mass (4) volume and hardness

15. Which property of the mineral diamond allows diamond powder to be used to shape gems for jewelry? (1) crystal shape (2) cleavage (3) luster (4) hardness

Rocks

A rock is any naturally formed solid on Earth or in any part of the universe. The definitions for rock and the individual rock types are not nearly as specific as those for a mineral. The reason is that rocks, except for the single-mineral ones, are mixtures of minerals, organic materials, glasses, and fragments of other rocks. A single-mineral rock is both a rock and a mineral and has a mineral's definite composition and properties.

In your study of rocks you will often have to refer to parts of the *Earth Science Reference Tables*. The Generalized Landscape Regions of New York State and the Generalized Bedrock Geology of New York State will allow you to see in which portions of New York State different rock types are found and the geologic ages (periods) of the rocks. The Rock Cycle in Earth's Crust diagram will allow you to quickly understand the relationships among the three major rock types and the processes that form them. The Scheme for Igneous Rock Identification, Scheme for Sedimentary Rock Identification, and Scheme for Metamorphic Rock Identification will provide you with the basic individual rock properties and how to identify most of the important rock types.

Rock Types

Rocks are classified into three categories—sedimentary, igneous, and metamorphic—based on the three methods of rock formation. Sometimes metamorphic and igneous rocks are grouped together as nonsedimentary rocks. As a group, rocks are distinguished and identified based on their composition and the texture. The **texture** of a rock is not how rough it feels, but the size, shape, and arrangement of the materials the rock is composed of. The majority of rocks are composed of individual grains of minerals called **mineral crystals.** Rocks that are made of intergrown or interconnecting mineral crystals are called crystalline.

Sedimentary Rocks

Rocks that form from an accumulation of sediments derived from preexisting rocks and/or organic materials are **sedimentary rocks.** These rocks form by various processes that occur on or within the top few kilometers of Earth's crust.

FORMATION OF SEDIMENTARY ROCKS Most sedimentary rocks are made up of solid fragments or sediments, often called "clasts," that have been weathered or eroded from older rocks. After formation, erosional agents, such as running water, glaciers, wind, ocean waves, currents, and gravity transport sediments to new locations under water or on land. Most sedimentary rocks form under large bodies of water, such as lakes, seas, and oceans, where the sediments are usually deposited in horizontal layers. Some methods of formation of sedimentary rocks are cementation, compaction, chemical action, and organic processes.

Cementation Often the clasts, such as sand, silt, and pebbles, are cemented together in a process called cementation. This happens as the sediments lose water and the dissolved minerals in the pores of sediments precipitate out, forming crystalline mineral material. Minerals, such as calcite, quartz, and hematite, are the common cements that glue the solid sediments together. Cementation can happen alone or in combination with other processes to form the clastic sedimentary rocks, such as siltstone and conglomerate. A **clastic sedimentary rock** is one that is largely composed of solid sediments, such as the sand in sandstone, or the tiny pieces of clay in shale.

Compaction Crustal movements and the weight of overlying water and sediments compress, or compact, sediments. This causes a reduction in volume due to the loss of pore space and water. The process is called compaction. While some sedimentary rocks, such as shale and coal, may form only by compaction, most of the clastic or fragmental rocks form due to compaction and cementation.

Chemical Action All natural liquid water on Earth contains dissolved minerals. In water, these dissolved minerals are called by various names—hardness in drinking water and salts in the sea—but they are all minerals that have been dissolved by chemical weathering. When these dissolved minerals precipitate, or drop out from the water, they form a crystalline mass of intergrown or

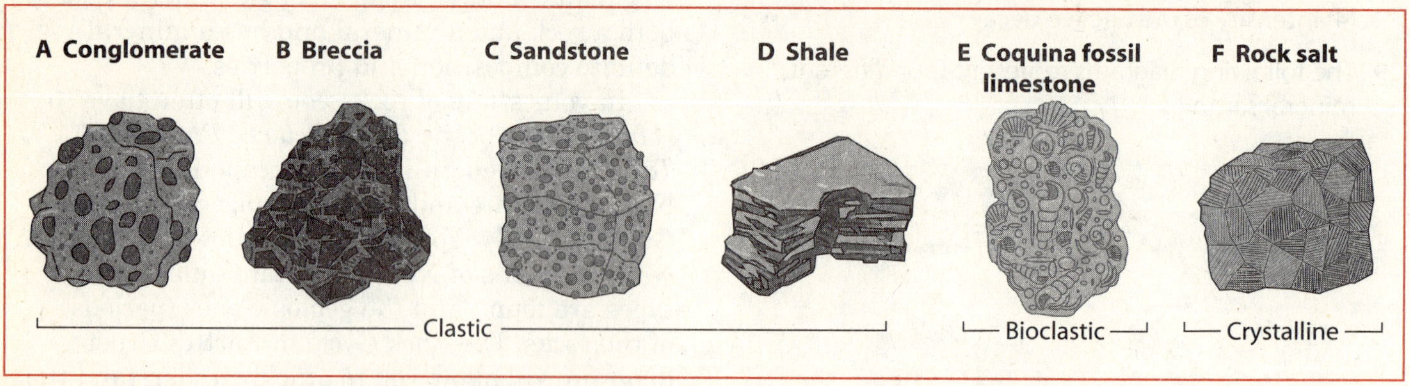

Figure 11-7. Some characteristics of sedimentary rocks: In diagrams A, B, and C the dark shading is sediment and the light color is the cement. **(A)** has unsorted sediments, mostly larger than sand, cemented together. **(B)** has sediments similar to conglomerate that are angular, not rounded. **(C)** has sorted sand-size sediments cemented together. **(D)** has compacted and sometimes cemented clay-sized sediments. **(E)** is an organic rock composed of shells cemented together. **(F)** is a chemical sedimentary rock of intergrown mineral crystals of the mineral halite.

interconnected mineral crystals called a **chemical sedimentary rock or evaporites.** This **precipitation** of minerals is the result of evaporation, saturation with dissolved minerals, or changes in temperature. Chemical sedimentary rocks or evaporites, are composed of interconnected crystals of just one mineral.

DIGGING DEEPER

Most precipitation of minerals occurs at the bottoms of lakes, seas, and oceans; but it also happens in soil, in caves—forming rock "icicles"— around mineral springs, and in homes—forming a rock film on water glasses and shower stalls.

Organic Processes When dissolved minerals are withdrawn from water by life forms, it is termed chemical—not mineral—precipitation. **Organic** means anything related to living organisms or to things that were alive. Any rock made by living organisms or mostly composed of materials from life forms is an organic, or **bioclastic sedimentary rock.** When a clam makes a shell, a coral makes a skeleton, or you make bones and teeth, chemicals are precipitated from water. Some organic sedimentary rocks, such as the limestone of a coral reef, are formed directly by chemical precipitation.

CHARACTERISTICS OF SEDIMENTARY ROCKS

Some features that distinguish sedimentary rocks from igneous and metamorphic rocks include the following:

- Most sedimentary rocks are clastic—composed of fragments called sediments, or clasts.
- The clasts or sediments are usually rounded particles because they have been moved by running water, wind, waves, or ocean currents.
- The clasts or sediments are often sorted into a small range of sizes because of the horizontal sorting described in Topic 10. (See Figure 10-4.)
- Some sedimentary rocks are organic and thus contain fossils. A **fossil** is any evidence of former life.
- One of the most distinguishing characteristics of sedimentary rocks is the beds or strata—parallel layers of the sediments in the rock. These beds are often seen in hand specimens, but are even more obvious outside where sedimentary rocks are exposed at Earth's surface in places such as road cuts or stream valleys.

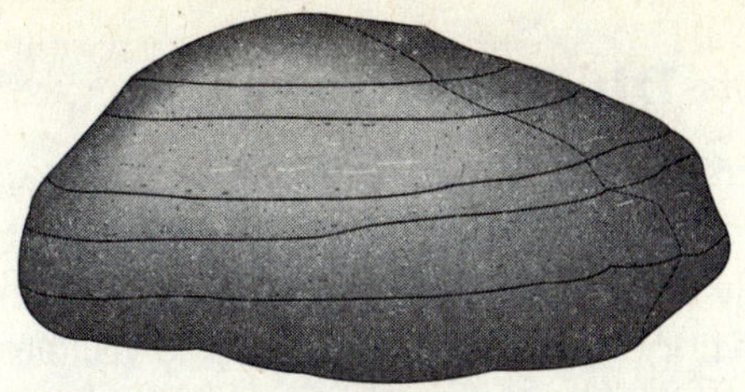

Figure 11-8. Layers in a sedimentary rock: Layers—called beds or strata—result from changes, often minor, in the types of sediment deposited at different times.

- Sedimentary rocks often contain features that indicate they formed at Earth's solid surface. Some features might be mud cracks, rain drop impressions, or ripple marks that formed on the top of a sand dune or at the bottom of the ocean. Other features might be fossils that indicate an earlier Earth's surface.
- The chemical sedimentary rocks are not composed of sediments or clasts, but are composed of interconnected mineral crystals of one mineral variety.

IDENTIFYING SEDIMENTARY ROCKS The more common clastic, or fragmental, sedimentary rocks are distinguished largely on the basis of their sediments—clay, silt, sand, or larger sediment sizes. If the sediments of a conglomerate are mostly rounded, the rock is regular conglomerate. If the sediments are mostly angular the rock is breccia. In basic rock identification, it makes no difference what type of mineral or rock fragments are in a clastic rock—texture is the main factor used for classification and identification. For example, if a rock is composed of sand size sediments, the rock is sandstone. Refer to the chart Scheme for Sedimentary Rock Identification in the *Earth Science Reference Tables.*

This chart also shows that nonclastic rocks are predominantly composed of one mineral. Therefore, you treat most of them as if they are minerals. Bioclastic limestone is composed of $CaCO_3$, so it will easily fizz in dilute hydrochloric acid. If the rock only fizzes after some of it is powdered, then the rock is likely dolostone. For example, rock salt is composed of the mineral halite and rock gypsum is composed of the mineral gypsum.

DIGGING DEEPER

To measure sand or larger sediments in a rock, a ruler can be placed on or next to the rock. If you can't easily see the sediments, but the rock feels gritty to a fingernail, then it is most likely siltstone. You can use a magnifying glass or microscope to see silt-sized sediments. If you can't see the sediments under a normal microscope, or if the rock does not feel gritty, it is likely shale.

Igneous Rocks

Rocks that form when natural, molten (liquid) rock-forming material cools and turns into a solid are **igneous rocks.** Liquid rock material beneath Earth's solid surface is called **magma.** When magma comes out onto or above Earth's solid surface, it becomes lava. If Earth was largely molten in its earliest stages of formation, then igneous rocks were the first rocks to form on Earth.

When magma solidifies beneath Earth's solid surface, it forms rocks called **intrusive** (plutonic) **igneous rocks.** The bodies, or masses, of these rocks can range from finger size up to the size of one or more of our states, such as Vermont. These bodies are called intrusions. Figure 11-9 illustrates many of the types of intrusions from the thin dikes and sills, common as light and dark streaks in eastern New York State road cuts, to massive batholiths. It is believed that most intrusions form, within Earth's solid outer layer—the lithosphere.

When lava solidifies on or above Earth's solid surface, the result is **extrusive igneous rock.** Extrusive, or volcanic rocks, form landscape features called extrusions. The two most common extrusions are lava flows and volcanoes. (See Figure 11-9.)

FORMATION OF IGNEOUS ROCKS All igneous rocks are the result of solidification—the change from a liquid to a solid. Most igneous rocks are produced as a result of the type of solidification called crystallization. Crystallization results when molten lava or magma cools and forms a solid composed of intergrown mineral crystals—a crystalline rock. Some igneous rocks that form at or above Earth's surface cool so fast that mineral crystals don't have a chance to form. The result is a type of solid called glass. In glass there is no pattern or arrangement of the atoms, therefore the substance is non-crystalline.

CRYSTAL SIZES AND GLASSES The size of the crystals in an igneous rock depends on the conditions in which the rock formed. The immediate cause of the difference in the size of the crystals, or lack of crystals, is the time in which the cooling

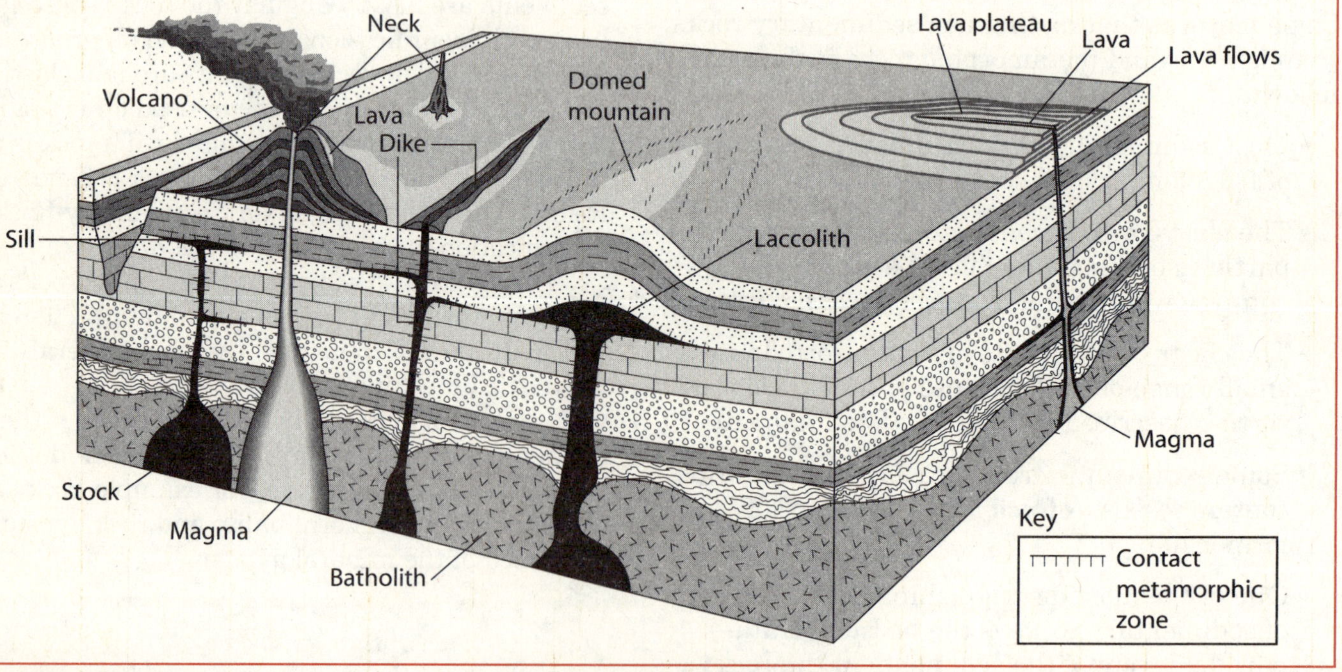

Figure 11-9. Examples of igneous intrusions and extrusions: The types of intrusions are batholith, stock, laccolith, sill, and dike. (Except for sill, you need not know the names of the various types of intrusions.) Note that both the volcano and the laccolith intrusion have formed mountains.

takes place. Generally, the longer the time of cooling, the larger the crystals become. However, the cooling time itself depends on the temperature and pressure of the environment, and the composition of the magma or lava. Generally, molten rock low in silica (SiO_2) content or high in water content will take longer to cool. The pressure and temperature deep within the lithosphere are very high, and therefore magma cools slowly—over many thousand years. The result is rocks with large or coarse crystals easily visible to the human eye.

The temperature and pressure at or near Earth's surface are much lower, and the lava there cools much more quickly, forming fine-grained rocks with small crystals, not easily seen with the unaided human eye. If the cooling is very fast (usually seconds to hours), a glassy rock with no or few mineral crystals forms. Most lava flows and volcanoes are composed of rocks with small mineral crystals or no crystals in them.

TEXTURES OF IGNEOUS ROCKS The texture of igneous rocks depends upon the size of the mineral crystals, presence of glassy material, and presence of rounded pores, or openings. These features are related to the cooling time of the magma and lava and the environment in which the rock formed. Rocks having crystals easily seen with the unaided eye called coarse-textured rocks. These coarse-textured rocks, such as granite and gabbro, almost always form within the lithosphere and are intrusive. Some pegmatite intrusive rocks have a very coarse texture, with some individual mineral crystals measuring over a meter in diameter.

Associated with lava flows and volcanoes, most of the extrusive igneous rocks have a fine texture of crystals smaller than one millimeter in size. To clearly see mineral crystals in fine-textured rocks, such as basalt and rhyolite, magnification is required. Many of the extrusive igneous rocks of lava flows and volcanoes have rounded openings in them caused by lava solidifying around trapped expanding gases. The rocks with these openings are said to have a vesicular texture. These openings or pockets are like the pores in the foam produced when you shake up a carbonated soft drink. Basalt with many pores in it is called scoria, or vesicular basalt. Volcanic glass with pores in it includes vesicular basaltic glass and pumice. See Figure 11-10 for illustrations of the textures of igneous rocks. Also see the Scheme for Igneous Rock Identification in the *Earth Science Reference Tables*.

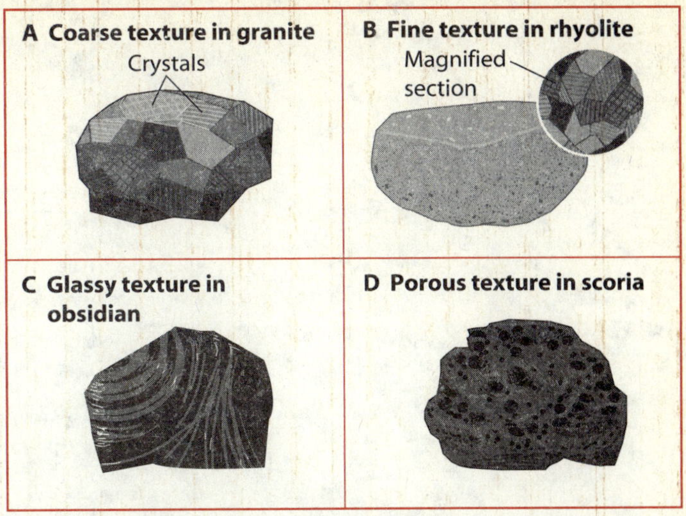

Figure 11-10. Textures of igneous rocks: (A) Slow cooling of magma results in coarse-sized crystals easily visible to the unaided eye. **(B)** Fast cooling of lava at Earth's surface results in fine-sized crystals not easily visible without magnification. A magnified section resembles the texture in **(A)**. **(C)** If lava cools very rapidly, a rock with a glassy texture of no minerals will form. **(D)** In fine—or glassy—textured rocks or extrusive igneous rocks, there is often a mixed, porous vesicular texture. The pores are due to expanding gas forming bubbles in the lava as it solidified.

IDENTIFICATION OF IGNEOUS ROCKS Igneous rocks are identified largely on the basis of texture (very coarse, coarse, fine, glassy, or vesicular) and percent mineral composition. In the coarse-grained rocks you can often identify the minerals easily because they are visible to the unaided eye. You match the percents of what you observe in the rock with the chart of percent by volume of the surface of a rock. With a strong enough microscope the same thing can be done with the fine-textured rocks. For aid in using the percent mineral composition chart see Figure 11-11 on the next page.

When microscopes are not available, both rock density and rock color can be a guide to the mineral composition. Refer to the Scheme for Igneous Rock Identification in the *Earth Science Reference Tables*. This scheme shows a method of classifying and identifying igneous rocks. The color of an igneous rock is more like shade, or tone—it is the overall lightness or darkness of the total rock—not the actual color. Light-colored rocks are on the left side of the scheme and dark-colored rocks are on the right. Nonvesicular igneous rocks range from 2.7 g/cm^3 for low-density rocks (on the left side) to approximately 3.4 g/cm^3 for high-density rocks (on the right side). Thus both density and color hint at mineral composition when the minerals can't be seen, or if identification of the minerals is uncertain.

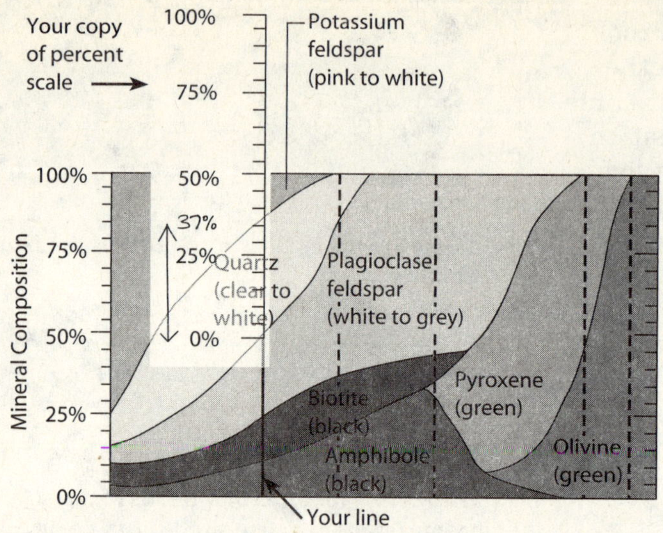

Figure 11-11. How to use the Scheme for Igneous Rock Identification in the *Earth Science Reference Tables* to measure the percent of mineral composition: Refer to the identification scheme in the *Earth Science Reference Tables* as you proceed. To obtain a percent for a particular part of the chart, do the following: (1) Copy the scale from the left side of the chart. (2) Draw a vertical line on the chart in pencil through the "t" in the word "quartz," as shown. (3) To compute the mineral composition of quartz, place your copy of the percent scale on the chart along your drawn line, as shown. (4) Align the 0% of your copy of the scale with the bottom of the quartz portion of the chart and read the percent at the top part of the mineral. The art shows quartz at about 37%. (5) Repeat this procedure for each of the other minerals. The values for the other minerals are: amphibole 10%, biotite 13%, plagioclase feldspar 24%, and potassium feldspar 14%. Usually you can obtain answers within 2–3 percent of the actual value.

The percent mineral composition divides igneous rocks on the diagram into vertical columns of igneous rock families usually named for the coarse-grained member. An example would be the granite family, which includes granite, pegmatite, rhyolite, vesicular rhyolite, pumice, and some obsidian. The left, or granite side, of the chart is also the felsic side, which indicates a high aluminum (Al) and silicon (Si) content compared to the peridotite and dunite side on the right, which is mafic. Mafic rocks are higher in iron (Fe) and magnesium (Mg) and lower in silicon and aluminum. Note that all the minerals listed on this chart are silicates, with the oxygen-silicon tetrahedron as the basic component of the minerals' atomic structure.

Metamorphic Rocks

Rocks that form from changes in previously existing rocks due to heat, pressure, and/or mineral fluids without weathering or melting are **metamorphic rocks.** The previously existing rocks can be sedimentary, igneous, or other metamorphic rocks. The process of forming metamorphic rocks is called **metamorphism.** The changes that produce metamorphic rocks occur within the lithosphere, usually many kilometers deep. The changed rocks resulting from metamorphism are often less porous and more dense than the original rocks. They also have larger mineral crystals, and often have a layering of mineral crystals called **foliation.**

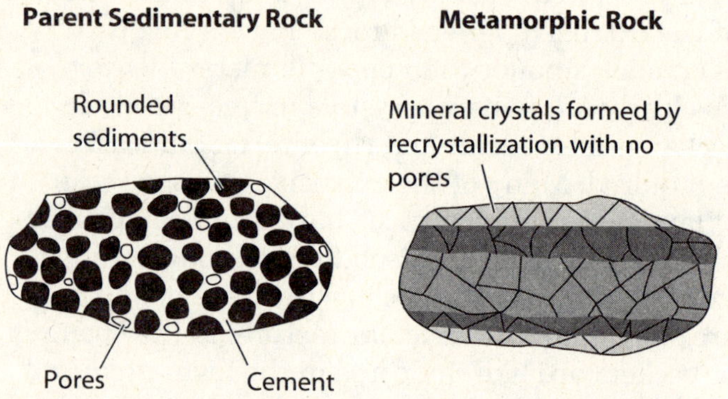

Figure 11-12. Metamorphic rock formed from sedimentary rock by recrystallization: Under the influence of heat and pressure, minerals in this clastic sedimentary rock combine by recrystallization to form mineral crystals of a coarse-texture (banded) crystalline rock. Recrystallization increases the rock's density by reducing the pore spaces.

FORMATION OF METAMORPHIC ROCKS

When metamorphism occurs the previously existing rocks, called parent rocks, are usually recrystallized. Recrystallization is the process of increasing the size of the mineral crystals or rock clasts and/or changing the mineral composition without melting. Under high heat and pressure conditions deep within the lithosphere, atoms can move small distances and become rearranged with changes in mineral composition without true melting, resulting in recrystallization. The various types of metamorphism are described in the sections that follow.

CONTACT METAMORPHISM When older rocks come in contact with the magma of an intrusion or lava of an extrusion, the heat and mineral fluids of the liquid rock alter the older rock in a process called **contact metamorphism.** Figure 11-13 shows the details of contact metamorphism. In a contact metamorphic zone there is a progression from igneous rocks, to metamorphic rocks, to the parent rocks, often without clear separations. At contact metamorphic zones, metamorphic rocks, such as hornfel, some marbles, and some quartzites are formed. Because there is mostly heat and not much directional pressure, the rocks formed by contact metamorphism usually don't have foliation.

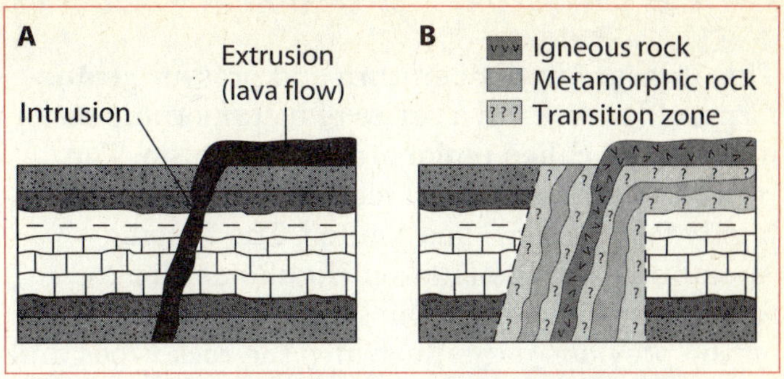

Figure 11-13. Transition of rock types in contact metamorphic zones: In diagram A, molten rock has flowed up through a crack in sedimentary rock to the surface, forming an intrusion below the surface and an extrusion (lava flow) on the surface. In the contact zone between the original local rock and the intrusion or extrusion, there is a blending of rock type from sedimentary, through metamorphic to igneous.

REGIONAL METAMORPHISM Sections of the lithosphere called plates may be hundreds of kilometers in width and tens of kilometers in depth. During the convergence (collision) of these plates, rocks are subjected to the high temperatures and pressures associated with a great thickness of overlying rocks and sediments and the pressures resulting from the collisions. These colliding plates often result in mountain building, which will be discussed in more detail in Topic 12. The closer an area is to the boundary of colliding plates, the greater the increase in temperature and pressure.

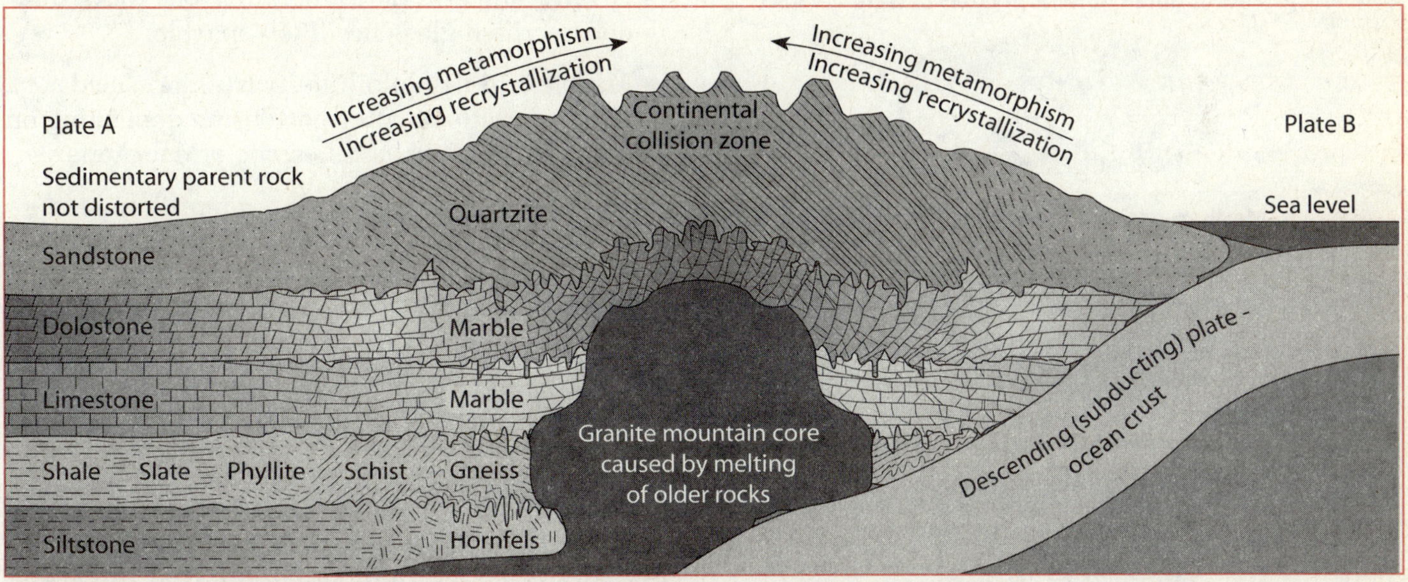

Figure 11-14. Conditions and rocks of regional metamorphism: This extremely ideal view shows two continents that have collided producing a young mountain range. The rocks become increasingly more deformed towards the center of the continents' zone of collision. As heat and pressure increase towards the collision zones, older sedimentary rocks progressively recrystallize into metamorphic rocks. With shale as the parent rock, a series of metamorphic rocks are formed as heat and pressure increases—shale → slate → phyllite → schist → gneiss. In the center of the mountain range, the heat and pressure have become so great that the rocks melted to form a granite intrusion of igneous rock.

This increase in temperature and pressure transforms older rocks to a series of metamorphic rocks in a process called **regional metamorphism.** The rocks formed by regional metamorphism are often highly folded (bent) and faulted. The Taconic Mountains and the Hudson Highlands show features of regional metamorphism. Figure 11-14 on the previous page illustrates the rock types and features formed by regional metamorphism.

TEXTURES OF METAMORPHIC ROCKS

Metamorphic rocks have two major types of textures—foliated and nonfoliated. Foliated rocks have layers of mineral crystals that have formed by recrystallization under directional pressures associated with regional metamorphism. These rocks are composed of two or more minerals and are made of interconnected mineral crystals. The three types of foliations are shown in Figure 11-15.

Nonfoliated metamorphic rocks are not layered because the minerals are not flat, or sheet-like, and/or the rocks were not subjected to a directional pressure. These rocks are composed of interconnected mineral crystals. Two single-mineral, crystalline rocks are quartzite, whose parent rock was pure quartz sandstone, and marble, whose parent rock is either limestone or dolostone.

Figure 11-15. Foliation in metamorphic rocks: There are three types of foliations. Metamorphic foliations are very thin in the variety shown in diagram A. The slightly thicker foliations have mineral crystals that are reasonably visible with various minerals blending together in one foliation as in diagram B. Extreme recrystallization in diagram C has resulted in a separation (segregation) of minerals into broad bands of different color-sometimes called banding.

IDENTIFICATION OF METAMORPHIC ROCKS

Similar to sedimentary and igneous rocks, metamorphic rocks are classified and identified based on composition and texture. If the rock has foliations that are thin, if it breaks into smooth layers, and if the mineral crystals are not easily visible, the rock is slate, or phyllite (if the surface is shiny). If the rock is foliated, the mineral crystals are just clearly visible, and the rock has a high-mica mineral content, then the rock is schist. If the rock has coarse foliations—banded—and the mineral crystals are easy to see and distinguish, then the rock is gneiss. Refer to the Scheme for Metamorphic Rock Identification in the *Earth Science Reference Tables*.

In the nonfoliated metamorphic rocks the composition of the rock is usually the key to identification and classification.

- A rock that looks something like sedimentary conglomerate, but whose crystallized pebbles are stretched out and broken through, is probably a meta-conglomerate.
- A grainy single-mineral rock that easily scratches glass—because the quartz mineral content is much harder than glass—is quartzite.
- Grainy single-mineral metamorphic rocks that don't scratch glass are likely marble.
- Marble will fizz in dilute hydrochloric acid either directly or after powdering depending on if the parent rock is limestone or dolostone.
- Hornfels are very difficult to identify unless you know the specimen was collected next to an intrusion or under an extrusion.

Environment of Rock Formation

The type of environment in which a rock formed is inferred from its composition, structure, and texture. The following are four examples of such inferences:

Large thick areas of rock salt in western New York State led to the inference that there was a large area of salty water in the past that has evaporated. This in turn suggests that the same could happen today in isolated seas of salt water exposed to a hot dry climate, such as the Mediterranean Sea or Black Sea.

The bent and twisted rock structure of surface metamorphic rocks in the Adirondack

Mountains of northern New York State suggests that this region once experienced one or more mountain-building periods. See Landscape Regions of New York State and Their Characteristics on the inside back cover of this book. The rock structure also suggests that much uplift and erosion has occurred to expose rocks that formed deep beneath the surface.

If the sediments in a clastic sedimentary rock are sharp and angular, such as those in breccia, it can be inferred that the rock was formed near where weathering produced the sediments. This is because any long-distance transporting would have rounded the sediments. A piece of igneous rock from a lava flow that has many large mineral crystals mixed with fine-grained crystals may indicate that solidification of magma had begun below the surface to produce the larger crystals. Then the lava with the large crystals erupted onto Earth's surface where the rest of the liquid solidified into rock.

The Rock Cycle

The **rock cycle** is a model used to show how the rock types—sedimentary, igneous, and metamorphic—are interrelated. It also shows the process that produces each rock type. Two examples of the rock cycle are Figure 11-16 and the Rock Cycle in Earth's Crust in the *Earth Science Reference Tables.* Some of the major concepts about the rock cycle are described in the following paragraphs.

ROCKS CHANGE TYPES Any rock type can change into any other rock type. Thus a specimen of any rock type can have materials in it that were once part of any other rock type. Examples include the solid sediments in sedimentary rocks, inclusions of older rocks in igneous rocks, or distorted fossils and sediments in metamorphic rocks.

NO PREFERRED DIRECTION OF MOVEMENT There is no preferred direction of movement of material in the rock cycle for any one mass of chemical elements. Any one piece of mass can stay in one place for any length of time, or it can follow any one of the paths of the rock cycle. As an example, there are sedimentary rocks over 3 billion years old that have remained largely unaltered. On the other hand a newly formed rock from a volcano can be eroded off by a hiker's boot, blown into evaporating water of a mineral spring, and become a sediment in a sandstone.

NO EXACT POINT OF SEPARATION There is often no exact point of separation between the rock types. One example is the progression from igneous rock, to contact metamorphic rock, to the older intruded rock associated with contact metamorphism (See Figure 11-13 on page 201.) In areas of regional metamorphism there are zones many miles wide where identification of the rock type is debatable. Whether the rock is sedimentary shale or metamorphic slate, slate or schist, schist or gneiss, and even metamorphic gneiss or igneous granite is questionable.

DRIVING FORCES The input of energy from Earth's interior, insolation from the sun, and gravity are the driving forces of the rock cycle. These forces create the processes of uplift, erosion and weathering, pressure, and melting.

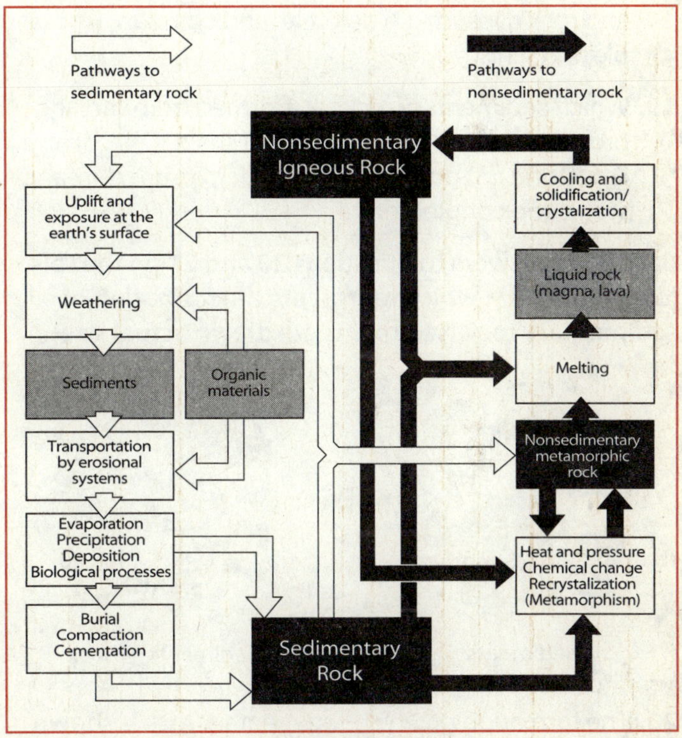

Figure 11-16. The rock cycle: Some of the processes by which one type of rock can be changed to another.

Review Questions

16. When dilute hydrochloric acid is placed on the sedimentary rock limestone and the metamorphic rock marble, a bubbling reaction occurs with both. What would this indicate? (1) The minerals of these rocks are similar. (2) Heat and pressure have changed the molecular structure of these two rocks. (3) The physical properties of these two rocks are identical. (4) The two rocks originated at the same location.

17. What do most igneous, sedimentary and metamorphic rocks have in common? (1) They are formed from molten material. (2) They are produced by heat and pressure. (3) They are composed mostly of minerals. (4) They are found mostly in distinct layers.

18. Which pair of rocks could each be composed of only one mineral? (1) dunite and rock salt (2) peridotite and rock gypsum (3) dolomite and obsidian (4) marble and schist

19. Which sedimentary rocks are formed from mostly organic matter? (1) rock salt and shale (2) coal and limestone (3) dolostone and rock gypsum (4) sandstone and conglomerate

Base your answers to questions 20 and 21 on the following diagram which represents the formation of a sedimentary rock (sediments are drawn actual size).

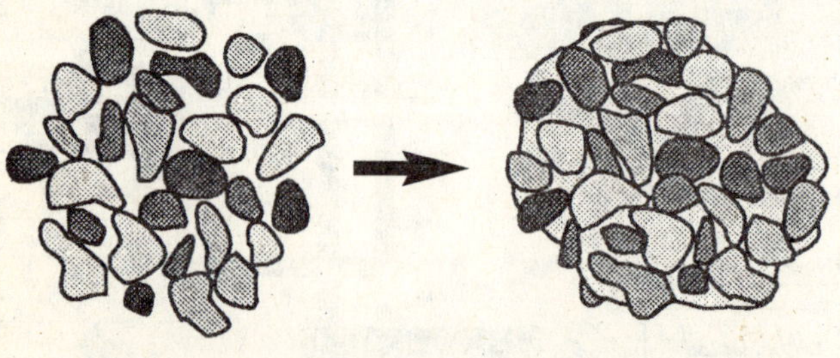

20. The formation of which sedimentary rock is shown in the diagram?

21. Which two processes formed this rock? (1) folding and faulting (2) melting and solidification (3) compaction and cementation (4) heating and application of pressure

22. Which would most likely occur during the formation of igneous rocks? (1) compression and cementation of sediments (2) recrystallization of unmelted material (3) solidification of molten materials (4) evaporation and precipitation of dissolved sediments

23. Which graph best shows the relationship between the size of the crystals in an igneous rock and the length of time it has taken the rock to solidify?

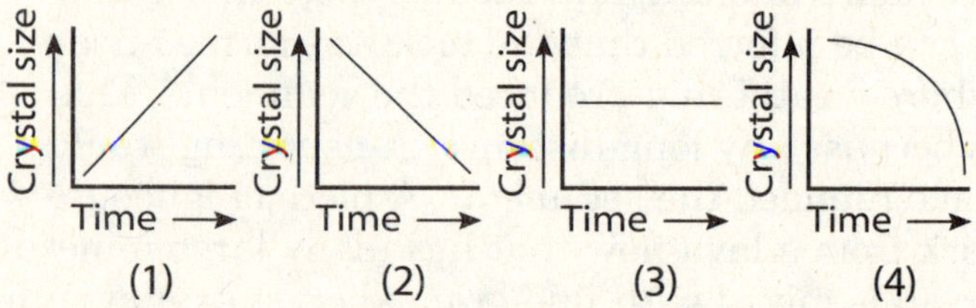

24. Which diagram below shows an area in which fine-grained igneous rocks are most likely to be found?

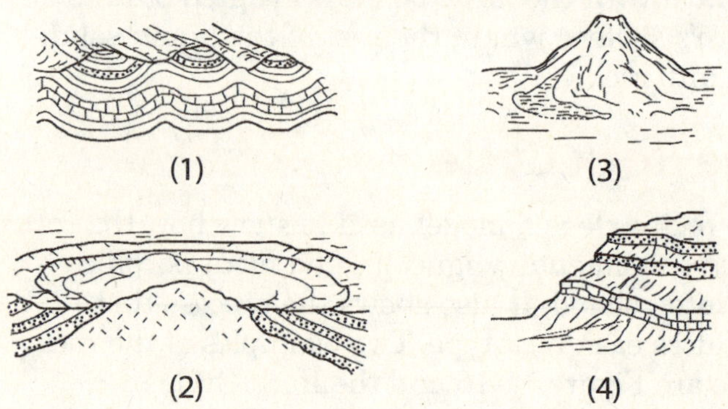

25. The green sand found on the shores of the Hawaiian Island volcanoes most probably consists of the mineral (1) quartz (2) olivine (3) biotite mica (4) potassium feldspar

26. Generally as the percentage of felsic minerals in a rock increases, the rock's color will become (1) darker and its density will decrease (2) lighter and its density will increase (3) darker and its density will increase (4) lighter and its density will decrease

27. The diagram below represents a cross section of a coarse-grained igneous rock (drawn to true scale). This rock is most likely (1) rhyolite (2) scoria (3) basalt (4) granite

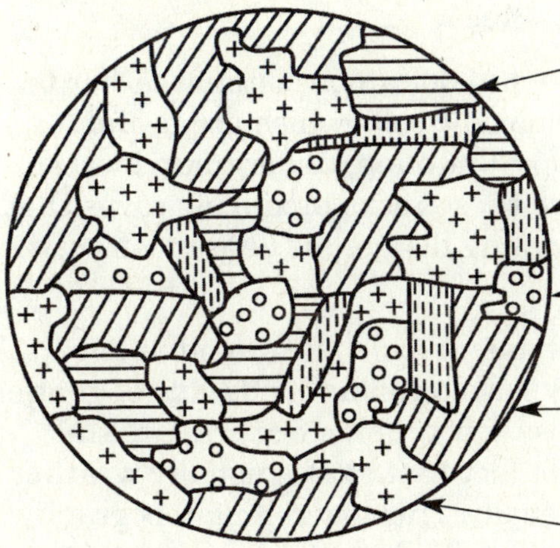

28. After collecting samples of igneous rocks, a student wishes to classify them as either intrusive or extrusive. Which characteristic of the samples might be the most useful to use?

29. A fine-grained rock has the following mineral composition: 50 percent potassium feldspar (orthoclase), 26 percent quartz, 13 percent plagioclase feldspar, 8 percent biotite mica, and 3 percent of the amphibole hornblende. The rock would most likely be (1) granite (2) rhyolite (3) gabbro (4) basalt

30. Metamorphic rocks form as the direct result of (1) precipitation from evaporating water (2) melting and solidification of magma (3) erosion and deposition of soil particles (4) heat and pressure causing changes in existing rock

31. What is the main difference between metamorphic rocks and most other rocks? (1) Most metamorphic rocks contain only one mineral. (2) Many metamorphic rocks have an organic composition. (3) Many metamorphic rocks exhibit foliation and distortion of structure. (4) Most metamorphic rocks contain a high amount of oxygen and silicon.

32. The regional metamorphism of a sandstone rock will cause the rock (1) to be melted (2) to recrystallize into smaller rock fragments (3) to become denser (4) to occupy a greater volume

33. Slate is formed by the (1) deposition of feldspars and micas (2) foliation of schist (3) metamorphism of shale (4) folding and faulting of gneiss

34. Which rock is composed of materials that show the greatest variety of rock origins? (1) a limestone composed of coral fragments cemented together by calcium carbonate (2) a conglomerate composed of pebbles of granite, siltstone, and gneiss (3) a very fine-grained basalt with sharp edges (4) a sandstone composed of rounded grains of quartz and feldspars

35. What ancient environment is the most likely inferred by the large rock salt deposits in the Syracuse, New York area?

36. Which statement about a large bedrock exposure of granite and gneiss is most likely correct? (1) a number of volcanoes is nearby (2) the granite and gneiss are the result of lava flows (3) the rocks were never under water (4) a great deal of erosion has taken place at this location

37. Which statement about inorganic rocks is true? (1) all inorganic rocks are formed from other rocks (2) all inorganic rocks except igneous rocks are formed from other rocks (3) all inorganic rocks except metamorphic rocks are formed from other rocks (4) all inorganic rocks except igneous and metamorphic rocks are formed from other rocks

Mineral Resources

Except for energy from the sun and the few things people might use from meteorites and similar objects, all things that people need come from Earth. These things, such as water to drink, air to breathe, plants to eat and use for lumber, and animals to provide clothing and milk products, are types of natural resources. Earth materials—including minerals, rocks, and fossils fuels—are grouped together as **mineral resources.** In a lifetime, each person in the United States, on the average, is responsible for the consumption of some $3\frac{1}{2}$ million pounds of minerals, rocks, and mineral fuels that are extracted from the crust of Earth.

Mineral Resources Are Nonrenewable

Some types of natural resources are renewable—which means they are replaced, or can be replaced, by Earth or sun processes at rates similar to the rates at which humans use them. Renewable natural resources include drinking water, trees, soil, oxygen, fish, and electromagnetic energy from the sun. Mineral resources are nonrenewable natural resources. This means that once the minerals are extracted—by mining or drilling—they are gone and will not be replaced at rates comparable to human life spans. Thus, there is a limited supply of the mineral resources, such as oil, gold, copper ore, sulfur, pure white marble, and other similar resources.

MEMORY JOGGER

You may recall that an ore is a rock or mineral deposit that can supply enough of a needed material to make it worthwhile to mine or drill for it from Earth's crust. Most minerals are mined to obtain one or more of the chemical elements the mineral contains.

Rock Properties and Humans

Humans often use rocks because of the characteristics of the whole rock. Some examples include the following:

- Slate is impermeable and cleaves along foliations to produce thin flat pieces that can be used for roofing and chalkboards.
- Basalt resists crushing, so it is used as a base under roads and railroad tracks.
- Coal can burn releasing much heat energy.
- The natural pore space and low density of pumice make it useful for building insulation.
- Granite and quartzite are very resistant to weathering because of their nonporous composition, thus they are used as building stone.

Rock Properties and Land Usage

The type of rock that underlies the landscape of an area has many effects on how people use the land. Some examples follow:

- Limestones and dolostones often weather to produce nonacidic soils which are useful for certain crops; but often, underground water can dissolve these rocks producing caves that collapse forming dangerous depressions called sinkholes.
- Regions having much shale are often low in topography with a rolling landscape of gentle slopes. These topographic features are the result of the fact that shales usually weather and erode easily. This low rolling topography makes it easy to build homes and other human constructions such as roads. The high populations and many transportation facilities of the Hudson-Mohawk Lowlands of eastern New York State are often built on shale.
- Certain rock types withstand tremendous weight without crumbling or flowing under pressure. These types of rocks, such as gneiss and granite, occurring near Earth's surface allow for the safe construction of high-rise buildings, such as in New York City.

Mineral Properties and Humans

The physical properties of a mineral often determine how people use the mineral. Some examples of minerals and their useful properties are listed below:

- Quartz, when put under even minor amounts of pressure, will vibrate in a very regular fashion, which makes it ideal to use in the inexpensive yet very accurate quartz watches.
- Many minerals with a high hardness are used in jewelry and as abrasives in sandpapers and drilling operations. Diamond, corundum (ruby when red and sapphire when blue), the garnets, and quartz (amethyst when purple and citrine when golden yellow) are used in great volumes in the jewelry and abrasive industries. Mines in the Adirondack Mountains of New York State are one of the largest producers of garnet in the world.

- Graphite is soft and has a black streak that makes it very useful as a main ingredient of pencil "lead." The lead is largely graphite, not the element lead.
- Talc is very soft and has one direction of cleavage, which makes it feel smooth or greasy. For these reasons it is used in cosmetics and many baby and foot powders. Portions of the Adirondack Mountains have talc mines.
- Hematite's red color has proven to be useful to humans for many thousands of years. It is the red color of cave paintings, many cosmetics, and most red paints.
- The use(s) of many minerals is listed in the chart "Properties of Common Minerals" in the *Earth Science Reference Tables*.

Global Distribution of Mineral Deposits

As shown in Figure 11-17 many of the more important elements required by our society are found in ores located outside of the United States. As the population of the United States and the rest of the world continues to grow, demand for the limited rock and mineral resources becomes more and more competitive. At present the United States has to import almost 100 percent of the ores for aluminum, tungsten, cobalt, manganese, and graphite—plus over 50 percent of the ores for tin, zinc, nickel, and chromium. These materials are crucial for the production of much of our advanced industrial and military equipment.

Fossil Fuels

Fossil fuels—which include oil (or petroleum), natural gas, coal, and oil shale—are the result of compaction and organic chemical changes in large deposits of organic sediments. These sediments are the remains of dead plants and animals.

Figure 11-18A on the next page indicates that at present three nonrenewable fossil fuels provide 90 percent of our energy needs. At present the United States can provide a large proportion of our coal and natural gas demands, though our reserves are dwindling fast. Over 50 percent of the oil used in the United States is imported. New York State produces small amounts of oil and natural gas, but has no commercial coal deposits.

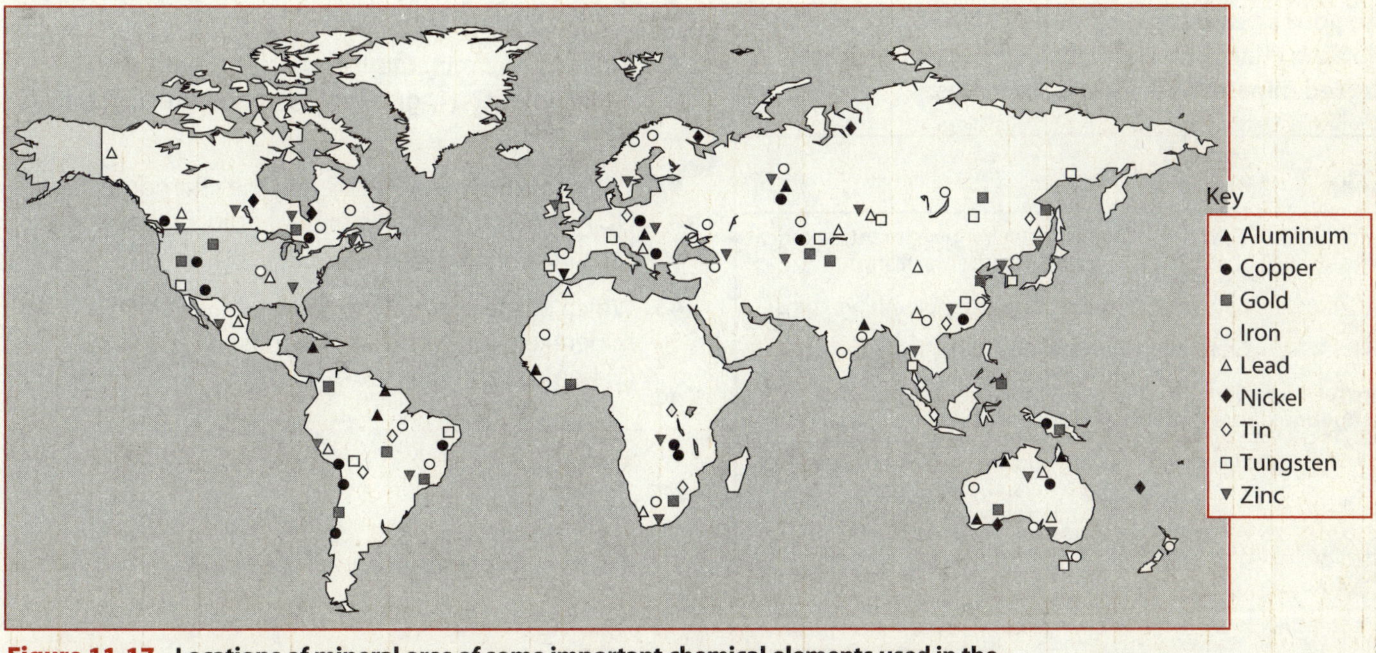

Figure 11-17. Locations of mineral ores of some important chemical elements used in the United States: The map shows that the United States must import much of the rock and mineral sources of aluminum, nickel, tin, tungsten, and zinc.

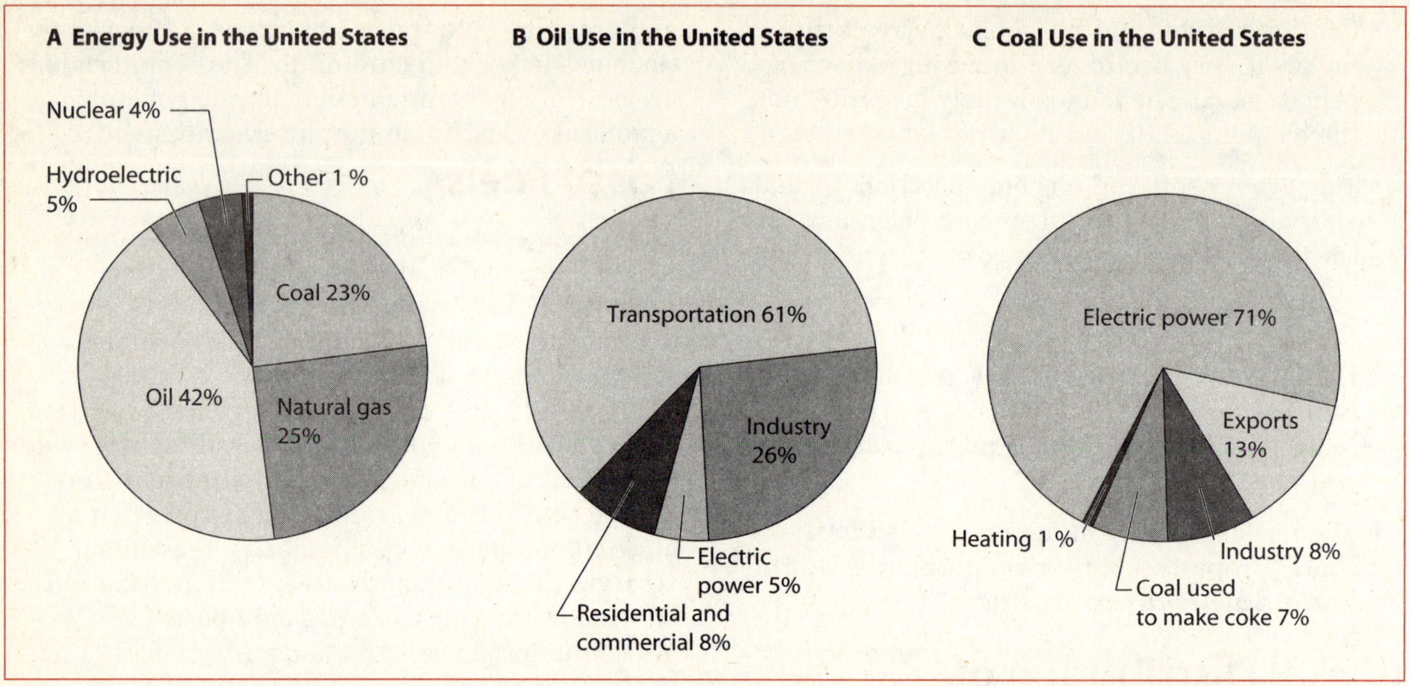

Figure 11-18. Pie graphs of sources of energy and uses of two fossil fuels in the United States: Diagram A lists the source of the energy we presently use in our nation. 90 percent of the energy comes from the nonrenewable fossil fuels-oil (petroleum), natural gas, and coal. Diagrams B and C illustrate the major uses of oil and coal in our society. Much of the oil is used in cars, trucks, trains, and airplanes (transportation), while much of the coal is burned to generate electricity.

Review Questions

38. Which geological resource in New York State resulted from glaciation? (1) coal and oil deposits (2) sand and gravel deposits (3) iron and zinc ores (4) garnet and quartz crystals

Base your answers to questions 39 through 43 on the following chart. The chart shows information about selected mineral and energy resources.

Group	Mineral Resource	Uses
Elements	Gold	coins, jewelry, investment, electrical conductors, dental fillings
	Copper	electrical wiring, plumbing, coins
	Graphite	lubricants, pencil "lead"
Mineral Compounds	Hematite (ore of iron)	construction, motor vehicles, machinery parts
	Halite	food additive, melting of ice, water softeners, chemicals
	Feldspar	abrasives (sandpaper), jewelry
Fuels	Coal	heating, electric generation plants, plastics and other synthetic chemicals
	Petroleum	automobile fuel, lubricants, plastics and other synthetic chemicals, medicines, heating

39. Name two resources that would last longer if people used public transportation.

40. What is the primary source of all the resources listed in the chart ? (1) recycled and discarded waste materials (2) deposits within Earth's crust (3) substances extracted from ocean water (4) meteorites that came from outer space

41. Which of the minerals listed below contain only one element? (1) graphite (2) hematite (3) halite (4) garnet

42. Which of the minerals listed in the chart does NOT have a metallic luster? (1) gold (2) copper (3) graphite (4) halite

43. Which of the materials from the chart is NOT a mineral or NOT made of minerals? (1) copper (2) hematite (3) halite (4) coal

44. The following table shows some properties of rocks.

Rock	Properties
Granite	Hard, coarse crystalline, long-lasting
Pumice	Low density, very porous
Sandstone	Porous, can be infiltrated by water
Slate	Hard, cleaves into thin sheets

Which rock would be the best one to use to face a government building or to build a monument?

45. Wood and coal are both organic natural resources. Explain why coal is considered a fossil fuel, while wood is not.

Questions for Regents Practice

Part A

1. Which of the following pairs of rocks usually contains only one mineral?

(1) rock gypsum and marble

(2) sandstone and conglomerate

(3) quartzite and schist

(4) dunite and gabbro

2. Which of the following properties is most useful in mineral identification?

(1) hardness

(2) color

(3) size

(4) texture

3. Which property of minerals is illustrated by the peeling of biotite mica into thin flat sheets?

(1) fracture

(2) cleavage

(3) a low hardness

(4) a weak streak

4. The main difference between sedimentary, metamorphic, and igneous rocks is the

(1) means by which they are located

(2) conditions under which they are formed

(3) minerals of which they are composed

(4) location in which they are found

5. Which rock was formed by the compaction and cementation of particles 0.07 centimeters in diameter?

(1) conglomerate

(2) sandstone

(3) shale

(4) siltstone

6. Dolostone and rock gypsum are formed by the processes of

(1) melting and solidification

(2) evaporation and precipitation

(3) erosion and deposition of clastic fragments

(4) weathering and metamorphism

7. Which would most likely cause molten rock material to become glassy igneous rock?

(1) cooling over a long period of time

(2) cooling under high pressure

(3) cooling on Earth's surface

(4) cooling at great depth within the crust

8. An igneous rock, which has crystallized deep below Earth's surface, has the following approximate composition: 70 percent pyroxene (augite), 15 percent plagioclase feldspar, and 15 percent olivine. What is the name of this igneous rock?

(1) granite

(2) rhyolite

(3) gabbro

(4) basalt

Use the following diagram to answer questions 9 and 10. The diagram shows an igneous rock intrusion in sedimentary rock layers.

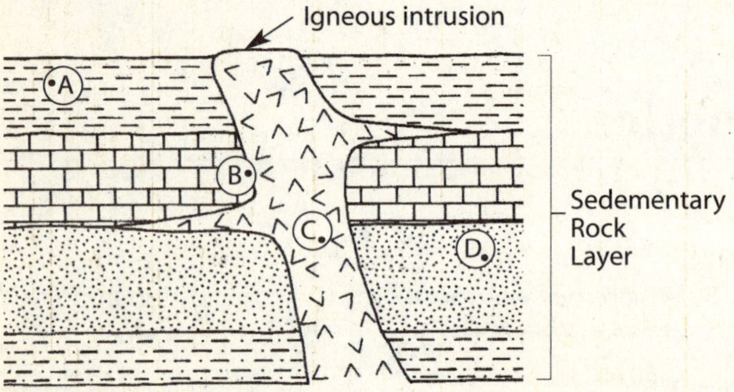

9. At which point would there most likely be contact metamorphic rock?

(1) A (2) B (3) C (4) D

10. Which rock type would most likely be located at the contact between rocks C and D?

(1) metaconglomerate

(2) gneiss

(3) marble

(4) quartzite

11. Which actual-size diagram best represents a sample of the metamorphic rock gneiss?

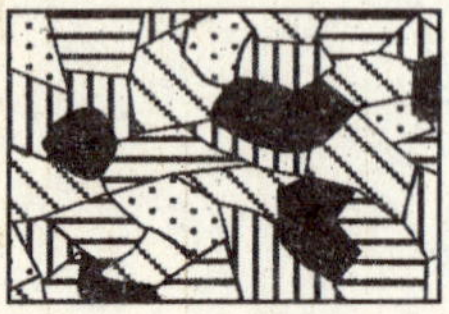

(1)

(3)

(2)

(4)

12. In which parts of New York State would you most likely find large amounts of bedrock formed by regional metamorphism?

(1) Atlantic Coastal Plain and Newark Lowlands

(2) Hudson Highlands and Adirondack Mountains

(3) Tug Hill Plateau and Allegheny Plateau

(4) Erie-Ontario Lowlands and the Catskills

13. Which type(s) of rock can be the source of deposited sediments?

(1) igneous and metamorphic rocks, only

(2) metamorphic and sedimentary rocks, only

(3) sedimentary rocks, only

(4) igneous, metamorphic, and sedimentary rocks

14. A certain igneous rock is composed of large mineral grains. This suggests that the rock formed

(1) on the surface, under high pressure, and at a rapid rate of cooling

(2) on the surface at high temperature, and at a slow rate of cooling

(3) deep underground under high pressure, at high temperature, and at a rapid rate of cooling

(4) deep underground under high pressure, at high temperature, and at a slow rate of cooling

15. Which characteristic would indicate that a rock was formed from sediments deposited in shallow water near shore rather than in deep water?

(1) hardness (2) dark color (3) a large grain size (4) a large amount of cement

Part B

Base your answers to questions 16 and 17 on the diagram below and the *Earth Science Reference Tables*. The diagram shows the elements found in four minerals.

	O	Si	Al	Fe	Ca	Na	C
Quartz	■	■					
Feldspar	■	■	■		■	■	
Olivine	■	■		■	■		
Diamond							■

■ = element present

16. Which of the minerals in the diagram has the greatest variety of elements in it? [1]

17. Which of the elements listed in the chart is second in abundance, by mass, in Earth's crust? [1]

Follow these directions for questions 18 through 23. The following numbered diagrams represent mineral specimens. Using these diagrams write the name of the mineral which is best described by each of the statements. (If a mineral has cleavage, the diagram illustrates it.)

1 Colorless

2 Colorless

3 Colorless to white

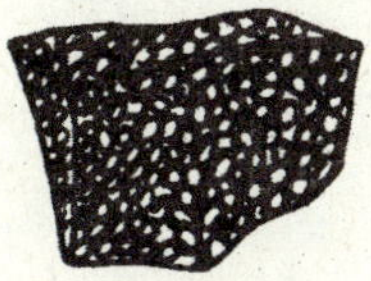

4 Red

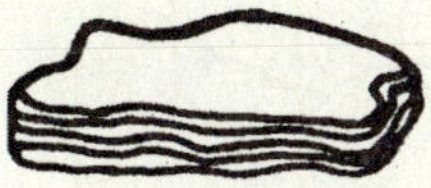

5 Black

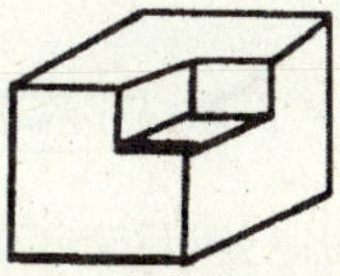

6 Gray

18. Diagram 1 is a mineral that easily bubbles when exposed to dilute acids. [1]

19. Diagram 2 is a very hard mineral that has a curved fracture. [1]

20. Diagram 3 shows intergrown crystals of this salty tasting mineral. [1]

21. Diagram 4 is an ore of iron with a red streak. [1]

22. Diagram 5 is a soft mineral with cleavage that forms thin flexible sheets. [1]

23. Diagram 6 is an ore of lead that is soft and has a metallic luster. [1]

Base your answers to questions 24 through 28 on the following diagram. The diagram represents a profile view of exposed rock layers. The layers are labeled A through H.

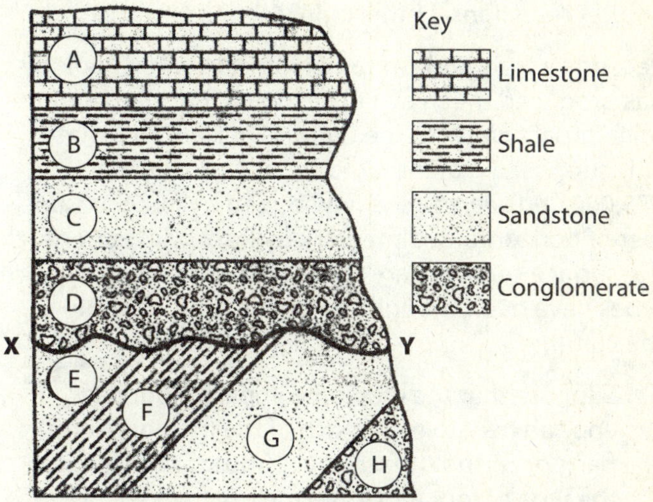

24. State the range of particle sizes of the sediments that formed rock layer C. [1]

25. State two ways in which the composition of rock layer A differs from the composition of rock layer B. [2]

26. State a method by which rock layer A could have formed. [1]

27. Based on information in the diagram, state a reason why you would choose to use rock from layer A instead of rock from layers C or D for a tombstone or statue. [1]

28. State the name of the sediment that was compacted to form rock unit B. [1]

---

29. Describe two conditions that can result in the metamorphosis of a rock. [2]

30. If an igneous rock layer is composed of vesicular andesite, identify three types of minerals that could be found in sand weathered from the rock layer [1]

Part C

The following paragraph provides information about a meteorite impact in New York State. Use data from the paragraph and your knowledge of earth science to answer questions 31 through 34.

Recently a New York State Geological Survey geologist has produced much evidence indicating a large meteorite impact crater in the Catskills of New York State. This supposed crater that formed approximately 375,000,000 years ago is now buried by hundreds of feet of horizontal sedimentary rock. By drilling beneath the surface sedimentary rock, samples of various rock types have been brought up from in and around the crater.

31. Suppose that the meteorite impact produced enough heat to melt much of the meteorite and parts of Earth's surface at the impact site. Describe the type of rock and its texture that would form from the melted meteorite and Earth rocks. [2]

32. Further out from the center of the crater, the meteorite impact caused high temperatures and very high pressure, but the temperatures were not high enough to melt the rocks. What type of rock would form in this environment? [1]

33. At the time of the impact, solid angular rock fragments, mostly larger than sand grains, were hurled far out from the crater. Describe how these fragments could form a sedimentary rock and suggest the most likely name of this sedimentary rock. [2]

34. State the geologic eon, era, and period when this supposed impact crater was formed. [2]

---

35. You observe that a sample of mineral has many smooth sides or surfaces. The sample has not been cut, sanded, or otherwise smoothed by people. Describe two ways in which the parallel smooth sides of the mineral specimen were most likely formed. Then describe an experiment that could help you determine which of the two ways formed the smooth sides on this mineral specimen. [3]

36. Two students were given nearly identical samples of the same mineral and were asked to identify the mineral. Student A said the mineral was graphite because it had a silver color and had a low hardness. Student B said the mineral was hematite because of its silver color and low hardness. Describe two mineral tests that could be used to determine the correct identification of the mineral. [2]

Base your answers to questions 37 and 38 on the rock cycle diagram below.

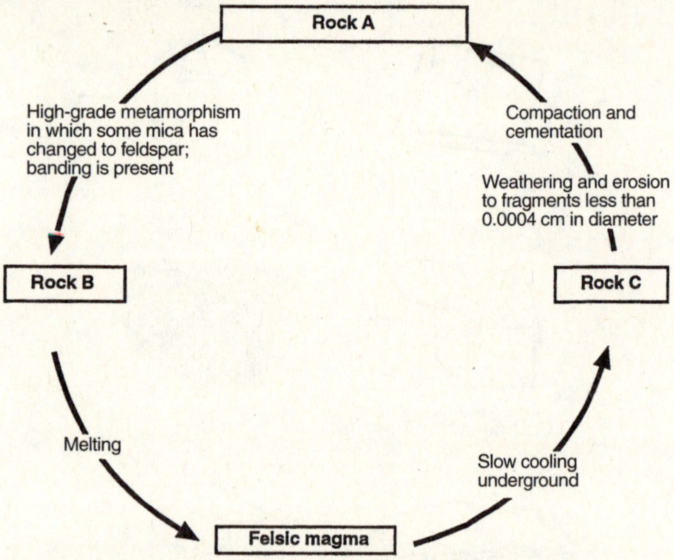

37. State the specific names of rocks *A*, *B*, and *C* in the diagram. Do *not* write the terms "sedimentary," "igneous," and "metamorphic." [3]

38. State *one* condition or process that would cause the high-grade metamorphism of rock *A*. [1]

---

39. A family wants to use rock materials as flooring in the entrance of their new house. They have narrowed their choice to granite or marble. Which of these rocks is more resistant to the physical wear of foot traffic and explain why this rock is more resistant. [2]

40. The mineral Wollastonite forms during the intense metamorphism of a sandy limestone. The expression below shows part of the process that results in the formation of wollastonite.

		Metamorphism		
$CaCO_3$ +	SiO_2	→	$CaSiO_2$ +	CO_2
Mineral 1	Mineral 2		Wollastonite	Carbon dioxide

a Name the *two* minerals involved in the formation of wollastonite. [1]

b What *two* conditions normally cause intense metamorphism? [1]

Earth's Dynamic Crust and Interior

TOPIC 12

VOCABULARY		
asthenosphere	island arc	P-waves
continental crust	lithosphere	seismic wave
convergent plate boundary	lithospheric plate	subduction
crust	mantle	S-waves
divergent plate boundary	mid-ocean ridge	tectonic plate
earthquake	Moho	transform plate boundary
epicenter	oceanic crust	tsunami
faulted	ocean trench	uplifted
folded	original horizontality	volcanic eruption
hot spot	outer core	volcano
inner core	plate	young mountains
	plate tectonic theory	

Much evidence indicates that Earth's crust is constantly undergoing change. You know that the crust moves because you have witnessed or read about earthquakes and the eruption of volcanoes. You also may have heard about measurements that indicate that mountains are "growing" and that the landmasses are changing their positions on Earth's surface.

Evidence of past movements of the crust is more indirect. Scientists must "read" the characteristics of rocks to learn about these past events. Thus, scientists have learned to understand Earth's lithosphere, surface features, and plate movements as well as the properties of Earth's interior. The **lithosphere** is the layer of rock that forms the outer shell at the top of Earth's interior. Earth's **crust** is the upper portion of the lithosphere. Note that whatever is stated in this chapter about the crust usually applies to the lithosphere as well.

Small Scale Crustal Changes

Much of the evidence for past movements of the crust is based on the concept of **original horizontality**. This concept assumes that sedimentary rocks and some extrusive igneous rocks, such as lava flows, form in horizontal layers parallel to Earth's surface. The layers of sedimentary rocks and extrusive igneous rocks are called strata or beds. Therefore, most strata found in positions other than horizontal are thought to have been deformed by crustal movement. (See Figure 12-1 on the next page.)

Rock layers (strata) that no longer show their original horizontality are called deformed layers. Some of the types of deformed layers are folded, tilted, and faulted. **Folded** rock layers are bent or curved. Tilted rock layers are slanted or tipped. **Faulted** rock layers are offset or displaced along a type of crack called a fault. A fault is a crack in a mass of rock along which there has been displacement, shifting, or movement of the rock layers on each side of the crack.

Displaced marine (ocean) rocks as well as rocks with fossils are often found in sedimentary rock hundreds or thousands of meters above sea level. A fossil is any evidence of former life. These displaced rocks and fossils indicate that the land has been raised up, or **uplifted,** to its present position. A lowering of sea level could not have changed the location of the fossils because it is believed that sea level has only ever varied a few hundred meters. On the other hand, fossils from shallow water and from land have been found many hundreds of meters below sea level, which indicates a sinking or lowering of part of Earth's crust.

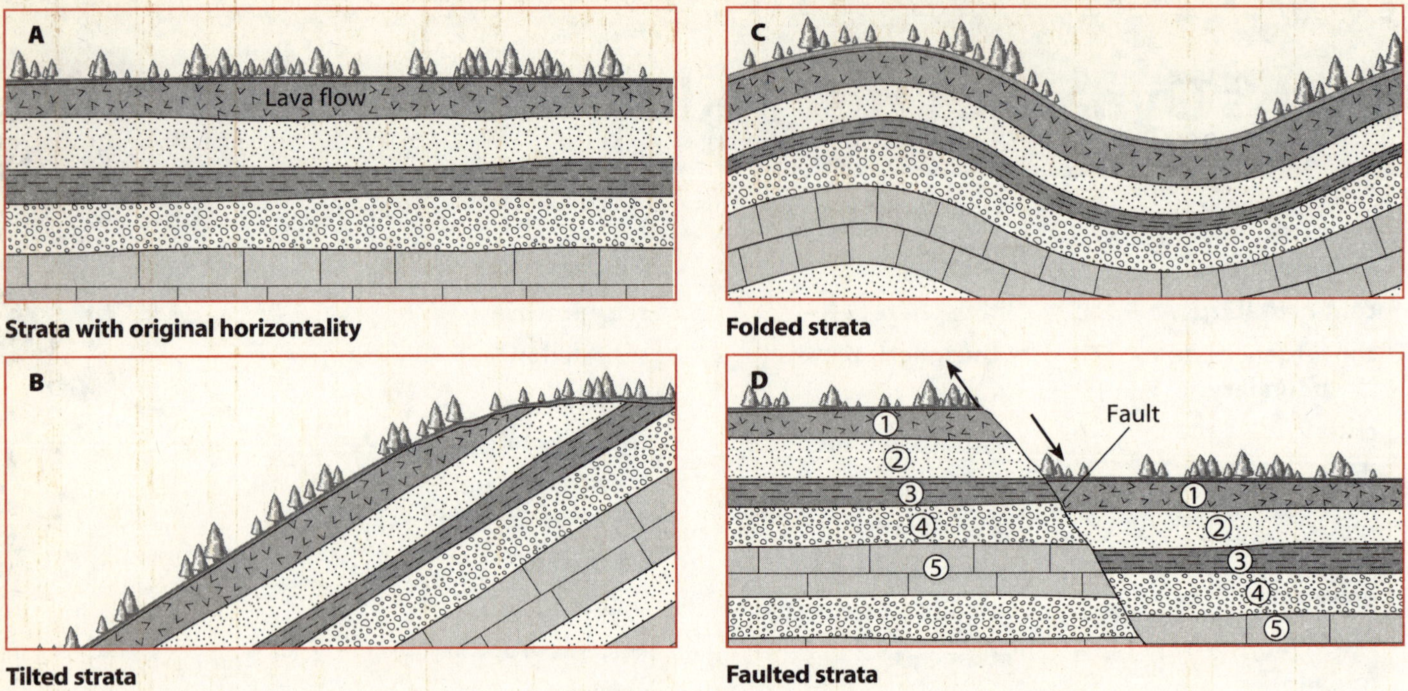

Figure 12-1. Strata, illustrating original horizontality and three types of deformed strata: The numbers in diagram D indicate strata that were originally continuous. The arrows show the direction of relative movement along the fault.

Review Questions

1. The following diagram represents a vertical cross section of sedimentary rock layers that have not been overturned. Which principle best supports the conclusion that these layers have undergone extensive movement since deposition?

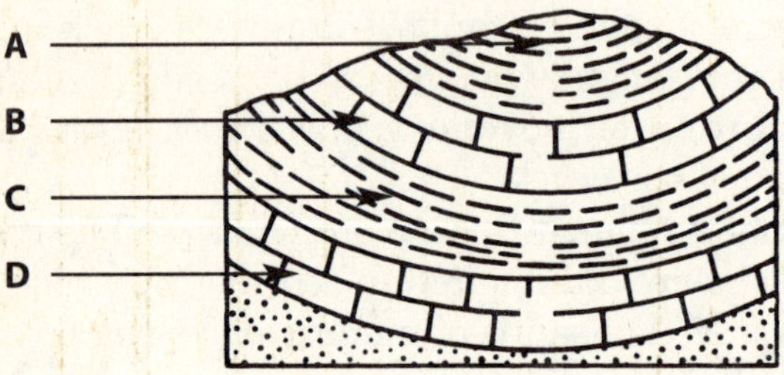

(1) Sediments are deposited with the youngest layers on top. (2) Sediments are deposited in horizontal layers. (3) Rock layers are older than igneous intrusions. (4) Sediments containing the remains of marine fossils are deposited above sea level.

2. The following diagrams show cross sections of exposed bedrock. Which cross section shows the least evidence of crustal movement?

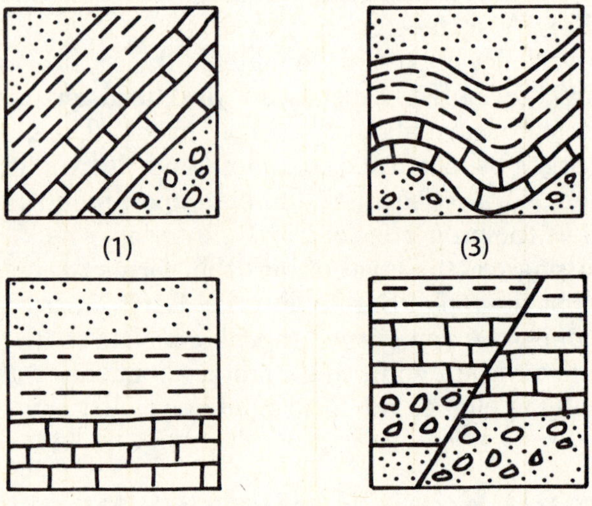

3. Draw a diagram of folded strata. Explain a geological event that might account for this deformation.

4. The following diagram represents a cross section of a portion of Earth's crust. What do these tilted rock layers suggest?

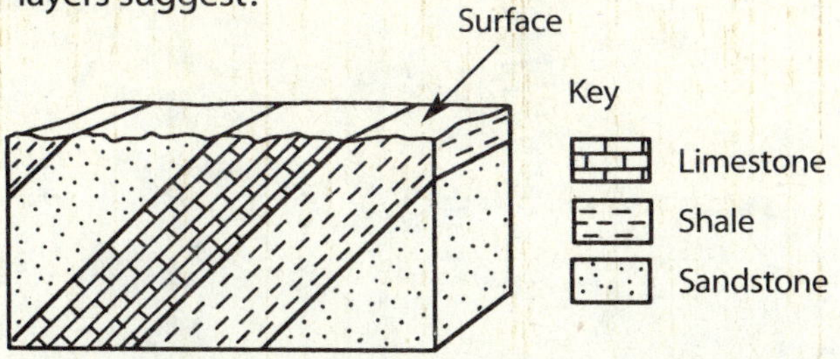

(1) This area remained fairly stable since the sediments were deposited. (2) The sediments were deposited at steep angles and then became rock. (3) Metamorphism followed the deposition of the sediments. (4) Crustal movement occurred sometime after the sediments were deposited.

5. A line of former beaches along a coast, all 50 meters above sea level, is evidence of (1) present erosion (2) the present melting of polar ice caps (3) land uplift (4) a decrease in the deposition of marine fossils

6. While hiking with a friend in New York State, you discover marine fossils in sedimentary strata at high elevations. What is the most logical explanation for their presence?

7. Recent measurements of elevation in New York State indicate that the land is slowly rising in the Adirondack Mountain region. Which statement best explains this change? (1) The Adirondack Mountains are in a region of crustal uplift. (2) The Adirondack Mountains are in a zone of few earthquakes. (3) The rocks in the Adirondack Mountains are younger than those in other regions of New York. (4) The gravitational attraction of the moon is greater in the higher Adirondack region.

8. Shallow-water fossils are found in rock layers deep beneath the ocean floor. This suggests that (1) shallow-water organisms always migrate to deeper water to die (2) parts of the ocean floor have been uplifted (3) parts of the ocean floor have sunk (4) the surface water cooled off, killing the organisms

Earthquakes and Igneous Activity, Including Volcanoes

An **earthquake** is a natural rapid shaking of the lithosphere caused by the release of energy stored in rocks. Most earthquakes are caused by the movement along faults, or faulting. Some earthquakes are associated with the movements of magma within the lithosphere and with volcanic eruptions. During an earthquake the potential energy stored in the rocks is given off as **seismic waves,** or earthquake waves. As shown in Figure 12-2, the earthquake starts at the focus, from which the waves are emitted. The location on Earth's land or water surface directly above the focus is called the **epicenter.** You often see epicenters plotted on maps of Earth's surface.

Scientists measure and record earthquake waves by using a seismograph—an instrument that shows how Earth shakes from the seismic waves. A seismograph records (on paper or another medium) "wiggles"—amplitudes of the wave—that represent Earth's shakings. The recording is called a seismogram. (See Figure 12-3.)

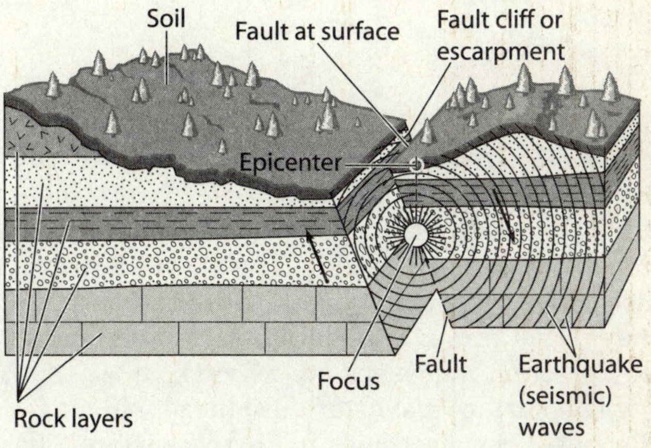

Figure 12-2. Terminology of an earthquake: Note the displacement of the rock layers along the fault. Also note that the epicenter is on Earth's surface directly above the focus.

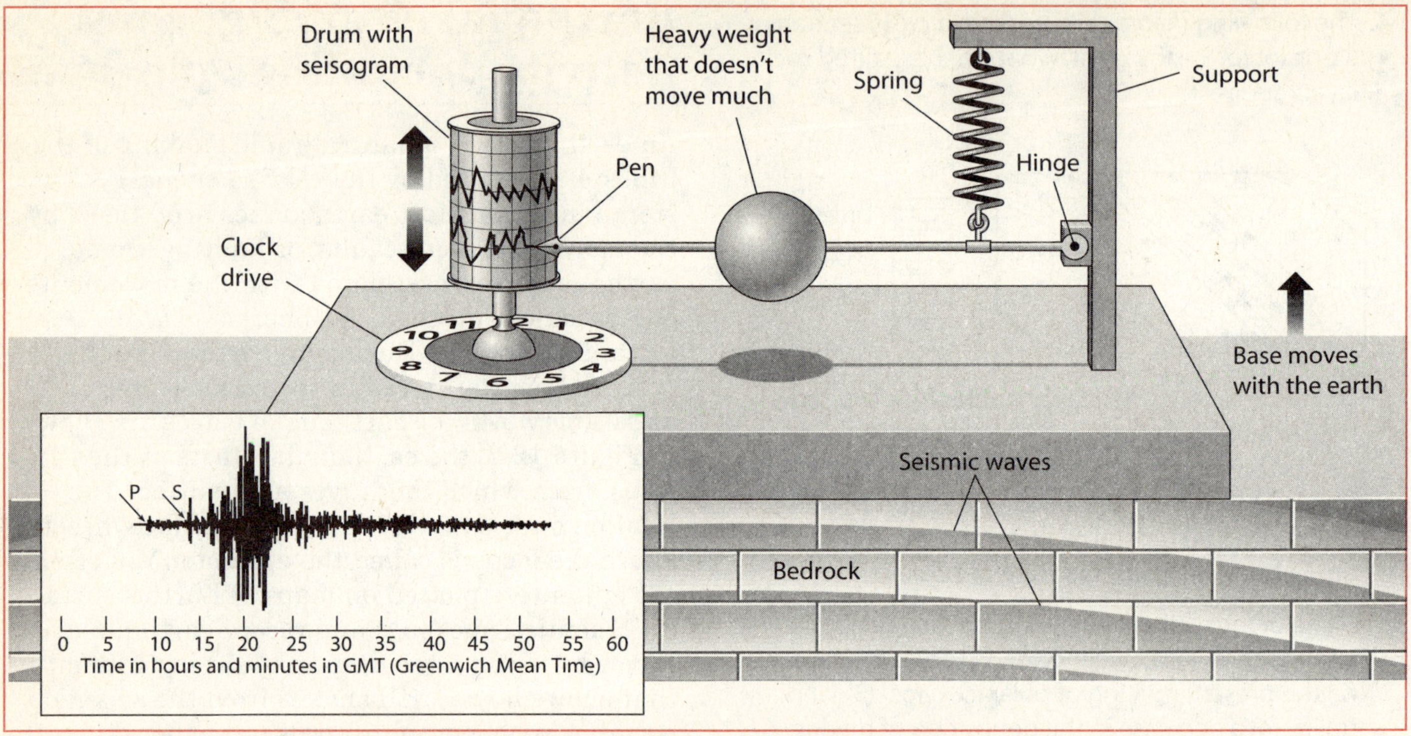

Figure 12-3. A simple seismograph and a seismogram: As Earth shakes during an earthquake, most of the seismograph moves up and down. Since the heavy weight resists movement because of momentum, the attached pen "writes" wiggles on a seismogram (shown in the insert). This seismogram has the record of the arrival times of P-waves (P) and S-waves (S) from an earthquake. P-waves arrive first because they are fastest.

Earthquake Waves

Seismic waves can be grouped into three categories —P-waves, S-waves, and surface waves. **P-waves**, also called primary waves, cause the particles they travel through to vibrate in the direction the waves are moving. The slower **S-waves,** also called secondary waves, cause the particles they travel through to vibrate at right angles to the direction the waves are moving.

The surface waves are produced when a P-wave or S-wave comes to Earth's surface. These waves only travel around Earth's solid surface and cause much of the surface shaking and damage of an earthquake.

PROPERTIES OF EARTHQUAKE WAVES

The different types of seismic waves have different properties depending on the materials they pass through:

- In any one material, P-waves travel faster than all other types of seismic wave. S-waves are the second fastest. When an earthquake occurs, the P-waves will reach a seismograph before the S-waves do. (See Figures 12-3 and 12-4.)
- The velocity of seismic waves in Earth depends on the properties of the materials they are passing through. Generally, the more dense and rigid a material is, the greater the velocity of the waves. As seismic waves pass from a material of one density to a material of higher or lower density, the waves are refracted, or bent.
- Within the same material, an increase in pressure increases the velocity of the seismic waves.
- P-waves will pass through solids, liquids, and gases. S-waves will only pass through solids.
- Some P-waves and S-waves are reflected by, or bounce off, dense rock layers within Earth. This property is often used to locate valuable rock and mineral resources within Earth.

LOCATION OF AN EPICENTER Epicenters are located by using the velocity differences between the P-waves and S-waves. Since P-waves move faster than S-waves, the farther an observer is from an epicenter, the larger the time interval between the arrival of the P-waves and S-waves. The distance to the epicenter is determined by comparing the interval with the graph data. (See Figure 12-4.)

To find the position of the epicenter, at least three seismograph locations must be used, and the epicenter distance must be calculated for each. For each of the three locations, the epicenter distance is then used as a radius, and circles are drawn on a globe or a map, as shown for locations A, B, and C in Figure 12-5. The place where all three circles intersect is the epicenter of the earthquake. Observations at one seismograph location provides only the distance to the epicenter, not the location or direction.

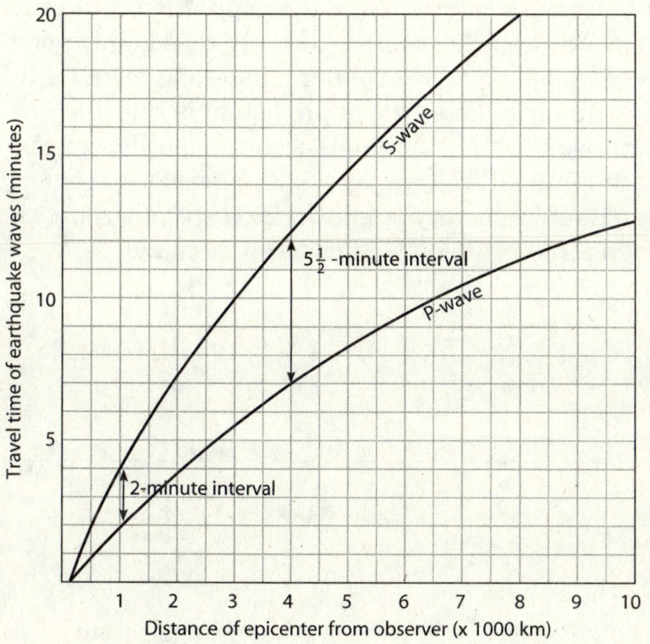

Figure 12-4. Travel times of P-waves and S-waves: The two curves in this graph show the time required for P-waves and S-waves to travel a given distance from the epicenter of an earthquake. Since the S-waves travel more slowly than the P-waves, the S-waves take longer to reach an observer. For example, if the observer is 1000 kilometers from the epicenter, the P-waves arrive 2 minutes after the earthquake occurs and the S-waves arrive 4 minutes after the occurrence. There is thus a 2-minute interval between observation of the P-waves and observation of the S-waves. At 4000 kilometers from the epicenter, it takes 7 minutes for the P-waves to arrive and 12 $\frac{1}{2}$ minutes for the S-waves; the time interval between them is thus 5 $\frac{1}{2}$ minutes. The graph can be used to find the distance from the epicenter if the time interval between the arrival of the two waves is known. A similar diagram can be found in the *Earth Science Reference Tables*.

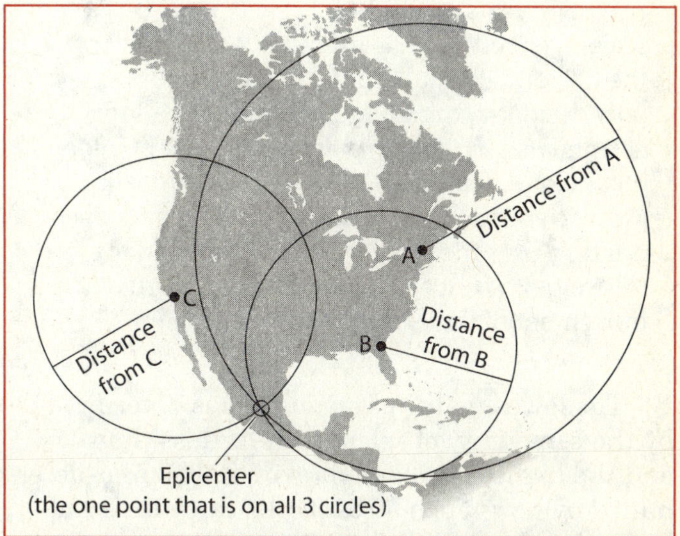

Figure 12-5. Locating the epicenter of an earthquake: Seismograph observations at three locations that are widely spaced are used. The distance to the epicenter is determined for each station, and a circle with this radius is then drawn around the station. The one point that lies on all three circles must be the epicenter.

FINDING THE ORIGIN TIME OF EARTHQUAKES

The time an earthquake originated can be determined from the epicenter distance and seismic-wave travel time. The farther an observer is from the epicenter, the longer it takes the seismic waves to travel to the observation point. For example, suppose the observer is 4000 kilometers from the epicenter. Figure 12-4 shows that it took the P-wave 7 minutes to arrive; thus the earthquake occurred 7 minutes earlier than the time at which the P-waves were observed on the seismograph. Try this example: If the S-wave first arrived at a station at 10 hr:12 min:30 sec and the seismograph is 5500 kilometers away from the epicenter, when did the earthquake occur?

MAGNITUDE OF EARTHQUAKES The strength of an earthquake can be measured in various ways. An earthquake intensity scale can be used to measure the various effects of an earthquake on humans and the types of damage the event causes. Generally, the closer to the epicenter, the greater the damage; that is, the intensity for an earthquake varies with distance.

Scientists most often use an earthquake magnitude scale to measure the strength of earthquakes. Magnitude scales use the height of the wiggles on seismograms to infer the total amount of energy released by an earthquake.

DIGGING DEEPER

For many years, the Richter Scale was used, but the scale—named in 1935 after Charles F. Richter of the California Institute of Technology—was really only designed for small-to moderate-strength earthquakes in California and would not provide meaningful data for the strongest earthquakes. Most scientists now use a magnitude scale, which not only uses the height of the wiggles on the seismograms but also considers the length of the movement of the earthquake.

The lowest value of magnitude is determined by the sensitivity of seismographs (less than 1), and the highest value (approximately 9.5) is determined by how much stress rocks can withstand before breaking. Figure 12-6 shows an example of one magnitude scale and the number of earthquakes that occur each year.

Earthquakes as a Natural Hazard

Earthquakes can cause great damage, injury, and death. What causes all the damage, death, and injuries during an earthquake? Earth doesn't open up and "swallow" buildings and people; the shaking ground causes a series of events, as illustrated in Figure 12-7 on the next page. Most injuries and fatalities are caused by parts of buildings falling on people and by the other related events.

EARTHQUAKE PREDICTION Scientists have been trying to successfully predict earthquakes for many years without a favorable outcome, except for long-term predictions. If we could predict earthquakes by a few weeks or even days, the population could then be evacuated. At present, the only reasonably good predictions have been general and long-term for a certain area. This type of prediction is possible because most earthquakes occur in specific zones. (See Figure 12-10 on page 222.) Therefore, if enough earthquake history is known for a specific zone, then a reasonable prediction can be made.

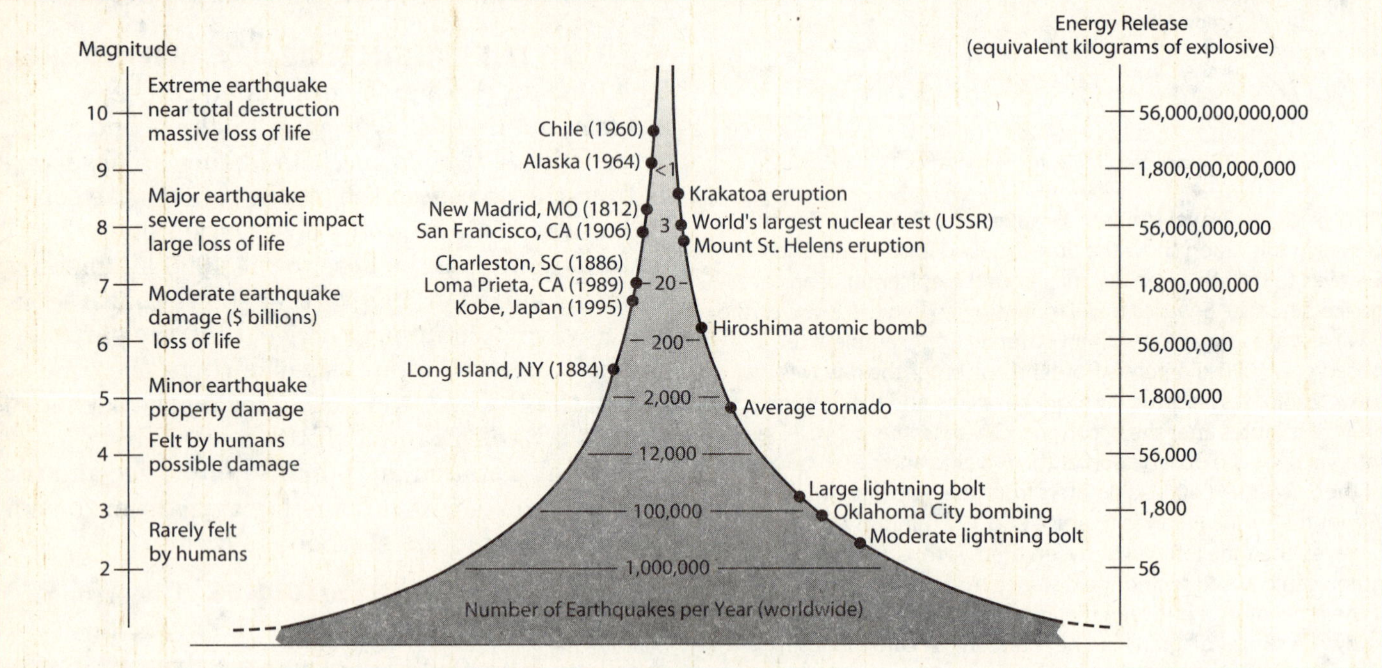

Figure 12-6. Earthquake magnitude scale: This scale provides one value for each earthquake that is an estimate of the total energy the event emits. Each whole-number value equals between 31 and 32 times more energy than the next lower number.

EMERGENCY PLANNING FOR EARTHQUAKES

Even though current methods of earthquake prediction are not of much value, the amount of damage, death, and injury from earthquakes can be greatly reduced with proper planning.

On an individual level, try to remember "drop, cover, and hold." Upon the first signs of shaking from what you believe to be an earthquake, don't go far, but drop down under a strong object (a strong desk or table, for example). Turn away from windows, and use one hand to cover your eyes. With the other hand, hold onto the strong object you are under. Do not try to run out of a building until the shaking is over. Most earthquakes last only 10 to 30 seconds, and it takes most of that time just to react.

Earthquake drills in the home, at work, and at school can help people to protect themselves, as can having emergency supplies at home, at work, in cars, and at school.

On a community level, proper planning of building sites and proper building construction techniques can greatly reduce the chance of death and injury during an earthquake. An example of important planning includes inspecting the soil and bedrock to ensure that new buildings are constructed on solid ground. Another example of proper planning is retrofitting older buildings to make them safer, such as bolting buildings to their foundations and cross-bracing walls.

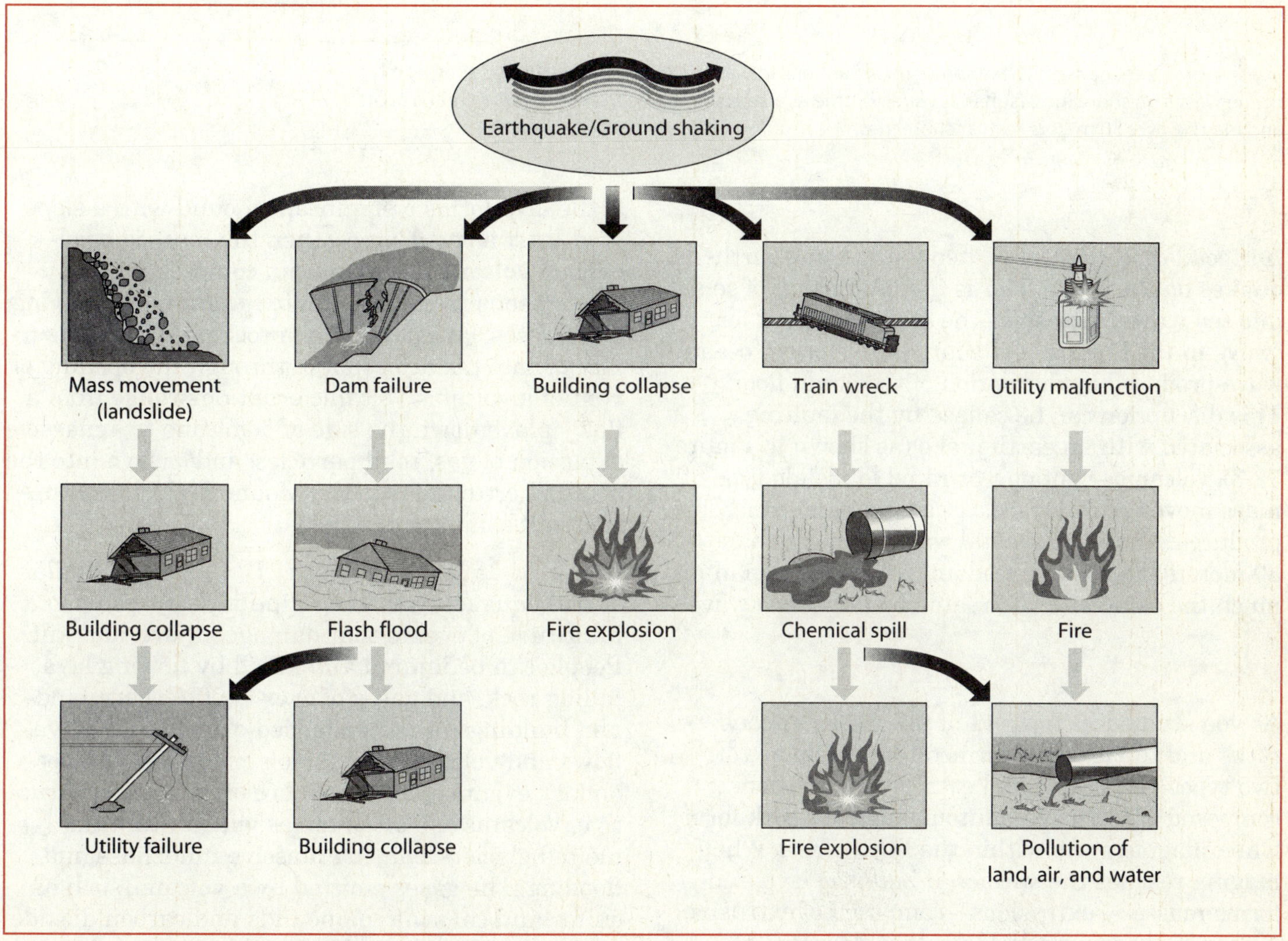

Figure 12-7. A cascade of earthquake disasters: This diagram illustrates many of the ways that earthquakes result in property damage, injury, and death.

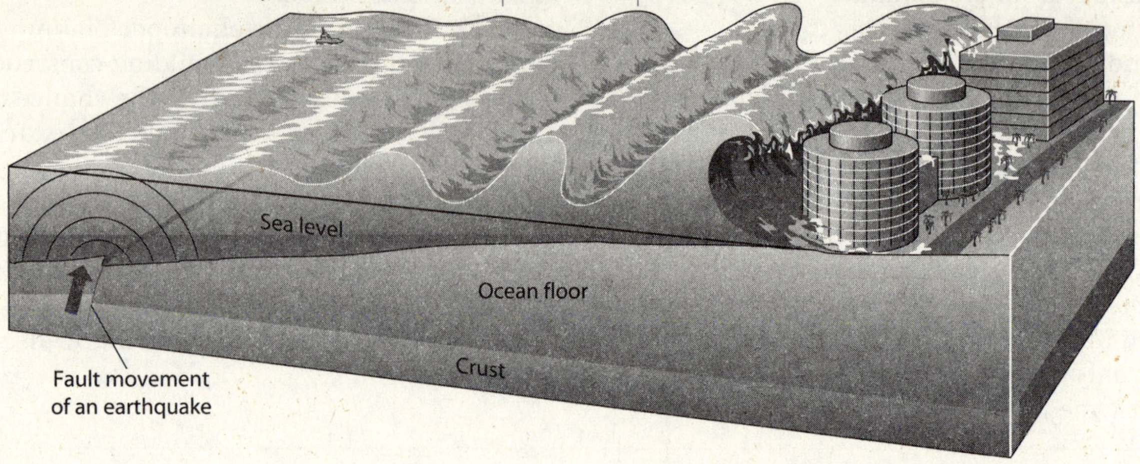

Figure 12-8. Tsunamis: These sometimes giant ocean waves can cause destruction similar to hurricane waves on shorelines. Tsunamis usually come as a series of waves from 10 to 60 minutes apart and may be over 30 meters high at shorelines.

Tsunami

A special condition associated with strong earthquakes on the ocean floor is the production of seismic sea waves. **Tsunami**—the Japanese word for "wave in the harbor"—is a large wavelength ocean wave produced by disruption of the ocean floor. This disruption can be caused by the faulting associated with an earthquake (as shown in Figure 12-8), volcanic eruptions, or rapid landslide type of mass movement. In rare cases a tsunami can produce destructive coastal waves of more than 30 meters. The shoreline damage from a tsunami is much the same as that created by hurricane waves.

Igneous Activity and Volcanoes

As you studied in Topic 11 of this book, igneous rocks and the features formed from them are of two types—intrusive and extrusive. Intrusions composed of intrusive (plutonic) igneous rock form when magma stays within the lithosphere. When magma reaches the surface, it becomes lava and forms masses—extrusions—composed of extrusive (volcanic) igneous rock. If the lava flows out of cracks and is very fluid, a feature called a "lava flow" forms. Successive lava flows can form large features like the Columbia Lava Plateau, which covers much of Oregon, Washington, and Idaho. (See Figure 12-9, Diagram D on the next page.)

If the lava forms a significant mound with a slope of at least a few degrees, then the mass is a volcano. A **volcano** is a mountain composed of extrusive igneous rocks. A **volcanic eruption** is the giving off of gases, lava, and/or lava rock onto Earth's surface or into the atmosphere through the opening or vent of a volcano. Volcanic eruptions range from a flow of lava down the side of a volcano to a massive explosion of gas, solid particles, and/or lava into the atmosphere similar to the Mount St. Helens eruption in Washington in 1980.

VOLCANOES AS A NATURAL HAZARD Similar to earthquakes, volcanic eruptions can result in a multitude of events that damage, injure, and kill. People can be injured and killed by flowing lava, falling rock, and gases of over 1000°C. Farmland and buildings are often buried or burned by lava flows and volcanic ash during eruptions. The secondary events of a volcano are often more destructive. Volcanic ash often mixes with water from melted glaciers, causing massive mudslides and flooding. The gases emitted by a volcano, such as sulfur and chlorine compounds and carbon dioxide, can cause immediate death or long-term lung damage. In massive eruptions, volcanic ash is ejected into the stratosphere, where it has been known to cool Earth by blocking insolation reaching Earth's surface.

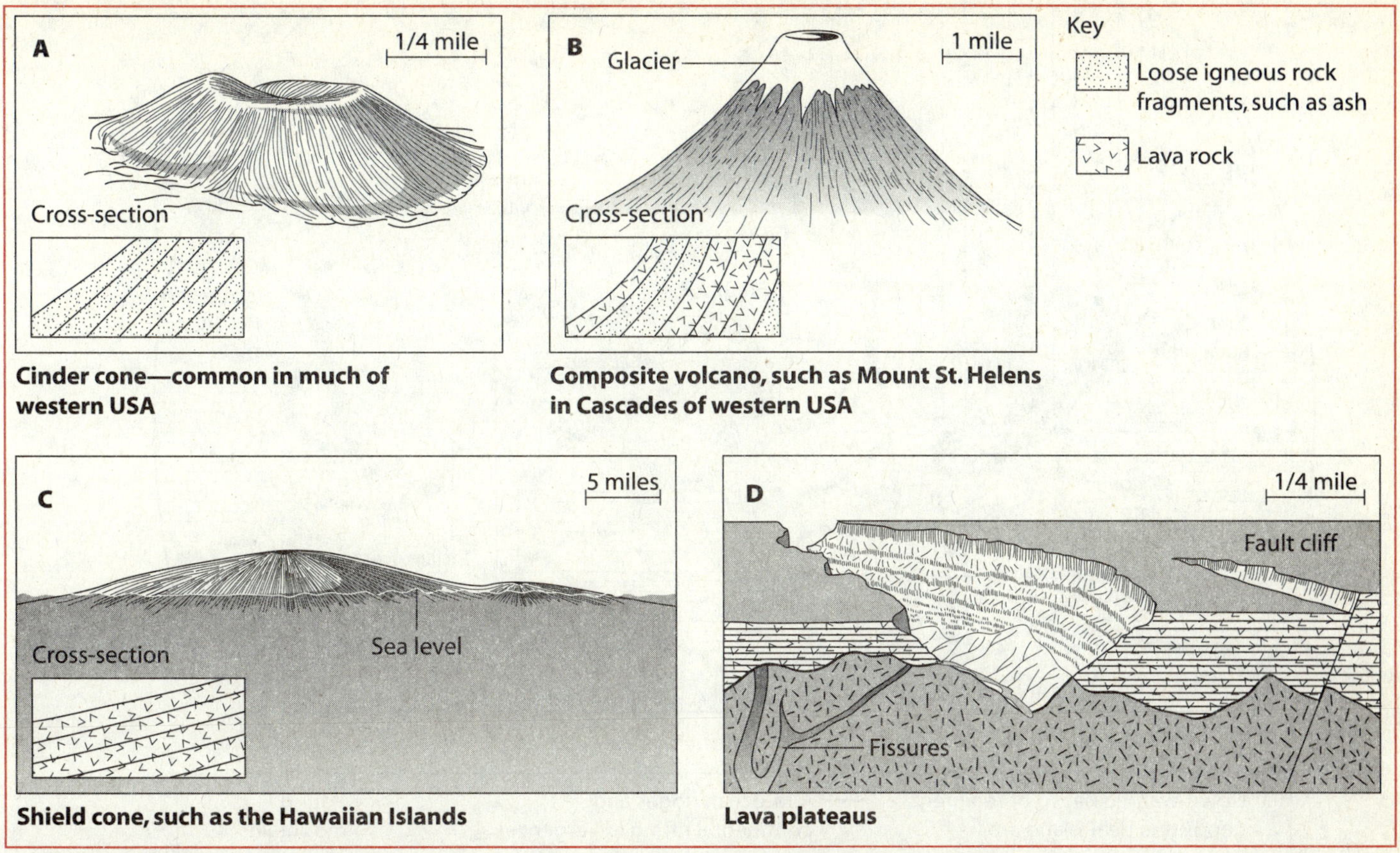

Figure 12-9. Volcanic mountains and plateaus: Volcanic mountains are composed of two types of igneous materials. One is lava rock formed from flows of liquid lava. The other is loose rock fragments—from ash to boulder size—formed by the exploding of lava and rock fragments into the air and their eventual settling to form a mountain. A volcanic plateau is relatively high landscape composed of horizontal layers of lava rock formed by many lava flows emitted from fissures (cracks) in the crust.

PREDICTION OF VOLCANIC ERUPTIONS There has been some success in predicting volcanic eruptions in sufficient time for people to escape from an area. Though most eruptions go unpredicted, some of the more dangerous volcanoes in populated regions are monitored by using a series of tools. Satellites measure infrared energy from the upper atmosphere and report on the increasing heat from rising magma. Tilt meters measure increasing slopes as the volcano inflates with magma. Elevation, benchmarks, latitude and longitude measurements, and topographic maps indicate the increases in elevation and width common before eruptions. When magma moves upward before volcanic eruptions, it pushes rock out of the way, causing hundreds of earthquakes. Using the three-seismograph method to determine epicenters, scientists can "watch" magma rise and can provide eruption predictions within an hour of an actual eruption. With predictions such as these, people have enough early warning to develop emergency action plans, including rescue routes.

Zones of Crustal Activity

Zones of frequent crustal activity can be located on Earth's surface. Major areas of uplift, sinking, earthquakes, and volcanic eruptions are often found together and are associated with features like young continental mountains, ocean trenches, island arcs, and mid-ocean ridges (mountain ranges in ocean areas). Note that the regions surrounding the Pacific Ocean contain a majority of these related events and features, and are thus referred to as the "Ring of Fire." Many of these associations are illustrated in Figure 12-10 on the next page. The fact that these activities and features are clustered hints at their common origins, which will be discussed later in this topic.

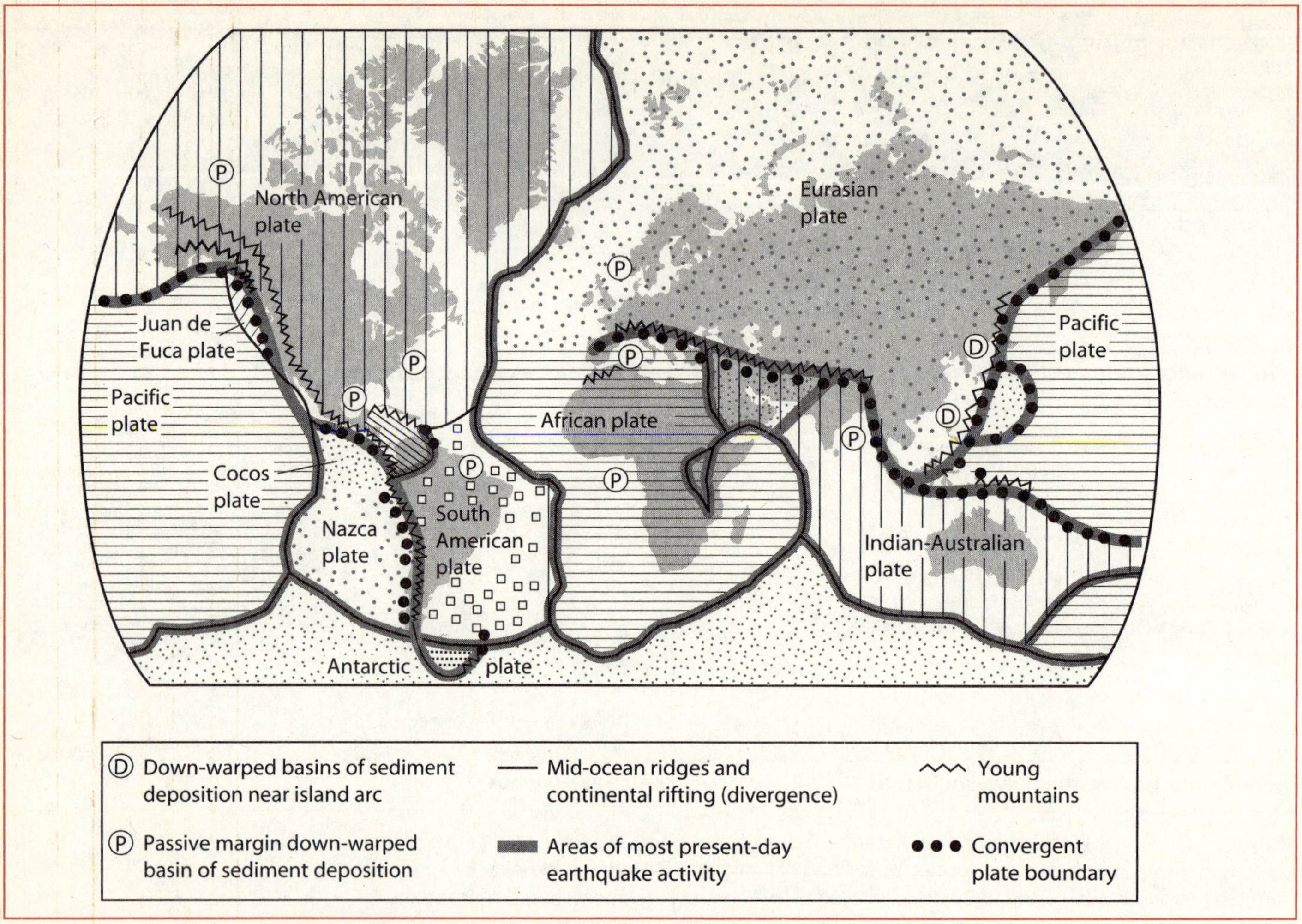

Figure 12-10. Zones of current crustal activity and Earth's major plates: Note that the borders of the plates are the locations of most current crustal activity, which includes volcanic eruptions and earthquakes. Most of the world's active volcanoes are located where mid-ocean ridges or young mountains are indicated on the map.

Review Questions

9. The landscape shown in the following diagram is an area of frequent earthquakes.

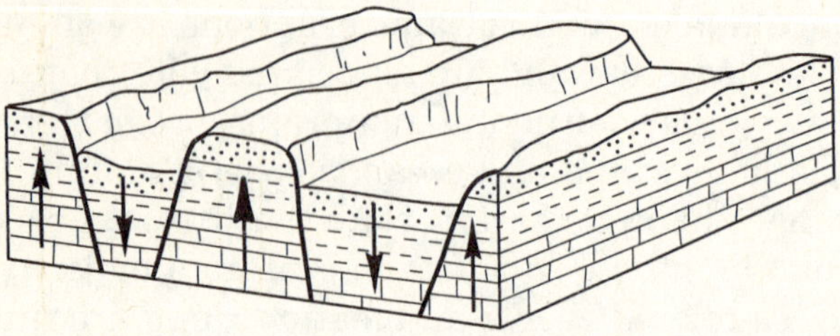

This landscape provides evidence for (1) converging convection cells within the rocks of the mantle (2) density differences in the rocks of the mantle (3) movement and displacement of the rocks of the crust (4) differential erosion of hard and soft rocks of the crust

10. Which statement about Earth's crust in New York State is best supported by the many faults found in the crust? (1) The crust has moved in the geologic past. (2) The crust has been inactive throughout the geologic past. (3) New faults will probably not develop in the crust. (4) An earthquake epicenter has not been located in the crust.

11. Where do most earthquakes originate? (1) within Earth's outer core (2) along specific belts within the crust (3) randomly across the entire Earth's surface (4) evenly spaced along the Moho interface

12. Recent volcanic activity in different parts of the world supports the inference that volcanoes are located mainly in (1) the centers of landscape regions (2) the central regions of the continents (3) zones of crustal activity (4) zones in late stages of erosion

13. Earthquakes generate primary waves (P-waves) and secondary waves (S-waves). Compared to the speed of secondary waves in a given Earth material, the speed of primary waves is (1) always slower (2) always faster (3) always the same (4) sometimes faster and sometimes slower

The following diagram illustrates how the observatories in Pasadena, California; Chicago, Illinois; and Washington, D.C. locate the epicenter of an earthquake. Base your answers to questions 14 through 17 on the diagram.

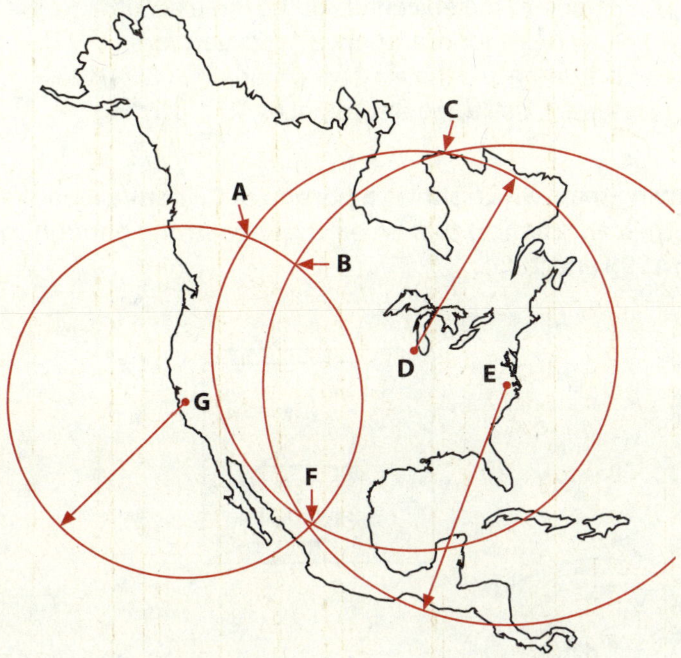

14. The epicenter of the earthquake is located nearest (1) A (2) B (3) C (4) F

15. The separation in time between the arrival of primary and secondary waves is (1) greatest at station G (Pasadena) (2) greatest at station D (Chicago) (3) greatest at station E (Washington, D.C.) (4) the same at all stations

16. If the method illustrated by the diagram is used to locate the epicenter of an earthquake, it appears unlikely for an individual observatory operating independently to determine the (1) direction to the epicenter (2) distance to the epicenter (3) distance to an epicenter under the ocean (4) interval between the initial and subsequent seismic waves

17. The time lapse between the arrival of the P-waves and S-waves on ONE recording of seismic waves can be used to determine the (1) magnitude of the earthquake (2) exact location of the focus (3) exact location of the epicenter (4) distance to the epicenter

Base your answers to questions 18 through 21 on the following diagram. The diagram represents a cross-section of Earth, showing the paths of earthquake waves from a single earthquake source. Recording stations for seismic waves are located on Earth's surface at points A through F, and they are all located in the same time zone.

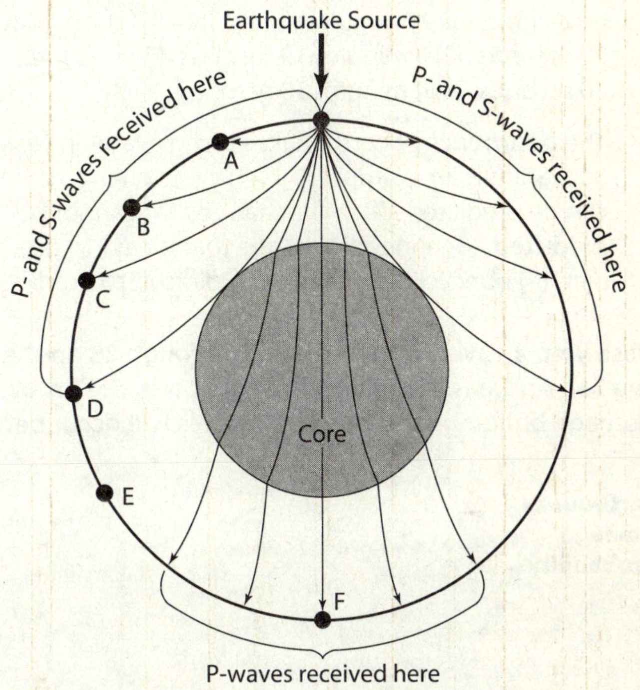

18. At which station is the distance in time between the arrival of P-waves and S-waves the greatest? (1) A (2) B (3) C (4) D

19. What explanation do scientists give for the reason that station F did NOT receive S-waves? (1) Earth's inner core is so dense that S-waves cannot pass through. (2) Earth's outer core is liquid, which does not allow S-waves to pass. (3) S-waves do not have enough energy to pass completely through Earth. (4) S-waves are absorbed by Earth's crust.

20. Seismic recording station D is 7700 kilometers from the epicenter. If the P-wave arrived at this station at 2:15 P.M., at approximately what time would the earthquake most likely have occurred? (1) 1:56 P.M. (2) 2:00 P.M. (3) 2:04 P.M. (4) 2:08 P.M.

21. Station B recorded the arrival of P-waves at 2:10 P.M. and the arrival of S-waves at 2:15 P.M. Approximately how far is station B from the earthquake epicenter? (1) 1400 km (2) 2400 km (3) 3400 km (4) 4400 km

22. You record evidence of an earthquake on your seismograph. What information about the earthquake's epicenter can be determined from your seismograph alone?

23. A seismic station is 2000 kilometers from an earthquake epicenter. According to the *Earth Science Reference Tables*, how long does it take an S-wave to travel from the epicenter to the station? (1) 7 minutes 20 seconds (2) 5 minutes 10 seconds (3) 3 minutes 20 seconds (4) 4 minutes 10 seconds

24. The occurrence of earthquakes along a fault in New York State is an example of a (1) cyclic change that can be predicted (2) cyclic change that cannot be predicted (3) noncyclic change that is easy to predict (4) noncyclic change that is difficult to predict

25. The following diagrams represent seismographic traces of three disturbances—A, B, and C—recorded by the same seismograph.

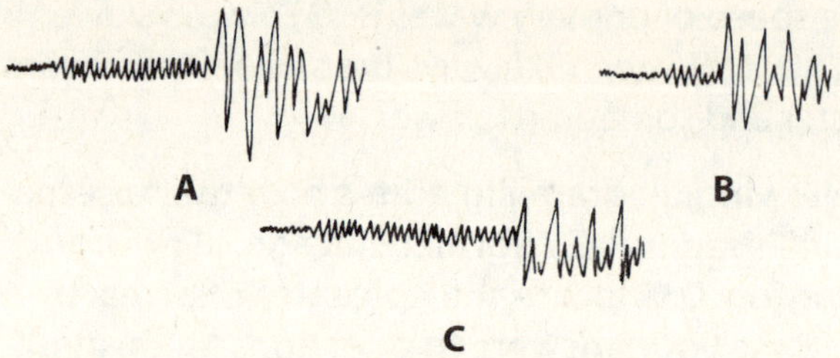

The traces indicate that the distance from the station to the epicenters of the three disturbances is (1) least for disturbance A (2) least for disturbance B (3) least for disturbance C (4) the same for all three disturbances

Base your answers to questions 26 through 28 on the following map, which shows a portion of California along the San Andreas Fault zone. The map gives the probability (percent chance) that an earthquake strong enough to damage buildings and other structures will occur between 1998 and 2024.

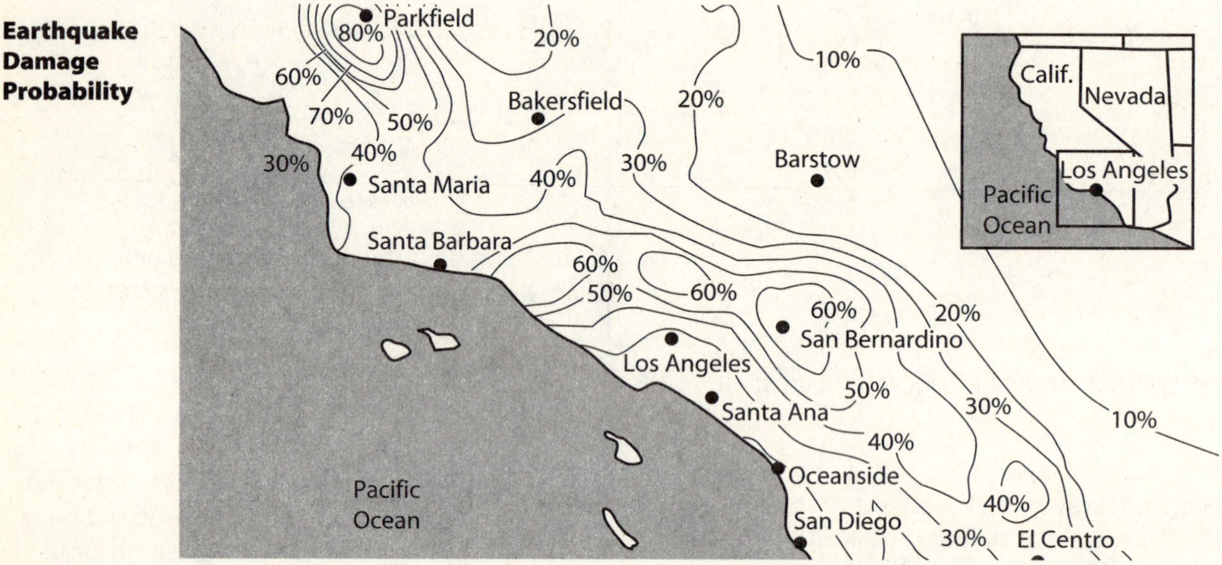

26. Which city has the greatest danger of damage from an earthquake? (1) Parkfield (2) San Diego (3) Santa Barbara (4) San Bernardino

27. Give a compass direction to indicate which part of California, shown on the map, has the least chance of being damaged by an earthquake.

28. Which map represents the most likely location of the San Andreas Fault line?

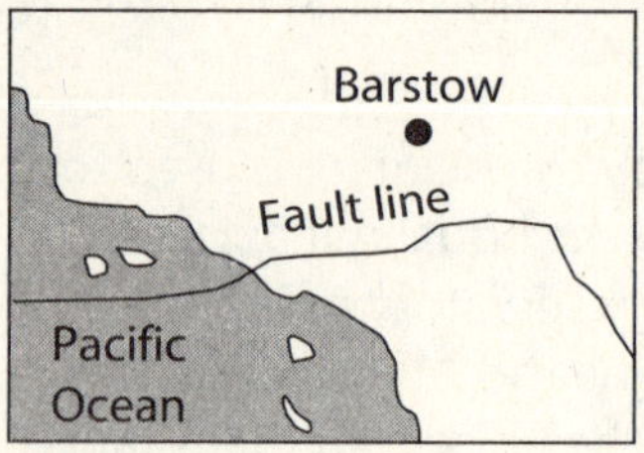

(1)

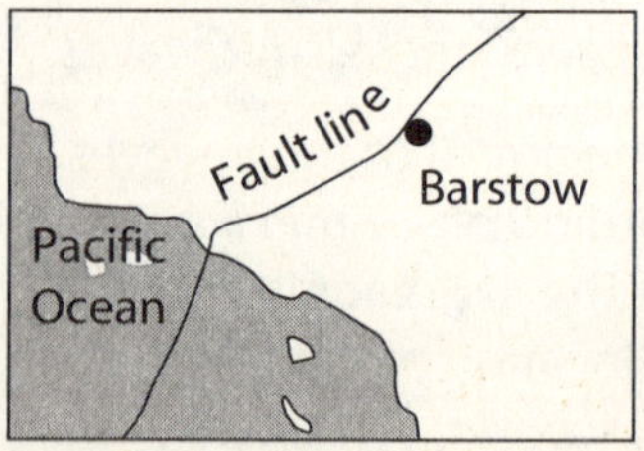

(2)

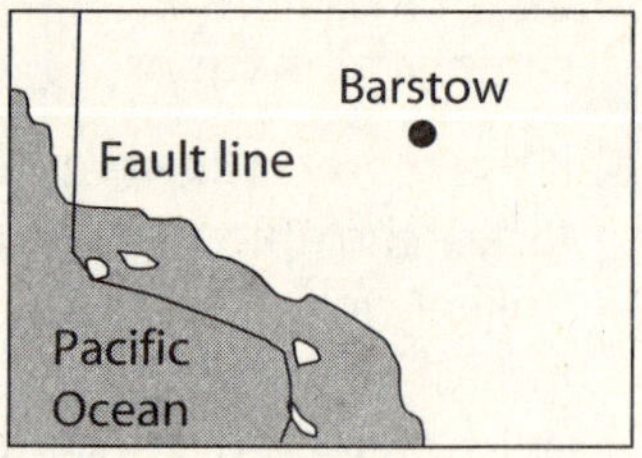

(3)

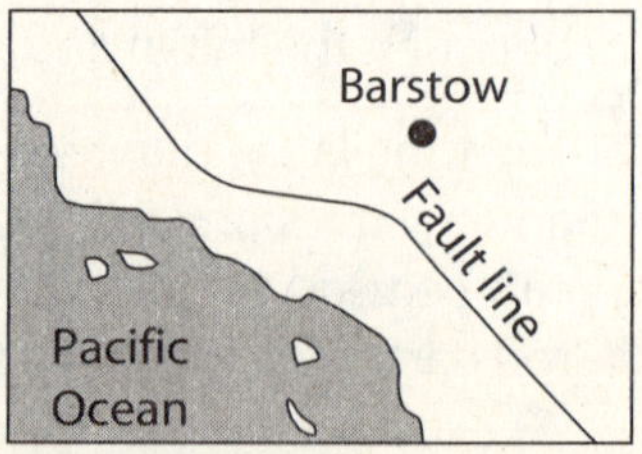

(4)

Base your answers to questions 29 through 33 on the following map. The map shows the time of arrival of seismic sea waves (tsunamis) from locations throughout the Pacific Ocean to the Hawaiian Islands. For example, an earthquake at any location on the line labeled "6 h" could produce a tsunami that would arrive in Honolulu 6 hours later.

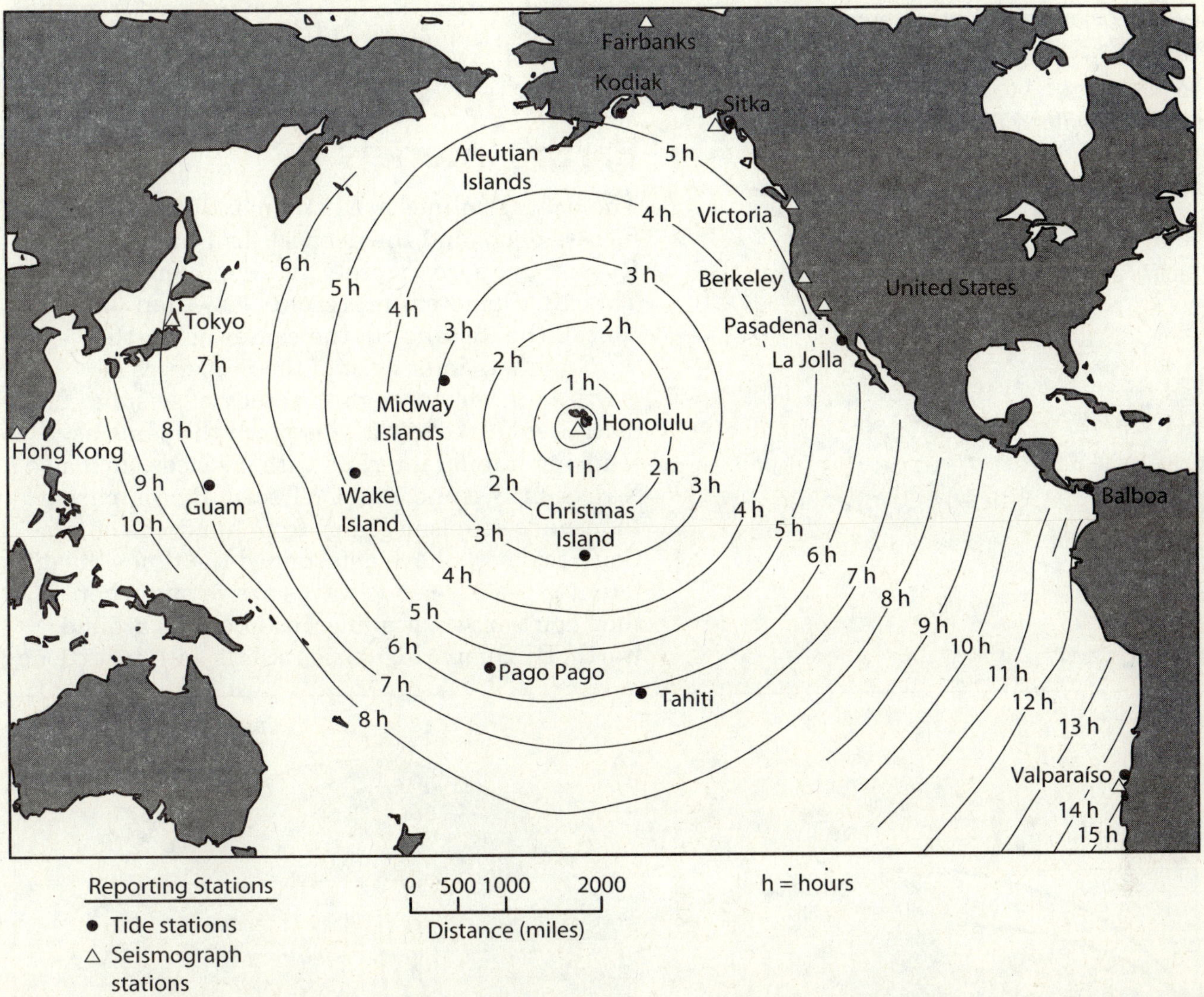

29. From which location would a tsunami arrive at Honolulu in the shortest time? (1) Kodiak (2) La Jolla (3) Tahiti (4) Wake Island

30. Which reporting station is NOT located on the Pacific plate? (1) Hong Kong (2) Midway (3) Honolulu (4) Tahiti

31. Approximately how fast do seismic sea waves travel from Midway to Honolulu, Hawaii? (1) 31 miles per hour (2) 500 miles per hour (3) 1500 miles per hour (4) 3000 miles per hour

32. The most likely reason for collecting this Pacific Ocean data is that Honolulu is (1) the site of frequent major eruptions (2) an equal distance from many seismic stations (3) surrounded by earthquake-prone zones (4) the site of a major tide station

33. Locations in this region where major earthquakes occur are most closely associated with areas of (1) volcanic eruptions (2) P-wave absorption (3) magnetic field reversal (4) mid-continental folding

34. Which information would be most useful for predicting the occurrence of an earthquake at a particular location? (1) elevation (2) climate (3) seismic history (4) number of nearby seismic stations

35. The diagram below represents a shoreline in New York State; several general features have been labeled.

Past movements along the fault line most likely caused the formation of (1) the sand bar (2) tsunami (3) ocean currents along the shore (4) tides

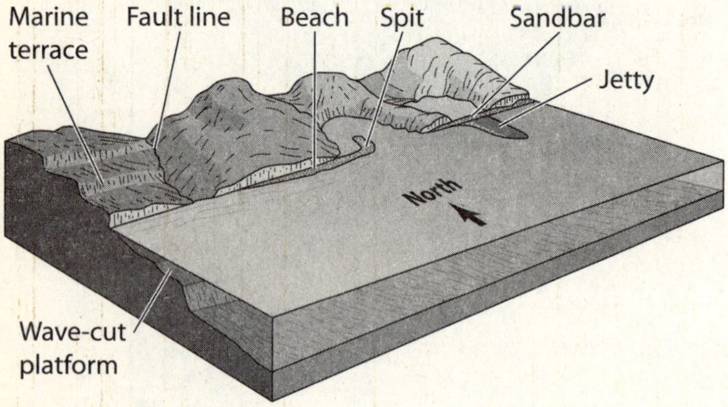

A Model of Earth's Interior

Getting information about Earth's interior from direct observation is a difficult task. Therefore, most of its properties have been inferred from the study of earthquake waves.

Methods of Studying Earth's Interior

The deepest mines are little more than 3.5 kilometers deep, and the deepest drill hole (in Russia) is approximately 12 kilometers deep. This 12 kilometers represents less than 0.2 percent of the distance to the center of Earth.

Scientists infer most of the properties of Earth's interior through the study of seismic (earthquake) waves. This is much the same as studying human interiors with methods such as X-rays, ultrasound, and CAT scan. By comparing thousands of seismograms from hundreds of locations, scientists have determined the times the different types of seismic waves arrive and when they don't arrive at numerous locations around Earth. By comparing these records with tests done

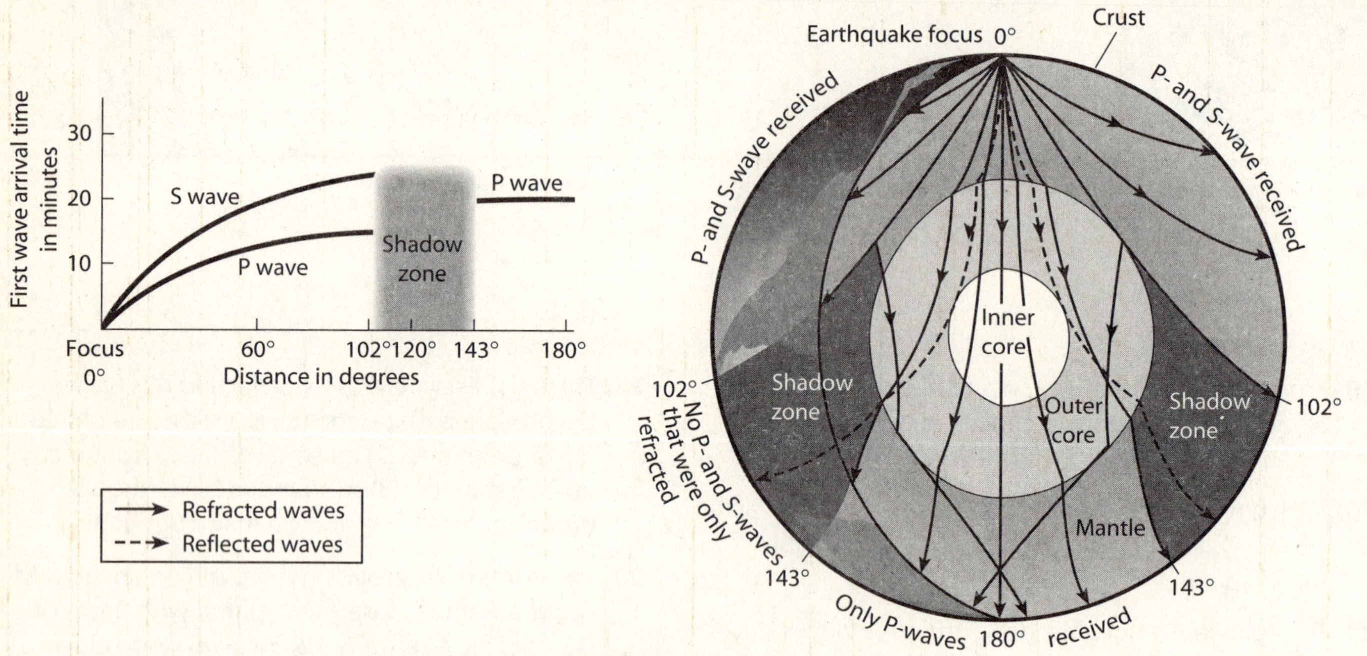

Figure 12-11. Arrival times and paths of seismic waves moving through Earth's interior: Analysis of the times seismic waves arrive has resulted in interpretations of reflections and refractions of these waves in Earth's interior. This analysis has resulted in the standard model of the zones of Earth's interior shown here.

using explosives in the oil industry and tests with nuclear bombs, scientists discovered that seismic waves refract, reflect, change velocity, and become absorbed by various parts of Earth's interior. Figures 12-11 and 12-12 show the most common model of Earth's interior that has resulted from these seismic wave studies.

Zones of Earth

Our analysis of seismic waves indicates that Earth is composed of a number of zones, as shown in Figures 12-11 and 12-12. The **crust** is the outermost part of Earth below the atmosphere or hydrosphere. Though mostly solid rock, the crust includes the soil and eroded and weathered rock deposits. Below the crust is the mostly solid part called the **mantle**, Earth's thickest zone that contains approximately 80 percent of Earth's volume. The mantle is separated from the crust by a thin interface called the **Moho**. Below the Moho, the rocks become denser. The whole crust and the very uppermost portion of the mantle (called the rigid mantle) are together called the **lithosphere**. (See Figure 12-12.) The lithosphere is divided into sections called "plates," which you will study in the next section.

Below the lithosphere is another portion of the upper mantle called the **asthenosphere**. This zone was discovered because it was found that seismic waves decreased in velocity from approximately 100 to 700 kilometers below Earth's surface. The asthenosphere is believed to be a plastic-like portion of the upper mantle that is at least partly molten. Much of the magma and lava of the crust and lithosphere is thought to originate in this zone. Note on the temperature graph in Figure 12-12 that the melting point and temperature curve intersect, indicating this partial melting. The asthenosphere allows the plates of the lithosphere to move around Earth's surface and to move up and down. Below the asthenosphere is the majority of the mantle, called the stiffer mantle.

Beneath the mantle Earth's core is divided into two parts—the **inner core** and the **outer core**. Figure 12-11 shows the outermost dimension—by area—of the shadow zone of the outer core that, due to refraction, does not receive any refracted seismic waves. The bottom of the outer core is indicated by the waves reflected off the outside edge of the inner core. Figure 12-11 also provides evidence that the outer core is liquid: Because no S-waves pass through the zone, it can't be a solid, and because

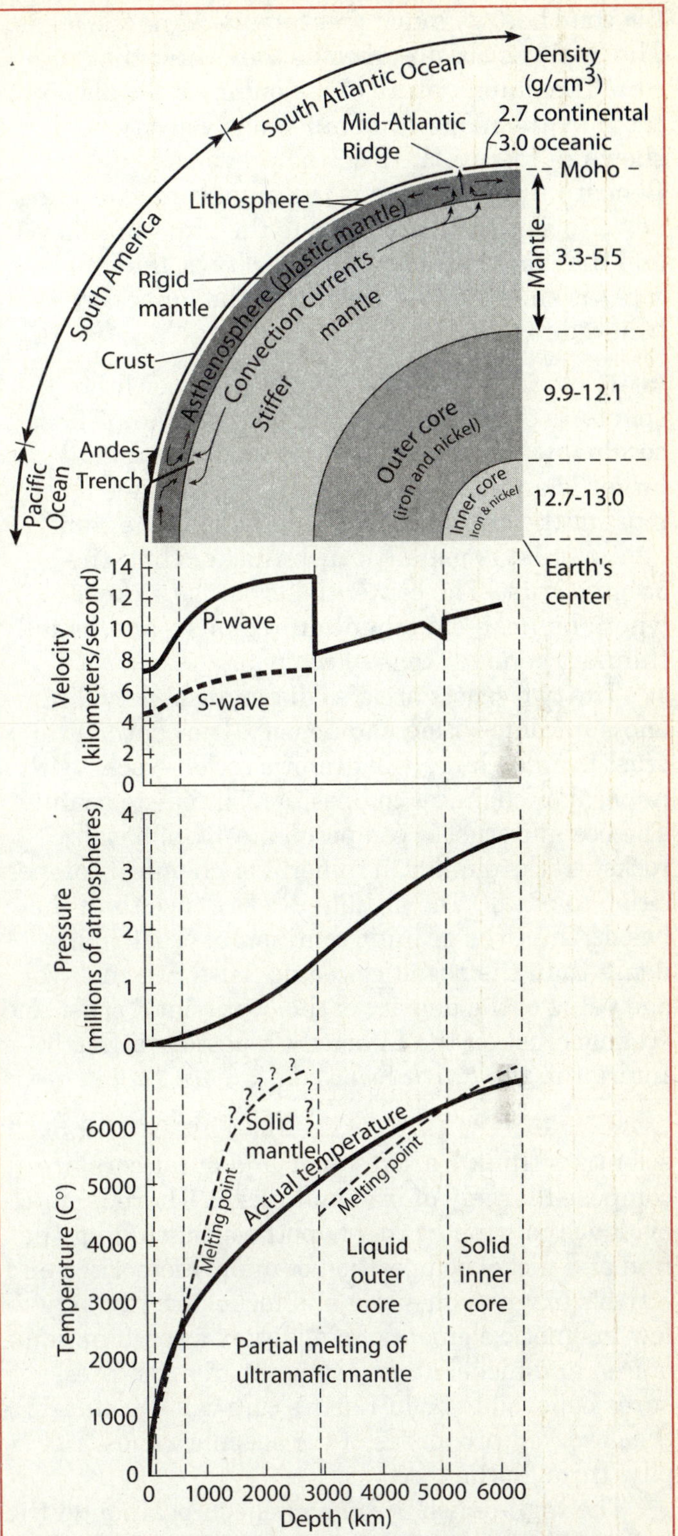

Figure 12-12. Properties and zones of Earth's interior: Analysis of the seismic waves and other studies have resulted in the inference about Earth's interior illustrated here. Note that most of Earth's interior is solid except for the outer core and portions of the asthenosphere, both of which are liquid. (See also the Inferred Properties of Earth's Interior in the *Earth Science Reference Tables.*)

the zone has too much pressure, it cannot be a gas. The S-waves are absorbed as they enter the outer core. The liquid nature of the outer core is also indicated by the major drop in P-wave velocity, as shown in Figure 12-12 on the previous page. Beneath the outer core is the inner core extending down to Earth's center. Due to the great pressures and the increase in P-wave velocity at the inner core (as shown in Figure 12-12), the inner core is believed to be a solid.

EARTH'S CRUST Earth's crust, or upper lithosphere, is divided into two major divisions. The **continental crust** makes up the continents and larger islands, and the **oceanic crust** makes up most of the crust beneath the oceans. The continental crust is usually much thicker than the oceanic crust. The contential crust is thickest where highest—in mountain regions—and usually thinnest beneath coastal regions.

The two crusts are also distinguished by differences in composition and density. The continental crust is made largely of granitic rocks—rocks with a mineral or chemical composition similar to granite. The oceanic crust is composed mostly of basaltic rocks—rocks similar in mineral or chemical composition to basalt. The granite is less dense than the basalt; thus, the granitic continental crust is less dense than the basaltic oceanic crust. See the average value of the density of the continental crust and oceanic crust on the Inferred Properties of Earth's Interior in the *Earth Science Reference Tables*.

COMPOSITION OF EARTH'S INTERIOR Many scientists think Earth's inner and outer core are composed largely of iron and nickel. There is much evidence to support this hypothesis, including the iron and nickel composition of many meteorites and Earth's magnetism—iron and nickel are two of a few magnetic elements. In addition, a combination of iron and nickel at the temperatures and pressures believed to be in Earth's core can account for the observed properties of the seismic waves that pass through the core.

The high-density iron-nickel composition of the core and the low-density composition of the crust indicate that the mantle must have a composition different from the crust and the core and an intermediate density.

Earth's crust is composed mostly of rocks and minerals, as illustrated on the various charts in the *Earth Science Reference Tables* depicting minerals, sedimentary rocks, metamorphic rocks, and igneous rocks. Generally, the crust is composed of low-density rocks with a mixture of granitic and basaltic compositions.

Review Questions

36. In developing a model of Earth's deep interior, most of the evidence was derived from (1) deep wells (2) mining operations (3) observation of other planets (4) seismic data

37. Through which zones of Earth do primary waves (P-waves) travel? (1) only the crust and mantle (2) only the mantle and outer core (3) only the outer and inner core (4) the crust, mantle, outer core, and inner core

38. According to the *Earth Science Reference Tables*, in which group are the zones of Earth's interior correctly arranged in order of increasing average density? (1) crust, mantle, outer core, inner core (2) crust, mantle, inner core, outer core (3) inner core, outer core, mantle, crust (4) outer core, inner core, mantle, crust

39. What evidence has been obtained concerning the existence of the Moho and Earth's mantle? (1) satellite images (2) well drillings (3) exposures in deep canyons like the Grand Canyon (4) refraction of earthquake waves

40. What is the relationship among density, temperature, and pressure inside Earth? (1) As depth increases, density, temperature, and pressure decrease. (2) As depth increases, density and temperature increase, but pressure decreases. (3) As depth increases, density increases, but temperature and pressure decrease. (4) As depth increases, density, temperature, and pressure increase.

41. The temperature of rock located 1000 kilometers below Earth's surface is approximately (1) 1000°C (2) 2600°C (3) 3300°C (4) 4300°C

42. In which parts of Earth's interior would melted or partially melted material be found? (1) stiffer mantle and inner core (2) stiffer mantle and outer core (3) crust and inner core (4) asthenosphere and outer core

43. An earthquake occurs in city A. Recordings on a seismograph in city B show only the presence of P-waves. City A and B are on opposite sides of Earth—180° apart. What does this information allow you to infer about the structure of Earth's interior?

44. How does the composition of the oceanic crust compare with the composition of the continental crust? (1) The oceanic crust is mainly limestone, while the continental crust is mainly sandstone. (2) The oceanic crust is mainly limestone, while the continental crust is mainly granitic. (3) The oceanic crust is mainly basaltic, while the continental crust is mainly sandstone. (4) The oceanic crust is mainly basaltic, while the continental crust is mainly granitic.

45. As one travels from an ocean shore to the interior of a continent, the thickness of Earth's crust generally (1) decreases (2) increases (3) remains the same

46. How does thickness and density of the continental crust compare to that of the oceanic crust? (1) The continental crust is thicker and less dense than the oceanic crust. (2) The continental crust is thicker and denser than the oceanic crust. (3) The continental crust is thinner and less dense than the oceanic crust. (4) The continental crust is thinner and denser than the oceanic crust.

47. The overall density of Earth is approximately 5.5 g/cm^3. The average density of Earth's crust is between 2.5 g/cm^3 and 3.0 g/cm^3. What does this suggest about the density of Earth's core?

48. The composition of some meteorites supports the inference that Earth's core is composed of (1) aluminum and calcium (2) iron and nickel (3) silicon and oxygen (4) magnesium and potassium

Plate Tectonics

People have always wondered about the origin of continents, mountain ranges, volcanoes, earthquakes, and the multitudes of other features and events. In the past, many legends, religious beliefs, and scientific theories have tried to explain Earth's features and events. Some of the older scientific theories include cooling and contraction of Earth, expansion of Earth, and continental drift. None of these earlier theories seems satisfactory for most of the scientific community today. Since the 1960s many new discoveries from the studies of ocean drilling, Earth's magnetism, satellite observations, and detailed analysis of rocks and fossils have lead to the plate tectonic theory. This theory has done for the earth sciences what evolution and genetics have done for the biological sciences. It has provided a unifying model to explain most, if not all, major features and events of Earth's lithosphere.

The Plate Tectonic Theory

The basic concept of the **plate tectonic theory** is that Earth's lithosphere is broken up into sections or pieces called **plates**—also called **lithospheric plates** and **tectonic plates**—and their movement and interaction produce major changes in Earth's surface. In this book the term "plates" will mostly be used.

These plates move about Earth's surface at a rate of a few centimeters per year (approximately the rate of fingernail growth). These plates can also move up and down—usually at rates of only millimeters per year—due to uplifting and sinking. The plates can move around and up and down because they are floating on the asthenosphere—a plastic-like layer of the mantle.

Refer to Figure 12-10 and the Tectonic Plates map in the *Earth Science Reference Tables* to become familiar with the names and locations of the various plates. You will observe that the plates usually don't follow continent or ocean boundaries. Note that the North American plate includes parts of Asia and North America; the Pacific, Arctic, and the Atlantic Oceans; and Greenland and part of Iceland. Westernmost California and Mexico are part of the Pacific plate. Most plates have continental and oceanic crust at their tops. A small number of plates are oceanic—that is, they have only oceanic crust at their tops, like the Nazca and Philippine plates.

DIGGING DEEPER

It should be understood that the continents don't drift or move on their own; they move as their associated plates move. Continental drift was a term used before common acceptance of plate tectonic theory. Continental drift means the continents by themselves are moving around Earth's solid surface. Figure 12-13 and the Inferred Position of Earth's Landmasses section of Geologic History of New York State in the *Earth Science Reference Tables* both show that the continents have drifted, but actually the continents only move as the plates move.

Figure 12-13. Changes in positions of Earth's landmasses from 700 million years ago to 250 million years into the future because of plate tectonics

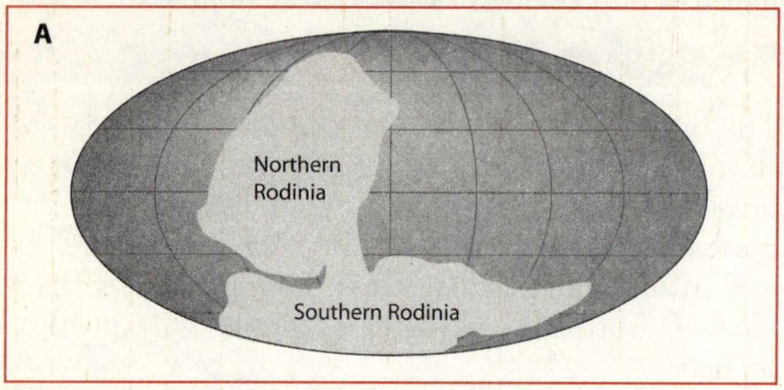

Earth around 700 millions years ago as supercontinent Rodinia is breaking up.

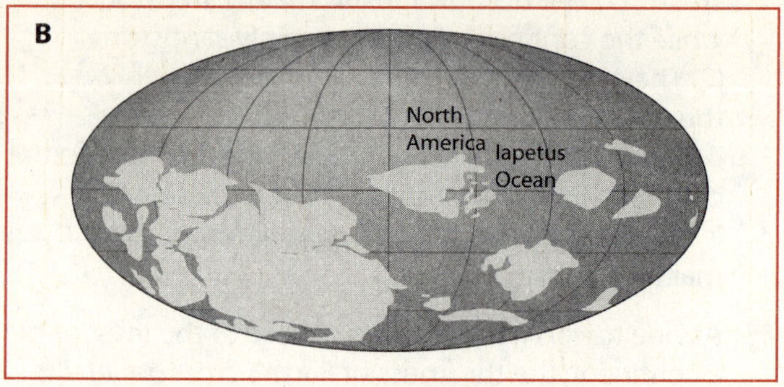

Earth around 550 million years after the break up of a supercontinent.

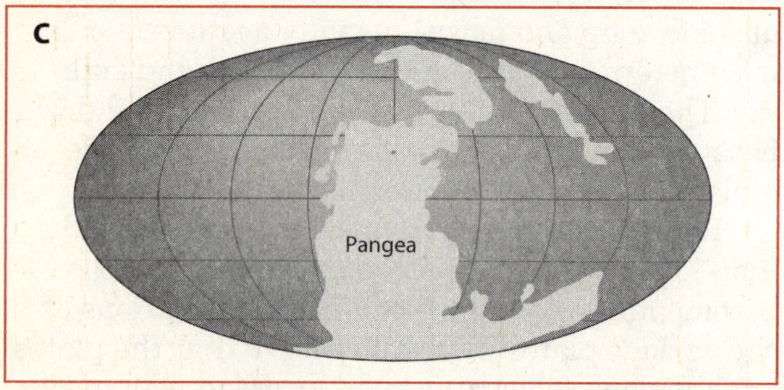

Earth around 250 million years ago when supercontinent Pangea is just beginning to break up.

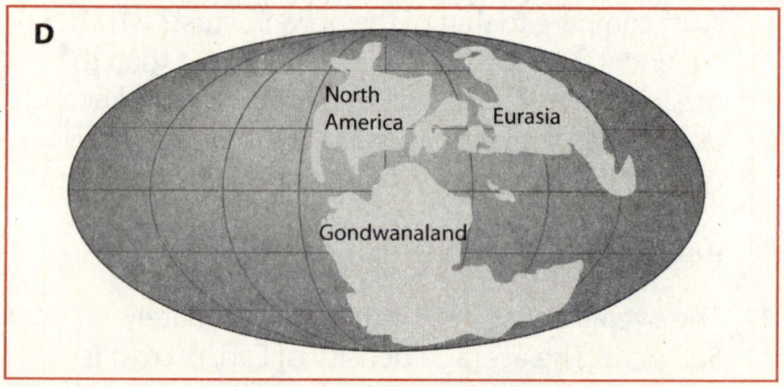

Earth around 135 millions years ago as Pangea continues to break up.

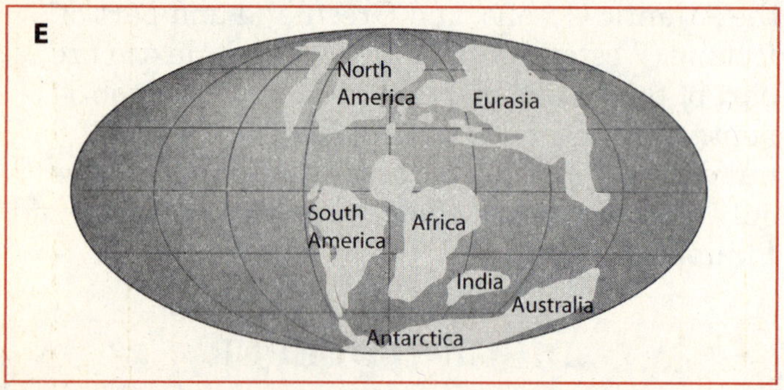

Earth around 100 million years ago with further break up of Pangea.

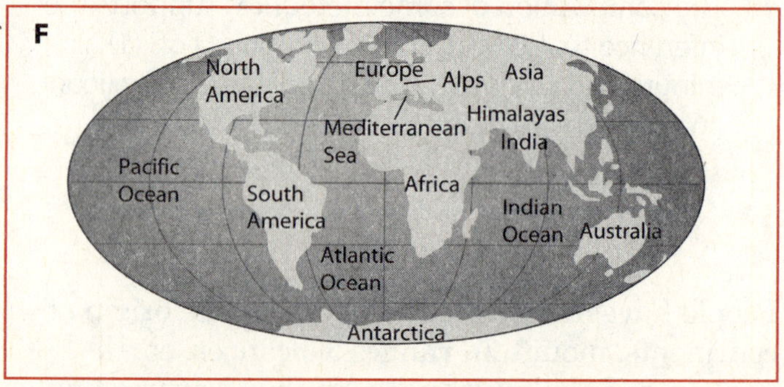

Earth at present.

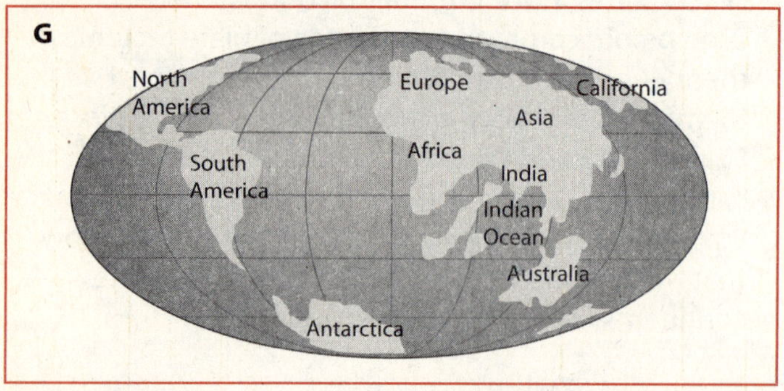

Earth 100 million years in future using present-day plate motions as a guide. Note bigger Atlantic Ocean and how Africa has split and collided with Europe.

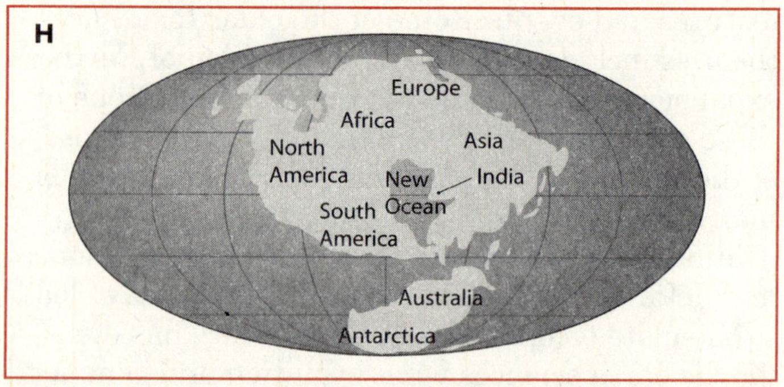

Using present-day plate motions as a guide, in approximately 250 million years, Earth may be close to another supercontinent.

At least three times, if not more, most of the large landmasses (continents and larger islands) have come together, forming supercontinents. Figure 12-13 shows how the supercontinent Rodinia formed approximately 800 million years ago and how the continents have split and formed since then. Note that diagrams G and H indicate the projected positions of landmasses in the future, based on the rates and directions the plates are moving in now.

The Three Types of Plate Boundaries

As the plates move, they interact in three ways—they can separate, collide, and slide by each other. Most of the major events involving Earth's crust—such as mountain building, earthquakes, and volcanic eruptions—occur at the boundaries where the plates interact. Each of the three types of plate boundaries has its own set of unique events and features, which are illustrated in Figure 12-14 on the next page.

DIVERGENT PLATE BOUNDARIES Where two plates separate, or diverge, the boundary is called a **divergent plate boundary**. At these locations the plates move apart and magma rises from below to fill in the separation, resulting in igneous intrusions and extrusions, such as lava flows and volcanoes. (See Figure 12-14B.) This divergence is sometimes called sea-floor spreading.

This magma and resulting lava form igneous rock that creates new crust and lithosphere, which are then split and divided in two by the divergence. The magma that forms in these regions is the result of divergence that lowers the confining pressure and melts rocks of the lower lithosphere and asthenosphere. The divergence also results in many earthquakes, most of which are shallow in depth. If the divergence is within the continental crust, the result is a continental rift valley of mountains created by faulting and much volcanic activity. (See Figure 12-14A.)

When the divergence is within oceanic crust, the faulting and volcanic activity result in a **mid-ocean ridge,** a mountain range at the bottom of the ocean that is composed mostly of volcanoes and lava flows. The mid-ocean ridges often have a central depression, or rift valley, as shown in Figures 12-14B.

CONVERGENT PLATE BOUNDARIES When two plates collide, or converge, the result is a **convergent plate boundary.** Much of the dramatic events and features of Earth's crust are created at these convergent boundaries. There are three varieties of convergent plate boundaries:

- both plates with oceanic crust on top
- both plates with continental crust on top
- one plate with oceanic crust and the other with continental crust on top

Where two plates with oceanic crust or oceanic and continental crust at their edges converge, the denser of the two plates sinks under the other plate in a process called **subduction.** As illustrated in Figure 12-14C and the Tectonic Plates map in the *Earth Science Reference Tables,* these regions of subduction of one oceanic plate under another result in ocean trenches and volcanic island arcs. The bending down of the subducting plate warps the crust, producing a long, steep, and narrow depression called an **ocean trench.** Some of these trenches are deeper below sea level than the highest mountains are above sea level. The subducting plate also results in magma bodies. Some of the magma breaks through Earth's solid surface and forms a series of volcanoes and volcanic islands—an **island arc.** The subducting process also results in a large amount of earthquake activity following the slope of the subducting plate into the bottom portions of the lithosphere. Very deep earthquakes only occur within subducting zones. The heat and pressure of the subduction causes large areas of crustal rocks to become metamorphosed in a process called "regional metamorphism," which was described in Topic 11 of this book.

If there is a continent near the two converging plates with oceanic crust, the crust at the margin of the continent may become bent down producing a down-warped ocean basin that is the site of much deposition of sediments. (See Figure 12-14C.) The sediments come from the eroding island arcs and the edge of the continents. Many of the rocks of eastern New York State formed in this type of down-warped basin before the formation of the supercontinent Pangea.

Where a plate with oceanic crust on top converges with a plate with continental crust on top, the denser oceanic crust subducts under the continental lithosphere. An ocean trench off the coast of

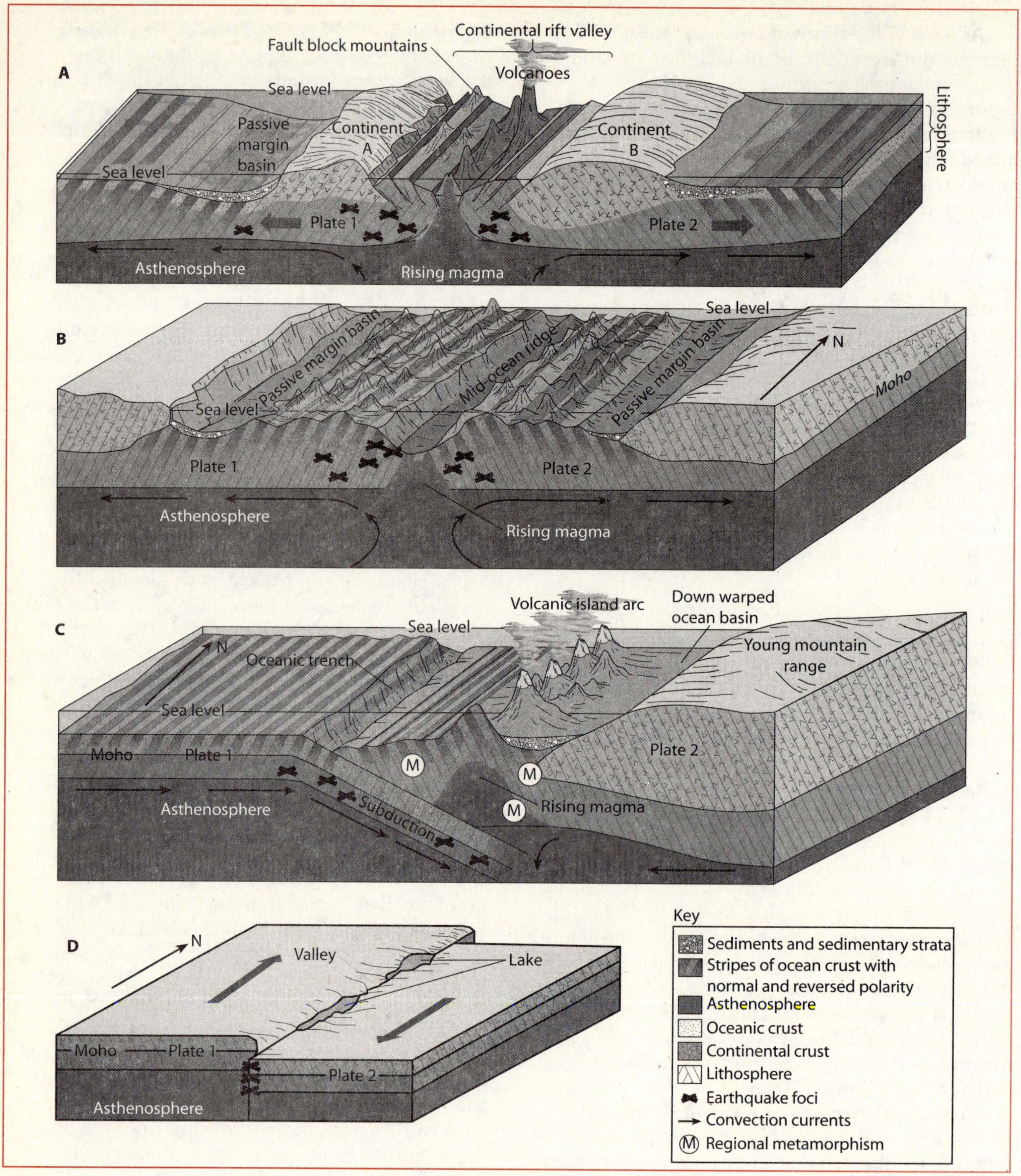

Figure 12-14. Types of plate boundaries: **(A)** Diverging plate boundaries in the continental crust result in a rift valley. **(B)** Diverging plate boundaries in the ocean result in a mid-ocean ridge, shallow-depth earthquakes, igneous intrusions, extrusions of lava flows, volcanoes, and passive margin basins in a down-warped part of crust. Plate 1 is moving west and plate 2 is moving east. **(C)** Converging plate boundaries result in an oceanic trench, volcanic island arcs, igneous intrusions, regional metamorphism, young mountains, and subduction of oceanic plate. Earthquake foci exist at various depths, indicating subductions. Plate 1 is moving east and plate 2 is moving west. **(D)** Transform plate boundaries result in many shallow-focus earthquakes and little to no igneous activity. Plate 1 is moving north and plate 2 is moving south.

a continent often forms from this event. The magma that forms from the subduction doesn't create an island arc because the continent is above the region of magma formation. What forms instead are relatively new mountain ranges called **young mountains.** Like the Cascades in western Oregon and Washington, these mountain ranges may be mostly volcanic. They may also be a combination of volcanic, faulted, and folded.

When two plates with continents at their edges converge, the two plate edges bunch up together, causing a great thickening of the crust and lithosphere. This bunching up creates the highest of young mountains—such as the Himalayas, where India is colliding with Asia.

Plate convergence that results in the growth of young mountain ranges is called orogeny. The term is also used to indicate the time when mountain building occurs. Refer to Important Geologic Events in New York within Geologic History of New York State in the *Earth Science Reference Tables,* and you will note the four orogenies (Grenville, Taconian, Acadian, and Appalachian, also known as the Alleghanian) that have resulted in mountains in New York State and surrounding regions.

TRANSFORM PLATE BOUNDARIES When two plates collide by sliding past each other, the boundary is called a **transform plate boundary.** The dragging of lithospheric rocks along the edges of the transform boundaries builds up much potential energy in rocks; this energy is eventually released as kinetic mechanical energy in the form of earthquakes. The San Andreas fault system in California is an example of a transform plate boundary between the North American and Pacific plates

Driving Forces of Plate Tectonics

While most scientists accept the major part of the plate tectonic theory, there is still much debate about what makes the plates move. The most accepted theory is that mantle convection currents drag or push the plates apart at places where plates diverge. The exact location of these convection currents is hotly debated. The energy source for these convection currents is the heat of Earth's interior, which causes hotter, less dense parts of the mantle to rise under diverging plates. Gravity pulls down the cooler, more dense regions of the mantle, causing falling convection currents in subduction areas. (See Figures 12-12 and 12-14 for an illustration of convection in the mantle.)

Hot Spots

One major aspect of lithosphere and crust that is difficut to explain by the conventional plate tectonic theory concerns major regions of volcanic activity in the interior parts of plates away from plate boundaries. These regions are called **hot spots**. Examples include the big island of Hawaii, the region around Yellowstone National Park in

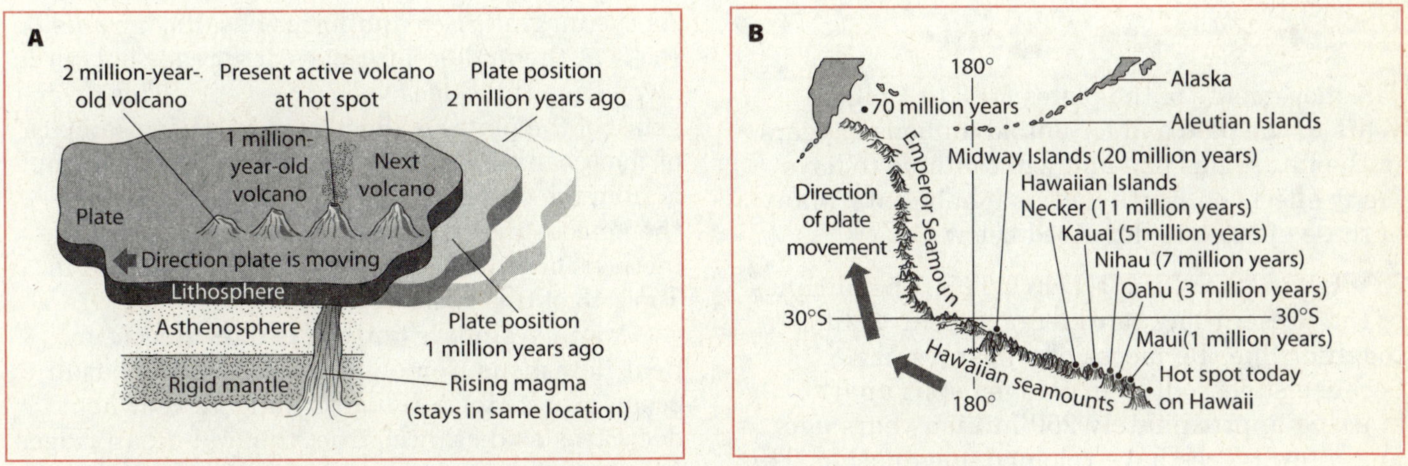

Figure 12-15. Hot spots, rising magma, and plate movements: It is believed that rising magma from the mantle melts through the plates of the lithosphere, forming volcanoes at hot spots. As the plate moves, new hot spots and volcanoes form, as shown in diagram A. The Hawaiian Islands' Emperor Seamount (underwater volcanoes) Trail was created by movements of the Pacific plate over an area of rising magma, as shown in diagram B. Note how the Pacific plate changed direction some 25 million to 30 million years ago.

Wyoming and Montana, and maybe even the Adirondack Mountains of New York State. See the hot spots on Tectonic Plates in the *Earth Science Reference Tables*.

Scientists have proposed that hot spots occur where rising magma from the lower mantle remains stationary for millions of years. When a plate moves over the rising magma, it melts its way to or near the surface to form a hot spot. Hot spots become regions of intrusive and extrusive volcanic activity that can build volcanoes and lava flows and push up regions of the crust to form mountains. Because the plate keeps moving over the rising magma, a series of volcanic mountains form for thousands of miles. These trails of hot-spot features can be used to infer past movements of the plates that moved over the hot spots. See Figure 12-15 on the previous page to help you understand rising magma and hot spots.

DIGGING DEEPER

An interesting theory related to hot spots is that the impact of an asteroid or comet on Earth could cause weak spots in the crust and mantle, allowing magma to rise and causing hot-spot regions. It has even been suggested that these asteroid- or comet-created areas of rising magma could cause the plates to begin moving by cracking the lithosphere and starting the rise of convection at new divergent plate boundaries.

Effects of Plate Tectonics

The movements of the plates for 3 to 4 billion years in the past and for untold millions of years in the future has had and will continue to have many effects on Earth and its inhabitants. Many of these effects are described below.

APPEARANCE OF CONTINENTS The outlines of the present-day continents appear to fit together like the pieces of a jigsaw puzzle because supercontinent Pangea split apart starting approximately 250 million years ago, and a few pieces have rejoined one another. This can be seen in Figure 12-13 and in the Inferred Positions of Earth's Landmasses section of the Geologic History of New York in the *Earth Science Reference Tables*.

FEATURES OF LANDMASSES Similarities in minerals, rocks, fossils, age, and structure features of mountain ranges are found at places where the continents and other landmasses may have fitted together in the past. These similar features on different landmasses indicate a commonality in age and origin. Today the continents are separated, and their respective life forms are often greatly different. The variation in life forms on the different land areas is a result of evolution. However, fossil evidence shows that in the past many plants and animals, such as the Glossopteris fossils of Figure 12-16, were the same throughout the world. Such a wide distribution of the same plants and animals probably could not have occurred unless the continents were connected. Figure 12-16 shows some of the now separated locations that have rocks, minerals, fossils, and mountain structures in common. These similarities suggest that the land areas were together when these features were formed, and that these areas have since separated as the plates moved.

As the plates move portions of Earth's surface to new locations, rocks are exposed to different climate conditions. The finding of rocks near the equator with evidence of glacial erosion and deposition and rocks near the poles with coal deposits of the same ages indicates a climate pattern of hot near the poles and cold near the equator. The probable answer to this dilemma is that plate movements have changed the locations of the landmasses.

AGE AND HEAT PATTERNS OF OCEANIC ROCKS As the magma rises and forms basaltic igneous rocks at the mid-ocean ridges, it spreads outwards away from the ridges with divergence. This divergence of the plates is illustrated by dating samples of the oceanic basaltic rocks. The farther the sample is from the center of a mid-ocean ridge, the older the igneous rock is. The rocks are dated using the methods of radioactive-decay dating explained in Topic 13 of this book.

Another similar pattern is indicated from heat flow measurements taken from the basaltic ocean crust. Measurements indicate that heat decreases as distances from the mid-ocean ridges increase. This makes sense because the hottest rocks should be closest to the magma and lava from which they formed.

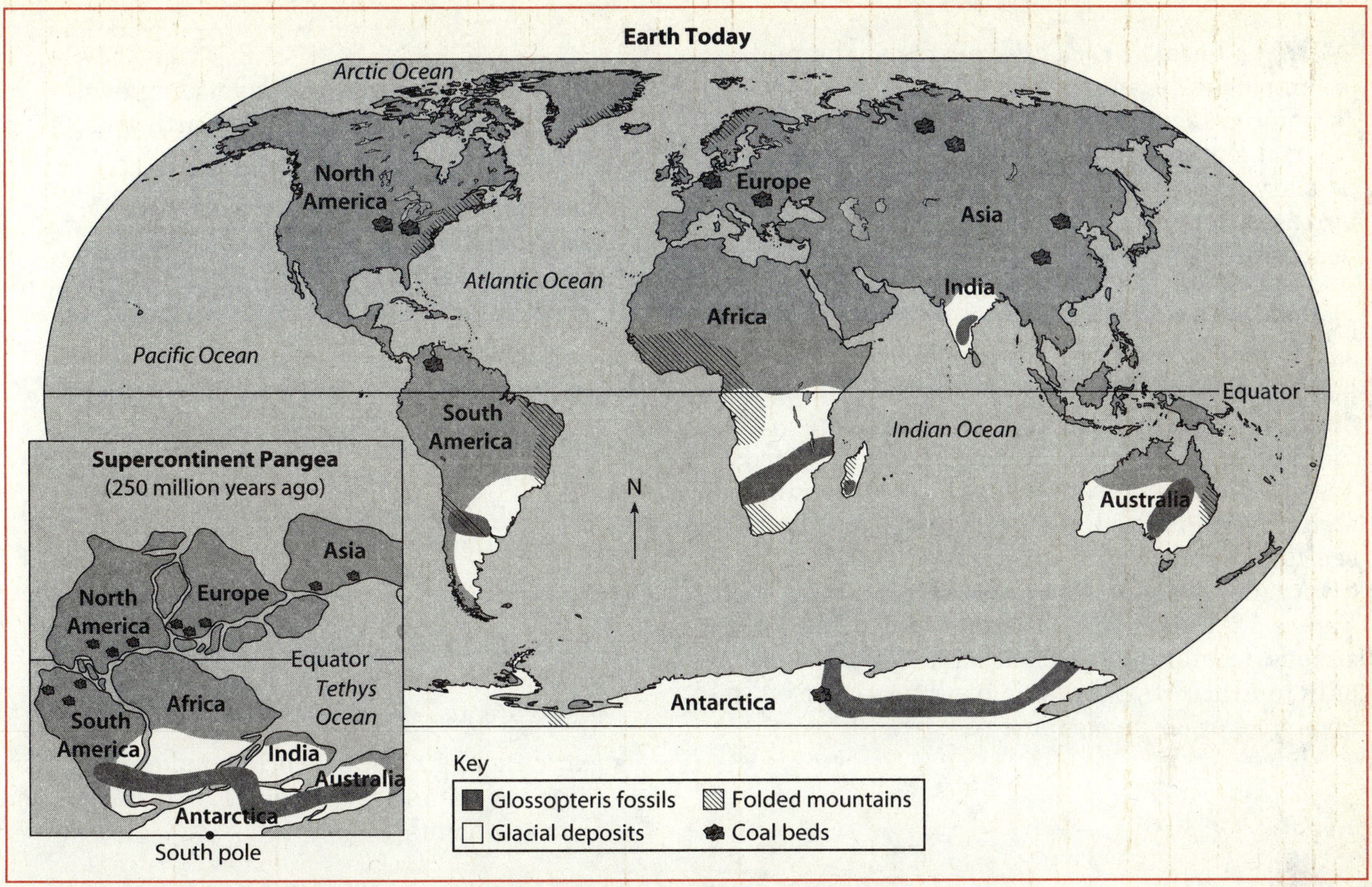

Figure 12-16. Changing features of landmasses due to plate movements: Many features of Earth's landmasses in rocks more than 250 million years old don't make sense unless the landmasses were in different locations approximately 250 million years ago when the supercontinent Pangea existed.

MAGNETIC PATTERNS OF OCEANIC BASALTIC ROCKS The divergence of the oceanic lithosphere and crust at mid-ocean ridges is shown by patterns in magnetism of the oceanic-basaltic rocks. The Earth's magnetic poles flip-flop in polarity (north changes to south and south changes to north) in periods of thousands of years in a process called reversal of Earth's magnetic polarity. The reasons for these magnetic reversals and the question of whether the magnetic reversals are cyclic or non-cyclic have not been answered. What is known is that Earth's magnetic field has reversed hundreds of times since the magnetic field's origin, probably billions of years ago. (See Figure 12-17.)

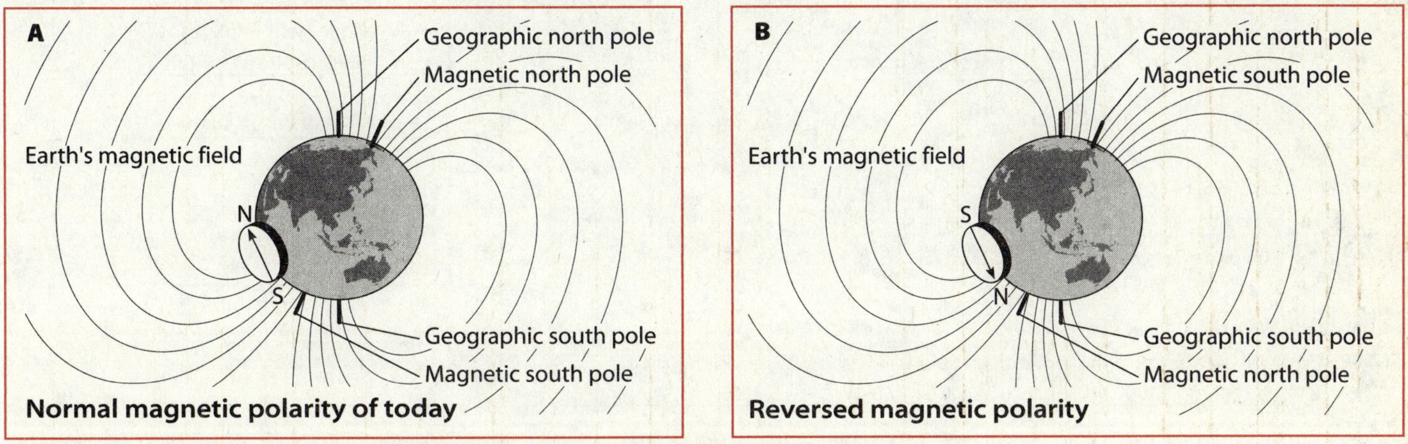

Figure 12-17. Normal and reversed polarity of Earth's magnetism: Earth's magnetic field flip-flops from normal polarity to reversed polarity in time spans of thousands of years.

When basaltic rock crystallizes at the mid-ocean ridges, its magnetic minerals are aligned; therefore, they record the particular polarity when the magma and lava solidify. Normal polarity is when magnetic north is near the north geographic pole and magnetic south is near the south geographic pole—as is true today. Reversed polarity is when magnetic north is near the south geographic pole and magnetic south is near the north geographic pole. It has been found that there is a pattern of corresponding stripes of basaltic rock on either side of the mid-ocean ridges of normal and reversed polarity. This evidence suggests that the corresponding stripes were formed at similar times, and the sea-floor spreading has separated them during divergence. These stripes are illustrated on Figure 12-14 on page 232. Note that these stripes are not a visual feature but a pattern only indicated by magnetic instruments used in airplanes or oceanographic research ships.

OTHER EFFECTS OF PLATE TECTONICS

The movement of the plates has had and will continue to have other effects on Earth and its inhabitants:

- environmental hazards, such as volcanic eruptions, earthquakes, and tsunamis
- changes in Earth's climate and weather, such as altering the locations of land and ocean areas and creating mountains to change wind patterns
- changes in the factors that cause our day-to-day weather by changing the distribution of land and ocean areas and other features, such as mountains and large continental regions
- the rock cycle
- the major landscape features of Earth—continents, ocean basins, mountains, plains, and plateaus
- exposing rocks to weathering and erosion that carve the details of the solid Earth surface and form sediments

Review Questions

Base your answers to questions 49 through 53 on the following diagram. The diagram is an earlier tectonic-plate model that represents one possible interpretation of the movements of Earth's rock surfaces according to the theory of plate tectonics. According to this interpretation, Earth's lithosphere consists of several large plates that are moving in relationship to one another. The arrows in the diagram show some of this relative motion of the plates. The diagram also shows the age of formation of the igneous rocks that make up the oceanic crust of the northern section of the Pacific plate.

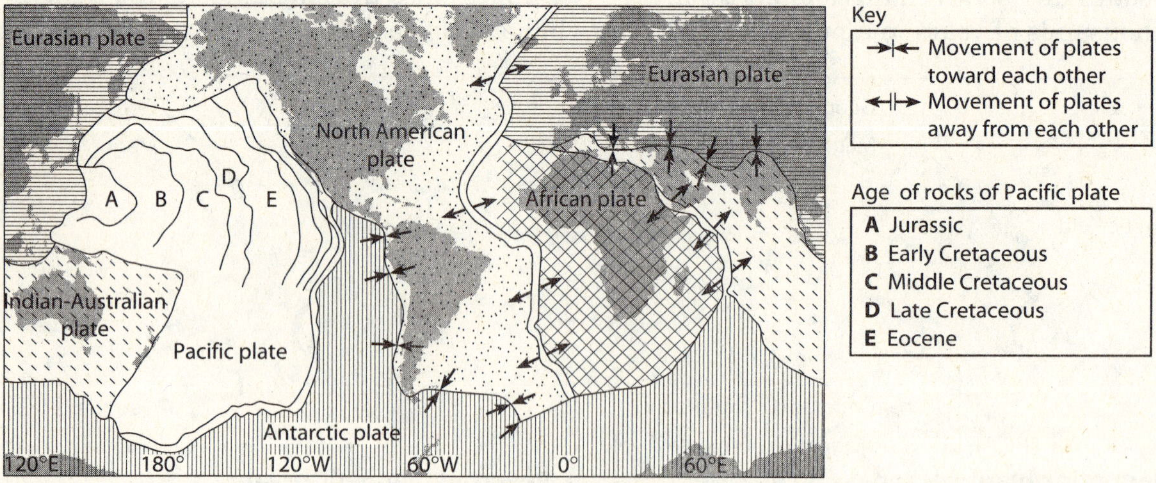

49. Which statement is best supported by the relative movement shown by the arrows in the diagram? (1) North America and South America are moving toward each other. (2) The Indian-Australian plate is moving away from the Eurasian plate. (3) The African plate and Eurasian plate are moving away from the North American plate. (4) The Antarctic plate is moving away from the North American plate.

50. The boundaries between all of these plates are best described as the sites of (1) frequent crustal activity (2) deep ocean depths (3) continental boundaries (4) magnetic age pattern

51. Which geologic structure is represented by the double line separating the North American plate from the African and Eurasian plates? (1) thick continental crust (2) thick layers of sediment (3) mid-ocean ridge (4) granitic igneous rock

52. Which provides the best explanation of the mechanism that causes these plates to move across Earth's surface? (1) convection currents in the mantle (2) faulting of the lithosphere (3) the spin of Earth on its axis (4) prevailing wind belts of the troposphere

53. The age of formation of the igneous rocks A, B, C, D, and E that make up the oceanic crust of the northern half of the Pacific plate suggests that this section of the Pacific plate is generally moving in which direction? (1) from north to south (2) from south to north (3) from west to east (4) from east to west

54. Scientists theorize that Africa and South America were once part of the same large continent. Cite two pieces of evidence that support this theory.

55. Which statement best supports the concept that continents have shifted position? (1) Basaltic rock is found to be progressively younger at increasing distances from a mid-ocean ridge. (2) Marine fossils are often found in deep-well drill cores. (3) The present continents appear to fit together as pieces of a larger landmass. (4) Areas of shallow-water seas tend to accumulate sediment, which gradually sinks

56. According to the *Earth Science Reference Tables*, during which geologic time period were the continents of North America, South America, and Africa closest together? (1) Tertiary (2) Cretaceous (3) Triassic (4) Ordovician

57. According to the theory of plate tectonics, the distance between two continents on opposite sides of a mid-oceanic ridge will generally (1) decrease (2) increase (3) remain the same

58. Describe two expected similarities between rock samples found equal distances from and on opposite sides of the Mid-Atlantic Ridge.

59. Which feature is commonly formed at a plate boundary where oceanic crust converges with continental crust? (1) a mid-ocean ridge (2) an oceanic trench (3) a transform fault (4) new oceanic crust

60. Evidence of subduction exists at the boundary between the (1) African and South American plates (2) Indian, Australian, and Antarctic plates (3) Pacific and Antarctic plates (4) Nazca and South American plates

61. Which of the following cross-sectional diagrams best represents a model for the movement of rock material below the crust along the mid-Atlantic ridge?

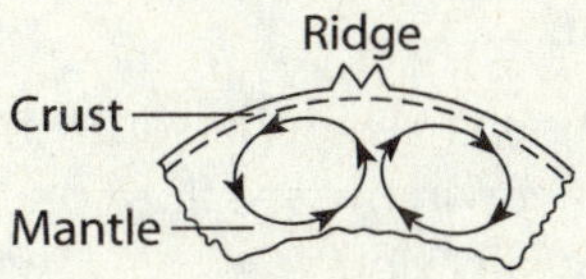

(1)

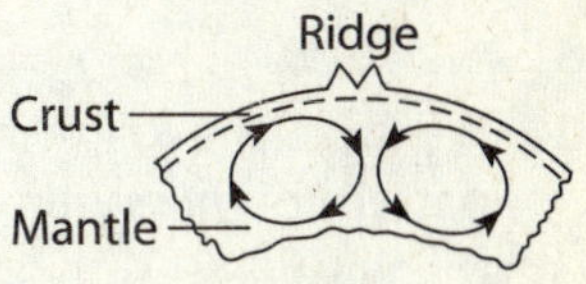

(3)

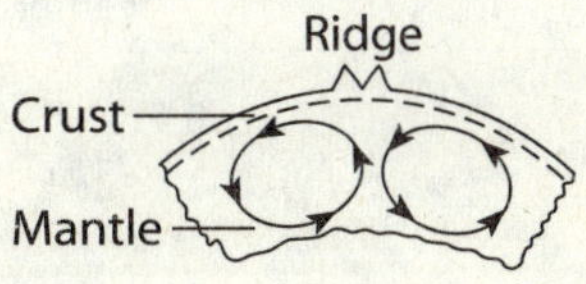

(2)

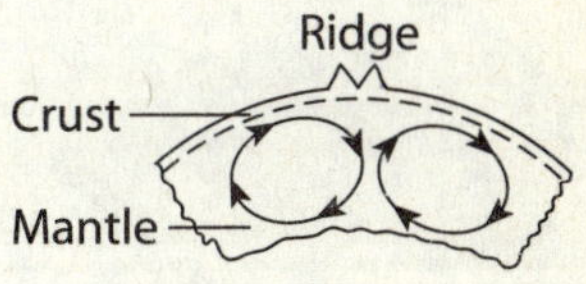

(4)

62. The following drawing represents the ocean floor between North America and Africa.

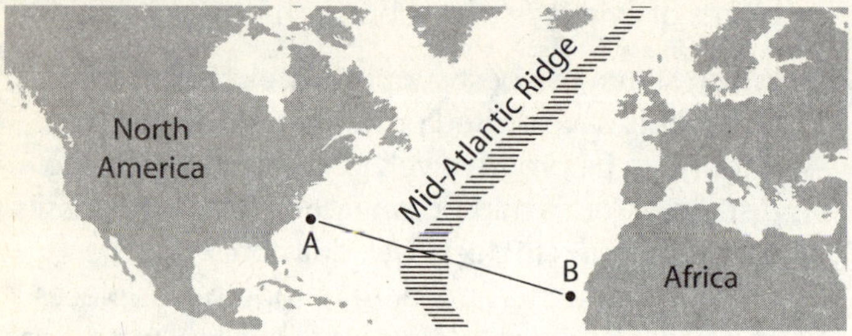

Which graph best represents the age of the bedrock in the ocean floor along line AB?

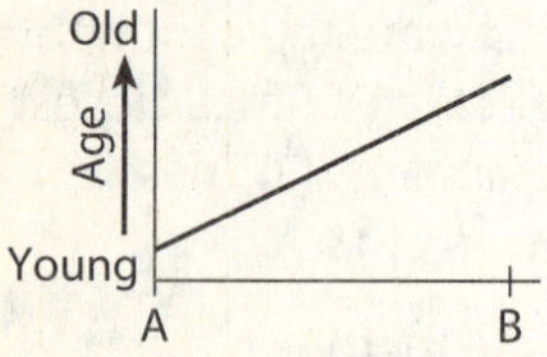

(1)

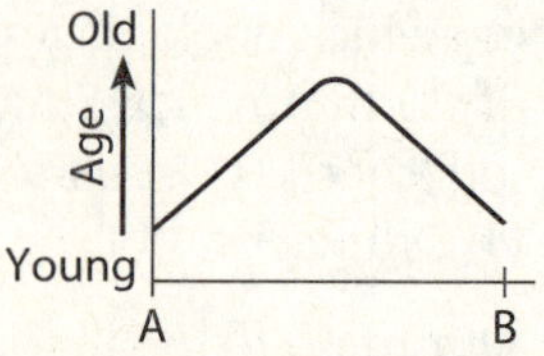

(2)

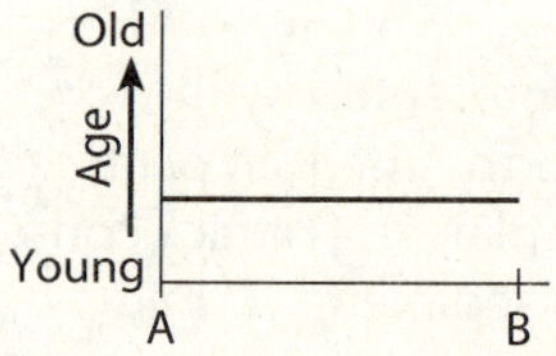

(3)

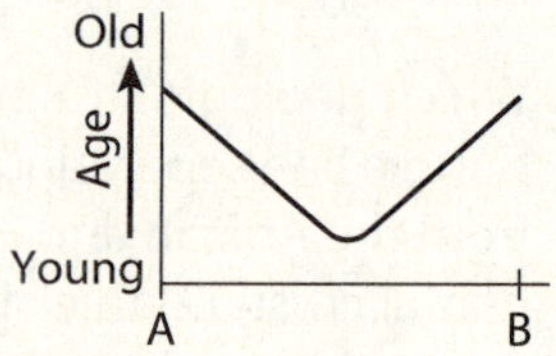

(4)

63. Igneous materials found along mid-ocean ridges contain magnetic particles that show reversal of magnetic orientation. This is evidence that (1) volcanic activity has occurred constantly throughout history (2) Earth's magnetic poles have exchanged their positions (3) igneous materials are always formed beneath oceans (4) Earth's crust does not move

Base your answers to questions 64 through 68 on the following diagram. The diagram shows the magnetic orientation of igneous rock on the sea floor on the east side of a mid-ocean ridge. The pattern on the west (left) side of the ridge has been omitted. The age of the igneous rock and its distance from the ridge center are shown.

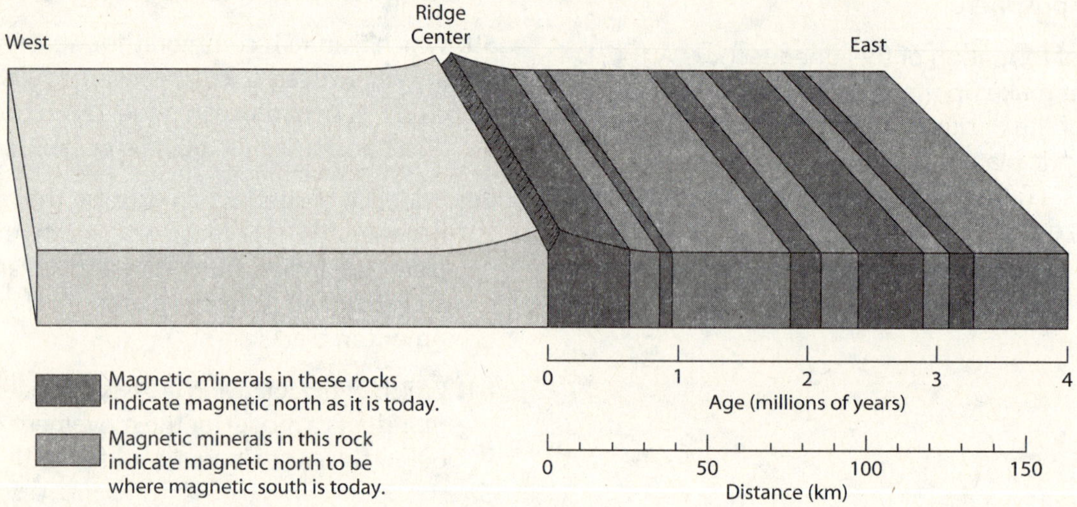

64. Which of the following diagrams best represents the pattern of magnetic orientation in the sea floor on the west (left) side of the ocean ridge?

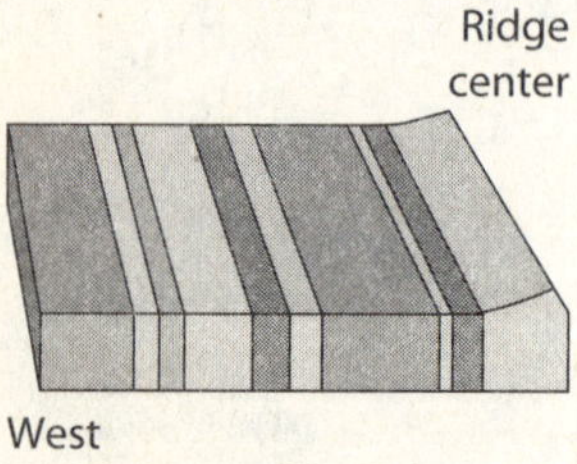

(1)

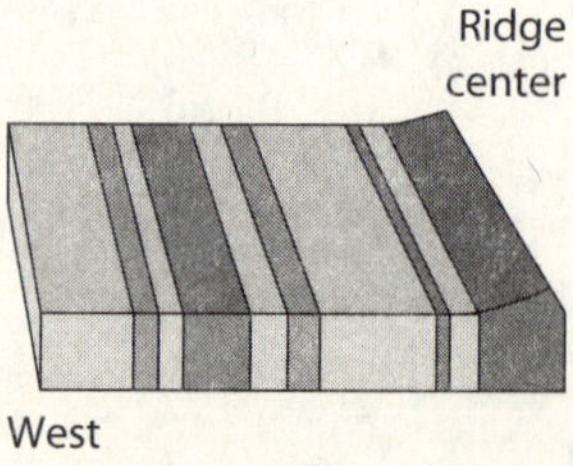

(2)

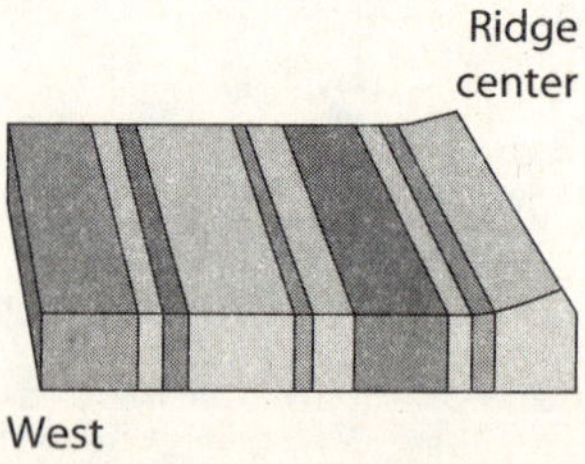

(3)

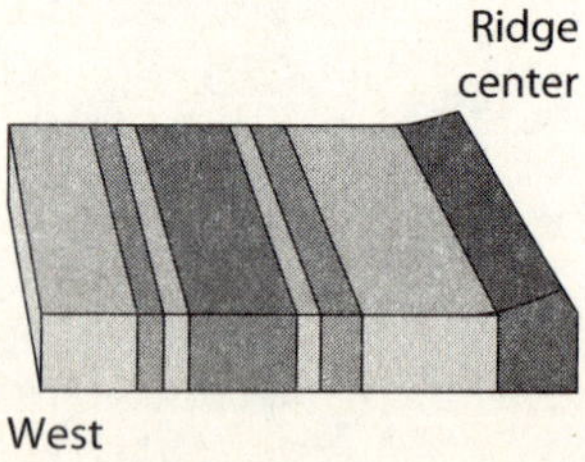

(4)

65. According to the diagram, what is the approximate rate of sea-floor spreading? (1) 1 km/million years (2) 2 km/million years (3) 40 km/million years (4) 50 km/million years

66. Which inference can best be made from the diagram? (1) The orientation of Earth's magnetic field has reversed with time. (2) The size of the continents has changed with time. (3) The elevation of sea level has changed with time. (4) The amount of fossil material preserved in the igneous rock has changed with time.

67. The crustal material on both sides of the ridge indicates that the tectonic plates are (1) diverging (2) converging (3) not moving (4) moving parallel to the ridge

68. As the distance from the center of the ridge increases, the age of the rocks (1) decreases (2) increases (3) remains the same (4) increases and decreases in a cyclic pattern

Base your answers to questions 69 through 73 on the following diagrams. Diagram I is a map showing the location and bedrock age of some of the Hawaiian Islands. Diagram II is a cross section of an area of Earth that illustrates a stationary magma source (rising magma and hot spot) and the process that could have formed the islands.

I

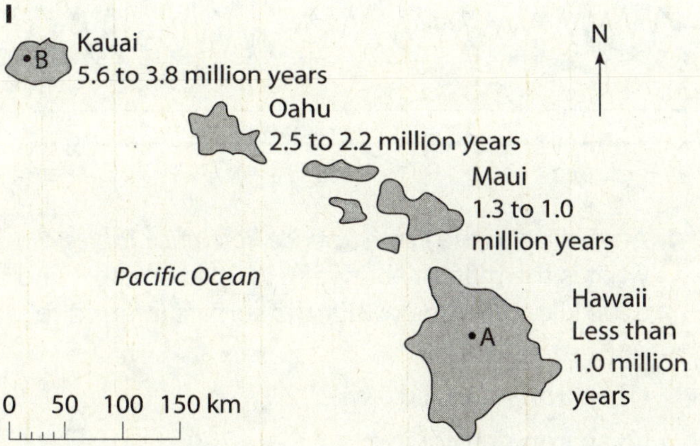

II

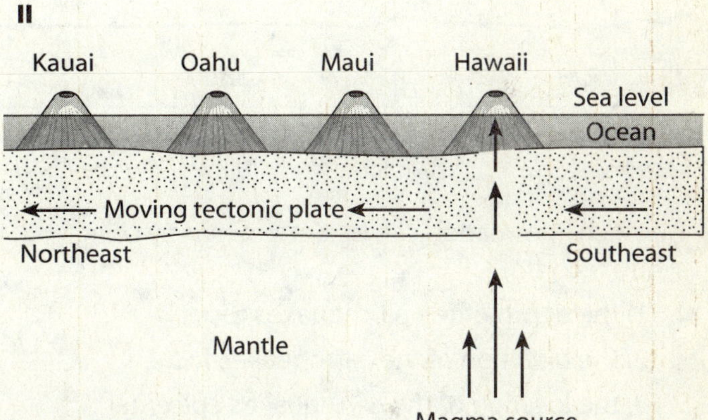

(Not drawn to scale.)

69. If each island formed as the tectonic plate moved over the magma source in the mantle, as shown in diagram II, where would the next volcanic island form? (1) northwest of Kauai (2) northeast of Hawaii (3) southeast of Hawaii (4) between Hawaii and Maui

70. Compared to the continental crust of North America, the oceanic crust in the area of the Hawaiian Islands is probably (1) thinner and similar in composition (2) thinner and different in composition (3) thicker and similar in composition (4) thicker and different in composition

71. Volcanic activity like that which produced the Hawaiian Islands is usually closely correlated with (1) nearness to the center of a large ocean (2) sudden changes in Earth's magnetic field (3) frequent major changes in climate (4) frequent earthquake activity

72. Which of the Hawaiian Islands has the greatest probability of a volcanic eruption? (1) Kauai (2) Oahu (3) Maui (4) Hawaii

73. Which of the following graphs best represents the ages of the Hawaiian Islands, comparing them from point A to point B?

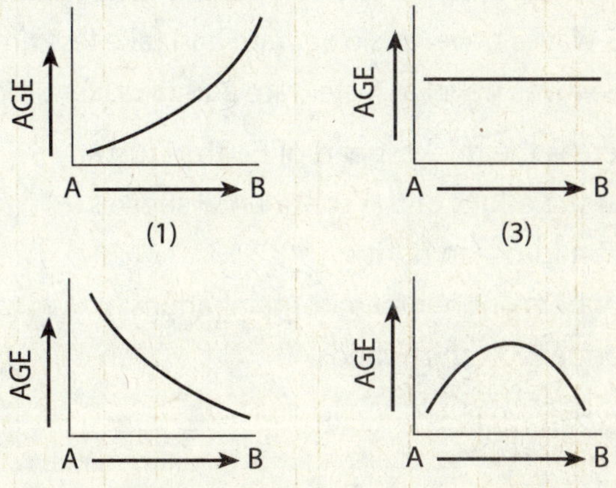

74. What do mid-ocean ridges and hot spots have in common? (1) They are associated with rising magma. (2) They are always associated with present-day plate boundaries. (3) They commonly are associated with earthquakes of great depth. (4) Neither is associated with plate motions.

75. Which observation provides the strongest evidence for the inference that convection cells exist within Earth's mantle? (1) Sea level has varied in the past. (2) Marine fossils are found at elevations high above sea level. (3) Displaced rock strata are usually accompanied by earthquakes and volcanoes. (4) Heat-flow readings vary at different locations in Earth's crust.

Questions for Regents Practice

Part A

1. The following is a record of earthquake waves from a seismic station.

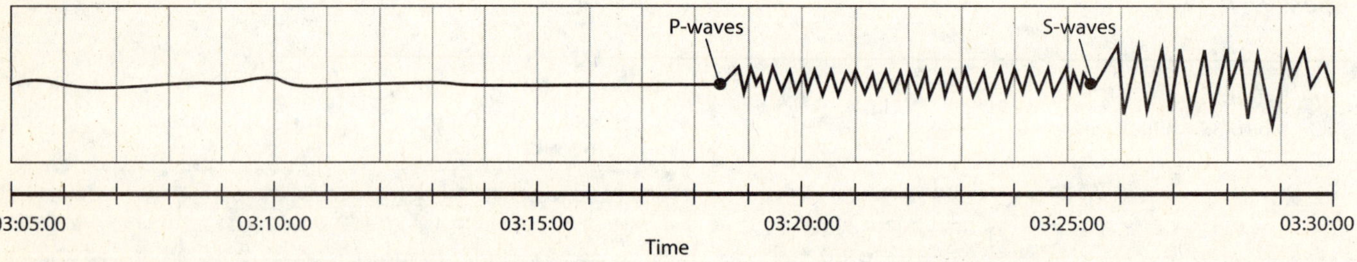

Which information can be determined by using this record?

(1) the depth of the earthquake's focus

(2) the direction of the earthquake's focus

(3) the location of the earthquake's epicenter

(4) the distance to the earthquake's epicenter

2. Use the *Earth Science Reference Tables* to decide which statement best describes the relationship between the travel rates and travel times of earthquake P-waves and S-waves from the focus of an earthquake to a seismic station.

(1) P-waves travel at a slower rate and take less time.

(2) P-waves travel at a faster rate and take less time.

(3) S-waves travel at slower rate and take less time.

(4) S-waves travel at faster rate and take less time.

3. Where is the thickest part of Earth's crust?

(1) at the edges of the continental shelves

(2) at mid-ocean ridges

(3) under continental mountain ranges

(4) under volcanic islands

4. According to the *Earth Science Reference Tables*, what is the material between 2900 kilometers and 5200 kilometers below Earth's surface inferred to be?

(1) an iron-rich solid

(2) an iron-rich liquid

(3) a silicate-rich solid

(4) a silicate-rich liquid

5. An earthquake's P-waves traveled 4800 kilometers and arrived at a seismic station at 5:10 P.M. At approximately what time did the earthquake occur?

6. The theory of plate tectonics suggests that the

(1) continents moved due to changes in Earth's orbital speed

(2) continents moved due to the Coriolis effect caused by Earth's rotation

(3) present-day continents of South America and Africa are moving toward each other

(4) present-day continents of South America and Africa once fit together like puzzle pieces

7. What is one of the main reasons volcanic eruptions can be predicted in enough time to allow people to escape?

(1) Many volcanoes emit X-rays before eruptions.

(2) Most volcanoes form a new crater before major eruptions.

(3) Most volcanoes swell in volume before major eruptions.

(4) Most animals escape from an area before most eruptions.

8. What is the best evidence that earthquakes and crustal movements are associated?

(1) associations of earthquake belts and young mountain ranges

(2) association of earthquake belts and belts of old metamorphic rocks

(3) studying faults in lava flows

(4) samples of rocks and fluids from deep drill holes

Base your answers to questions 9 through 11 on the following map.The map shows a mid-ocean ridge and trenches in the Pacific Ocean.

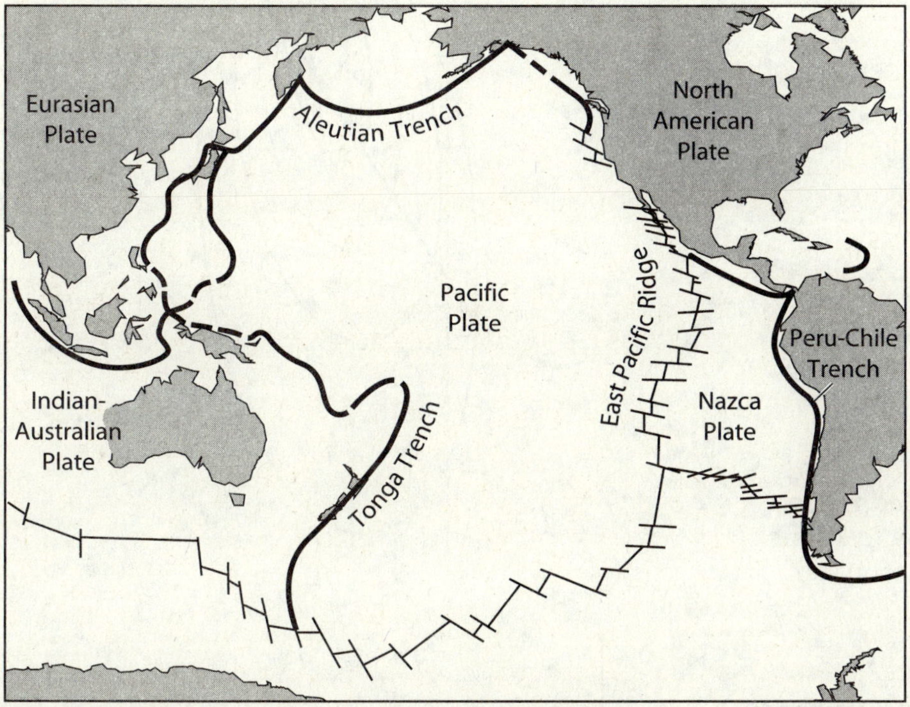

9.Movement of the plates shown in the diagram is most likely caused by

(1) the revolution of Earth

(2) the erosion of Earth's crust

(3) shifting of Earth's magnetic poles

(4) convection in Earth's mantle

10. The crust at the mid-ocean ridge is composed mainly of

(1) granite (3) shale

(2) basalt (4) limestone

11. Mid-ocean ridges such as the East Pacific Ridge are best described as

(1) mountains containing folded sedimentary rocks

(2) plateaus containing fossils of present-day marine life

(3) sections of the ocean floor that contain the youngest oceanic crust

(4) sections of the ocean floor that are the remains of a submerged continent

12. At which depth below Earth's surface is the density most likely 10.5 grams per cubic centimeter?

(1) 1500 km (3) 3500 km
(2) 2000 km (4) 6000 km

13. What happens to P-waves and S-waves from a crustal earthquake when the waves reach Earth's core?

(1) S-waves are transmitted through the outer core, but P-waves are not transmitted.

(2) P-waves are transmitted through the outer core, but S-waves are not transmitted.

(3) Both P-waves and S-waves are transmitted through the outer core.

(4) Neither P-waves nor S-waves are transmitted through the outer core.

14. At a depth of 2000 kilometers, the temperature of the stiffer mantle is inferred to be

(1) 6500°C (3) 3500°C
(2) 4200°C (4) 1500°C

15. Tsunamis can be directly caused by

(1) offshore surface ocean currents
(2) gravitational effects of the moon
(3) underwater earthquakes
(4) underwater lava flows at mid-ocean ridges

16. A large earthquake occurred at 45° N, 75° W on September 5, 1994. According to the *Earth Science Reference Tables,* which location in New York State was closest to the epicenter of the earthquake?

(1) Buffalo (3) Albany
(2) Massena (4) New York City

17. Oxygen is the most abundant element by volume in Earth's

(1) inner core
(2) crust
(3) hydrosphere
(4) troposphere

18. Which geologic event occurred most recently?

(1) initial opening of the Atlantic Ocean
(2) formation of the Hudson Highlands
(3) formation of the Catskill delta
(4) collision of North America and Africa

Part B

Base your answers to questions 19 through 21 on the following map. The star symbol represents a volcano located on the mid-Atlantic ridge in Iceland. The isolines represent the thickness, in centimeters, of volcanic ash deposited from an eruption of this volcano. Points A and B represent locations in the area.

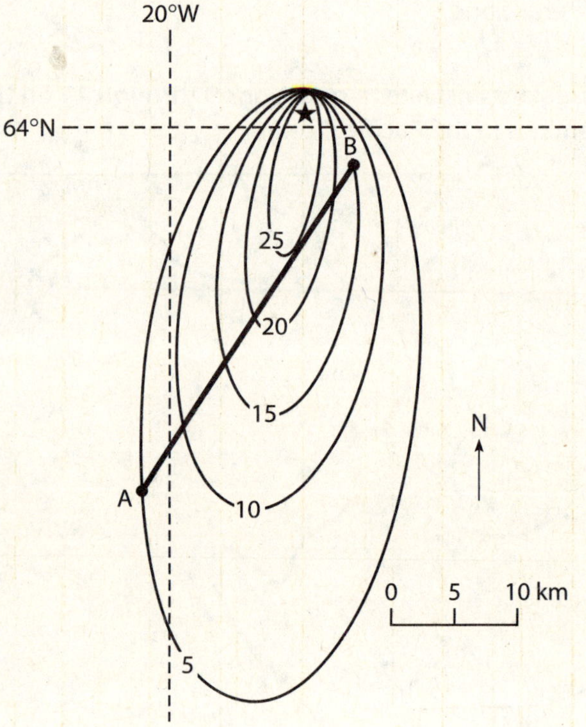

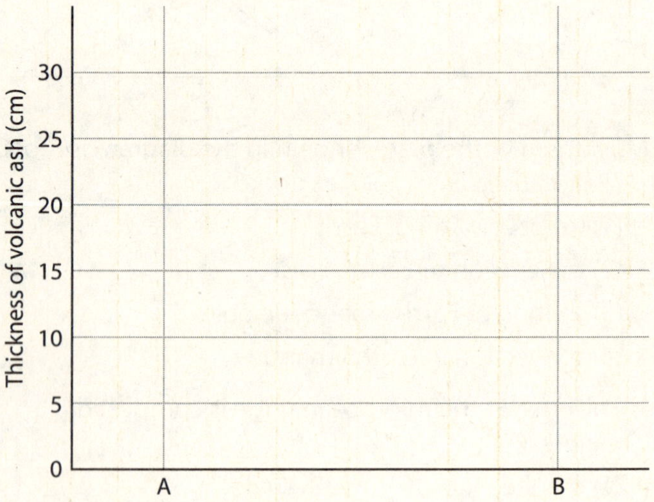

19. On a grid like the one shown on the previous page and using the directions below, construct a profile of the ash thickness between point A and point B.

a. Plot the thickness of the volcanic ash along line AB by marking with a dot each point where an isoline is crossed by line AB. [2]

b. Connect the dots to complete the profile of the thickness of the volcanic ash. [1]

20. State one factor that could have produced this pattern of deposition of the ash. [1]

21. State why volcanic eruptions are likely to occur in Iceland. [1]

Base your answers to questions 22 through 24 on the following map and the information given. The map shows the location of major islands and coral reefs in the Hawaiian Island chain. Their ages are given in millions of years.

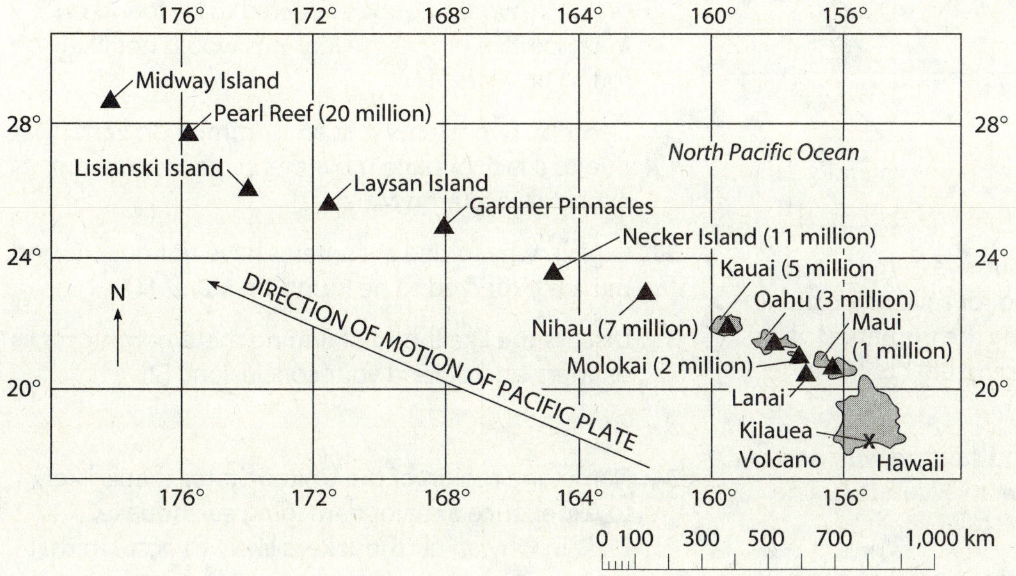

The islands of the Hawaiian chain formed from the same source of molten rock. The movement of the Pacific plate over the Hawaiian hot rising magma created a trail of extinct volcanoes that makes up the Hawaiian Islands. The island of Hawaii (lower right) has active volcanoes, including Mauna Loa and Kilauea. Kilauea is considered to be almost directly over the stationary source of rising magma.

22. Expressed in kilometers, approximately how far has the Pacific plate moved since Necker Island was located over the rising magma at X? [1]

23. What is the latitude and longitude of Lisianski Island? [1]

24. State the name of a specific rock type that would be common on the surface around Kilauea volcano. (An example of a specific rock type is sandstone, NOT sedimentary rock.) [1]

Base your answers to questions 25 through 28 on the map below.

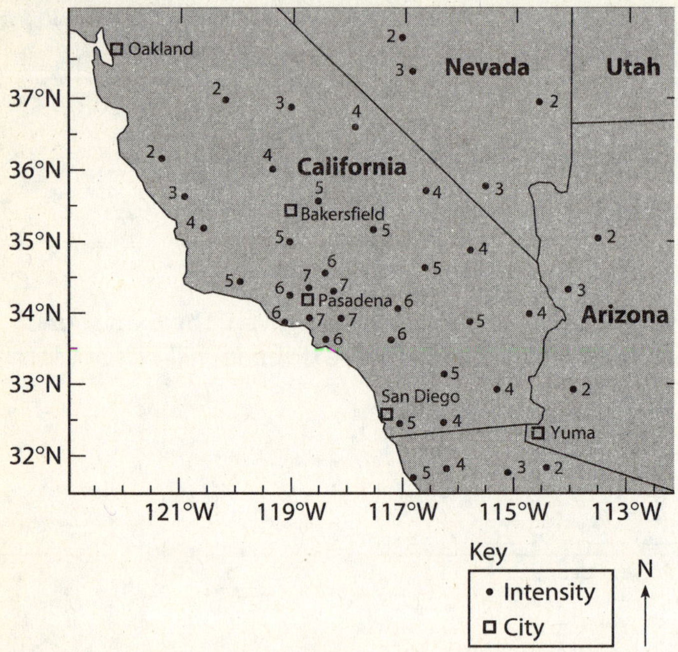

An earthquake occurred in the southwestern United States. Intensity units were plotted for several locations on a map such as the following. (As the numerical value of intensities increases, the damaging effects of the earthquake waves also increases.)

25. Using an interval of 2 units and starting with an isoline representing 2 units, draw an accurate isoline map of earthquake intensity. [4]

26. State the name of the city that is closest to the earthquake epicenter. [1]

27. Identify the most likely cause of earthquakes that occur in the area shown on the map. [1]

28. A newspaper article indicates that a major earthquake is expected in the local area. State three actions individuals could take to increase safety or reduce injury from an earthquake [3]

Part C

Use the following information and your knowledge of earth science to answer questions 29 through 33.

Studies of the planet Mars from 1965 to the present have allowed scientists to infer that like Earth, Mars has or had hot spots; but unlike Earth there is no evidence of plates or movement of plates.

29. Describe a landscape feature composed of igneous rocks that you would expect to find on Mars and explain why you would expect to find it there. [2]

30. Identify four large (hundreds of miles in length) types of landscape features that, although common on Earth, would not be expected to be found on Mars. Explain why these features would not likely be found on Mars. [4]

31. Describe two events that are common on Earth that, due to a lack of plate movements, you would not expect to occur on Mars. [2]

32. Explain why folded mountains have not been found, nor are expected to be found, on Mars. [1]

33. Discuss the likelihood of finding metamorphic rocks on Mars, and defend your conclusion. [2]

34. Name one region of the United States that is likely to experience a major damaging earthquake. Explain why an earthquake is likely to occur in that region. [1]

Interpreting Geologic History

TOPIC 13

VOCABULARY		
absolute age	inclusion	radioactive dating
bedrock	index fossil	radioactive decay
carbon-14 dating	intrusion	species
correlation	isotope	unconformity
extrusion	organic evolution (theory of)	uranium-238
fossil	outgassing	volcanic ash
geologic time scale	principle of superposition	
half-life		

The composition, structure, position, and fossil content of Earth's rocks provide information about Earth's geologic history. Evidence in Earth's rocks dates back approximately 4 billion years. In the *Earth Science Reference Tables,* carefully study the Geologic History of New York State and the Generalized Bedrock Geology of New York State. A bedrock, or geologic, map shows the rock types that underlie a specific area and their age.

Relative Dating of Rocks and Events

Relative dating is the determination of the age of a rock or event in relation to the age of other rocks or events. The following descriptions explain how the methods of relative dating make it possible to determine the relative age of a rock or event. Note that relative age differs from **absolute age,** which refers to the actual age of a rock or an event in years.

Principle of Superposition

When observing layers of sedimentary rock, or some extrusive igneous rocks, geologists usually infer that the bottom layer is the oldest. Each overlying layer is then progressively younger, so that the top layer is the youngest, or most recently formed. This inference is called the **principle of superposition,** and it is the basis for methods of relative dating. (See Figure 13-1.) Superposition relates to the original horizontality of deposited sediments or of volcanic rock layers, a concept that is discussed in Topic 12.

Exceptions to the principle of superposition sometimes occur when changes in Earth's crust

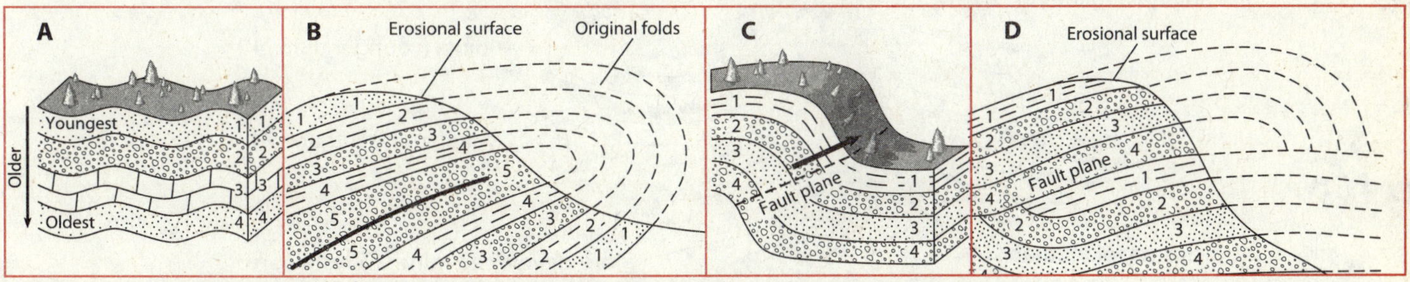

Figure 13-1. The principle of superposition and possible exceptions: (A) Normally, a rock layer is younger than the layers below it. **(B)** Overturned folds can result in an exception to the principle. Layer 1 at the lower right is actually a continuation of layer 1 at the upper left. **(C)** and **(D)** Movement of layers along an overthrust fault can result in an exception to the principle. Layers numbered alike are the same age. The arrow in diagram C shows the direction of overthrusting.

cause the deformation of rock layers. Some exceptions are illustrated in Figure 13-1. For example, when there are overturns in folded rock layers, or when movement along faults thrusts older rock layers over younger layers, the principle of superposition no longer holds true.

Dating Intrusions, Extrusions, and Inclusions

When molten rock—magma—squeezes into preexisting rocks and crystallizes, it forms an igneous rock body called an **intrusion,** as shown in Figure 13-2. Intrusions can vary in thickness from centimeters to hundreds of kilometers. An intrusion is younger, in relative age, than any rock it cuts through.

When molten rock—lava—flows on Earth's surface and solidifies, it forms a mass of igneous rock called an **extrusion** as shown in Figure 13-2. Extrusions include lava flows and volcanoes. The extrusion is younger than any rocks beneath it, but will be older than any rocks that may later form on top of it.

An **inclusion** is a body of older rock within igneous rock. Often when magma rises towards Earth's surface, pieces of the rock that the magma is intruding—pushing through—will fall into the magma. Usually these pieces of older rock will melt to become part of the magma. However, if the temperature of the magma is lower, as occurs when the magma is about to solidify, the older body will not melt. The result is the formation of an inclusion as shown in Figure 13-3. Some of Earth's oldest dated rocks have inclusions in them; thus, scientists know that even older rocks exist. However, they may not know how much older these rocks are.

MEMORY JOGGER

You may recall from Topic 11 that contact metamorphism occurs when magma or lava comes in contact with older bedrock and alters the minerals and/or sediments by recrystallization. This process results in a younger metamorphic rock with a new texture and/or mineral composition.

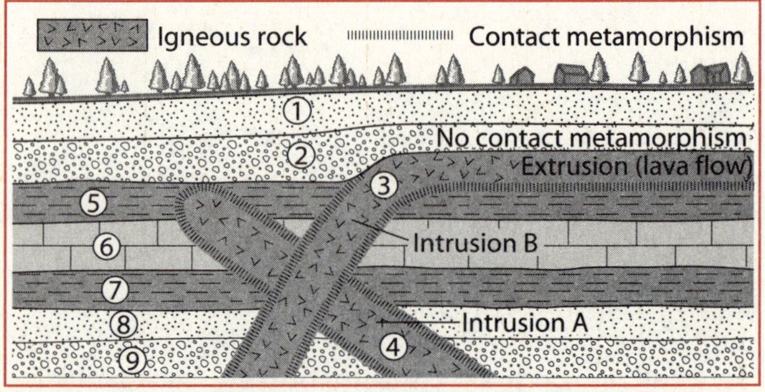

Figure 13-2. Relative ages of intrusions and extrusions: Intrusion A is younger than the layers it intrudes. Intrusion B is younger than intrusion A because it cuts intrusion A. The layers above the extrusion must be younger than the extrusion because of the absence of contact metamorphism along their common boundary. The rocks are numbered in order of increasing age.

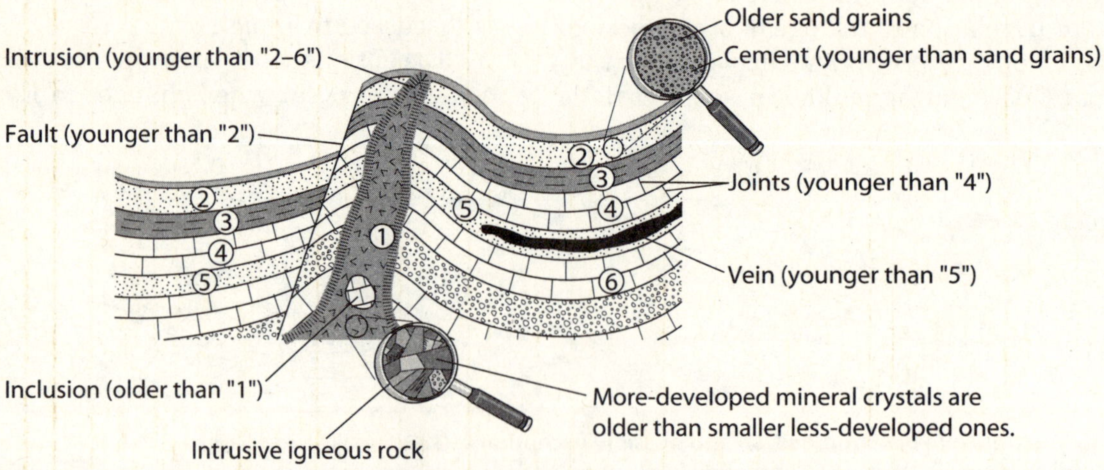

Figure 13-3. Relative ages of structural and internal features of rocks: Folds, faults, joints, and veins are younger than the rocks in which they occur. Sediments such as sand grains and inclusions in igneous rocks are older than the rocks in which they occur.

Dating Rock Features

Many of the structural features within rocks provide evidence to use in relative dating. For example, a rock is older than any fault, joint, tilting, or fold that appears in it. The reason is that rocks form without these features, and are folded, tilted, or cracked afterward.

Rock characteristics, as shown in Figure 13-3 on the previous page, can also be used in relative dating. In sedimentary rocks, the sediments are older than the rock itself, while the cement is younger. In igneous rocks, individual mineral crystals vary in age because they form at different temperatures reached by the magma as it cools and hardens over thousands or millions of years. Older mineral crystals are generally larger and show a well-developed crystal shape.

Rocks may also contain mineral deposits called veins. A vein forms when a watery mineral solution fills a crack or permeable zone in the rock. Like an intrusion, a vein is younger than the rock around it. Veins often contain valuable natural resources such as gold, silver, and lead ores.

Review Questions

1. According to the Generalized Bedrock Geology of New York State in the *Earth Science Reference Tables,* what is the geologic age of the bedrock found at the surface at 43° 30′N latitude by 75° 00′ W longitude? (1) Devonian (2) Cambrian (3) Early Ordovician (4) Middle Proterozoic

2. Which area in New York State is located on rock formations that contain large amounts of salt deposits? (1) Syracuse (2) Long Island (3) New York City (4) Old Forge

3. What is the age of the surface bedrock at Niagara Falls, New York? (1) Devonian (2) Silurian (3) Ordovician (4) Cambrian

4. Suppose you are examining a section of sedimentary rock. What evidence could you observe that would support the principle of superposition?

5. The following diagram represents a cross-sectional view of a portion of Earth's crust showing sedimentary rock layers that have not been overturned. The letters identify the specific layers. Which rock layer is probably the oldest? (1) A (2) B (3) C (4) D

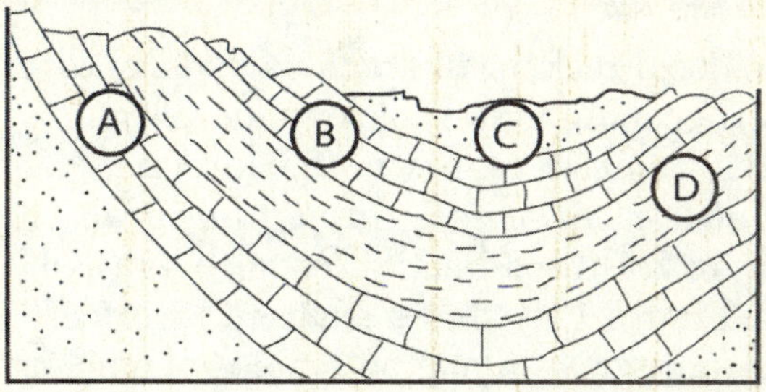

6. The following diagram shows a portion of Earth's crust. What is the relative age of the igneous rock? (1) It is older than the limestone but younger than the shale. (2) It is younger than the limestone but older than the shale. (3) It is older than both the limestone and the shale. (4) It is younger than both the limestone and the shale.

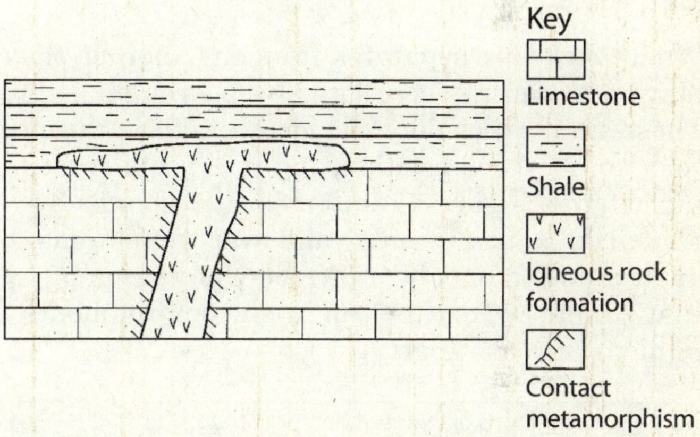

7. The following diagram represents exposed bedrock. Which geologic event occurred last? (1) the intrusion of A (2) the fault along line B (3) the fold at C (4) the deposition of gravel at D

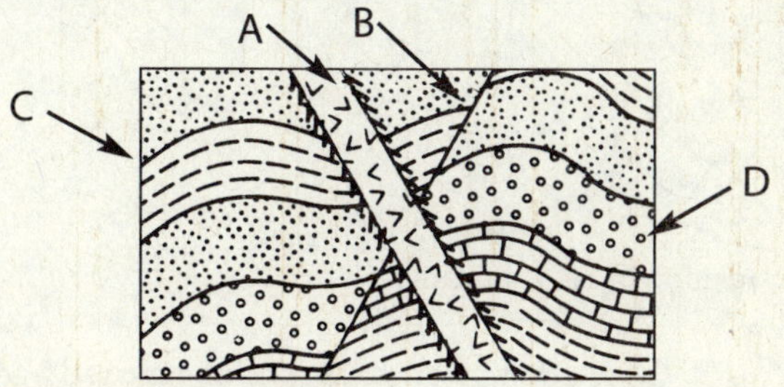

Correlation

The process of **correlation** makes it possible to show that rocks or geologic events from different places are the same or similar in age. Correlation is an important tool in unraveling the sequence of geologic events in an area. Correlation is also useful in finding certain mineral resources, such as fossil fuels, which are found in rocks of a specific age. Some of the methods of correlation are described below.

Correlation by Exposed Bedrock

An area's local rock, or **bedrock,** is usually covered by soil, other loose materials, or human-built structures. Sometimes, however, bedrock is exposed at Earth's surface through natural processes or by human activities such as road-building. Figure 13-4 shows exposed layers of bedrock—outcrops—along both sides of a river valley. Where bedrock is exposed, correlation can be accomplished by directly following the continuity of the rock layers. Direct observation of rock layers is especially easy in arid regions where little soil or vegetation covers the outcrops.

Correlation by Similarities in Rocks

Where rocks are separated from one another, they may be tentatively correlated by their similarities. These similarities may include overall appearance, color, mineral composition, and rock sequence, as indicated in Figure 13-4. Correlation of rocks by similarity is usually only valid over small areas, and even then may be incorrect. One reason is that similar rocks can form in like environments millions of years apart.

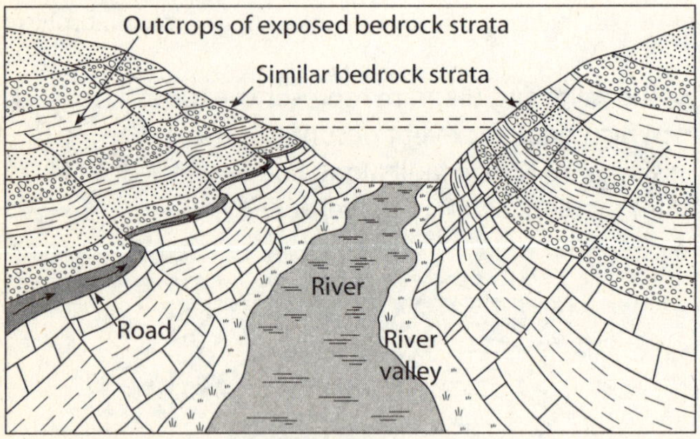

Figure 13-4. Correlation by observations of exposed bedrock and rock similarities: A geologist could follow and correlate bedrock layers in this exposed hillside by walking along the road on the left side of the river valley. The geologist might also be able to correlate layers on opposite sides of the valley by observing similarities in color, texture, and sequence of rock layers.

Correlation by Use of Index Fossils

One of the best methods of correlation involves the use of particular fossils or groups of fossils. **Fossils** are the remains or evidence of former living things—such as bones, shells, footprints, or organic compounds (such as DNA). With minor exceptions, fossils are found exclusively in sedimentary rocks. They are rarely found in igneous and metamorphic rocks because fossils are usually destroyed by the melting of igneous rocks and by the heat and pressure associated with the formation of metamorphic rocks.

The fossils used in correlation are called **index fossils.** As shown in Figure 13-5, the fossil remains of a particular type of life form may become useful as an index fossil if two conditions are met. First, the life form must have lived over a wide geographical area—so that its fossil remains will have a large horizontal distribution throughout rocks formed around the same time. Second, the life form must have existed for a relatively short time,

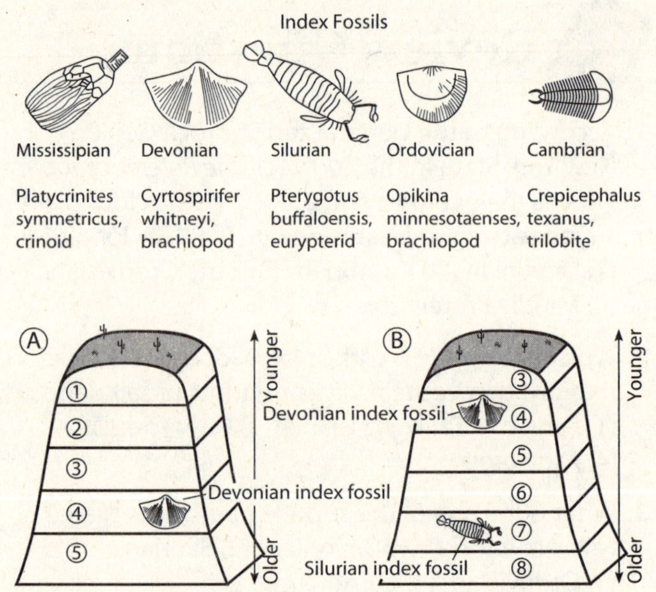

Figure 13-5. Correlation of rock layers by means of index fossils: The layers are in order from younger to older. In diagram A, a fossil of an organism known to have lived only during the Devonian Period is found in rock layer 4. This layer was therefore deposited during Devonian times. The layers above it are younger, and those below it are older. In diagram B, a similar fossil is found in one of the layers. This is therefore also a Devonian layer. A second index fossil, of the Silurian Period, is found in layer 7. This indicates that layers 5 and 6 are not younger than the Devonian and not older than the Silurian layers.

so that its fossil remains have a small vertical distribution in the rock layers in which the fossils occur.

MEMORY JOGGER

You may recall that most sedimentary rocks form from compacted and cemented sediments. In contrast, igneous rocks form directly from molten magma or lava, while metamorphic rocks form when rocks are subjected to heat and pressure.

Correlation by Volcanic Ash and Meteorite Deposits

During volcanic eruptions, **volcanic ash** consisting of sand-sized and clay-sized particles of extrusive igneous rock is shot into the air. In large eruptions, volcanic ash scatters over wide areas of Earth's surface. The scattered ash then settles among other sediments in many different environments. Since ash from each volcanic eruption has a unique mineralogical, or chemical, composition, specific ash deposits can be detected in rock layers. Such ash deposits are useful in correlation because they are widely distributed and represent a small interval of time. Deposits of volcanic ash serve as specific age markers in rocks and glacial ice that may be thousands of kilometers apart.

MEMORY JOGGER

You may recall from the *Earth Science Reference Tables* that sand particles measure between 0.0006 cm and 0.2 cm in diameter and that clay particles measure between 0.00001 cm and 0.0004 cm.

In a similar fashion, rock particles and debris produced by the impact of an asteroid, comet, or meteorite can cover large parts of Earth's surface. For example, at the end of the Mesozoic Era approximately 65 million years ago, a large comet or asteroid hit Mexico's Yucatan peninsula. Debris from the impact scattered over much of Earth's surface and can be found on all continents and in the sediments of most ocean areas. The debris, which included particles of the comet or asteroid, formed identifiable deposits which can be used in correlation.

Review Questions

8. The simplest way to correlate exposed rock layers in the same general vicinity when they contain no fossils is by (1) examining intrusions (2) radioactive dating (3) following the continuity of the layers (4) tracing a fault

9. The following diagrams represent the rock layers and fossils found at four widely separated areas of exposed bedrock.

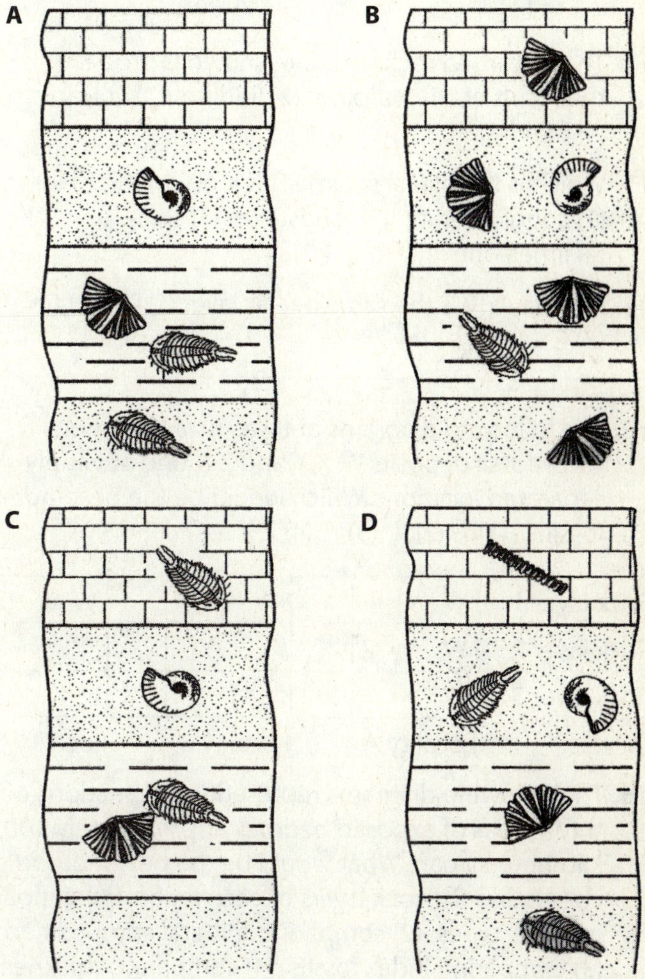

Which fossil appears to be the best index fossil?

(1) (2) (3) (4)

10. Describe two essential characteristics of an index fossil.

11. In a conglomerate rock sample that is composed of limestone particles cemented together by calcite, what is the oldest part of the rock sample?

Base your answers to questions 12 through 14 on the following diagram.

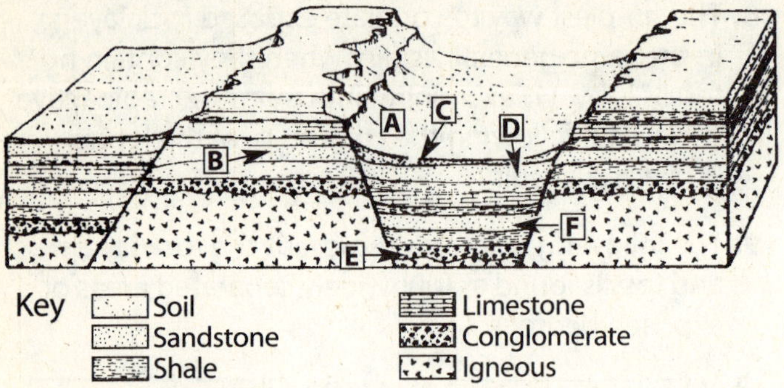

12. The disturbed rock structure shown is probably the result of (1) warping (2) folding (3) volcanism (4) faulting

13. Which is the oldest sedimentary rock in this diagram? (1) conglomerate (2) sandstone (3) shale (4) limestone

14. Rock layer B is the same age as layer (1) C (2) D (3) E (4) F

15. The following diagrams of three profile sections show fossil deposits W, X, Y, and Z found at widely separated locations. Which would be the best index fossil? (1) W (2) X (3) Y (4) Z

	Locality A	Locality B	Locality C
Rock layer 1	W	W	W Z
Rock layer 2	W Z	Y	Z
Rock layer 3	W X	X	X Z

16. The following diagrams represent cross sections of three areas of exposed bedrock approximately 100 kilometers apart. What would be the best method of correlating the rock layers of each area? (1) comparing rock types (2) comparing mineral composition (3) comparing index fossils (4) comparing thickness of rock layers

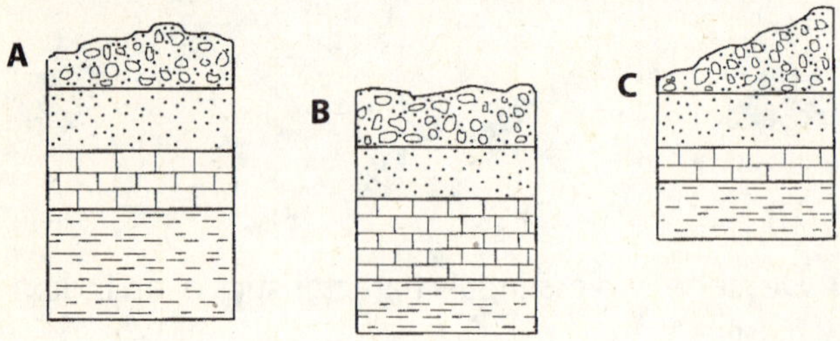

17. Why might it be more likely for a fossil to be found in a sedimentary rock than in either an igneous or a metamorphic rock?

18. Why can layers of volcanic ash found between other rock layers often serve as good geologic time markers? (1) Volcanic ash usually occurs in narrow bands around volcanoes. (2) Volcanic ash usually contains index fossils. (3) Volcanic ash usually contains the radioactive isotope carbon-14. (4) Volcanic ash usually is rapidly deposited over a large area.

19. According to the *Earth Science Reference Tables,* which rock is most likely the oldest? (1) conglomerate containing the tusk of a mastodont (2) shale containing trilobite fossils (3) sandstone containing fossils of flowering plants (4) siltstone containing dinosaur footprints

20. The best basis for concluding that a certain layer of shale rock in New York State was deposited at the same time as one in California is that both (1) are the same distance below the surface (2) contain similar fossil remains (3) are sedimentary rocks (4) have the same chemical composition

Geologic History From the Rock Record

As life forms on Earth constantly evolve, or change over time, some life forms exist or are dominant only during specific intervals of geologic time. Thus, fossils in rocks can be used to order an area's geologic events according to relative age. For example, the rock record shows that dinosaurs existed in a long interval called the Mesozoic Era. However, certain types of dinosaurs existed only for shorter time intervals. In general, fossils in rock layers serve to establish the relative age of the rock layers.

Geologic Time Scale

Mainly on the basis of changing fossil evidence, geologists have been able to divide geologic time into divisions. From longest to shortest, these divisions of time are called <u>eons</u>, <u>eras</u>, <u>periods</u>, and <u>epochs</u>. A model of these divisions of geologic time is called the **geologic time scale**. The Geologic

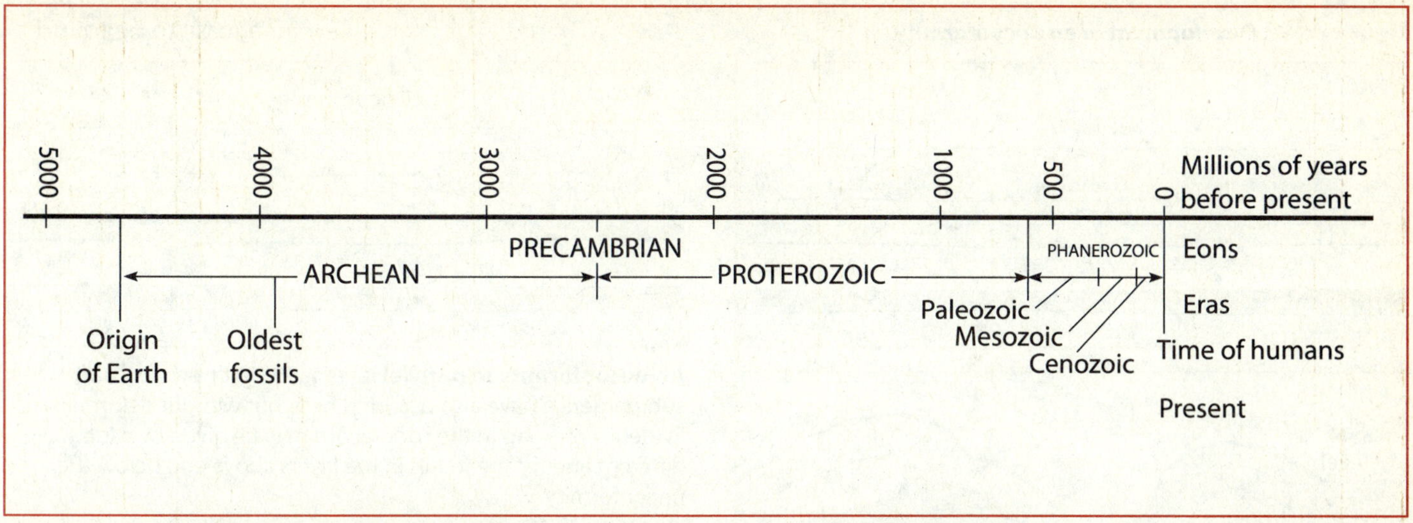

Figure 13-6. Timeline of Earth's geologic time: Life has existed for a large percentage of Earth's history. However, humans have existed for a comparatively short time (only 0.04 percent of Earth's history).

History of New York State in the *Earth Science Reference Tables* represents the geologic time scale as a chart, while Figure 13-6 represents it as a timeline. Note that eons, eras, and other divisions of geologic time are not exact units of time such as hours or minutes. For example, in Figure 13-6, which shows the relative lengths of eons and eras, no two eras represent the same amount of time.

The Precambrian is a "super" eon composed of the earliest Archean and Proterozoic eons and represents approximately 88 percent of all geologic time. Precambrian fossils are rare and difficult to identify because the earliest known life forms were small and had no hard parts. Also, most Precambrian rocks have been either buried by more recent rocks, eroded away, or converted to metamorphic rocks or magma. To become familiar with the major divisions of geologic time, review the Geologic History of New York in the *Earth Science Reference Tables*. Also, learn to recognize the names of life forms associated with particular eons, eras, periods, and epochs.

Unconformities

In attempting to read the rock record, geologists often find evidence of buried eroded surfaces called **unconformities.** One example is shown in Figure 13-7 on the next page. An unconformity in the rock record of an area indicates that at some time in its geologic history, uplift occurred. This uplift exposed the rocks to weathering and erosion, which removed part of the rock record. A later sinking of land or rise in sea level caused the area to be covered by water. New sediments were then deposited on the eroded land surface, producing the unconformity. Most unconformities show a lack of parallelism between the older rock layers beneath the erosion surface and the younger rock layers above. The lack of parallelism occurs because the older rock layers were folded or tilted during uplift.

The presence of an unconformity means that parts of the rock record are missing—like missing pages or chapters in a history book. Although there is a gap in the record of geologic time, an unconformity is still useful in relative dating. The rocks above an unconformity are younger than the unconformity, and the rocks below it considerably older. An unconformity also usually correlates with a time of mountain building, called an orogeny, and the associated plate convergence. In the *Earth Science Reference Tables*, the times of orogenies are listed as Important Geologic Events in New York in the Geologic History of New York State.

Figure 13-7. Development of an unconformity

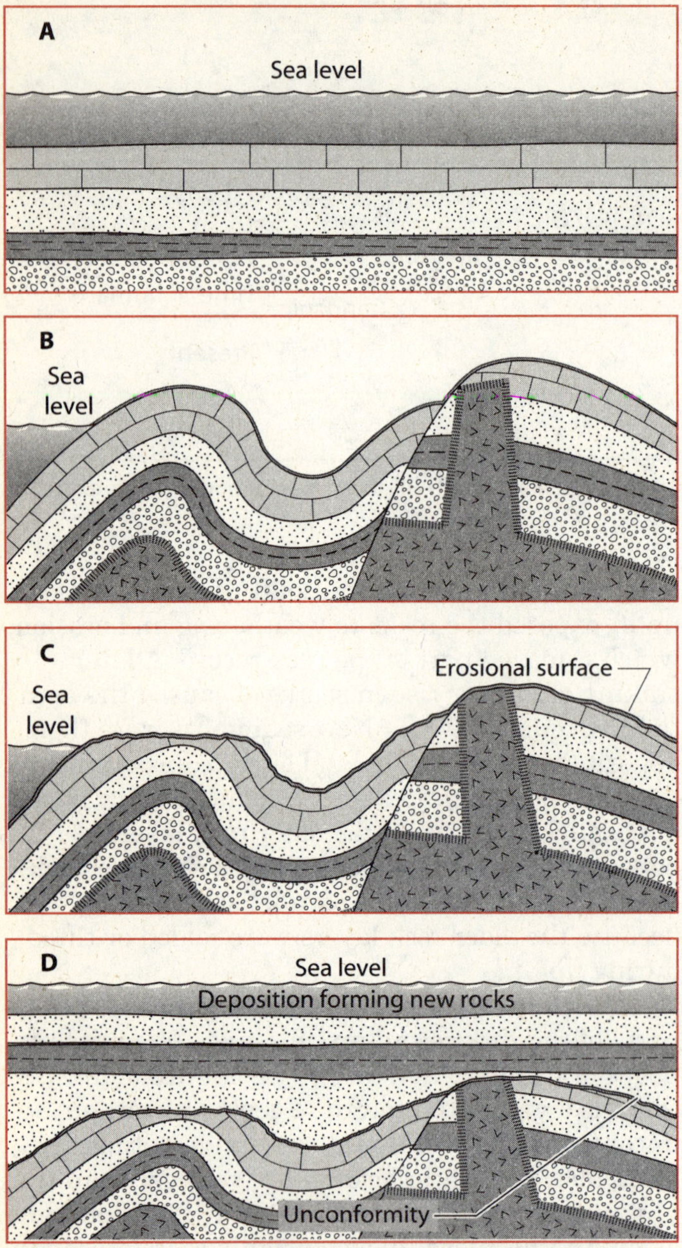

Events producing an unconformity with lack of parallelism **(A)** Deposition forms sedimentary rock layers. **(B)** Crustal deformation and uplift occur. **(C)** Weathering and erosion remove part of the rock record. **(D)** Submergence (moved below sea level) and new deposition result in an unconformity along the now buried erosional surface.

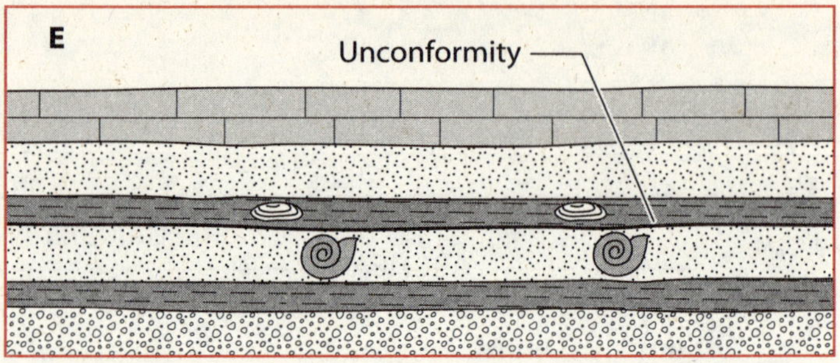

An unconformity in parallel layers **(E)** Uplift, erosion, and submergence have also occurred here, but without deformation. Evidence of a gap in the rock record may be given by widely different ages of the fossils in the layers above and below the unconformity.

Uniformity of Process

One of the basic principles geologists use to interpret geologic history is the uniformity of process, which implies that "the present is the key to the past." This principle assumes that geologic processes happening today also occurred in the past and that much of the rock record can be interpreted by observing present geologic processes. Uniformity of process does not mean that different processes could not have happened or that past geologic processes always occurred at the same rate as they do today.

As an example of uniformity of process, suppose that a geologist in tropical Africa finds a conglomerate sedimentary rock layer. The sediments in the rock are unsorted, have a wide range of sizes, and have some partly rounded shapes with various scratches. Based on glacial processes known today, the geologist can infer that this conglomerate layer was the result of a past glaciation.

Uniformity of process implies that the chemical, physical, geological, and biological characteristics of nature remain the same throughout time. Thus, present observations can not only help geologists interpret Earth's past, but also make certain predictions about its future. By gathering sufficient data, it becomes possible to accurately predict certain natural disasters, such as volcanic eruptions, earthquakes, landslides, storms, and floods. Coupled with appropriate emergency procedures when needed, such predictions can help minimize death, injury and property loss.

Review Questions

21. You are searching for fossils in the surface bedrock in Elmira, New York. Why are you unlikely to find dinosaur fossils at this location?

22. During which time was the majority of the exposed bedrock in New York State deposited? (1) Precambrian (2) Mesozoic (3) Cenozoic (4) Paleozoic

23. Rocks containing fossils of the earliest land plants could most likely be found in New York State bedrock near (1) Syracuse (2) Oswego (3) Ithaca (4) Old Forge

24. Approximately how long ago were the Taconic Mountains uplifted? (1) 540 million years ago (2) 470 million years ago (3) 310 million years ago (4) 120 million years ago

25. According to the Geologic History of New York State in the *Earth Science Reference Tables*, what is the estimated age of Earth as a planet in millions of years? (1) 540 (2) 4000 (3) 4600 (4) 5000

26. The geologic time scale has been subdivided into a number of time units largely on the basis of (1) fossil evidence (2) rock thickness (3) rock types (4) radioactive dating

27. Which line is the best representation of the relative duration of each of the geologic time intervals?

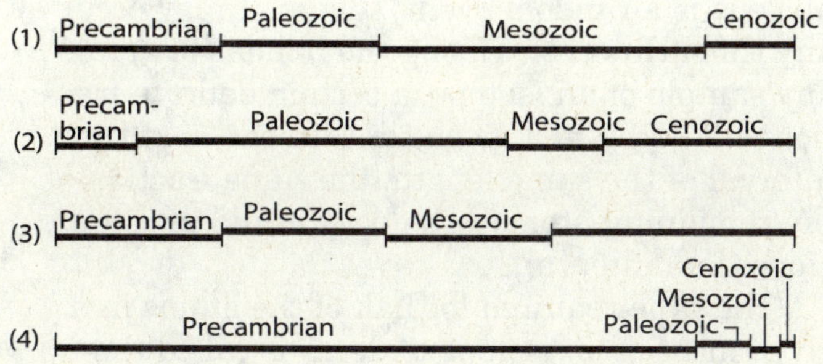

(1) 1 (2) 2 (3) 3 (4) 4

Base your answers to questions 28 through 32 on the following diagram.

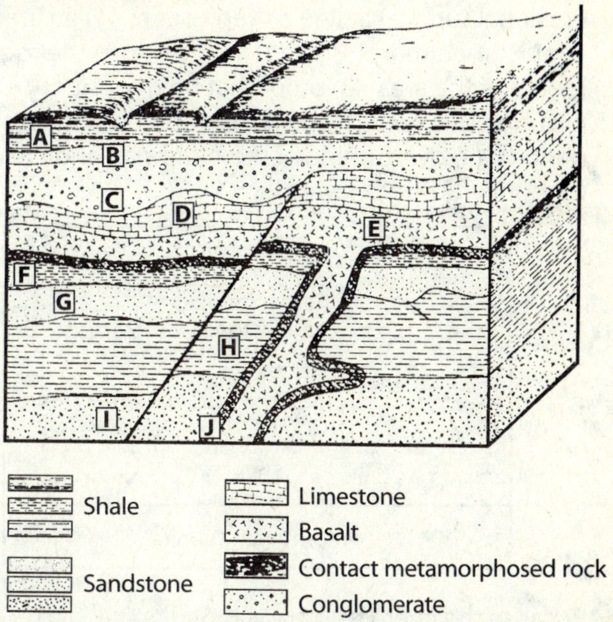

28. Which is the most recently formed rock? (1) A (2) B (3) C (4) D

29. An unconformity is located between (1) A and B (2) B and C (3) C and D (4) F and G

30. The conclusion that the limestone layer D is younger than the basalt layer E is supported by evidence of (1) faulting (2) contact metamorphism (3) igneous intrusion (4) fossils

31. The last event before faulting occurred was the formation of (1) C (2) D (3) E (4) J

32. Which must have most recently preceded the formation of layer D? (1) faulting (2) orogeny (3) intrusion (4) uplift

33. An unconformity between two sedimentary layers is most likely produced by (1) the deposition of gravel followed by the deposition of sand and silt (2) continuous sedimentation in a deep basin over a long period (3) uplift followed by extensive erosion, submergence, and deposition (4) a period of extrusive vulcanism followed by another period of extrusive vulcanism

34. The following diagram shows a cross section of part of Earth's crust. What does the unconformity (buried erosional surface) at line XY represent? (1) an area of contact metamorphism (2) a time gap in the rock record of the area (3) proof that no deposition occurred between the Cambrian and Carboniferous periods (4) overturning of the Cambrian and Carboniferous rock layers

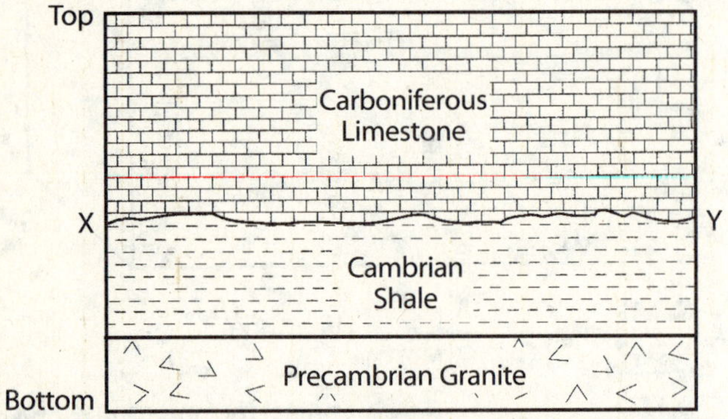

35. Because the chemical, physical, geological, and biological characteristics of nature remain the same throughout time, we can interpret the past and (1) prevent natural disasters (2) identify unconformities (3) predict natural hazard events (4) determine the age of bedrock

36. Fossils are rarely found in Precambrian rocks because, at that time (1) no life existed (2) few life forms had hard parts (3) few sedimentary rocks formed (4) a meteorite impact destroyed most fossils

37. "The present is the key to the past." Explain what this statement means in the context of geologic history.

Absolute Dating of Rocks Using Radioactive Decay

The principle of superposition and methods of correlation help to determine the relative age of rock layers, but do not give their absolute age. Absolute age is expressed in specific units, such as years before the present. One method of absolute dating is the counting of annual layers of glacial lake sediments, which is similar to tree-ring dating. However, the primary method of absolute dating is described in the text that follows.

Isotopes and Radioactive Decay

An element is a substance consisting of atoms that are chemically alike. Most elements exist in several varieties called **isotopes.** The difference between one isotope of an element and another is in the mass of its atoms. For example, the mass of an atom of the most common isotope of carbon is 12 units. This isotope, carbon-12, is usually just called, "carbon," while another isotope in which the atoms have a mass of 14 units is called carbon-14.

Almost all the mass of an atom is concentrated in a central region called the nucleus (plural, nuclei). The nuclei of the atoms of many isotopes are unstable or radioactive. They emit particles and electromagnetic energy in a process called **radioactive decay,** nuclear decay or nuclear fission thus changing into atoms of other isotopes and elements. The nucleus remaining after radioactive decay may also be unstable and decay further over time. Eventually, a stable isotope—one that is not radioactive or does not undergo radioactive decay—forms.

URANIUM-238 One of the most important radioactive isotopes used in dating rocks is **uranium-238,** the isotope of uranium whose atoms have a mass of 238 units. The nuclei of its atoms pass though a series of radioactive decays, eventually producing atoms of lead-206, a stable isotope of the element lead.

HALF-LIFE The decay of any individual atomic nucleus is a random event; that is, it may occur at any time. However, among the billions of atoms in any sample of an isotope, a certain definite fraction will decay in a given time. In the next time interval of the same length, the same fraction of the remaining atoms will decay as did in the previous time interval.

The time required for half of the atoms in a given mass of an isotope to decay is called the **half-life** of the isotope. In a given sample of a radioactive isotope, half of the original atoms will have decayed to other isotopes at the end of one half-life period, and half will remain unchanged. At the end of the next half-life period, half of these remaining atoms will have decayed, leaving one-fourth of the original atoms unchanged. This halving of the number of unchanged atoms during successive half-life periods continues indefinitely.

Each radioactive isotope has its own characteristic half-life. The half-life of an isotope is not affected by any environmental factors, such as temperature, pressure, or involvement in chemical reactions. The half-life is also not affected by the amount, mass, or volume of the sample. By the principle of uniformity of process, it can be assumed that the half-life of a given isotope has remained the same throughout Earth's history. Half-lives for different isotopes vary widely—from fractions of seconds to billions of years. Table 13-1 shows some examples. For more information, see Physical Constants—Radioactive Decay Data in the *Earth Science Reference Tables.*

Table 13-1. Radioactive Decay Data

Radioactive Isotope	Disintegration	Half-life in Years (exponential/ordinal)
Carbon-14	$C^{14} \longrightarrow N^{14}$	$5.7 \times 10^3 = 5700$
Potassium-40	$K^{40} \rightarrow Ar^{40}$, $K^{40} \rightarrow Ca^{40}$	$1.3 \times 10^9 = 1{,}300{,}000{,}000$
Uranium-238	$U238 \longrightarrow Pb^{206}$	$4.5 \times 10^9 = 4{,}500{,}000{,}000$
Rubidium-87	$Rb^{87} \longrightarrow Sr^{87}$	$4.9 \times 10^{10} = 49{,}000{,}000{,}000$

Radioactive Dating

The half-life of a radioactive isotope can be used, along with the ratio between the amount of the original isotope and the amount of its decay-product, to estimate the absolute age of a rock sample. This method is called **radioactive dating.** For example, consider a rock formed with uranite mineral grains containing uranium-238 and no lead-206. As time passed, the uranium-238 would slowly change to lead-206—its stable decay product—at its fixed half-life rate. At the end of one half-life period (4.5 billion years for uranium-238), half the uranium-238 atoms would have changed to atoms of lead-206. If a rock is found today with this 1:1 ratio of uranium-238 atoms to lead-206 atoms, it can be concluded that the rock formed 4.5 billion years ago. If the ratio between uranium-238 and lead-206 is higher than 1:1, the rock must have formed less than 4.5 billion years ago.

The age corresponding to any particular ratio of uranium-238 to lead-206 can be calculated mathematically. The curves in Figure 13-8 show how the percentage of uranium-238 decreases and the percentage of lead-206 increases with time in a given rock. The shape of the curves for any other radioactive isotope and its stable decay product would be similar. However, the time intervals would differ due to the different length of the half-life period.

CARBON-14 DATING Radioactive elements with long half-lives, such as uranium-238, are used to date rocks that are hundreds of millions, or even billions, of years old. Over such long periods, any isotope with a short half-life would have decayed to such an extent that the remaining amounts would be immeasurable.

Some radioactive isotopes with short half-lives are useful for dating rocks and organic remains of relatively recent origin. One such useful isotope is carbon-14, with a half-life of 5700 years. **Carbon-14 dating**—also called radiocarbon dating—can be used to date rocks and organic remains, such as the bones of a mastodont, up to approximately 70,000 years in age. See the Geologic History of New York State in the *Earth Science Reference Tables.*

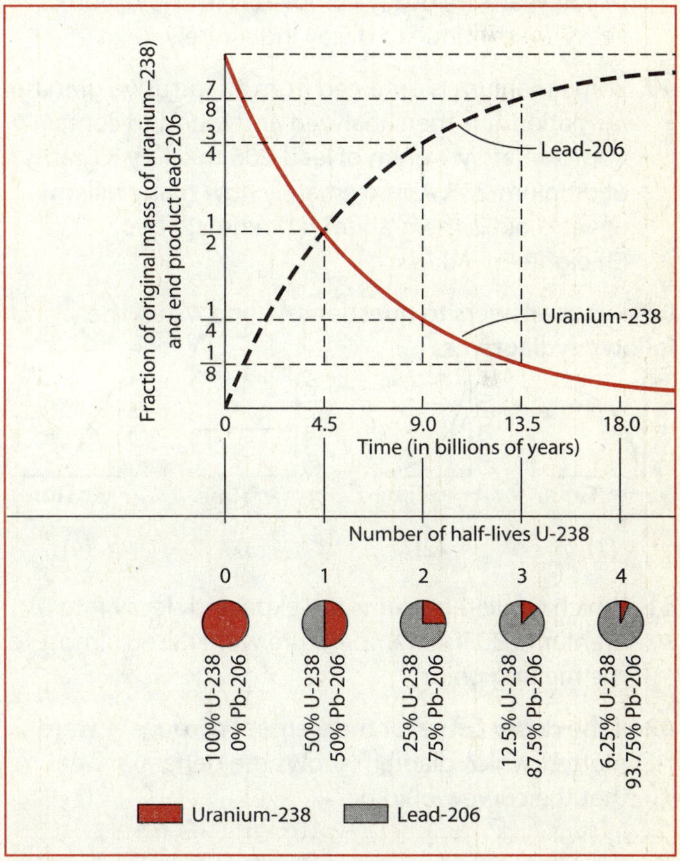

Figure 13-8. Curves showing radioactive decay of uranium-238 and formation of the stable decay product lead-206: After one half-life (4.5 billion years), half, or 50 percent, of the original uranium atoms have become lead atoms. By five half-lives, only 3.125 percent of the uranium atoms are left, and there are 96.875 percent lead atoms.

Review Questions

Base your answers to questions 38 through 40 on the following graph, which shows the radioactive decay of uranium-238.

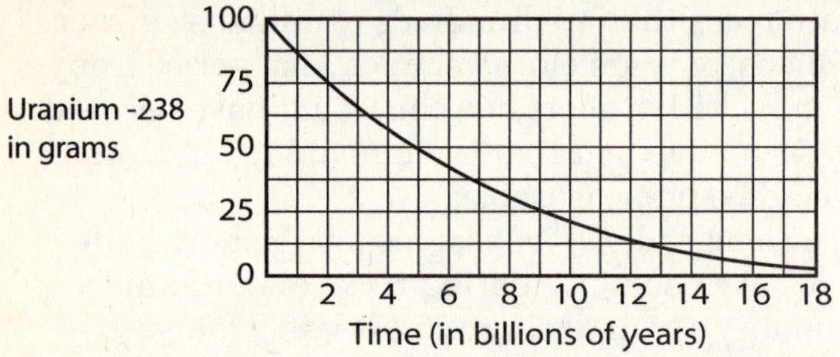

38. The half-life of uranium-238 as shown on this graph is closest to (1) 2.5 billion years (2) 3.5 billion years (3) 4.5 billion years (4) 5.5 billion years

39. Because of the half-life nature of the decay process, uranium-238 will theoretically (1) decay completely before 4.5 billion years (2) decay completely in 18 billion years (3) decay completely in 200 billion years (4) continue to decay indefinitely

40. Some uranium is obtained from an intrusive granite formation. It is then analyzed and found to contain approximately 1 gram of lead-206 to every 3 grams of uranium-238. Approximately how many billions of years old is the granite? (1) nine (2) two (3) eighteen (4) four

Base your answers to questions 41 and 42 on the following diagrams.

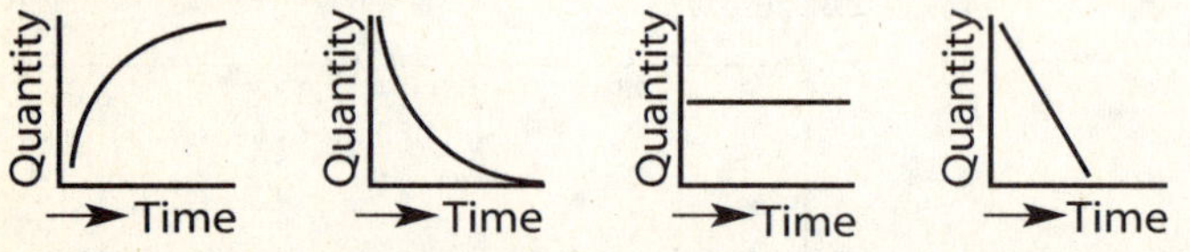

(1) (2) (3) (4)

41. Which curve represents the expected decay rate of uranium-238 if its temperature were raised almost to the melting point?

42. If the decay curve for the element carbon-14 were plotted, which diagram shows the general shape that the curve would be?

43. The half-life of a particular radioactive substance (1) decreases as pressure on it increases (2) decreases as its mass decreases (3) increases as the temperature increases (4) is independent of mass, temperature, and pressure

44. If a specimen of a given radioactive substance is reduced in size, its half-life (1) decreases (2) increases (3) remains the same

45. Which radioactive substance shown on the following graph has the longest half-life? (1) A (2) B (3) C (4) D

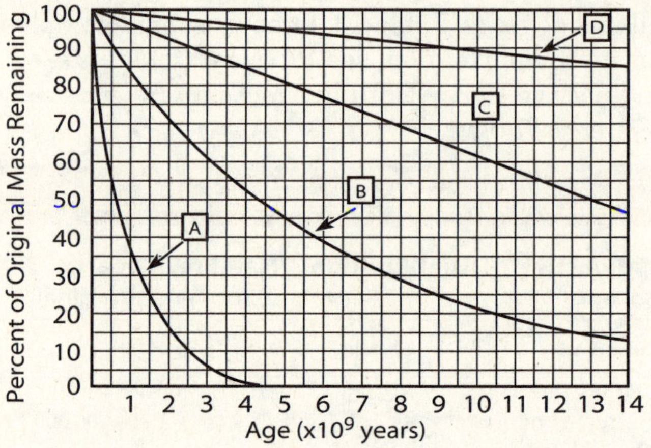

46. Why is carbon-14 usually NOT used to accurately date objects more than 50,000 years old? (1) Carbon-14 has a relatively short half-life, and too little carbon-14 is left after 50,000 years. (2) Carbon-14 has a relatively long half-life and not enough carbon-14 has decayed after 50,000 years. (3) Carbon-14 has been introduced as an impurity in most materials older than 50,000 years. (4) Carbon-14 has only existed on Earth during the last 50,000 years.

47. Carbon-14 is often used to date fossils from the Holocene Epoch. Why would carbon-14 be less useful in dating an Ordovician fossil?

48. The radioactive isotope K^{40} decays to Ar^{40}. The diagram below shows a model of the relative amounts of K^{40} and Ar^{40} remaining after one half-life.

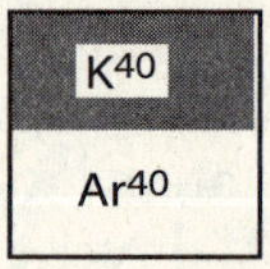

Which of the following diagrams best illustrates the relative amounts of K^{40} and Ar^{40} remaining after two half-lives?

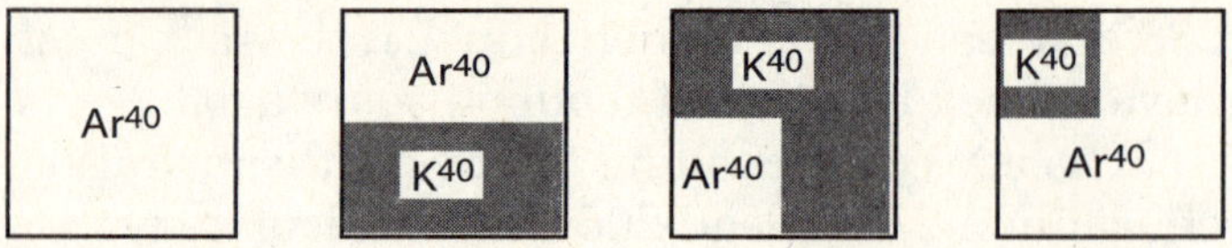

(1) (2) (3) (4)

49. Uranium-238 rather than carbon-14 is used to date most Earth materials because uranium-238 (1) was more abundant when Earth formed (2) has a longer half-life (3) decays at a constant rate (4) is easier to collect and test

Evolution of Earth and Life

For a long time, people have wondered about Earth's history and how life on Earth came to be. From the study of fossils and rock layers, scientists have been able to gain much knowledge of Earth's past.

Variations in Fossils and Environments

The fossil record, preserved in sedimentary rocks, shows that a wide variety of life forms have lived in Earth's changing environments over time. Most of these life forms are now extinct. Since the chances of fossilization are low, and the percent of all sedimentary rocks studied for fossils is minuscule, most forms of past life probably have not been identified.

Comparisons of fossils with similar life forms alive today make it possible to infer facts about Earth's past environments, as shown in Figure 13-9. For example, fossils of ancient corals have been found in Devonian period rocks of the Allegheny Plateau in western New York State. As most present-day corals live in shallow, warm ocean waters, it can be inferred that these rocks were formed in a shallow, warm-water ocean environment. The Devonian/Mississippian map, at the far right of the table Geologic History of New York State in the *Earth Science Reference Tables,* shows that New York State was much nearer to the equator at that time and thus had a warmer climate.

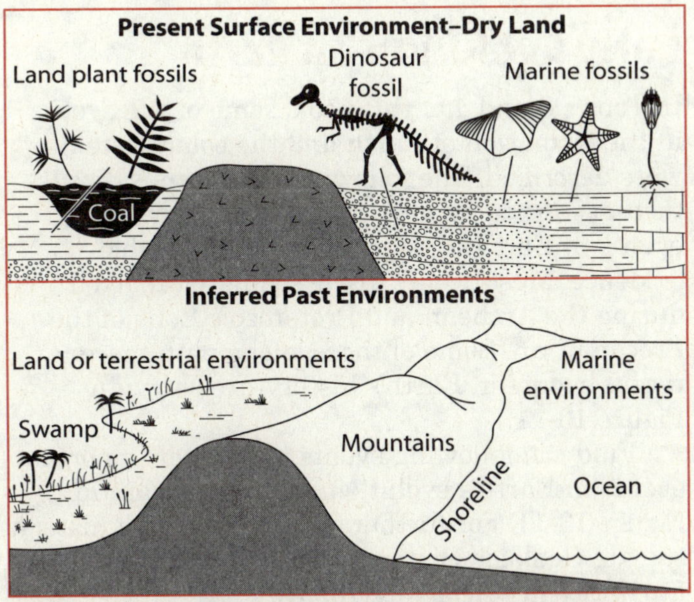

Figure 13-9. Reconstruction of past environments from fossil evidence and rock properties: From the existence of dinosaur fossils, marine fossils, and changes in sedimentary rock type, the location of the former shoreline and ocean can be inferred. From the existence of plant fossils and coal, the location of the former swamp can be inferred.

ENVIRONMENT EVOLUTION AND PLATE TECTONICS A major reason for changes in Earth's environments over geologic time has been the movements of plates and their associated landmasses. In the *Earth Science Reference Tables* these movements are shown in a series of maps in the column Inferred Positions of Earth's Landmasses in the Geologic History of New York State. As the plates and their landmasses move, their latitude changes. The result is a change in climate. For example, the climate of a landmass which has moved closer to one of Earth's poles will become colder. Similarly, the climate of a landmass which has moved closer to the equator will become warmer.

Plate movements also result in other environmental changes. For example, plate convergence can result in continent-to-continent collisions that form long, high mountain ranges such as the Himalayas. Such mountain ranges can produce local climate changes, as discussed in Topic 8. Also, rising magma at rapidly diverging plate boundaries can lead to a rise in the ocean floor, pushing ocean water onto the landmasses.

ENVIRONMENTAL EVOLUTION AND ROCK TYPES As Earth's environments changed over time, specific rocks and minerals formed, as well as deposits of fossil fuels. For example, hot and humid conditions during the Carboniferous Period of the Paleozoic Era and an abundance of land above sea level resulted in widespread swamps. The worldwide coal deposits of today were formed from these swamps. During parts of the Silurian Period of the Paleozoic Era, plate movements in hot and dry environments led to the isolation of many ocean areas. Thus isolated, the waters of these ocean areas evaporated— precipitating evaporite rocks and mineral deposits, such as salt and gypsum. Deposits of salt and gypsum are found in western New York State.

Fossils and the Evolution of Life

A **species** of living things is a group of organisms which are similar enough to be able to interbreed and produce fertile young. However, not all members of a species are exactly alike. For example, in humans and in different species of dogs one sees that many variations in each species exist. When variations in a species are graphed, a bell-shaped curves often forms, as shown in Figure 13-10.

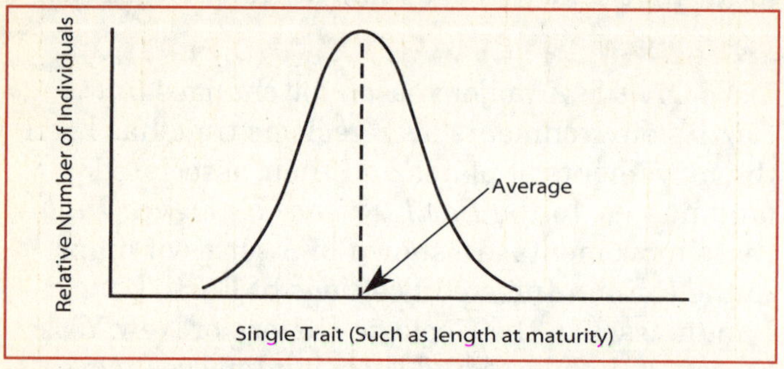

Figure 13-10. Variations in a single trait within a species: In any population of a species, each trait usually has an average value. In most individuals, the trait will be close to the average. In a few, it will be much below average; in a similar number, it will be much above average. If the departure from average in either direction has an advantage for survival and can be inherited, the average will shift in this favorable direction in later generations.

The theory of **organic evolution** states that life forms change through time. As environmental conditions change, variations within a species give certain individuals a greater chance of surviving and reproducing. Genetic material coding for these variations will be passed on to the offspring, and the favorable variations will be preserved. At the same time, individuals with unfavorable variations will gradually die out or become extinct. After a long period of time, individuals with many accumulated variations can lose the ability to interbreed with earlier varieties of the species. These individuals then become a new species.

The fossil record provides evidence for the theory of organic evolution. Fossils from adjacent intervals of geologic time sometimes show a gradual transition from an older species to a newer one.

Evidence in the fossil record also shows that evolution does not always occur at the same rate. Instead, there are times of rapid extinctions and subsequently rapid evolution of new species (punctuated evolution). Some of these times of rapid extinctions and evolution correspond to major divisions in the geologic time scale.

Rapid Evolution of Life After an Impact Event

An impact event, such as the collision of a comet or asteroid with Earth, may cause catastrophic environmental changes leading to rapid extinctions and evolution. Such an event probably occurred approximately 65 million years ago, at the interface of the Mesozoic and Cenozoic eras. At this time, a mass extinction occurred— about 70 percent of all existing species became extinct, including many species of dinosaurs. This mass extinction ended the Mesozoic Era and the dominance of reptiles on land and in the oceans. The Cenozoic Era began, and mammals became the dominant life forms on land.

Evidence exists that about 65 million years ago, Earth collided with one or more comets or asteroids many kilometers in diameter. This impact event created huge tsunami, burned much of Earth's vegetation, and hurled millions of tons of dust-sized aerosols into the stratosphere. The aerosols remained in the stratosphere for many years, blocking out much of the insolation from the sun. With the amount of solar energy reaching Earth's surface drastically reduced, surface temperatures cooled. Photosynthesis also declined, reducing the growth of plants and algae which form the basis of the food supply for most living things. These environmental changes led to a mass extinction, and a chance for previously less successful life forms, such as mammals, to thrive and evolve. There is evidence that similar impact events occurred at the ends of the Triassic, Permian, and Devonian Periods of the geologic time scale.

Early Evolution on Earth

In Topics 3 and 4 of this book, some of the events in the formation of Earth and the solar system were described. The following paragraphs discuss further inferences about some of the earliest events in Earth's history, based on available evidence. Most events discussed here occurred during the Archean and Proterozoic Eons of the Precambrian. Some of these, along with more recent events in Earth's history, are shown in Figure 13-11.

Find the following events, and other important events in Earth's evolution, on the diagram in Figure 13-11. For further details of Earth's history, see the Geologic History of New York State in the *Earth Science Reference Tables*.

- Evidence from the radioactive decay dates of moon rocks and meteorites shows that the planet Earth probably formed approximately 4.6 billion years ago.

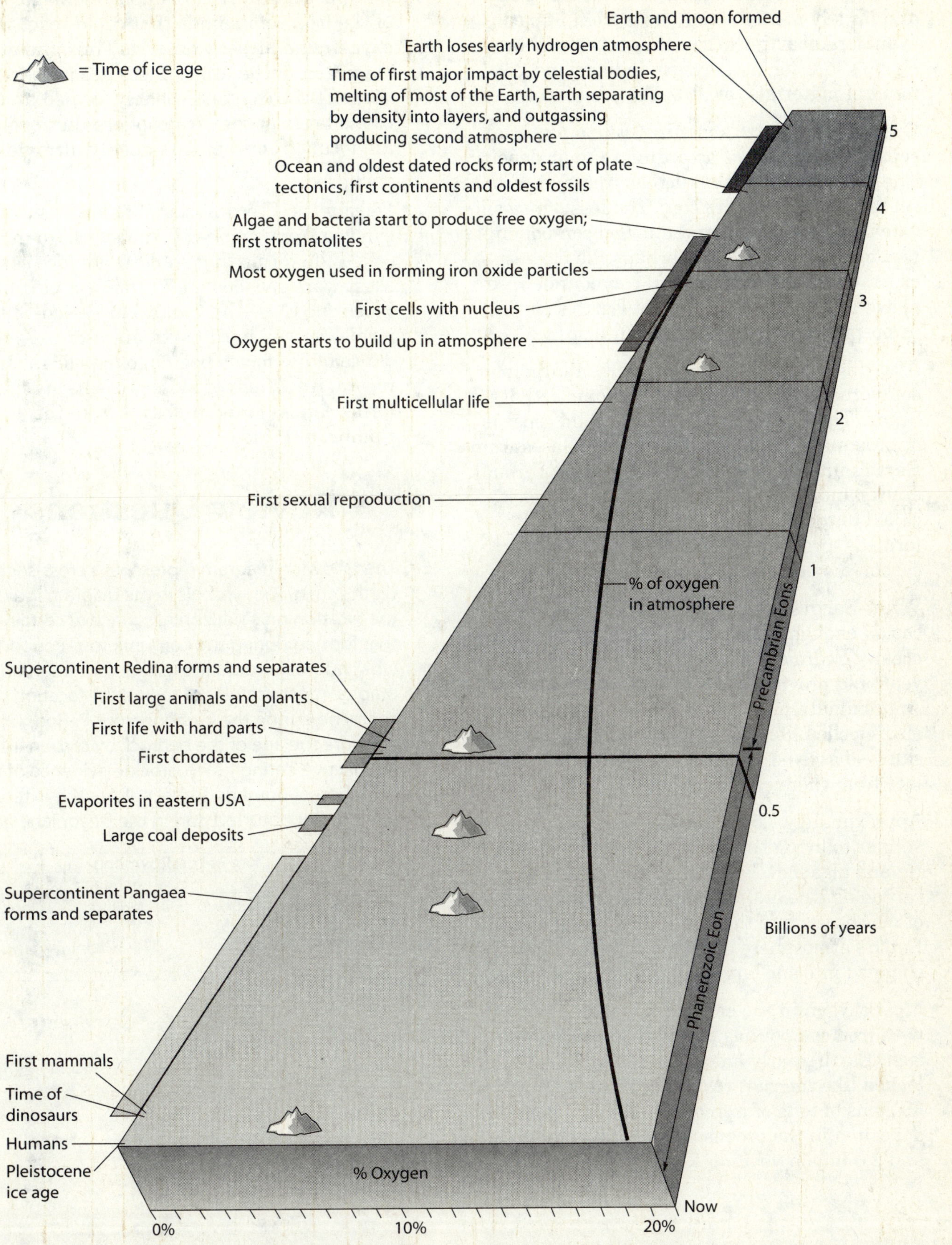

Figure 13-11. History of Earth: The diagram shows some highlights of the evolution of Earth and life.

- During its early formation, Earth heated up and largely melted due to heat from impact events, radioactive decay of isotopes within Earth's interior, and gravity-induced movement of materials towards Earth's center.
- During early Earth's melting, Earth materials separated into zones according to their density. The densest materials, such as iron and nickel, settled to form Earth's core. The less dense silicates, and other rock and mineral components, became the mantle and perhaps the earliest crust. Gaseous elements and compounds rose above Earth's surface. Most of Earth's earliest atmosphere probably escaped into space.
- After hundreds of million of years, a largely solid crust formed, and plate tectonic activity began. The earliest evidence of a solid crust is approximately 4.2 billion years old. Gases from Earth's interior seeped out of the crust through cracks and volcanic eruptions in a process called **outgassing.** A second atmosphere formed—probably composed of water vapor, carbon dioxide, nitrogen, and other gases.
- When Earth's atmosphere and surface had cooled enough, water precipitated to form the oceans. Sedimentary rocks more than 4 billion years old provide evidence of this early ocean. Approximately 3.8 billion years ago, or earlier, single-celled life forms were present. Ocean salts—dissolved minerals—began to accumulate from chemical weathering of Earth's crust.
- Approximately 3.5 billion years ago, life forms similar to living stromatolites (colonies of bacteria and algae) evolved. These life forms used carbon dioxide and released free oxygen, some of which rose into the atmosphere. Over time, Earth's atmosphere of mostly carbon dioxide changed into one largely of nitrogen and oxygen.
- Most oxygen in the early atmosphere reacted with iron compounds forming iron oxide, or rust. Earth's early landmasses may have looked like the rust-colored surface of Mars. Millions of tons of iron oxide minerals were deposited in the oceans between 3.5 billion and 2.8 billion years ago.
- By approximately 2.8 billion years ago, most of the iron compounds that could react with oxygen had already done so. Thus, the amount of oxygen in the atmosphere began to increase. Some of the oxygen probably formed an early ozone layer in the stratosphere that protected life forms from the sun's deadly ultraviolet radiation.
- During the Precambrian, life forms slowly evolved from single-celled bacteria to multi-celled life forms that could reproduce sexually. Most were soft-bodied. In the first part of the Paleozoic Era—the Cambrian Period—life forms evolved hard parts, and extensive evidence of life forms begins to appear in the rock record. Ancestors of most present-day life forms can be found in fossils from the Cambrian Period.

Review Questions

50. The following diagram represents a cross-section of Earth's crust. The symbols in the diagram indicate the location on a horizontal surface of certain fossils that formed during the Carboniferous Period. For what purpose would the fossil information on the map be most useful? (1) to find the location of the shoreline during the Carboniferous Period (2) to measure the age of the bedrock by carbon-14 radioactive dating (3) to provide evidence of the evolution of humans (4) to indicate the extent of folding that occurred during the Devonian Period

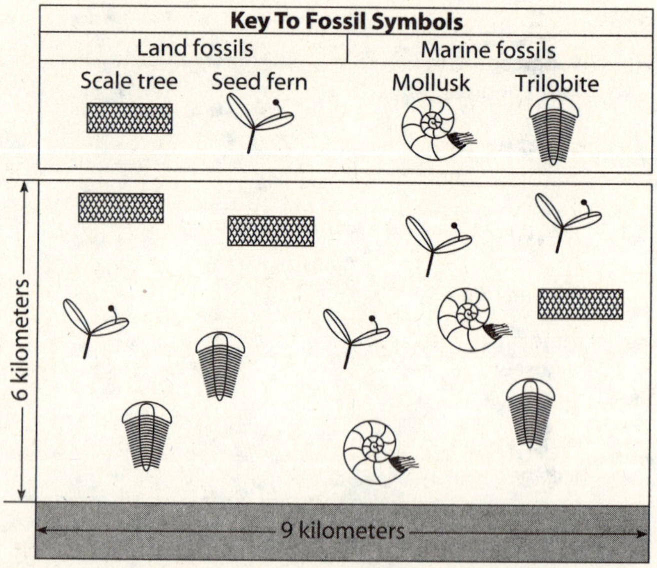

51. Shark and coral fossils are found in the rock record of certain land areas. What does the presence of these fossils indicate about those areas? (1) They have undergone glacial deposition. (2) They were once covered by thick vegetation. (3) They have undergone intense metamorphism. (4) They were once covered by shallow seas.

52. Approximately 3.5 billion years ago photosynthetic organisms evolved. What dramatic change in Earth's atmosphere resulted primarily from this event?

53. The changes observed in the fossil record from the Precambrian to the Cenozoic Era best provide evidence of (1) sublimation (2) radioactive decay (3) organic evolution (4) planetary motion

54. Which statement is false? (1) Two members of one species must be exactly alike. (2) The members of a species can vary in physical features. (3) The members of a species can interbreed and produce fertile young. (4) A species may exist on Earth for millions of years.

55. The following diagram represents a site in western New York State. The remains of a mastodont were found in sediments from a swamp located above limestone bedrock. Fossils of Paleozoic coral were discovered in the limestone beneath the swamp. Observations made at this site provide evidence that (1) the local climate and environment have changed dramatically (2) fossil coral provided food for mastodonts (3) North America has drifted to the east from Europe and Africa (4) mastodonts and coral were alive at the same time

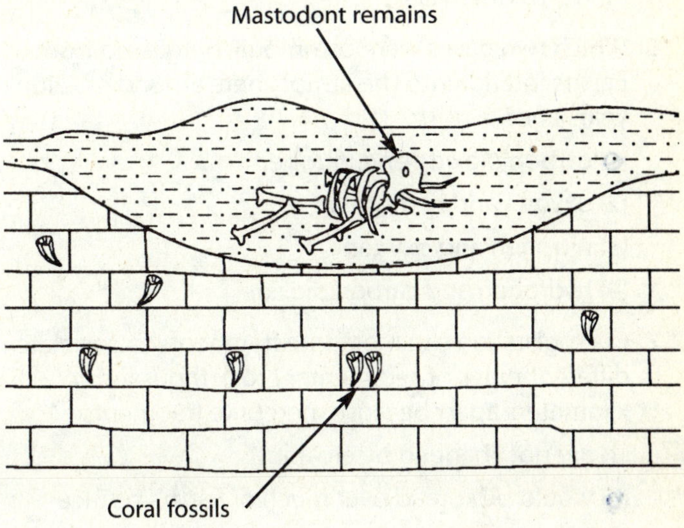

(Not drawn to scale.)

Questions for Regents Practice

Part A

1. Which of the following landscape features of New York State were most recently formed?

(1) the Finger Lakes

(2) the Catskills

(3) the Palisades sill

(4) the rocks of the Adirondack Mountains

2. Trilobite fossils from different time periods show small changes in appearance. These observations suggest that the changes may be the result of

(1) evolutionary development

(2) a variety of geologic processes

(3) periods of destruction of the geologic record

(4) the gradual disintegration of radioactive substances

3. From the study of fossils, what can be inferred about most species of plants and animals that have lived on Earth?

(1) They are still living today.

(2) They are unrelated to modern life forms.

(3) They existed during the Cambrian Period.

(4) They have become extinct.

4. An igneous rock contains one-half of its original amount of potassium-40. According to the *Earth Science Reference Tables,* the age of the igneous rock is closest to

(1) 0.7×10^9 years

(2) 1.3×10^9 years

(3) 2.1×10^9 years

(4) 2.6×10^9 years

5. Mesozoic rocks and fossils found in Australia would most likely match Mesozoic rocks and fossils found in
(1) Europe
(2) Antarctica
(3) the Atlantic Ocean
(4) North America

6. Which two gases were commonly outgassed from Earth's interior into the atmosphere around 4 billion years ago?
(1) carbon dioxide and nitrogen
(2) oxygen and hydrogen
(3) nitrogen and oxygen
(4) hydrogen and carbon dioxide

7. Geologists can use meteorite fragments to correlate different types of sedimentary rock thousands of kilometers apart because meteorite fragments
(1) are not changed by erosion
(2) would be spread over much of Earth's surface
(3) are all formed from the same minerals
(4) are often found in layers of sedimentary rock

8. What event in the evolution of life, around 540 million years ago, began an extensive fossil record on Earth?
(1) flowering plants evolved
(2) the extinction of armored fish
(3) animals evolved hard parts
(4) modern coral groups evolved

9. Trilobite fossils were recently discovered in Himalayan Mountain bedrock. During which geologic period could these organisms have lived?
(1) Tertiary
(2) Cretaceous
(3) Triassic
(4) Cambrian

10. What is the geologic age of the surface bedrock of most of the Allegheny Plateau landscape region in New York State?
(1) Cambrian
(2) Devonian
(3) Silurian
(4) Ordovician

11. When did the Jurassic Period end?
(1) 65 million years ago
(2) 142 million years ago
(3) 163 million years ago
(4) 206 million years ago

12. According to the *Earth Science Reference Tables,* which geologic event is associated with the Grenville Orogeny?
(1) the formation of the ancestral Adirondack Mountains
(2) the advance and retreat of the last continental ice sheet
(3) the separation of South America from Africa
(4) the initial opening of the Atlantic Ocean

13. Present-day corals live in warm, shallow ocean water. Which inference is best supported by the discovery of Ordovician-age corals in the surface bedrock of western New York State?
(1) Western New York State was covered by a warm, shallow sea during Ordovician time.
(2) Ordovician-age corals lived in the forest of western New York State.
(3) Ordovician-age corals were transported to western New York State by cold, freshwater streams.
(4) Western New York was covered by a continental ice sheet that created coral fossils during the Ordovician period.

14. Unconformities (buried erosional surfaces) provide evidence that
(1) many life forms have become extinct
(2) faults are older than the rock in which they are found
(3) part of the geologic record has been destroyed
(4) metamorphic rocks have formed from sedimentary rocks

15. Based on studies of fossils found in subsurface rocks near Buffalo, New York, scientists have inferred that the climate of this area during the Ordovician Period was much warmer than the present climate. Which statement best explains this change in climate?
(1) The sun emitted less sunlight during the Ordovician Period.
(2) Earth was farther from the sun during the Ordovician Period.

(3) The North American continent was nearer to the equator during the Ordovician Period.

(4) Many huge volcanic eruptions occurred during the Ordovician Period.

Part B

Base your answers to questions 16 through 19 on the following diagrams. Columns A and B represent two widely separated areas of exposed bedrock. The symbols show the rock types and the locations of fossils found in the rock layers. The rock layers have not been overturned.

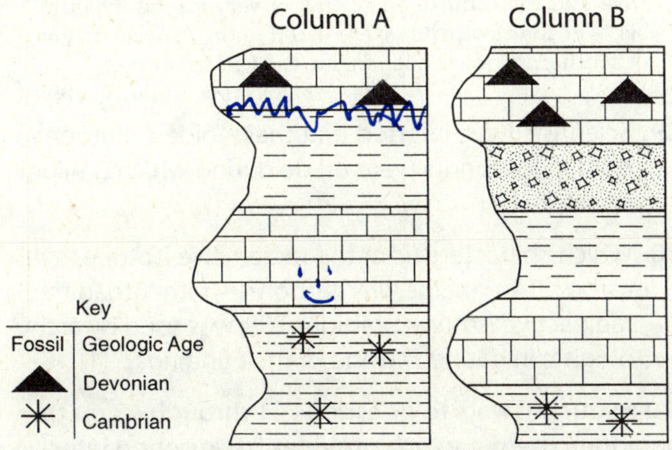

16. State one method of correlating rock layers found in the exposed bedrock represented by Column A with rock layers found in the exposed bedrock represented by Column B. [1]

17. An unconformity (buried erosion surface) exists between two layers in the exposed bedrock represented by Column A. Identify the location of the unconformity by drawing a thick wavy line ∿∿ at the correct position on column A.

18. State the evidence that limestone is the most resistant layer in these areas of exposed bedrock. [1]

19. State the oldest possible age, in millions of years, for the fossils in the siltstone layer. [1]

---

20. In what era and period of geologic time were much of the present-day coal deposits formed? [1]

21. What produced most of the oxygen presently found in Earth's atmosphere? [1]

22. How can divergence of lithospheric plates affect the evolution of life on a continent? [1]

23. Why do large masses of the evaporite sedimentary rocks—rock salt and rock gypsum—exist in Silurian rocks of western New York State but are not forming there today? [1]

24. What are two events associated with the impact of a comet or an asteroid with Earth that would cause a major extinction of life? [2]

25. Why did most of the iron that existed near Earth's surface around 4.5 billion years ago end up in Earth's core? [1]

26. Compare the age of the Earth and our solar system to the age of the most distant galaxies. [1]

27. While exploring a stream a student found a rock containing a trilobite fossil. Name the most likely type of rock this student found. [1]"

28. If a rock contains a trilobite fossil in what geologic era did the rock most likely form? [1]

Part C

Base your answers to questions 29 through 32 on the geologic cross section below. Radioactive dating indicates that the granite intrusion is 279 million years old and the vesicular basalt is 260 million years old. The rock layers have not been overturned.

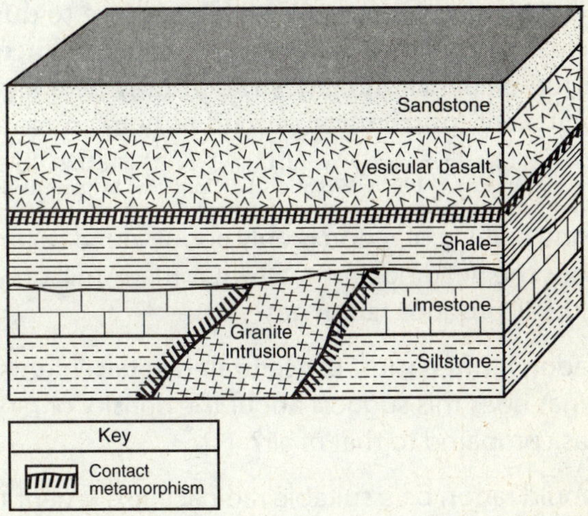

29. List the six rock units in the order from the oldest to the youngest. [1]

30. During which goelogic time period did the shale layer form? [1]

31. The granite intrusion caused part of the limestone layer to undergo metamorphism. What metamorphic rock would most likely be found in this zone of contact metamophism? [1]

32. Between which two rock layers in the diagram is there most likely an unconformity? [1]

33. Why can life, as we presently know it, exist on Earth and not on other planets in our solar system? [1]

34. In natural exposed bedrock, how is it possible that a meter-wide piece of sandstone can be found within an igneous rock such as granite? [1]

Read the following article about radon. Use your knowledge of earth science and the information in the following article to answer questions 35–38.

Radon (gas) gained public attention in 1984 when a worker in a Pennsylvania nuclear power plant set off radiation alarms—not when he left work, but as he entered. His clothing and hair were contaminated with radon decay products. Investigation revealed that his basement at home had a radon level 2800 times the average level in indoor air. The home was located along a geological formation known as the Reading Prong—a mass of uranium-bearing rock that runs from near Reading, Pennsylvania, to near Trenton, New Jersey.

Radon is continually generated by the gradual decay of uranium. Because uranium has a half-life of about 4.5 billion years, radon will be with us forever. Radon itself decays, having a half-life of only about four days. Its decay products (except lead 206) are all radioactive solids that adhere to dust particles, many of which we inhale. Steadily accumulating evidence indicates radon may be a significant cause of lung cancer, second only to smoking.

From *Earth: An Introduction to Physical Geology* by Edward J. Tarbuck and Frederick K. Lutgens.

35. If the half-life of radon is only about four days, why does the Reading Prong, which is many million years old, still produce radon gas? [1]

36. Radon is often most concentrated in basements. What does this suggest about the density of radon gas compared to that of air? [1]

37. Would radon be a suitable radioactive element to use in radioactive dating? Explain. [2]

38. If you were a homeowner on the Reading Prong, what might you do to ensure the lowest possible concentration of radon gas in your home? [1]

Base your answers to questions 39 and 40 in part on the newspaper article below.

Ancient human footprints found

PARIS—In the darkness of an underground cave lined with prehistoric paintings, French scientists believe they have discovered the oldest footprints of humans in Europe.

Embedded in damp clay, the imprints, slightly more than 8 inches long, appear to be those of a boy, 8 or 10 years old, who was walking barefoot between 25,000 and 30,000 years ago, prehistorians said Wednesday.

They said the dates are only hypothetical because there is no precise way to determine when the markings were made. But Michel-Andre Garcia, one prehistorian who has studied the site, said that the carbon datings in the cave and the context make this "a very strong hypothesis." The four footprints were found in the Ardeche region of southern France, deep inside the Chauvet cave.

—*Times Union,* June 10, 1999

39. Scientists have inferred that these "oldest" European human footprints were made during which geologic epoch? [1]

40. Which characteristic of the radioactive isotope carbon-14 explains why carbon-14, rather than the radioactive isotope uranium-238, was used by archeologists in dating the age of their findings? [1]

Base your answers to questions 41 through 43 on the paragraph below, which provides background information regarding recent fossil discoveries in Canada.

Scientific evidence indicates that the earliest mammals may have evolved approximately 225 million years ago from an ancient reptile group called the therapsids. For millions of years afterward, early mammals and therapsids coexisted until the therapsids apparently became extinct 165 million years ago. However, geologists have recently found a fossil they believe to be a therapsid that is only 60 million years old. They found the fossil, which they have named *Chronoperates paradoxus* (paradoxical time-wanderer), near Calgary in Canada. This find suggests that for 105 million years after the apparent extinction of the therapsids, a few of the reptiles continued to live in a narrow geographic range in Canada.

41. According to fossil evidence, during which geologic period did the earliest mammals appear on Earth? [1]

42. Explain briefly why *Chronoperates paradoxus* would not be a good index fossil. [1]

43. State one method geologists could have used to determine that *Chronoperates paradoxus* lived 60 million years ago. [1]

Landscape Development and Environmental Change

TOPIC 14

VOCABULARY		
escarpment	plain	stream drainage pattern
landscape	plateau	uplifting forces
landscape region	ridges	
mountain		

Landscapes, or topography, are the features of Earth's surface at the interfaces between the atmosphere, or hydrosphere, and the top of the lithosphere. Landscapes also exist on all other solid celestial bodies. Some of the characteristics of landscapes are the slope of the land, shape of the landscape features, stream drainage patterns, stream slope, soil properties, and evidences of human occupation such as cities and mines. A **stream drainage pattern** is the shape of the stream courses in an area as viewed from the sky. (See Figure 14-8 on page 274.) Landscape features range in size from a raindrop impression in mud to a continent.

MEMORY JOGGER

You may recall from Topic 2 that the lithosphere is the layer of rock that forms Earth's solid outer shell. The hydrosphere is the layer of liquid water that rests on the lithosphere. The atmosphere is a mixture of gases that surrounds the rest of Earth.

Measuring Landscape Characteristics

The shape and slope of the land, the stream drainage patterns, and some soil features can be measured by using actual observations or models such as contour maps, aerial photographs, satellite images, and the types of landscape maps and drawings shown in this topic. The use of these tools has shown that landscape features such as mountains, valleys, and steam drainage patterns have distinctive shapes by which they can be identified. These models can also be interpreted to show the changes in landscapes through time, for example, shoreline erosion, deforestation, retreat of glaciers, and the advance of deserts.

GRADIENT AND PROFILE One measurable characteristic of land and streams described in Topic 2 is the gradient, or slope. Also described in Topic 2 are profiles that show the elevation, shape, and slope of landscape features. (See Figure 14-1.)

MOUNTAINS, PLATEAUS, AND PLAINS On the basis of gradient, elevation, and rock structure, landscapes are divided into three major types—mountains, plateaus, and plains. (See Figure 14-1.)

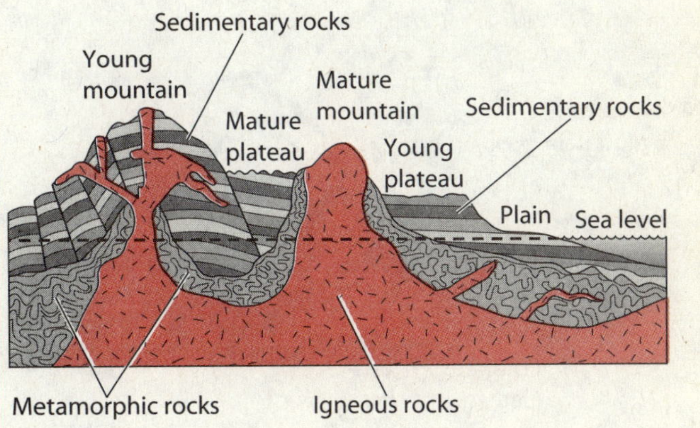

Figure 14-1. Profiles and structures of the three major types of landscapes—mountains, plateaus, and plains

A **mountain** is an area of high elevation, compared to the surrounding area or sea level, that usually has regions of steep gradient and many changes in slope. Its peak is usually thousands of feet higher than its base. Internally, mountains are characterized by distorted rock structures such as faults, folds, or volcanic rocks. Mountains are usually composed of much igneous rock and metamorphic rock.

There are many types of mountains, and it is their origin that determines their type. Some examples include faulted mountains, folded mountains, and volcanic mountains. (See Figure 14-6 on page 272.) The origin of these mountain types was discussed in Topic 12.

A **plateau** is also an area of high elevation, but it has undistorted horizontal rock structure and often a more level slope or gradient than mountains have. A plateau may have steep slopes because of valleys cut by streams and/or glaciers. Examples include the Grand Canyon in the Colorado Plateau or the many valleys of the Catskill Plateau of New York. Plateaus are composed of horizontal strata of sedimentary rocks or layers of extrusive igneous lava flows.

A **plain,** usually composed of sedimentary rocks, has a low elevation and generally level surface with little change in slope. One example is the Atlantic Coastal Plain of Long Island, New York. Some earth scientists consider old mountain areas that have been eroded to low elevations a type of plain. Composed of distorted metamorphic rocks, parts of the Manhattan Prong landscape region around New York City are an example of this type of eroded mountain or plain.

Landscape Regions

Landscape characteristics—elevation, bedrock structure, stream drainage patterns, soil characteristics, and amount of slope—occur in combinations that form identifiable areas called **landscape regions.**

CONTINENTAL LANDSCAPE REGIONS Any continental land mass has several landscape regions that can be identified by the surface characteristics. The landscape regions of the contiguous United States are shown in Figure 14-2. The landscape regions for New York State are shown in Figure 14-3 and on the inside back cover. You should also carefully study the New York State landscape regions in the *Earth Science Reference Tables.*

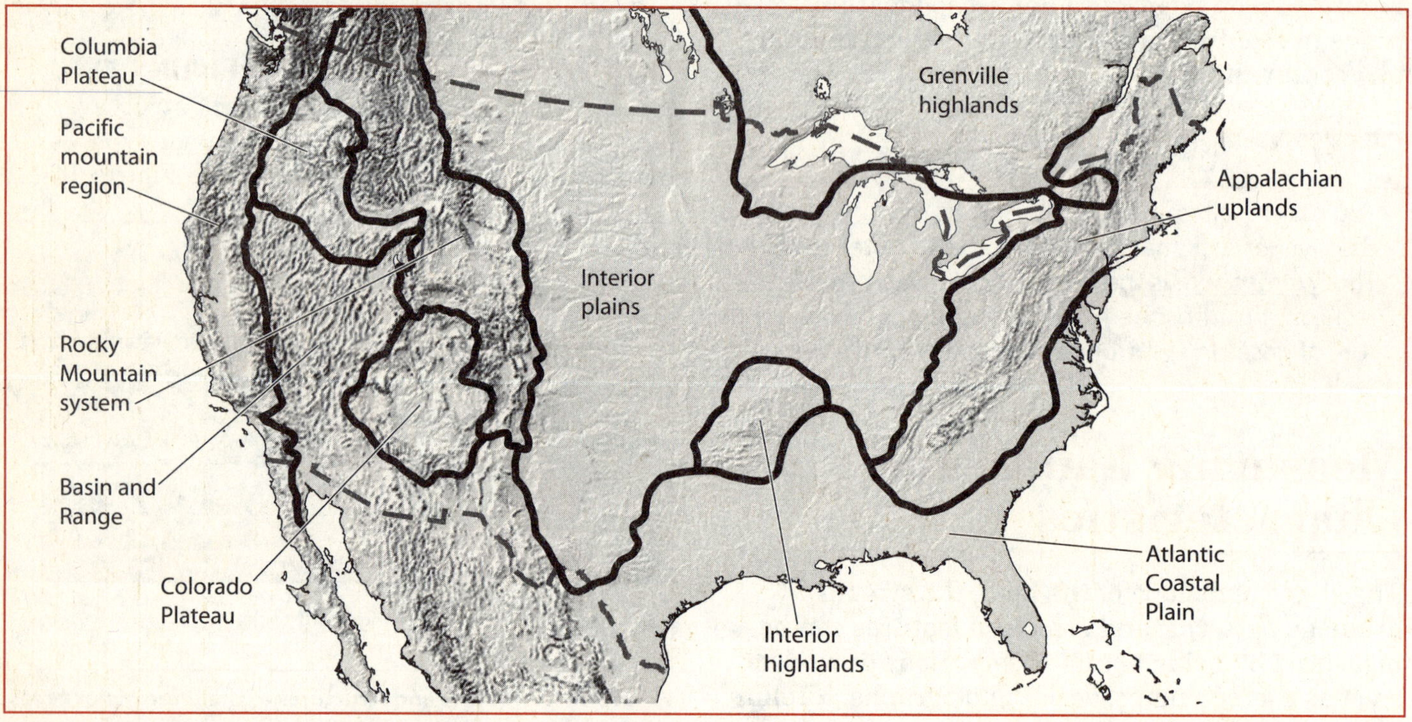

Figure 14-2. Major landscape regions of the contiguous United States

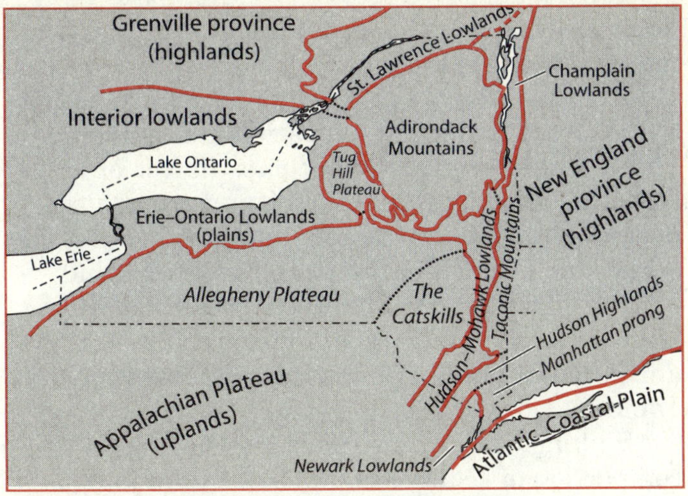

Figure 14-3. Generalized landscape regions of New York State

LANDSCAPE BOUNDARIES The boundaries between landscape regions are usually well defined, as shown in Figures 14-2 and 14-3 for the United States and New York State. Landscape boundaries usually consist of characteristics of the landscape that have been brought about by changes in the structure of rocks. Such characteristics include mountain edges, fault zones, ridges, cliffs, changes in the amount of slope, or river courses that follow the direction of the structural change.

Review Questions

1. Why would you expect to find distorted rock structures in mountain landscapes?

2. The major landscape regions of the United States are identified chiefly on the basis of (1) similar surface characteristics (2) similar climatic conditions (3) nearness to major mountain regions (4) nearness to continental boundaries

3. In which New York State landscape region is 44° N latitude and 74°30′ W longitude found? (1) Hudson-Mohawk Lowlands (2) St. Lawrence Lowlands (3) Adirondack Mountains (4) the Catskills

4. Which New York State landscape region has the lowest elevation, the most level land surface, and is composed primarily of Cretaceous through Pleistocene unconsolidated sediments? (1) the Hudson-Mohawk Lowlands (2) the Atlantic Coastal Plain (3) the Champlain Lowlands (4) the Erie-Ontario Lowlands

5. Boundaries between landscape regions in New York State are best described as being (1) straight lines running north and south (2) invisible on the surface (3) usually well defined (4) generally unchanging

6. The boundaries between landscape regions are usually indicated by sharp changes in (1) bedrock structure and elevation (2) weathering rate and method of deposition (3) soil associations and geologic age (4) stream discharge rate and direction of flow

Factors of Landscape Development

Two groups of forces form and change landscapes—the uplifting forces and the leveling, or destructional forces. Uplifting forces are also called tectonic forces because they are the results of Earth's plate movements.

Uplifting and Leveling Forces

Uplifting forces originate beneath or within Earth's lithosphere. They displace and form new rock material to raise the land, build mountains, and cause the growth of continental regions. These uplifting forces include volcanic action, earthquakes, and the many other movements associated with plate interactions. These forces obtain their energy from Earth's interior as the result of decay of radioactive materials and from the residual heat from Earth's formation. It is inferred that most of the heat is distributed to the lithosphere by convection currents within the mantle.

MEMORY JOGGER

You may recall from Topic 5 that a convection current transfers heat energy by circular movements caused by differences in density.

Most leveling forces operate at or near Earth's solid surface. These forces lower and level out Earth's surface by breaking down the rocks of landmasses, transporting sediments along Earth's surface, and depositing sediments. The leveling forces include weathering, erosion, deposition, and sinking. The source of energy for the leveling forces is energy from the sun—insolation—and gravity.

Leveling forces, such as erosion and weathering, are always at work at all areas of Earth's solid surface, but the uplifting forces may or may not be operating at the same time. When both groups of forces are present, the landscape will be uplifted or lowered, depending on which forces are dominant (operating at a faster rate).

If the uplifting forces are dominant, level land of low elevation may be changed into mountains with steep hillslopes and high elevations. With higher elevations, streams will gain more energy to erode, glaciers may form, and higher lands will alter wind patterns. These uplifting forces are most dominant at plate boundaries where earthquakes, volcanic actions, and the growth of young mountains are going on.

In areas where the leveling forces are acting alone or are dominant, elevation is decreasing, and the land eventually becomes flatter or smoother in slope. Weathering and erosion are destroying the rocks faster than uplift (if any) is occurring. Deposition by running water, wind, gravity, glaciers, and shoreline processes contributes to the leveling and forms the many depositional landscape features discussed in Topic 10.

Sinking of Earth's surface may be due to the compaction of sediments under the influence of gravity. The withdrawal of fluids such as water and petroleum can also contribute to the lowering of the land.

Figure 14-4. Stages of landscape development

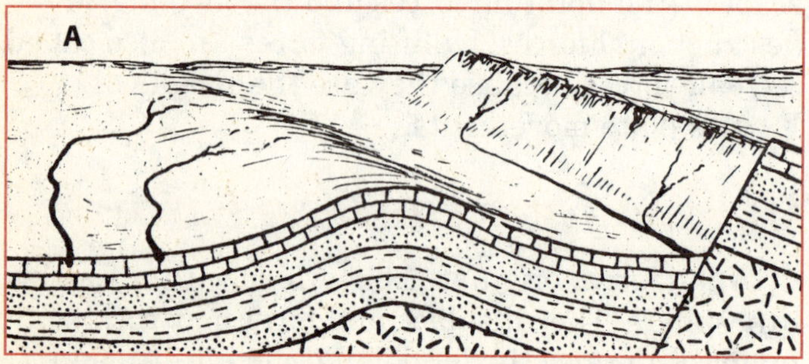

Youth Uplifting forces are dominant, causing folding and faulting, and the formation of mountains with high elevations and steep slopes.

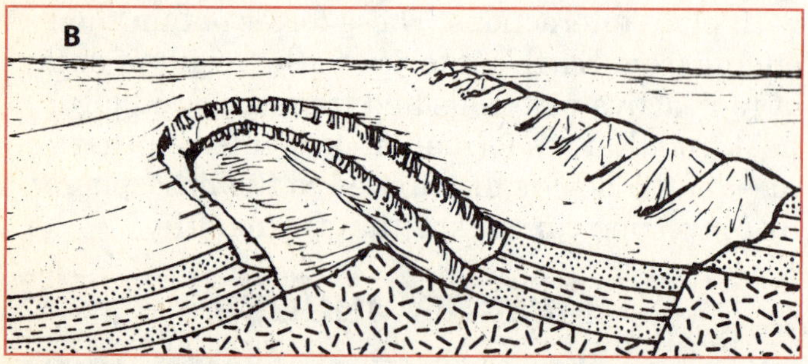

Maturity Leveling forces are dominant, creating a rugged landscape with lower elevations.

Old Age Leveling forces are still dominant but less effective because low elevations and gentle slopes provide little potential energy.

Time and Landscape Stages

The features of a landscape at any one time are partly the result of the length of time the uplifting and leveling processes have acted on the rocks and loose materials of Earth's surface. An area often experiences a time when the uplifting forces are dominant, followed by a long time when the leveling forces are dominant. When this happens, an area's landscape goes through stages of development, with each stage having characteristic conditions.

Figure 14-4 is a model of the stages of landscape development in a mountainous region of a humid climate. In stage A—youth—the uplifting gives the leveling forces such as gravity, running water, and glaciers a large amount of potential energy, and the rate of leveling becomes great. As the uplifting forces decrease in intensity, the leveling forces carve away the land, creating a rugged landscape of lower elevation characteristic of stage B—maturity. If the leveling forces continue to be dominant for millions of years, the landscape will become progressively smoother in slope and lower in elevation. When the former mountain region is smooth enough to resemble a plain, it is considered to be in stage C—old age.

Although landscape development is a continuous process leading to a gradual change in the landscape, at any one time the landscape condition reflects a state of balance among many environmental factors. This is a condition of dynamic equilibrium. If a change occurs in any of the factors, a change in the landscape features will occur until a new equilibrium is established.

Effects of Climate on Landscape Development

The rate of development and the characteristics of the landscape in an area are greatly influenced by temperature and moisture conditions, which in turn affect the type and amount of vegetation. Any change in an area's climate will alter the rate of development and the conditions of the landscape. On the other hand, changes in landscape sometimes cause changes in climate. For example, the growth of a new mountain range will block air mass movement and change the precipitation and temperature patterns. In turn, more landscape changes will occur in response to the climate change. The effect of mountains on climate is described in Topic 8. One example of this effect is that increased elevations may result in alpine glaciers that will carve rugged, sharp landscapes.

HILLSLOPES IN ARID AND HUMID CLIMATES

The steepness of hillslopes is partly affected by the balance between production of sediments by weathering and the removal of these sediments by erosion. In arid climates, there is little vegetation to hold the sediments in place on steeper slopes or at the bases of slopes, so wind and running water can rapidly carry the sediments away. This rapid removal of sediments causes many arid regions to be characterized by steep slopes and sharp and angular landscape features common in the southwestern United States. Arid regions are also characterized by sand dunes and bedrock sandblasted and sculptured by wind-blown sand. (See Figure 14-5A.)

In humid climates, sediments on hillslopes and at the bases of hillslopes are better held in place by more abundant vegetation; therefore, areas with humid climates are characterized by the smoother and more rounded landscape features common in the eastern United States. The surfaces of the humid landscapes are more likely to be soil and thick vegetation, not the rough and irregular bedrock surfaces of an arid region. The greater rate of chemical weathering of bedrock in humid climates also contributes to their rounded landscapes. (See Figure 14-5B.)

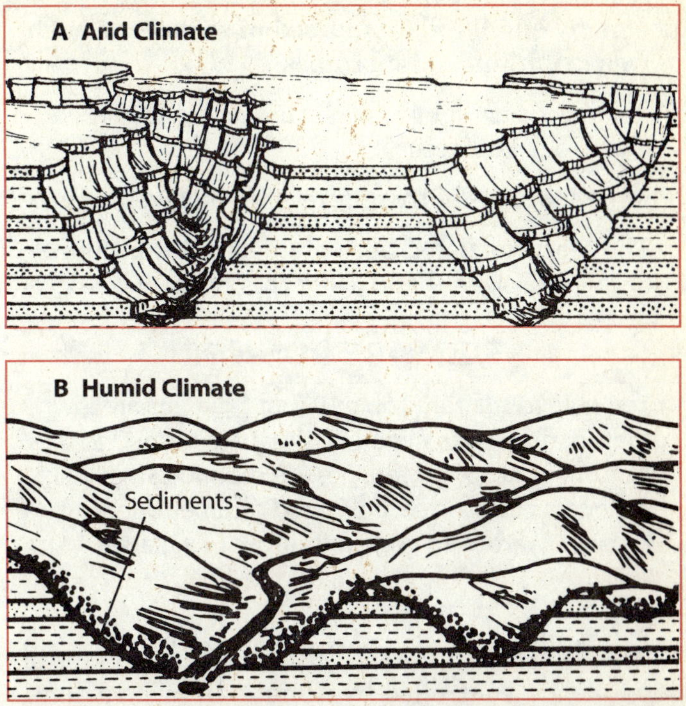

Figure 14-5. Landscape development of a plateau as affected by climate: The rock types and rock structure are assumed to be similar in both diagrams. Diagram A illustrates the steep and angular slopes characteristic of an arid climate. Note that the more resistant sandstone (shown by dots) results in the steepest slopes. Diagram B illustrates the smoother, more rounded landscapes of a humid climate, such as the Tug Hill and Catskill plateaus of New York State.

GLACIATION AND LANDSCAPES In climates where glaciers exist or have existed in recent times (such as the Pleistocene ice ages—see the Geologic History of New York State in the *Earth Science Reference Tables),* the landscape will show much evidence of glacial erosion and deposition. Some of the landscape features of glaciation are

- mountaintops and steep slopes without much soil
- transported soil covering large areas
- soil with a large range of particle size—clay to boulders—even at the surface
- wide valleys with U-shaped profiles—as compared to a stream's V-shaped profile
- many lakes, including finger lakes and kettle lakes
- disrupted—not well connected—stream drainage patterns

- many small hills composed of sediments such as drumlins and moraines
- polish, parallel grooves, and scratches on bedrock surfaces
- the features of the glaciers themselves

The features listed here are described in more detail in Topics 9 and 10 of this book.

MEMORY JOGGER

You may recall from Topic 10 that drumlins are low, long, narrow streamlined hills that are formed at the bottom of glaciers. Moraines are mounds, small hills, or sheets of unsorted, unlayered sediment deposited directly from an edge or bottom of a glacier.

STREAM AND CLIMATE Some characteristics of streams are controlled by climate. In arid regions, most streams, being temporary or intermittent, are without water for much of the time; many streams in humid regions are permanent, having discharge most or all of the time.

Internal drainage is common in arid regions and occurs when runoff is channeled into basins not connected by streams to the oceans. Examples of such basins include the Great Salt Lake region in Utah, Death Valley in California, and other basins in Nevada. Internal drainage exists in arid regions because the streams have not had time to carve interconnecting valleys for drainage to the oceans. Also characteristic of arid regions are land deltas located at the base of steep slopes where streams rapidly lose velocity and deposit sediment.

SOILS AND CLIMATE One of the most important factors in determining soil characteristics—composition, porosity, permeability, and particle size—is climate. Soils in arid regions are often thin or nonexistent because there is little or no vegetation to help make the soil and hold in place. The soil that does exist is often very sandy because smaller sediments have been blown away. Arid soils often contain many minerals not commonly found in the soil of humid regions, where infiltration and runoff would dissolve and carry minerals away. If the climate is hot and humid, like many areas near the equator, the soil will be infertile because chemical weathering and infiltration rapidly remove organic material and mineral nutrients, leaving the soil lacking these components.

Review Questions

7. Which New York State landscape region is best represented by the block diagram below? (1) Allegheny Plateau (2) Adirondack Mountains (3) Atlantic Coastal Plain (4) Erie-Ontario Lowlands

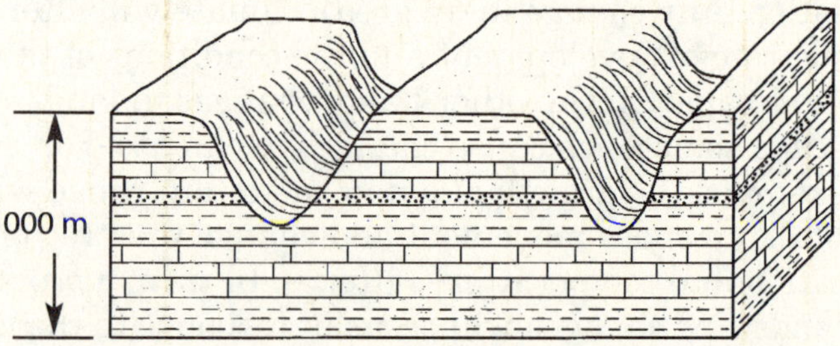

8. The Adirondacks and the Catskills are both areas of high elevation. Why are the Adirondacks classified as a mountain landscape, whereas the Catskills are classified as a plateau landscape?

9. Landscape regions in which leveling forces are dominant over uplifting forces are often characterized by (1) volcanoes (2) mountain building (3) low elevations and gentle slopes (4) high elevations and steep slopes

10. In a certain area, the processes of uplift and leveling are in dynamic equilibrium. If this condition continues, what will be the overall long-term effect on the elevation of the area?

11. Which change would occur in a landscape region where uplifting forces are dominant over leveling forces? (1) topographic features becoming smoother with time (2) no dynamic equilibrium over time (3) streams decreasing in velocity with time (4) hillslopes increasing in steepness with time

12. Which characteristics of a landscape region would provide the best information about the stage of development of the landscape? (1) the age and fossil content of the bedrock (2) the types of hillslopes and the stream patterns (3) the amount of precipitation and the potential evapotranspiration (4) the type of vegetation and the vegetation growth rate

13. The three diagrams below represent the same location at different times. What is the best explanation for the differences in appearance of this location?

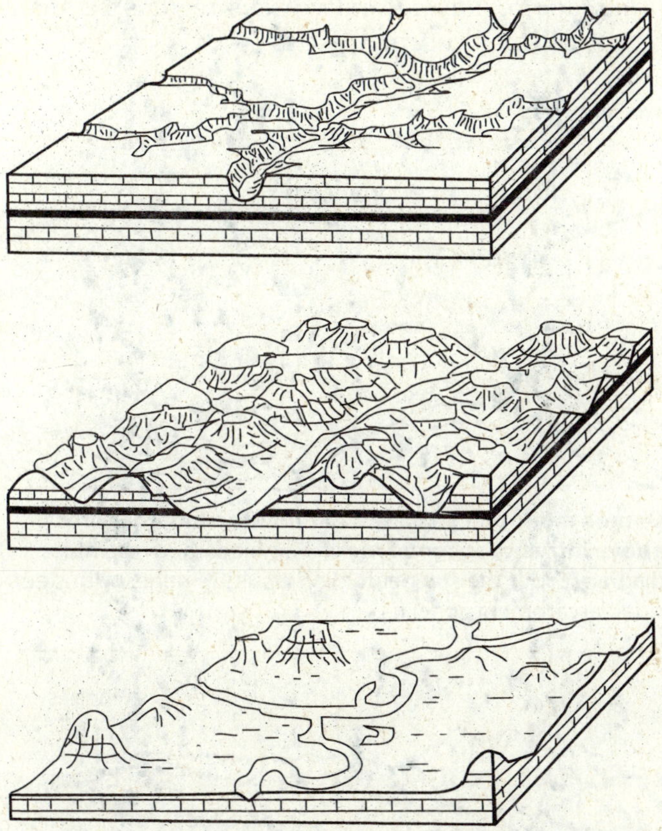

(1) This region was exposed to leveling forces for different lengths of time. (2) This region has different underlying bedrock structures. (3) This region developed in different climate regions. (4) This region has the same type of underlying rock layers.

14. Which factor is most important in determining the evolution of a landscape? (1) surface topography (2) plant cover (3) climate (4) development of drainage

15. The diagrams below represent geologic cross sections from two widely separated regions.

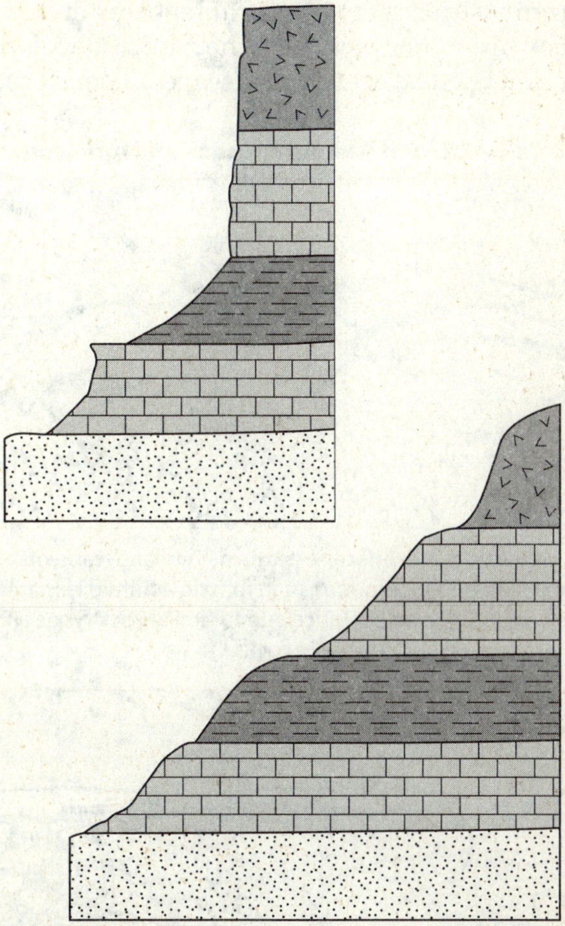

The layers of rock appear similar, but the hillslopes and shapes are different. These differences are most likely the result of (1) volcanic eruptions (2) earthquake activity (3) age of bedrock (4) climate variations

16. List two landscape characteristics you would expect to find in an area which has undergone glacial erosion.

17. Contrast the cross-sectional profiles of a glacial valley and a stream valley.

18. Characteristics such as composition, porosity, permeability, and particle size are usually used to describe different types of (1) hillslopes (2) stream drainage patterns (3) soils (4) ridges

Bedrock

The composition and structural features of bedrock are major factors in the rate of development and the features of landscapes. Therefore, the primary reason New York State has several landscape regions is that various regions have different bedrock characteristics.

Figure 14-6. Effect of bedrock structure and rock types on landscape features

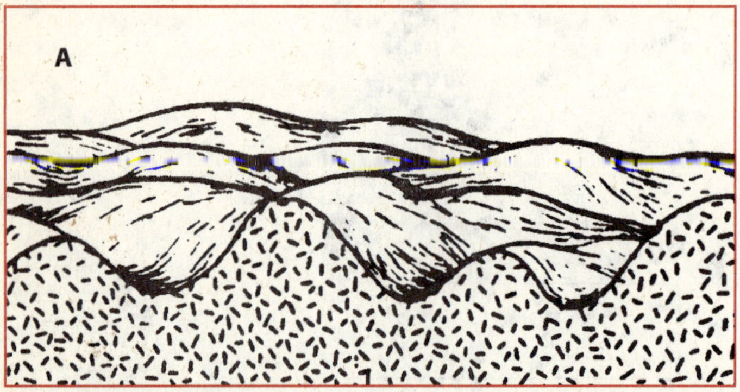

Random mountain landscape pattern This landscape of random rounded mountaintops can form in a humid climate with little difference in rock resistance due to similar rock types and no major structural distortion of the rocks.

Horizontal undistorted sedimentary rocks of a plateau This landscape of an arid region has uniform elevations, with steep V-shaped valleys cut by streams. Note the sandstone layers result in steeper cliffs because the sandstone is a more resistant rock.

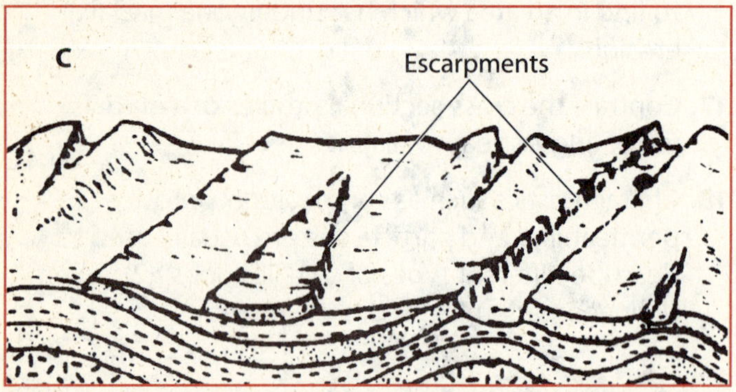

Gently folded sedimentary strata of varying resistance Note that the resistant sandstone forms steep escarpments, or cliffs, in this mountain landscape.

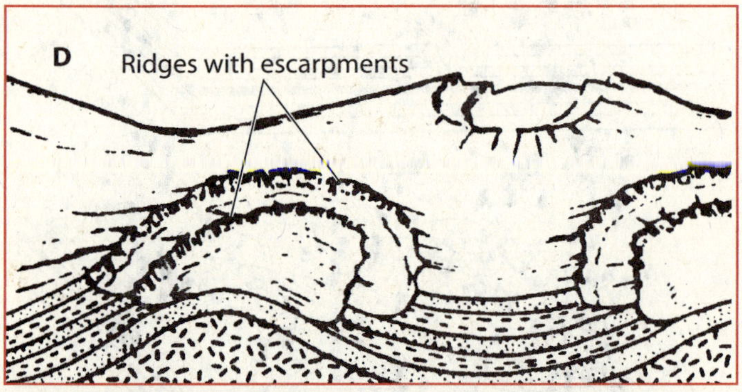

Domed mountain structure of folded sedimentary rocks above intrusive igneous rocks This landscape resembles diagram C, but the steeper folding results in ridges with steep-sided escarpments or cliffs.

Complex bedrock structure This mountainous region consists of faulted, folded, and intruded rocks of all three rock types. Some of the more-resistant sedimentary rock layers form ridges of varying elevations and slopes.

Fault block mountain in igneous rocks This landscape of ridges of varying elevations is caused by faulting.

Rock Types and Hillslopes

In any one climate, different rock types have varying degrees of resistance to weathering and erosion. If all the rocks in an area exposed to the surface have about the same resistance, the landscape will be controlled by structural features such as faults, folds, and joints and by uplifting and leveling forces. (See diagram E in Figure 14-6 on the previous page.) If there are no special structural features or rock resistance differences, the landscape features will be random in location and without rapid changes in hillslopes. (See diagram A in Figure 14-6.)

If the rocks exposed at the surface have different degrees of resistance, the rocks will weather and erode at varying rates. For example, sandstone layers will generally be more resistant than shale in areas of sedimentary rock and will weather and erode away more slowly. The result will be a marked difference in slope between layers of different resistance, often with steeper slopes called **escarpments,** or cliffs, along the edge of the more-resistant layers. (See the escarpments in diagram C and small cliffs in diagram B associated with the sandstone layers in Figure 14-6.)

In landscapes of folded, faulted, igneous, or metamorphic rocks, resistant rocks form long narrow uplands called **ridges.** Ridges form the more-resistant rocks of the higher portions of the southern part of the Hudson-Mohawk Lowlands landscape region of southeastern New York. This region is often called the ridge and valley landscape region. (See the ridges in diagrams D, E, and F of Figure 14-6.) Figure 14-7 illustrates many of the various landforms that can develop because of differences in rock resistance.

STRUCTURAL FEATURES AND HILLSLOPES

Types of rock structure—such as horizontality, folds, joints, and faults—have a major effect on landscapes. If an area's rocks have many faults or joints, then the forces of weathering and erosion will find it easier to break up and remove the rocks, resulting in a lower area like a valley. A study of the diagrams in Figure 14-6 will show the influence of horizontal rocks, folded rocks, and faulted rocks on the landscape appearance. Generally, the greater the variety of structural features found in an area, the more varied the changes in hillslopes and types of landscape features created, as shown in diagram E of Figure 14-6.

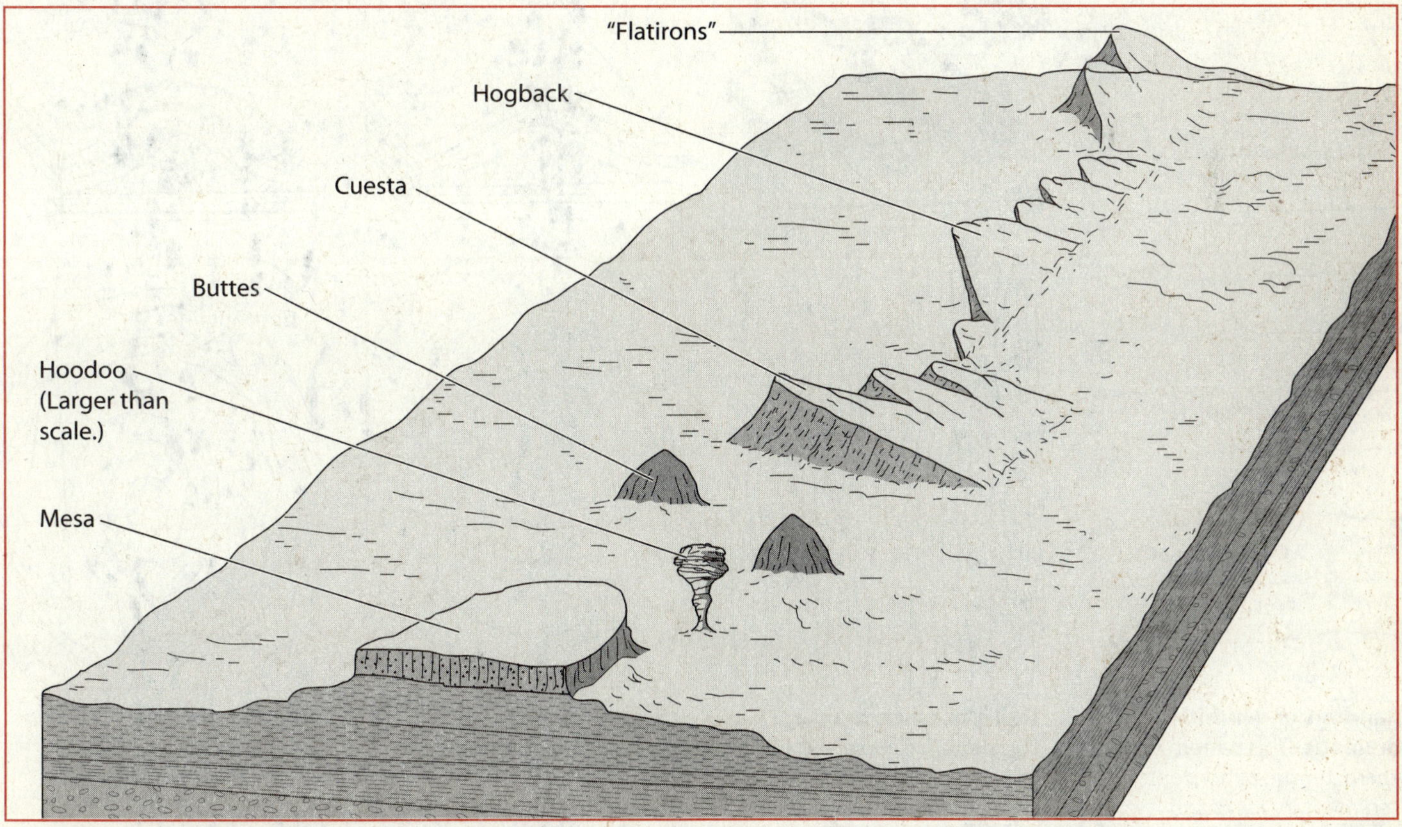

Figure 14-7. Some types of escarpments and ridges: Individual features shown here are only examples of resistant rocks in landscapes. You do NOT need to learn the names of these features.

STREAMS AND BEDROCK CHARACTERISTICS

The direction, drainage pattern, features, and gradient of streams are often directly related to the resistance and structure of underlying bedrock. In horizontal rock layers with little difference in resistance, the streams will develop a random pattern and have few changes in gradient as shown in diagram A of Figure 14-8. When the rocks in an area have varying degrees of resistance or there are folded, domed, jointed, or faulted rocks, the stream drainage patterns and directions of the streams will be at least partly, if not totally, controlled by these rock features. See diagrams B, C, and D of Figure 14-8. In such cases, there are often sharp changes in gradient, as indicated by waterfalls and rapids flowing over ridges and escarpments.

SOILS AND ROCK COMPOSITION The different compositions of the soil are due in part to differences in the underlying bedrock, or in the transported sediment from which the soil formed. A soil formed from a sedimentary rock such as sandstone, will be sandier than soil formed from an igneous rock such as basalt. Even though most of New York's soil has been transported by glaciers or streams, there is still a close correlation in soil properties and rock types, as shown on a geologic map such as the Generalized Bedrock Geology of New York State in the *Earth Science Reference Tables*.

Review Questions

19. The primary reason several landscape regions have formed in New York State is that the various regions have different (1) climates (2) latitudes (3) soil characteristics (4) bedrock characteristics

Figure 14-8. Effect of bedrock structure and varying rock resistance on stream drainage patterns

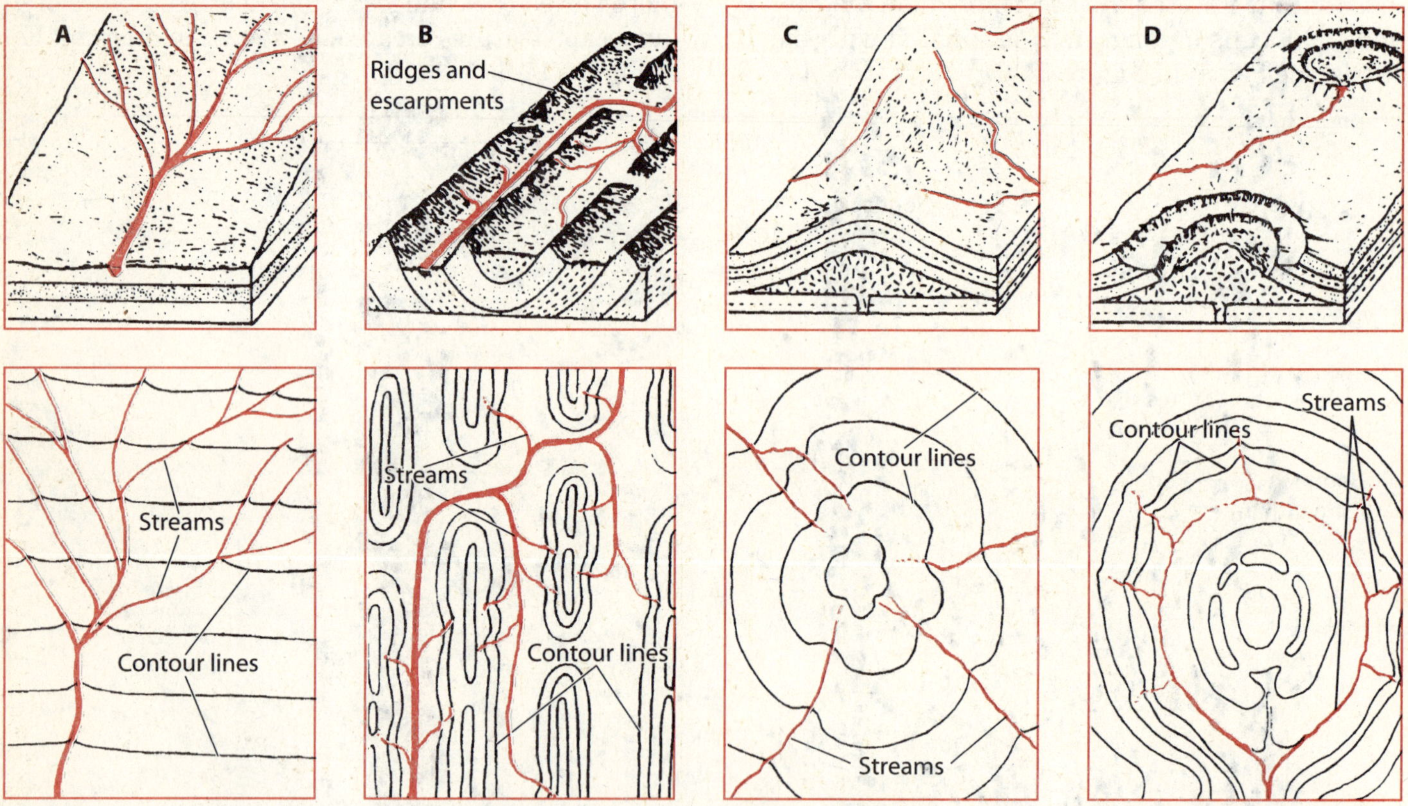

Random, or dendritic, drainage This pattern is characteristic of horizontal sedimentary rocks and lava flows with little difference in rock resistance.

Trellis, or block, drainage This pattern is observed in folded rocks with much difference in resistance, and also in faulted and jointed rock.

Radial drainage This pattern occurs in areas of domed structure (such as some volcanoes) with little difference in rock resistance.

Annular drainage This pattern of concentric circles is found in areas of domed structure with much difference in rock resistance.

20. Which cross-sectional diagram below best represents a landscape region that resulted from faulting?

(1)

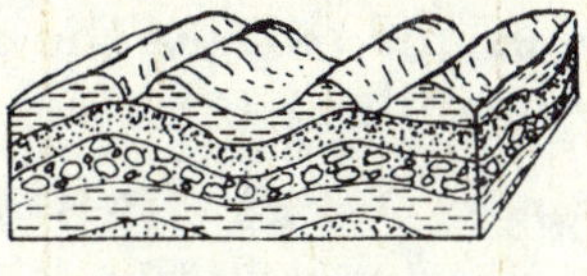
(3)

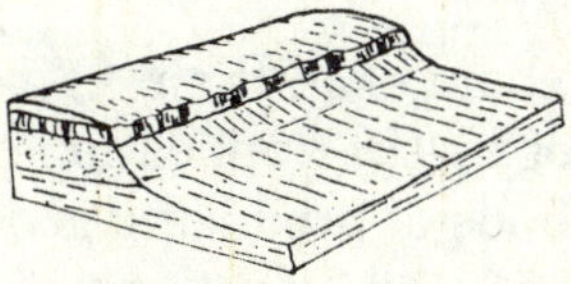
(2)

(4)

21. In the cross section of the hill shown below, which rock units are probably most resistant to weathering?

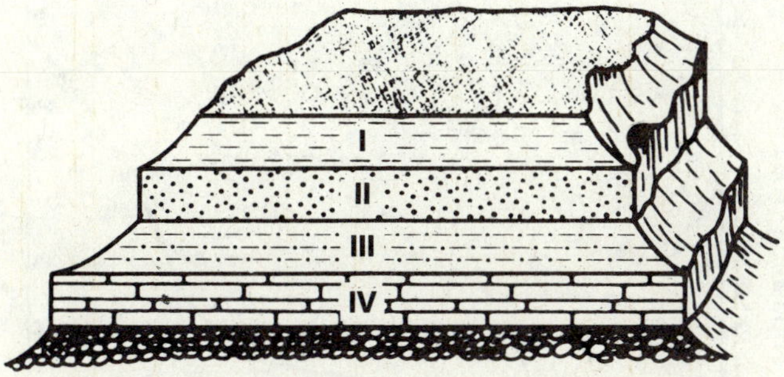

(1) I and II (2) II and III (3) I and III (4) II and IV

22. How was the valley shown in the diagram below most likely formed?

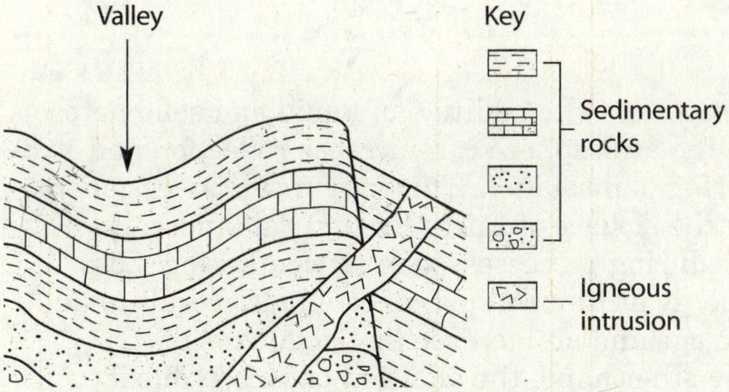

(1) by the deposition of sediments (2) by the extrusion of igneous material (3) by the faulting of rock layers (4) by the folding of rock layers

23. Which of the following stream patterns is most characteristic of horizontal rock structure?

(1)

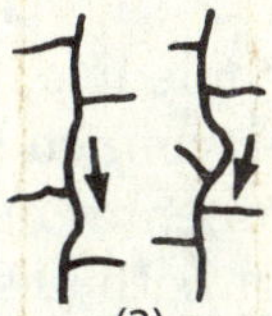
(3)

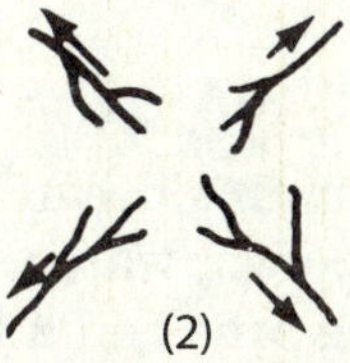
(2)

(4)

Base your answers to questions 24 and 25 on the diagram below of a section of Earth's crust.

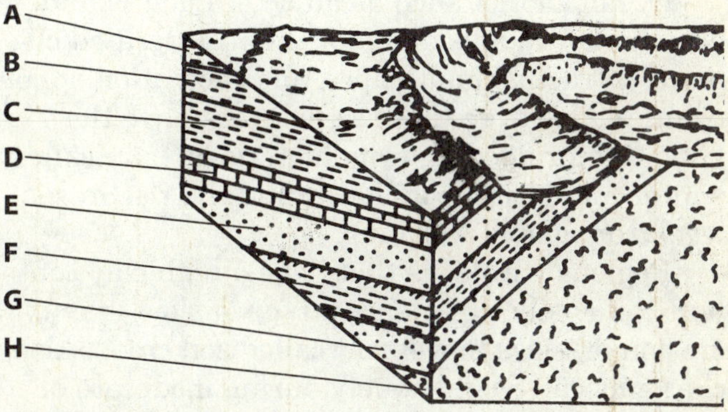

24. Which kind of stream pattern would most likely be found on the type of landscape shown in the diagram?

(1)

(3)

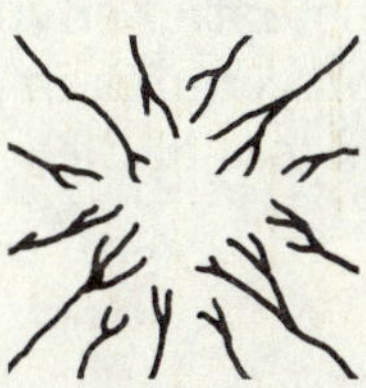
(2)

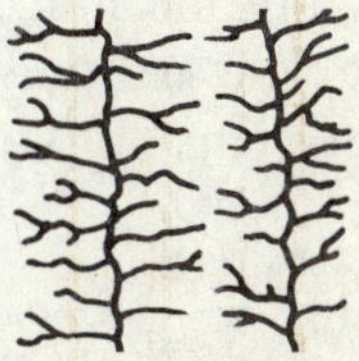
(4)

25. According to the surface landscape development indicated in the diagram, which rock type or types are most resistant to weathering and erosion in this environment? (1) rock E (2) rock H (3) rocks C and F (4) rocks D and G

People, Environmental Change, and Landscapes

How people live, work, play, and keep safe is in part determined by landscape factors that include environmental hazards. People also affect the environment both positively and negatively.

Effects of Landscape Conditions on People

Many people have settled near streams because of the ease of transportation, hydropower, water supply, and recreation. However, some areas on a flood plain could have dangerous flooding. People have made the choice to live in these flood-prone areas, but they risk losing their lives or possessions.

Ocean shores, such as much of Long Island, New York, are appealing for their moderated climate, attractive landscapes, and recreation opportunities. However, these same areas have the potential for flooding from high tides, hurricanes, winter mid-latitude cyclones, and the rise in sea level from global warming.

Regions with soils that have a high clay content can easily experience a type of slow gravity erosion or mass movement called soil creep. Also, clay soils can easily liquefy during moderate or strong earthquakes. Constructing roads and buildings on ridges and hillslopes having soil with a high clay content can result in mass movements such as landslides. Building on bedrock that is too steep or on rock that is highly fractured can also result in landslides. These mass movements can cause harm to people and to buildings.

Proper environmental planning, coupled with laws and zoning limitations, can reduce potential future damage to buildings as well as minimize human injury or death. If it is impractical to remove humans from an area of potential environmental hazard, planning escape procedures in advance can help avoid injury and loss of life.

Environmental Change of the Landscape by People

People have greatly affected the environment, including the landscape. They have cut down forests and plowed up the land, allowing the soil to be eroded. They have carved up or smoothed out the land for mining and construction of roads, buildings, and airports. They have polluted the land with their trash. They have added chemicals to the land, air, and water that aid weathering. They have even made "new" land by filling in lakes and parts of the ocean with their garbage in landfills.

LANDSCAPE POLLUTION AND POPULATION Landscape pollution is generally greatest in areas of high population density. Figure 14-9 is a graph showing that the human population has been increasing at a very rapid rate.

People also cause rapid changes in their environment because of technology—the application of scientific discoveries to methods of producing goods and services. Many changes caused by actions of people have resulted in the extinction of wildlife, loss of natural habitats, and destruction of landscapes that were used for farming and grazing.

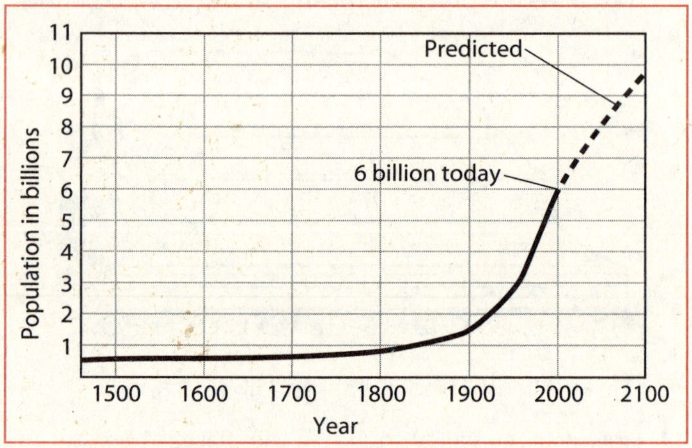

Figure 14-9. **Estimated human population of the world (projected to the year 2100)**

ATMOSPHERIC POLLUTION AND LANDSCAPE CHANGE The addition of liquid and solid aerosols to the atmosphere can increase reflection and scattering of insolation. This may result in less energy at the surface of Earth for natural landscape-producing processes, such as weathering. Less energy at Earth's surface could also result in an ice age and an increase in glacial erosion. On the other hand, the addition of carbon dioxide, water vapor, and methane to the atmosphere can increase the greenhouse effect and thus cause an increase in energy at Earth's surface for landscape change. Such changes may include an increase in arid environments and a decrease in glacial environments. Many industrial and individual activities add substances to the atmosphere that cause greater amounts of cloud coverage than would naturally occur. Increased cloud coverage can increase the amount of precipitation, which in turn affects the amounts of stream erosion. The addition of sulfur and nitrogen compounds to the atmosphere, largely by burning fossil fuels and through mineral refining, has significantly increased the acid content of precipitation. This acid precipitation has increased chemical weathering of rocks, including many stone buildings, and has caused the destruction of animal species and large areas of forest. One prime example of air pollution and landscape change is the modern urban micro-environment, as shown in Figure 14-10.

Environmental Conservation

Natural resources—such as soil for agriculture, land for home sites, and pure water for consumption, irrigation, manufacturing processes, and recreation—can be conserved by careful planning. Reclamation of polluted environments, reduction in production of environmental pollutants, and the choice to use alternative resources and recycle our waste products are some ways to conserve natural resources. This can result in preservation of natural resources and minimal environmental harm while still providing the resources needed for human use.

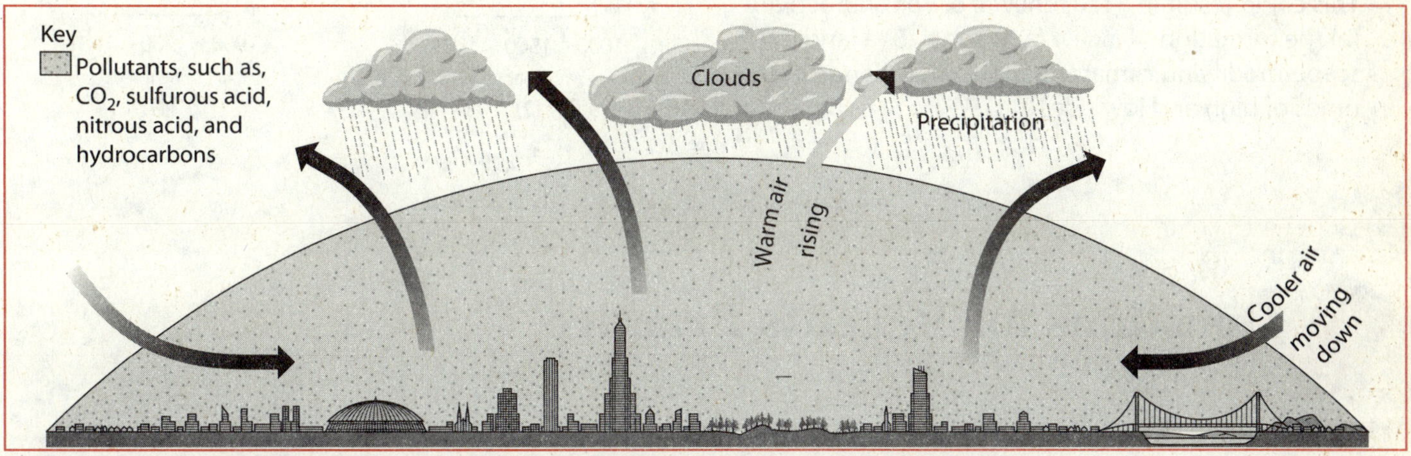

Figure 14-10. An urban microenvironment: A city or urban area is an example of how humans have altered, or polluted, the atmosphere and landscapes. The changes that have directly altered the environment include the addition of pollutants and the constructions of objects such as buildings and roads. The concentration of humans and buildings cause heat to be given off that increases the air temperature above the urban area. The result is an atmosphere that is hotter, more smoggy, less windy, more humid, cloudier, more foggy, and less transparent. Other results include more precipitation and a higher acid precipitation.

Review Questions

26. The diagram at the right shows a New York State highway road cut.

The water well has probably gone dry because (1) crustal uplift has altered the landscape (2) human activities have altered the landscape (3) the climate has become more humid (4) the bedrock porosity has changed

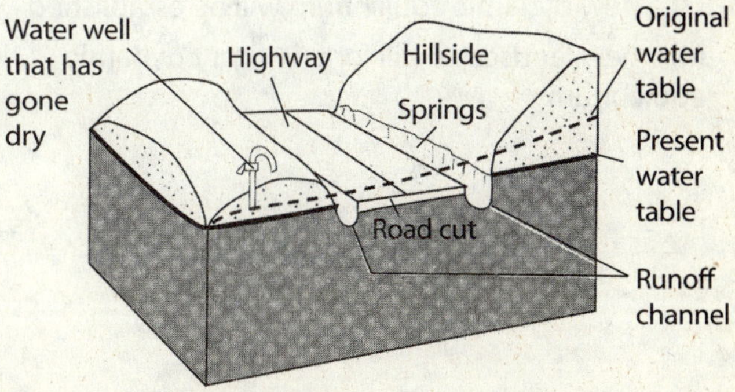

27. Humans can cause rapid changes in the environment, sometimes producing catastrophic events. Which statement below is the best example of this concept? (1) Mountainside highway construction causes a landslide. (2) Lightning causes a forest fire. (3) Shifting crustal plates cause an earthquake. (4) Changing seasonal winds cause flooding in an area.

28. As population increases, the amount of alteration of the landscape by human activity will most likely (1) decrease (2) increase 3) remain the same

29. The addition of pollutants to the atmosphere may change the rate of energy absorption and radiation. This could result in (1) a renewal of volcanic activity (2) the formation of a new fault zone (3) a landscape-modifying climate change (4) a change in the times of high and low tides

30. Which graph best represents human population growth?

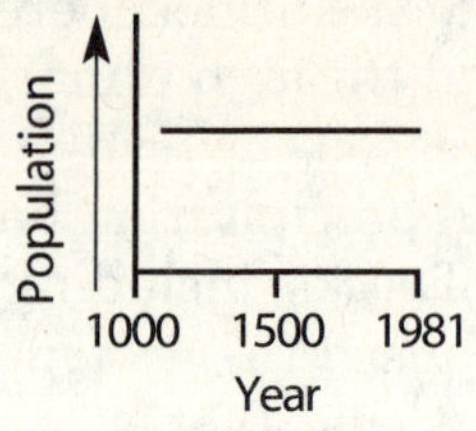

(1)

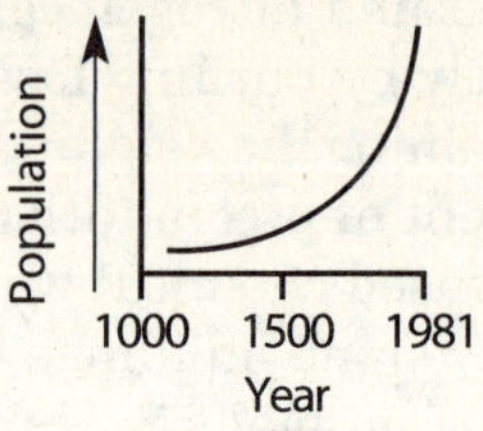

(3)

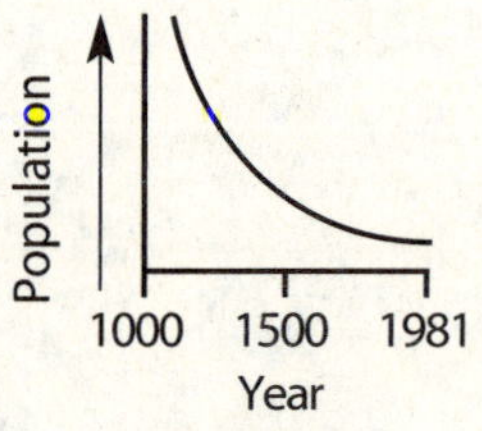

(2)

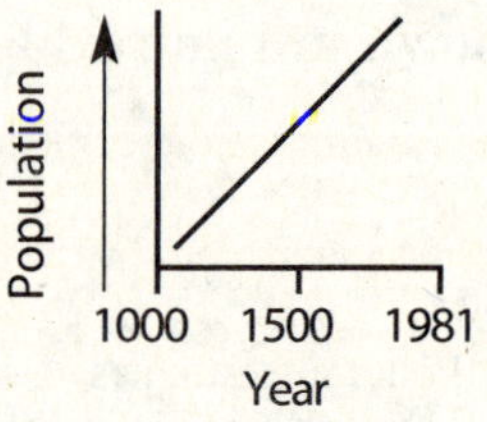

(4)

Questions for Regents Practice

Part A

1. Models of hillslopes and landform shapes can be made with the use of

(1) isobars

(2) contour lines

(3) isotherms

(4) parallels

2. If a landscape is in dynamic equilibrium and uplift occurs,

(1) there can never again be dynamic equilibrium in the area

(2) the dynamic equilibrium will remain the same

(3) a new dynamic equilibrium will be established

(4) a new landscape will form without dynamic equilibrium

3. If a new mountain range were to form where the Mississippi River is today, the landscape east of this new mountain range would become more angular and less rounded because the climate would become

(1) drier and warmer

(2) drier and cooler

(3) moister and warmer

(4) moister and cooler

Refer to the diagram below to answer questions 4 through 9.

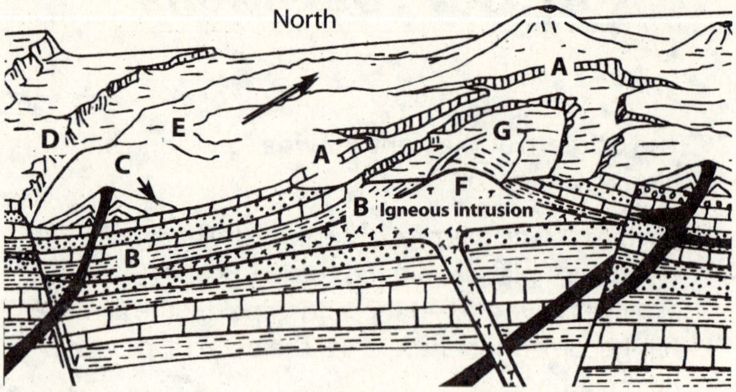

4. The climate of this area is most likely
(1) arid
(2) humid
(3) hot
(4) cold

5. The cliffs or escarpments around location A are most directly the result of
(1) igneous intrusion
(2) resistant rock layers
(3) the different ages of the rock
(4) movement along faults

6. What can be said about rock layer B compared to rock layer F?
(1) B is younger
(2) B is more resistant
(3) B is less resistant
(4) B is lower because of its environment of deposition

7. How would you classify the geologic forces producing the landscapes at locations C and G?
(1) At both places uplifting forces are dominant.
(2) At both places leveling forces are dominant.
(3) At C uplifting is dominant, and at G leveling is dominant.
(4) At C leveling is dominant, and at G uplifting is dominant.

8. What immediate effects would increased rainfall have on the rate of leveling of this landscape?
(1) increase
(2) decrease
(3) remain the same

9. The primary reason for the cliff at D is
(1) volcanic action
(2) resistant rock layers
(3) movement along a fault
(4) joints in the rock layers

---

10. What is the most practical way to control pollution?
(1) slowing the rate of technological growth
(2) decreasing the population
(3) careful planning of environmental usage
(4) more government control of industry

11. The most practical way individuals can help fight pollution is by
(1) becoming aware of the problems and thus changing their attitudes and actions
(2) recycling paper
(3) using fewer products produced by technology
(4) changing their lifestyles to conform to that of nineteenth-century rural America

12. Which New York State landscape region is mostly a direct result of the last ice age?
(1) Newark Lowlands
(2) Taconic Mountains
(3) Atlantic Coastal Plain
(4) Hudson-Mohawk Lowlands

13. A landscape region with only intermittent (not permanent) streams and no drainage connected to the ocean would most likely have what type of climate?
(1) arid
(2) humid
(3) hot and wet
(4) cool and wet

14. Which New York State landscape region is composed mostly of intensely metamorphosed surface bedrock?
(1) Hudson-Mohawk Lowlands
(2) Tug Hill Plateau
(3) Hudson Highlands
(4) Allegheny Plateau

15. If global warming continues to result in higher ocean levels, the increased ocean wave and current erosion will have the greatest effect on what New York State landscape region?
(1) Newark Lowlands
(2) Atlantic Coastal Plain
(3) Erie-Ontario Lowlands
(4) Hudson Highlands

Part B

Base your answers to questions 16 through 18 on the information below.

A mountain is a landform with steeply sloping sides whose peak is usually thousands of feet higher than its base. Mountains often contain a great deal of igneous and metamorphic rock and have distorted rock structures caused by faulting and folding of the crust.

A plateau is a broad, level area at a high elevation. It usually has an undistorted, horizontal rock structure. A plateau may have steep slopes as a result of erosion.

16. State why marine fossils are not usually found in the bedrock of the Adirondack Mountains. [1]

17. State the agent of erosion that is most likely responsible for shaping the Catskill Plateau so that it physically resembles a mountainous region. [1]

18. State the approximate age of the surface bedrock of the Catskills. [1]

Base your answers to questions 19 through 21 on the map below and your knowledge of earth science. The map shows houses, asphalt roads, drilled water wells and the amount of insecticide in parts per million (ppm) for each well except at house Q. The insecticide was spread on the lawn of one of the houses and seeped into the ground water.

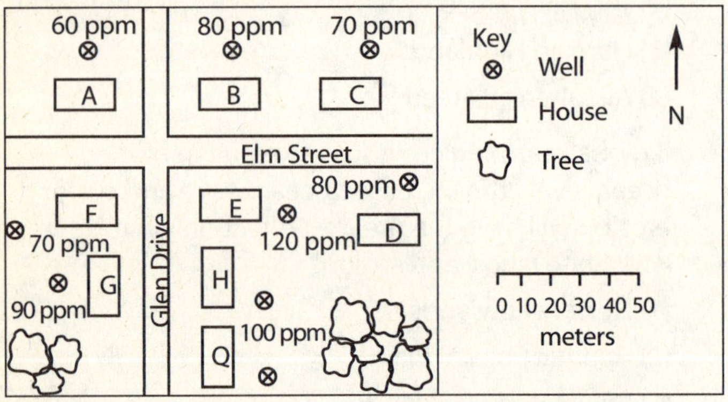

19. Write the letter of the house that is the probable source of the insecticide. [1]

20. State a reasonable value for the concentration of insecticide in ppm at house Q. [1]

21. These houses were recently built on land that was originally forest. Indicate how the amount of runoff from precipitation compares today with that of the original forest. Then state a reason for this change in runoff. [2]

Part C

The following article, though fictitious, represents events that have happened in Washington State along the Pacific coast. Use your knowledge of earth science and the information in the following article to answer questions 22 through 25.

The death toll from a mudslide in the town of Coastal, Washington, rose to over 50 today, and local authorities began discussing whether to allow people to move back into this housing development, to only use the land for seaside recreation, or to declare the land off limits. Because the forecast is calling for another two weeks of heavy rain on top of what has already fallen in this area, local emergency management officials feel that there is a real threat of more devastating mudslides.

Many of the 70 or so homes that were destroyed or severely damaged had been built on or just behind a sea cliff that has been declared unstable by local and state authorities. Since the houses were built over ten years ago, the owners, under the pre-existing laws, were allowed to remain in their homes on the cliff.

Local geologists indicate that one of the major reasons for the danger of additional mudslides is the fact that the soil is partly composed of clay layers formed by ash from eruptions of the volcanoes in the Cascade Mountains to the east. Another factor is related to new home construction a few miles inland, which has resulted in both deforestation and disruption of the local stream drainage.

The emergency management and other government officials now have to make the tough decisions: Should people be evicted from their homes in nearby areas with similar landscape? Should people be allowed to rebuild in the mudslide area? Should the area be zoned just as park land? Should the area be declared unsafe for any human activity?

22. What landscape and weather conditions resulted in these mudslides at Coastal, Washington? [2]

23. What three landscape features would you expect geologists and government officials to investigate to determine future land use in this part of Coastal, Washington? [3]

24. What feature of the Tectonic Plate Map in the *Earth Science Reference Tables* might affect the use of land in this part of Washington State? [1]

25. Describe an actual experiment geologists could conduct at this landscape site to help determine future mudslide potential. [2]

Appendix 1: Earth Science Reference Tables

The University of the State of New York • THE STATE EDUCATION DEPARTMENT • Albany, New York 12234 • www.nysed.gov

Earth Science Reference Tables

PHYSICAL CONSTANTS

Radioactive Decay Data

RADIOACTIVE ISOTOPE	DISINTEGRATION	HALF-LIFE (years)
Carbon-14	$C^{14} \rightarrow N^{14}$	5.7×10^{3}
Potassium-40	$K^{40} \rightarrow Ar^{40}$, $K^{40} \rightarrow Ca^{40}$	1.3×10^{9}
Uranium-238	$U^{238} \rightarrow Pb^{206}$	4.5×10^{9}
Rubidium-87	$Rb^{87} \rightarrow Sr^{87}$	4.9×10^{10}

Specific Heats of Common Materials

MATERIAL	SPECIFIC HEAT (calories/gram • C°)
Water: solid	0.5
Water: liquid	1.0
Water: gas	0.5
Dry air	0.24
Basalt	0.20
Granite	0.19
Iron	0.11
Copper	0.09
Lead	0.03

Properties of Water

Energy gained during melting	80 calories/gram
Energy released during freezing	80 calories/gram
Energy gained during vaporization	540 calories/gram
Energy released during condensation	540 calories/gram
Density at 3.98°C	1.00 gram/milliliter

EQUATIONS

Percent deviation from accepted value
$$\text{deviation (\%)} = \frac{\text{difference from accepted value}}{\text{accepted value}} \times 100$$

Eccentricity of an ellipse
$$\text{eccentricity} = \frac{\text{distance between foci}}{\text{length of major axis}}$$

Gradient
$$\text{gradient} = \frac{\text{change in field value}}{\text{distance}}$$

Rate of change
$$\text{rate of change} = \frac{\text{change in field value}}{\text{time}}$$

Density of a substance
$$\text{density} = \frac{\text{mass}}{\text{volume}}$$

2001 EDITION

This edition of the Earth Science Reference Tables should be used in the classroom beginning in the 2000–2001 school year. The first examination for which these tables will be used is the January 2001 Regents Examination in Earth Science.

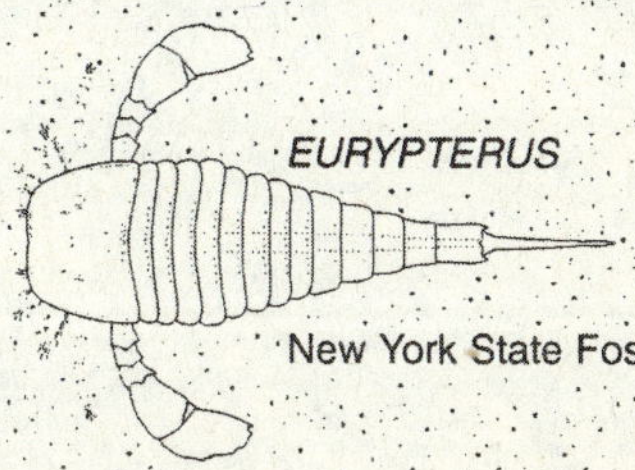

EURYPTERUS

New York State Fossil

Generalized Landscape Regions of New York State

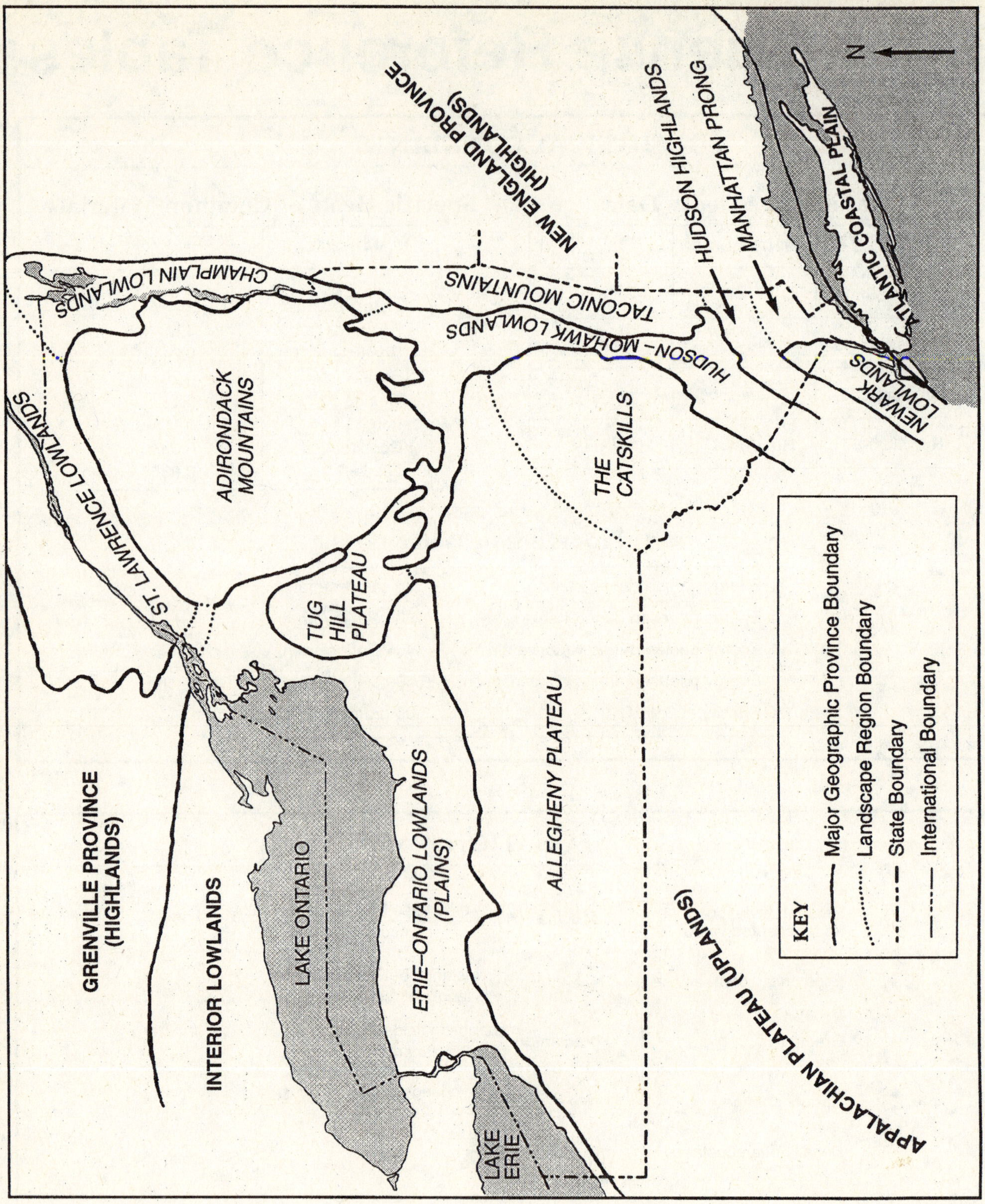

Generalized Bedrock Geology of New York State

modified from
GEOLOGICAL SURVEY
NEW YORK STATE MUSEUM
1989

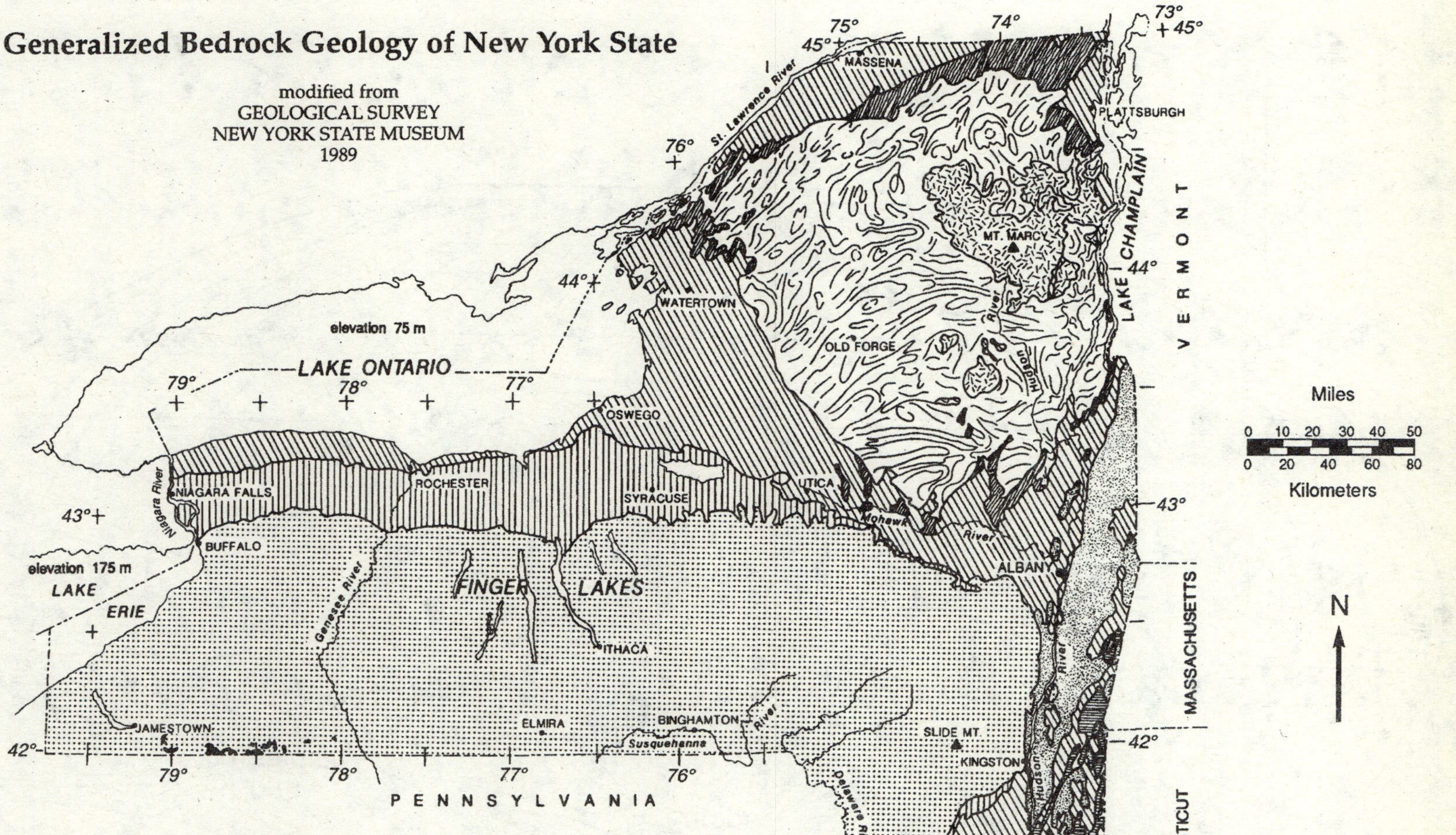

GEOLOGICAL PERIODS AND ERAS IN NEW YORK

Dominantly Sedimentary Origin

- CRETACEOUS, TERTIARY, PLEISTOCENE (Epoch) weakly consolidated to unconsolidated gravels, sands, and clays
- LATE TRIASSIC and EARLY JURASSIC conglomerates, red sandstones, red shales, and diabase (in Palisades Sill)
- PENNSYLVANIAN and MISSISSIPPIAN conglomerates, sandstones, and shales
- DEVONIAN, SILURIAN } limestones, shales, sandstones, and conglomerates *Silurian also contains salt, gypsum, and hematite.*
- ORDOVICIAN, CAMBRIAN } limestones, shales, sandstones, and dolostones

Dominantly Metamorphosed Rocks

- CAMBRIAN and EARLY ORDOVICIAN sandstones and dolostones *Moderately to intensely metamorphosed east of the Hudson River.*
- CAMBRIAN and ORDOVICIAN (undifferentiated) quartzites, dolostones, marbles, and schists *Intensely metamorphosed; includes portions of the Taconic Sequence and Cortlandt Complex.*
- TACONIC SEQUENCE sandstones, shales, and slates *Slightly to intensely metamorphosed rocks of CAMBRIAN through MIDDLE ORDOVICIAN ages.*

Intensely Metamorphosed Rocks (regional metamorphism about 1,000 m.y.a.)

- MIDDLE PROTEROZOIC gneisses, quartzites, and marbles *Lines are generalized structure trends.*
- MIDDLE PROTEROZOIC anorthositic rocks

Surface Ocean Currents

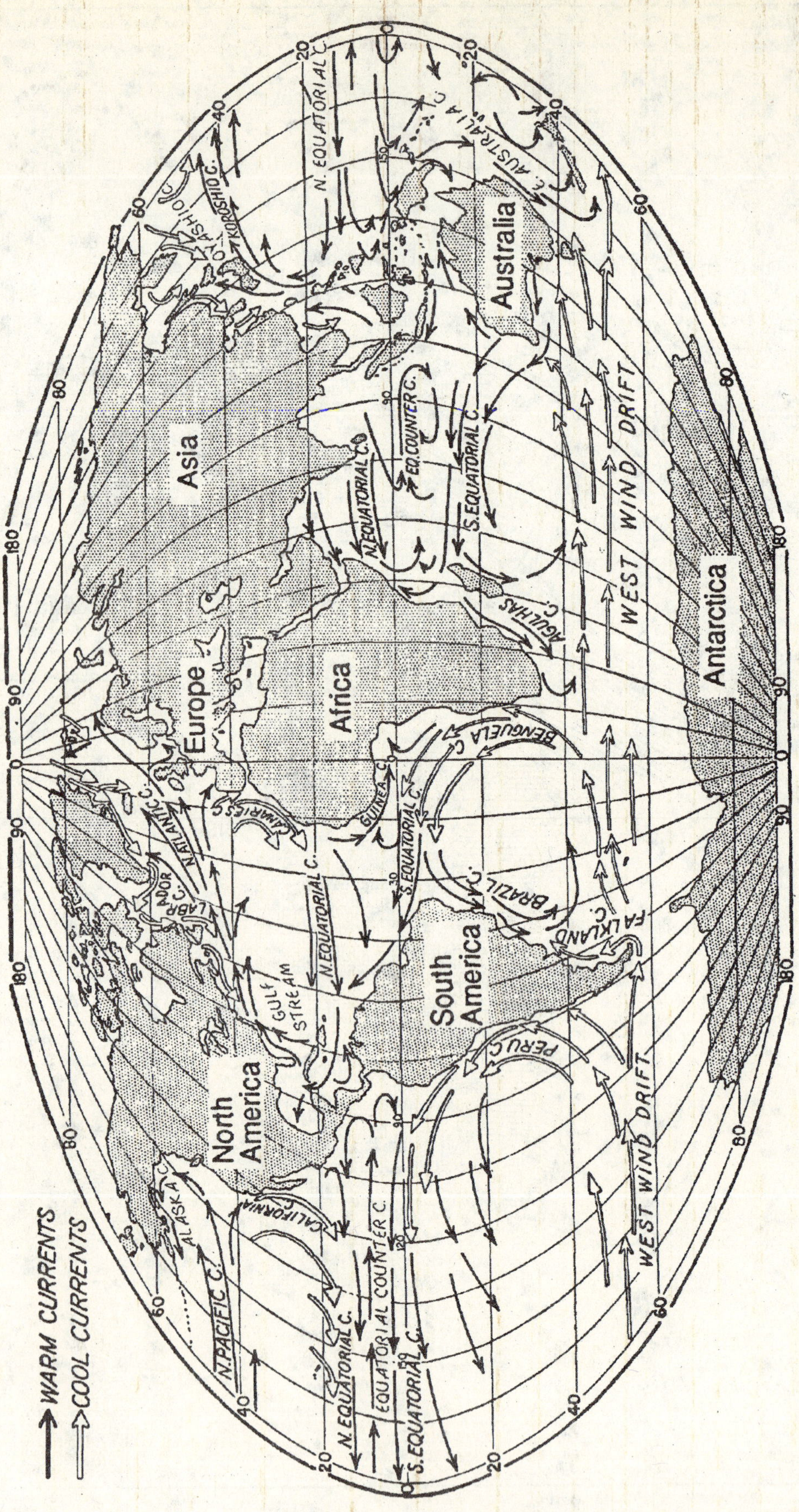

Tectonic Plates

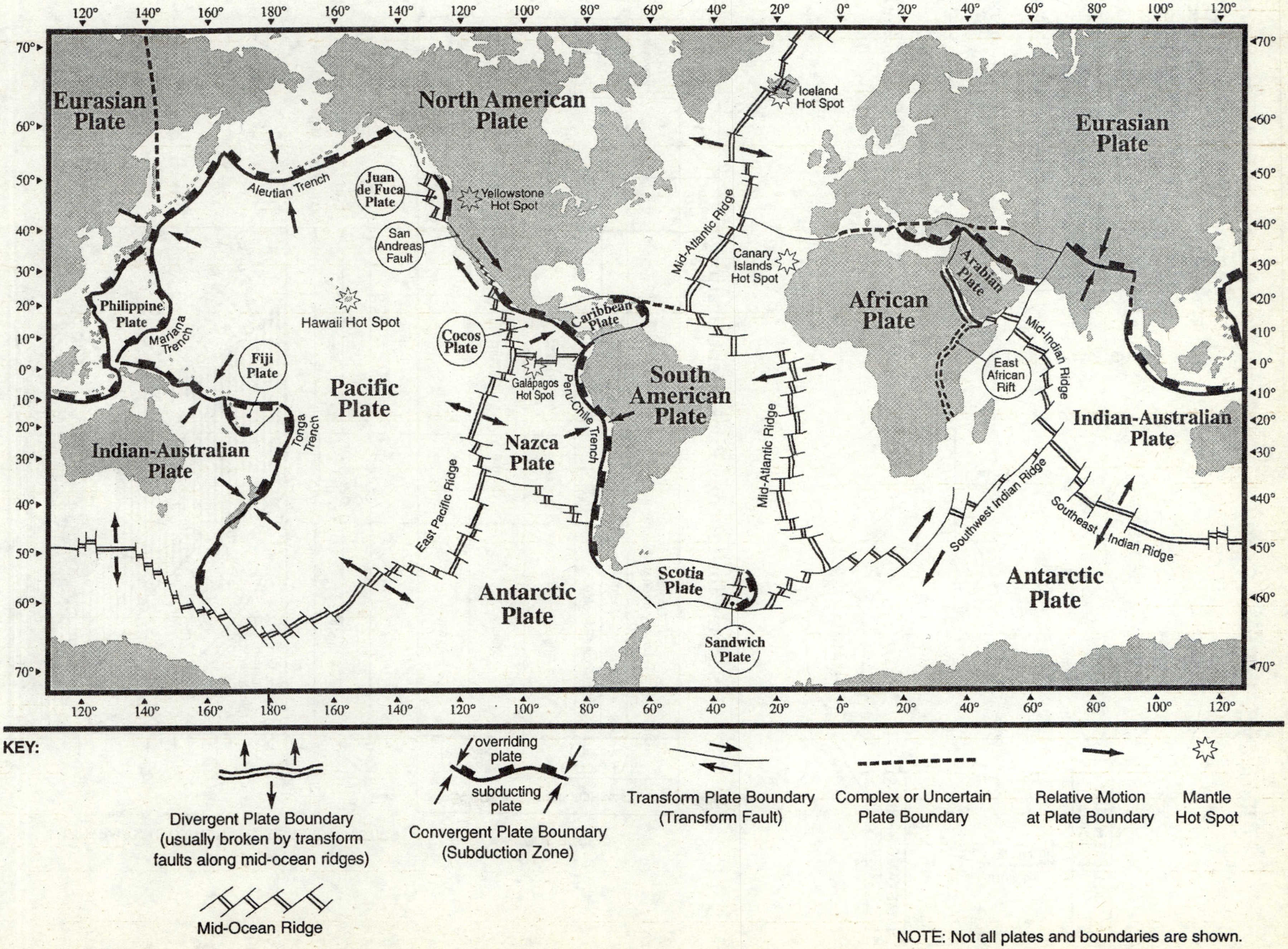

NOTE: Not all plates and boundaries are shown.

Rock Cycle in Earth's Crust

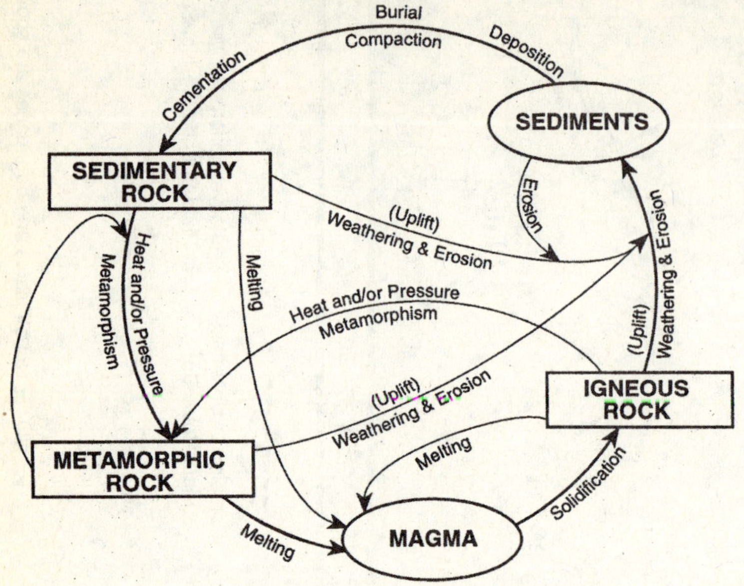

Relationship of Transported Particle Size to Water Velocity

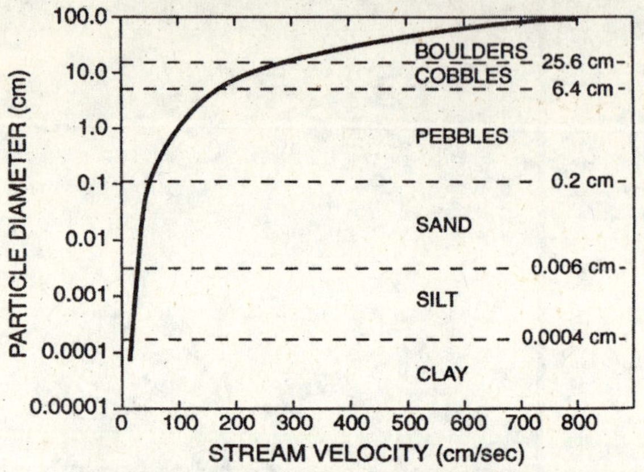

*This generalized graph shows the water velocity needed to maintain, but not start, movement. Variations occur due to differences in particle density and shape.

Scheme for Igneous Rock Identification

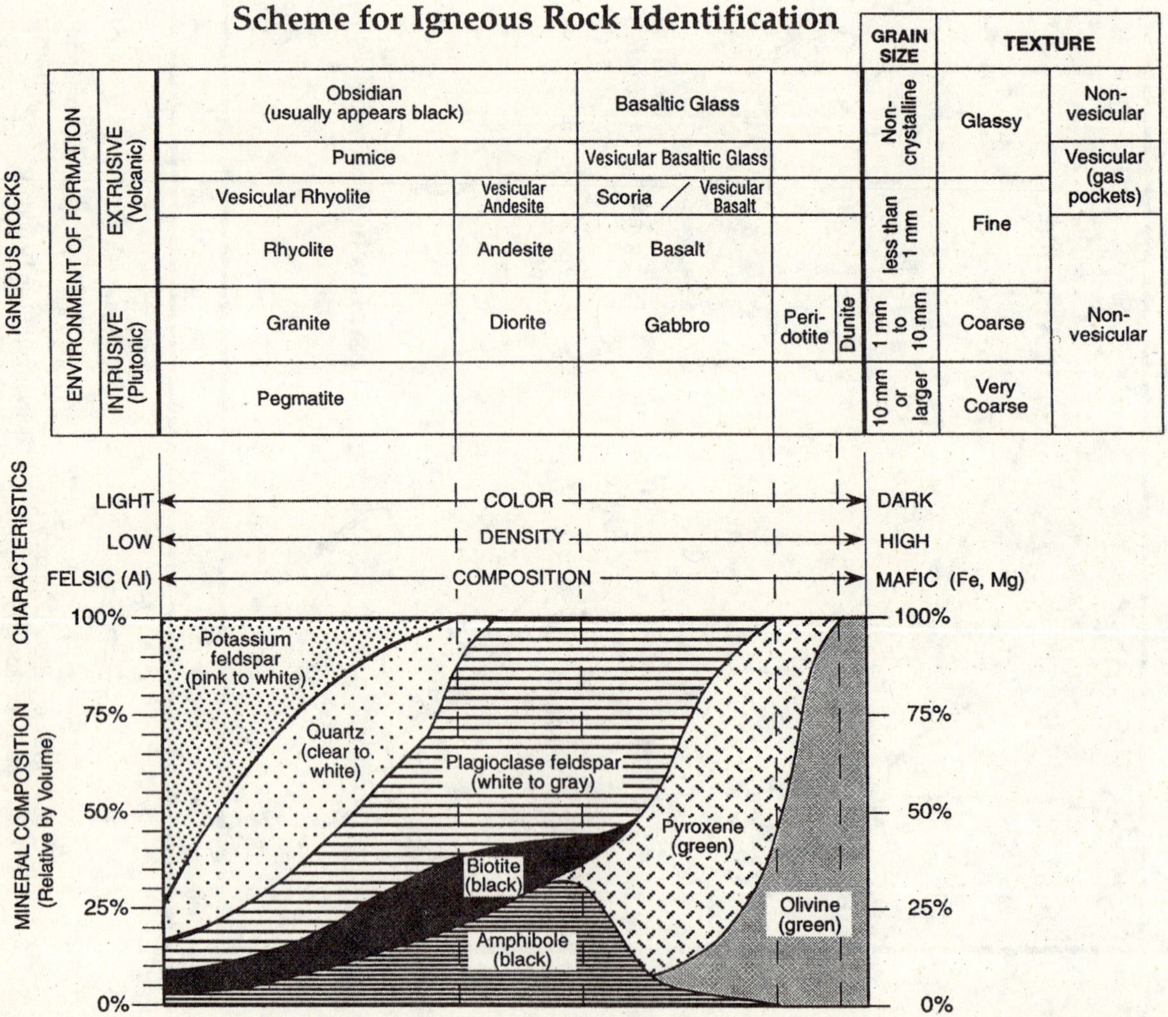

Scheme for Sedimentary Rock Identification

INORGANIC LAND-DERIVED SEDIMENTARY ROCKS					
TEXTURE	**GRAIN SIZE**	**COMPOSITION**	**COMMENTS**	**ROCK NAME**	**MAP SYMBOL**
Clastic (fragmental)	Pebbles, cobbles, and/or boulders embedded in sand, silt, and/or clay	Mostly quartz, feldspar, and clay minerals; may contain fragments of other rocks and minerals	Rounded fragments	Conglomerate	
			Angular fragments	Breccia	
	Sand (0.2 to 0.006 cm)		Fine to coarse	Sandstone	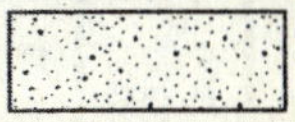
	Silt (0.006 to 0.0004 cm)		Very fine grain	Siltstone	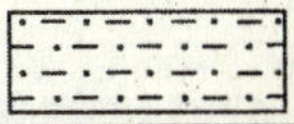
	Clay (less than 0.0004 cm)		Compact; may split easily	Shale	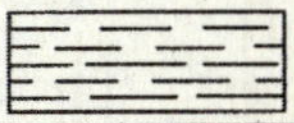

CHEMICALLY AND/OR ORGANICALLY FORMED SEDIMENTARY ROCKS					
TEXTURE	**GRAIN SIZE**	**COMPOSITION**	**COMMENTS**	**ROCK NAME**	**MAP SYMBOL**
Crystalline	Varied	Halite	Crystals from chemical precipitates and evaporites	Rock Salt	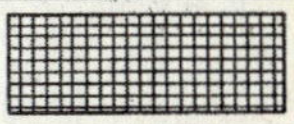
	Varied	Gypsum		Rock Gypsum	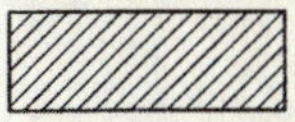
	Varied	Dolomite		Dolostone	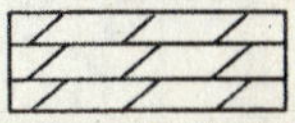
Bioclastic	Microscopic to coarse	Calcite	Cemented shell fragments or precipitates of biologic origin	Limestone	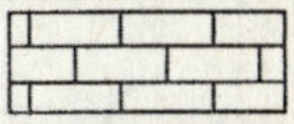
	Varied	Carbon	From plant remains	Coal	

Scheme for Metamorphic Rock Identification

TEXTURE		GRAIN SIZE	COMPOSITION	TYPE OF METAMORPHISM	COMMENTS	ROCK NAME	MAP SYMBOL
FOLIATED	MINERAL ALIGNMENT	Fine	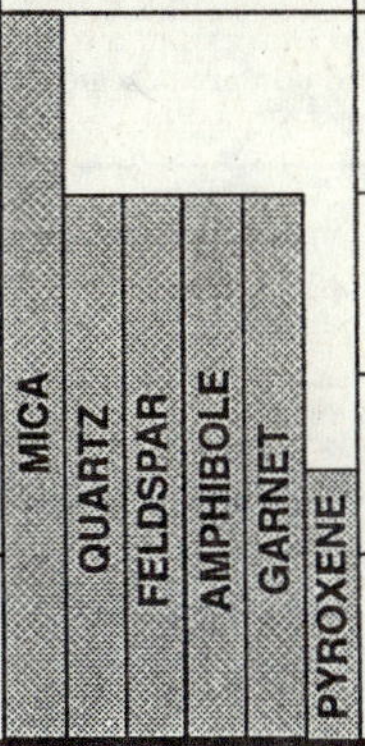	Regional	Low-grade metamorphism of shale	Slate	
		Fine to medium		(Heat and pressure increase with depth)	Foliation surfaces shiny from microscopic mica crystals	Phyllite	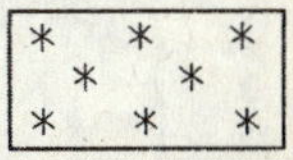
					Platy mica crystals visible from metamorphism of clay or feldspars	Schist	
	BAND-ING	Medium to coarse			High-grade metamorphism; some mica changed to feldspar; segregated by mineral type into bands	Gneiss	
NONFOLIATED		Fine	Variable	Contact (Heat)	Various rocks changed by heat from nearby magma/lava	Hornfels	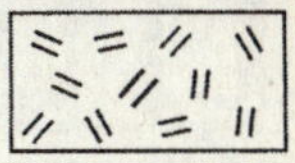
		Fine to coarse	Quartz	Regional or Contact	Metamorphism of quartz sandstone	Quartzite	
			Calcite and/or dolomite		Metamorphism of limestone or dolostone	Marble	
		Coarse	Various minerals in particles and matrix		Pebbles may be distorted or stretched	Metaconglomerate	

GEOLOGIC HISTORY

(Fossils not drawn to scale)

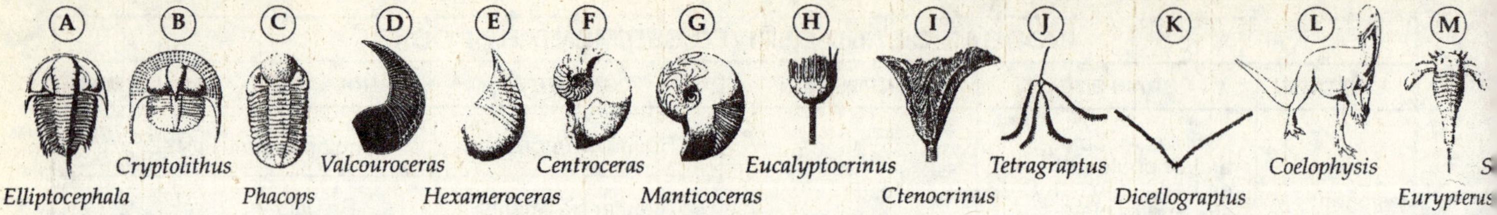

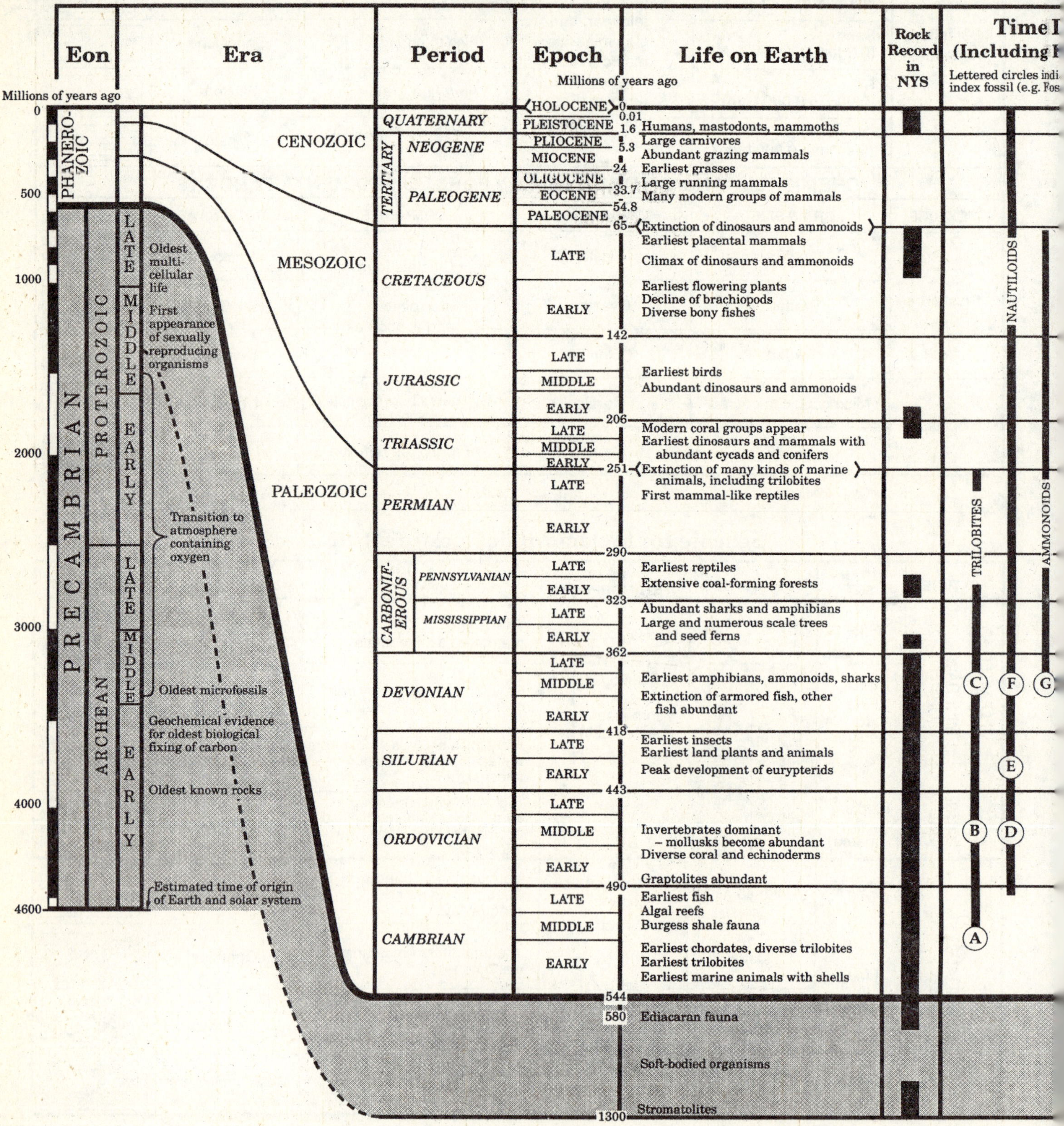

OF NEW YORK STATE

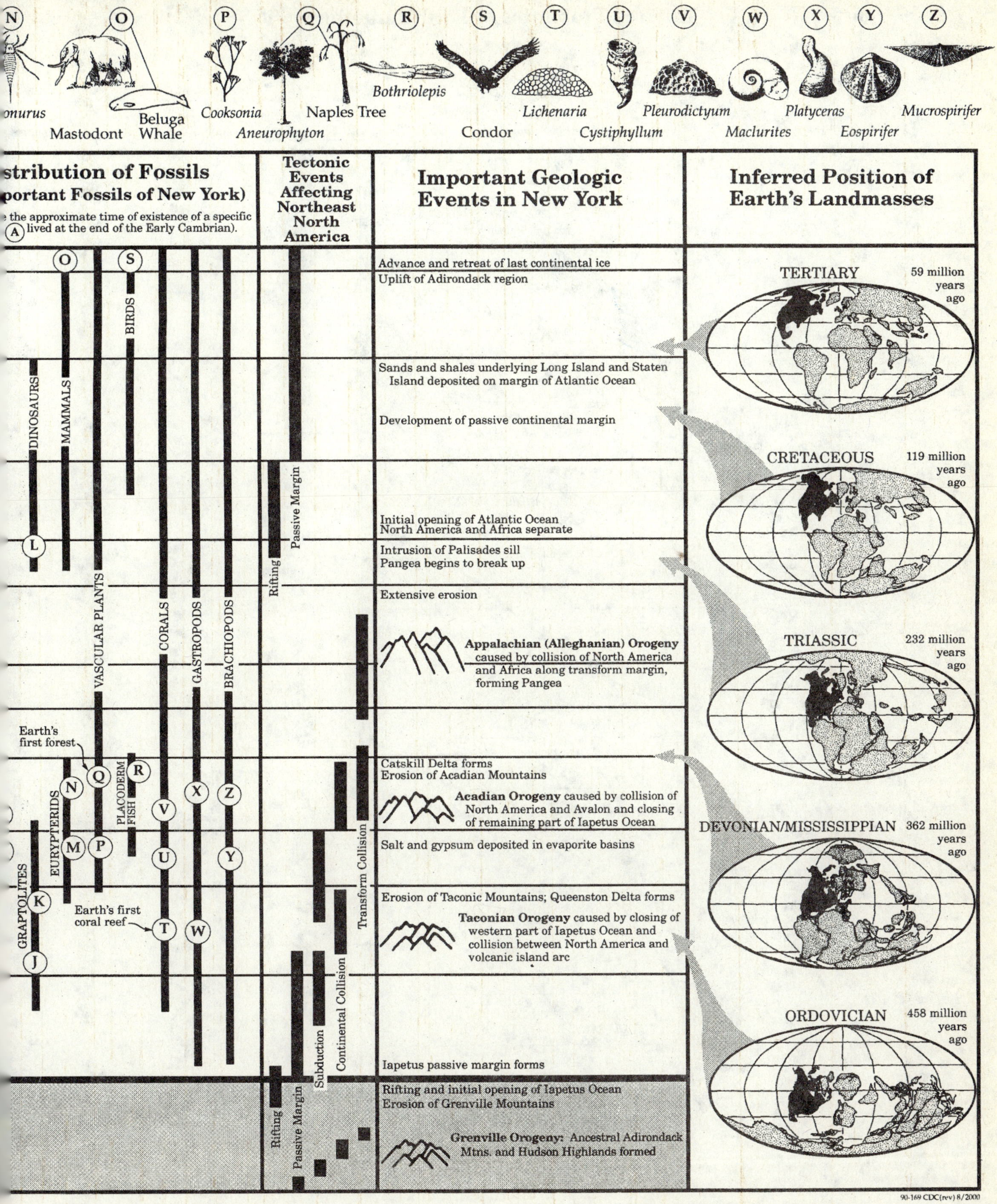

Inferred Properties of Earth's Interior

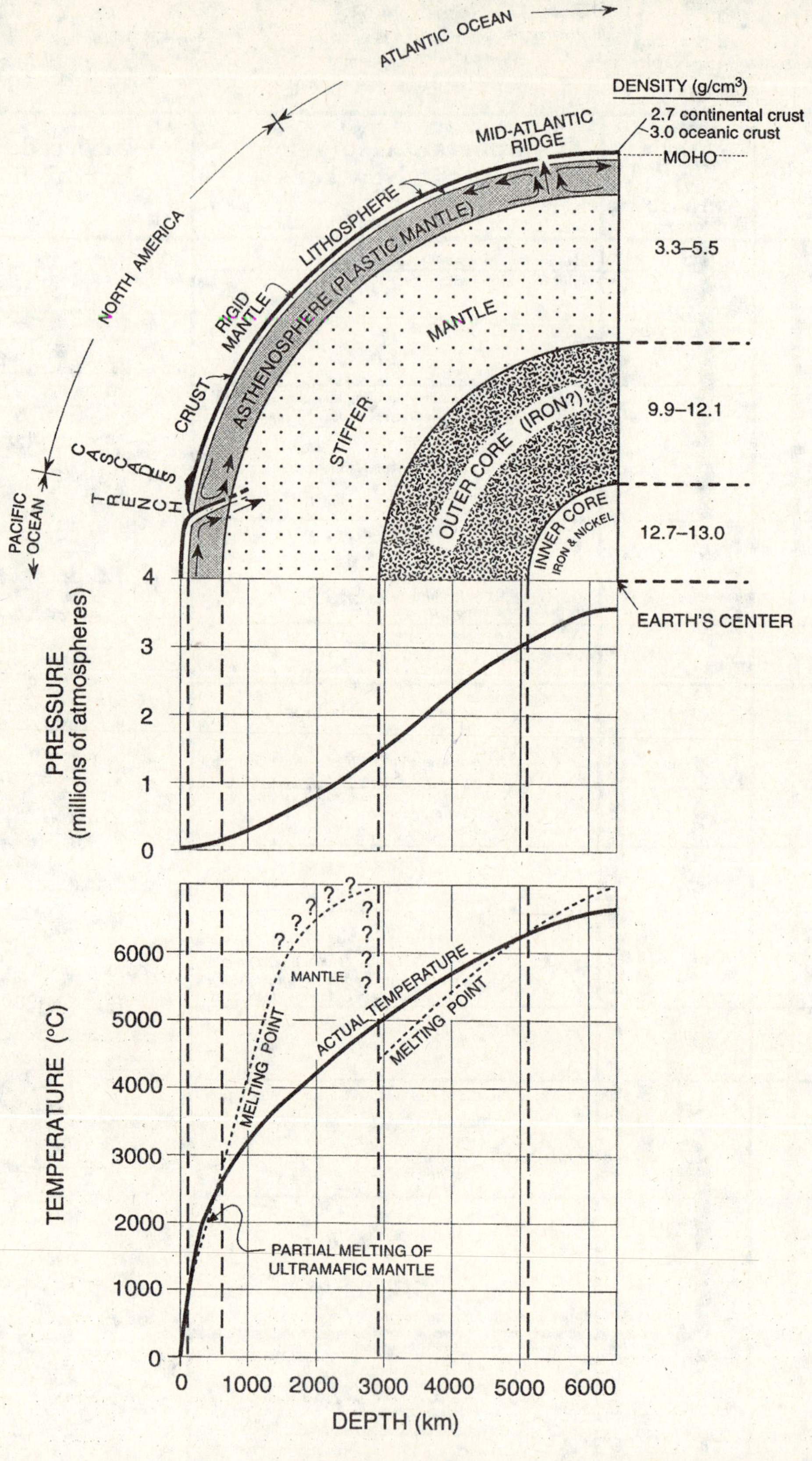

Average Chemical Composition of Earth's Crust, Hydrosphere, and Troposphere

ELEMENT (symbol)	CRUST		HYDROSPHERE	TROPOSPHERE
	Percent by Mass	Percent by Volume	Percent by Volume	Percent by Volume
Oxygen (O)	46.40	94.04	33.0	21.0
Silicon (Si)	28.15	0.88		
Aluminum (Al)	8.23	0.48		
Iron (Fe)	5.63	0.49		
Calcium (Ca)	4.15	1.18		
Sodium (Na)	2.36	1.11		
Magnesium (Mg)	2.33	0.33		
Potassium (K)	2.09	1.42		
Nitrogen (N)				78.0
Hydrogen (H)			66.0	
Other	0.66	0.07	1.0	1.0

Earthquake P-wave and S-wave Travel Time

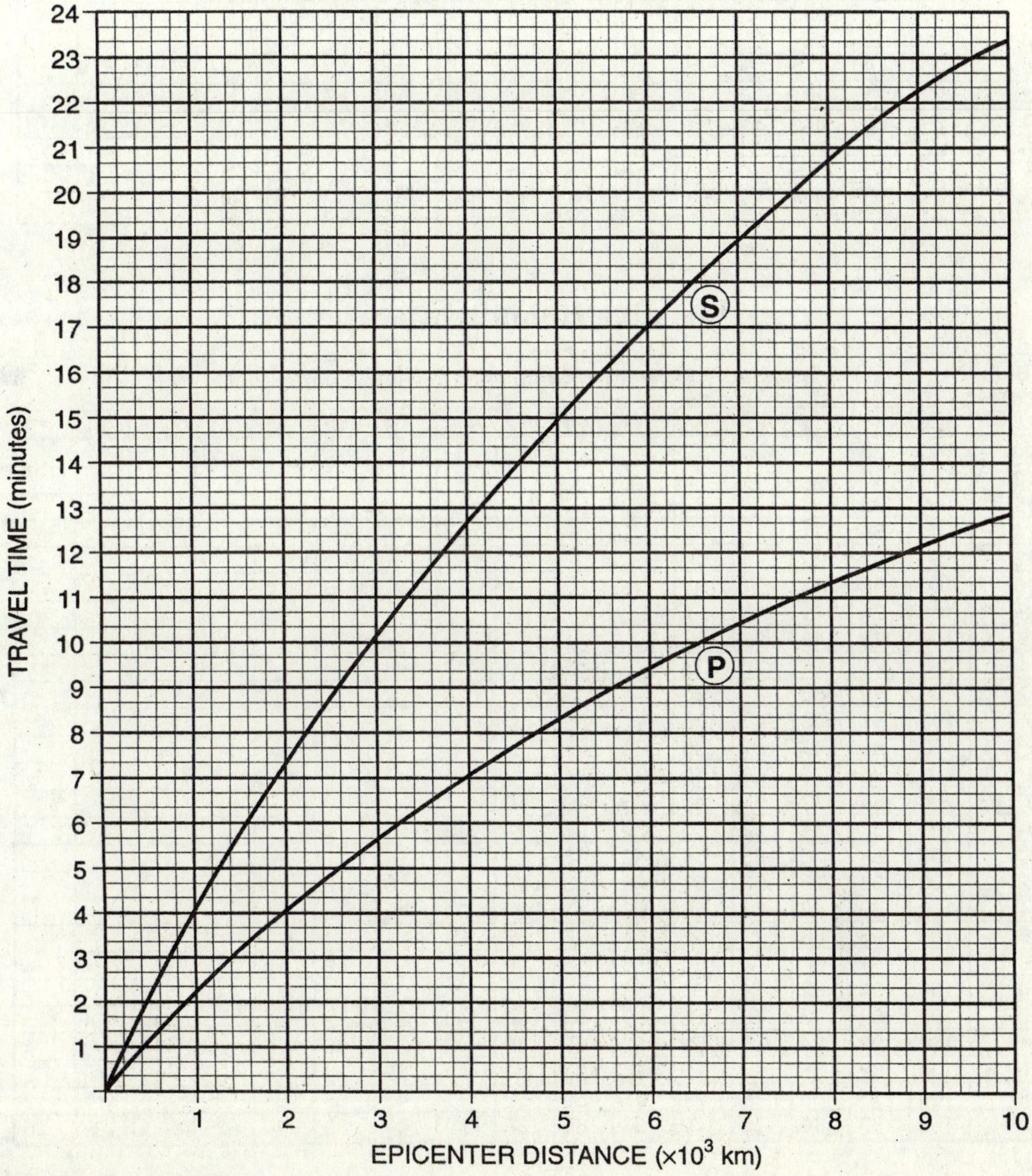

Dewpoint Temperatures (°C)

Dry-Bulb Temperature (°C)	Difference Between Wet-Bulb and Dry-Bulb Temperatures (C°)															
	0	1	2	3	4	5	6	7	8	9	10	11	12	13	14	15
−20	−20	−33														
−18	−18	−28														
−16	−16	−24														
−14	−14	−21	−36													
−12	−12	−18	−28													
−10	−10	−14	−22													
−8	−8	−12	−18	−29												
−6	−6	−10	−14	−22												
−4	−4	−7	−12	−17	−29											
−2	−2	−5	−8	−13	−20											
0	0	−3	−6	−9	−15	−24										
2	2	−1	−3	−6	−11	−17										
4	4	1	−1	−4	−7	−11	−19									
6	6	4	1	−1	−4	−7	−13	−21								
8	8	6	3	1	−2	−5	−9	−14								
10	10	8	6	4	1	−2	−5	−9	−14	−28						
12	12	10	8	6	4	1	−2	−5	−9	−16						
14	14	12	11	9	6	4	1	−2	−5	−10	−17					
16	16	14	13	11	9	7	4	1	−1	−6	−10	−17				
18	18	16	15	13	11	9	7	4	2	−2	−5	−10	−19			
20	20	19	17	15	14	12	10	7	4	2	−2	−5	−10	−19		
22	22	21	19	17	16	14	12	10	8	5	3	−1	−5	−10	−19	
24	24	23	21	20	18	16	14	12	10	8	6	2	−1	−5	−10	−18
26	26	25	23	22	20	18	17	15	13	11	9	6	3	0	−4	−9
28	28	27	25	24	22	21	19	17	16	14	11	9	7	4	1	−3
30	30	29	27	26	24	23	21	19	18	16	14	12	10	8	5	1

Relative Humidity (%)

Dry-Bulb Temperature (°C)	Difference Between Wet-Bulb and Dry-Bulb Temperatures (C°)															
	0	1	2	3	4	5	6	7	8	9	10	11	12	13	14	15
−20	100	28														
−18	100	40														
−16	100	48														
−14	100	55	11													
−12	100	61	23													
−10	100	66	33													
−8	100	71	41	13												
−6	100	73	48	20												
−4	100	77	54	32	11											
−2	100	79	58	37	20	1										
0	100	81	63	45	28	11										
2	100	83	67	51	36	20	6									
4	100	85	70	56	42	27	14									
6	100	86	72	59	46	35	22	10								
8	100	87	74	62	51	39	28	17	6							
10	100	88	76	65	54	43	33	24	13	4						
12	100	88	78	67	57	48	38	28	19	10	2					
14	100	89	79	69	60	50	41	33	25	16	8	1				
16	100	90	80	71	62	54	45	37	29	21	14	7	1			
18	100	91	81	72	64	56	48	40	33	26	19	12	6			
20	100	91	82	74	66	58	51	44	36	30	23	17	11	5		
22	100	92	83	75	68	60	53	46	40	33	27	21	15	10	4	
24	100	92	84	76	69	62	55	49	42	36	30	25	20	14	9	4
26	100	92	85	77	70	64	57	51	45	39	34	28	23	18	13	9
28	100	93	86	78	71	65	59	53	47	42	36	31	26	21	17	12
30	100	93	86	79	72	66	61	55	49	44	39	34	29	25	20	16

Temperature

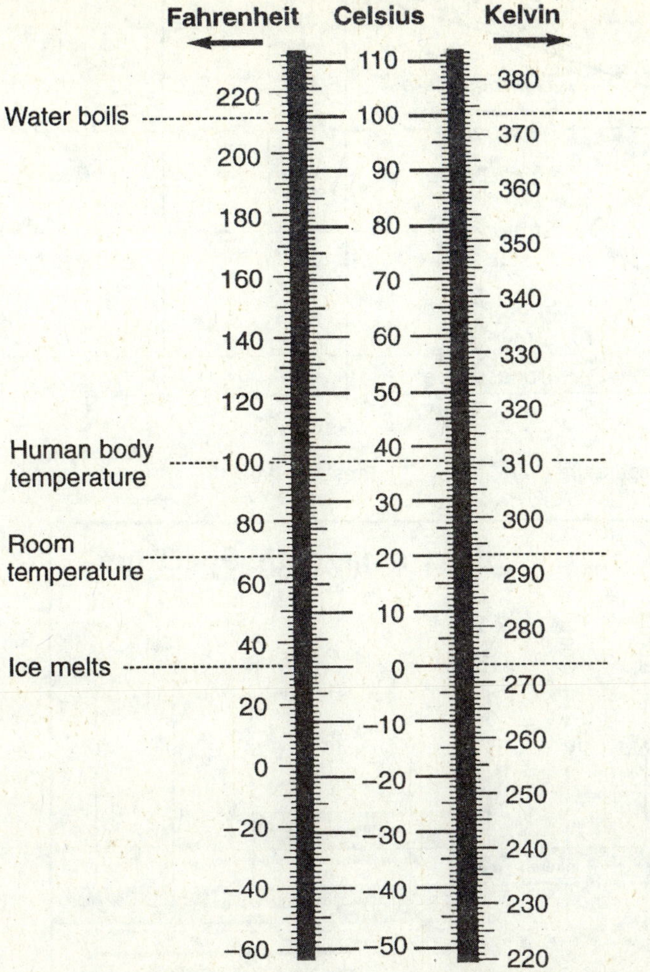

Pressure

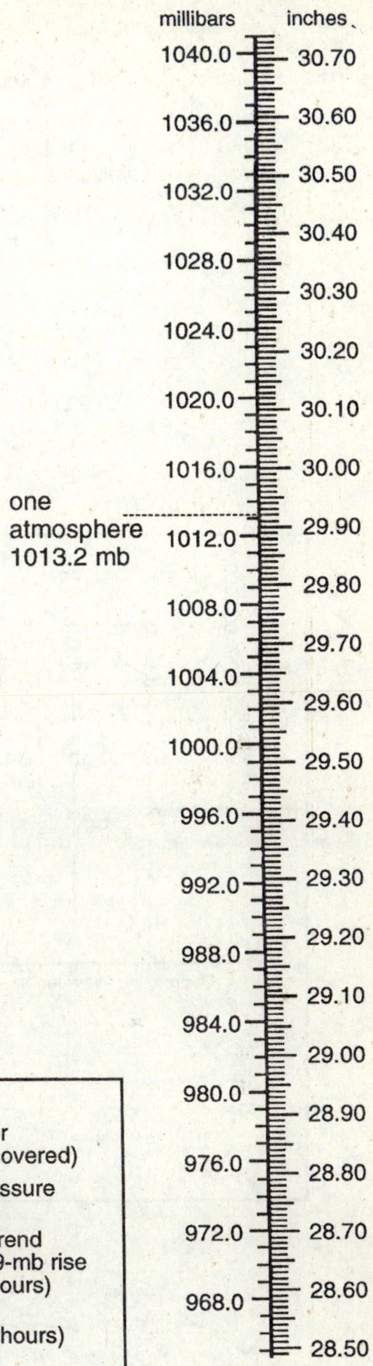

Weather Map Symbols

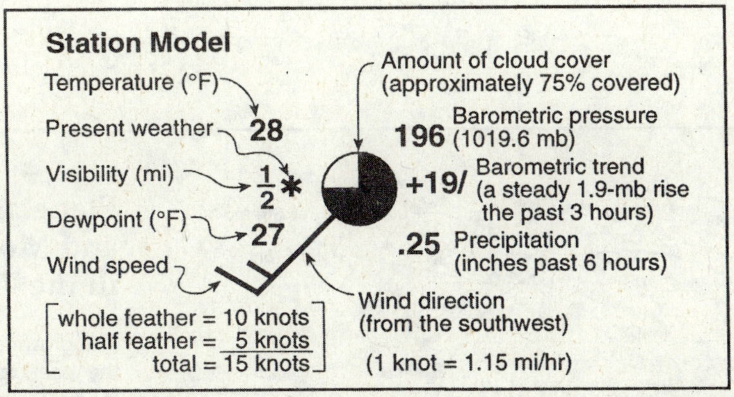

Present Weather

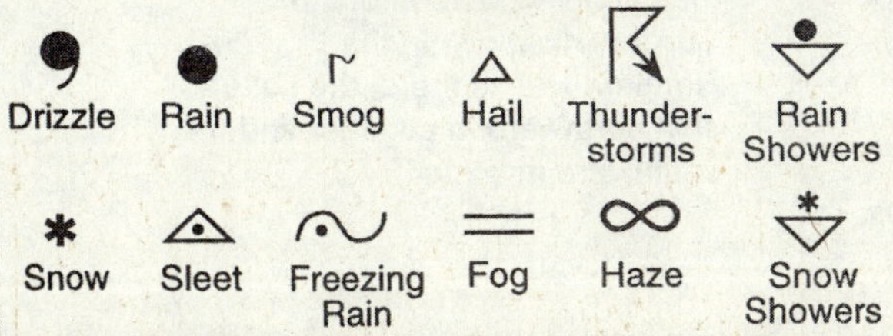

Air Masses

- cA continental arctic
- cP continental polar
- cT continental tropical
- mT maritime tropical
- mP maritime polar

Front Symbols

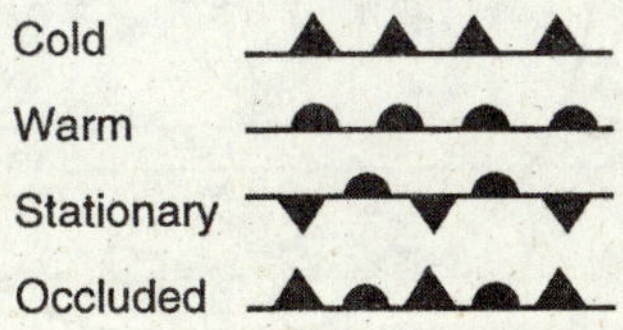

Hurricane

Selected Properties of Earth's Atmosphere

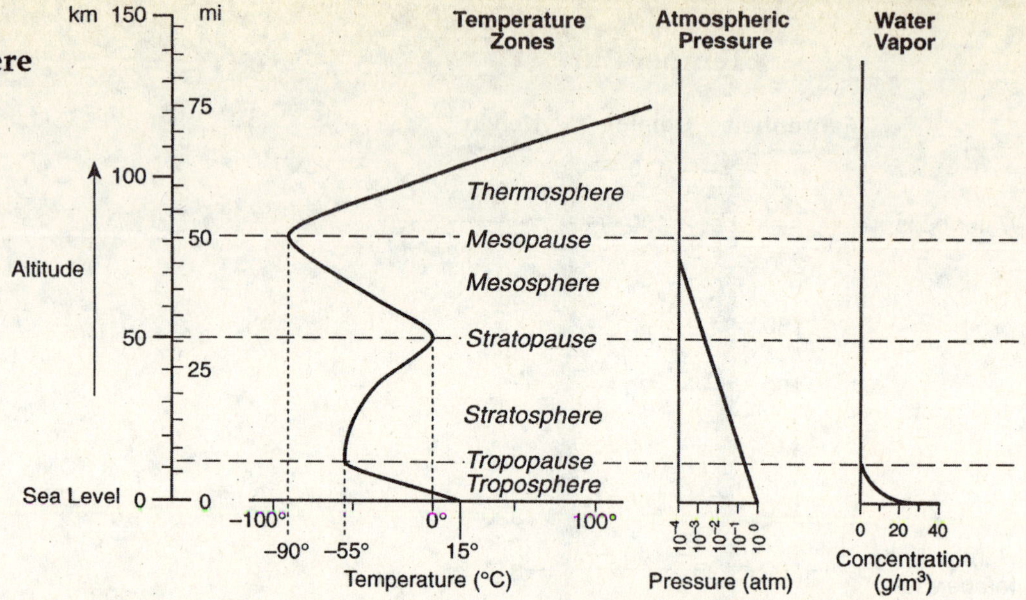

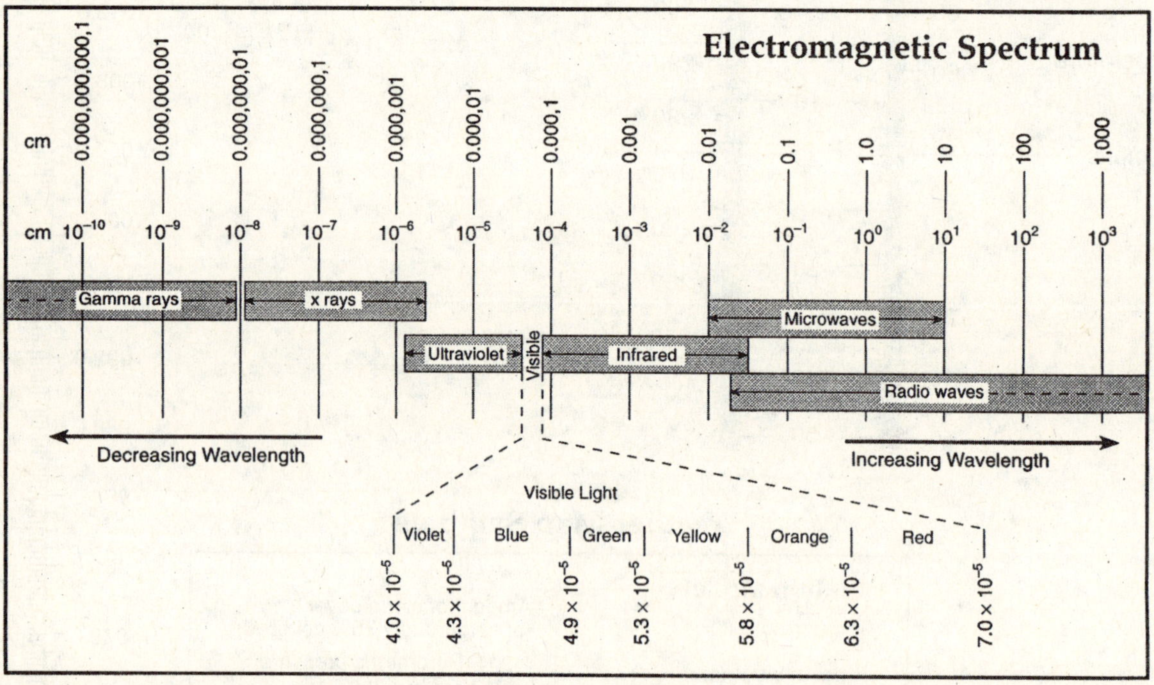

Planetary Wind and Moisture Belts in the Troposphere

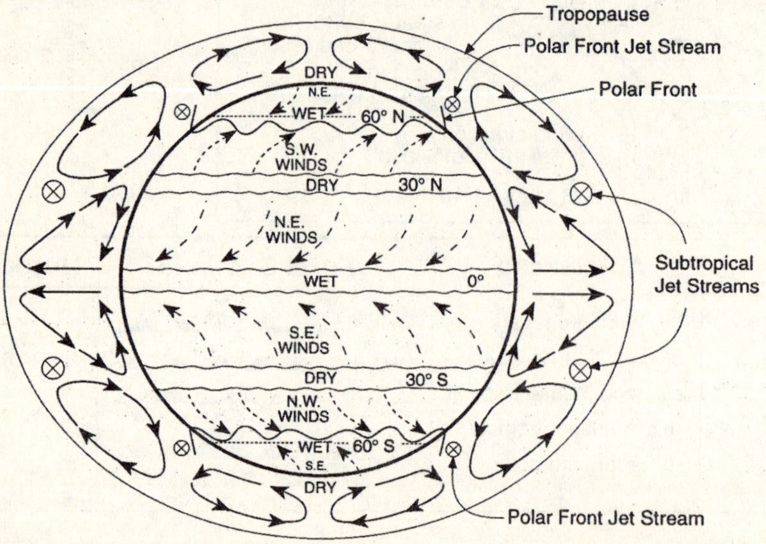

The drawing to the left shows the locations of the belts near the time of an equinox. The locations shift somewhat with the changing latitude of the Sun's vertical ray. In the Northern Hemisphere, the belts shift northward in summer and southward in winter.

Luminosity and Temperature of Stars

(Name in italics refers to star shown by a ⊕)

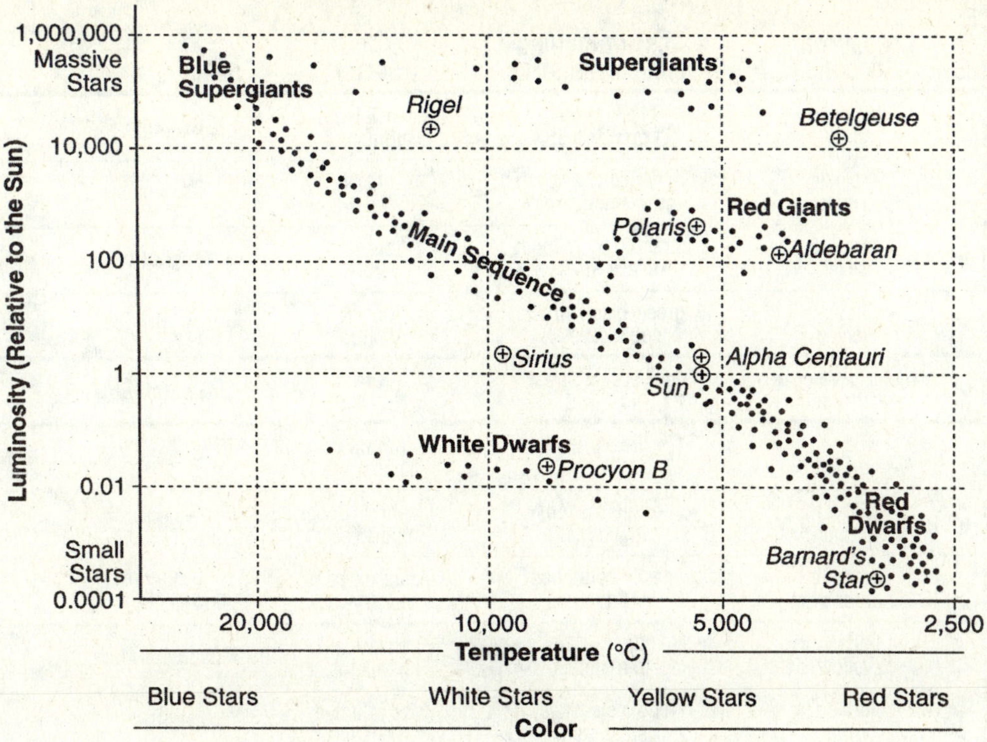

Luminosity is the brightness of stars compared to the brightness of our Sun as seen from the same distance from the observer.

Solar System Data

Object	Mean Distance from Sun (millions of km)	Period of Revolution	Period of Rotation	Eccentricity of Orbit	Equatorial Diameter (km)	Mass (Earth = 1)	Density (g/cm^3)	Number of Moons
SUN	—	—	27 days	—	1,392,000	333,000.00	1.4	–
MERCURY	57.9	88 days	59 days	0.206	4,880	0.553	5.4	0
VENUS	108.2	224.7 days	243 days	0.007	12,104	0.815	5.2	0
EARTH	149.6	365.26 days	23 hr 56 min 4 sec	0.017	12,756	1.00	5.5	1
MARS	227.9	687 days	24 hr 37 min 23 sec	0.093	6,787	0.1074	3.9	2
JUPITER	778.3	11.86 years	9 hr 50 min 30 sec	0.048	142,800	317.896	1.3	16
SATURN	1,427	29.46 years	10 hr 14 min	0.056	120,000	95.185	0.7	18
URANUS	2,869	84.0 years	17 hr 14 min	0.047	51,800	14.537	1.2	21
NEPTUNE	4,496	164.8 years	16 hr	0.009	49,500	17.151	1.7	8
PLUTO	5,900	247.7 years	6 days 9 hr	0.250	2,300	0.0025	2.0	1
EARTH'S MOON	149.6 (0.386 from Earth)	27.3 days	27 days 8 hr	0.055	3,476	0.0123	3.3	—

Properties of Common Minerals

LUSTER	HARD-NESS	CLEAVAGE	FRACTURE	COMMON COLORS	DISTINGUISHING CHARACTERISTICS	USE(S)	MINERAL NAME	COMPOSITION*
Metallic Luster	1–2	✔		silver to gray	black streak, greasy feel	pencil lead, lubricants	**Graphite**	C
Metallic Luster	2.5	✔		metallic silver	very dense (7.6 g/cm^3), gray-black streak 	ore of lead	**Galena**	PbS
Metallic Luster	5.5–6.5		✔	black to silver	attracted by magnet, black streak	ore of iron	**Magnetite**	Fe_3O_4
Metallic Luster	6.5		✔	brassy yellow	green-black streak, cubic crystals 	ore of sulfur	**Pyrite**	FeS_2
Either	1–6.5		✔	metallic silver or earthy red	red-brown streak	ore of iron	**Hematite**	Fe_2O_3
Nonmetallic Luster	1	✔		white to green	greasy feel	talcum powder, soapstone	**Talc**	$Mg_3Si_4O_{10}(OH)_2$
Nonmetallic Luster	2		✔	yellow to amber	easily melted, may smell	vulcanize rubber, sulfuric acid	**Sulfur**	S
Nonmetallic Luster	2	✔		white to pink or gray	easily scratched by fingernail	plaster of paris and drywall	**Gypsum** (Selenite)	$CaSO_4{\bullet}2H_2O$
Nonmetallic Luster	2–2.5	✔		colorless to yellow	flexible in thin sheets 	electrical insulator	**Muscovite Mica**	$KAl_3Si_3O_{10}(OH)_2$
Nonmetallic Luster	2.5	✔		colorless to white	cubic cleavage, salty taste 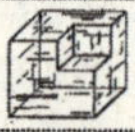	food additive, melts ice	**Halite**	NaCl
Nonmetallic Luster	2.5–3	✔		black to dark brown	flexible in thin sheets 	electrical insulator	**Biotite Mica**	$K(Mg,Fe)_3$ $AlSi_3O_{10}(OH)_2$
Nonmetallic Luster	3	✔		colorless or variable	bubbles with acid 	cement, polarizing prisms	**Calcite**	$CaCO_3$
Nonmetallic Luster	3.5	✔		colorless or variable	bubbles with acid when powdered	source of magnesium	**Dolomite**	$CaMg(CO_3)_2$
Nonmetallic Luster	4	✔		colorless or variable	cleaves in 4 directions	hydrofluoric acid	**Fluorite**	CaF_2
Nonmetallic Luster	5–6	✔		black to dark green	cleaves in 2 directions at 90° 	mineral collections	**Pyroxene** (commonly Augite)	$(Ca,Na)(Mg,Fe,Al)$ $(Si,Al)_2O_6$
Nonmetallic Luster	5.5	✔		black to dark green	cleaves at 56° and 124° 	mineral collections	**Amphiboles** (commonly Hornblende)	$CaNa(Mg,Fe)_4(Al,Fe,Ti)_3$ $Si_6O_{22}(O,OH)_2$
Nonmetallic Luster	6	✔		white to pink	cleaves in 2 directions at 90°	ceramics and glass	**Potassium Feldspar** (Orthoclase)	$KAlSi_3O_8$
Nonmetallic Luster	6	✔		white to gray	cleaves in 2 directions, striations visible	ceramics and glass	**Plagioclase Feldspar** (Na-Ca Feldspar)	$(Na,Ca)AlSi_3O_8$
Nonmetallic Luster	6.5		✔	green to gray or brown	commonly light green and granular	furnace bricks and jewelry	**Olivine**	$(Fe,Mg)_2SiO_4$
Nonmetallic Luster	7		✔	colorless or variable	glassy luster, may form hexagonal crystals 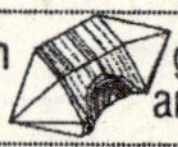	glass, jewelry, and electronics	**Quartz**	SiO_2
Nonmetallic Luster	7		✔	dark red to green	glassy luster, often seen as red grains in NYS metamorphic rocks	jewelry and abrasives	**Garnet** (commonly Almandine)	$Fe_3Al_2Si_3O_{12}$

*Chemical Symbols: Al = aluminum, C = carbon, Ca = calcium, Cl = chlorine, F = fluorine, Fe = iron, H = hydrogen, K = potassium, Mg = magnesium, Na = sodium, O = oxygen, Pb = lead, S = sulfur, Si = silicon, Ti = titanium

✔ = dominant form of breakage

Appendix 2: Strategies for Answering Test Questions

This appendix provides strategies to help you answer various types of questions on the Regents Examination for Physical Setting: Earth Science. Strategies are provided for answering multiple-choice and constructed-response questions as well as for questions based on diagrams, data tables, and graphs and questions that use the *Earth Science Reference Tables*.

Strategies for Multiple-Choice Questions

Multiple-choice questions will likely account for more than 50 percent of the Regents Examination for Physical Setting: Earth Science. Part A is comprised totally of multiple-choice questions, and Part B includes some. Therefore, it is important to be good at deciphering multiple-choice questions. Here are a few helpful strategies. For any one question, not all strategies will need to be used. The numbers are provided for reference, not to specify an order (except for Strategies 1 and 2).

1. Always read the entire question, but wait to read the choices. (See Strategy 4).
2. Carefully examine any data tables, diagrams, photographs or relevant part(s) of the *Earth Science Reference Tables* associated with the question.
3. Underline key words and phrases in the question that signal what you should be looking for in the answer. This will make you read the question more carefully. This strategy applies mostly to questions with a long introduction.
4. Try to think of an answer to the question before looking at the choices given. If you think you know the answer, write it on a separate piece of paper before reading the choices. Next, read all of the choices and compare them to your answer before making a decision. Do not select the first answer that seems correct. If your answer matches one of the choices, and you are quite sure of your response, you are probably correct. Even if your answer matches one of the choices, carefully consider all of the answers because the obvious choice is not always the correct one. If there are no exact matches, re-read the question and look for the choice that is most similar to your answer.
5. Eliminate any choices that you know are incorrect. Lightly cross out the numbers for those choices on the exam paper. Each choice you can eliminate increases your chances of selecting the correct answer.
6. If the question makes no sense after reading through it several times, leave it for later. After completing the rest of the exam, return to the question. Something you read on the other parts of the exam may give you some ideas about how to answer this question. If you are still unsure, go with your best guess. There is no penalty for guessing, but answers left blank will be counted as wrong. If you employ your best test-taking strategies, you just may select the correct answer.

Strategies for Constructed-Response Questions

Some questions in Part B and all questions in Part C of the Regents Examination for Physical Setting: Earth Science require a constructed response. No matter which type of answer is requested, the following strategies will help you write constructed responses.

1. Always read through the entire question.
2. Underline key words and phrases in the question that signal what you should be looking for in the answer. This will make you read the question more carefully.
3. Look over the *Earth Science Reference Tables* for any helpful information related to the question.

Using Key Words to Direct Your Answers

Key Word	What Direction Your Answers Should Take
Analyze	• Break the idea, concept, or situation into parts, and explain how they relate. • Carefully explain relationships, such as cause and effect.
Discuss	• Make observations about the topic or situation using facts. • Thoroughly write about various aspects of the topic or situation.
Describe	• Illustrate the subject using words. • Provide a thorough account of the topic. • Give complete answers.
Explain	• Clarify the topic of the question by spelling it out completely. • Make the topic understandable. • Provide reasons for the outcome.
Define	• State the exact meaning of topic or word. • Explain what something is or what it means.
Compare	• Relate two or more topics with an emphasis on how they are alike. • State the similarities between two or more examples.
Contrast	• Relate two or more topics with an emphasis on how they are different. • State the differences between two or more examples.
Design	• Plan an experiment or component of an experiment. Map out your proposal, being sure to provide information about all of the required parts.
State	• Express in words. • Explain or describe using at least one fact, term, or relationship.

4. Write a brief outline, or at least a few notes to yourself, about what should be included in the answer.

5. Pay attention to key words that indicate how to answer the question and what you need to say in your answer. Several of these words are very common. For example, you might be asked to discuss, describe, explain, define, compare, contrast, or design. The table above lists key words and directions for your answers.

6. When you write your answer, don't be so general that you are not really saying anything. Be very specific. You should use the correct terms and clearly explain the processes and relationships. Be sure to provide details, such as the names of processes, names of structures, and, if it is appropriate, how they are related. If only one example or term is required, do not give two or more. If one is correct and the other is wrong, your answer may be marked wrong.

7. If a question has two or three parts, answer each part separately. This will make it easy for the person scoring your paper to find all of the information. When writing your answer, don't shortchange one part of the question by spending too much time on another part.

8. Note that you will not lose points for incorrect grammar, spelling, punctuation, or poor penmanship. However, such errors and poor penmanship could impair your ability to make your answer clear to the person scoring your paper. If that person cannot understand what you are trying to say, you will not receive the maximum number of points.

Strategies for Questions Based on Diagrams

Both multiple-choice and extended-response questions frequently include diagrams or pictures. Usually the diagrams provide information needed to answer the question. The diagrams may be realistic, or they may be schematic. Schematic drawings show the relationships among parts and sometimes the sequence in a system. Follow these steps:

1. First study the diagram and think about what the diagram shows you. Be sure to read any information, such as titles or labels, that go with the diagram.
2. Read the question. Follow the strategies for either multiple-choice or constructed-response questions listed previously.

Strategies for Questions Based on Data Tables

Most data tables contain information that summarizes a topic. A table uses rows and columns to condense information and to present it in an organized way. Rows are the horizontal divisions going from left to right across the table, while columns are vertical divisions going from top to bottom. Column headings name the type of information included in a table. Sometimes different categories of information are listed down the left-hand column of the table. See the sample data table, and identify the kind of information in the columns and rows. When answering a question with a data table, use the following strategies.

1. Find the title of the table. It is usually located across the top.
2. Determine the number of columns in the table and their purpose.
3. Determine the number of rows and their purpose.
4. Read across the rows and down the columns to determine what the relationships are.
5. Now you are ready to read the question with the data table. Answer the question by using the suggested strategies for multiple-choice or constructed-response questions listed previously.

Table 1 Planet Data

Planet	Average Distance From the Sun (AU)	Average Orbital Speed (km/s)
Mercury	0.4	48.0
Venus	0.7	35.0
Earth	1.0	30.0
Mars	1.5	24.0
Jupiter	5.2	13.0
Saturn	9.6	10.0
Uranus	19.0	7.0
Neptune	30.0	5.1
Pluto	39.0	4.7

Strategies for Questions Based on Graphs

Graphs represent relationships in a visual form that is easy to read. Three different types of graphs commonly used on science Regents Examinations are line graphs, bar graphs, and circle graphs. Line graphs are the most common, and they show the relationship between two changing quantities, or variables. When a question is based on any of the three types of graphs, the information you need to correctly answer the question can usually be found on the graph.

When answering a question that includes a graph, first ask yourself these questions:

- What information does the graph provide?
- What are the variables?
- What seems to happen to one variable as the other changes?

After a careful analysis of the graph, use the appropriate strategies for multiple-choice or constructed-response questions.

Use of the Earth Science Reference Tables To Help Answer Questions

In recent Regents Examinations between 20% to 35% of the questions have involved the use of the *Earth Science Reference Tables*. You should become thoroughly familiar with all details of these tables. Sometimes the questions will specifically refer you to the reference tables, but most often you will be expected to know what information is included within the reference tables. Listed below are some of the ways these reference tables are used in Regents Examination questions.

- to find a specific fact, such as the composition of the Earth's inner core
- to find the relationship between two facts on different parts of the reference tables, such as the latitude and longitude of the center of the Tug Hill plateau
- using an equation on the reference tables to solve a problem, such as a rate of change of an event
- graphing or recognizing the correct graph of data on the reference tables such as the percent by mass of the elements in the Earth's crust
- decoding a graphic symbol in a question such as present weather or air mass type on a weather map
- performing a procedure using part of the reference tables, such as determining the distance of a location to an epicenter using the Earthquake P-wave and S-wave Travel Time graph
- interpretation of data on the reference tables, such as the temperature of white dwarf stars

Appendix 3: Earth Science Performance Test

About 15% of the Earth Science Regents test grade is derived from a performance test designed to evaluate various laboratory and classroom procedures refined during the yearly course of study. The test is designed to measure some of the standards outlined in the Physical Setting/Earth Science Core Curriculum of the New York State Education Department. This performance test is normally given about two weeks before the rest of the regents exam. The test will be divided into four groups of tasks that can be completed in forty to forty-five minutes. As of this time it has not been decided which group of tasks will be included, but they will likely be four of the six stations listed below. This performance test will be given as part of the June 2007 Earth Science Regents. What is included on performance tests before and after June 2007 may or may not be similar to the descriptions that follow.

Station 1: Mineral and Rock Identification

"Using a mineral identification kit, the student will determine the properties of a mineral and will use those properties to identify that mineral from a flowchart. Using rock identification charts from the *Earth Science Reference Tables* and the characteristics observed in each rock sample, the student will also name an igneous, a sedimentary, and a metamorphic rock."

It is suggested that students study the sections in this book on rock and mineral identification on pages 192–194, 197, 199–200, and 202 to be prepared for this task.

Station 2: Locating an Earthquake Epicenter

"Using seismic data, the P-wave and S-wave travel time graph from the *Earth Science Reference Tables,* a safe drawing compass, and a map, the student will determine the location of an earthquake epicenter"

It is suggested that students study the section in this book on location of an epicenter on page 217 to be prepared for this task.

Station 3: Atmospheric Moisture

"Using dry-bulb and wet-bulb thermometers, the generalized graph for determining cloud-base altitude, and the dewpoint and relative humidity tables and the temperature conversion chart from the *Earth Science Reference Tables,* the student will determine dewpoint and relative humidity of the air and the possible altitude of a cloud base. The student will record data on a weather map station model."

The Generalized Graph for Determining Cloud Base Altitude and the diagram of How to Use the Graph for Determining Cloud Base Altitude are shown on the next page.

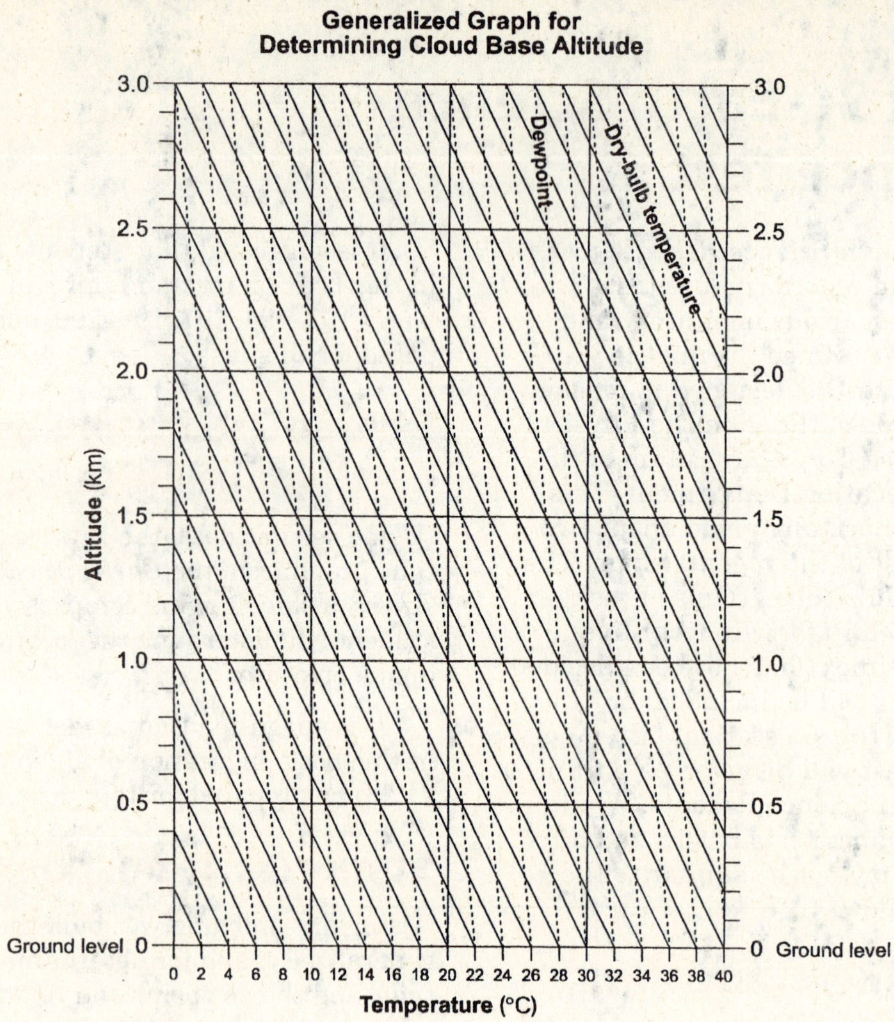

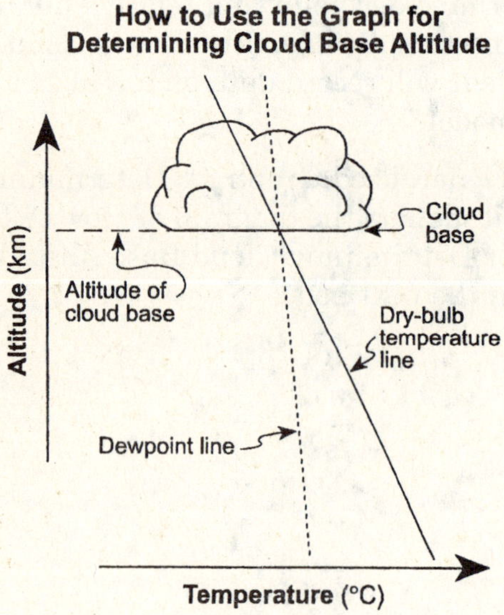

It is suggested that students study the sections in this text on determining dewpoint, relative humidity and calculating cloud base altitude plus understanding the weather map station model on pages 118–119, 121 and 131 to be prepared for this task.

Station 4: Density of Fluids

"Using graduated cylinders containing different fluids, an electronic balance, a magnifier or hand lens and a calculator, the student will determine the mass, volume and density of the fluids and will apply this information to demonstrate an understanding of relative density."

It is suggested that students study the sections in this text on density on pages 3 and 4 to be prepared for this task.

Station 5: Data Collecting, Graphing, and Predicting

"Using a column of fluid, three sizes of spherical particles of the same density, a stopwatch, and a calculator, the student will determine the average settling time for each of the three sizes of particles. The student will construct a line graph of average settling time vs. particle diameter and predict the settling time for another particle of a given diameter."

Station 6: Constructing and Analyzing an Elliptical Orbit

"Using two pins, a looped string, a metric ruler, and a calculator, the student will construct an ellipse, determine its eccentricity, and relate this information to our solar system."

It is suggested that students study the sections on planet revolution and eccentricity of planet orbits on pages 42–43 in this text to be prepared for this task.

Doing Well On The Performance Test

How well you perform on this test is determined by the accuracy and precision of your answers. You are encouraged to carefully read the directions given with the test and to record your answers neatly and accurately. Be especially careful to answer the questions with the degree of accuracy asked for, not less and not more. As an example, if the directions ask for an answer to the nearest 0.1 (tenth) of a gram, don't just give a whole number or extend your answer to the nearest 0.01 (hundredth). Follow the directions. Another common problem is using or recording the wrong units, such as using inches instead of centimeters or millimeters instead of centimeters.

Regents Examinations

The New York Regents Examinations that follow are provided so that you can practice taking an Earth Science Regents Examination. All of the tests included in this book cover the key concepts of the new Physical Setting: Earth Science core curriculum. You will find that the content of this book provides specifically what you need to review for the Physical Setting: Earth Science Regents Examination.

The best way to use these examinations is to take the entire test after you have reviewed the course content. Use the tests to determine if you have reviewed enough to do well on the Regents Examination and to determine where further review will be most helpful.

Do not look up any information or answers while you take the examinations. Answer each question just as you would during a real test. As you take the examination, use the margin of the paper to note any question where you are just guessing. Leave the more difficult questions for last, but be sure to answer each question. Every point counts so do not skip over a long question that is only worth a point or two. A long question could be easier than it looks and may make the difference between an A or a B or between passing or failing.

When you finish, have your teacher score your Examination and help you determine the areas where you need the most work. Also review the "guesses" you noted in the margin to find out what you need to study to ensure that you will be able to answer similar questions on the next test. Once you have determined your weaknesses, you can focus your review on those topics in this book.

Reviewing the areas where you know the least will give you the best chance of improving your final score. Spending time on areas where you are doing quite well will not produce much improvement in your total score, but it is still important if time permits.

Physical Setting Earth Science June 19, 2005

Part A

Answer all questions in this part.

Directions (1–35): For *each* statement or question, write on your separate answer sheet the *number* of the word or expression that, of those given, best completes the statement or answers the question. Some questions may require the use of the *Earth Science Reference Tables*.

1 Which list of three planets and Earth's Moon is arranged in order of increasing equatorial diameter?

(1) Earth's Moon, Pluto, Mars, Mercury
(2) Pluto, Earth's Moon, Mercury, Mars
(3) Mercury, Mars, Earth's Moon, Pluto
(4) Mars, Mercury, Pluto, Earth's Moon

2 If Earth's axis were tilted 35° instead of 23.5°, the average temperatures in New York State would most likely

(1) decrease in both summer and winter
(2) decrease in summer and increase in winter
(3) increase in summer and decrease in winter
(4) increase in both summer and winter

3 Which star has a higher luminosity and a lower temperature than the Sun?

(1) *Rigel* (3) *Alpha Centauri*
(2) *Barnard's Star* (4) *Aldebaran*

4 Starlight from distant galaxies provides evidence that the universe is expanding because this starlight shows a shift in wavelength toward the

(1) red-light end of the visible spectrum
(2) blue-light end of the visible spectrum
(3) ultraviolet-ray end of the electromagnetic spectrum
(4) gamma-ray end of the electromagnetic spectrum

5 On which day of the year would the intensity of insolation at Kingston, New York, most likely be greatest?

(1) March 21 (3) September 23
(2) June 21 (4) December 21

6 The diagram below represents the elliptical orbit of a moon revolving around a planet. The foci of this orbit are the points labeled F_1 and F_2.

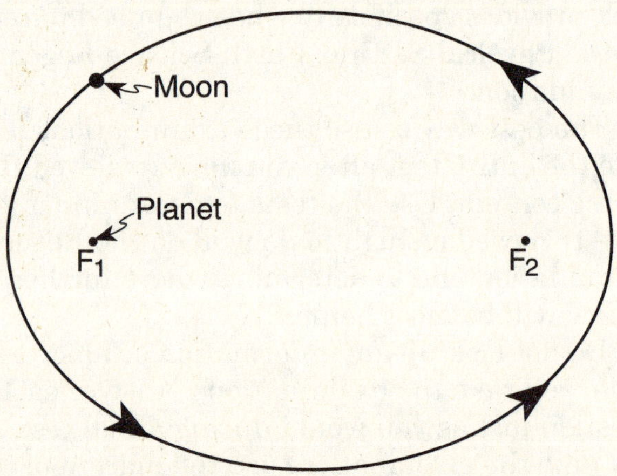

(Drawn to scale)

What is the approximate eccentricity of this elliptical orbit?

(1) 0.3 (3) 0.7
(2) 0.5 (4) 1.4

7 The coldest climates on Earth are located at or near the poles primarily because Earth's polar regions

(1) receive mostly low-angle insolation
(2) receive less total yearly hours of daylight
(3) absorb the greatest amount of insolation
(4) are usually farthest from the Sun

8 Compared to an inland location, a location on an ocean shore at the same elevation and latitude is likely to have

(1) cooler winters and cooler summers
(2) cooler winters and warmer summers
(3) warmer winters and cooler summers
(4) warmer winters and warmer summers

9 The diagram below represents a Foucault pendulum swinging freely for 8 hours.

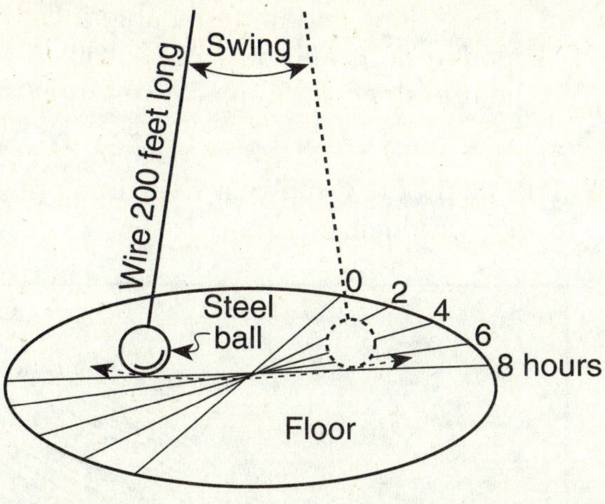

The Foucault pendulum appears to gradually change its direction of swing due to Earth's

(1) orbit around the Sun
(2) curved surface
(3) tilted axis
(4) spin on its axis

10 The diagram below shows the noontime shadows cast by a student and a tree.

If the time is solar noon and the student is located in New York State, in what direction is the student facing?

(1) north
(2) south
(3) east
(4) west

11 The diagram below represents the Moon in its orbit, as viewed from above Earth's North Pole. Position 1 represents a specific location of the Moon in its orbit.

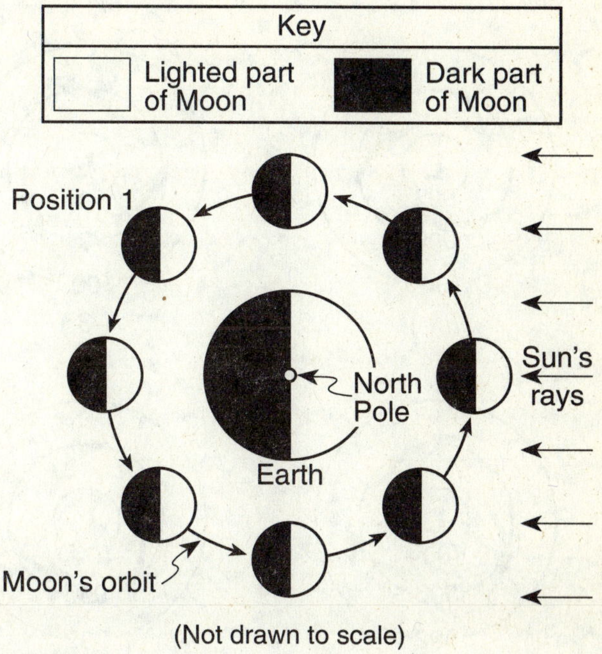

Which phase of the Moon will be seen from Earth when the Moon is at position 1?

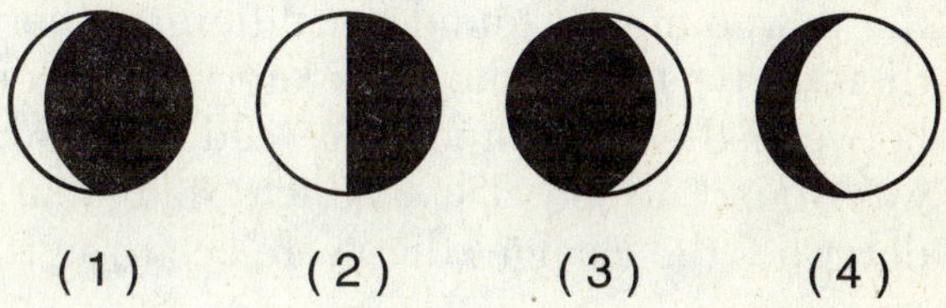

12 During nighttime cooling, most of the energy radiated by Earth's oceans into space is

(1) ultraviolet rays
(2) gamma rays
(3) visible light rays
(4) infrared rays

13 Landscapes with horizontal bedrock structure, steep slopes, and high elevations are classified as

(1) plateau regions
(2) plain regions
(3) lowland regions
(4) mountain regions

14 Which map view best represents the pattern of isobar values, in millibars, and the pattern of wind flow, shown by arrows, at Earth's surface surrounding a Northern Hemisphere low-pressure center?

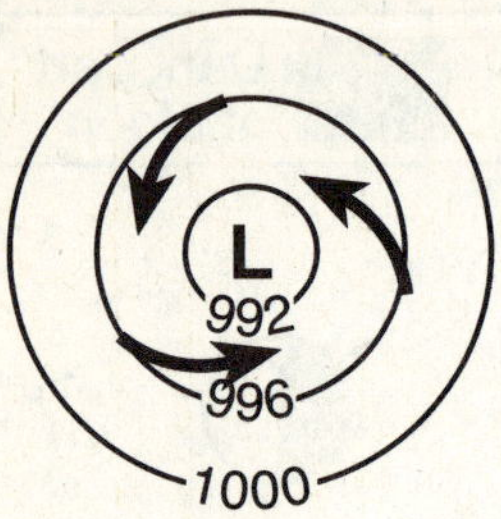

(1)

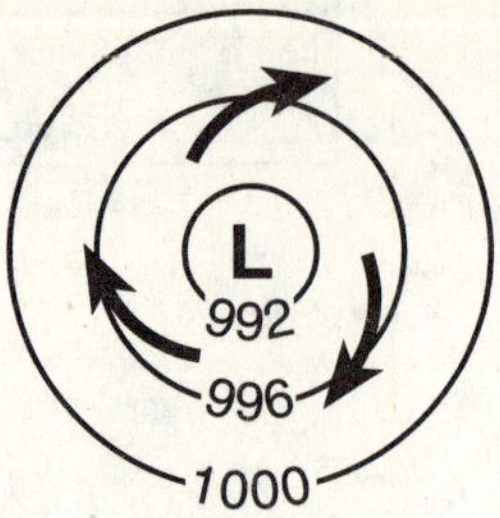

(3)

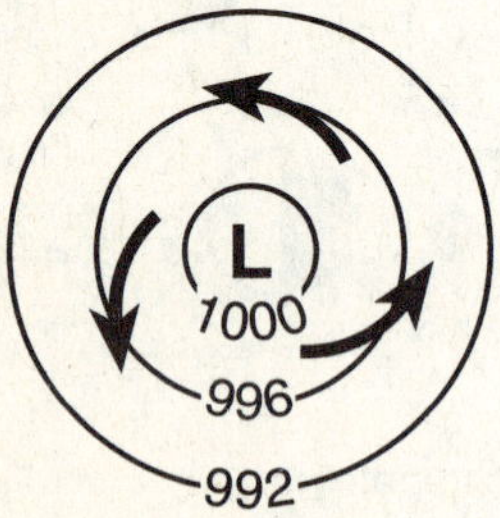

(2)

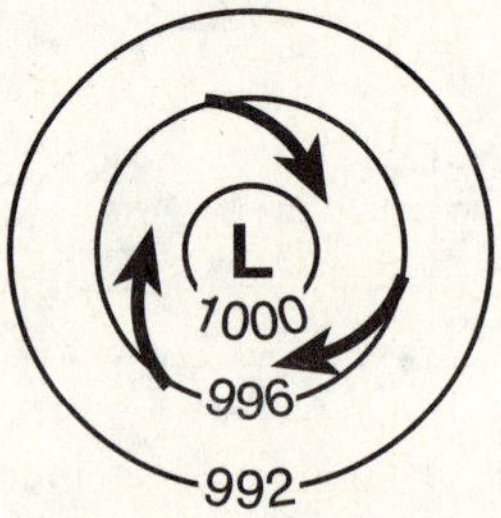

(4)

15 An observer measured the air temperature and the dewpoint and found the difference between them to be 12°C. One hour later, the difference between the air temperature and the dewpoint was found to be 4°C. Which statement best describes the changes that were occurring?

(1) The relative humidity was decreasing and the chance of precipitation was decreasing.
(2) The relative humidity was decreasing and the chance of precipitation was increasing.
(3) The relative humidity was increasing and the chance of precipitation was decreasing.
(4) The relative humidity was increasing and the chance of precipitation was increasing.

16 Which two ocean currents are both warm currents that primarily flow away from the equator?

(1) Guinea Current and Labrador Current
(2) Brazil Current and Agulhas Current
(3) Alaska Current and Falkland Current
(4) Canaries Current and Gulf Stream Current

17 Which surface soil conditions allow the most infiltration of rainwater?

(1) steep slope and permeable soil
(2) steep slope and impermeable soil
(3) gentle slope and permeable soil
(4) gentle slope and impermeable soil

18 The map below shows a river emptying into an ocean, producing a delta.

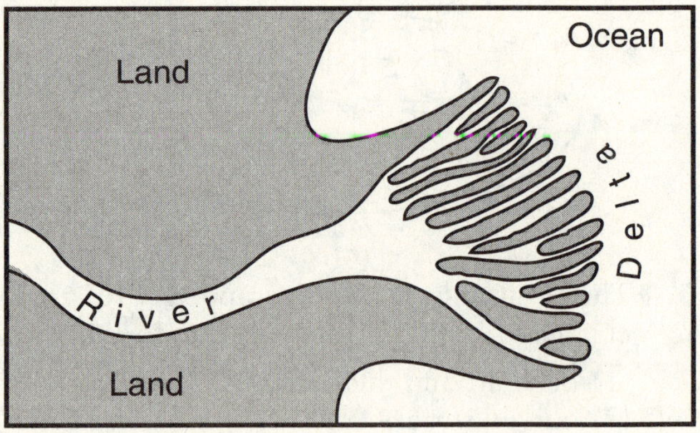

Which graph best represents the relationship between the distance from the river delta into the ocean and the average size of sediments deposited on the ocean floor?

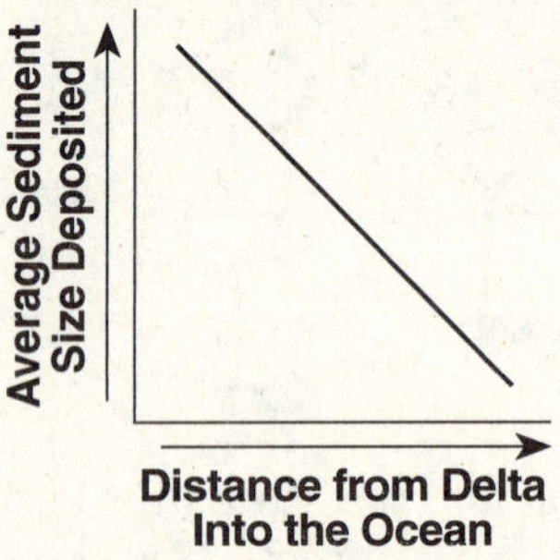

(1)

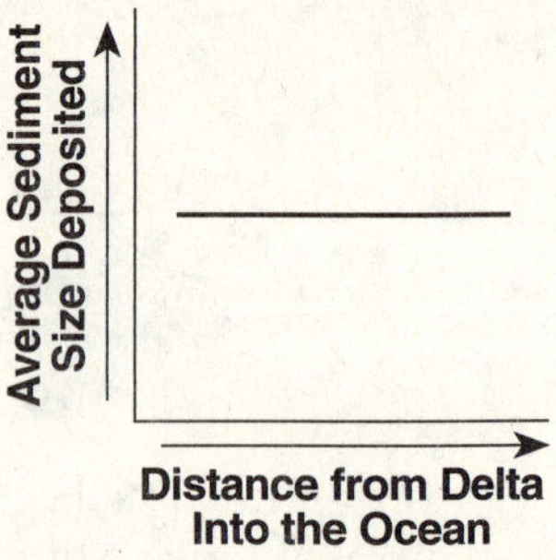

(3)

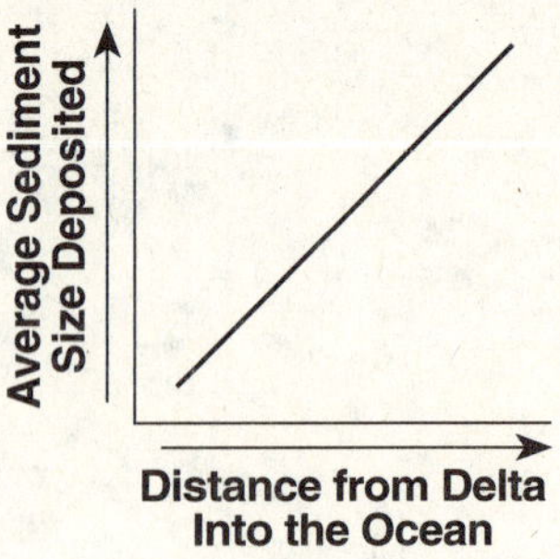

(2)

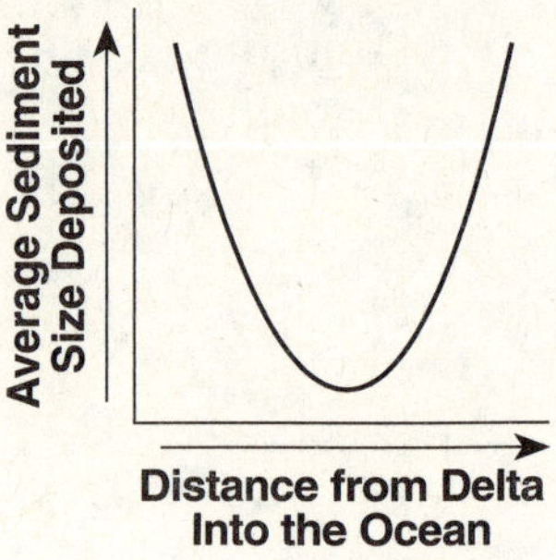

(4)

19 Which process could lead directly to the formation of pumice rock?

(1) precipitation of minerals from evaporating seawater
(2) metamorphism of unmelted rock material
(3) deposition of quartz sand
(4) explosive eruption of lava from a volcano

20 The rock shown below has a foliated texture and contains the minerals amphibole, quartz, and feldspar arranged in coarse-grained bands.

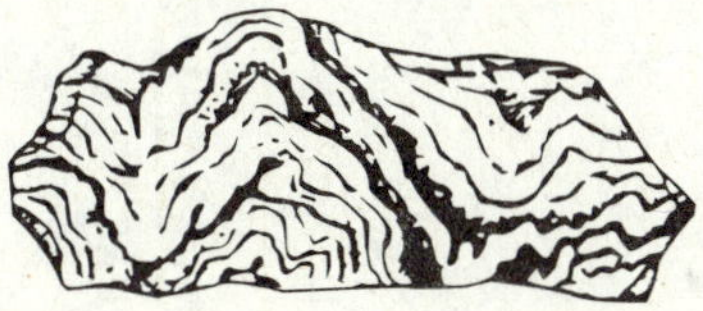

Which rock is shown?

(1) slate (3) gneiss
(2) dunite (4) quartzite

21 The photograph below shows the intergrown crystals of a pegmatite rock.

(Actual size)

Which characteristic provides the best evidence that this pegmatite solidified deep underground?

(1) low density (3) felsic composition
(2) light color (4) very coarse texture

22 Which igneous rock, when weathered, could produce sediment composed of the minerals potassium feldspar, quartz, and amphibole?

(1) gabbro (3) andesite
(2) granite (4) basalt

Base your answers to questions 23 and 24 on the photograph below. The photograph shows several broken samples of the same colorless mineral.

23 Which physical property of this mineral is most easily seen in the photograph?

(1) fracture (3) streak
(2) hardness (4) cleavage

24 Which mineral is most likely shown in the photograph?

(1) quartz (3) galena
(2) calcite (4) halite

---

25 The large coal fields found in Pennsylvania provide evidence that the climate of the northeastern United States was much warmer during the Carboniferous Period. This change in climate over time is best explained by the

(1) movements of tectonic plates
(2) effects of seasons
(3) changes in the environment caused by humans
(4) evolution of life

26 What is the inferred temperature at the boundary between Earth's stiffer mantle and outer core?

(1) 2,500°C (3) 5,000°C
(2) 4,500°C (4) 6,200°C

27 Which color of the visible spectrum has the *shortest* wavelength?

(1) violet (3) yellow
(2) blue (4) red

28 The diagram below shows the interaction of two tectonic plates.

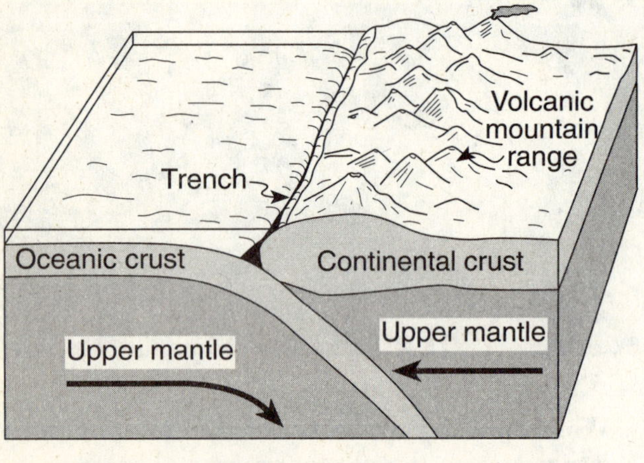

(Not drawn to scale)

The type of plate boundary represented in the diagram most likely exists between the

(1) Antarctic Plate and the African Plate
(2) Antarctic Plate and the Indian-Australian Plate
(3) South American Plate and the Nazca Plate
(4) South American Plate and the African Plate

29 When 1 gram of liquid water at 0° Celsius freezes to form ice, how many total calories of heat are lost by the water?

(1) 1 (3) 80
(2) 0.5 (4) 540

30 The shore of which New York State body of water has large amounts of metamorphic bedrock exposed at the surface?

(1) western shore of Lake Champlain
(2) eastern shore of Lake Erie
(3) southern shore of Long Island Sound
(4) southern shore of Lake Ontario

31 Which graph best shows the radioactive decay of carbon-14?

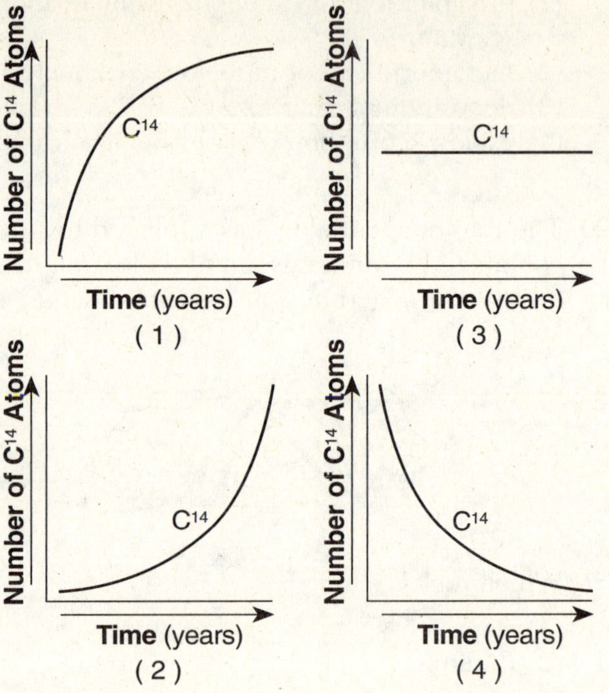

32 Approximately what percentage of the estimated age of Earth does the Cenozoic Era represent?

(1) 1.4% (3) 11.9%
(2) 5.0% (4) 65.0%

33 The geologic cross section below shows the geologic age of two rock layers separated by an unconformity.

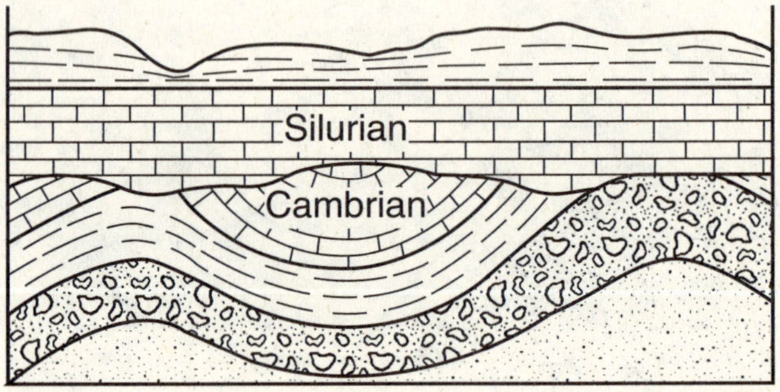

The unconformity at the bottom of the Silurian rock layer indicates a gap in the geologic time record. What is the *minimum* time, in millions of years, shown by the gap?

(1) 13 (3) 101
(2) 47 (4) 126

34 The diagram below represents a map view of a stream drainage pattern.

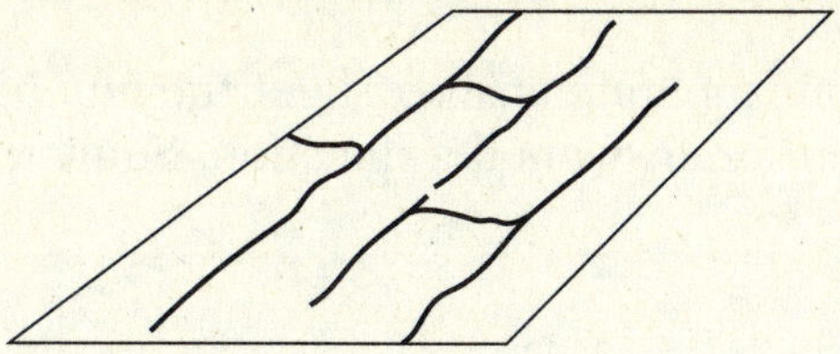

Which underlying bedrock structure most likely produced this stream drainage pattern?

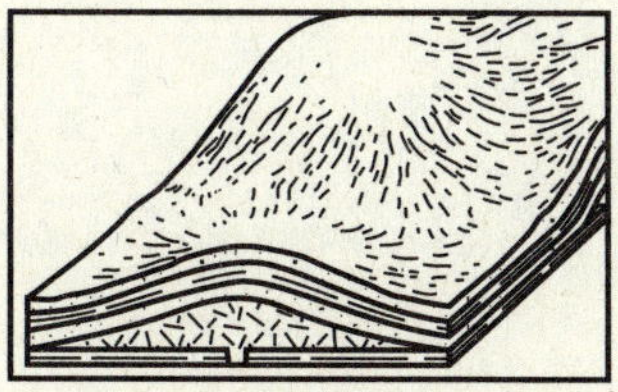

(1)

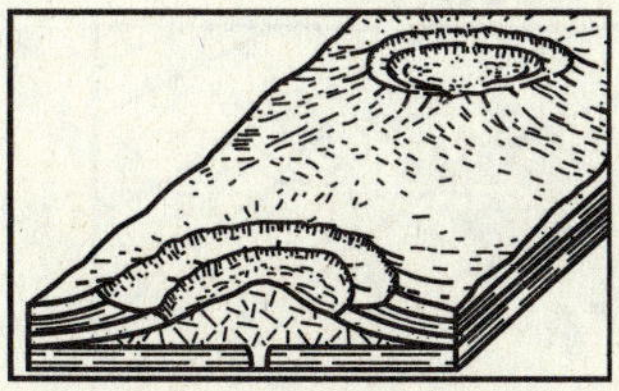

(3)

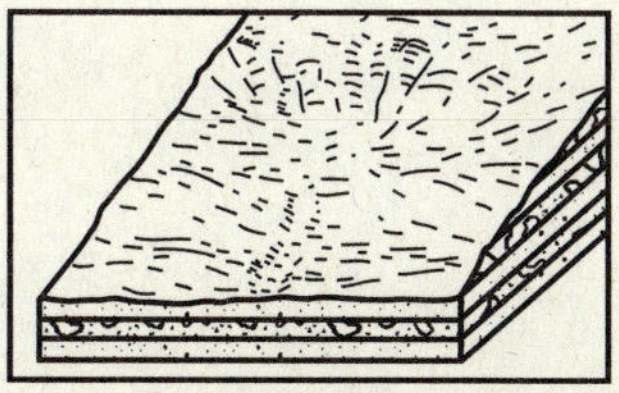

(2)

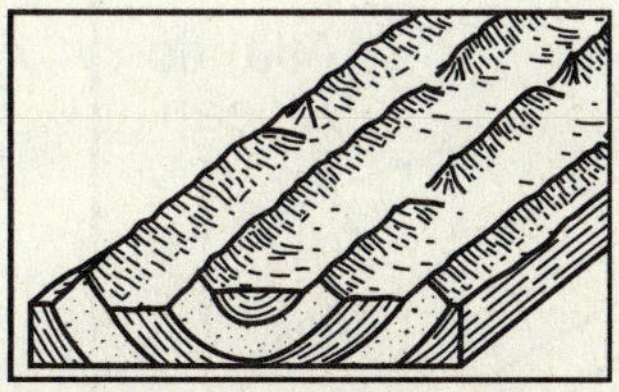

(4)

35 Compared to a maritime tropical air mass, a maritime polar air mass has a

(1) higher temperature and more water vapor
(2) higher temperature and less water vapor
(3) lower temperature and more water vapor
(4) lower temperature and less water vapor

Part B–1

Answer all questions in this part.

Directions (36–50): For *each* statement or question, write on your separate answer sheet the *number* of the word or expression that, of those given, best completes the statement or answers the question. Some questions may require the use of the *Earth Science Reference Tables*.

Base your answers to questions 36 through 38 on the map below, which shows sea-level air pressure, in millibars, for a portion of the eastern coast of North America. Points *A, B, C,* and *D* are sea-level locations on Earth's surface.

Sea-Level Air Pressures

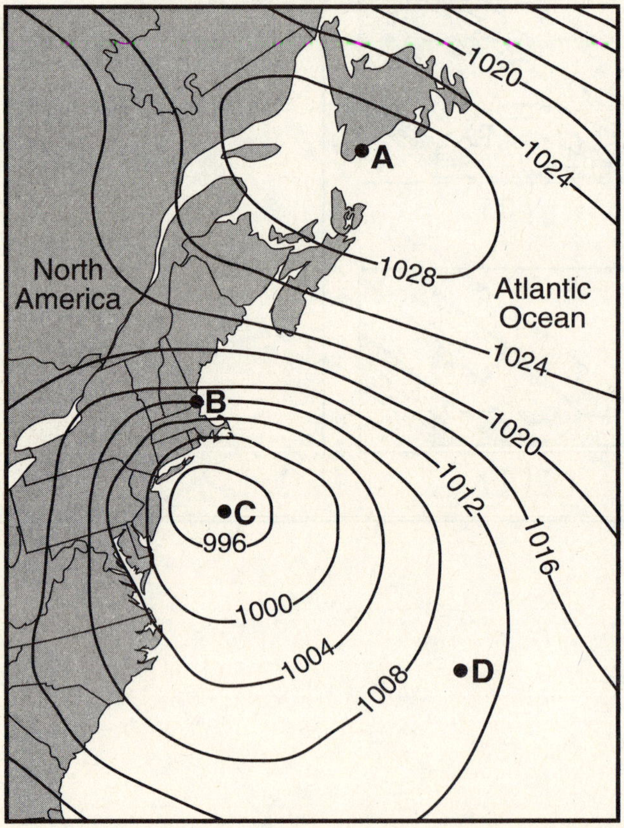

36 Which weather instrument was used to measure the air pressures?

(1) thermometer
(2) wind vane
(3) sling psychrometer
(4) barometer

37 Which location most likely recorded the highest wind speed?

(1) *A*
(2) *B*
(3) *C*
(4) *D*

38 The air pressure recorded at point *D* was most likely

(1) 1014 mb
(2) 1012 mb
(3) 1010 mb
(4) 1006 mb

Base your answers to questions 39 through 41 on the newspaper article shown below and on your knowledge of Earth science.

Fossilized Jellyfish Found in Wisconsin

Fossil hunters have unearthed the largest collection of fossilized jellyfish ever discovered, including the largest fossilized jellyfish ever found.

The remains of soft-bodied animals such as jellyfish are relatively rare because they don't have bones, fossil dealer Dan Damrow, James W. Hagadorn of the California Institute of Technology and Robert H. Dott Jr. of the University of Wisconsin at Madison noted in describing the find in the journal Geology.

About a half-billion years ago, during the Cambrian period, the quarry in Mosinee, Wis., where the deposits were found was a small lagoon. The jellyfish apparently died when they were washed up by a freak tide or storm, the researchers said. The jellyfish remains were probably preserved because of a lack of erosion from sea water and wind, and a lack of scavengers, the researchers concluded.

"It is very rare to discover a deposit which contains an entire stranding event of jellyfish," Hagadorn said. "These jellyfish are not just large for the Cambrian, but are the largest jellyfish in the entire fossil record."

Washington Post, January 2002

39 These fossilized jellyfish were most likely discovered in which type of rock?

(1) sandstone
(2) granite
(3) pumice
(4) slate

40 Which two marine organisms most likely lived at the same time as these jellyfish?

(1) crinoids and dinosaurs
(2) ammonoids and placoderm fish
(3) brachiopods and gastropods
(4) amphibians and eurypterids

41 Which evidence would lead scientists to suspect that a tide or storm had washed up these jellyfish on a beach?

(1) Primitive life existed on land 500 million years ago.
(2) The rock containing the jellyfish fossils has distorted crystal structure.
(3) Treeroot fossils appear to have been pitted and folded.
(4) Large ripple marks were found in the fossil-containing rock layers.

Base your answers to questions 42 through 44 on the topographic map below. Points *A*, *B*, *C*, *D*, and *X* represent locations on the map. Elevations are measured in feet.

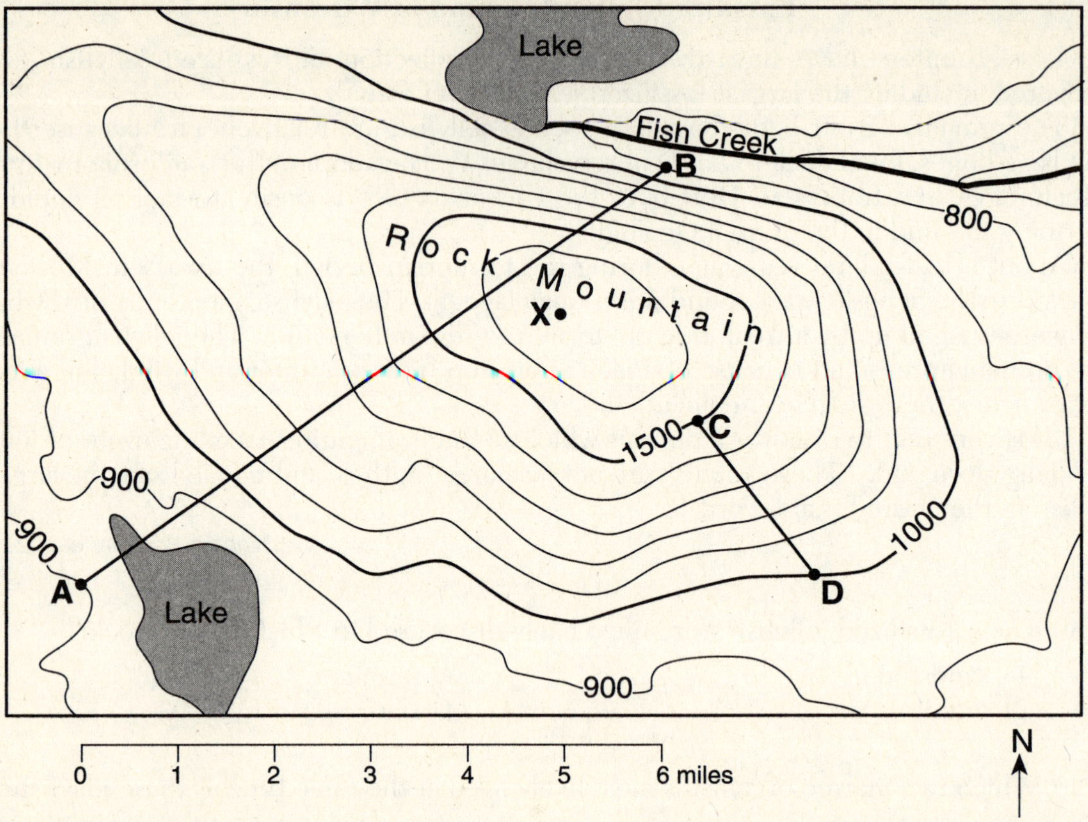

42 What is the highest possible elevation of point *X* on Rock Mountain?

(1) 1,599 ft
(2) 1,600 ft
(3) 1,601 ft
(4) 1,699 ft

43 What is the average gradient of the slope along straight line *CD*?

(1) 100 ft/mi
(2) 250 ft/mi
(3) 500 ft/mi
(4) 1,000 ft/mi

44 Which cross section best represents the profile along straight line *AB*?

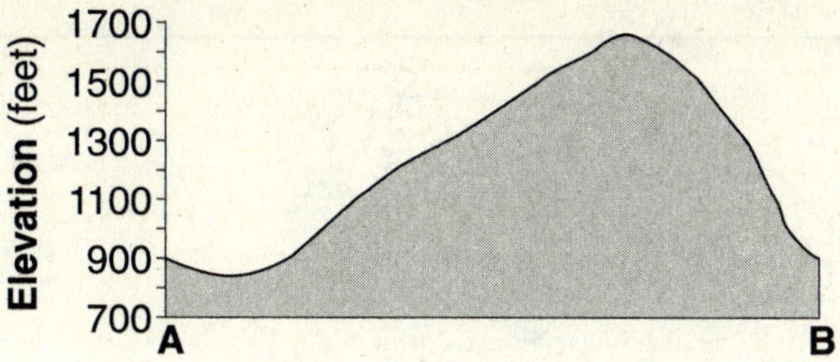

(1)

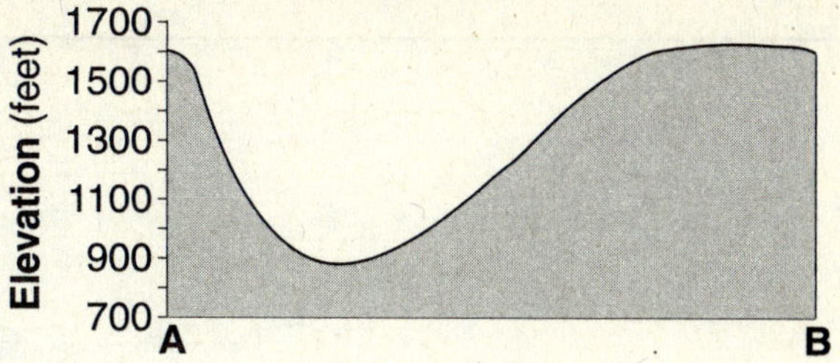

(3)

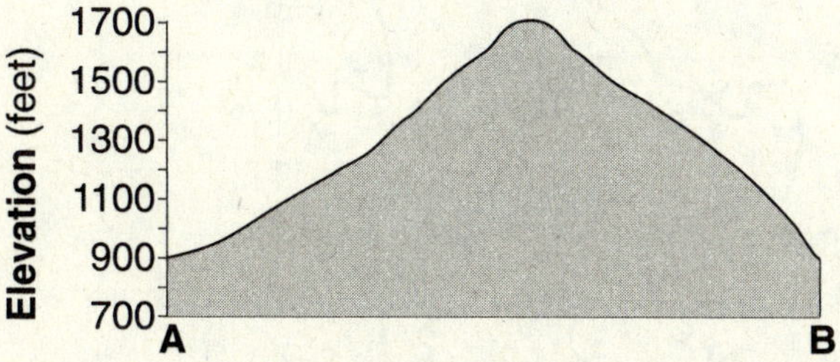

(2)

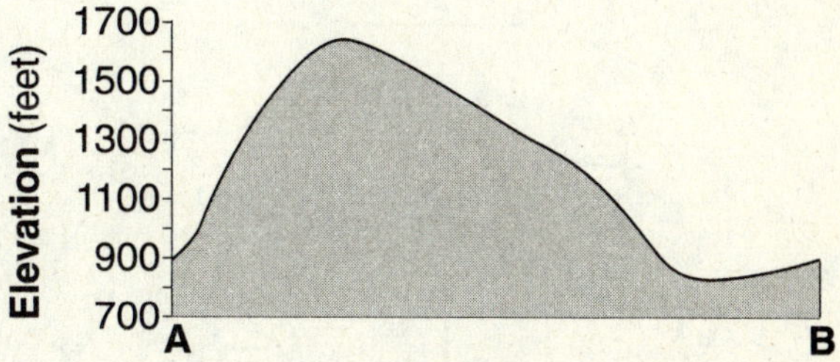

(4)

Base your answers to questions 45 through 47 on the map below, which shows watershed regions of New York State.

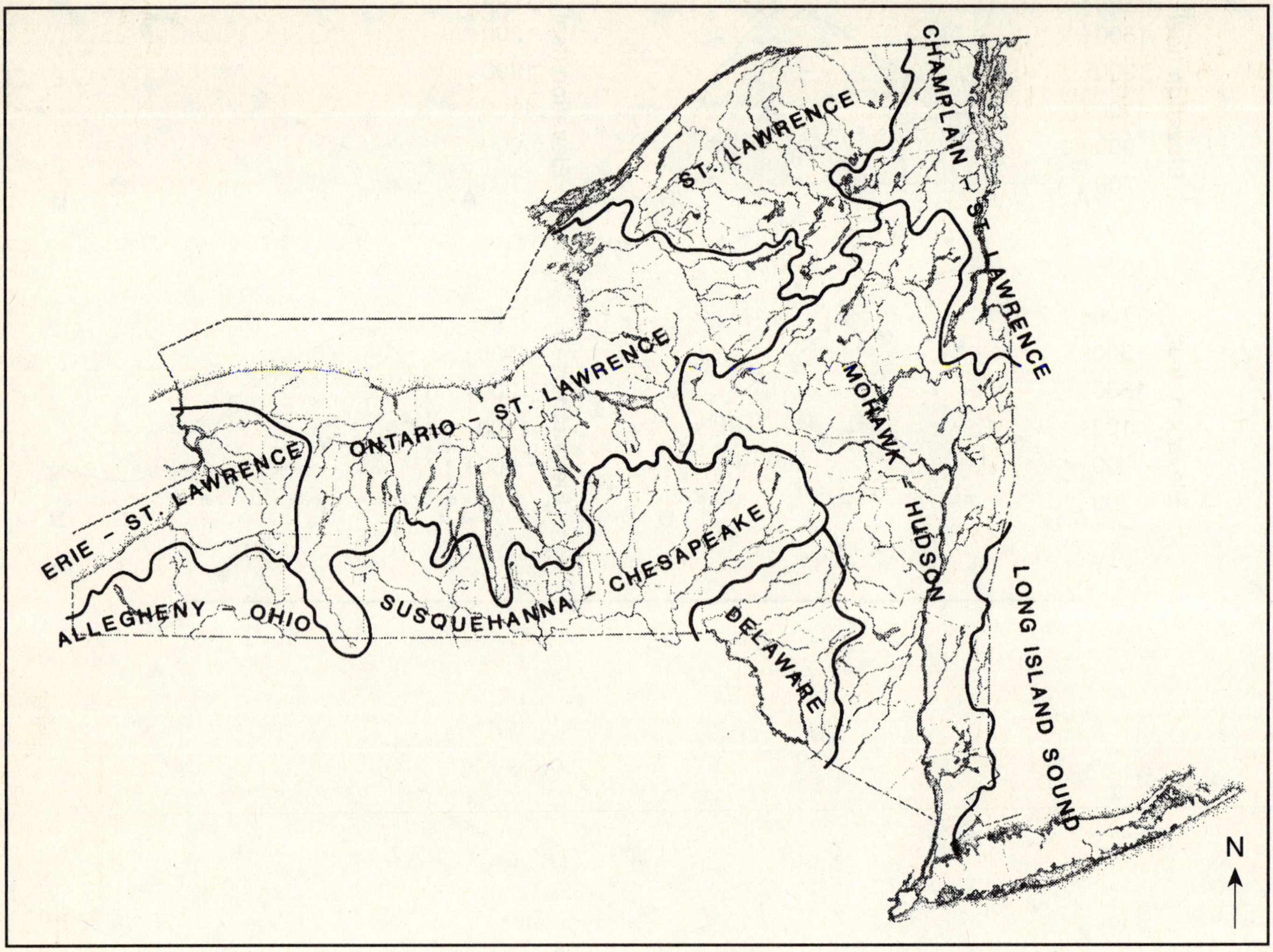

45 On which type of landscape region are both the Susquehanna-Chesapeake and the Delaware watersheds located?

(1) plain
(2) plateau
(3) mountain
(4) lowland

46 In which watershed is the Genesee River located?

(1) Ontario-St. Lawrence
(2) Susquehanna-Chesapeake
(3) Mohawk-Hudson
(4) Delaware

47 Most of the surface bedrock of the Ontario-St. Lawrence watershed was formed during which geologic time periods?

(1) Precambrian and Cambrian
(2) Ordovician, Silurian, and Devonian
(3) Mississippian, Pennsylvanian, and Permian
(4) Triassic, Jurassic, and Cretaceous

Base your answers to questions 48 through 50 on the diagram below, which represents an exaggerated view of Earth revolving around the Sun. Letters *A*, *B*, *C*, and *D* represent Earth's location in its orbit on the first day of each of the four seasons.

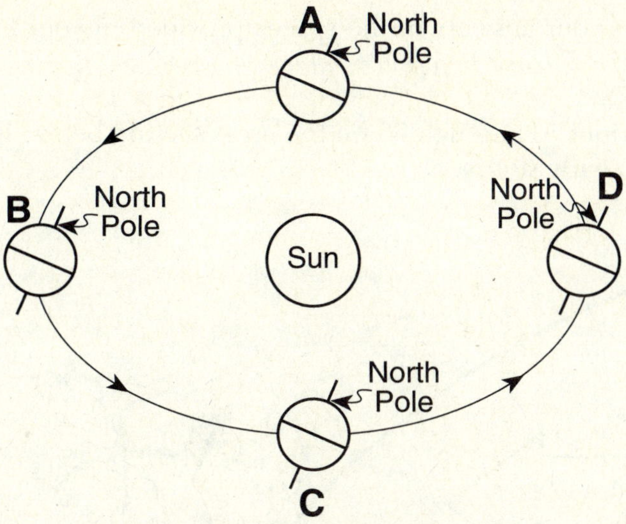

(Not drawn to scale)

48 Which location in Earth's orbit represents the first day of fall (autumn) for an observer in New York State?

(1) *A*
(2) *B*
(3) *C*
(4) *D*

49 Earth's rate of revolution around the Sun is approximately

(1) 1° per day
(2) 360° per day
(3) 15° per hour
(4) 23.5° per hour

50 Which observation provides the best evidence that Earth revolves around the Sun?

(1) Stars seen from Earth appear to circle *Polaris*.
(2) Earth's planetary winds are deflected by the Coriolis effect.
(3) The change from high ocean tide to low ocean tide is a repeating pattern.
(4) Different star constellations are seen from Earth at different times of the year.

Part B–2

Answer all questions in this part.

Directions (51–65): Record your answers in the spaces provided in your answer booklet. Some questions may require the use of the *Earth Science Reference Tables.*

Base your answers to questions 51 through 53 on the cross section below, which shows a typical cold front moving over New York State in early summer.

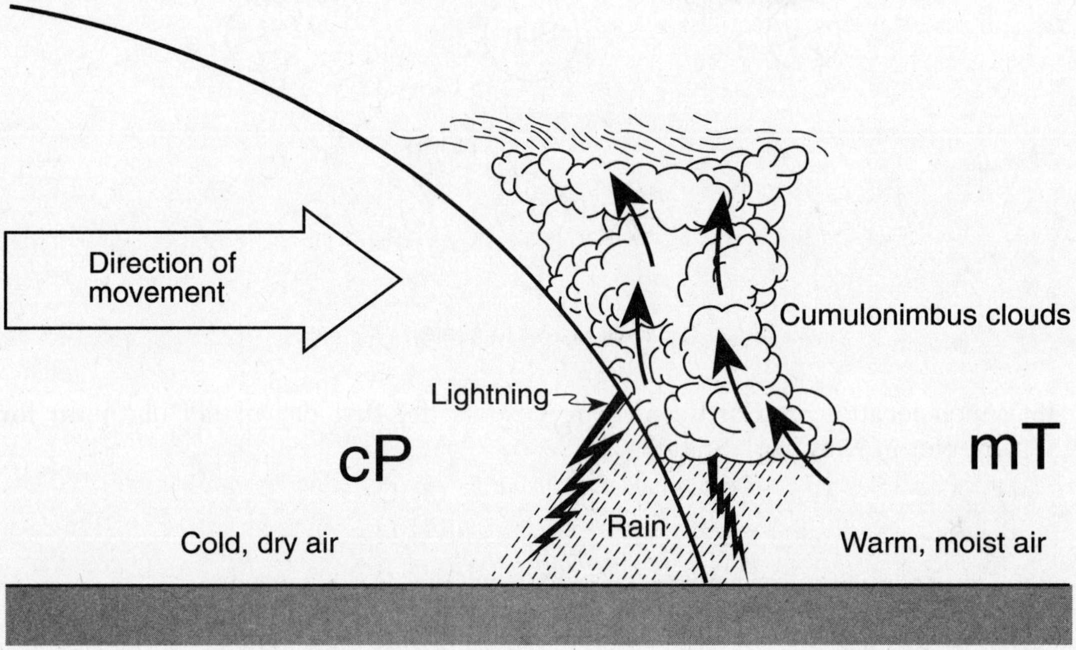

51 Explain why the warm, moist air is rising at the frontal boundary. [1]

52 State *one* process that causes clouds to form in this rising air. [1]

53 Central Canada was the geographic source region for the cP air mass shown in the cross section. Identify the most likely geographic source region for the mT air mass shown in the cross section. [1]

Base your answers to questions 54 through 57 on the map below and on your knowledge of Earth science. The map shows the location of the epicenter, ⓧ, of an earthquake that occurred on April 20, 2002, about 29 kilometers southwest of Plattsburgh, New York.

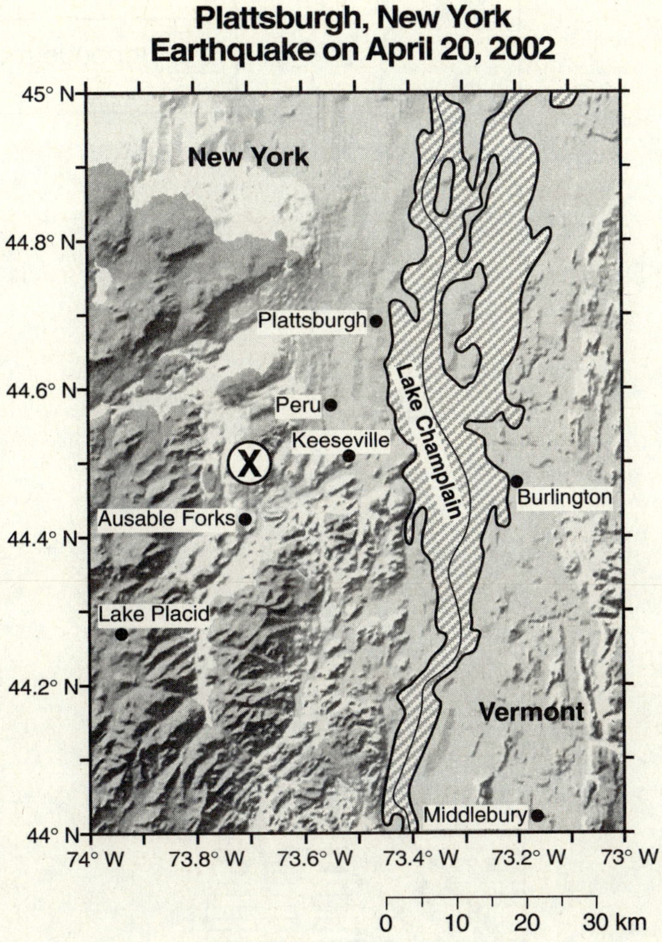

54 State the latitude and longitude of this earthquake epicenter. Express your answers to the *nearest tenth of a degree* and include the compass directions. [1]

55 What is the *minimum* number of seismographic stations needed to locate the epicenter of an earthquake? [1]

56 Explain why this earthquake was most likely felt with greater intensity by people in Peru, New York, than by people in Lake Placid, New York. [1]

57 A seismic station located 1,800 kilometers from the epicenter recorded the *P*-wave and *S*-wave arrival times for this earthquake. What was the difference in the arrival time of the first *P*-wave and the first *S*-wave? [1]

Base your answers to questions 58 through 60 on the cross section of the bedrock outcrop shown below and on your knowledge of Earth science. Index fossils found in some of the rock units are shown. The rock units are labeled I through IX.

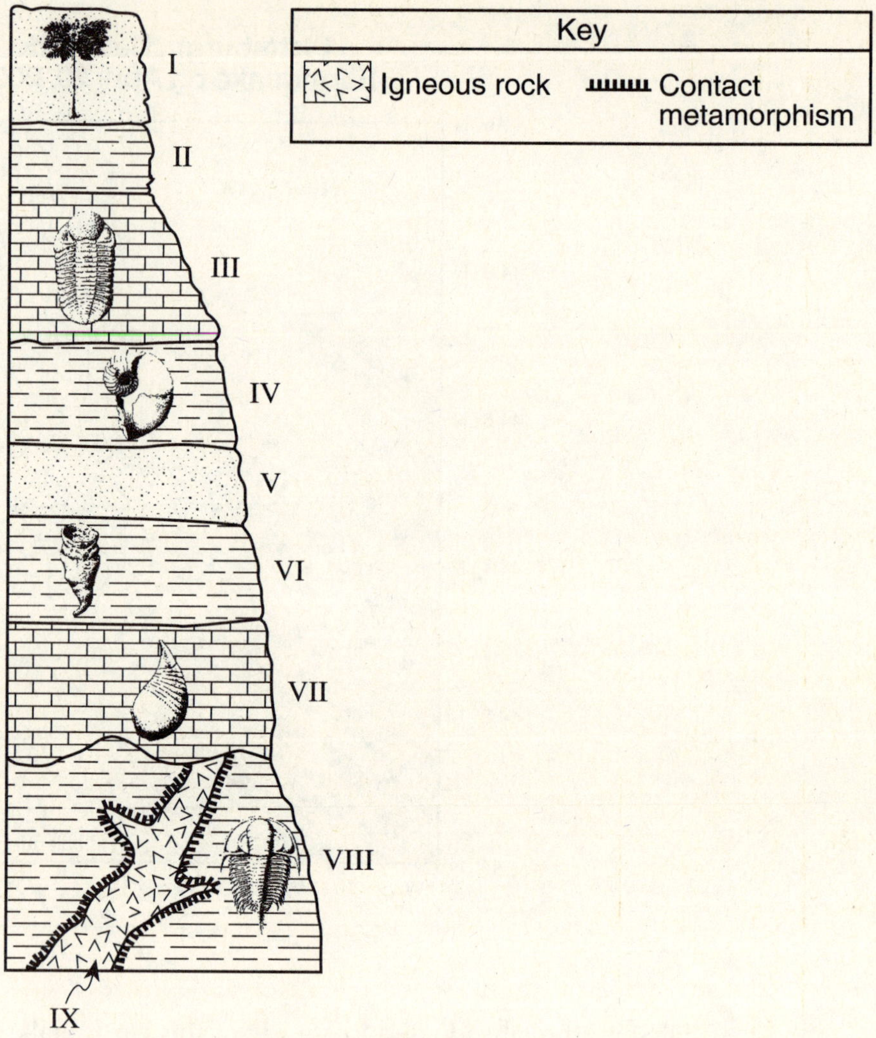

58 In the space provided *in your answer booklet,* number the relative age of rock units VII, VIII, and IX from 1 to 3, with number 1 indicating the oldest rock and number 3 indicating the youngest rock. [1]

59 The fossil shown in rock unit VIII is a member of an extinct group of fossils. State *two* other index fossils that are also members of the same group of extinct fossils. [1]

60 Based on the fossils shown in the limestone and shale layers, state the type of environment in which these sedimentary rocks were deposited. [1]

Base your answers to questions 61 through 63 on the block diagrams below, which show three types of streams with equal volumes.

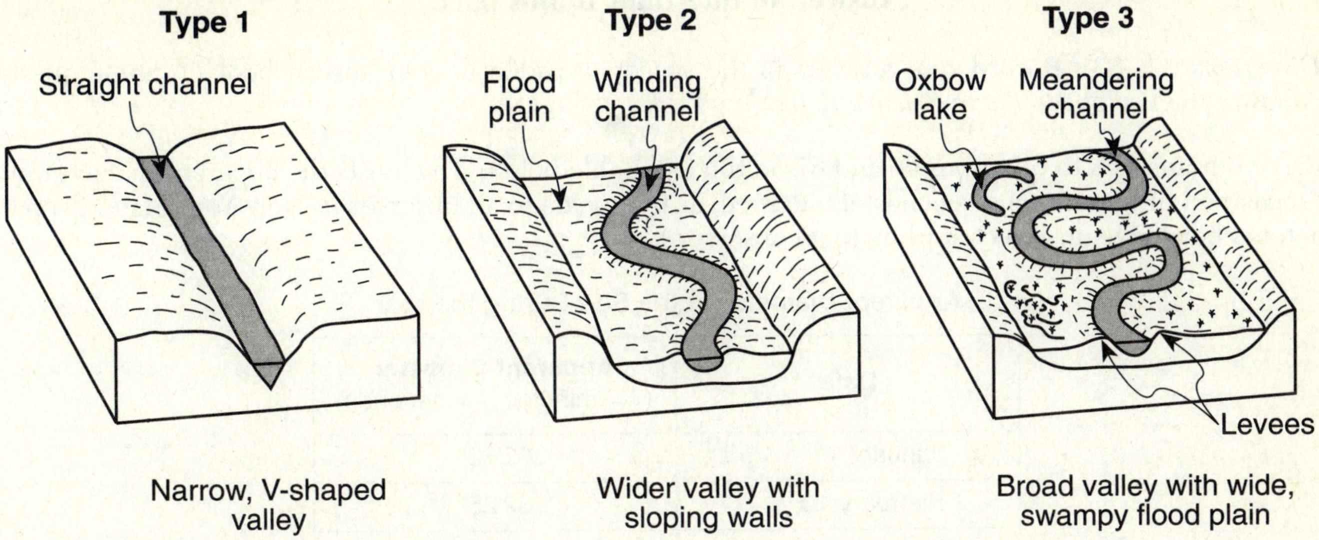

61 Explain how the differences between the type 1 and type 3 stream channels indicate that the average velocities of the streams are different. [1]

62 Explain why the outside of the curve of a meandering channel experiences more erosion than the inside of the curve. [1]

63 Explain how the cobbles and pebbles that were transported by these streams became smooth and rounded in shape. [1]

Base your answers to questions 64 and 65 on the station model below, which shows the weather conditions at Rochester, New York, at 4 p.m. on a particular day in June.

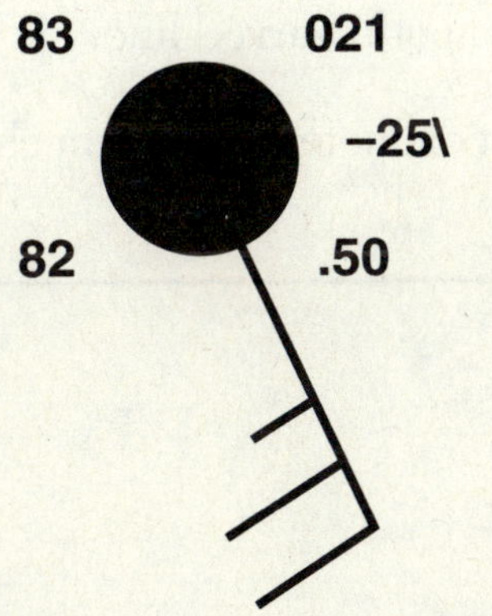

64 What was the actual barometric pressure, according to the station model, to the *nearest tenth of a millibar*? [1]

65 The winds shown by this station model were blowing from which compass direction and at what wind speed? [1]

Part C

Answer all questions in this part.

Directions (66–83): Record your answers in the spaces provided in your answer booklet. Some questions may require the use of the *Earth Science Reference Tables.*

Base your answers to questions 66 and 67 on the data table below, which lists the apparent diameter of the Sun, measured in minutes and seconds of a degree, as it appears to an observer in New York State. (Apparent diameter is how large an object appears to an observer.)

Apparent Diameter of the Sun During the Year

Date	Apparent Diameter (' = minutes " = seconds)
January 1	32'32"
February 10	32'25"
March 20	32'07"
April 20	31'50"
May 30	31'33"
June 30	31'28"
August 10	31'34"
September 20	31'51"
November 10	32'18"
December 30	32'32"

66 On the grid provided *in your answer booklet,* graph the data shown on the table by marking with a dot the apparent diameter of the Sun for *each* date listed and connecting the dots with a smooth, curved line. [2]

67 Explain why the apparent diameter of the Sun changes throughout the year as Earth revolves around the Sun. [1]

____

Base your answers to questions 68 through 71 on the passage below and on your knowledge of Earth science.

Watching the Glaciers Go

Mountain glaciers and ice caps in tropical areas of the world are melting fast and may vanish altogether by the year 2020. That was the chilling news last year from Lonnie Thompson, a geologist at Ohio State University's Byrd Polar Research Center who has been studying icy areas near the equator in South America, Africa, and the Himalayas for two decades.

It doesn't take a glacier scientist to see the changes. In 1977, when Thompson visited the Quelccaya ice cap in Peru, it was impossible not to notice a schoolbus-size boulder stuck in its grip. When Thompson returned in 2000, the rock was still there but the ice wasn't — it had retreated far into the distance.

Most scientists believe the glaciers are melting because of global warming — the gradual temperature increase that has been observed with increasing urgency during the past decade. Last year a panel of the nation's top scientists, the National Research Council, set aside any lingering skepticism about the phenomenon, concluding definitively that average global surface temperatures are rising and will continue to do so.

"Watching the Glaciers Go,"
Popular Science, vol. #7, January 2002

68 State *one* greenhouse gas that is an excellent absorber of infrared radiation and may be responsible for global warming. [1]

69 Describe the arrangement of sediment deposited directly from glaciers. [1]

70 Some glaciers currently exist near Earth's equator due to the cold, snowy climate of certain locations. Which type of landform exists where these glaciers occur? [1]

71 Describe *one* action humans could take to reduce the global warming that is melting the Quelccaya ice cap. [1]

Base your answers to questions 72 through 75 on the map and cross section of the Finger Lakes Region shown below and on your knowledge of Earth science.

Finger Lakes Region of New York State

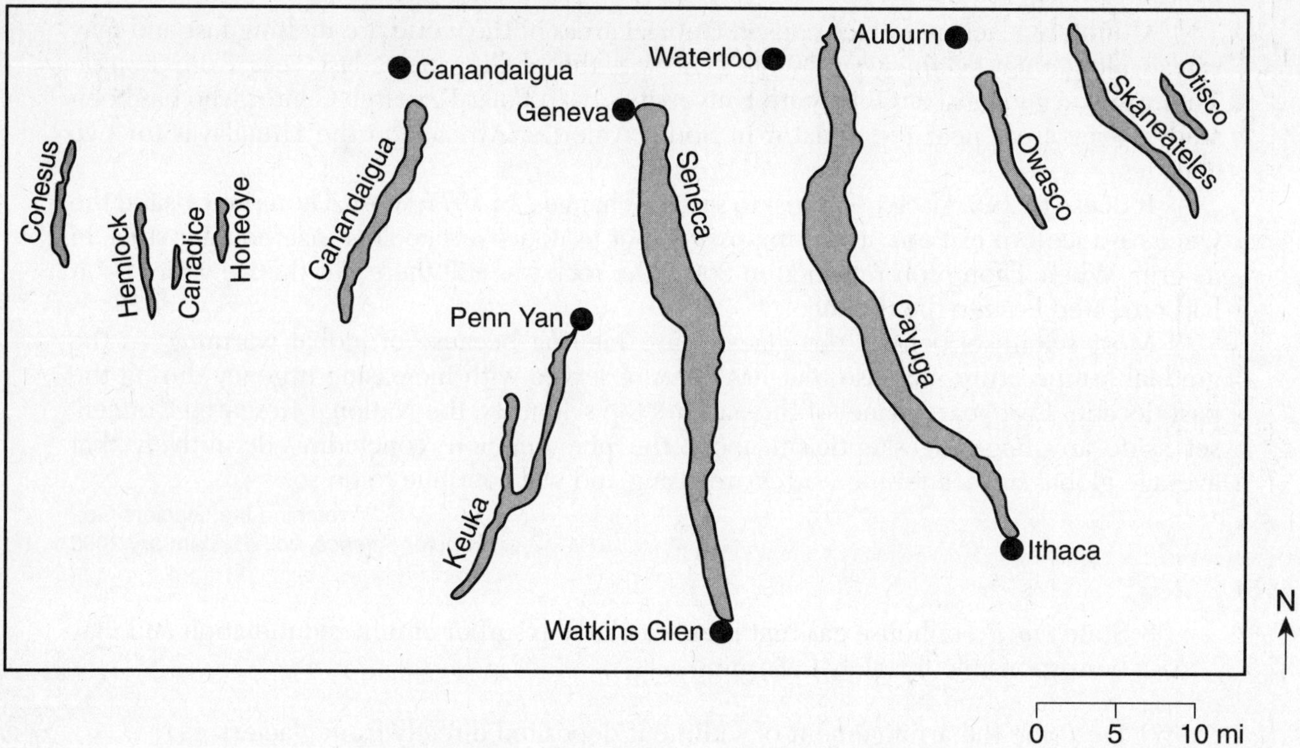

Elevation and Depth of the Finger Lakes of New York State (measured in feet)

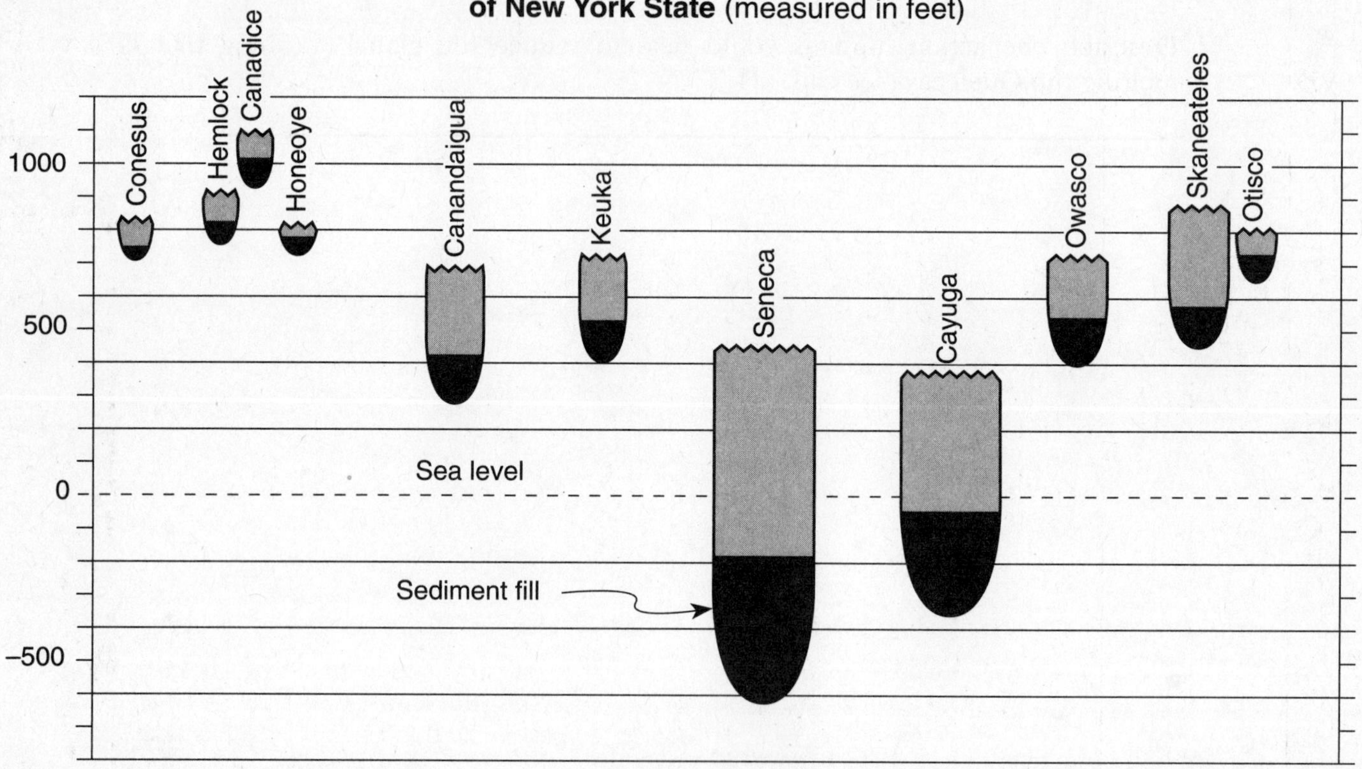

72 According to the cross section, how thick from top to bottom is the sediment fill in Seneca Lake? [1]

73 State *one* possible explanation for the north-south orientation of the Finger Lakes. [1]

74 During some winters, a few of the Finger Lakes remain unfrozen even though the land around the lakes is frozen. Explain how the specific heat of water can cause these lakes to remain unfrozen. [1]

75 Identify *two* processes that normally occur to form the type of surface bedrock found in the Finger Lakes Region. [1]

Base your answers to questions 76 through 79 on the diagram provided in your answer booklet, which shows observations made by a sailor who left his ship and landed on a small deserted island on June 21. The diagram represents the apparent path of the Sun and the position of *Polaris,* as observed by the sailor on this island.

76 On the diagram provided *in your answer booklet,* draw an arrow on the June 21 path of the Sun to show the Sun's direction of apparent movement from sunrise to sunset. [1]

77 The sailor was still on the island on September 23. On the diagram provided *in your answer booklet,* draw the Sun's apparent path for September 23, as it would have appeared to the sailor. Be sure your September 23 path indicates the correct altitude of the noon Sun and begins and ends at the correct points on the horizon. [2]

78 Based on the sailor's observations, what is the latitude of this island? Include the units and the compass direction in your answer. [1]

79 The sailor observed a 1-hour difference between solar noon on the island and solar noon at his last measured longitude onboard his ship. How many degrees of longitude is the island from the sailor's last measured longitude onboard his ship? [1]

Base your answers to questions 80 through 83 on the world map shown below and on your knowledge of Earth science. Letters *A* through *H* represent locations on Earth's surface.

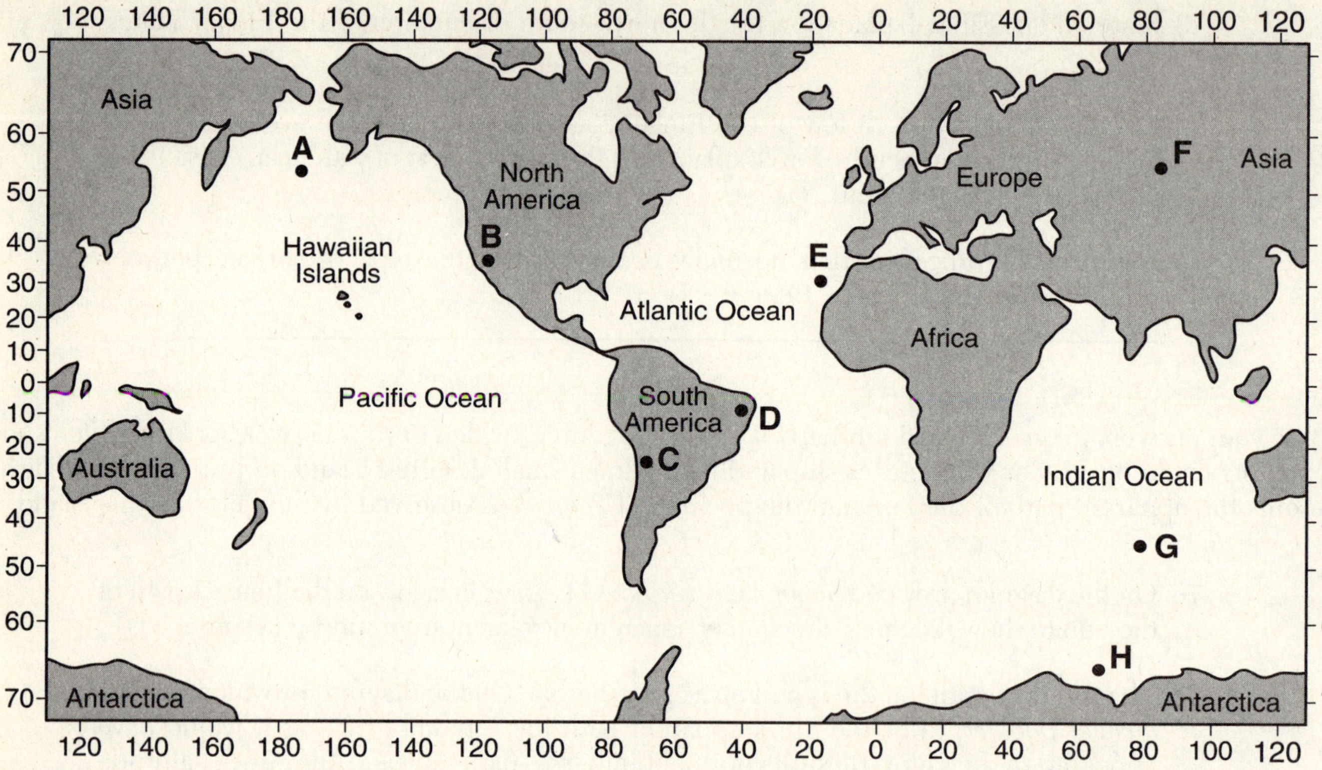

80 Explain why most earthquakes that occur in the crust beneath location *B* are shallower than most earthquakes that occur in the crust beneath location *C*. [1]

81 Explain why location *A* has a greater probability of experiencing a major earthquake than location *D*. [1]

82 Explain why a volcanic eruption is more likely to occur at location *E* than at location *F*. [1]

83 Explain why the geologic age of the oceanic bedrock increases from location *G* to location *H*. [1]

Part A

Answer all questions in this part.

Directions (1–35): For *each* statement or question, write on your separate answer sheet the *number* of the word or expression that, of those given, best completes the statement or answers the question. Some questions may require the use of the *Earth Science Reference Tables*.

1 Which planet's orbit around the Sun is most nearly circular?

(1) Mercury
(2) Neptune
(3) Pluto
(4) Venus

2 How many times will the Sun's perpendicular rays cross Earth's Equator between March 1 of one year and March 1 of the next year?

(1) 1
(2) 2
(3) 3
(4) 4

3 Which star's surface temperature is closest to the temperature at the boundary between Earth's mantle and core?

(1) *Sirius*
(2) *Rigel*
(3) the Sun
(4) *Betelgeuse*

4 Which graph best represents the relationship between soil particle size and the rate at which water infiltrates permeable soil?

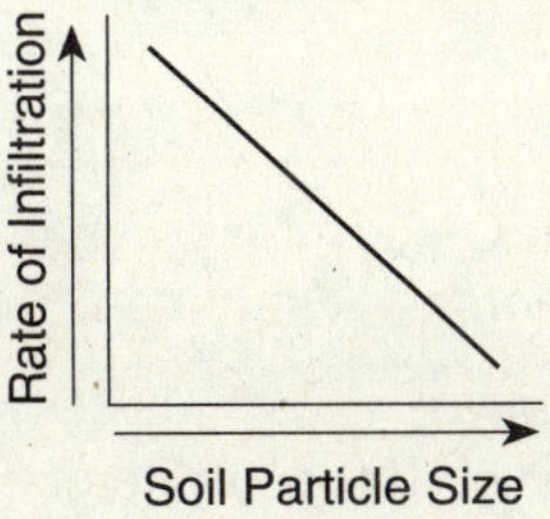

(1)

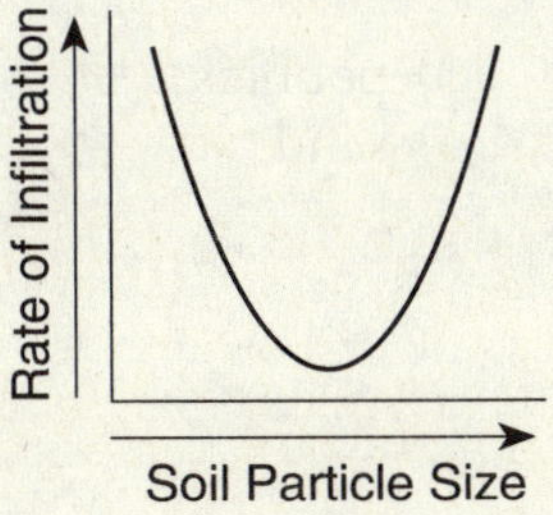

(2)

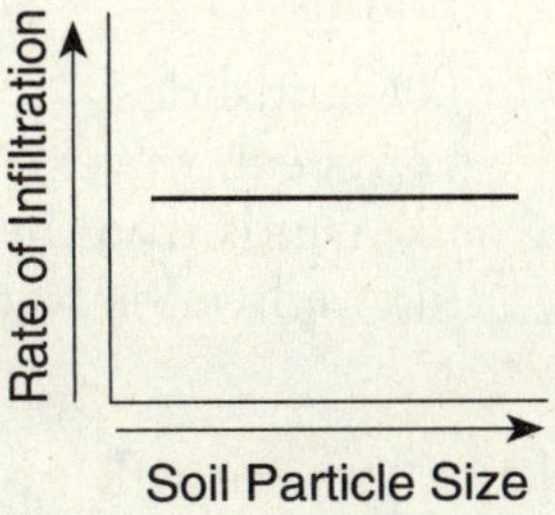

(3)

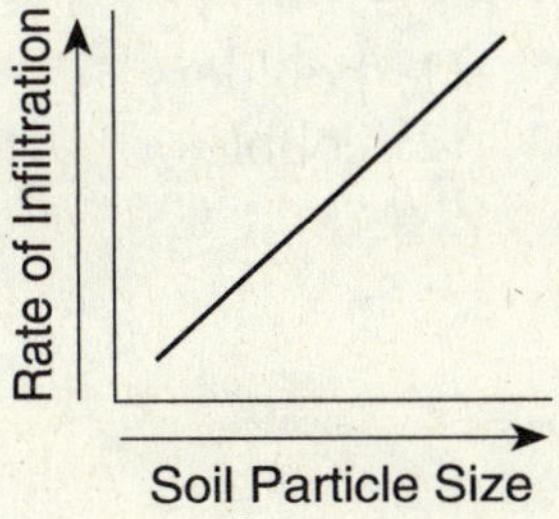

(4)

5 The diagram below shows a standard spectrum compared to a spectrum produced from a distant star.

Standard Spectrum

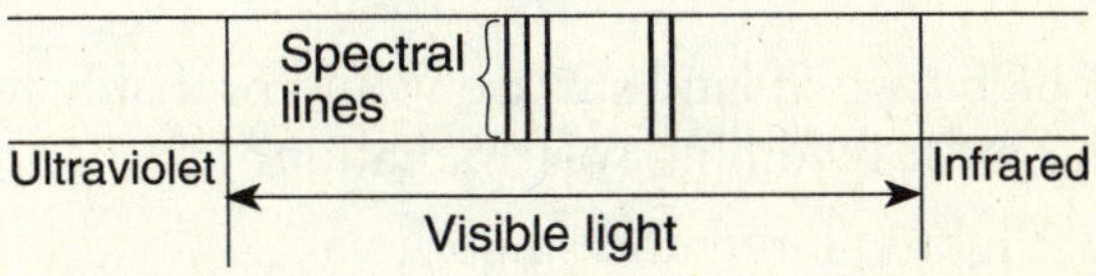

Spectrum from Distant Star

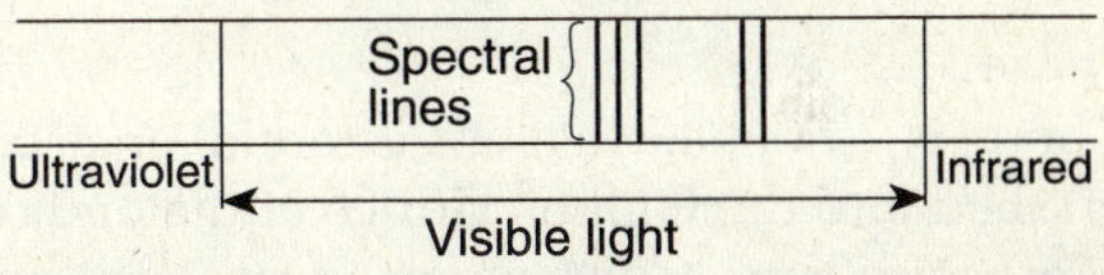

Which conclusion can be made by comparing the standard spectrum to the spectrum produced from this distant star?

(1) The star's spectral lines have shifted toward the ultraviolet end of the spectrum and the star is moving toward Earth.
(2) The star's spectral lines have shifted toward the ultraviolet end of the spectrum and the star is moving away from Earth.
(3) The star's spectral lines have shifted toward the infrared end of the spectrum and the star is moving toward Earth.
(4) The star's spectral lines have shifted toward the infrared end of the spectrum and the star is moving away from Earth.

6 Scientists are concerned about the decrease in ozone in the upper atmosphere primarily because ozone protects life on Earth by absorbing certain wavelengths of

(1) x-ray radiation
(2) ultraviolet radiation
(3) infrared radiation
(4) microwave radiation

7 It is inferred that during the early Archean Era the atmosphere of Earth contained water vapor, carbon dioxide, nitrogen, and other gases in small amounts. These gases probably came from

(1) precipitation of groundwater
(2) volcanic eruptions
(3) evaporation of Paleozoic oceans
(4) convection currents in the mantle

8 Which ocean current flows northeast along the eastern coast of North America?

(1) Gulf Stream (3) California
(2) North Equatorial (4) Labrador

9 Which type of land surface would probably reflect the most incoming solar radiation?

(1) light colored and smooth
(2) light colored and rough
(3) dark colored and smooth
(4) dark colored and rough

10 There is evidence that an asteroid or a comet crashed into the Gulf of Mexico at the end of the Mesozoic Era. Consequences of this impact event may explain the

(1) extinction of many kinds of marine animals, including trilobites
(2) extinction of ammonoids and dinosaurs
(3) appearance of the earliest birds and mammals
(4) appearance of great coal-forming forests and insects

11 What is the approximate location of the Canary Islands hot spot?

(1) 32° S 18° W (3) 32° N 18° W
(2) 32° S 18° E (4) 32° N 18° E

12 How long would it take for the first *S*-wave to arrive at a seismic station 4,000 kilometers away from the epicenter of an earthquake?

(1) 5 min 40 sec (3) 12 min 40 sec
(2) 7 min 0 sec (4) 13 min 20 sec

13 Which New York State river flows generally southward?

(1) St. Lawrence River (3) Genesee River
(2) Niagara River (4) Hudson River

14 The map below shows the stream drainage patterns for a region of Earth's surface. Points *A*, *B*, *C*, and *D* are locations in the region.

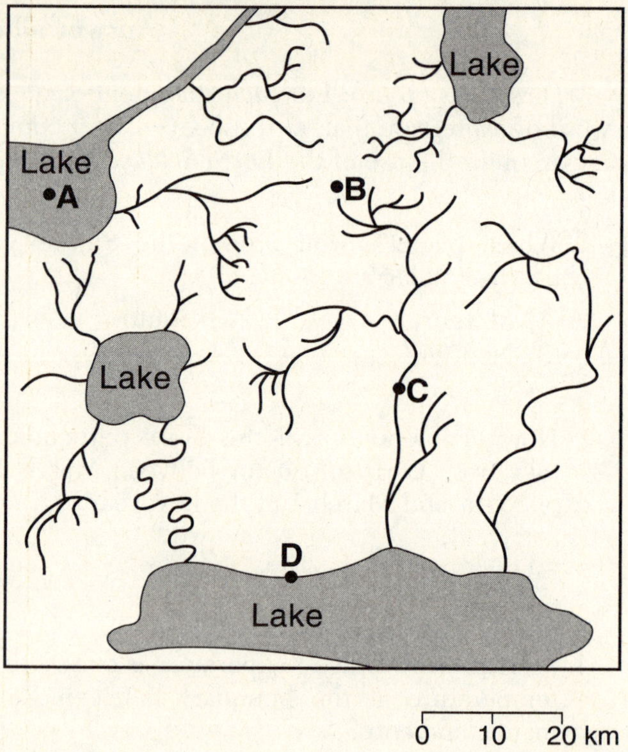

The highest elevation most likely exists at point

(1) *A* (3) *C*
(2) *B* (4) *D*

15 Outwash plains are formed as a result of deposition by

(1) landslides
(2) ocean waves
(3) winds from hurricanes
(4) meltwater from glaciers

16 A stream with a velocity of 100 centimeters per second flows into a lake. Which sediment-size particles would the stream most likely deposit first as it enters the lake?

(1) boulders (3) pebbles
(2) cobbles (4) sand

17 The cross section below shows a soil profile.

This soil was formed primarily by

(1) erosion by glaciers
(2) erosion by running water
(3) capillarity and human activity
(4) weathering and biological activity

18 Which type of rock most likely contains fossils?

(1) scoria
(2) gabbro
(3) schist
(4) shale

19 In which New York State landscape region is most of the surface bedrock composed of metamorphic rock?

(1) Adirondacks
(2) Catskills
(3) Erie-Ontario Lowlands
(4) Newark Lowlands

20 A human fingernail has a hardness of approximately 2.5. Which two minerals are *softer* than a human fingernail?

(1) calcite and halite
(2) sulfur and fluorite
(3) graphite and talc
(4) pyrite and magnetite

21 The photograph below shows an escarpment (cliff) located in the western United States. The directions for north and south are indicated by arrows. A fault in the sedimentary rocks is shown on the front of the escarpment.

The photograph shows that the fault most likely formed

(1) after the rock layers were deposited, when the north side moved downward
(2) after the rock layers were deposited, when the north side moved upward
(3) before the rock layers were deposited, when the south side moved downward
(4) before the rock layers were deposited, when the south side moved upward

22 Which mountain range resulted from the collision of North America and Africa, as parts of Pangea joined together in the late Pennsylvanian Period?

(1) Appalachian Mountains
(2) Acadian Mountains
(3) Taconic Mountains
(4) Grenville Mountains

23 Which physical characteristic best describes the rock phyllite?

(1) glassy texture with gas pockets
(2) clastic texture with angular fragments
(3) bioclastic texture with cemented shell fragments
(4) foliated texture with microscopic mica crystals

24 The diagram below represents a part of the crystal structure of the mineral kaolinite.

Structure of Kaolinite

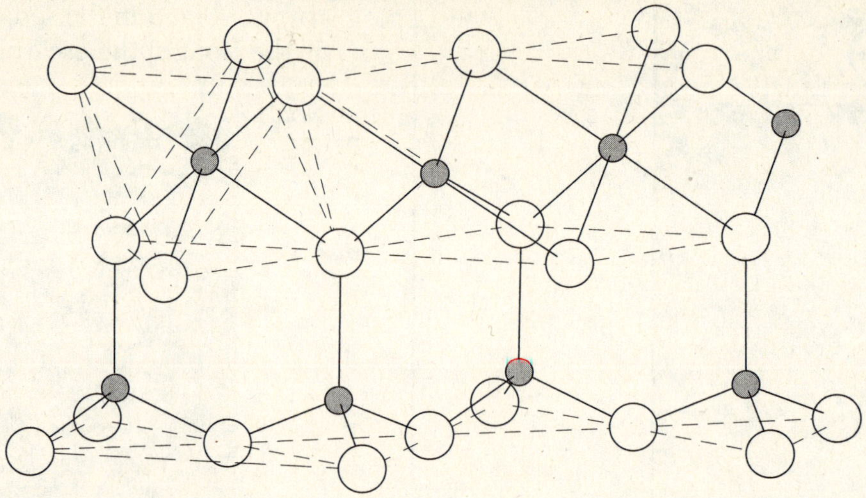

An arrangement of atoms such as the one shown in the diagram determines a mineral's

(1) age of formation
(2) infiltration rate
(3) physical properties
(4) temperature of formation

---

25 The map below shows the path of a river. The arrow shows the direction the river is flowing. Letters *A* and *B* identify the banks of the river.

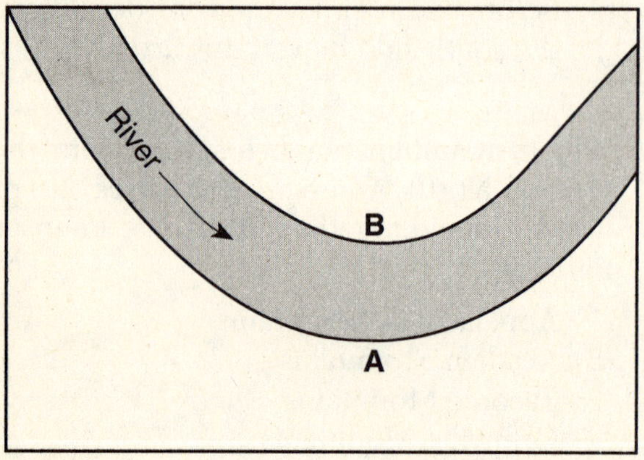

The water depth is greater near bank *A* than bank *B* because the water velocity near bank *A* is

(1) faster, causing deposition to occur
(2) faster, causing erosion to occur
(3) slower, causing deposition to occur
(4) slower, causing erosion to occur

26 Which home-building material is made mostly from the mineral gypsum?

(1) plastic pipes
(2) window glass
(3) drywall panels
(4) iron nails

27 The two most abundant elements by mass in Earth's crust are oxygen and

(1) potassium
(2) hydrogen
(3) nitrogen
(4) silicon

Base your answers to questions 28 through 30 on the diagram below, which represents the planetary wind and moisture belts in Earth's Northern Hemisphere.

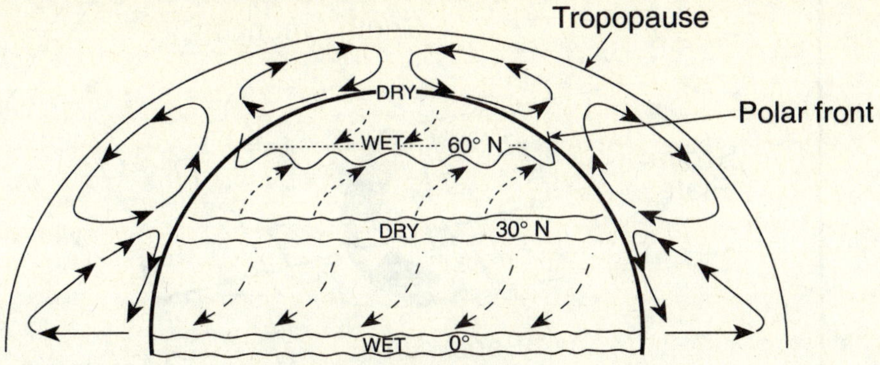

(Not drawn to scale)

28 The climate at 90° north latitude is dry because the air at that location is usually

(1) warm and rising
(2) warm and sinking
(3) cool and rising
(4) cool and sinking

29 The paths of the surface planetary winds are curved due to Earth's

(1) revolution
(2) rotation
(3) circumference
(4) size

30 The tropopause is approximately how far above sea level?

(1) 12 mi
(2) 12 km
(3) 60 mi
(4) 60 km

---

31 The diagram below shows four magnified block-shaped sandstone samples labeled *A*, *B*, *C*, and *D*. Each sandstone sample contains quartz grains of different shapes and sizes. The quartz grains are held together by hematite cement.

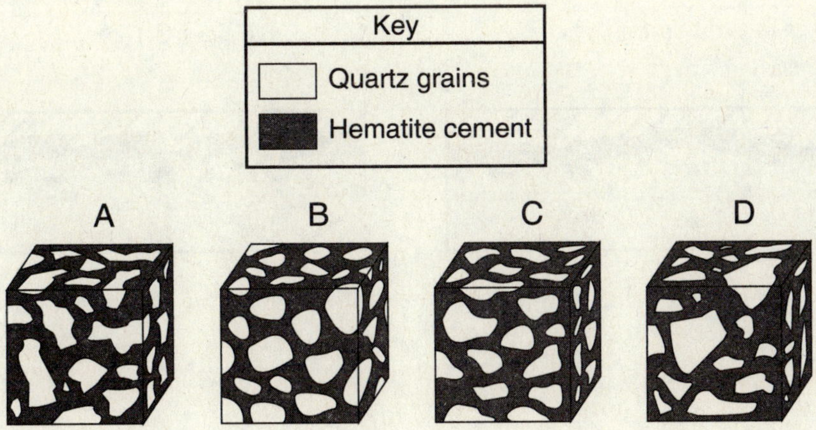

In which sample did the quartz grains undergo the most abrasion during erosional transport?

(1) *A*
(2) *B*
(3) *C*
(4) *D*

32 On the map below, line *AB* is drawn across several of Earth's tectonic plates in the South Atlantic Ocean.

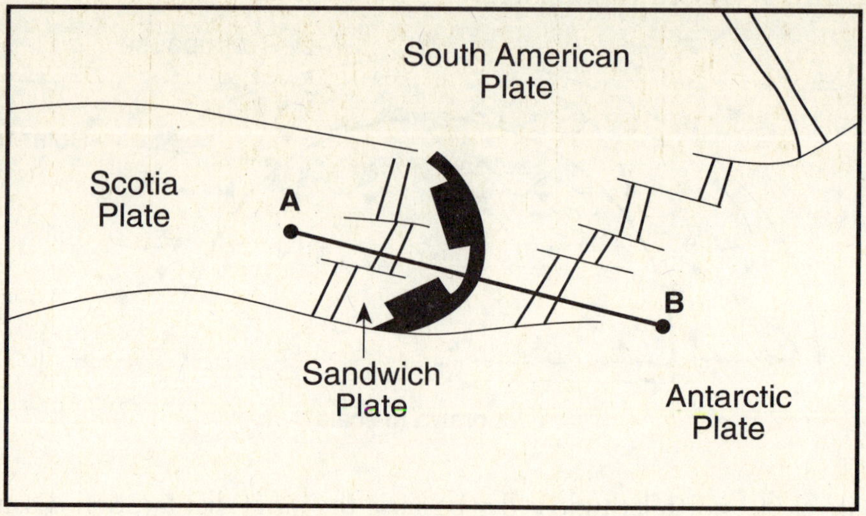

Which cross section best represents the plate boundaries and mantle movement beneath line *AB*?

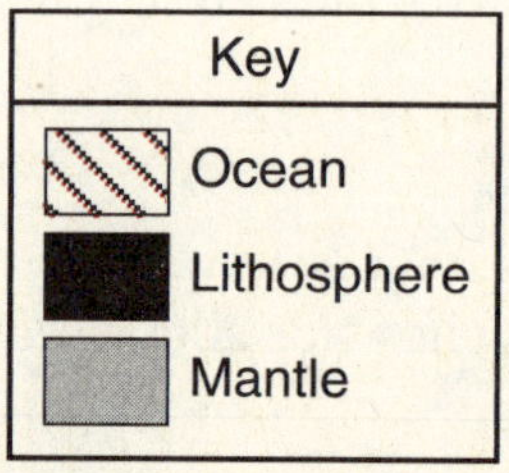

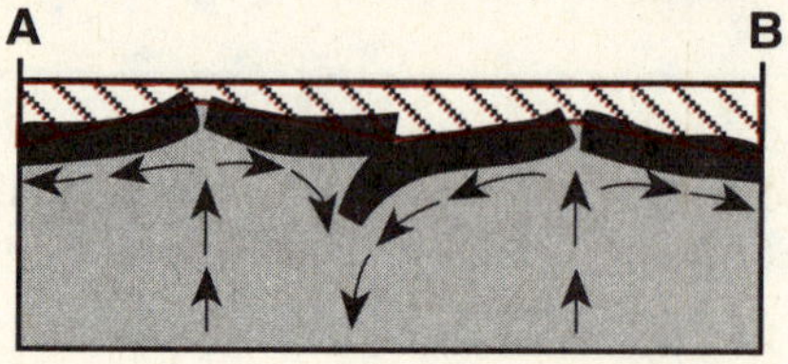

(1)

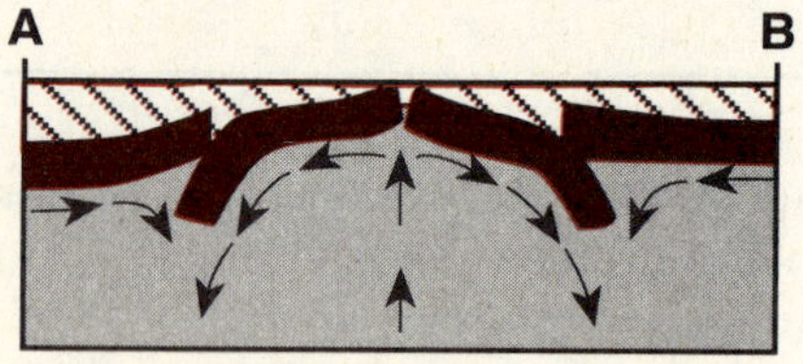

(3)

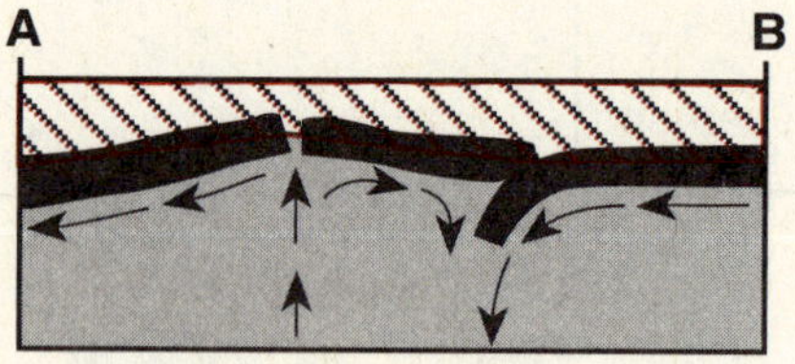

(2)

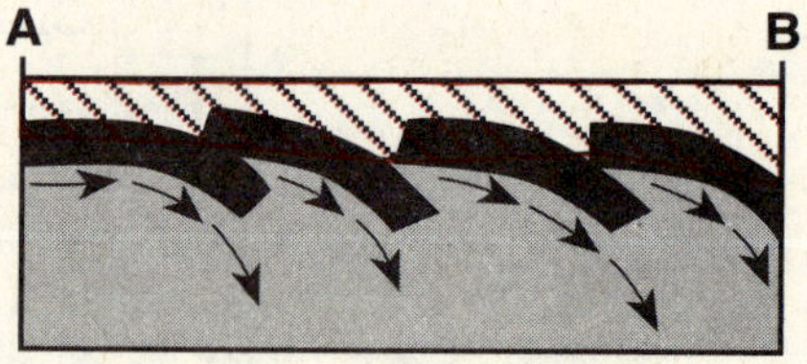

(4)

33 The diagrams below show four major types of fault motion occurring in Earth's crust. Which type of fault motion best matches the general pattern of crustal movement at California's San Andreas fault?

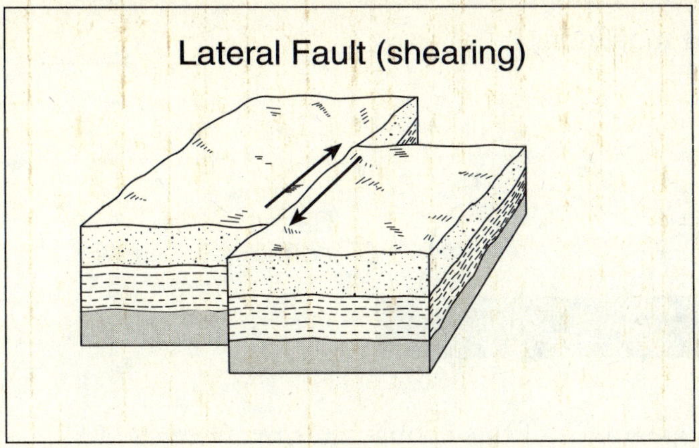

(1)

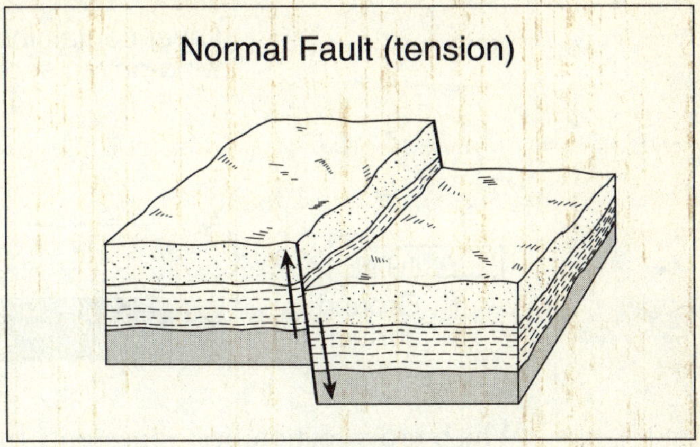

(3)

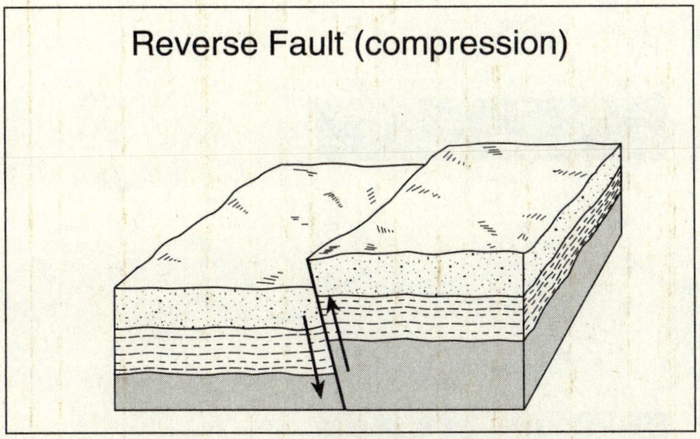

(2)

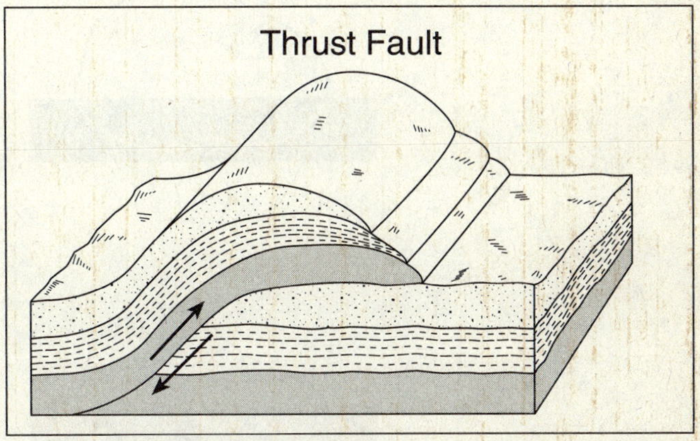

(4)

34 The cross section below shows the movement of wind-driven sand particles that strike a partly exposed basalt cobble located at the surface of a windy desert.

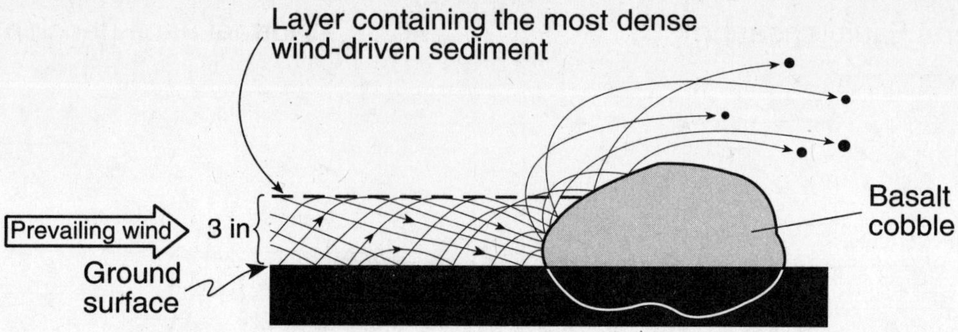

Which cross section best represents the appearance of this cobble after many years of exposure to the wind-driven sand?

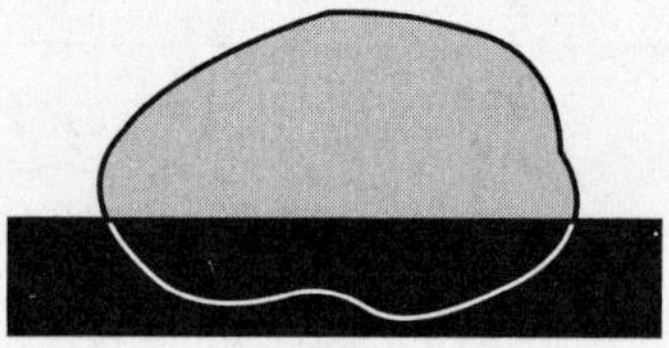

(1)

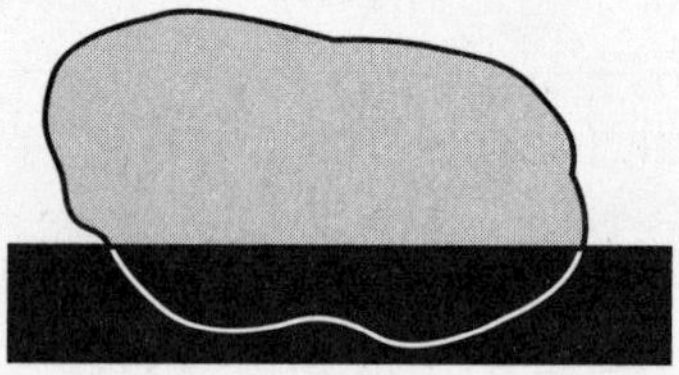

(3)

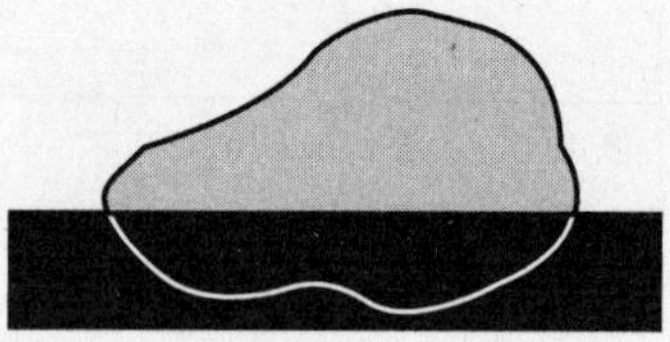

(2)

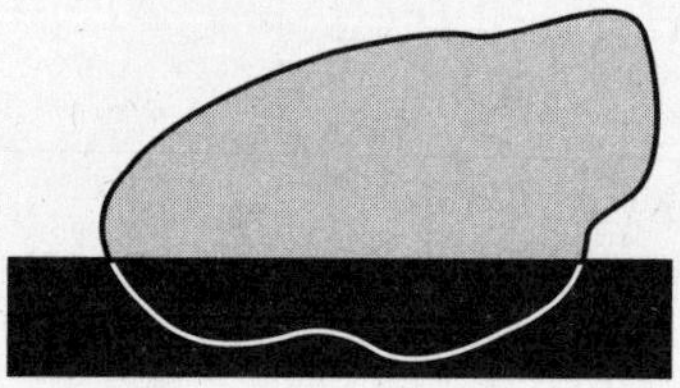

(4)

35 Each dot on the graph below shows the result of separate scientific studies of the relationship between the rates of erosion in regions of different relief. Relief is the local difference between the highest and the lowest elevations.

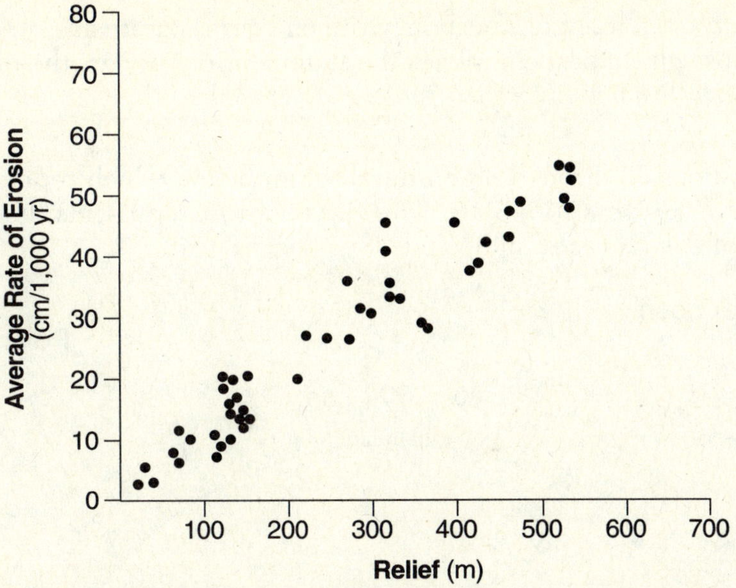

The results of these combined studies indicate that with each 100-meter increase in relief, the rate of erosion generally

(1) decreases at a rate of 10 cm/1,000 years
(2) decreases at a rate of 20 cm/1,000 years
(3) increases at a rate of 10 cm/1,000 years
(4) increases at a rate of 20 cm/1,000 years

Part B–1

Answer all questions in this part.

Directions (36–50): For *each* statement or question, write on your separate answer sheet the *number* of the word or expression that, of those given, best completes the statement or answers the question. Some questions may require the use of the *Earth Science Reference Tables*.

Base your answers to questions 36 through 38 on the diagram below, which represents a model of the sky (celestial sphere) for an observer in New York State. The curved arrow represents the Sun's apparent path for part of one day. The altitude of *Polaris* is also indicated.

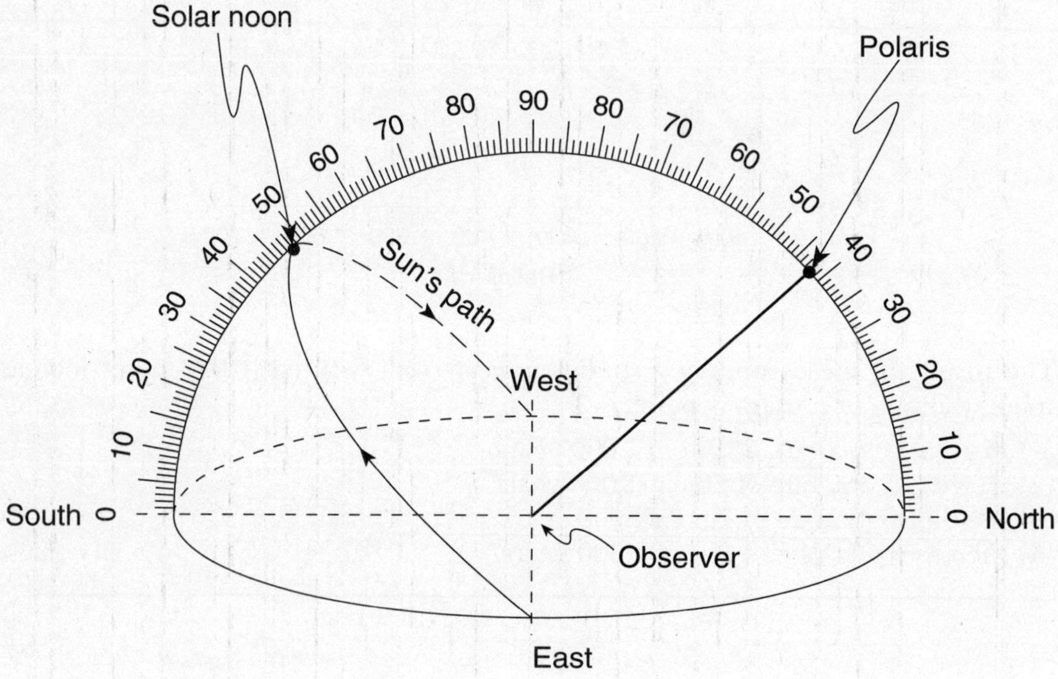

36 According to this diagram, what is the Sun's altitude at solar noon?

(1) 23.5° (3) 48°
(2) 42° (4) 90°

37 Where is this observer most likely located?

(1) Massena (3) Slide Mountain
(2) Oswego (4) Mt. Marcy

38 On which date could this observation of the Sun's apparent path have been made?

(1) March 21 (3) October 21
(2) July 21 (4) December 21

Base your answers to questions 39 through 42 on the weather map below. The map shows a low-pressure system and some atmospheric conditions at weather stations *A*, *B*, and *C*.

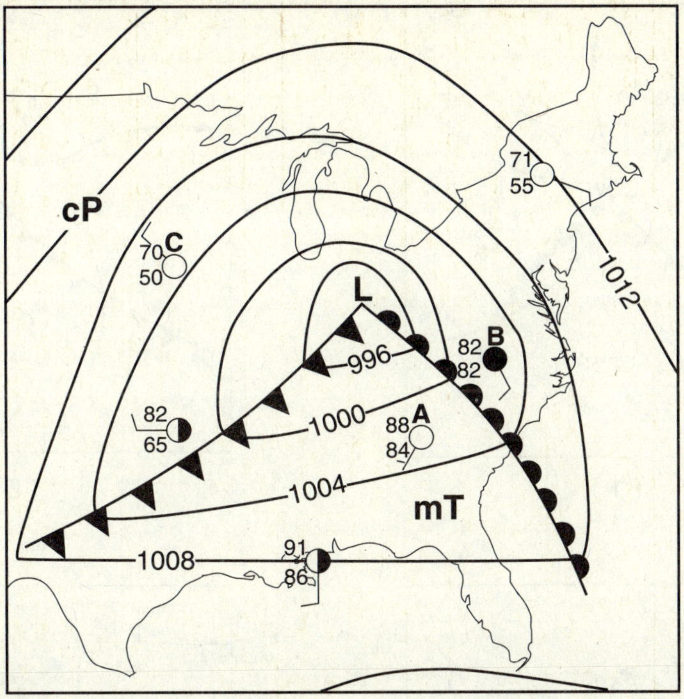

39 Which type of weather is usually associated with a cP air mass, as shown near weather station *C*?

(1) moist and cool
(2) moist and warm
(3) dry and cool
(4) dry and warm

40 Which cross section best represents the air masses, air movement, clouds, and precipitation occurring behind and ahead of the warm front located between stations *A* and *B*?

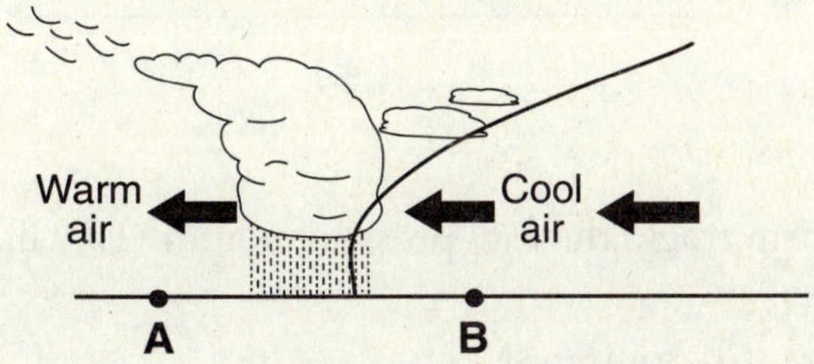

(1)

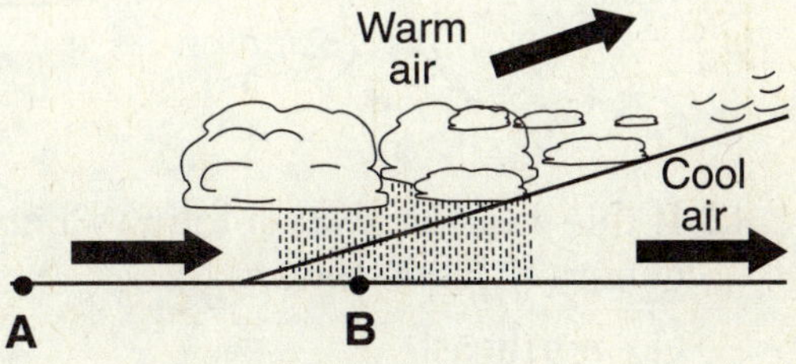

(3)

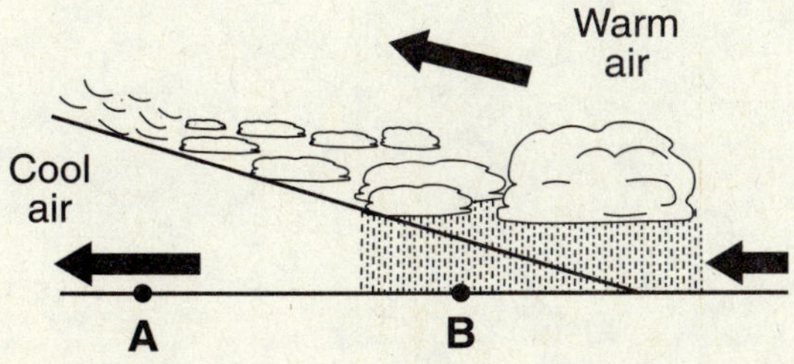

(2)

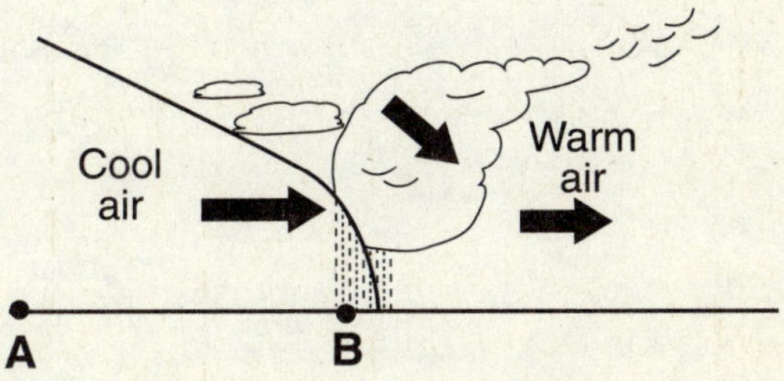

(4)

41 The arrows on which map best represent the direction of surface winds associated with this low-pressure system?

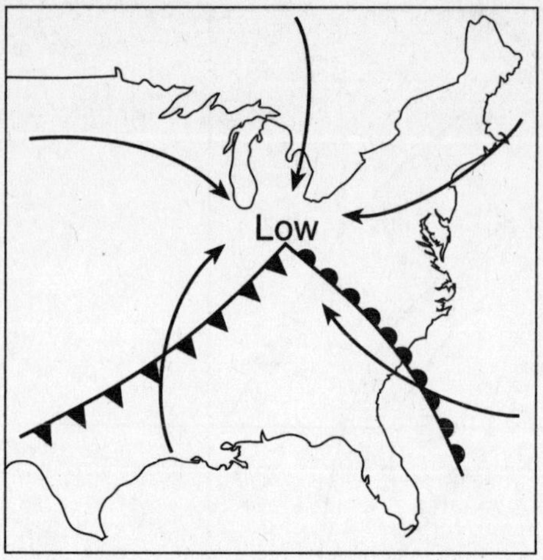

(1)

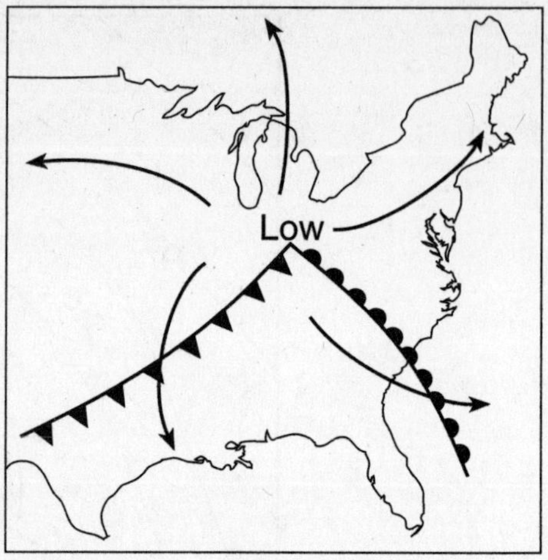

(3)

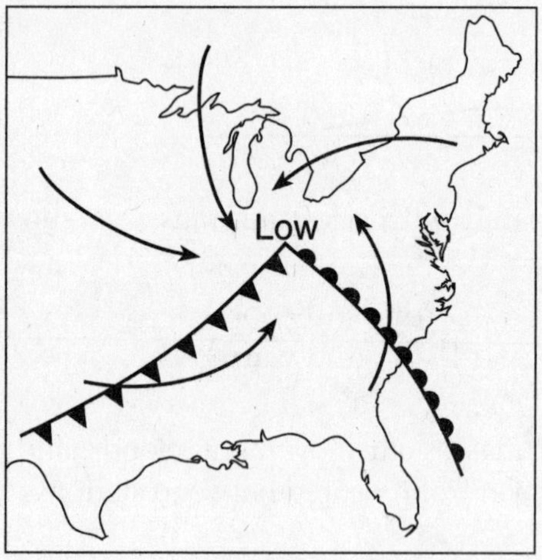

(2)

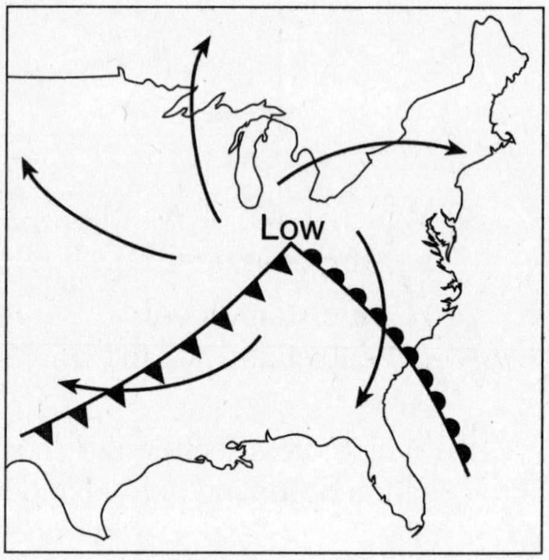

(4)

42 If this weather system follows a normal storm track, the low-pressure center (L) will generally move toward the

(1) northeast
(2) northwest
(3) southeast
(4) southwest

Base your answers to questions 43 and 44 on the bedrock cross section below. The cross section represents part of Earth's crust where natural gas, oil, and water have moved upward through a layer of folded sandstone and filled the pore spaces at the top of the sandstone layer.

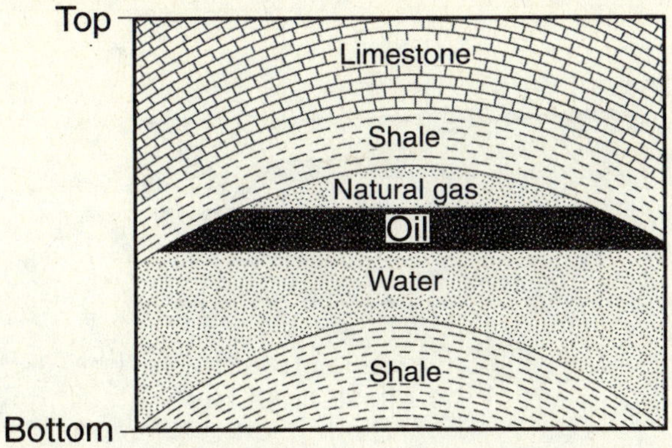

43 The final arrangement of the natural gas, oil, and water within the sandstone was caused by differences in their

(1) density
(2) specific heat
(3) relative age
(4) radioactive half-life

44 The natural gas, oil, and water are trapped within the top of the sandstone and do not move upward through the shale because, compared to the sandstone, the shale has

(1) lower permeability
(2) less foliation
(3) larger pore spaces
(4) larger particles

Base your answers to questions 45 through 47 on the map below, which shows the drainage basin of the Mississippi River system. Several rivers that flow into the Mississippi River are labeled. The arrow at location *X* shows where the Mississippi River enters the Gulf of Mexico.

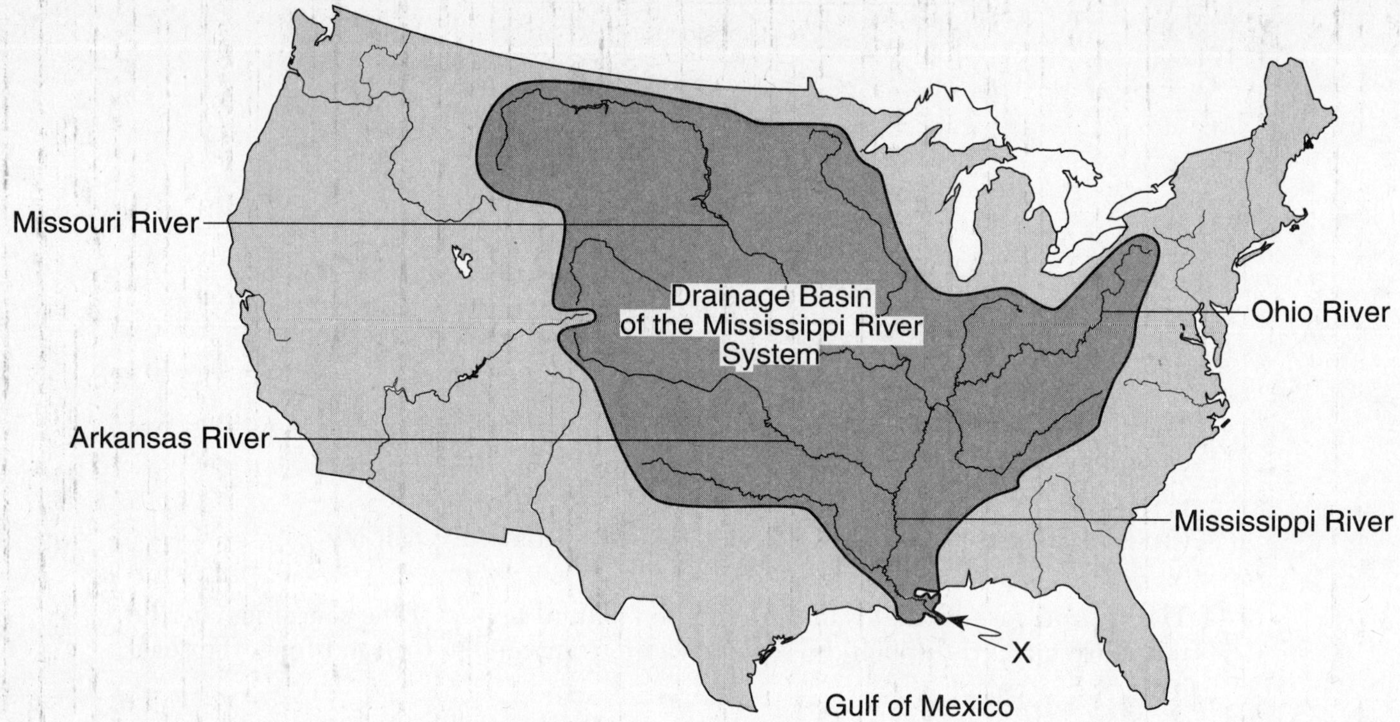

45 The entire land area drained by the Mississippi River system is referred to as a

(1) levee
(2) watershed
(3) meander belt
(4) floodplain

46 Sediments deposited at location *X* by the Missisippi River most likely have which characteristics?

(1) angular fragments arranged as mixtures
(2) rock particles arranged in sorted beds
(3) rocks with parallel scratches and grooves
(4) high-density minerals with hexagonal crystals

47 The structure formed by the deposition of sediments at location *X* is best described as a

(1) moraine
(2) tributary
(3) delta
(4) drumlin

Base your answers to questions 48 through 50 on the reading passage and the drawing below and on your knowledge of Earth science.

Fossil With Signs of Feathers Is Cited as Bird-Dinosaur Link

Paleontologists have discovered in China a fossil dinosaur with what are reported to be clear traces of feathers from head to tail, the most persuasive evidence so far, scientists say, that feathers predated the origin of birds and that modern birds are descendants of dinosaurs.

Entombed in fine-grained rock, the unusually well-preserved skeleton resembles that of a duck with a reptilian tail, altogether about three feet in length. Its head and tail are edged with the imprint of downy fibers. The rest of the body, except for bare lower legs, shows distinct traces of tufts and filaments that appear to have been primitive feathers. On the backs of its short forelimbs are patterns of what look like modern bird feathers.

Other dinosaur remains with what appear to be featherlike traces have been unearthed in recent years, but nothing as complete as this specimen, paleontologists said. Etched in the rock like a filigree decoration surrounding the skeleton are imprints of where the down and feathers appear to have been.

The 130-million-year-old fossils were found a year ago by farmers in Liaoning Province in northeastern China. After an analysis by Chinese and American researchers, the fossil animal was identified as a dromaeosaur, a small fast-running dinosaur related to velociraptor. The dinosaurs belonged to a group of two-legged predators known as advanced theropods . . .

excerpted from "Fossil With Signs of
Feathers Is Cited as Bird-Dinosaur Link"
John Noble Wilford
New York Times, April 26, 2001

The drawing below shows an artist's view of the dinosaur, based on the fossilized remains.

48 During which period of geologic time have paleontologists inferred that the feathered dinosaur mentioned in the passage existed?

(1) Cambrian
(2) Cretaceous
(3) Paleogene
(4) Permian

49 This feathered dinosaur is not considered an index fossil because it

(1) existed too long ago
(2) was preserved in ash
(3) was a land-dwelling animal
(4) was found in only one area

50 The reference to the bird-dinosaur link is most likely referring to the concept of

(1) plate tectonics
(2) evolution
(3) dynamic equilibrium
(4) recycling

Part B–2

Answer all questions in this part.

Directions (51–62): Record your answers in the spaces provided in your answer booklet. Some questions may require the use of the *Earth Science Reference Tables.*

Base your answers to questions 51 and 52 on the diagram provided *in your answer booklet.* The diagram shows a model of Earth's orbit around the Sun. Two motions of Earth are indicated. Distances to the Sun are given for two positions of Earth in its orbit.

51 On the diagram provided *in your answer booklet*, place an **X** on Earth's orbit to indicate Earth's position on May 21. [1]

52 Explain why New York State experiences summer when Earth is at its greatest distance from the Sun. [1]

____

Base your answers to questions 53 through 55 on the field map provided *in your answer booklet*. The map shows elevations, measured in feet, of a number of points in a certain geographic region. Contour lines have been drawn for the 100-foot and 120-foot elevations. Points *A* and *B* represent two spot elevations on the map.

53 On the map provided *in your answer booklet,* draw the 60-foot contour line. Make sure that the contour line extends to the edges of the map. [1]

54 Toward which general compass direction does Elma Creek flow? [1]

55 Calculate the gradient between points *A* and *B*. Label the answer with the correct units. [2]

____

56 On the weather map station model provided *in your answer booklet*, using the proper format, record the *six* weather conditions shown below. [2]

Wind: from the northwest
Wind speed: 10 knots
Barometric pressure: 1022.0 mb
Cloud cover: 50%
Visibility: 5 mi
Precipitation (in the past 6 hours): .45 in

Base your answers to questions 57 and 58 on the map and table below. The map shows the name and location of the volcanic peaks in the Cascade Mountain Range of the northwestern United States west of the Yellowstone Hot Spot. The table shows the major eruptions of each peak over the past 4,000 years.

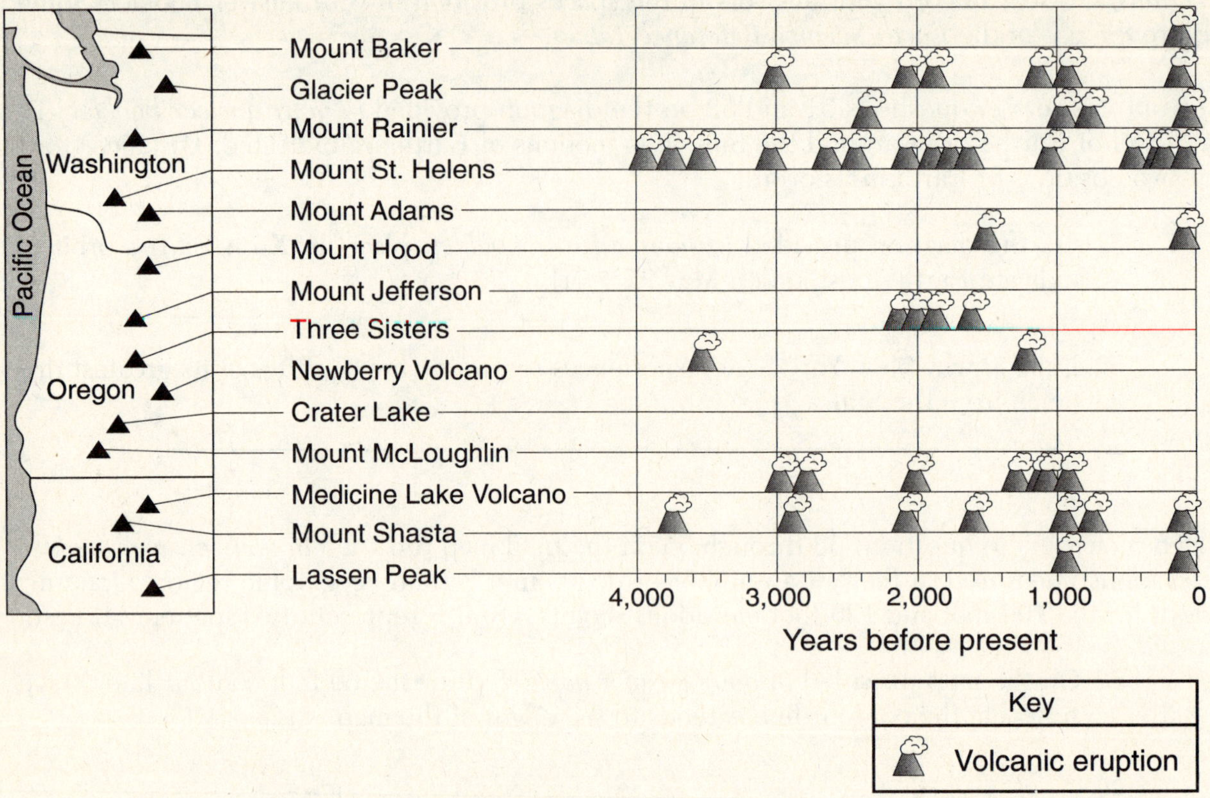

57 During which geologic epoch did the volcanic activity shown on the table occur? [1]

58 On the cross section provided *in your answer booklet,* place an arrow in the continental crust and an arrow in the oceanic crust to show the relative directions of plate movement. [1]

---

59 Which layer of Earth is composed of both the crust and the rigid mantle? [1]

Base your answers to questions 60 and 61 on the diagram below, which represents a model of the radioactive decay of a particular element. The diagram shows the decay of a radioactive element (▭) into the stable decay element (▭) after one half-life period.

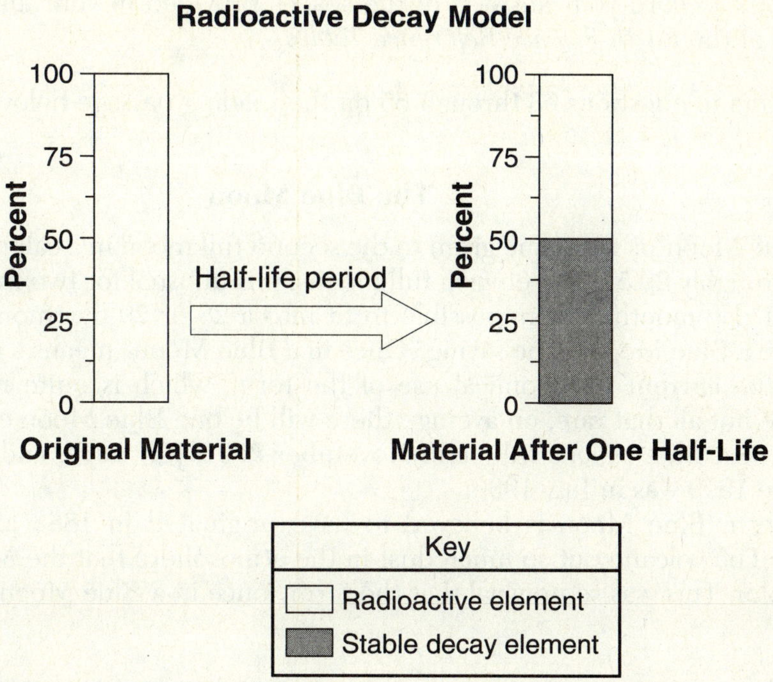

60 On the diagram provided *in your answer booklet,* shade in the amount of stable decay element present after the second half-life period. [1]

61 If the radioactive element in this model is carbon-14, how much time will have passed after one half-life? [1]

62 The diagram below shows warm, moist air moving off the ocean and over a mountain, causing precipitation between points 1 and 2.

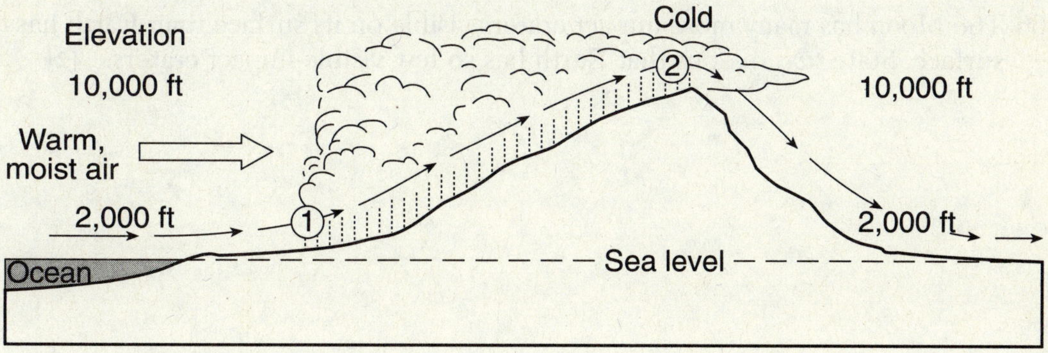

Describe *two* changes that occur to the warm, moist air between points 1 and 2 that would cause cloud formation. [2]

Part C

Answer all questions in this part.

Directions (63–80): Record your answers in the spaces provided in your answer booklet. Some questions may require the use of the *Earth Science Reference Tables.*

Base your answers to questions 63 through 65 on the reading passage below and on your knowledge of Earth science.

The Blue Moon

A "Blue Moon" is the name given to the second full moon in a calendar month. Because there are roughly 29.5 days between full moons, it is unusual for two full moons to "fit" into a 30 or 31 day month (and impossible to fit into a 28 or 29 day month, so February can never have a Blue Moon). The saying "Once in a Blue Moon" means a rare occurrence, and predates the current astronomical use of the term, which is quite recent. In fact, Blue Moons are not all that rare, on average there will be one Blue Moon every 2.5 years. After 1999, the next Blue Moons will be in November 2001; July 2004; and June 2007. The last one before 1999 was in July 1996.

The term Blue Moon is believed to have originated in 1883 after the eruption of Krakatoa. The volcano put so much dust in the atmosphere that the Moon actually looked blue in color. This was so unusual that the term "once in a Blue Moon" was coined.

"The Blue Moon"
David R. Williams
nssdc.gsfc.nasa.gov/planetary/lunar/blue_moon.html

63 Explain why a Blue Moon never occurs during the month of February. [1]

64 What is the greatest number of full-Moon phases, visible from Earth, that are possible in a span of 1 year? [1]

65 In the space provided *in your answer booklet,* draw the relative positions of Earth, the Moon, and the Sun, as viewed from space, so that a full-Moon phase would be visible to an observer on Earth. Label Earth, the Moon, and the Sun in your drawing. [1]

---

66 The Moon has many more impact craters visible on its surface than Earth has on its surface. State *two* reasons that Earth has so few visible impact craters. [2]

Base your answers to questions 67 and 68 on the map and data table shown below. The map shows some tectonic plates and the boundaries between them. Letters *A* and *B* are locations on Earth's surface. The data table shows the depth below Earth's surface of five earthquakes measured from location *A* toward location *B*.

Map

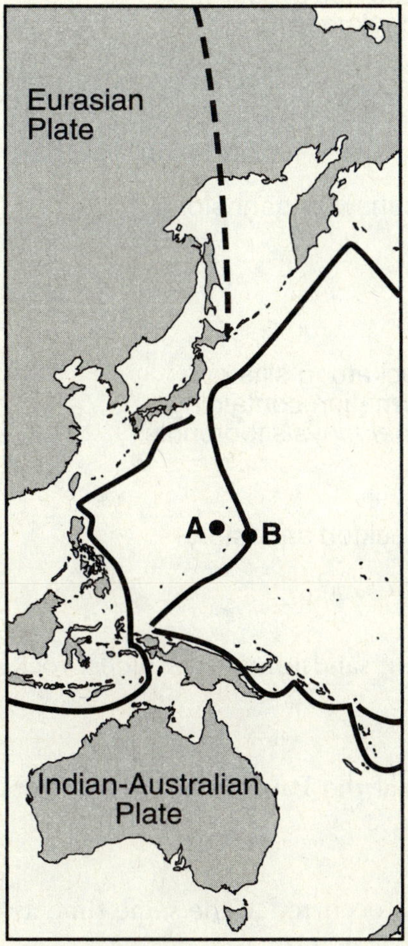

Data Table

Earthquake	Distance from Location *A* Toward Location *B* (km)	Depth Below Earth's Surface (km)
1	100	600
2	200	400
3	250	300
4	300	250
5	400	60

67 On the grid provided *in your answer booklet,* plot the depths of the *five* earthquakes from location *A* toward location *B*. [1]

68 Identify the type of plate boundary or geologic feature found at location *B*. [1]

____

Base your answers to questions 69 through 71 on the cross section below, which shows several rock formations found in New York State. The rock layers have not been overturned.

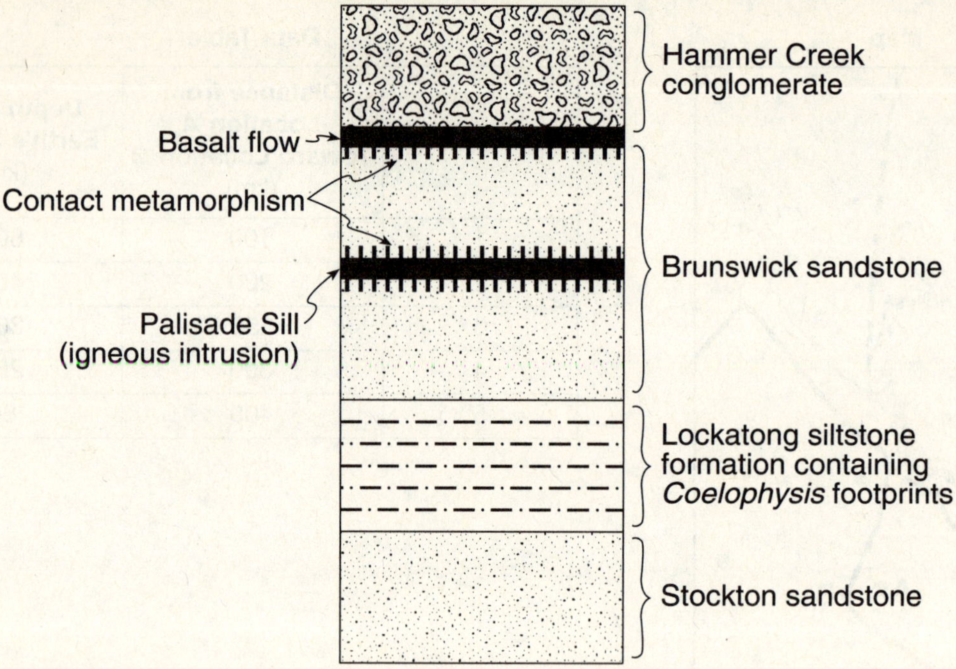

69 How does this cross section indicate that the Stockton sandstone is the oldest rock layer? [1]

70 State one piece of evidence that supports the fact that the Palisade Sill is younger than the Brunswick sandstone. [1]

71 State one tectonic event affecting North America that occurred at the same time as the Palisade Sill intrusion. [1]

Base your answers to questions 72 through 74 on the map below, the graph on the next page, and your knowledge of Earth science. The map shows the length of the growing season in New York State, expressed in days. The growing season is the average number of days between the last frost in spring and the first frost in fall. The graph line shows the relationship between the latitudes of Riverhead, New York; Albany, New York; and Massena, New York; and the length of the growing season at these three locations.

Length of Growing Season (in days)

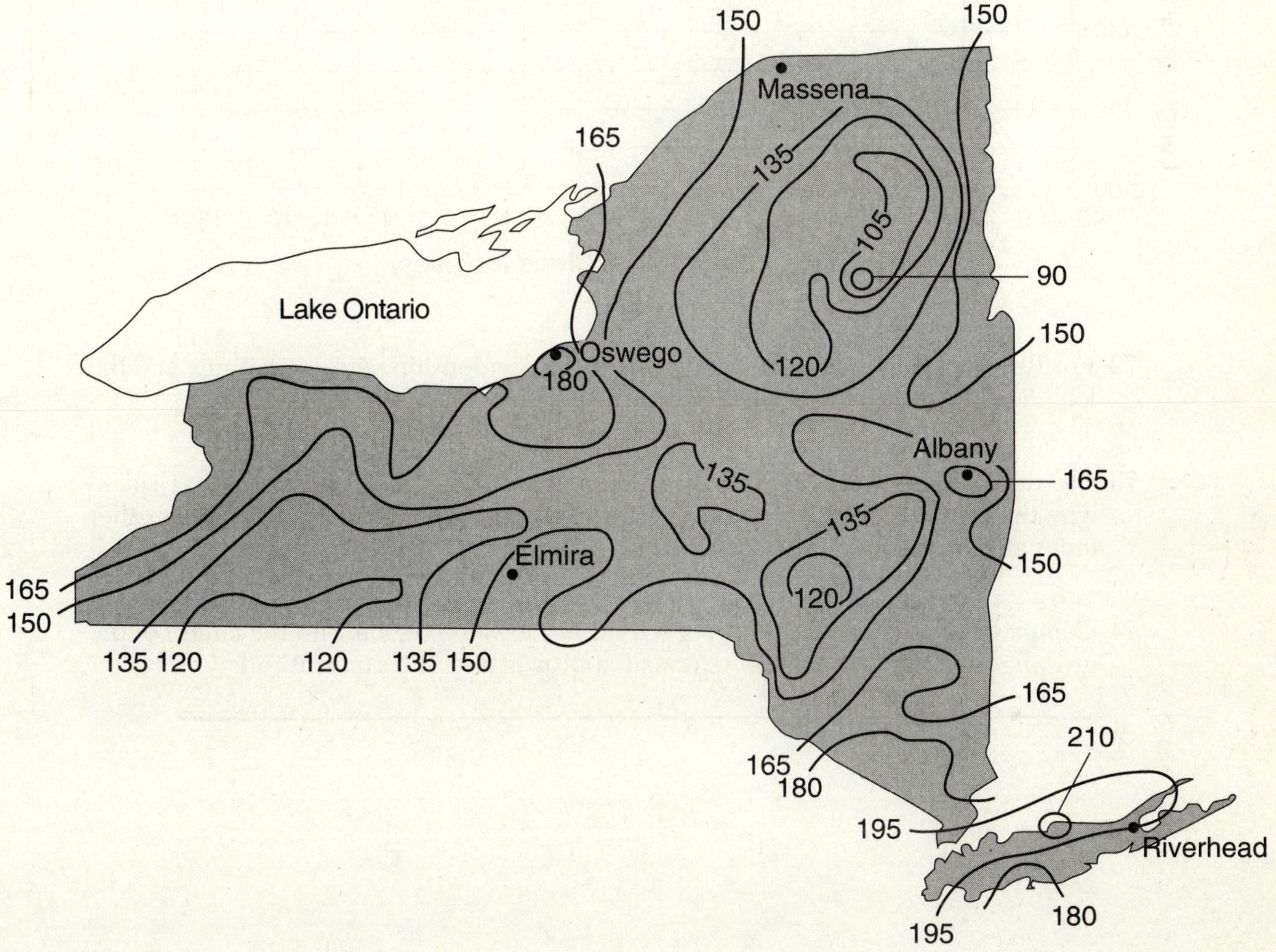

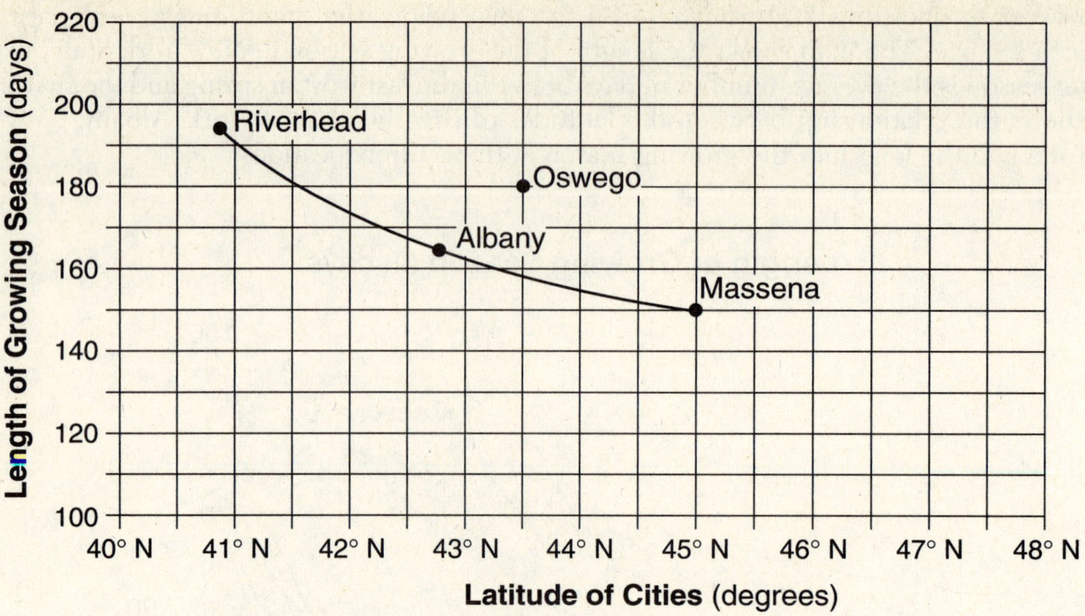

72 For Riverhead, Albany, and Massena, state the relationship between latitude and the length of the growing season shown by the graph. [1]

73 The data for Oswego, New York, have been plotted separately on the graph. Explain why the location of Oswego causes it to have a growing season longer than other cities at the same latitude. [1]

74 Compare the length of the growing season in a lowland region with the length of the growing season in a mountain region at approximately the same latitude. [1]

Base your answers to questions 75 through 77 on the cross section below and on your knowledge of Earth science. The cross section shows a portion of Earth's crust. The age, in millions of years, of each boundary between the different sedimentary rock layers is shown. The age of boundary *X* between the sedimentary rock and the metamorphic rock is not shown. Assume no overturning has occurred.

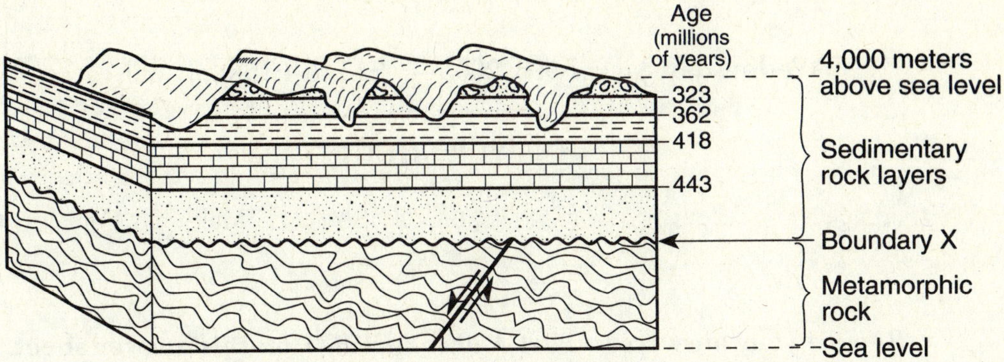

75 Identify the geologic feature represented by boundary *X*. [1]

76 Describe how the rock type below boundary *X* was formed. [1]

77 Identify by name one index fossil that existed when the limestone rock shown in the cross section was being formed. [1]

Base your answers to questions 78 through 80 on the diagram and table below. The diagram represents a felsic igneous rock. Letters *A*, *B*, and *C* represent three different minerals in the rock sample. The table describes the physical properties of minerals *A*, *B*, and *C* found in the igneous rock sample.

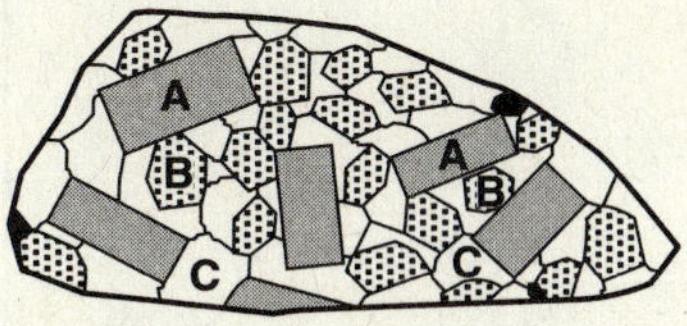

(Actual size)

Mineral	Key	Physical Properties
A		pink, cleaves in two directions at 90°
B		white, cleaves in two directions, striations visible
C		colorless or clear with a glassy luster

78 State the texture of this igneous rock. [1]

79 On the table provided *in your answer booklet*, state the names of minerals *A*, *B*, and *C*. [2]

80 State *two* processes responsible for the formation of an igneous rock. [1]

Tear Here

The University of the State of New York

REGENTS HIGH SCHOOL EXAMINATION

PHYSICAL SETTING
EARTH SCIENCE

Wednesday, January 26, 2005 — 1:15 to 4:15 p.m., only

ANSWER SHEET

Student .... Sex: ☐ Male ☐ Female Grade ....

Teacher .... School ....

Record your answers to Part A and Part B–1 on this answer sheet.

Part A

1	13	25 ....
2	14	26 ....
3	15	27 ....
4	16	28 ....
5	17	29 ....
6	18	30 ....
7	19	31 ....
8	20	32 ....
9	21	33 ....
10	22	34 ....
11	23	35 ....
12	24	

Part A Score

☐

Part B–1

36	44 ....
37	45 ....
38	46 ....
39	47 ....
40	48 ....
41	49 ....
42	50 ....
43	

Part B–1 Score

☐

Write your answers to Part B–2 and Part C in your answer booklet.

The declaration below should be signed when you have completed the examination.

I do hereby affirm, at the close of this examination, that I had no unlawful knowledge of the questions or answers prior to the examination and that I have neither given nor received assistance in answering any of the questions during the examination.

____

Signature

Tear Here

The University of the State of New York

REGENTS HIGH SCHOOL EXAMINATION

PHYSICAL SETTING EARTH SCIENCE

Wednesday, January 26, 2005 — 1:15 to 4:15 p.m., only

ANSWER BOOKLET

☐ Male

Student .... Sex: ☐ Female

Teacher ....

School .... Grade ....

Answer all questions in Part B–2 and Part C. Record your answers in this booklet.

☐ **Performance Test Score (Maximum Score: 23)**

Part	Maximum Score	Student's Score
A	35	
B–1	15	
B–2	15	
C	20	

Total Written Test Score (Maximum Raw Score: 85) ☐

Final Score (from conversion chart) ☐

Raters' Initials:

Rater 1 .... **Rater 2** ....

Part B–2

For Raters Only

51

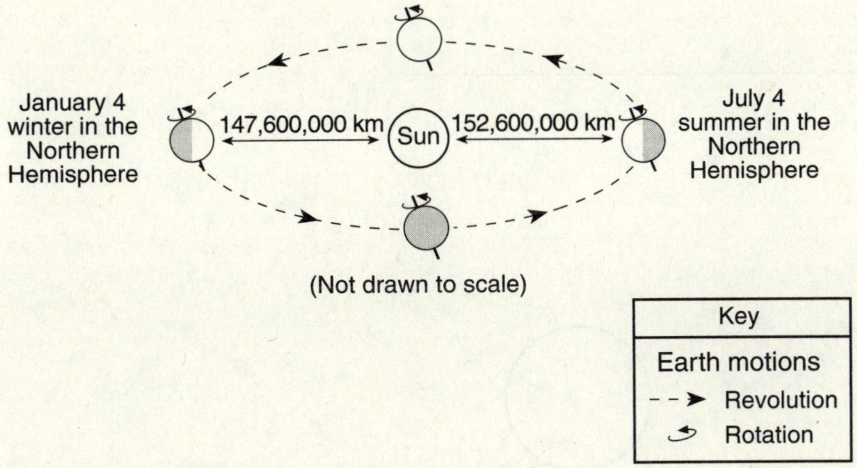

51 ☐

52 ____

____

____

52 ☐

[1]

For Raters Only

53

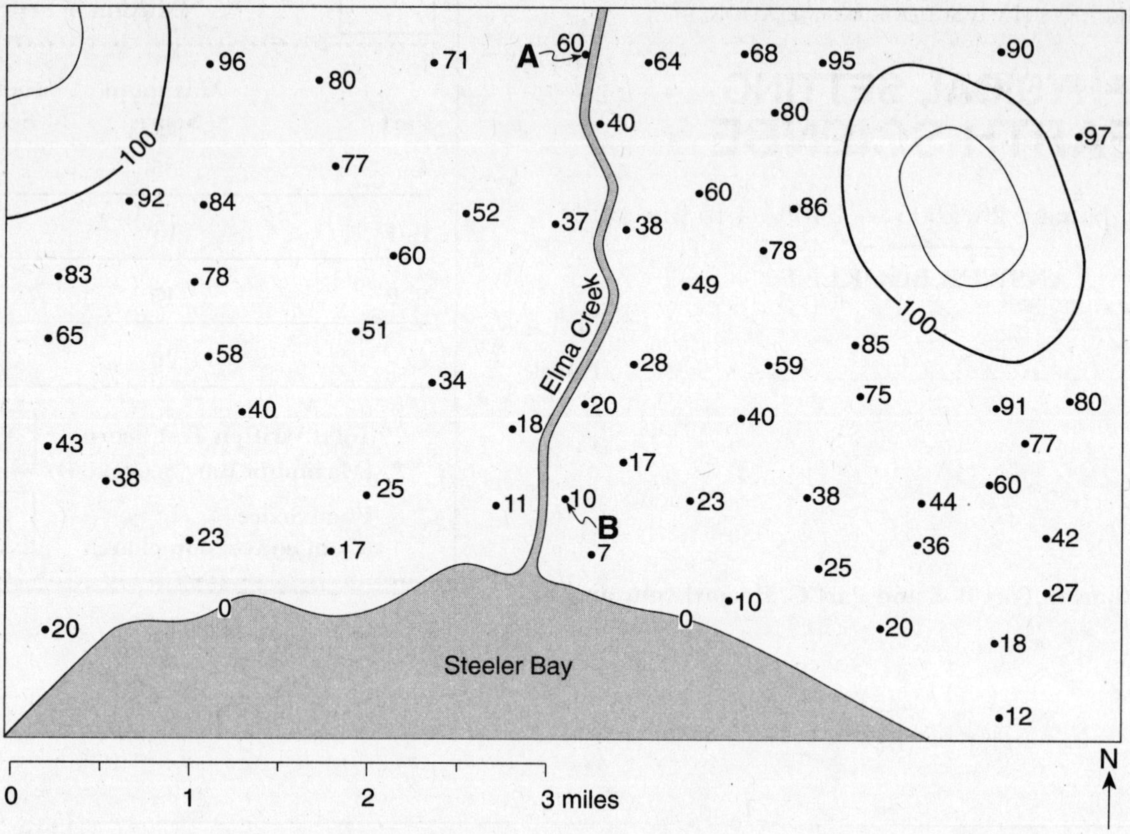

53 ☐

54 ______________________________ 54 ☐

55 Gradient = ______________________________ 55 ☐

56

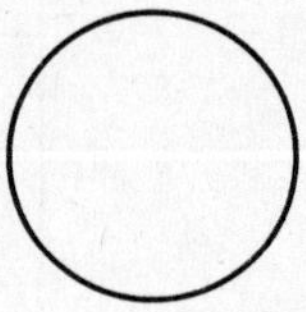

56 ☐

57 ______________________ **Epoch** 57 ☐

[2]

For Raters Only

58

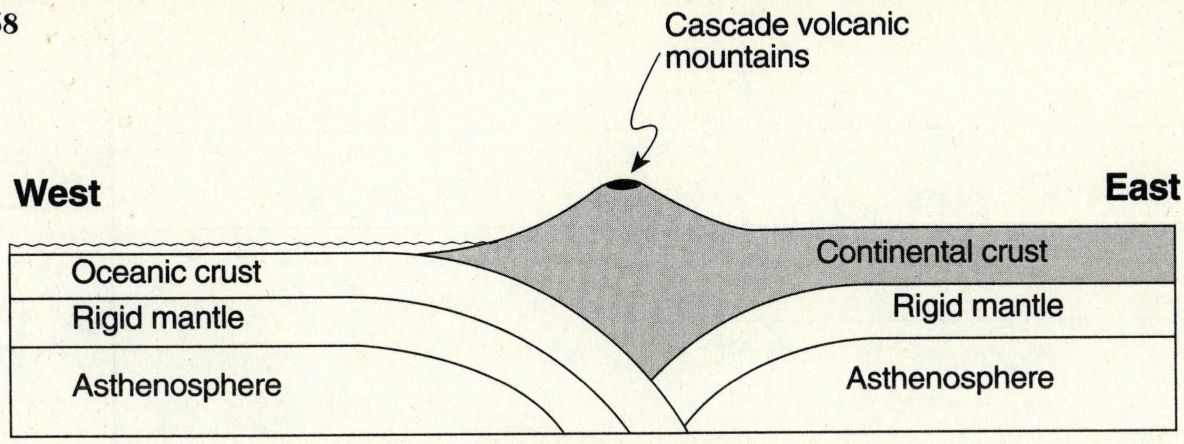

(Not drawn to scale)

58 ☐

59 ______________________________ 59 ☐

60

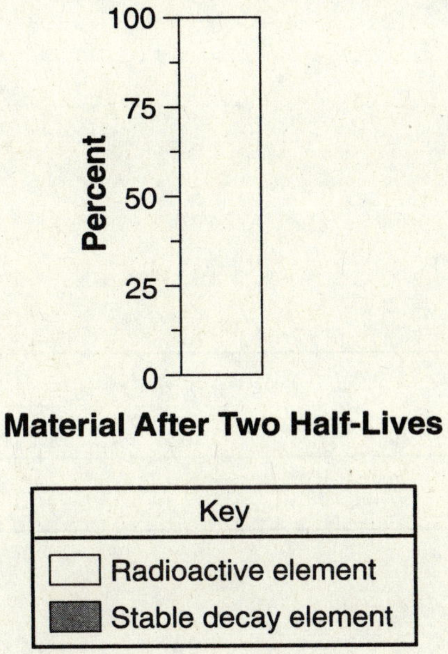

60 ☐

61 ______________________________ years 61 ☐

62 (1) ____

(2) ____ 62 ☐

Total Score for Part B–2

For Raters Only

Part C

63 ______________________________ 63 [ ]

____

64 ____________________ 64 [ ]

65

65 [ ]

66 (1) ______________________________ 66 [ ]

(2) ______________________________

[4]

For Raters Only

67

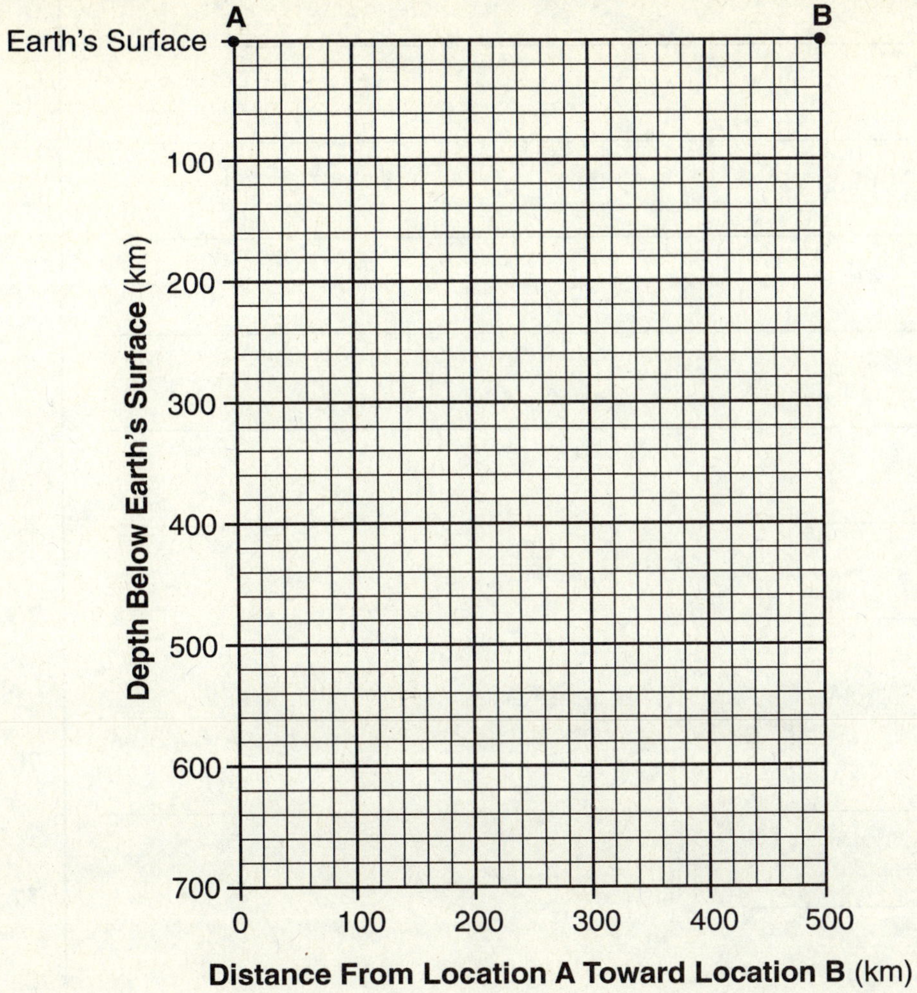

67

68 ______________________ 68

69 ______________________ 69

70 ______________________ 70

71 ______________________ 71

For Raters Only

72 ____ 72

____

73 ____ 73

____

74 ____ 74

____

75 ____________________ 75

76 ____ 76

____

77 ____________________ 77

78 ____________________ 78

79 79

Mineral	Name of Mineral
A	
B	
C	

80 ____________________ and ____________________ 80

Total Score for Part C

[6]

Physical Setting Earth Science August 17, 2004

Part A

Answer all questions in this part.

Directions (1–35): For *each* statement or question, write on your separate answer sheet the *number* of the word or expression that, of those given, best completes the statement or answers the question. Some questions may require the use of the *Earth Science Reference Tables*.

1 One complete cycle of the phases of the Moon takes approximately one

(1) day
(2) week
(3) month
(4) year

2 The diagram below shows the Moon at four positions in its orbit around Earth.

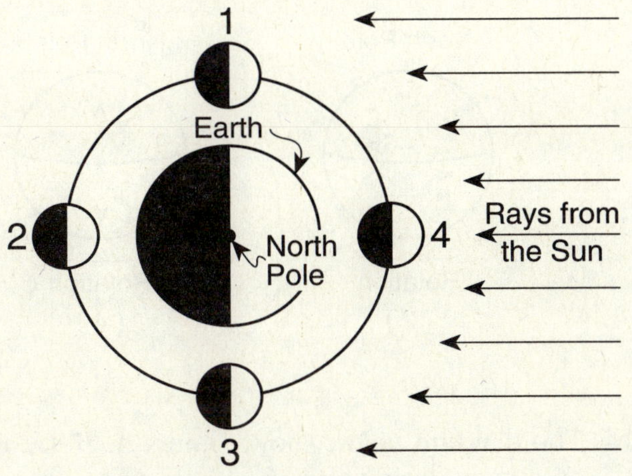

(Not drawn to scale)

An observer on Earth could see a solar eclipse when the Moon is at position

(1) 1
(2) 2
(3) 3
(4) 4

3 Which two stars have the most similar luminosity and temperature?

(1) *Betelgeuse* and *Barnard's Star*
(2) *Rigel* and *Betelgeuse*
(3) *Alpha Centauri* and the Sun
(4) *Sirius* and *Procyon B*

4 Which planet would float if it could be placed in water?

(1) Mercury
(2) Earth
(3) Saturn
(4) Pluto

5 The diagram below is a constructed ellipse. F_1 and F_2 are the foci of the ellipse.

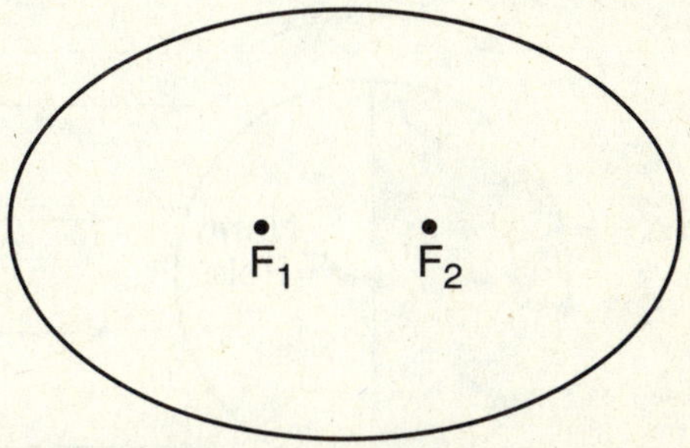

The eccentricity of this constructed ellipse is closest to the eccentricity of the orbit of which planet?

(1) Mercury
(2) Earth
(3) Saturn
(4) Pluto

6 The diagram below shows the noontime shadow cast by a vertical post located in New York State.

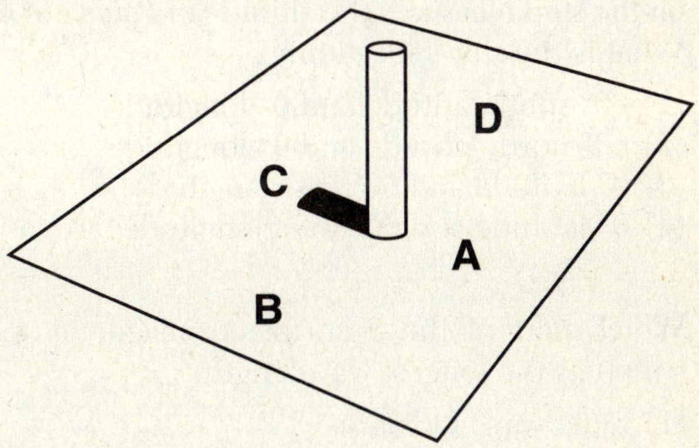

Which letter indicates a location west of the post?

(1) *A*
(2) *B*
(3) *C*
(4) *D*

7 On March 21, at which location in New York State would sunrise occur first?

(1) Riverhead
(2) Syracuse
(3) Buffalo
(4) Massena

8 Earth is farthest from the Sun during the Northern Hemisphere's summer, and Earth is closest to the Sun during the Northern Hemisphere's winter. During which season in the Northern Hemisphere is Earth's orbital velocity greatest?

(1) winter
(2) spring
(3) summer
(4) fall

9 The diagram below represents the direction of Earth's rotation as it appears from above the North Pole. Point *X* is a location on Earth's surface.

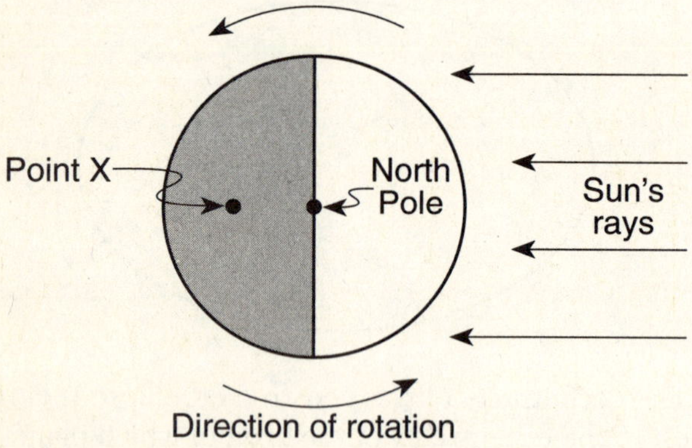

The time at point *X* is closest to

(1) 9 a.m.
(2) 12 noon
(3) 9 p.m.
(4) 12 midnight

10 As a ship crosses the Prime Meridian, an observer on the ship measures the altitude of *Polaris* at 60°. What is the ship's location?

(1) 60° south latitude and 0° longitude
(2) 60° north latitude and 0° longitude
(3) 0° latitude and 60° east longitude
(4) 0° latitude and 60° west longitude

11 Which part of the Sun's electromagnetic spectrum has the longest wavelength?

(1) radio wave radiation
(2) infrared radiation
(3) visible light radiation
(4) x-ray radiation

12 Which weather change is most likely indicated by rapidly falling air pressure?

(1) Humidity is decreasing.
(2) Temperature is decreasing.
(3) Skies are clearing.
(4) A storm is approaching.

13 Which diagram correctly shows how surface winds are deflected (curved) in the Northern and Southern Hemispheres due to Earth's rotation?

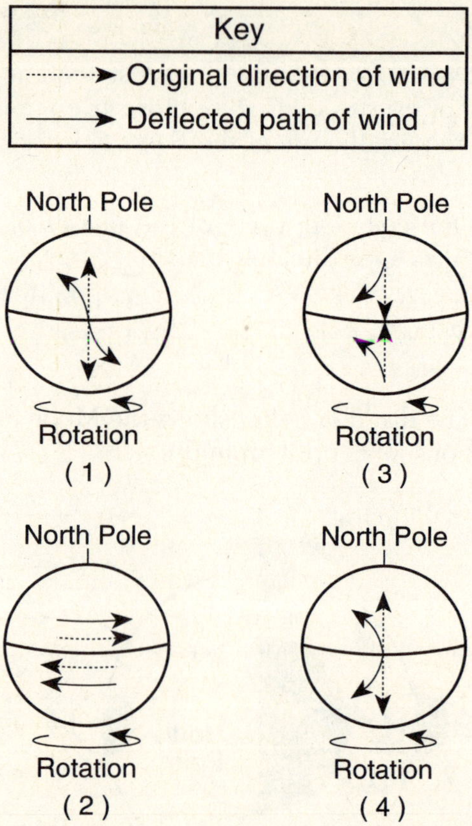

14 The diagram below shows points *A, B, C,* and *D* on a meandering stream.

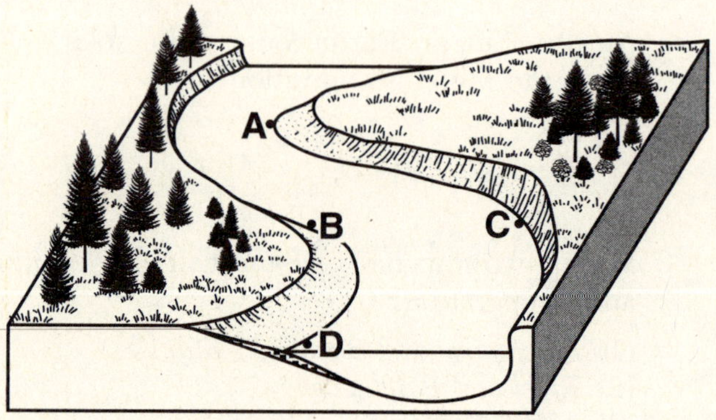

At which point does the greatest stream erosion occur?

(1) *A*
(2) *B*
(3) *C*
(4) *D*

15 The diagram below shows a greenhouse.

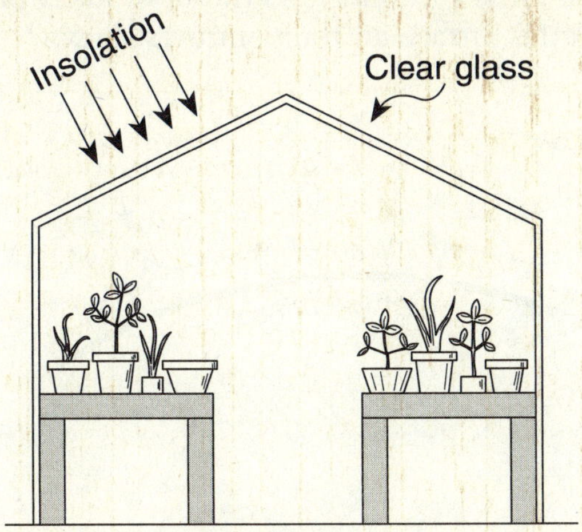

Greenhouse

What is the primary function of the clear glass of the greenhouse?

(1) The glass reduces the amount of insolation entering the greenhouse.
(2) The glass allows all wavelengths of radiation to enter and all wavelengths of radiation to escape.
(3) The glass allows short wavelengths of radiation to enter, but reduces the amount of long-wavelength radiation that escapes.
(4) The glass allows long wavelengths of radiation to enter, but reduces the amount of short-wavelength radiation that escapes.

16 The cross section below shows the prevailing winds that cause different climates on the windward and leeward sides of this mountain range.

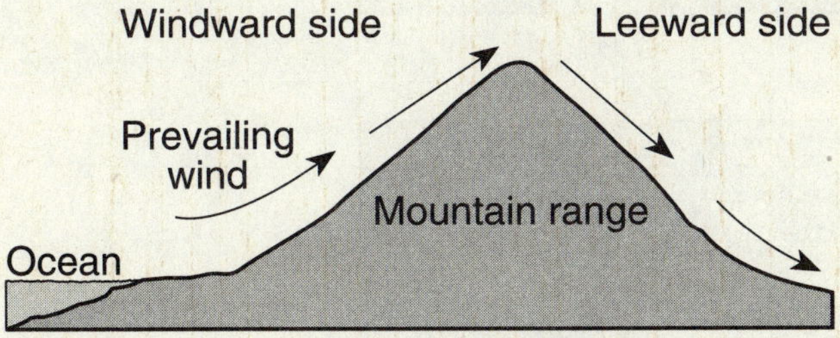

Compared to the climate conditions on the leeward side of this mountain range, the conditions on the windward side are usually

(1) cooler and wetter
(2) cooler and drier
(3) warmer and wetter
(4) warmer and drier

17 The California Ocean Current, which flows along the west coast of North America, is a

(1) cool current, flowing north
(2) cool current, flowing south
(3) warm current, flowing north
(4) warm current, flowing south

18 Which change at a particular location in a stream usually causes more sediments to be deposited at that location?

(1) decrease in stream velocity
(2) decrease in stream width
(3) increase in stream slope
(4) increase in stream discharge

19 Rocks are classified as igneous, sedimentary, or metamorphic based primarily on their

(1) texture
(2) crystal or grain size
(3) method of formation
(4) mineral composition

20 The internal atomic structure of a mineral most likely determines the mineral's

(1) color, streak, and age
(2) origin, exposure, and fracture
(3) size, location, and luster
(4) hardness, cleavage, and crystal shape

21 How are the minerals biotite mica and muscovite mica different?

(1) Biotite mica is colorless, but muscovite mica is not.
(2) Biotite mica contains iron and/or magnesium, but muscovite mica does not.
(3) Muscovite mica scratches quartz, but biotite mica does not.
(4) Muscovite mica cleaves into thin sheets, but biotite mica does not.

22 Which three minerals are most commonly found in the igneous rock granite?

(1) amphibole, calcite, and hematite
(2) amphibole, biotite mica, and gypsum
(3) plagioclase feldspar, pyroxene, and olivine
(4) plagioclase feldspar, potassium feldspar, and quartz

23 The cross sections below show different patterns of air movement in Earth's atmosphere. Air temperatures at Earth's surface are indicated in each cross section. Which cross section shows the most likely pattern of air movement in Earth's atmosphere that would result from the surface air temperatures shown?

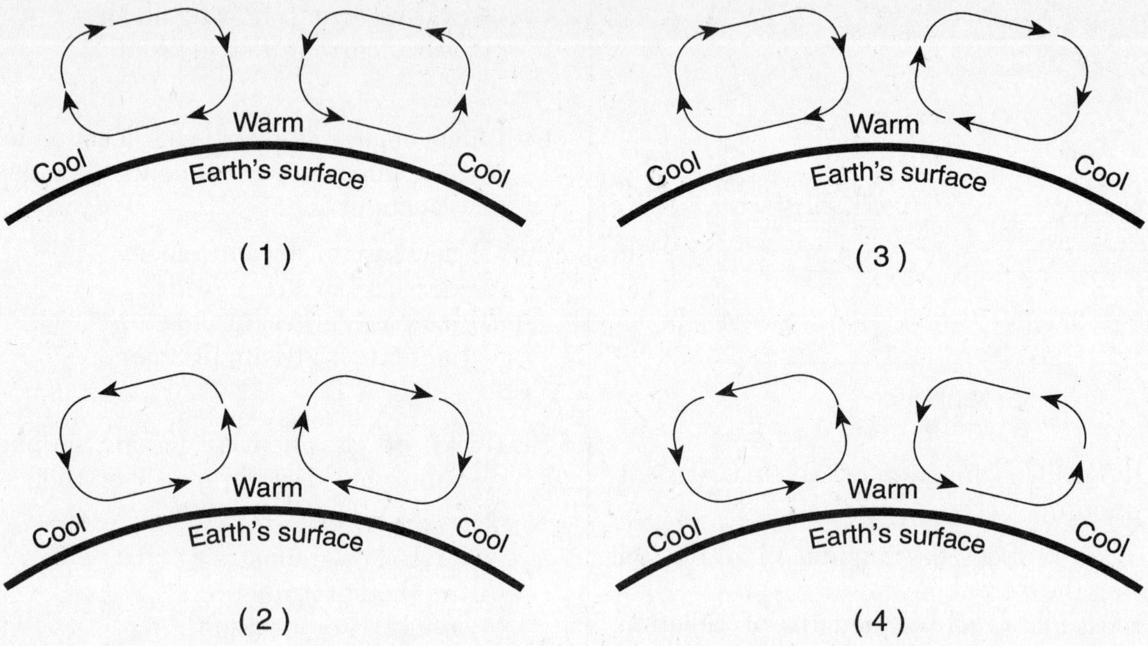

24 The diagrams below represent four permeable sediment samples. The sediments are composed of the same material, but differ in particle size and sorting. Which sediment sample will most likely have the fastest groundwater infiltration rate?

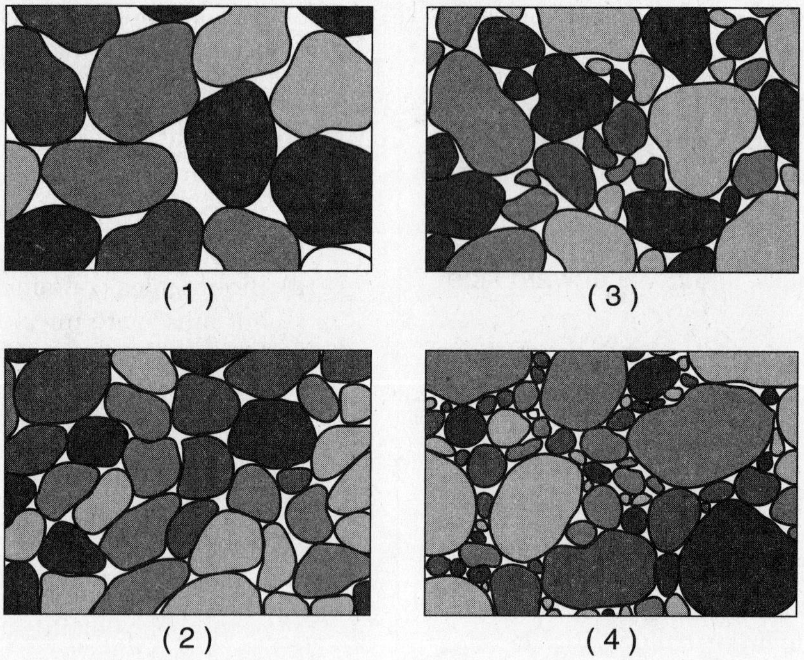

25 The diagram below represents three seismograms showing the same earthquake as it was recorded at three different seismic stations, *A*, *B*, and *C*.

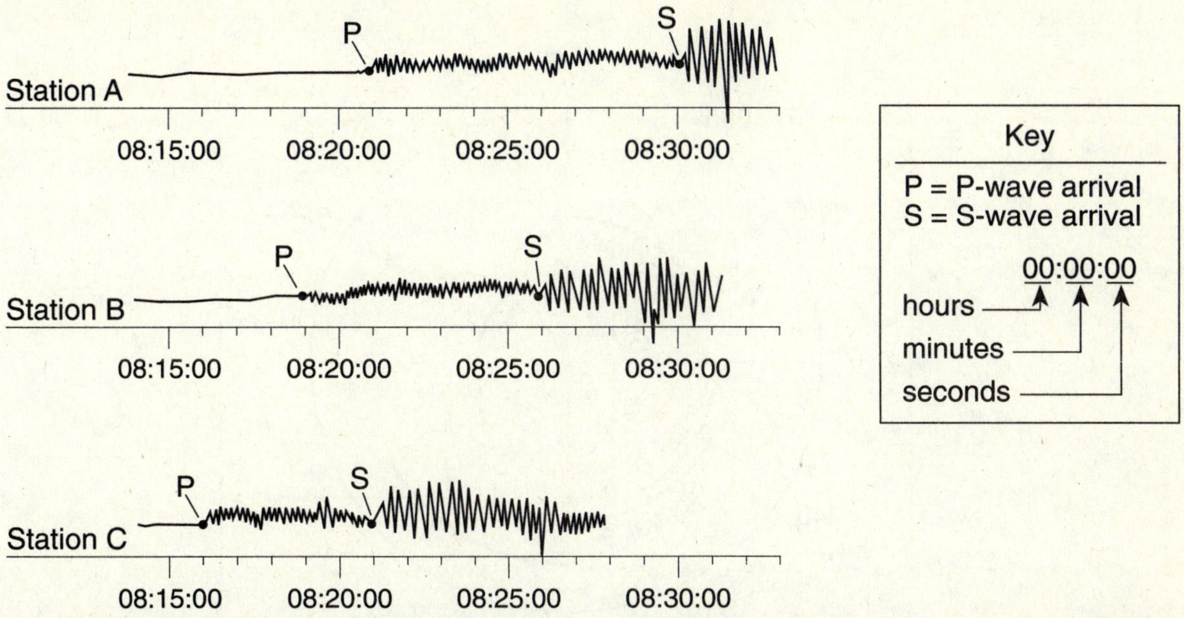

Which statement correctly describes the distance between the earthquake epicenter and these seismic stations?

(1) *A* is closest to the epicenter, and *C* is farthest from the epicenter.
(2) *B* is closest to the epicenter, and *C* is farthest from the epicenter.
(3) *C* is closest to the epicenter, and *A* is farthest from the epicenter.
(4) *A* is closest to the epicenter, and *B* is farthest from the epicenter.

26 A camera was placed outside at night and pointed directly at *Polaris* and several other stars. The lens was kept open and a time-exposure photograph was taken. The diagram below represents that photograph of *Polaris* and star trails, with an angular protractor to measure apparent motion.

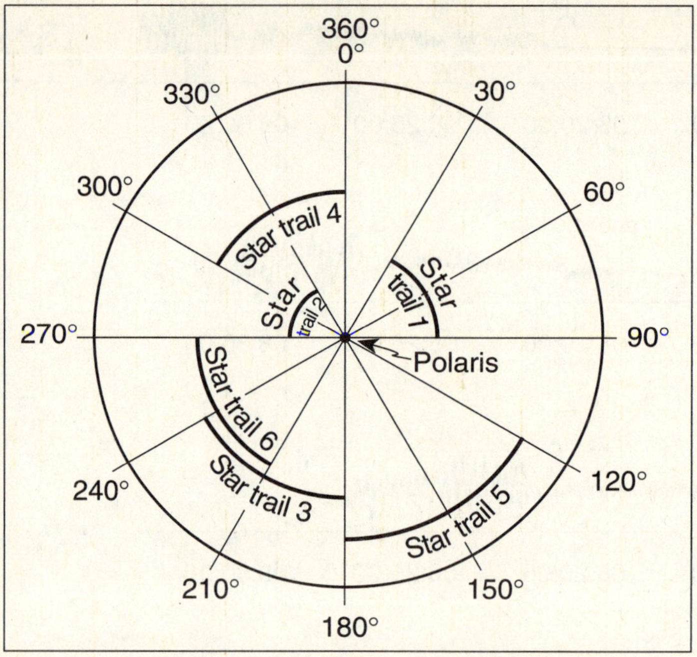

How many hours was the lens kept open to create the star trails in this photograph?

(1) 1 hour
(2) 6 hours
(3) 3 hours
(4) 4 hours

27 The sequence of bedrock cross sections below represents the same landscape region over a period of geologic time.

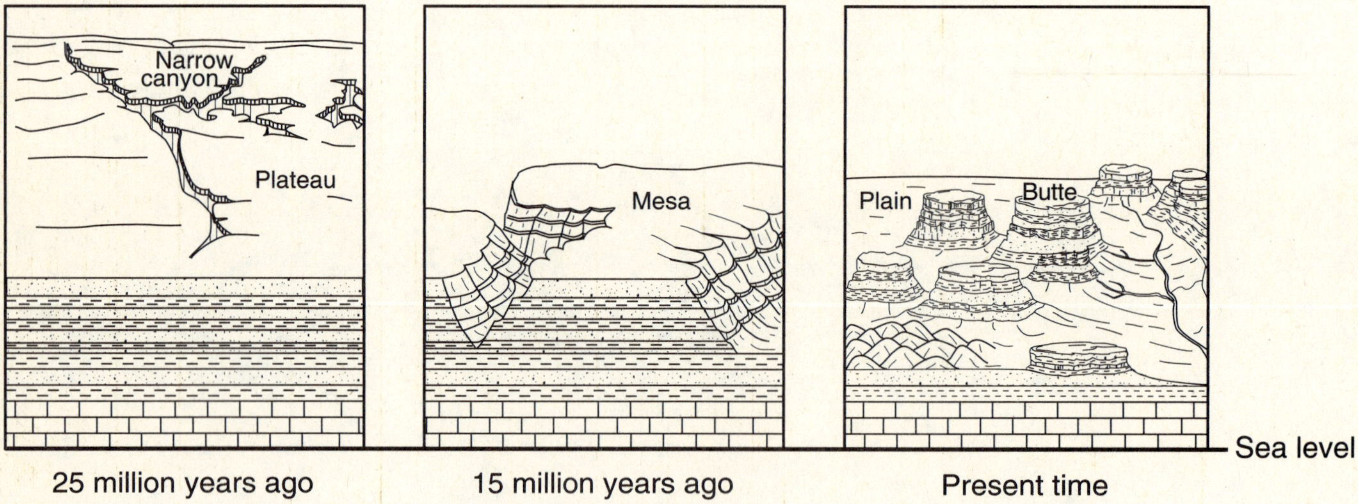

This sequence best represents

(1) an arid region that experienced mostly uplifting forces
(2) an arid region that experienced mostly erosional forces
(3) a humid region that experienced mostly uplifting forces
(4) a humid region that experienced mostly erosional forces

28 Which method of energy transfer is primarily responsible for energy being lost from Earth into space?

(1) conduction
(2) convection
(3) solidification
(4) radiation

29 When a continental crustal plate collides with an oceanic crustal plate, the continental crust is forced to move over the oceanic crust. What is the primary reason that the continental crust stays on top of the oceanic crust?

(1) Continental crust is less dense.
(2) Continental crust deforms less easily.
(3) Continental crust melts at higher temperatures.
(4) Continental crust contains more mafic minerals.

30 Earth's fossil record shows evidence that

(1) very few life-forms have become extinct
(2) life-forms existed on land before life-forms existed in water
(3) more complex life-forms probably have evolved from less complex life-forms
(4) older bedrock contains a great variety of life-forms, while younger bedrock contains less variety of life-forms

31 Large garnet mineral crystals are found in the metamorphic surface bedrock in which New York State landscape region?

(1) Catskills
(2) Adirondacks
(3) Erie-Ontario Lowlands
(4) Tug Hill Plateau

32 Bedrock outcrops *A* and *B* are located at two different locations along the Genesee River in western New York State. Rock layers 1, 2, and 4 are the same in both outcrops.

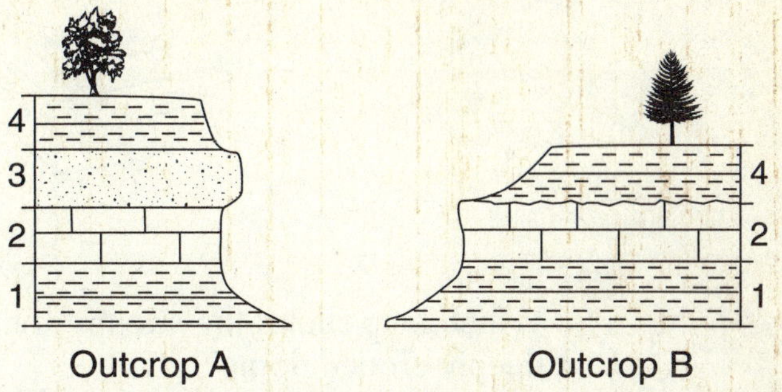

Which statement best explains why rock layer 3 is missing from outcrop *B*?

(1) A fault exists between outcrops *A* and *B*.
(2) Erosion created an unconformity between rock layers 2 and 4 in outcrop *B*.
(3) A volcanic eruption destroyed rock layer 3 in outcrop *B*.
(4) Metamorphism of outcrop *A* created rock layer 3.

33 Fossilized footprints of *Coelophysis* dinosaurs have been found in bedrock closest to which New York State location?

(1) New York City
(2) Old Forge
(3) Watertown
(4) Niagara Falls

34 A sample of wood that originally contained 100 grams of carbon-14 now contains only 25 grams of carbon-14. Approximately how many years ago was this sample part of a living tree?

(1) 2,850 years
(2) 5,700 years
(3) 11,400 years
(4) 17,100 years

35 The block diagram below represents a deeply eroded dome.

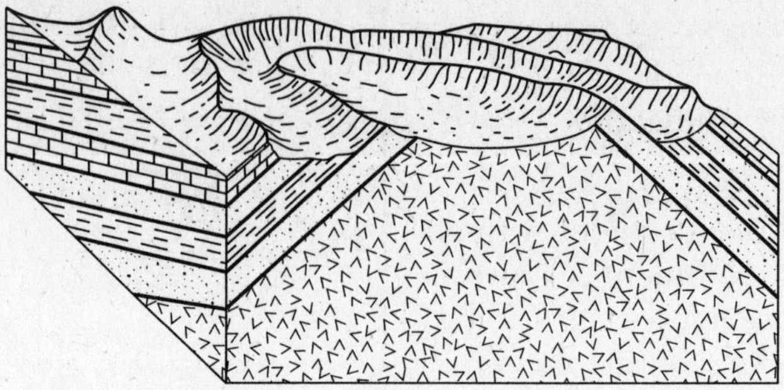

Which map shows the stream drainage pattern that would most likely develop on this deeply eroded dome?

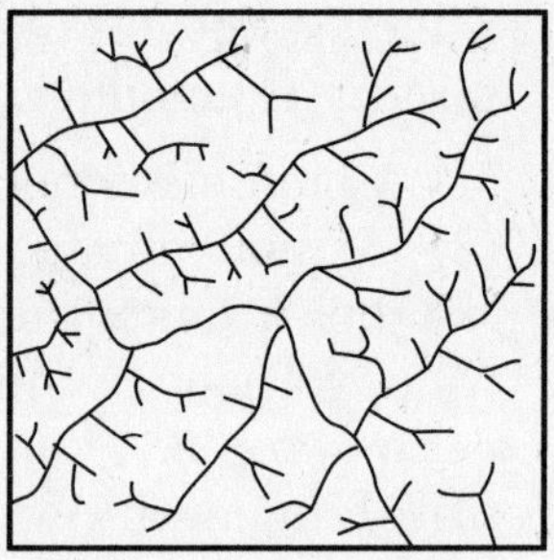

(1)

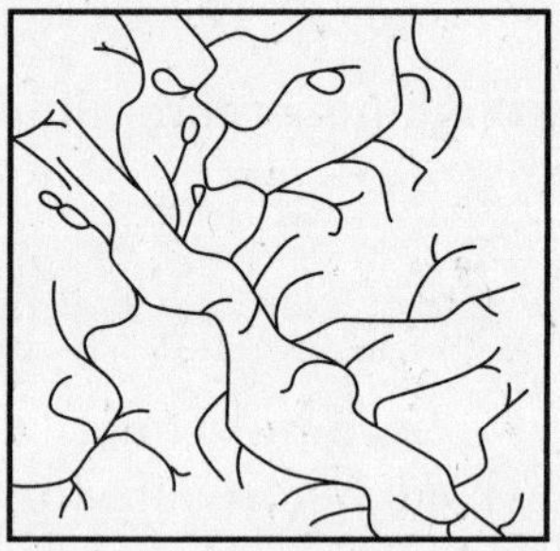

(3)

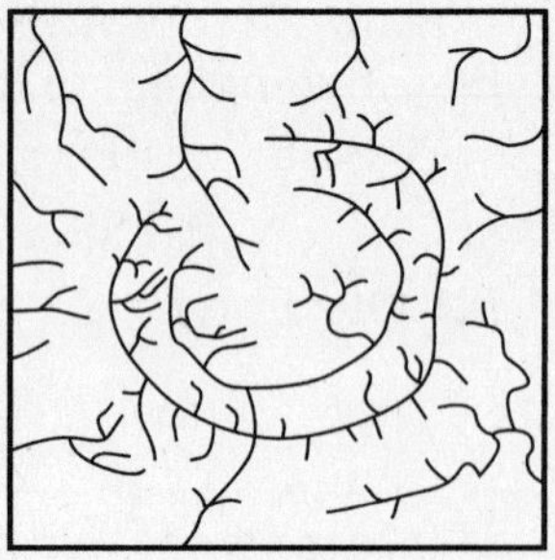

(2)

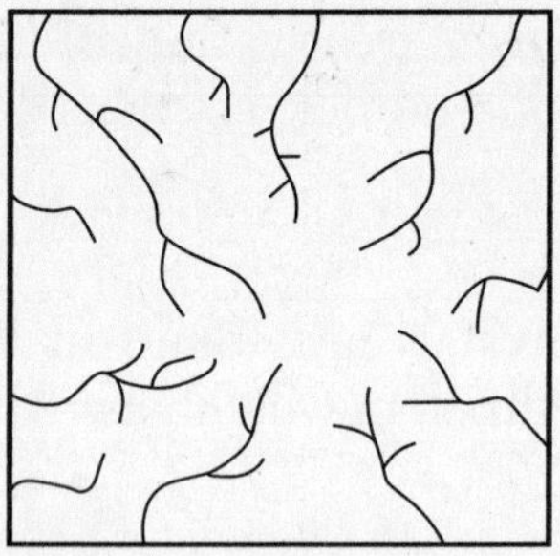

(4)

Part B–1

Answer all questions in this part.

Directions (36–50): For *each* statement or question, write on your separate answer sheet the *number* of the word or expression that, of those given, best completes the statement or answers the question. Some questions may require the use of the *Earth Science Reference Tables*.

Base your answers to questions 36 through 38 on the diagram below, which shows the altitude and apparent position of the noontime Sun, as seen from various latitudes on Earth on a particular day of the year. Letters *A* through *D* represent locations on Earth's surface.

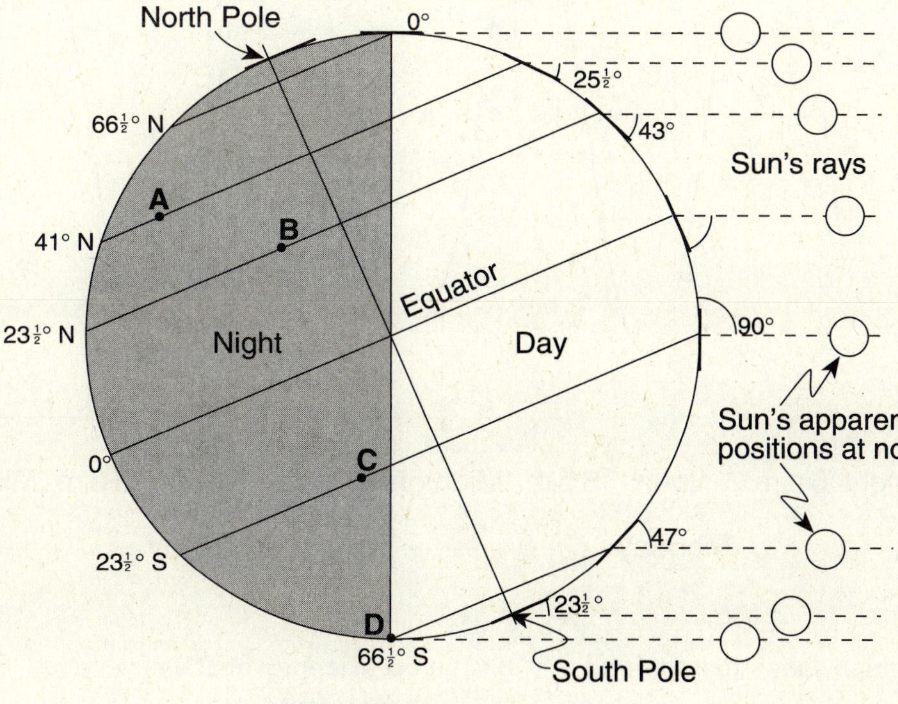

(Not drawn to scale)

36 Which lettered location will experience the *shortest* period of daylight during one Earth rotation on this day?

(1) *A* (3) *C*
(2) *B* (4) *D*

37 What is the altitude of the noontime Sun at the Equator on this date?

(1) $23\frac{1}{2}$° (3) $66\frac{1}{2}$°
(2) 43° (4) 90°

38 Which season will begin at 41° N latitude, three months after the date represented by this diagram?

(1) summer (3) winter
(2) fall (4) spring

Base your answers to questions 39 through 41 on the graph below, which shows the water levels of ocean tides measured in Boston, Massachusetts, for a 2-day period.

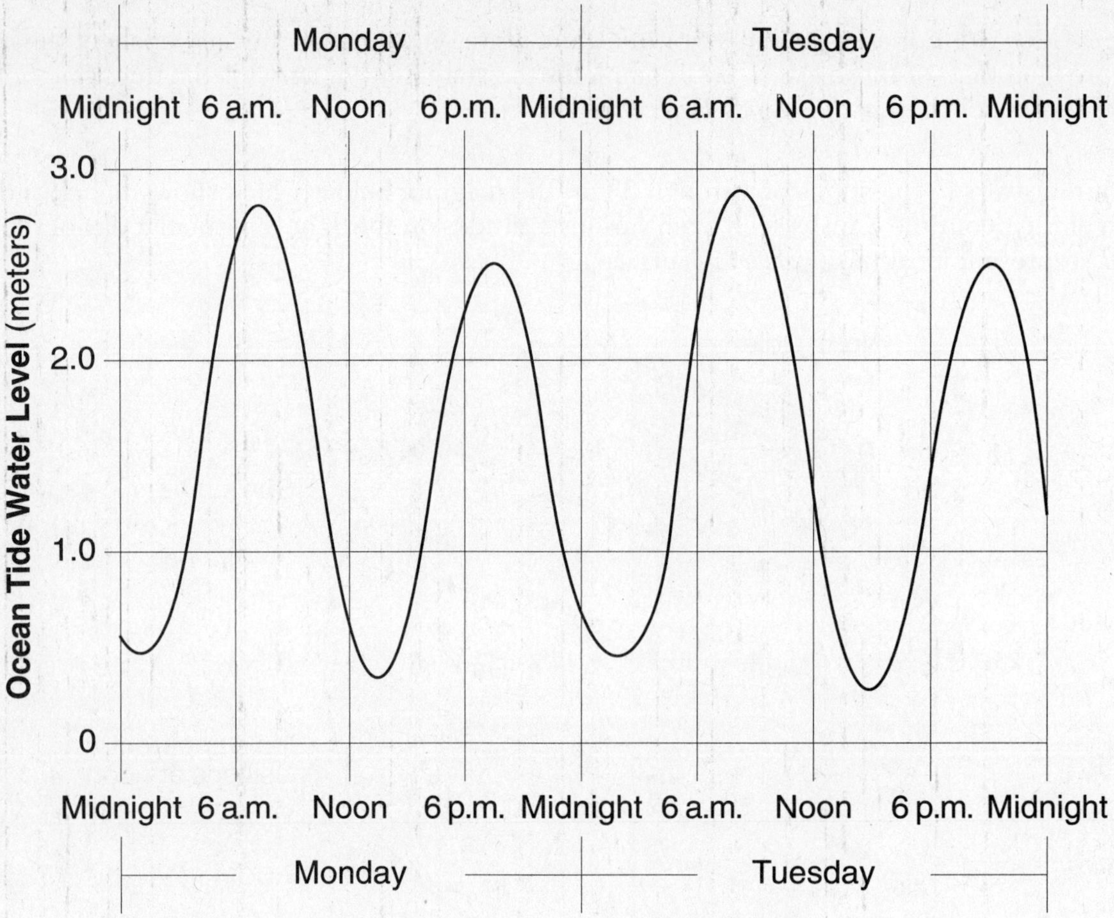

39 The graph shows that high tides at Boston occur approximately every

(1) 3.5 hours
(2) 6.0 hours
(3) 12.5 hours
(4) 16.0 hours

40 If the trends shown by the graph continue, which statement best describes the next low tide at Boston that is expected to occur on Wednesday?

(1) It will occur about 3 a.m. with a 0.4-meter water level.
(2) It will occur about 6 a.m. with a 0.6-meter water level.
(3) It will occur about 9 p.m. with a 2.6-meter water level.
(4) It will occur about 10 p.m. with a 2.8-meter water level.

41 The gravitational pull of the Moon has the greatest influence on the water levels of Earth's ocean tides. If the distance between the Moon and Earth were to *decrease* steadily for the week following the time shown on the graph, which water-level changes would be expected to occur?

(1) High tides would get higher and low tides would get lower.
(2) High tides would get lower and low tides would get higher.
(3) Both high tides and low tides would get higher.
(4) Both high tides and low tides would get lower.

Base your answers to questions 42 through 45 on the satellite image below, which shows a Northern Hemisphere hurricane, and on your knowledge of Earth science.

42 What is the usual surface wind pattern around the eye of Northern Hemisphere hurricanes?

(1) clockwise and outward
(2) clockwise and inward
(3) counterclockwise and outward
(4) counterclockwise and inward

43 Which air mass is normally associated with the formation of hurricanes?

(1) continental tropical
(2) maritime tropical
(3) continental polar
(4) maritime polar

44 Clouds form in the hurricane because the air is

(1) sinking, expanding, and cooling
(2) sinking, compressing, and warming
(3) rising, expanding, and cooling
(4) rising, compressing, and warming

45 When the eye of this hurricane reaches 43° N latitude, this hurricane will most likely be pushed by planetary winds toward the

(1) northwest
(2) northeast
(3) southwest
(4) southeast

Base your answers to questions 46 through 48 on the diagram below, which shows details of a section of a rift valley in the center of a mid-ocean ridge. The vertical lines in the diagram represent faults and fractures within the ocean floor bedrock.

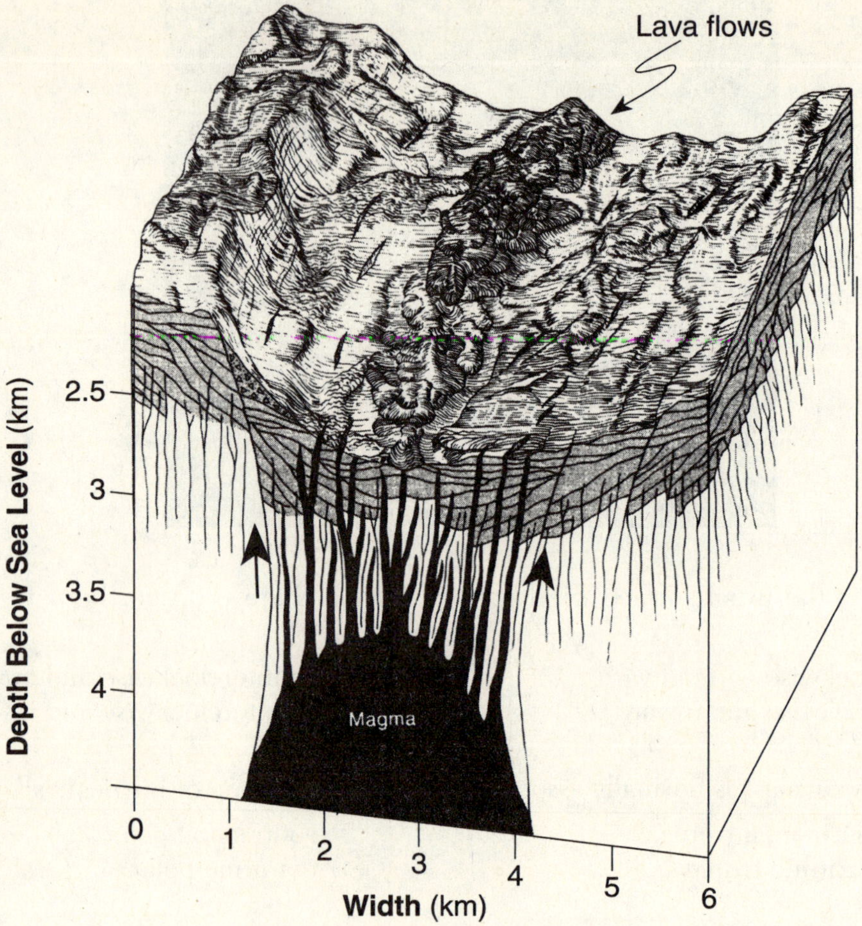

46 What will be the primary result of the continuation of the geologic processes indicated at this location?

(1) Earth's magnetic field will reverse direction.
(2) Continental crust will be forced downward.
(3) Earth's circumference will increase.
(4) New oceanic crust will form.

47 Which type of crustal plate boundary is shown in this diagram?

(1) divergent
(2) convergent
(3) universal
(4) transform

48 The dark-colored lava flows shown in the diagram were pushed from the magma chamber onto the surface of the ocean floor. Which characteristics are present in the solid rock that formed when the lava flows cooled?

(1) generally small grain size and mafic composition
(2) generally small grain size and felsic composition
(3) generally large grain size and mafic composition
(4) generally large grain size and felsic composition

Base your answers to questions 49 and 50 on the cross section below. The cross section shows a typical bedrock structure where oil and natural gas deposits are found.

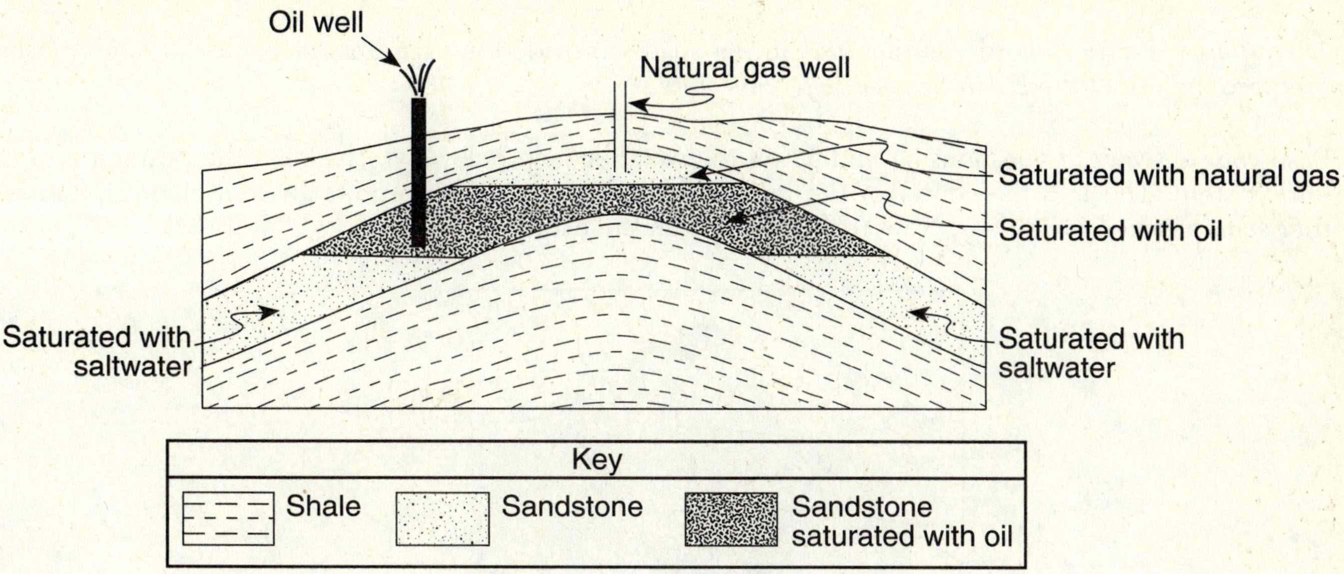

49 The natural gas, oil, and saltwater have formed layers at different levels in the same rock layer due to the

(1) principle of superposition
(2) principle of original horizontality
(3) differences in the density of the three materials
(4) differences in the geologic age of the three materials

50 According to the diagram, in which type of rock are these natural gas and oil deposits found?

(1) coarse-textured igneous rock
(2) foliated metamorphic rock
(3) porous clastic sedimentary rock
(4) intrusive crystalline sedimentary rock

Part B–2

Answer all questions in this part.

Directions (51–63): Record your answers in the spaces provided in your answer booklet. Some questions may require the use of the *Earth Science Reference Tables.*

Base your answers to questions 51 and 52 on the weather map below, which shows the position of a low-pressure system. The **L** is the center of the low. The shaded portion represents an area of precipitation. A weather station model for Albany, New York, is shown on the map.

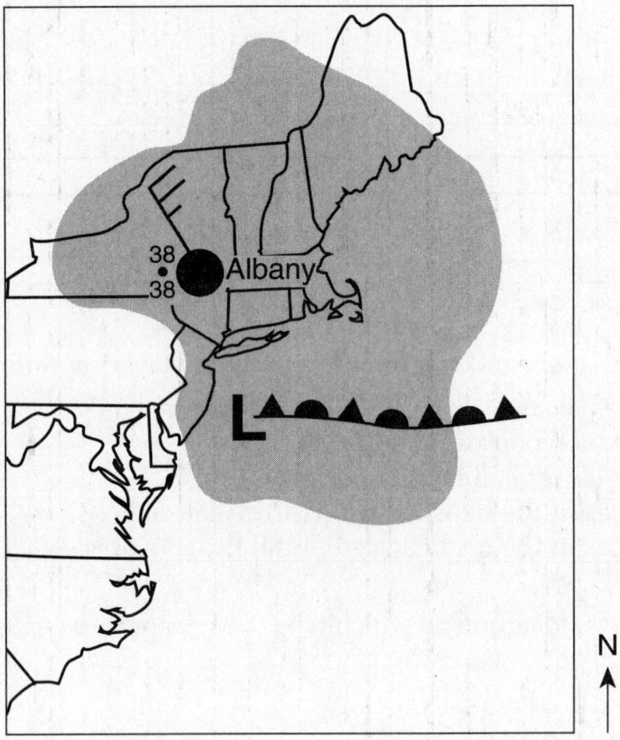

51 What type of front extends eastward from the low-pressure center? [1]

52 *In your answer booklet,* complete the weather data table for Albany, New York, based on the station model shown on the map. [2]

Base your answers to questions 53 and 54 on the diagram below, which shows a clear plastic tube containing water and a beaker containing a mixture of rounded quartz grains of different sizes.

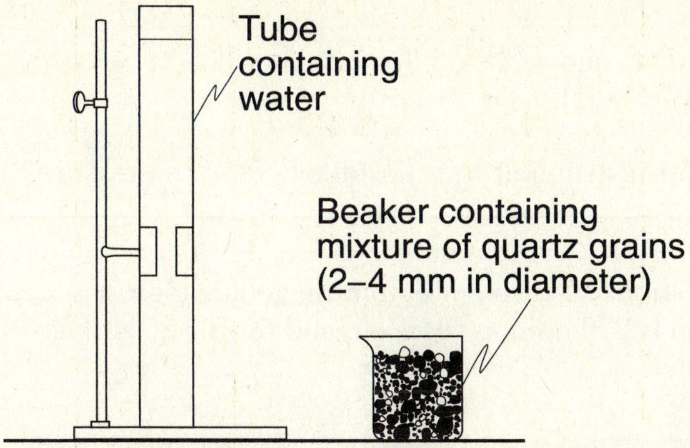

(Not drawn to scale)

53 When the rounded quartz grains are poured all at once into the tube, the grains will settle to the bottom of the tube. On the cross section provided *in your answer booklet*, draw the approximate grain sizes and pattern of arrangement of the rounded quartz grains at the bottom of the tube. [1]

54 The side-view diagram below shows the same mixture and amount of rounded quartz grains being poured all at once into a moving stream with a depth of 3 meters.

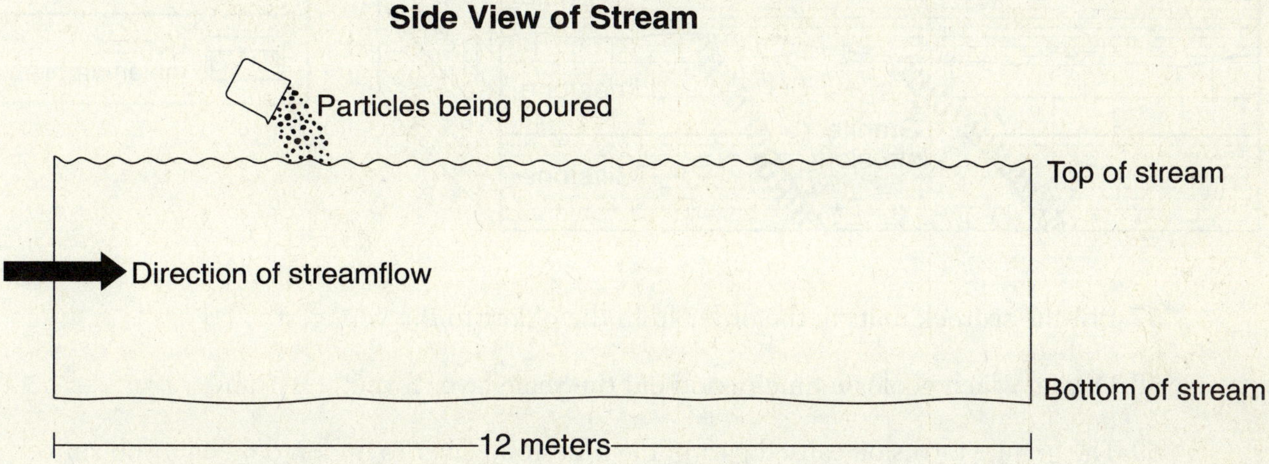

Describe the general location of the 2-mm-diameter rounded quartz grains compared to the 4-mm-diameter rounded quartz grains as they are transported and deposited downstream. [1]

____

Base your answers to questions 55 and 56 on the weather map provided *in your answer booklet,* which shows surface air-pressure readings, in millibars, at various locations in the United States and Canada. The 1020-millibar isobars have been drawn and labeled.

55 Draw the 1024- and 1028-millibar isobars on the weather map provided *in your answer booklet.* [2]

56 What weather instrument was most likely used to measure these air pressures? [1]

Base your answers to questions 57 through 60 on the geologic cross section below. Radioactive dating indicates that the granite intrusion is 279 million years old and the vesicular basalt is 260 million years old. The rock layers have not been overturned.

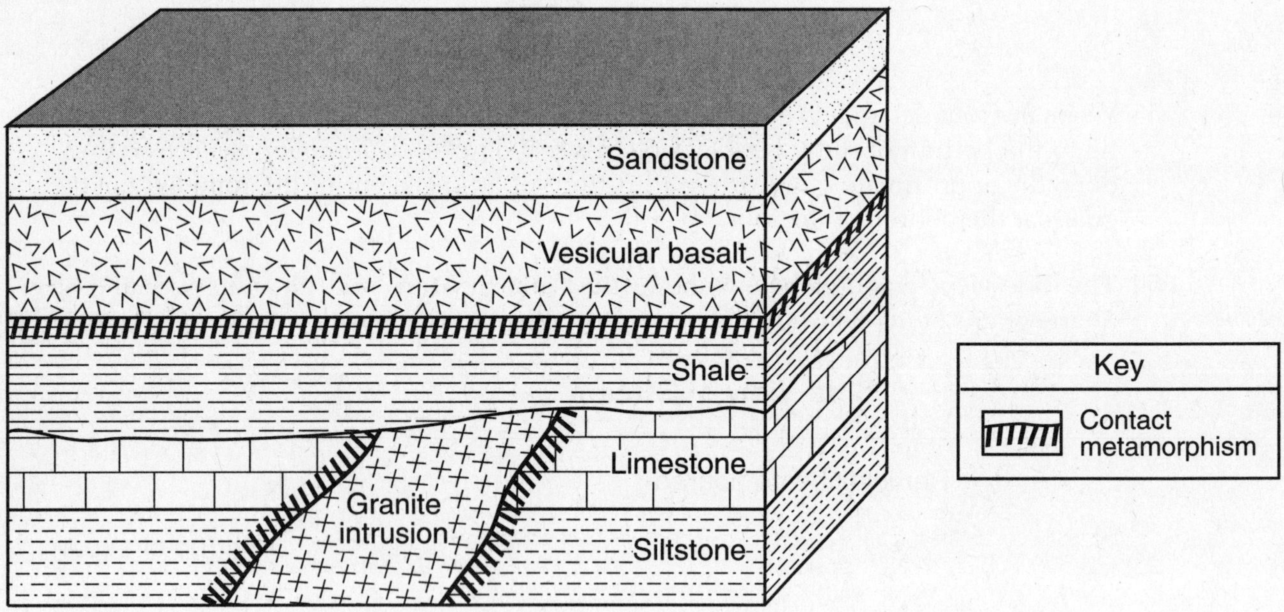

57 List the six rock units in the order from the oldest to the youngest. [1]

58 During which geologic time period did the shale layer form? [1]

59 The granite intrusion caused part of the limestone layer to undergo metamorphism. What metamorphic rock would most likely be found in this zone of contact metamorphism? [1]

60 Describe the rate of cooling that must occur for magma to form vesicular basalt. [1]

Base your answers to questions 61 through 63 on the information, table, and photographs below and on your knowledge of Earth science.

"Herkimer Diamonds"

Gem-quality "Herkimer Diamonds" are hexagonal-shaped quartz crystals found in some of the surface bedrock of Herkimer, New York. Herkimer is located at approximately 43° north latitude and 75° west longitude. The oldest of these gemstones are believed to be approximately 500 million years old. These quartz crystals are magnificent works of nature that have a natural diamondlike geometric shape formed when the quartz crystallized. Natural "Herkimer Diamonds" were not cut or shaped by humans. Due to their appearance, "Herkimer Diamonds" are commonly used in jewelry. These quartz crystals are not true diamonds.

Mineral Characteristics of "Herkimer Diamonds" (Quartz) and True Diamonds

Mineral	Color	Chemical Composition	Luster	Hardness	Dominant Form of Breakage
"Herkimer Diamond" (quartz)	Colorless or variable	SiO_2	Glassy	7	Fracture
True diamond	Colorless or variable	C	Glassy	10	Cleavage

Photographs of "Herkimer Diamonds" (Quartz)

61 List *two* mineral characteristics that differ between "Herkimer Diamonds" and true diamonds. [1]

62 State one use for "Herkimer Diamonds" (quartz), other than their use in jewelry. [1]

63 On the New York State map provided *in your answer booklet,* mark with a dot the location of Herkimer, New York. Draw a small circle around your dot (⊙) to make the dot easily seen. [1]

Part C

Answer all questions in this part.

Directions (64–77): Record your answers in the spaces provided in your answer booklet. Some questions may require the use of the *Earth Science Reference Tables.*

Base your answers to questions 64 and 65 on the topographic map shown below. Points *A, B, C,* and *D* are reference points on the map. Elevations are measured in meters.

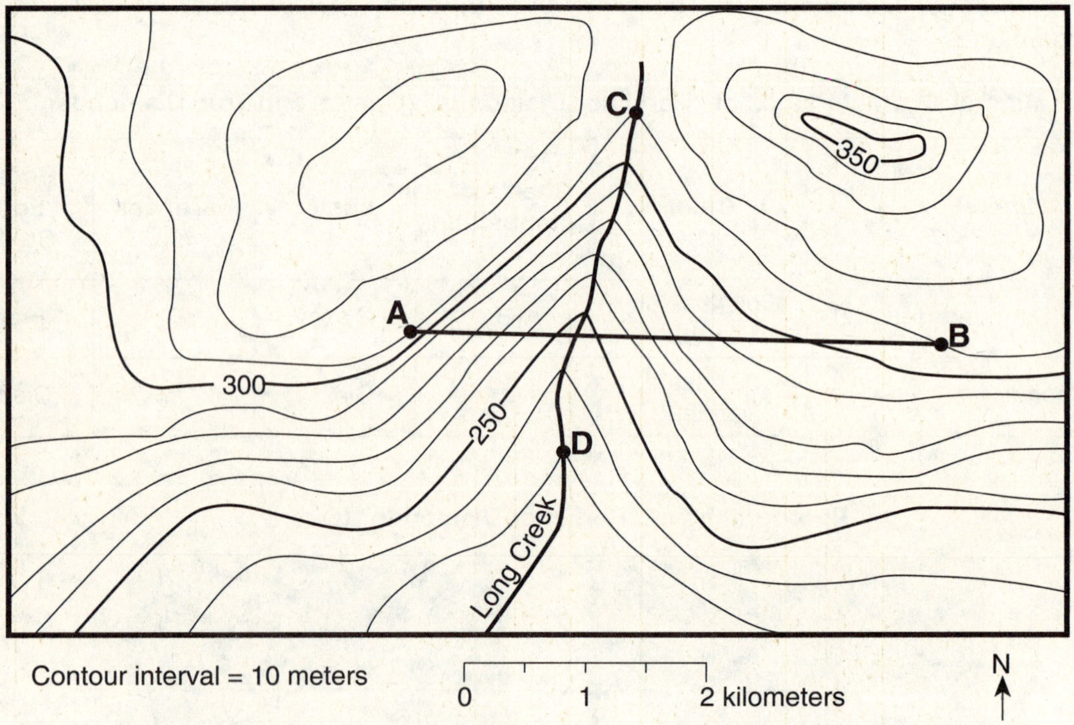

64 On the grid provided *in your answer booklet,* construct a topographic profile along line *AB,* by plotting a point for the elevation of *each* contour line that crosses line *AB* and connecting the points with a smooth, curved line to complete the profile. [2]

65 Calculate the gradient of Long Creek between points *C* and *D* and label the answer with the correct units. [2]

____

Base your answers to questions 66 and 67 on the map below, which shows one method of classifying Earth's surface into latitudinal climate belts. In the tropical climate belt, the average monthly temperatures never drop below 18°C. In the polar climate belts, the average monthly temperatures never rise above 10°C. The isotherms show the average monthly temperature of the coolest and warmest months. Effects of elevation have been omitted.

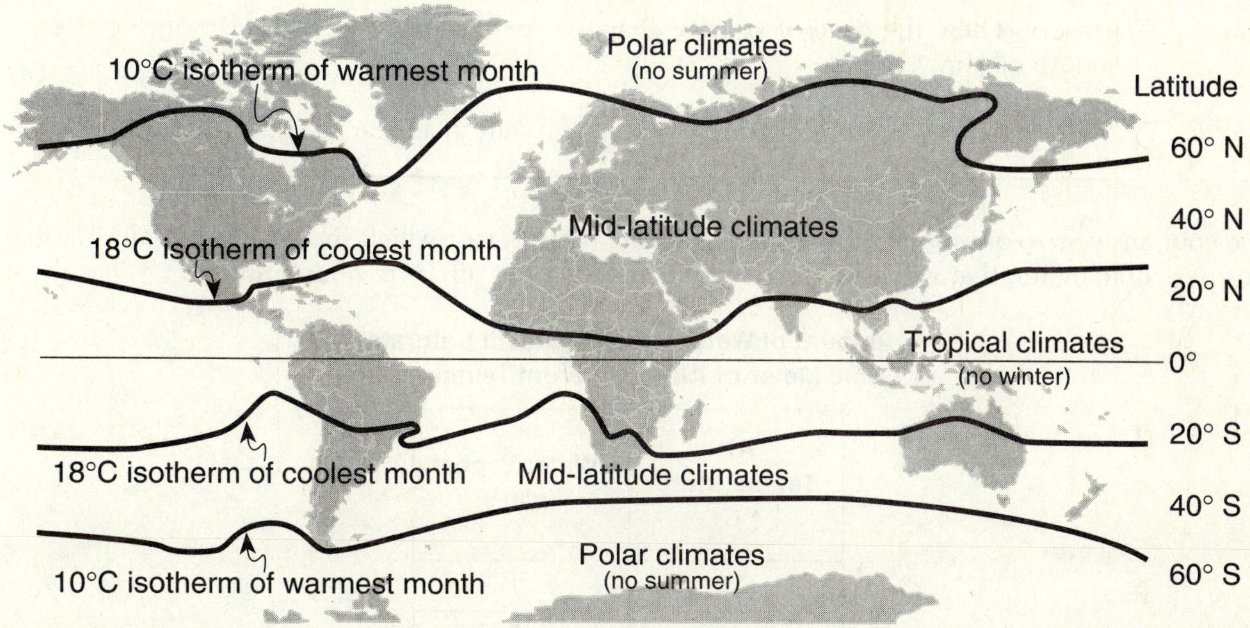

66 According to the isotherms on the map, locations in the mid-latitude climate belts have average monthly temperatures between what values? [1]

67 Describe a specific characteristic of insolation received in the tropical climate belt region that causes the average monthly temperature to remain warm all year. [1]

68 New York State's Adirondacks are classified as a mountain landscape region. Describe one bedrock characteristic and one land surface characteristic that were used to classify the Adirondacks as a mountain landscape region. [2]

69 The map provided *in your answer booklet* shows six source regions for different air masses that affect the weather of North America. The directions of movement of the air masses are shown. Using the standard two-letter air-mass symbols from the *Earth Science Reference Tables*, label the air masses by writing the correct symbol in each circle on the map. [2]

Base your answers to questions 70 and 71 on the information below and on your knowledge of Earth science.

Accurate observations of the Sun were made by a New York State observer. This person observed the time of sunrise and the position of sunrise along the eastern horizon for each day during the month of May.

70 Describe how the time of sunrise changed for the observer each day during the month of May. [1]

71 State the actual Earth motion that causes the Sun to appear to rise each day. [1]

Base your answers to questions 72 and 73 on the data table below, which shows the amount of water vapor, in grams per cubic meter, that will saturate 1 cubic meter of air at different temperatures.

Amount of Water Vapor That Will Saturate 1 Cubic Meter of Air at Different Temperatures

Air Temperature (°C)	Water Vapor (g/m³)
–20	1
–10	2
0	5
10	9
20	17
30	29
40	50

72 On the grid provided *in your answer booklet*, construct a line graph of the data, following the directions below.

a Place the name of the correct variable along the *y*-axis. Include the correct units. [1]
b Mark an appropriate numerical scale showing equal intervals along the *y*-axis. [1]
c Plot the amount of water that will saturate 1 cubic meter of air at the temperatures shown in the data table. Connect the points with a smooth, curved line. [1]

73 Describe the relationship between the air temperature and the amount of water vapor necessary to saturate the air. [1]

Base your answers to questions 74 through 77 on the information below and on your knowledge of Earth science.

In the 1930s, most scientists believed that Earth's crust and interior were solid and motionless. A small group of scientists were talking about "continental drift," which is the idea that Earth's crust is not stationary, but is constantly shifting and moving.

From seismic data, geophysical evidence, and laboratory experiments, scientists now generally agree that lithospheric plates move at the surface. Both Earth's surface and interior are in motion. Solid rock in the mantle can be softened and shaped when subjected to the heat and pressure within Earth's interior over millions of years.

Subduction processes are believed by many scientists to be the driving force of plate tectonics. At present, this theory cannot be directly observed and confirmed. The lithospheric plates have moved in the past and are still moving today. The details of why and how they move will continue to challenge scientists.

74 Earth's crust is described as "constantly shifting and moving." Give one example of geologic evidence that supports the conclusion that continents have drifted apart. [1]

75 The information given suggests that "subduction processes are the driving force of plate tectonics." Identify a specific location of a subduction zone on Earth. [1]

76 According to the *Earth Science Reference Tables,* at what inferred depth is *mantle* rock partially melted and slowly moving below the lithospheric plates? [1]

77 According to the geologic record, during which geologic time period did the lithospheric plates that made up Pangea begin to break up? [1]

Tear Here

The University of the State of New York

REGENTS HIGH SCHOOL EXAMINATION

PHYSICAL SETTING
EARTH SCIENCE

Tuesday, August 17, 2004 — 12:30 to 3:30 p.m., only

ANSWER SHEET

Student .... Sex: ☐ Male ☐ Female Grade ....

Teacher .... School ....

Record your answers to Part A and Part B–1 on this answer sheet.

Part A

1	13	25 ....
2	14	26 ....
3	15	27 ....
4	16	28 ....
5	17	29 ....
6	18	30 ....
7	19	31 ....
8	20	32 ....
9	21	33 ....
10	22	34 ....
11	23	35 ....
12	24	

Part A Score

☐

Part B–1

36	44 ....
37	45 ....
38	46 ....
39	47 ....
40	48 ....
41	49 ....
42	50 ....
43	

Part B–1 Score

☐

Write your answers to Part B–2 and Part C in your answer booklet.

The declaration below should be signed when you have completed the examination.

I do hereby affirm, at the close of this examination, that I had no unlawful knowledge of the questions or answers prior to the examination and that I have neither given nor received assistance in answering any of the questions during the examination.

____
Signature

Tear Here

The University of the State of New York

REGENTS HIGH SCHOOL EXAMINATION

PHYSICAL SETTING EARTH SCIENCE

Tuesday, August 17, 2004 — 12:30 to 3:30 p.m., only

ANSWER BOOKLET

Student . . . . Sex: ☐ Male ☐ Female

Teacher . . . .

School . . . . Grade . . . .

Answer all questions in Part B–2 and Part C. Record your answers in this booklet.

Performance Test Score (Maximum Score: 23) ☐

Part	Maximum Score	Student's Score
A	35	
B–1	15	
B–2	15	
C	20	

Total Written Test Score (Maximum Raw Score: 85) ☐

Final Score (from conversion chart) ☐

Raters' Initials:

Rater 1 . . . . **Rater 2** . . . .

Part B–2

For Raters Only

51 ______________________________ **front** — **51** ☐

52 **Weather Data Table for Albany** — **52** ☐

Relative humidity (%)	
Wind direction from	
Wind speed (knots)	
Present weather	

53 — **53** ☐

Cross Section of the Bottom of the Tube

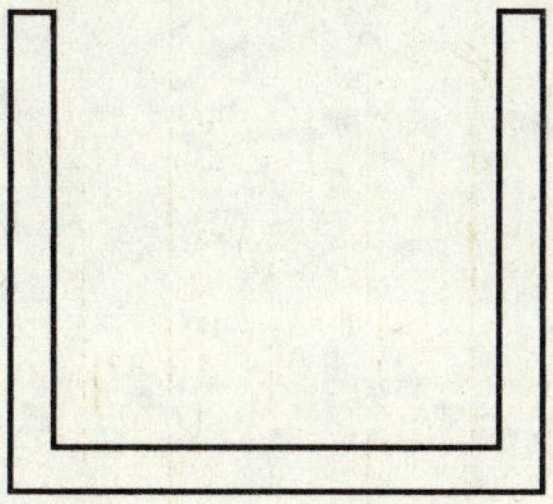

Key
Approximate size
○ 2-mm diameter
○ 3-mm diameter
○ 4-mm diameter

For Raters Only

54 ______________________________________ 54 ☐

____

55

Surface Air Pressures

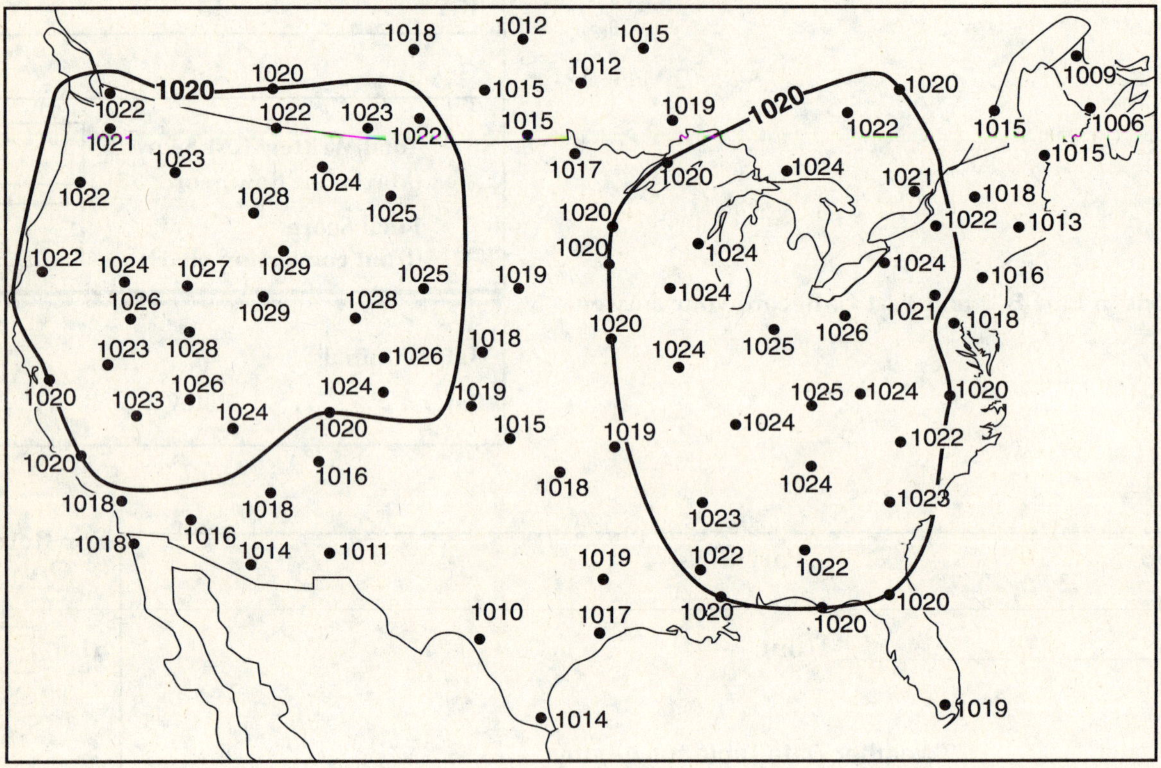

55 ☐

56 ______________________ 56 ☐

57 Oldest (1) ______________________ 57 ☐

(2) ______________________

(3) ______________________

(4) ______________________

(5) ______________________

Youngest (6) ______________________

For Raters Only

58 ________________ Period | 58 ☐

59 ________________ | 59 ☐

60 ________________ | 60 ☐

61 (1) ________________

(2) ________________ | 61 ☐

62 ________________ | 62 ☐

63 | 63 ☐

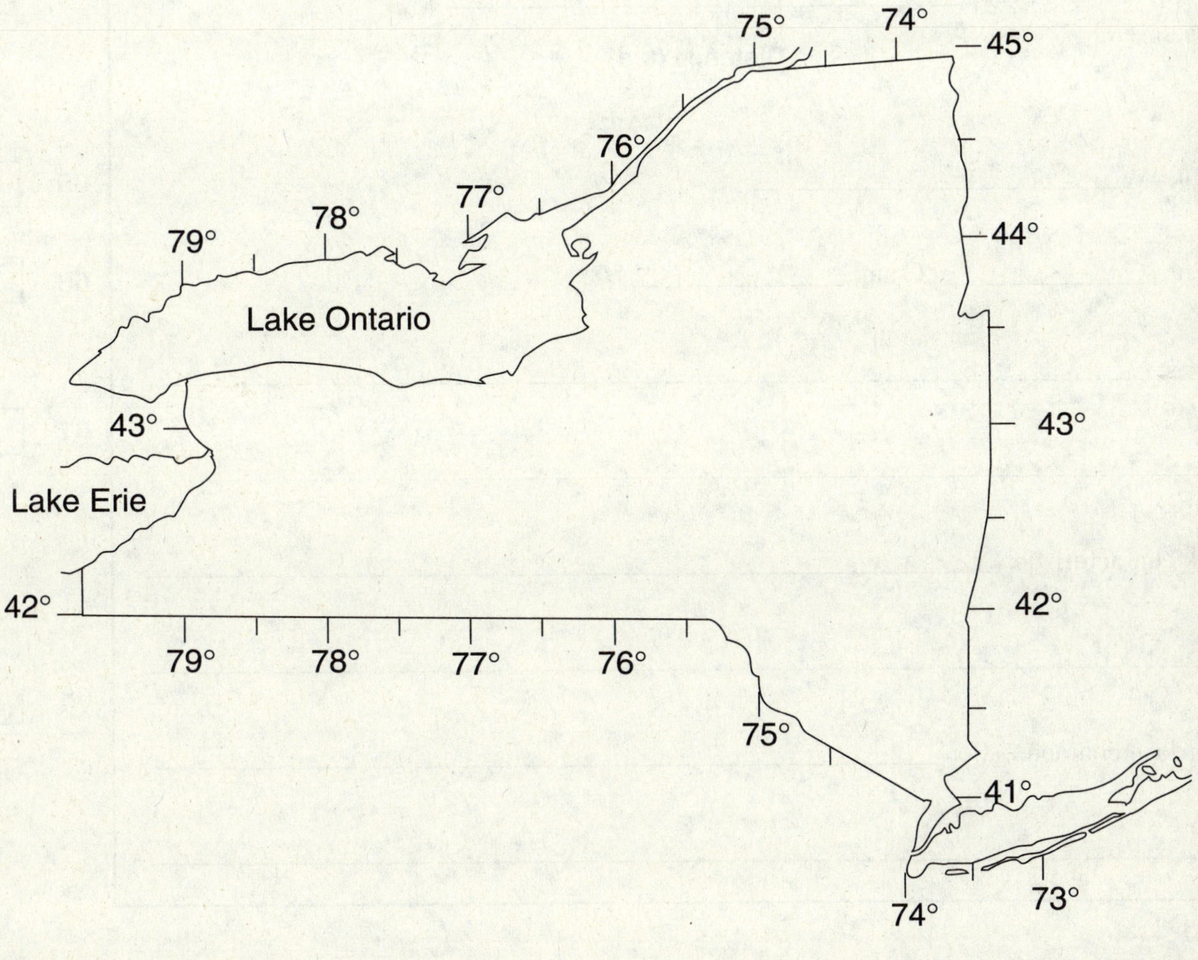

☐

Total Score for Part B–2

[OVER]

For Raters Only

Part C

64

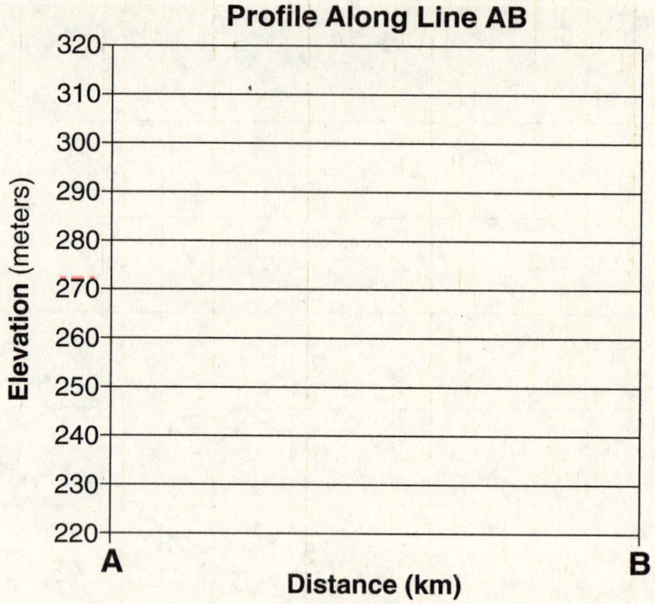

64 ☐

65 Gradient = ______________________ 65 ☐

66 Between ____________ **°C** and ____________ **°C** 66 ☐

67 ____

____ 67 ☐

68 Bedrock characteristic: ________________________________

____ 68 ☐

Land surface characteristic: ______________________________

____

[4]

For Raters Only

69

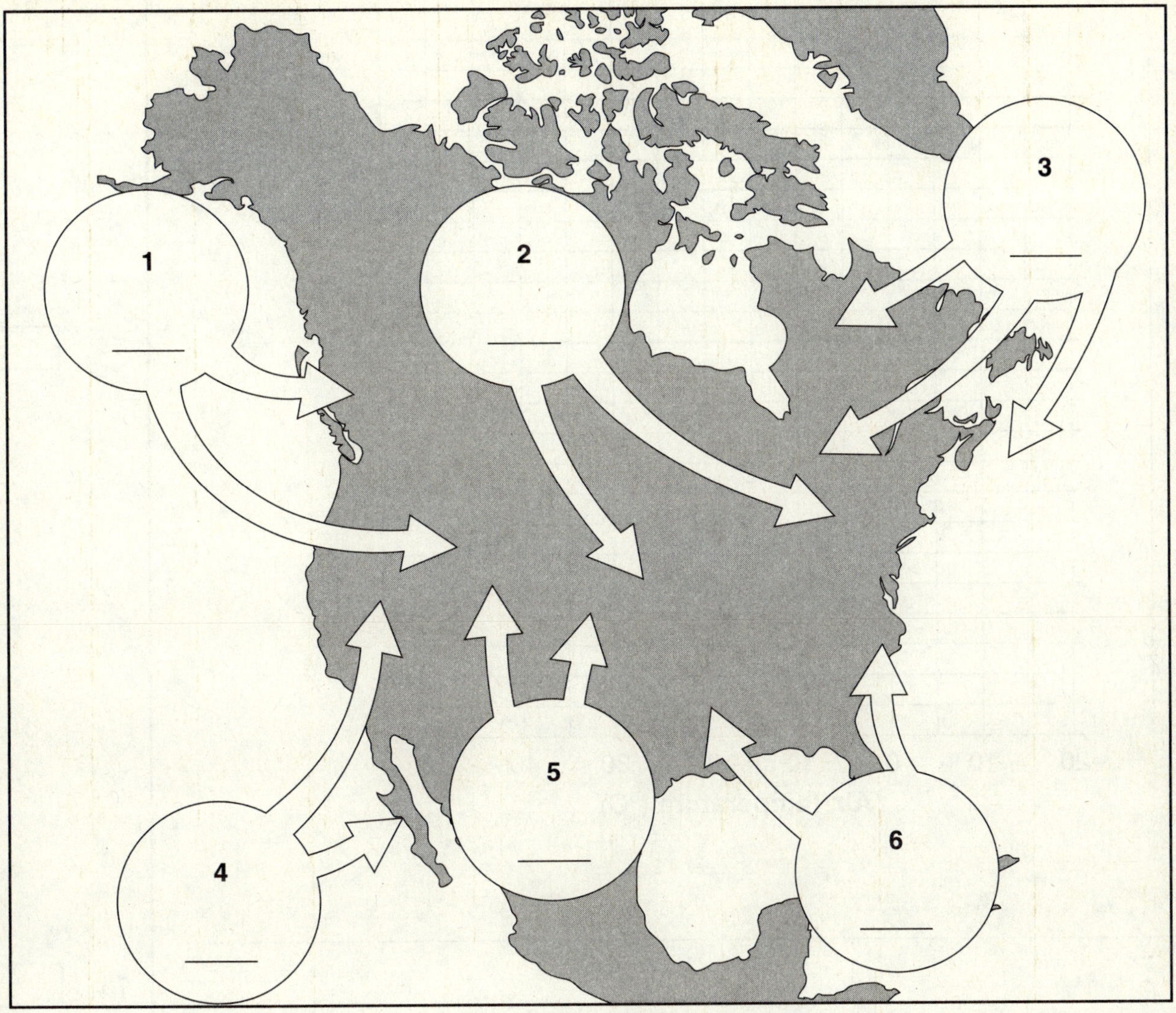

69 ☐

70 ____

____

70 ☐

71 ____

71 ☐

72

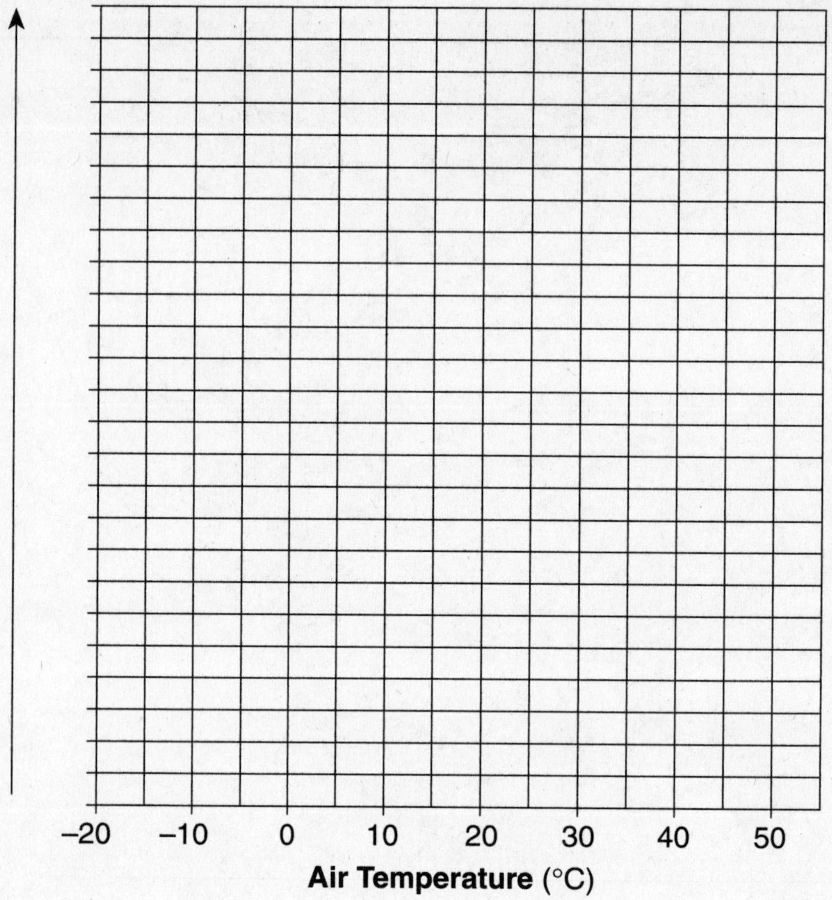

For Raters Only

72 ☐

73 ______________________________

____

73 ☐

74 ______________________________

____

74 ☐

75 ______________________________ 75 ☐

76 ____________________ km 76 ☐

77 ____________________ Period 77 ☐

Total Score for Part

Physical Setting Earth Science June 19, 2004

Part A

Answer all questions in this part.

Directions (1–35): For *each* statement or question, write on your separate answer sheet the *number* of the word or expression that, of those given, best completes the statement or answers the question. Some questions may require the use of the *Earth Science Reference Tables*.

1 The motion of a Foucault pendulum provides evidence of

(1) the Sun's rotation (3) Earth's rotation
(2) the Sun's revolution (4) Earth's revolution

2 Which form of electromagnetic radiation has a wavelength of 1.0×10^{-3} centimeter?

(1) ultraviolet (3) radio waves
(2) infrared (4) microwaves

3 The time required for the Moon to show a complete cycle of phases when viewed from Earth is approximately

(1) 1 day (3) 1 month
(2) 1 week (4) 1 year

4 Which planet has an orbital eccentricity most like the orbital eccentricity of the Moon?

(1) Pluto (3) Mars
(2) Saturn (4) Mercury

5 On June 21, where will the Sun appear to rise for an observer located in New York State?

(1) due west (3) north of due east
(2) due east (4) south of due east

6 Which statement best describes sediments deposited by glaciers and rivers?

(1) Glacial deposits and river deposits are both sorted.
(2) Glacial deposits are sorted, and river deposits are unsorted.
(3) Glacial deposits are unsorted, and river deposits are sorted.
(4) Glacial deposits and river deposits are both unsorted.

7 The diagram below shows four different chemical materials escaping from the interior of early Earth.

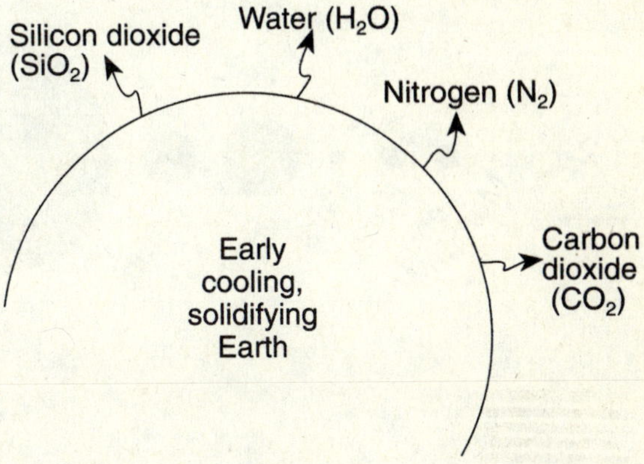

Which material contributed *least* to the early composition of the atmosphere?

(1) SiO_2 (3) N_2
(2) H_2O (4) CO_2

8 The diagram below shows a fossil found in the surface bedrock of New York State.

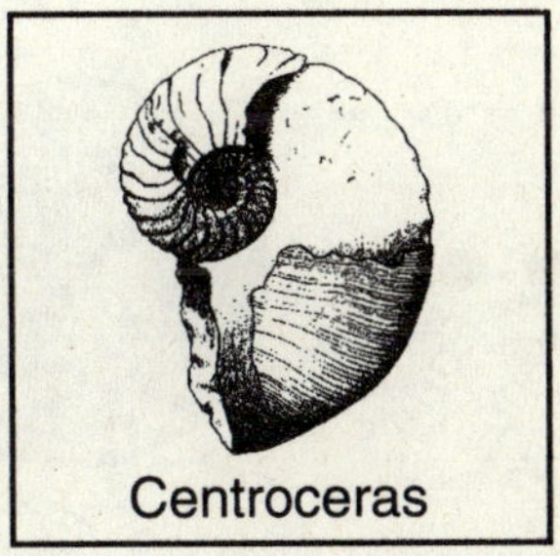

Which other fossil is most likely to be found in the same age bedrock?

(1) *Phacops* (3) *Coelophysis*
(2) condor (4) *Tetragraptus*

9 Soil composed of which particle size usually has the greatest capillarity?

(1) silt (3) coarse sand
(2) fine sand (4) pebbles

10 Which sequence correctly shows the relative size of the nine planets of our solar system?

(1) 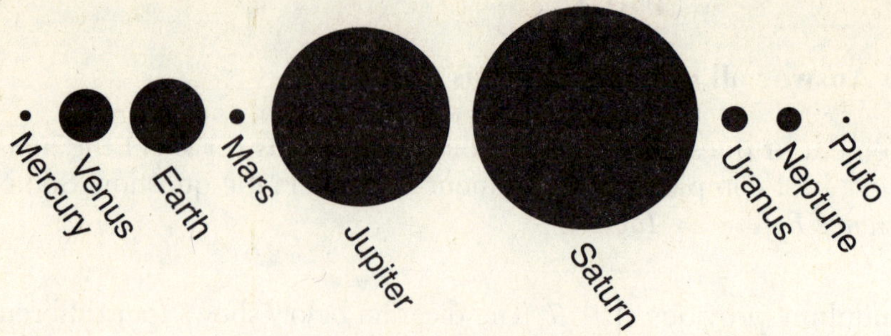

(2) 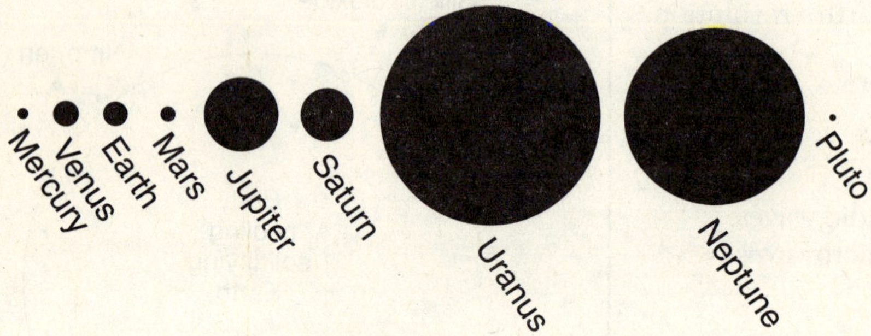

(3) 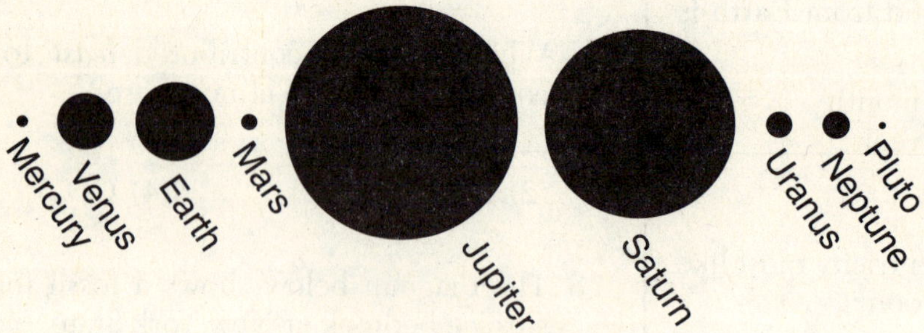

(4) 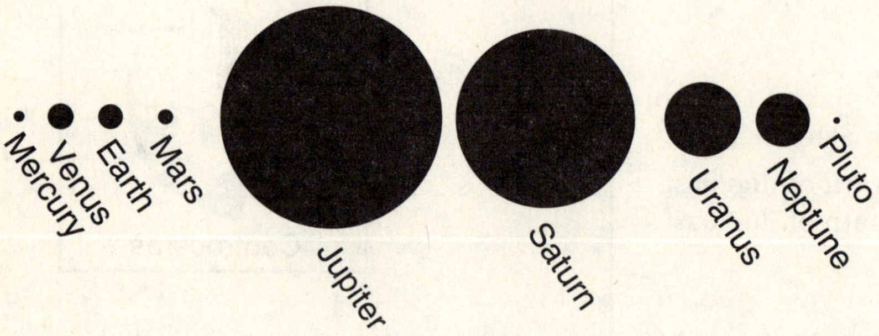

11 The graph below shows changes in the atmosphere occurring above typical air-mass source regions *A*, *B*, *C*, and *D*. Changes in air temperature and altitude are shown as the graphed lines. Changes in water-vapor content, in grams of vapor per kilogram of air, are shown as numbers on each graphed line.

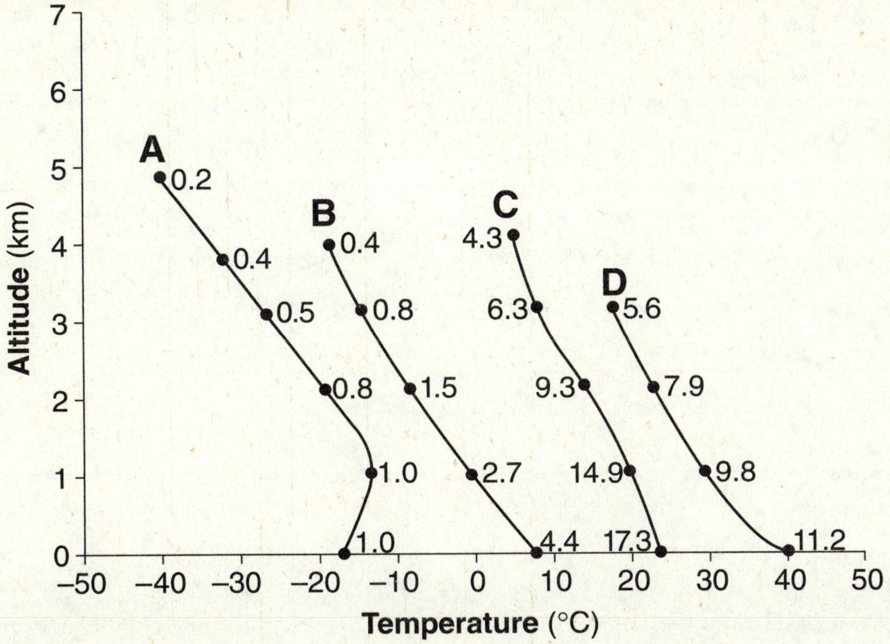

Which list best identifies each air-mass source region?

(1) *A* — cT, *B* — cP, *C* — mP, *D* — mT
(2) *A* — cP, *B* — mP, *C* — mT, *D* — cT
(3) *A* — mP, *B* — mT, *C* — cT, *D* — cP
(4) *A* — mT, *B* — cT, *C* — cP, *D* — mP

---

12 Earth's outer core and inner core are both inferred to be

(1) liquid
(2) solid
(3) composed of a high percentage of iron
(4) under the same pressure

13 Surface winds on Earth are primarily caused by differences in

(1) air density due to unequal heating of Earth's surface
(2) ocean wave heights during the tidal cycle
(3) rotational speeds of Earth's surface at various latitudes
(4) distances from the Sun during the year

14 Which nonfoliated rock forms only in a zone of contact metamorphism?

(1) conglomerate
(2) hornfels
(3) pegmatite
(4) quartzite

15 During a dry summer, the flow of most large New York State streams generally

(1) continues because some groundwater seeps into the streams
(2) increases due to greater surface runoff
(3) remains unchanged due to transpiration from grasses, shrubs, and trees
(4) stops completely because no water runs off into the streams

16 The density of Earth's crust is

(1) less than the density of the outer core but greater than the density of the mantle
(2) greater than the density of the outer core but less than the density of the mantle
(3) less than the density of both the outer core and the mantle
(4) greater than the density of both the outer core and the mantle

17 Which map best represents the direction of surface winds associated with the high- and low-pressure systems?

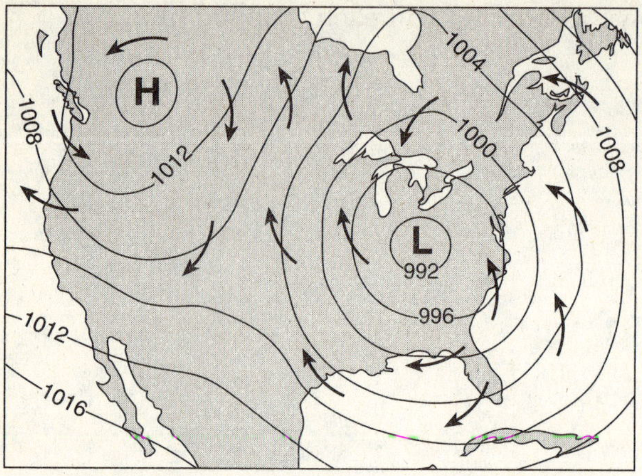

(1)

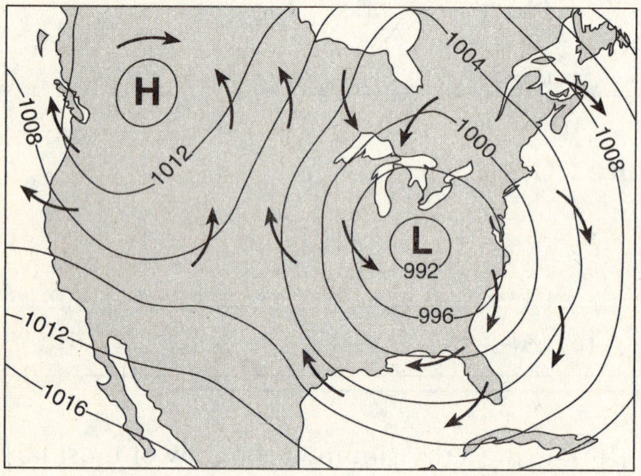

(2)

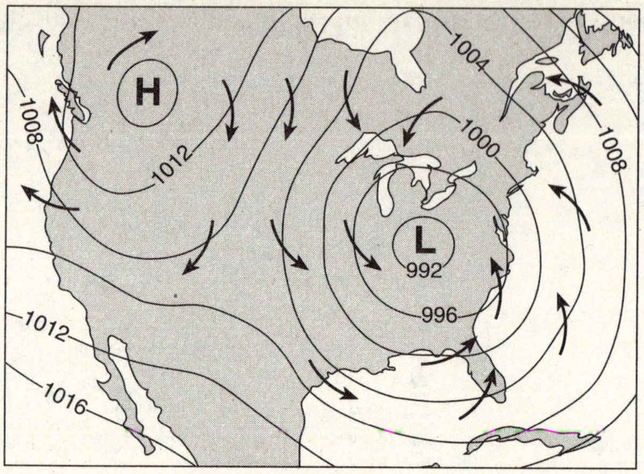

(3)

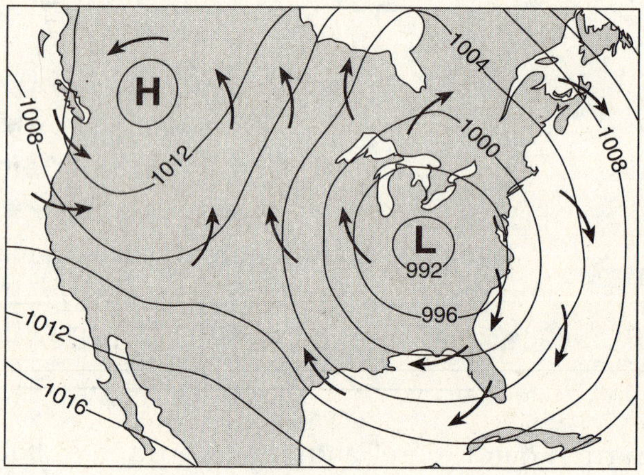

(4)

18 In each diagram below, the mass of the star is the same. In which diagram is the force of gravity greatest between the star and the planet shown?

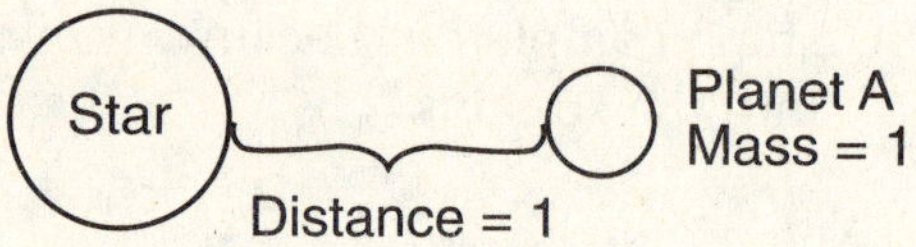

(1)

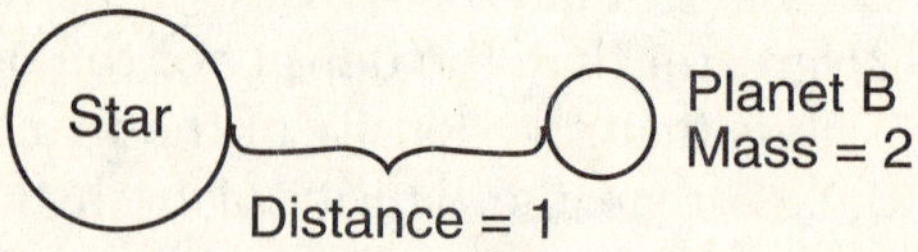

(2)

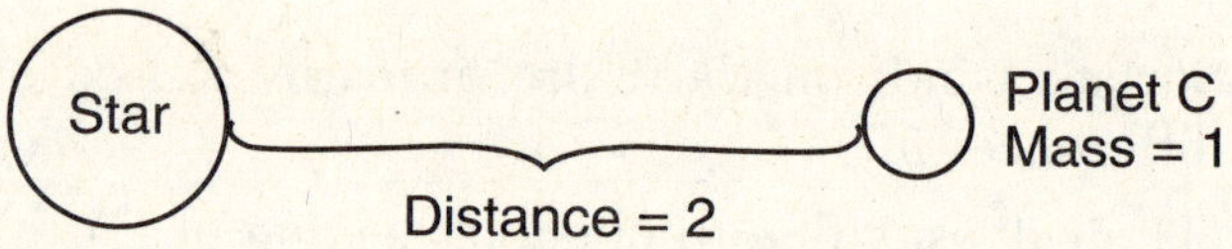

(3)

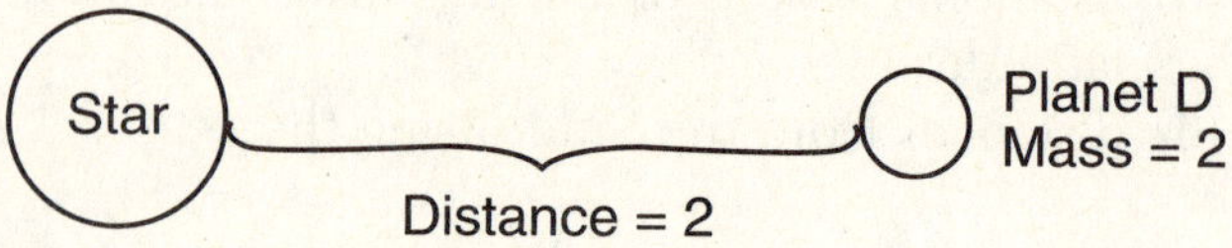

(4)

19 The cross section below shows rock layers that underwent crustal movement during an igneous intrusion in the Cretaceous Period.

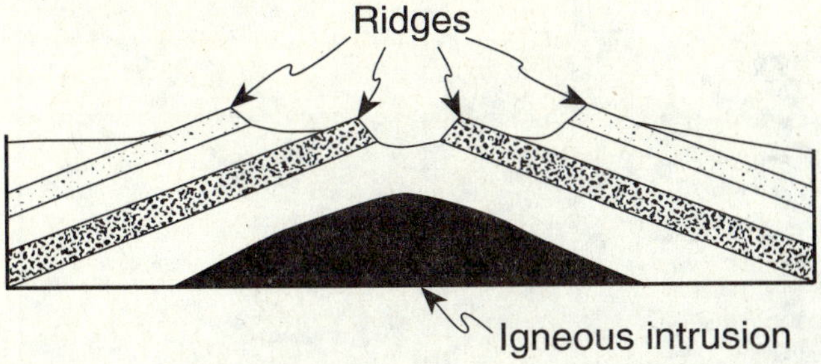

Which statement best describes the cause of the ridges shown?

(1) The rock layers were evenly weathered.
(2) Some rock layers were more resistant to weathering and erosion.
(3) The igneous intrusion flowed over the surface.
(4) More deposition occurred at the ridge sites after uplift.

20 The picture below shows a geological feature in the Kalahari Desert of southwestern Africa.

Which process most likely produced the present appearance of this feature?

(1) wind erosion
(2) volcanic eruption
(3) earthquake vibrations
(4) plate tectonics

21 Which group of organisms, some of which were preserved as fossils in early Paleozoic rocks, are still in existence today?

(1) brachiopods
(2) eurypterids
(3) graptolites
(4) trilobites

22 The diagram below shows the shadow cast by a telephone pole on March 21 at solar noon at a location in New York State.

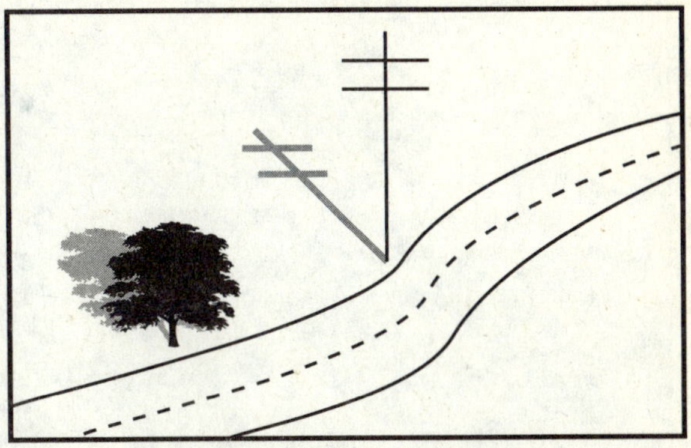

Shadow Cast on March 21

Which shadow was cast by the same telephone pole on June 21 at solar noon?

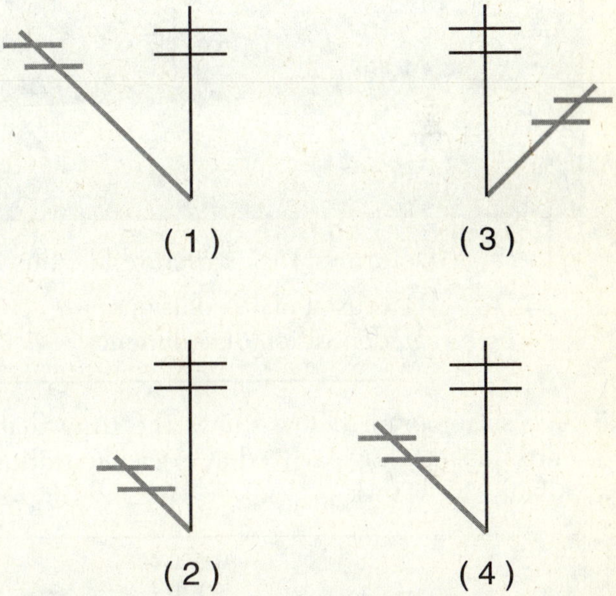

23 Which two New York State landscape regions are formed mostly of surface bedrock that is approximately the same geologic age?

(1) Manhattan Prong and Atlantic Coastal Plain
(2) Erie-Ontario Lowlands and Adirondack Mountains
(3) Adirondack Mountains and Allegheny Plateau
(4) Tug Hill Plateau and St. Lawrence Lowlands

24 The photograph below shows deformed rock structure found on Earth's surface.

Deformed rock structure like this is most often caused by

(1) crustal plate collisions
(2) deposition of sediments
(3) extrusion of magma
(4) glacial movement

---

25 The seismogram below shows the time that an earthquake *P*-wave arrived at a seismic station in Albany, New York.

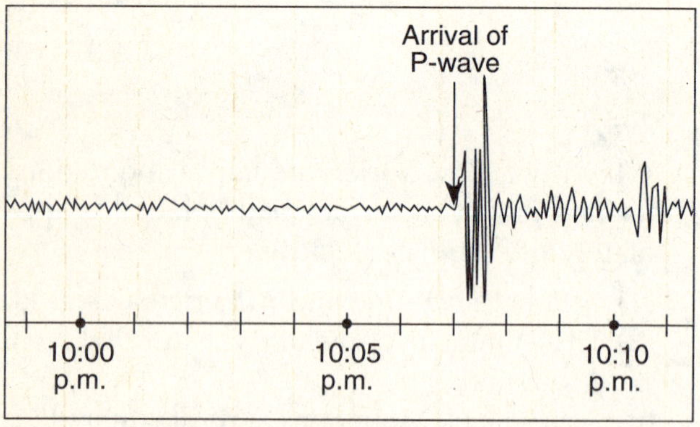

If the earthquake occurred at exactly 10:00 p.m., approximately how far from the earthquake epicenter was Albany, New York?

(1) 1,900 km
(2) 3,200 km
(3) 4,000 km
(4) 5,200 km

26 On each topographic map below, the straight-line distance from point *A* to point *B* is 5 kilometers. Which topographic map shows the steepest gradient between *A* and *B*?

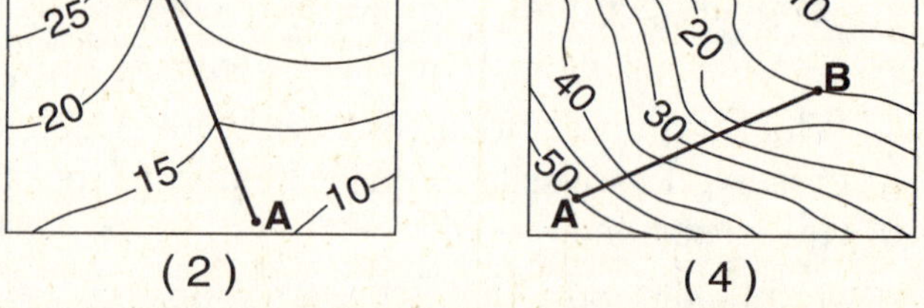

27 Which seismogram was recorded approximately 4,000 kilometers from an earthquake epicenter?

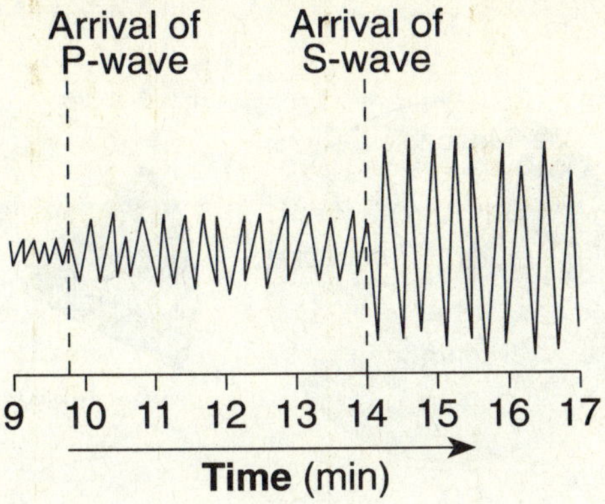

(1)

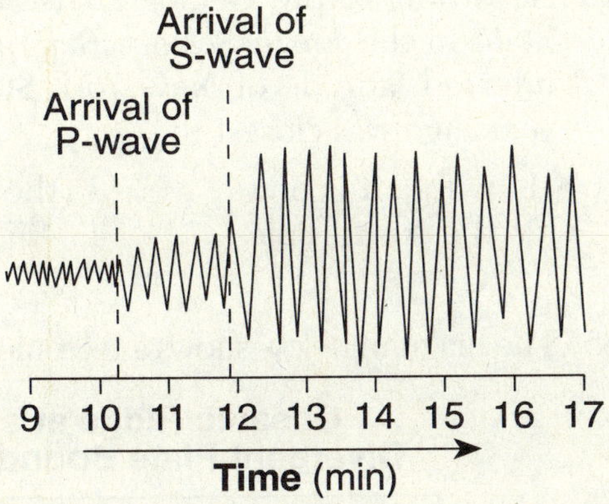

(2)

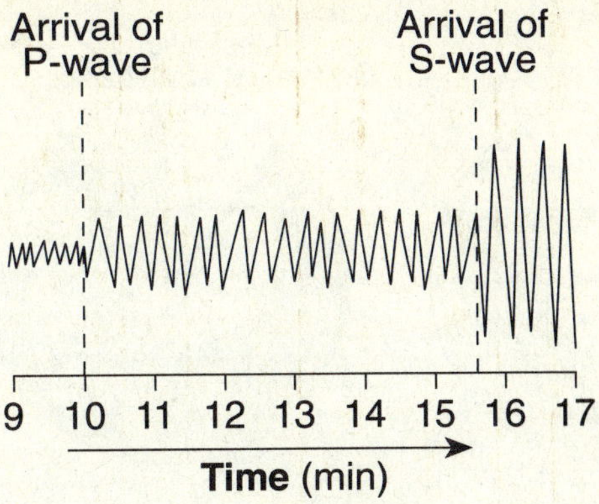

(3)

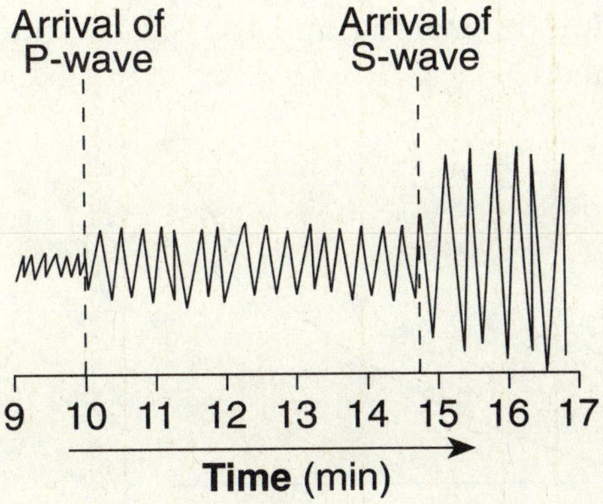

(4)

---

28 When the velocity of a stream suddenly *decreases,* the sediment being transported undergoes an increase in

(1) particle density
(2) erosion
(3) deposition
(4) mass movement

29 When granite melts and then solidifies, it becomes

(1) a sedimentary rock
(2) an igneous rock
(3) a metamorphic rock
(4) sediments

30 During the Permian Period, sedimentary bedrock in the Appalachian Region was subjected to high temperature and pressure. Calcite deposits that had existed in this environment would most likely have formed

(1) schist
(2) gabbro
(3) marble
(4) gneiss

31 The satellite photograph below shows a geologic feature composed of silt, sand, and clay.

The geologic feature shown in the photograph was primarily deposited by which agent of erosion?

(1) glaciers
(2) wind
(3) wave action
(4) running water

32 Which graph shows the relative duration of geologic time for the Precambrian, Paleozoic, Mesozoic, and Cenozoic time intervals?

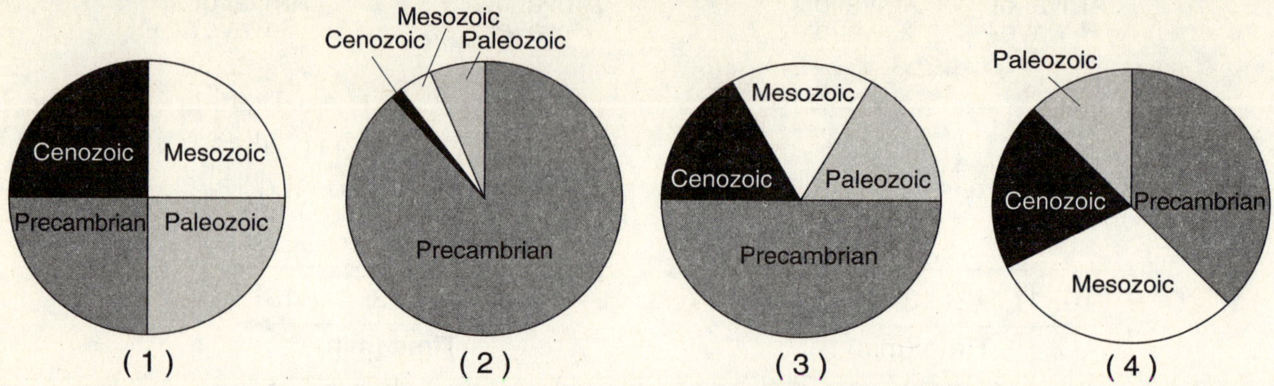

33 The graph below shows the relationship between the cooling time of magma and the size of the crystals produced.

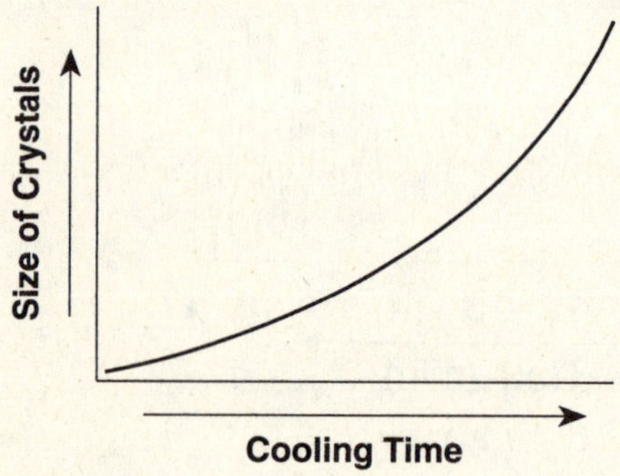

Which graph correctly shows the relative positions of the igneous rocks granite, rhyolite, and pumice?

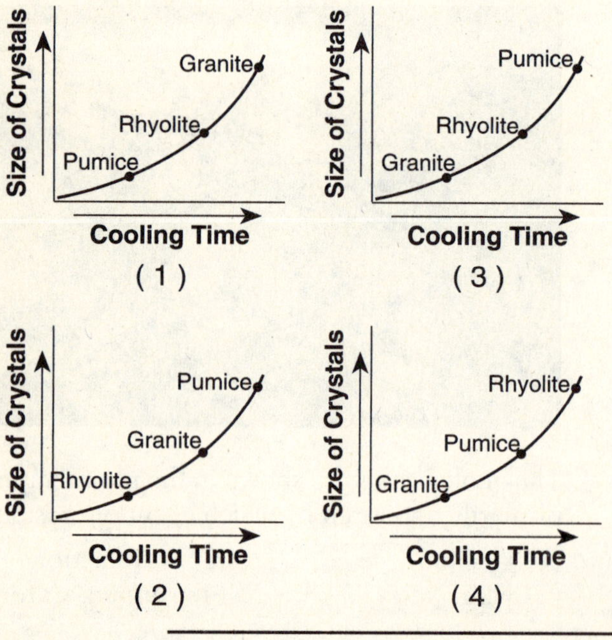

34 According to the Geologic History of New York State in the *Earth Science Reference Tables*, the inferred latitude of New York State 362 million years ago was closest to

(1) where it is now
(2) the North Pole
(3) the Equator
(4) 45° south

35 The diagram below shows a tectonic plate boundary.

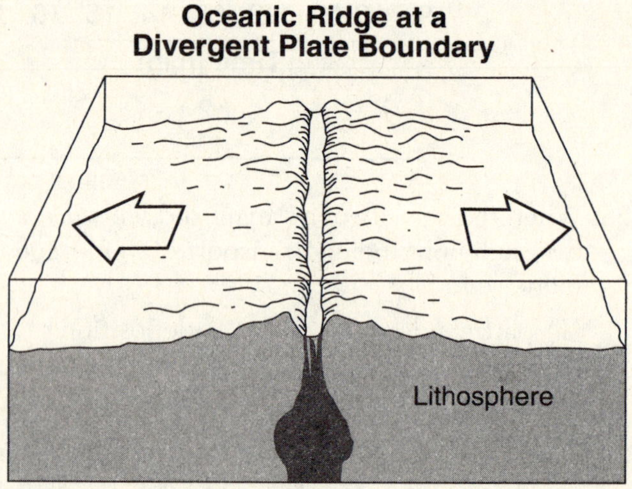

Which mantle hot spot is at a plate boundary like the one shown in this diagram?

(1) Hawaii Hot Spot
(2) Yellowstone Hot Spot
(3) Galapagos Hot Spot
(4) Canary Hot Spot

Part B–1

Answer all questions in this part.

Directions (36–50): For *each* statement or question, write on your separate answer sheet the *number* of the word or expression that, of those given, best completes the statement or answers the question. Some questions may require the use of the *Earth Science Reference Tables*.

Base your answers to questions 36 through 38 on the diagram below, which shows two possible sequences in the life cycle of stars, beginning with their formation from nebular gas clouds in space.

The Life Cycles of Stars

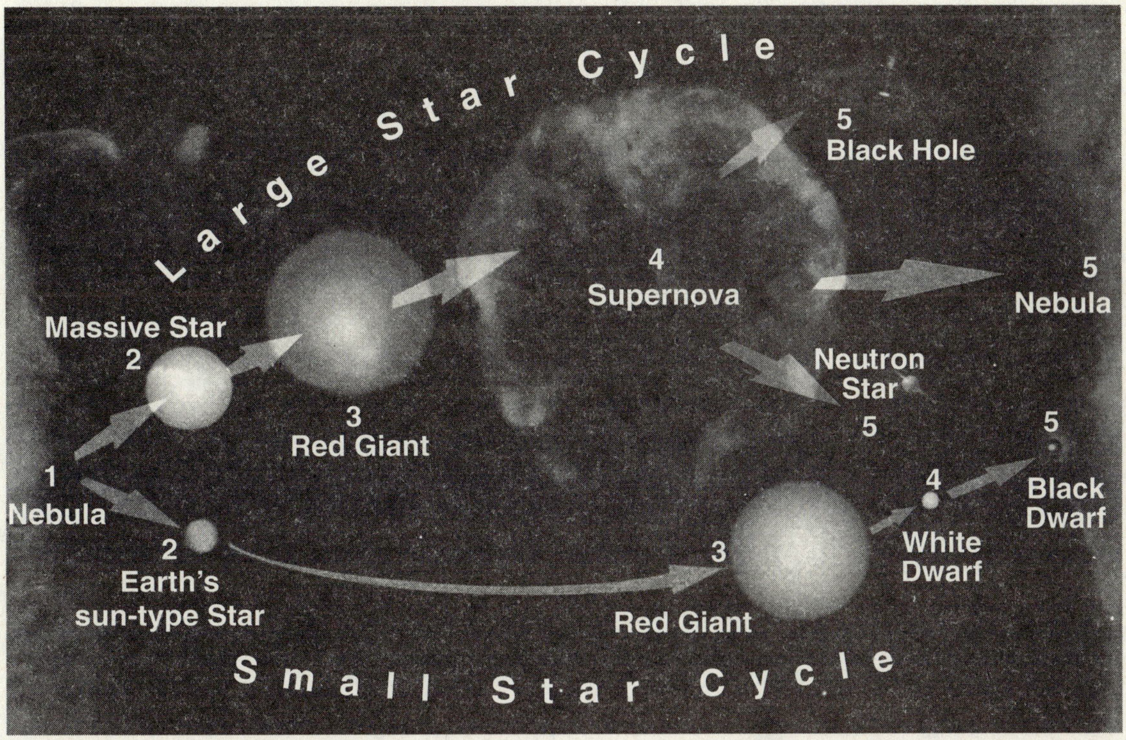

36 According to the diagram, the life-cycle path followed by a star is determined by the star's initial

(1) mass and size
(2) temperature and origin
(3) luminosity and color
(4) luminosity and structure

37 Stars like Earth's Sun most likely formed directly from a

(1) nebula
(2) supernova
(3) red giant
(4) black dwarf

38 According to the diagram, a star like Earth's Sun will eventually

(1) explode in a supernova
(2) become a black hole
(3) change into a white dwarf
(4) become a neutron star

Base your answers to questions 39 and 40 on the maps below, which show changes in the distribution of land and water in the Mediterranean Sea region that scientists believe took place over a period of 6 million years.

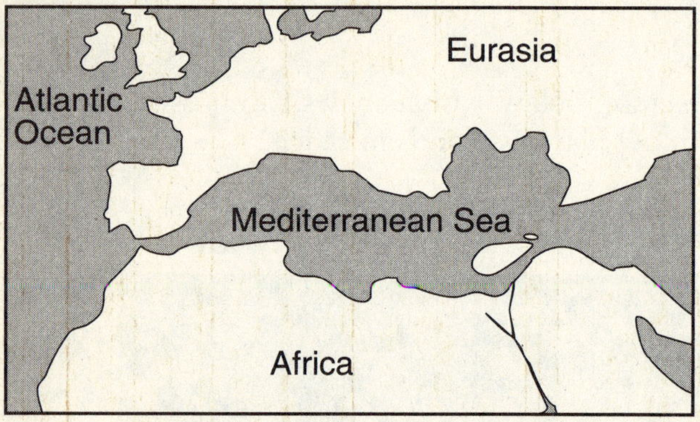

About 10 Million Years Ago

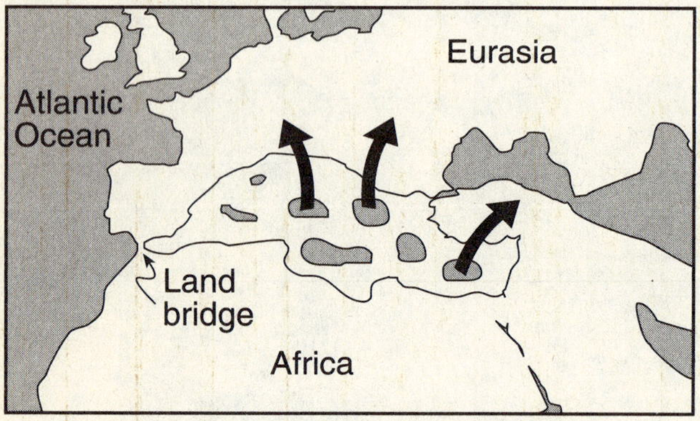

About 8 to 5.5 Million Years Ago
Evaporation from Mediterranean Sea

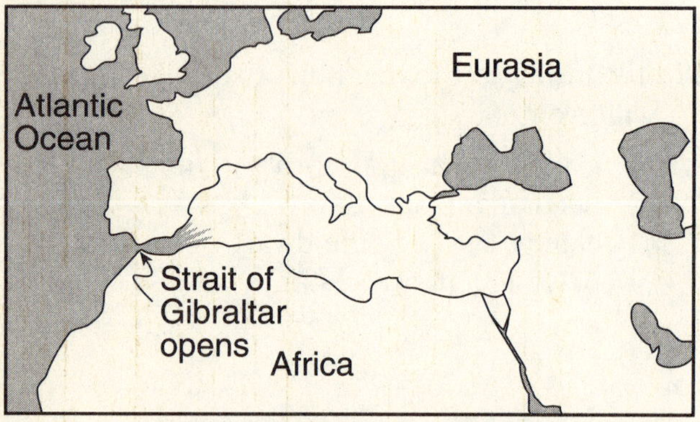

About 4 Million Years Ago
Mediterranean Sea Refills
with Atlantic Ocean Water

39 Which type of rock was precipitated from seawater as the Mediterranean Sea evaporated between 8 million years ago and 5.5 million years ago?

(1) rock salt
(2) basalt
(3) sandstone
(4) metaconglomerate

40 During which geologic time period did the changes shown in the maps take place?

(1) Cambrian
(2) Cretaceous
(3) Permian
(4) Neogene

Base your answers to questions 41 through 45 on the maps below. Points *A*, *B*, *C*, *X*, and *Y* are locations on the topographic map. The small map identifies the New York State region shown in the topographic map.

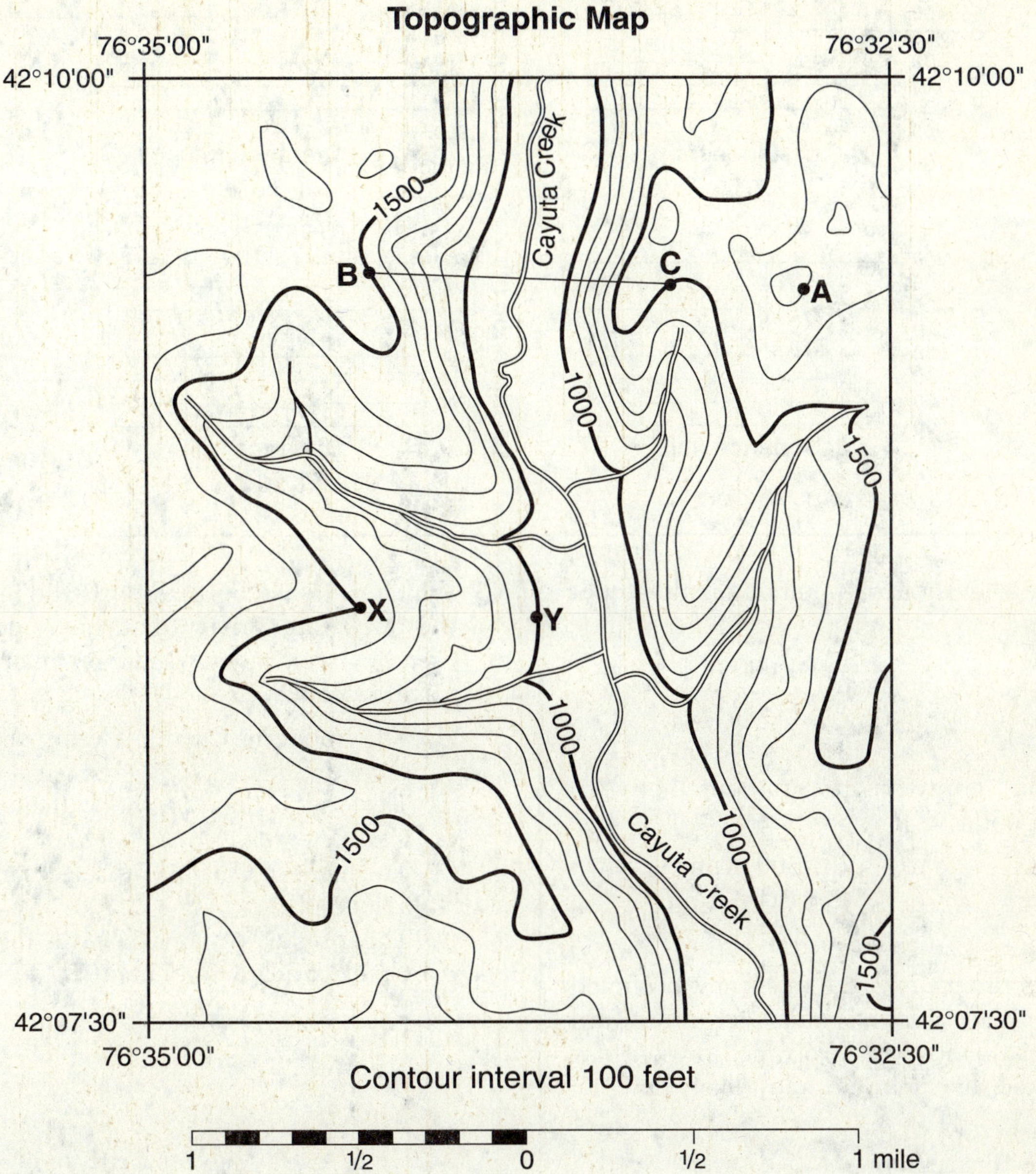

41 Which graph best represents the profile from point *B* to point *C*?

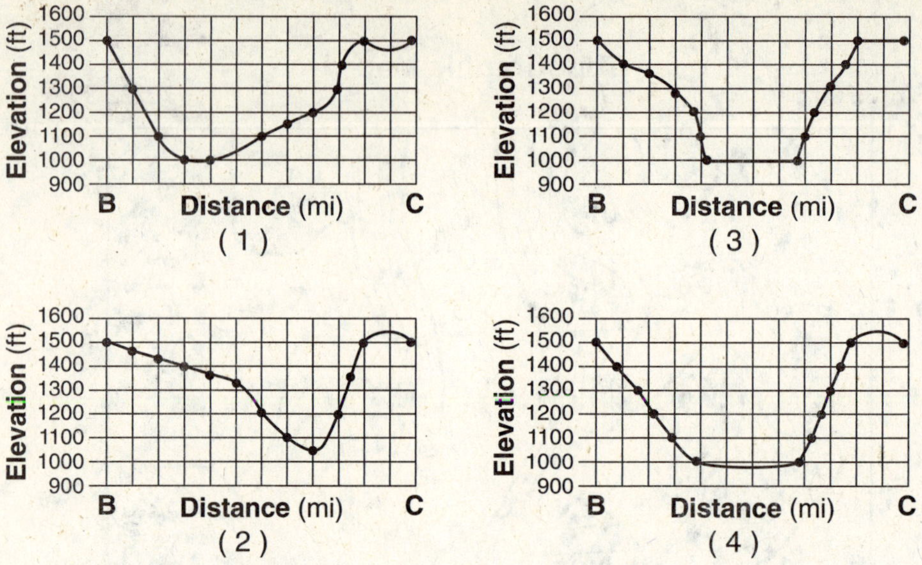

42 What is the elevation of point *A* on the topographic map?

(1) 1,700 ft
(2) 1,650 ft
(3) 1,600 ft
(4) 1,550 ft

43 What is the approximate gradient between point *X* and point *Y*?

(1) 100 ft/mi
(2) 250 ft/mi
(3) 500 ft/mi
(4) 1,000 ft/mi

44 At the end of the Ice Age, the valley now occupied by Cayuta Creek was a channel for southward flowing glacial meltwater. Into which present-day river valley did this meltwater most likely flow?

(1) Hudson River
(2) Genesee River
(3) Delaware River
(4) Susquehanna River

45 Which evidence best supports the inference that the meltwater river that once occupied the Cayuta Creek valley was larger than the modern Cayuta Creek?

(1) The modern Cayuta Creek occupies a V-shaped valley.
(2) The valley floor is wider than the modern Cayuta Creek.
(3) The modern Cayuta Creek lacks meanders and a flood plain.
(4) The tributary streams meet the modern Cayuta Creek at nearly right angles.

Base your answers to questions 46 through 50 on the two cross sections below, which represent the Pacific Ocean and the atmosphere near the Equator during normal weather (cross section *A*) and during El Niño conditions (cross section *B*). Sea surface temperatures (SST) are labeled and trade-wind directions are shown with arrows. Cloud buildup indicates regions of frequent thunderstorm activity. The change from normal sea level is shown at the side of each diagram.

Cross Section A: Normal Weather

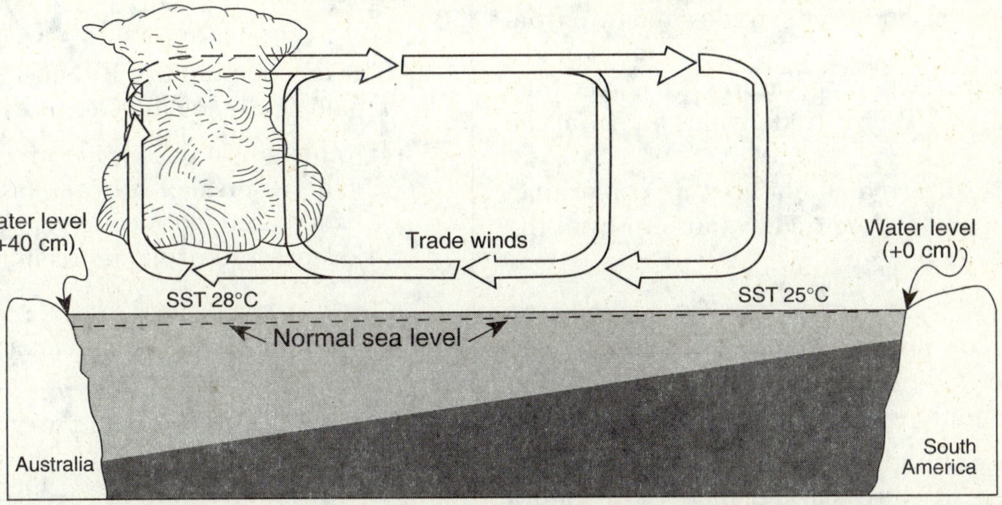

Cross Section B: El Niño Conditions

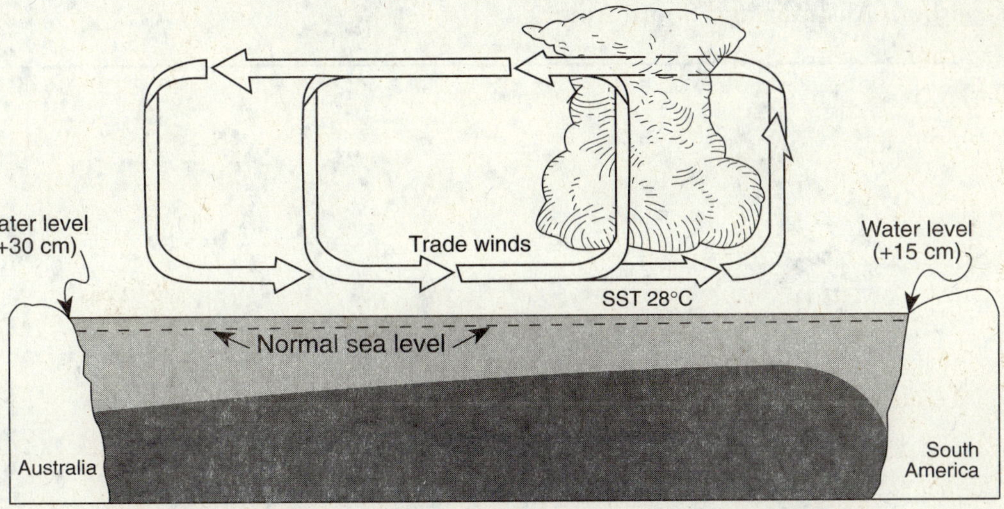

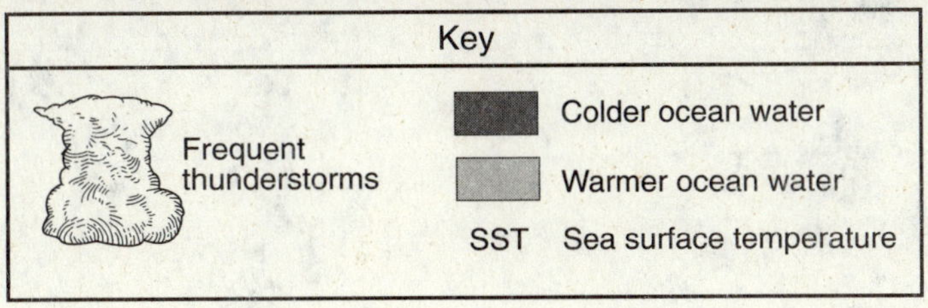

46 Which statement correctly describes sea surface temperatures along the South American coast and Pacific trade winds during El Niño conditions?

(1) The sea surface temperatures are warmer than normal, and Pacific trade winds are from the west.
(2) The sea surface temperatures are warmer than normal, and Pacific trade winds are from the east.
(3) The sea surface temperatures are cooler than normal, and Pacific trade winds are from the west.
(4) The sea surface temperatures are cooler than normal, and Pacific trade winds are from the east.

47 Compared to normal weather conditions, the shift of the trade winds caused sea levels during El Niño conditions to

(1) decrease at both Australia and South America
(2) decrease at Australia and increase at South America
(3) increase at Australia and decrease at South America
(4) increase at both Australia and South America

48 During El Niño conditions, thunderstorms increase in the eastern Pacific Ocean region because the warm, moist air is

(1) less dense, sinking, compressing, and warming
(2) less dense, rising, expanding, and cooling
(3) more dense, sinking, compressing, and warming
(4) more dense, rising, expanding, and cooling

49 The development of El Niño conditions over this region of the Pacific Ocean has caused

(1) changes in worldwide precipitation patterns
(2) the reversal of Earth's seasons
(3) increased worldwide volcanic activity
(4) decreased ozone levels in the atmosphere

50 Earth's entire equatorial climate zone is generally a belt around Earth that has

(1) high air pressure and wet weather
(2) high air pressure and dry weather
(3) low air pressure and wet weather
(4) low air pressure and dry weather

Part B–2

Answer all questions in this part.

Directions (51–64): Record your answers in the spaces provided in your answer booklet. Some questions may require the use of the *Earth Science Reference Tables.*

51 The atmospheric conditions at a given location are represented by the weather station model below.

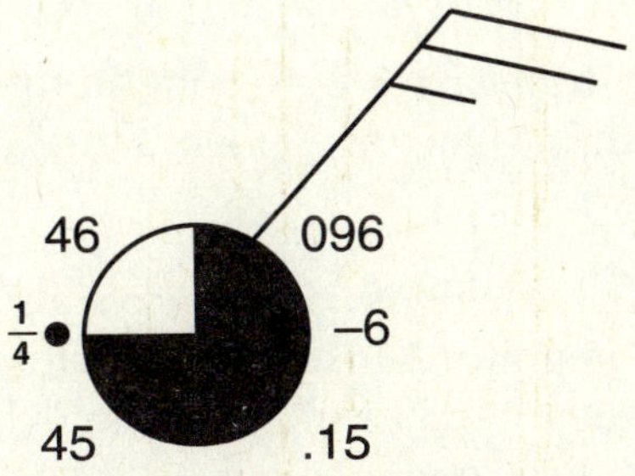

On the lines provided *in your answer booklet*, fill in the correct information for *each* variable listed, based on this weather station model. [2]

Base your answers to questions 52 through 54 on the diagram provided *in your answer booklet*, which represents the Sun's rays striking Earth at a position in its orbit around the Sun.

52 On the diagram provided *in your answer booklet,* neatly and accurately shade the area of Earth that is in darkness. [1]

53 On the diagram provided *in your answer booklet,* draw the line of latitude that is receiving the Sun's direct perpendicular rays on this date. [1]

54 What month of the year is represented by the diagram? [1]

---

55 The diagram provided *in your answer booklet* shows the Sun, the Moon, and Earth in line with one another in space. On the diagram, draw *two* dots (•) on the surface of Earth to indicate the locations where the highest ocean tides are most likely occurring. [1]

56 Using the "Luminosity and Temperature of Stars" graph in the *Earth Science Reference Tables*, list the five stars below in order of *decreasing* relative luminosity, with letter *a* being the brightest. [1]

Aldebaran, Betelgeuse, Polaris, Sirius, the Sun

Base your answers to questions 57 through 61 on the geologic cross section provided *in your answer booklet,* which represents an outcrop of various types of bedrock and bedrock features in Colorado.

57 On the cross section provided *in your answer booklet,* indicate with arrows the direction of movement on *both* sides of the fault. [1]

58 According to this cross section, what is the amount of vertical movement of the shale along the fault? Express your answer to the *nearest tenth of a meter.* [1]

59 Place the geologic events listed *in your answer booklet* in order by numbering them from oldest (1) to youngest (4). [1]

60 The shale and sandstone layers both contain fossilized leaves from the *Fagopsis* tree, an index fossil for the Oligocene Epoch. State a possible age for these rock layers, in million years. [1]

61 The vesicular basalt includes zircon crystals containing the radioactive isotope U-235, which disintegrates to the stable isotope Pb-207. The zircon crystals have 98.44% of the original U-235 remaining, and 1.56% has decayed to Pb-207. Based on the table below, how many half-lives have elapsed since the formation of these crystals? [1]

Percent of U-235 Remaining	Percent Decayed to Pb-207	Half-Lives Elapsed
99.22	0.78	$\frac{1}{64}$
98.44	1.56	$\frac{1}{32}$
96.88	3.12	$\frac{1}{16}$
93.75	6.25	$\frac{1}{8}$
87.50	12.5	$\frac{1}{4}$
75.0	25.0	$\frac{1}{2}$
50.0	50.0	1
37.5	62.5	$1\frac{1}{2}$
25.0	75.0	2
12.5	87.5	3
6.25	93.75	4

Base your answers to questions 62 through 64 on diagram 1 below and on diagram 2 *in your answer booklet,* which show some constellations in the night sky viewed by a group of students. Diagram 1 below shows the positions of the constellations at 9:00 p.m. Diagram 2 *in your answer booklet* shows their positions two hours later.

Diagram 1 — 9:00 p.m.

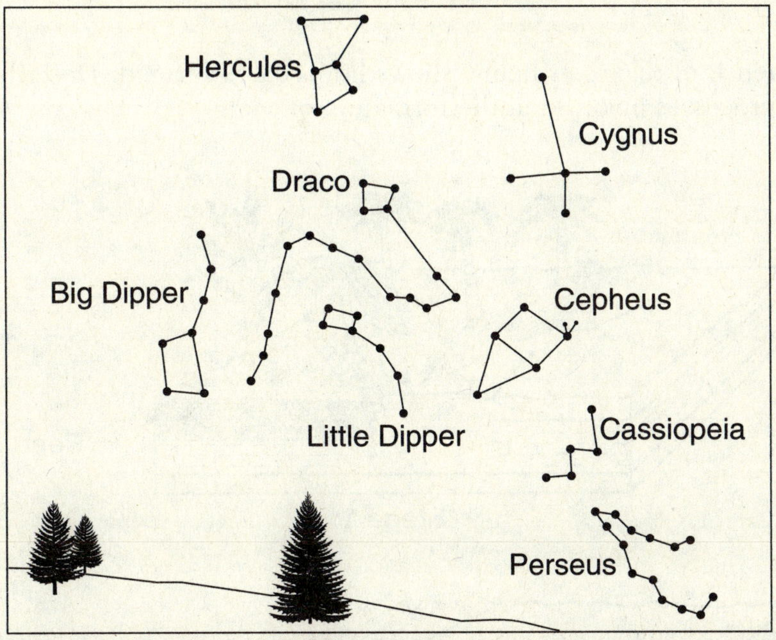

62 Circle *Polaris* on diagram 2 provided *in your answer booklet*. [1]

63 In which compass direction were the students facing? [1]

64 Describe the apparent direction of movement of the constellations Hercules and Perseus during the two hours between student observations. [1]

____

Part C

Answer all questions in this part.

Directions (65–81): Record your answers in the spaces provided in your answer booklet. Some questions may require the use of the *Earth Science Reference Tables.*

65 The sequence of diagrams below shows how coal is formed. Describe the material and *two* processes involved in the formation of coal. [2]

Stage 1

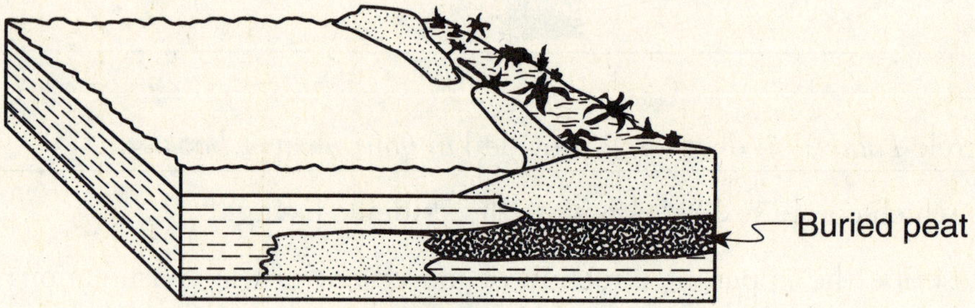

Stage 2

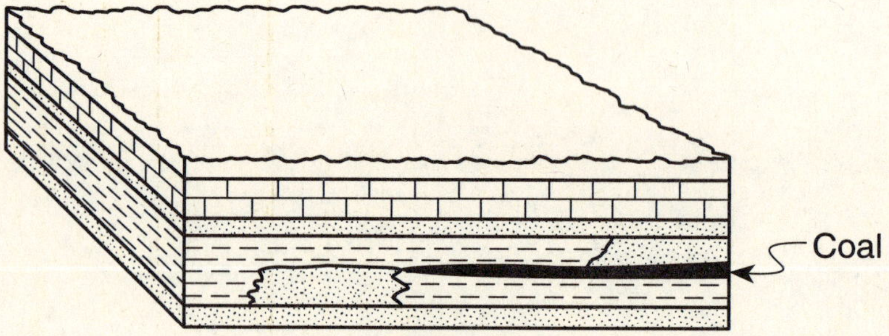

Stage 3

Base your answers to questions 66 and 67 on the table and graph below. The table labeled "Animal Key" shows symbols to represent various animal groups that exist on Earth. The graph shows inferred changes in Earth's average temperatures over the last 500 million years.

Animal Key

Letter	Picture	Animal Group
A		Birds
B		Fish
C		Amphibians
D		Mammals
E		Humans
F		Reptiles

Inferred Changes in Earth's Average Temperature

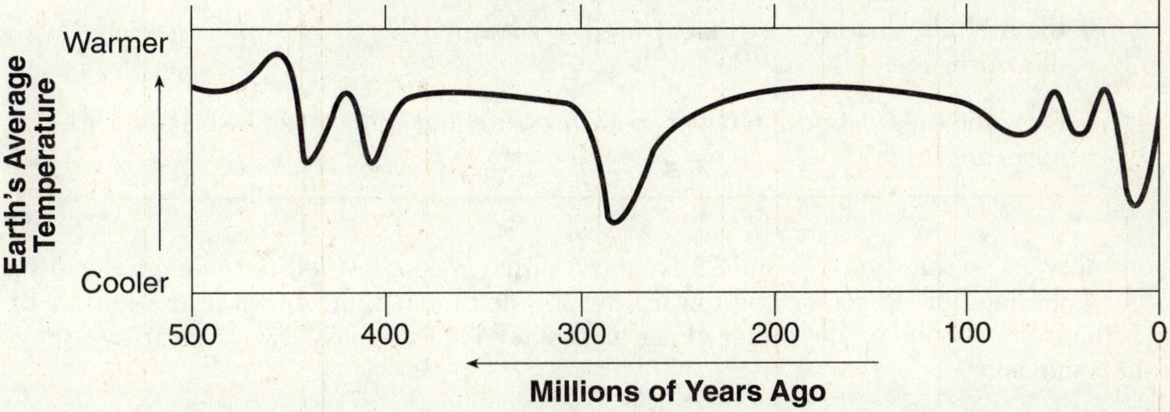

66 On the graph provided *in your answer booklet*, indicate when each of the life-forms in the table is believed to have first appeared on Earth by placing the letter for *each* animal group in the correct box. The correct location for earliest fish, letter *B*, has already been plotted above the graph. [2]

67 The two factors listed below could have caused the temperature variations shown on the graph. For *each* factor, state the effect that the increase described would have had on Earth's temperature, and explain why that temperature change would have taken place. [2]

Factors

A Increase in carbon dioxide (CO_2) and water vapor (H_2O gas) content of Earth's atmosphere

B Increase in volcanic ash in Earth's atmosphere

Base your answers to questions 68 through 71 on the data table below, which shows recorded information for a major Atlantic hurricane. Use the map provided *in your answer booklet* to answer questions 68 and 69.

Hurricane Data

Date	Time	Latitude	Longitude	Maximum Winds (knots)	Air Pressure (mb)
Sept. 10	11:00 a.m.	19° N	59° W	70	989
Sept. 11	11:00 a.m.	22° N	62° W	95	962
Sept. 12	11:00 a.m.	23° N	67° W	105	955
Sept. 13	11:00 a.m.	24° N	72° W	135	921
Sept. 14	11:00 a.m.	26° N	77° W	125	932
Sept. 15	11:00 a.m.	30° N	79° W	110	943

68 Using the latitude and longitude data in the table, place an **X** on the map provided *in your answer booklet* for *each* location of the hurricane during these 6 days. Connect all the **X**s with a solid line. [1]

69 Label the September 15 (9/15) position of the hurricane on the map. Starting from this plotted position on September 15, draw a dashed line on the map provided *in your answer booklet* to indicate the storm's most likely path for the next 5 days. [1]

70 Identify the weather instrument used to measure the air pressure associated with this hurricane. [1]

71 Describe the relationship between air pressure and wind speed associated with this hurricane. [1]

Base your answers to questions 72 and 73 on the weather map provided *in your answer booklet*, which shows a large white band of clouds moving toward the southeast. The line shown in the middle of the white cloud band is the frontal boundary between a cP air mass and an mT air mass. Two large arrows show the direction the front is moving.

72 On the frontal boundary line on the weather map provided *in your answer booklet,* draw the weather front symbol to represent the front moving toward the southeast. [1]

73 On the same weather map, place an **X** centered on the geographic region that was most likely the source of the warm, moist (mT) air mass. [1]

Base your answers to questions 74 through 79 on the reading passage and maps below and on your knowledge of Earth science. The enlarged map shows the location of volcanoes in Colombia, South America.

Fire and Ice — and Sluggish Magma

On the night of November 13, 1985, Nevado del Ruiz, a 16,200-foot (4,938 meter) snow-capped volcano in northwestern Colombia, erupted. Snow melted, sending a wall of mud and water raging through towns as far as 50 kilometers away, and killing 25,000 people.

Long before disaster struck, Nevado del Ruiz was marked as a trouble spot. Like Mexico City, where an earthquake killed at least 7,000 people in October 1985, Nevado del Ruiz is located along the Ring of Fire. This ring of islands and the coastal lands along the edge of the Pacific Ocean are prone to volcanic eruptions and crustal movements.

The ring gets its turbulent characteristics from the motion of the tectonic plates under it. The perimeter of the Pacific, unlike that of the Atlantic, is located above active tectonic plates. Nevado del Ruiz happens to be located near the junction of four plate boundaries. In this area an enormous amount of heat is created, which melts the rock 100 to 200 kilometers below Earth's surface and creates magma.

Nevado del Ruiz hadn't had a major eruption for 400 years before this tragedy. The reason: sluggish magma. Unlike the runny, mafic magma that makes up the lava flows of oceanic volcanoes such as those in Hawaii, the magma at this type of subduction plate boundary tends to be sticky and slow moving, forming the rock andesite when it cools. This andesitic magma tends to plug up the opening of the volcano. It sits in a magma chamber underground with pressure continually building up. Suddenly, tiny cracks develop in Earth's crust, causing the pressure to drop. This causes the steam and other gases dissolved in the magma to violently expand, blowing the magma plug free. Huge amounts of ash and debris are sent flying, creating what is called an explosive eruption.

Oddly enough, the actual eruption of Nevado del Ruiz didn't cause most of the destruction. It was caused not by lava but by the towering walls of sliding mud created when large chunks of hot ash and pumice mixed with melted snow.

Locator Plate Map **Enlarged Map Showing Volcanoes of Colombia**

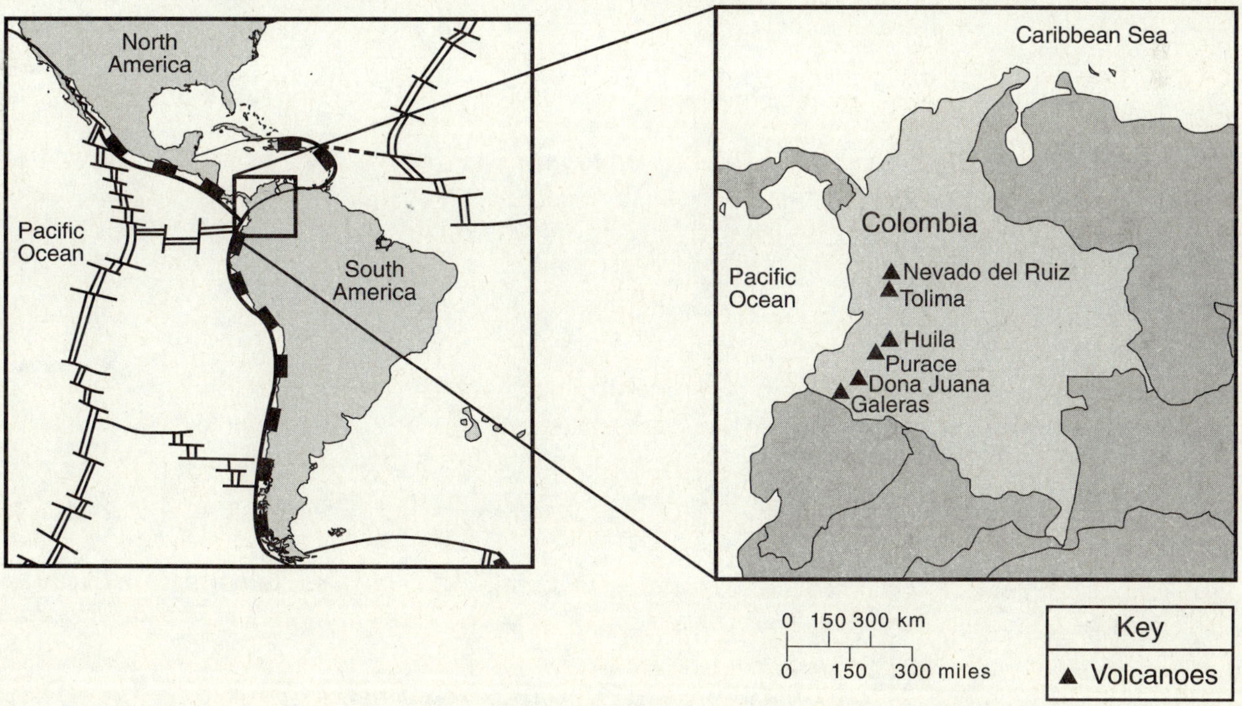

74 What are the names of the *four* tectonic plates located near the Nevado del Ruiz volcano? [1]

75 What caused most of the destruction associated with the eruption of Nevado del Ruiz? [1]

76 What caused the magma to expand, blowing the magma plug free? [1]

77 Vesicular texture is very common in igneous rocks formed during andesitic eruptions. Explain how this texture is formed. [1]

78 Why are eruptions of Nevado del Ruiz generally more explosive than most Hawaiian volcanic eruptions? [1]

79 Describe one emergency preparation that may reduce the loss of life from a future eruption of the Nevado del Ruiz volcano. [1]

GO RIGHT ON TO THE NEXT PAGE ➪

Base your answers to questions 80 and 81 on the cross section below. The cross section represents a part of Texas where weakly cemented sandstone is exposed at the surface. The mineral cement holding the sandstone grains together is calcite. Area *X* is a circular depression of loose sand that has been partially removed by prevailing winds. Sand dunes have developed downwind from depression *X*.

Present Day, Dry Climate

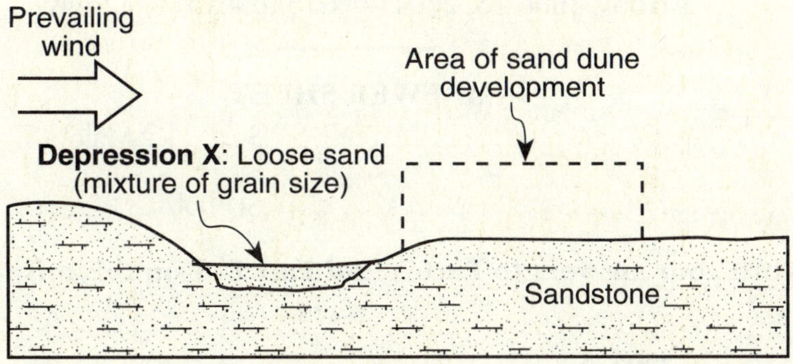

80 On the diagram of the area of sand dune development provided *in your answer booklet*, draw a sketch showing the general sideview of a sand dune formed by a wind blowing in the direction indicated. Your sketch should clearly show any variations in the slope of the sides of the dune. [1]

81 The cross section below shows this same area of Texas near the end of the last ice age when this area had a much wetter climate. More infiltration of rainwater was occurring at area *X*. Scientists infer that depression *X* was an area where slightly acidic rainwater collected and infiltrated into the sandstone.

Late Pleistocene, Wetter Climate

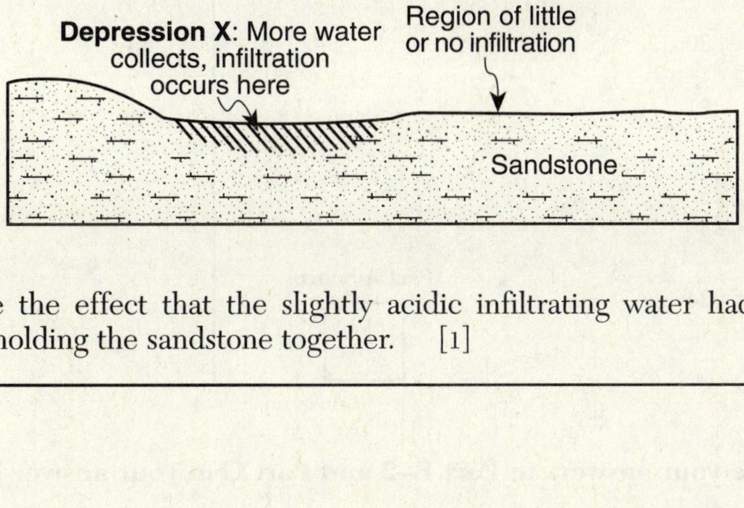

Describe the effect that the slightly acidic infiltrating water had on the calcite cement holding the sandstone together. [1]

____

Tear Here

The University of the State of New York

REGENTS HIGH SCHOOL EXAMINATION

PHYSICAL SETTING
EARTH SCIENCE

Friday, June 18, 2004 — 1:15 to 4:15 p.m., only

ANSWER SHEET

Student .... Sex: ☐ Male ☐ Female Grade ....

Teacher .... School ....

Record your answers to Part A and Part B–1 on this answer sheet.

Part A

1	13	25 ....
2	14	26 ....
3	15	27 ....
4	16	28 ....
5	17	29 ....
6	18	30 ....
7	19	31 ....
8	20	32 ....
9	21	33 ....
10	22	34 ....
11	23	35 ....
12	24	

Part A Score ☐

Part B–1

36	44 ....
37	45 ....
38	46 ....
39	47 ....
40	48 ....
41	49 ....
42	50 ....
43	

Part B–1 Score ☐

Write your answers to Part B–2 and Part C in your answer booklet.

The declaration below should be signed when you have completed the examination.

I do hereby affirm, at the close of this examination, that I had no unlawful knowledge of the questions or answers prior to the examination and that I have neither given nor received assistance in answering any of the questions during the examination.

____

Signature

Tear Here

The University of the State of New York

REGENTS HIGH SCHOOL EXAMINATION

PHYSICAL SETTING EARTH SCIENCE

Friday, June 18, 2004 — 1:15 to 4:15 p.m., only

ANSWER BOOKLET

Student .... Sex: ☐ Male ☐ Female

Teacher ....

School .... Grade ....

Answer all questions in Part B–2 and Part C. Record your answers in this booklet.

☐ **Performance Test Score (Maximum Score: 23)**

Part	Maximum Score	Student's Score
A	35	
B–1	15	
B–2	15	
C	20	

Total Written Test Score (Maximum Raw Score: 85) ☐

Final Score (from conversion chart) ☐

Raters' Initials:

Rater 1 .... **Rater 2** ....

For Raters Only

Part B–2

51 Air pressure: __________ **mb**

Air temperature: ________ **°F**

Amount of precipitation during last six hours: ________ **inch(es)**

Cloud cover: ________ **%**

Present weather: ______________________

51 ☐

[1] [OVER]

For Raters Only

52 and 53

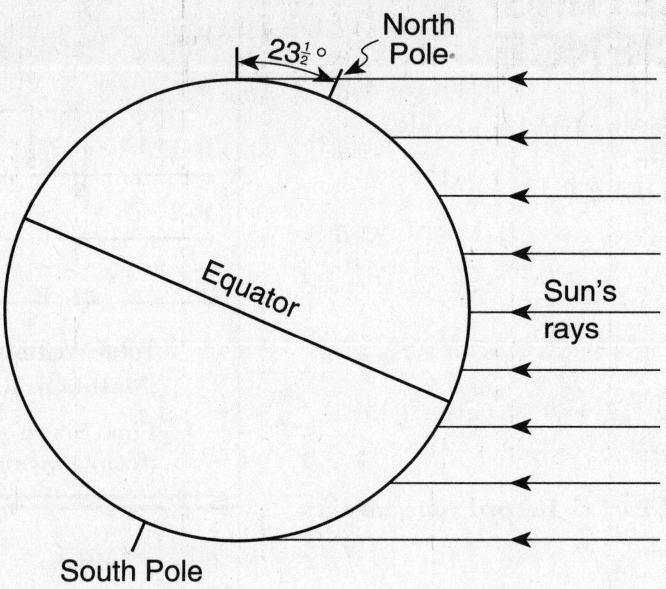

52 ☐

53 ☐

54 ______________________________ 54 ☐

55

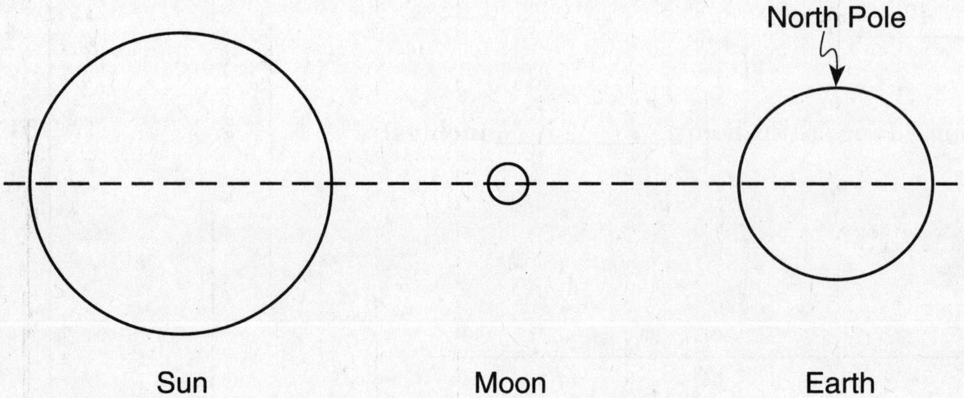

(Not drawn to scale)

55 ☐

For Raters Only

56 Brightest: (*a*) ____________________

(*b*) ____________________

(*c*) ____________________ 56 [ ]

(*d*) ____________________

Least bright: (*e*) ____________________

57

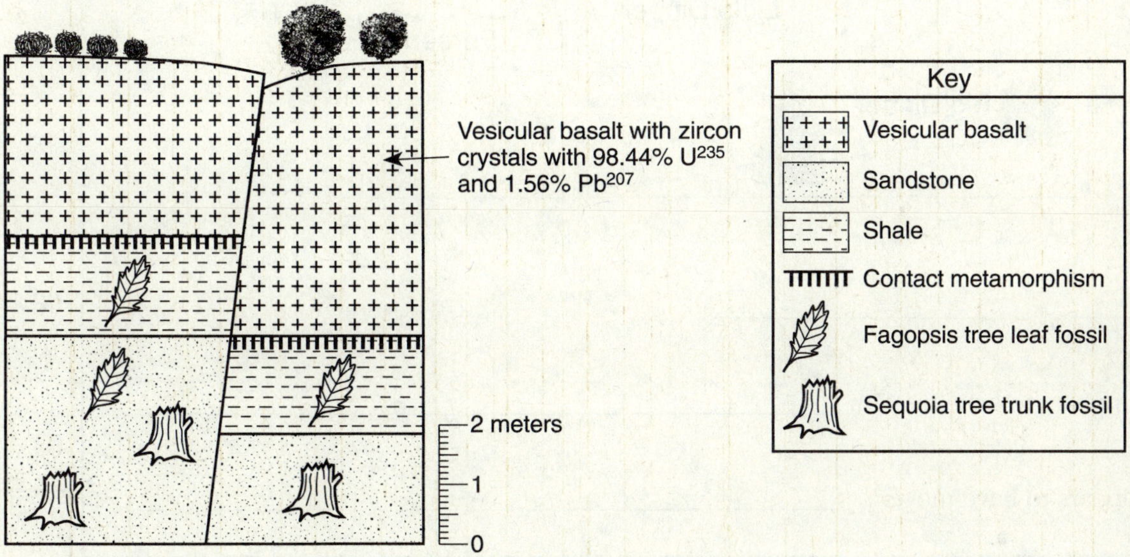

57 [ ]

58 ____________________ **meter(s)** 58 [ ]

59 _____ The fault was formed.

_____ The shale was deposited.

_____ The vesicular basalt was formed. 59 [ ]

_____ The sandstone was deposited.

60 ____________________ **million years** 60 [ ]

61 ____________________ **half-lives** 61 [ ]

62

Diagram 2 — 11:00 p.m.

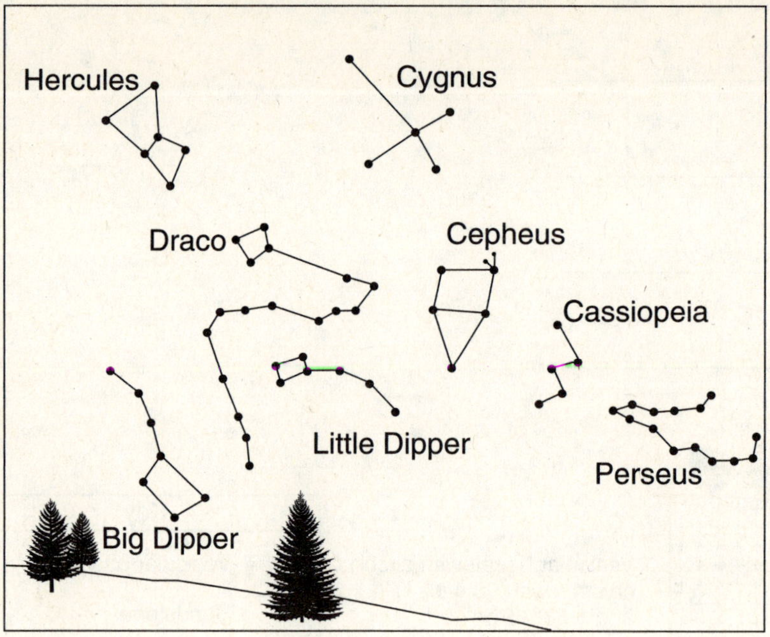

63 ________________________________

64 Hercules appears to have moved: ________________________________

Perseus appears to have moved: ________________________________

For Raters Only

62 ☐

63 ☐

64 ☐

Total Score for Part B–

[4]

For Raters Only

Part C

65 Material: ______________________ 65

Processes: ______________ and ______________

66

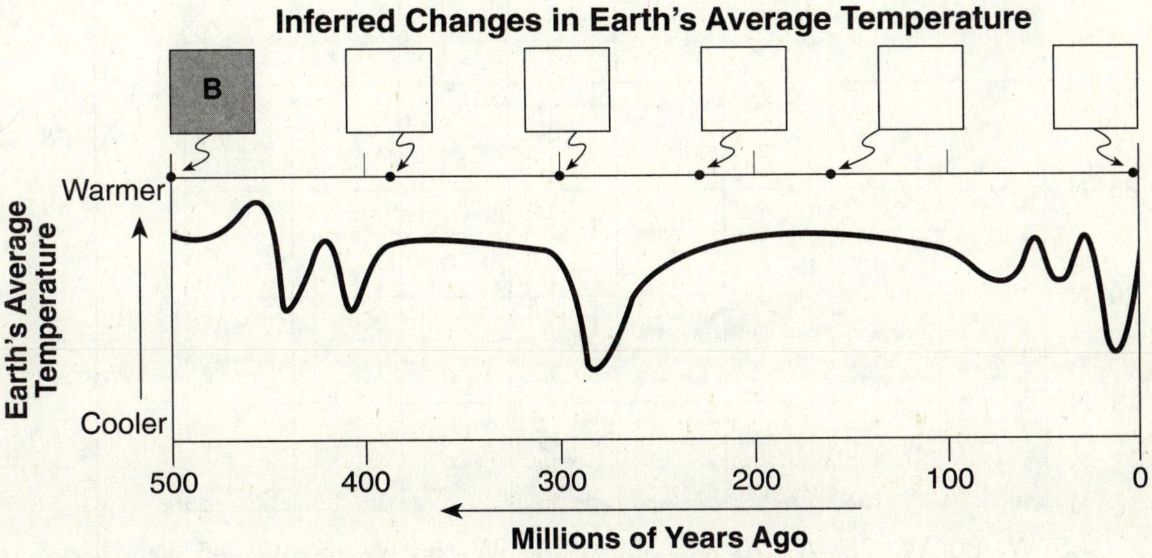

66

67 Factor *A*:

Effect on Earth's temperature: ______________

Why temperature changes: ______________

______________ 67

Factor *B*:

Effect on Earth's temperature: ______________

Why temperature changes: ______________

____

	For Raters Only

68 and **69**

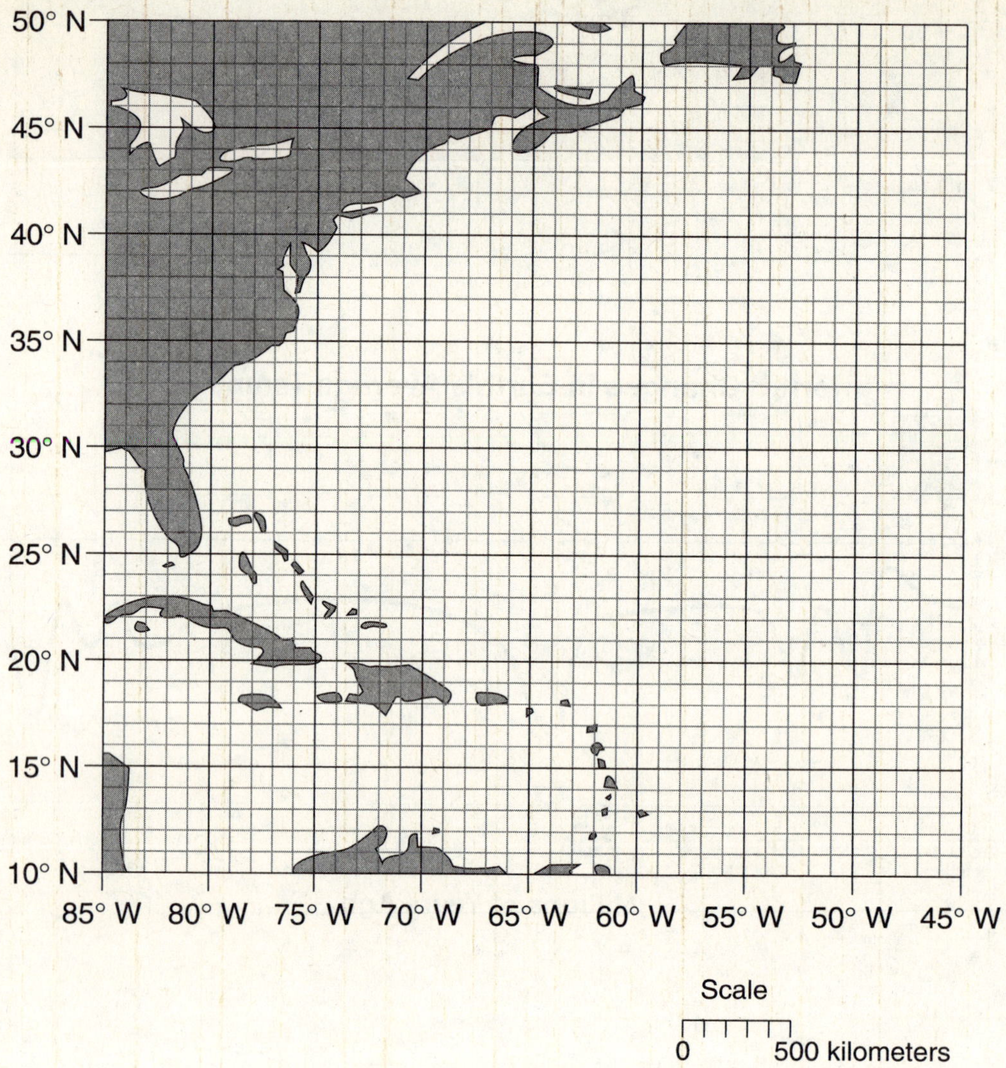

68 ☐

69 ☐

70 ______________________

71 ______________________

____

70 ☐

71 ☐

[6]

For Raters Only

72 and **73**

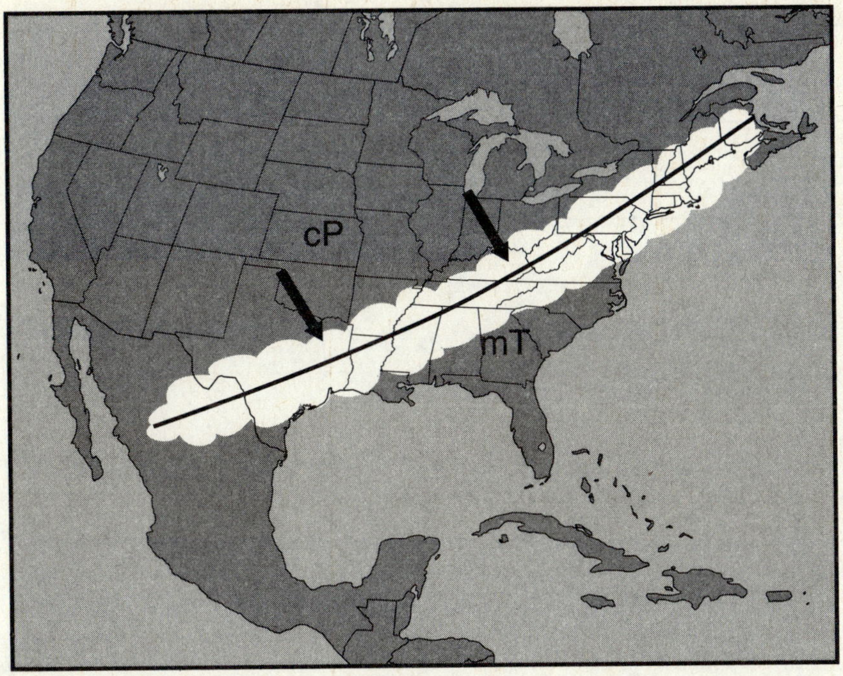

72 ☐

73 ☐

74 (1) ____________________

(2) ____________________

(3) ____________________

(4) ____________________

74 ☐

75 ____________________

____

75 ☐

76 ____________________

____

76 ☐

77 ____________________

____

77 ☐

78

79

80

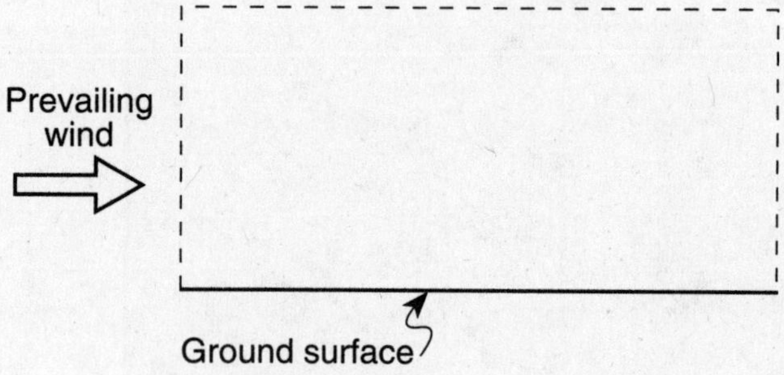

81

For Raters Only

78

79

80

81

Total Score for Part

[8]

Physical Setting Earth Science January 27, 2004

Part A

Answer all questions in this part.

Directions (1–35): For *each* statement or question, write on your separate answer sheet the *number* of the word or expression that, of those given, best completes the statement or answers the question. Some questions may require the use of the *Earth Science Reference Tables*.

1 Which graph best represents the force of gravity between Earth and the Sun during one revolution of Earth around the Sun?

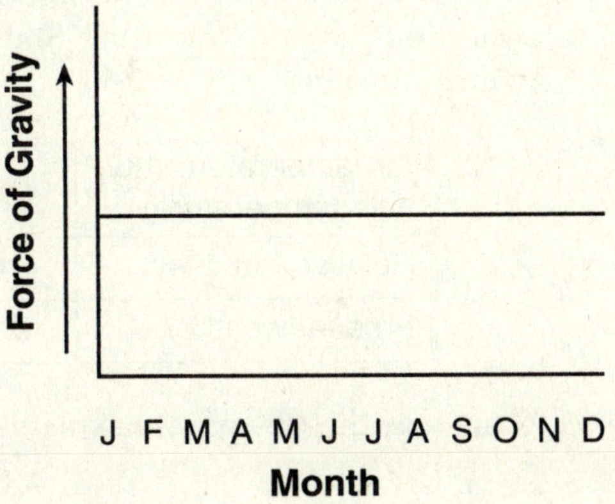

(1)

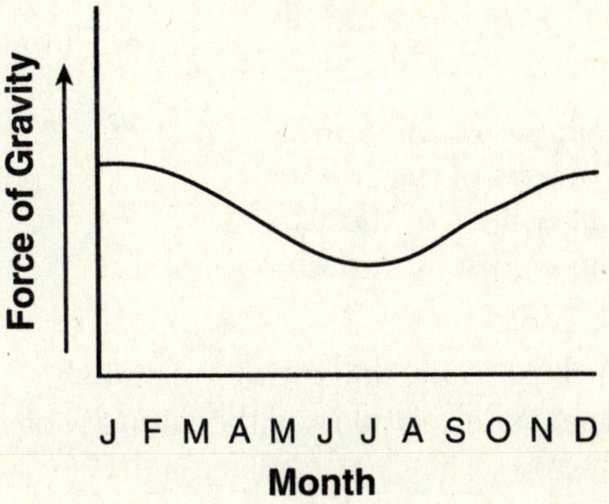

(3)

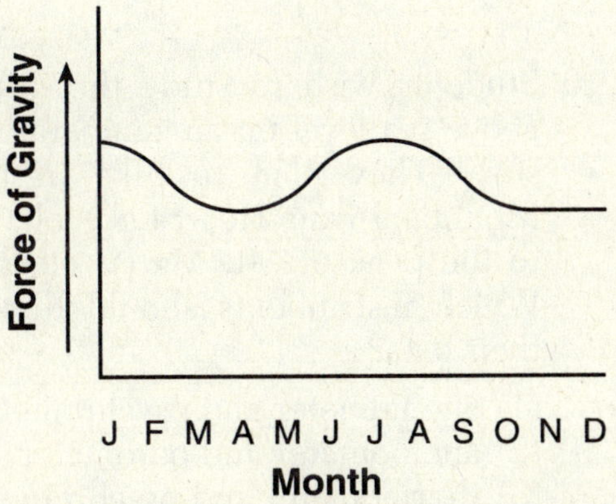

(2)

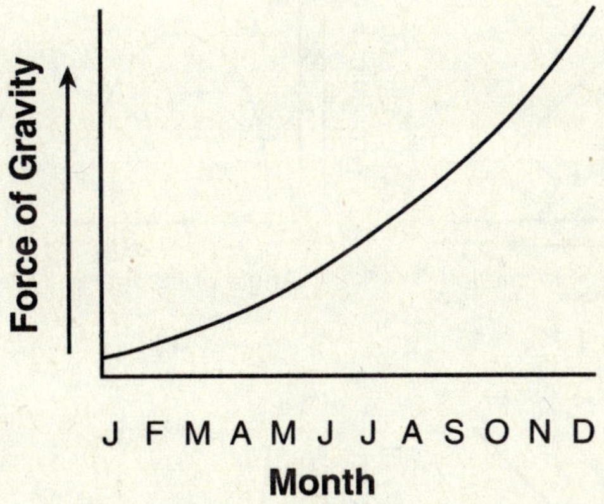

(4)

2 Which celestial feature is largest in actual size?

(1) the Moon
(2) Jupiter
(3) the Sun
(4) the Milky Way

3 The best evidence that Earth rotates is provided by the

(1) location of mid-oceanic ridge volcanoes and the distribution of index fossils
(2) movement of Foucault pendulums and the Coriolis effect on air movement
(3) pattern of changing seasons and the depth of meteor impacts
(4) rate of uranium-238 decay and changes in atmospheric composition

4 How do Jupiter's density and period of rotation compare to Earth's?

(1) Jupiter is less dense and has a longer period of rotation.
(2) Jupiter is less dense and has a shorter period of rotation.
(3) Jupiter is more dense and has a longer period of rotation.
(4) Jupiter is more dense and has a shorter period of rotation.

5 What is represented by the diagram below?

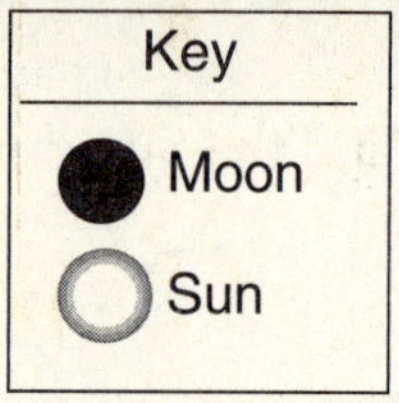

(1) changing phases of the Sun
(2) changing phases of the Moon
(3) stages in an eclipse of the Sun
(4) stages in an eclipse of the Moon

6 Which graph best represents the relationship between the angle of insolation and the intensity of insolation?

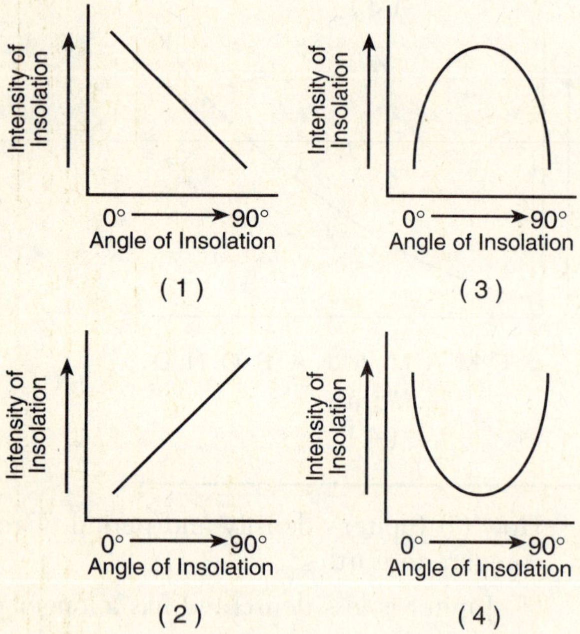

7 A square meter of surface of which of these natural areas would most likely absorb the most insolation during a clear day?

(1) a fast-moving river
(2) a dark-green forest
(3) a beach with white sand
(4) a snow-covered field

8 Which observable change would occur in New York State if Earth's rate of rotation were one-half its present rate?

(1) The Sun would rise in the southwest each day.
(2) The length of a day would be longer.
(3) The time needed to complete a cycle of Moon phases would be greater.
(4) The seasonal changes would not occur.

9 The data below represent some of the weather conditions at a New York State location on a winter morning.

Air temperature (dry-bulb temperature)	0°C
Relative humidity	81%
Present weather	snow

What was the dewpoint at this time?

(1) 1°C
(2) 2°C
(3) –3°C
(4) –5°C

10 Students wish to study the effect of elevation above sea level on air temperature and air pressure. They plan to hike in the Adirondack Mountains from Heart Lake, elevation 2,179 feet, to the peak of Mt. Marcy, elevation 5,344 feet. Which instruments should they use to collect their data?

(1) anemometer and psychrometer
(2) anemometer and barometer
(3) thermometer and psychrometer
(4) thermometer and barometer

11 Weather-station measurements indicate that the dewpoint temperature and air temperature are getting farther apart and that air pressure is rising. Which type of weather is most likely arriving at the station?

(1) a snowstorm
(2) a warm front
(3) cool, dry air
(4) maritime tropical air

12 The diagram below represents a naturally occurring geologic process.

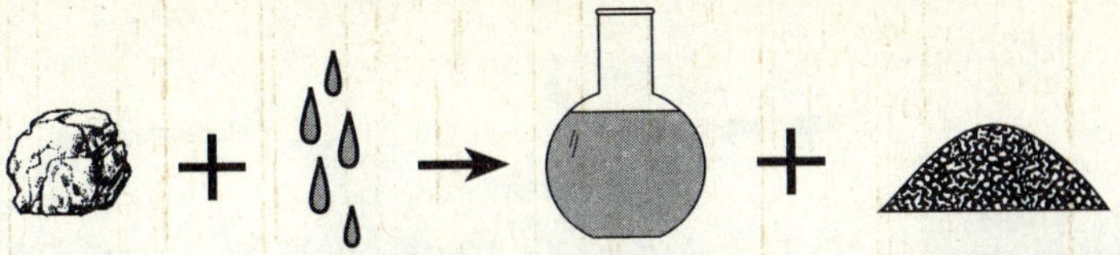

Which process is best illustrated by the diagram?

(1) cementation
(2) erosion
(3) metamorphism
(4) weathering

---

13 The graph below shows the average number of days each year that thunderstorms occur at different latitudes on Earth.

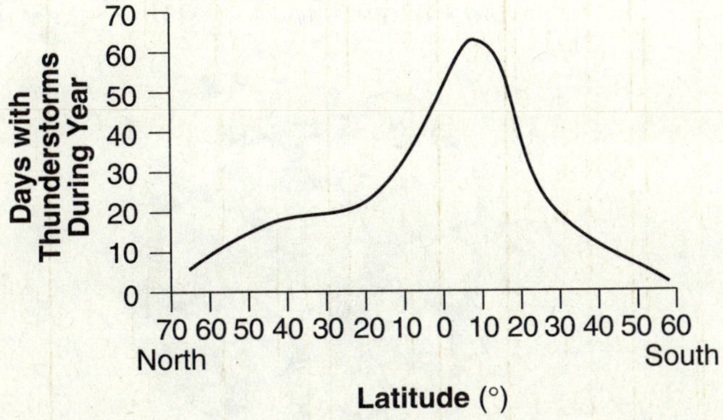

According to the graph, what is the approximate number of days each year that thunderstorms occur at locations along the 40° N parallel of latitude?

(1) 8 days
(2) 18 days
(3) 24 days
(4) 32 days

14 One result of a large volcanic eruption is that surface air temperatures decrease over a sizable region of Earth. This phenomenon occurs because volcanic eruptions usually *decrease* the

(1) transparency of the atmosphere
(2) number of dust particles entering the atmosphere
(3) amount of moisture in the atmosphere
(4) reflection of sunlight within the atmosphere

15 When rainfall occurs, the rainwater will most likely become surface runoff if the land surface is

(1) sandy
(2) impermeable
(3) covered with grass
(4) nearly flat

16 Fossil pollen has been recovered from sediments deposited in late-Pleistocene lakes. The pollen's geologic age can most accurately be measured by using

(1) rubidium-87
(2) potassium-40
(3) oxygen-18
(4) carbon-14

17 Andrija Mohorovičić discovered the interface between the crust and the mantle that is now named for him. His discovery of the "Moho" was based on analysis of

(1) landscape boundaries
(2) continental coastlines
(3) erosional surfaces
(4) seismic waves

18 Which phrase best describes coal?

(1) low density, mafic
(2) chemical precipitate
(3) organic plant remains
(4) glassy texture, volcanic

19 Which mineral will scratch glass (hardness = 5.5), but not pyrite?

(1) gypsum
(2) fluorite
(3) orthoclase
(4) quartz

20 The observed difference in density between continental crust and oceanic crust is most likely due to differences in their

(1) composition
(2) thickness
(3) porosity
(4) rate of cooling

21 The diagrams below show gradual stages 1, 2, and 3 in the development of a river delta where a river enters an ocean.

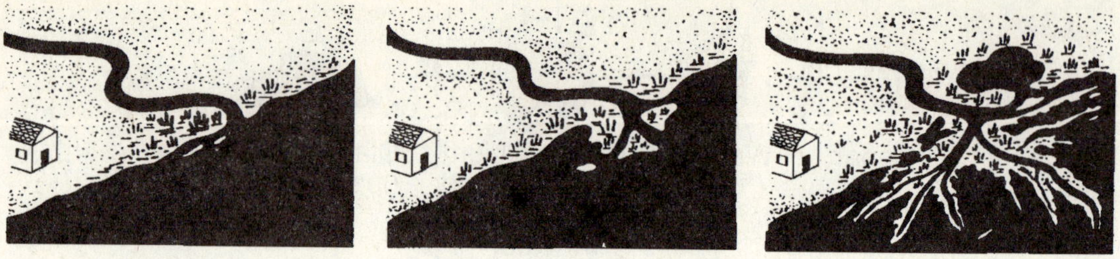

Stage 1 Stage 2 Stage 3

Which statement best explains why the river delta is developing at this site?

(1) The rate of deposition is less than the rate of erosion.
(2) The rate of deposition is greater than the rate of erosion.
(3) Sea level is slowly falling.
(4) Sea level is slowly rising.

---

22 According to plate tectonic theory, during which geologic time interval did the continents of North America and Africa separate, resulting in the initial opening of the Atlantic Ocean?

(1) Mesozoic Era
(2) Paleozoic Era
(3) Proterozoic Eon
(4) Archean Eon

23 Which rock most probably formed directly from lava cooling quickly at Earth's surface?

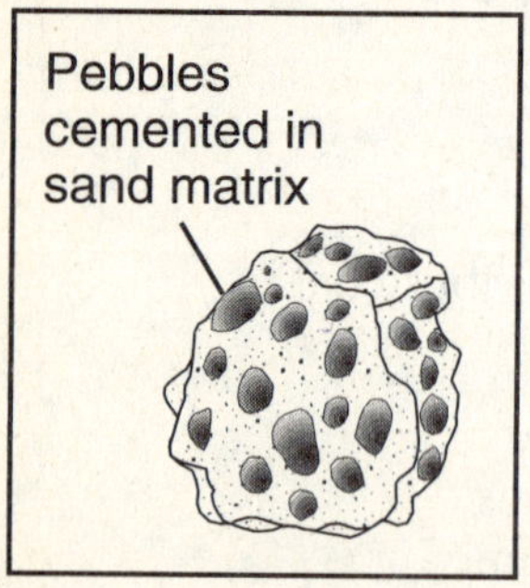

(1)

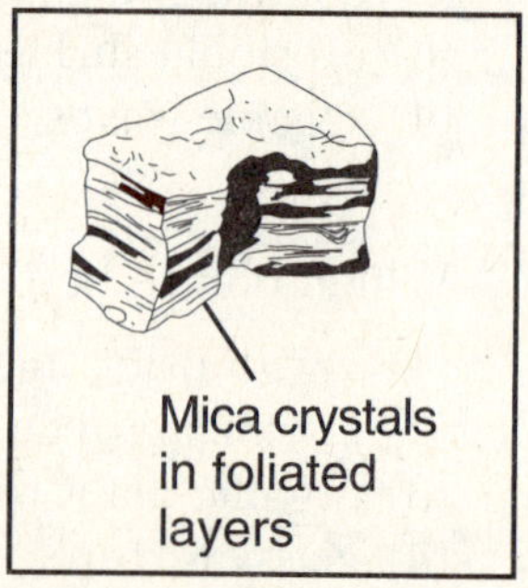

(3)

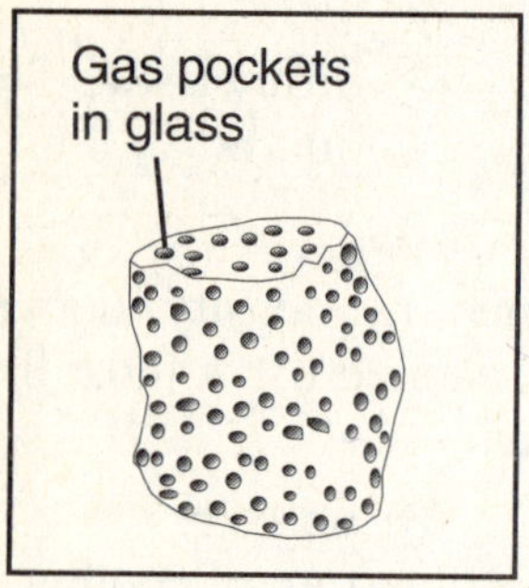

(2)

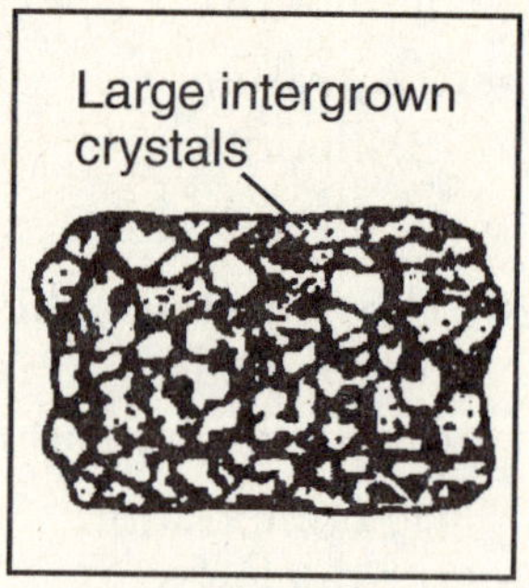

(4)

24 The diagram below shows a sedimentary rock sample.

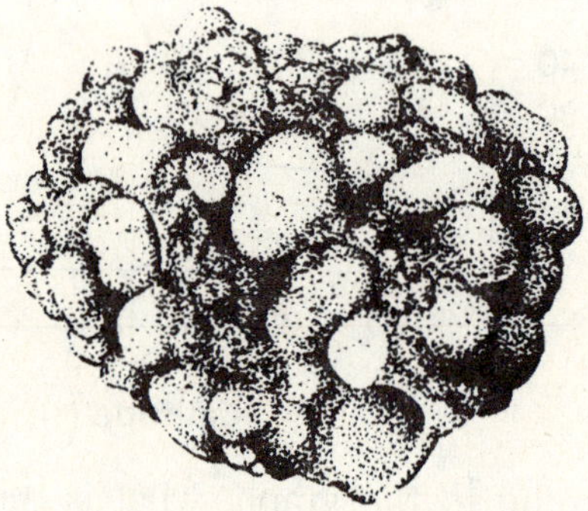

(Shown actual size)

Which agent of erosion was most likely responsible for shaping the particles forming this rock?

(1) mass movement
(2) wind
(3) glacial ice
(4) running water

25 Near which location in New York State would a geologist have the greatest chance of finding dinosaur footprints in the surface bedrock?

(1) 41° 10' N latitude, 74° W longitude
(2) 42° 10' N latitude, 74° 30' W longitude
(3) 43° 30' N latitude, 76° W longitude
(4) 44° 30' N latitude, 75° 30' W longitude

26 Geologic cross sections *A* through *F* shown below represent different stages in the development of one part of Earth's crust over a long period of geologic time.

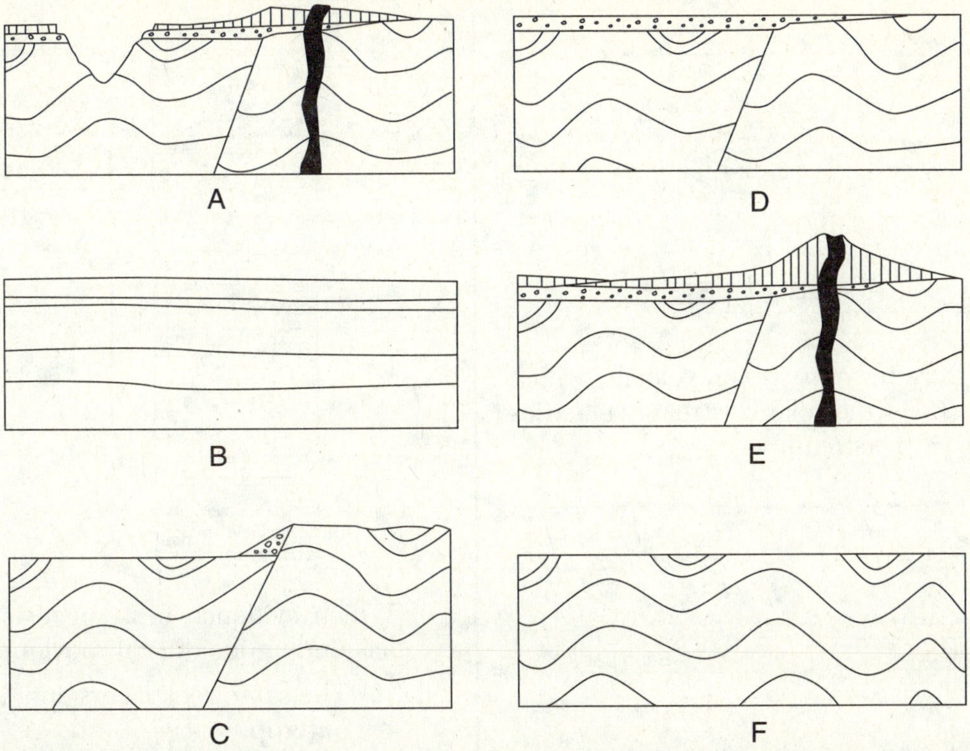

What is the correct order of development from the original (oldest) stage to the most recent (youngest) stage?

(1) $B \rightarrow D \rightarrow C \rightarrow F \rightarrow A \rightarrow E$
(2) $B \rightarrow F \rightarrow C \rightarrow D \rightarrow E \rightarrow A$
(3) $E \rightarrow A \rightarrow D \rightarrow F \rightarrow C \rightarrow B$
(4) $E \rightarrow A \rightarrow F \rightarrow C \rightarrow D \rightarrow B$

---

27 The list below shows characteristics that vary from place to place on Earth.

a Radioactive substances
b Bedrock structures
c Duration of insolation
d Hillslopes
e Stream patterns
f Atmospheric composition

Observations and measurements of which three characteristics would be most useful in describing landscapes?

(1) *a*, *b*, and *c*
(2) *b*, *c*, and *f*
(3) *b*, *d*, and *e*
(4) *d*, *e*, and *f*

28 Uranium-238 that crystallized at the same time Earth formed has undergone approximately how many half-lives of radioactive decay?

(1) one half-life
(2) two half-lives
(3) three half-lives
(4) four half-lives

29 The photograph below shows a piece of halite that has been recently broken.

Which physical property of halite is demonstrated by this pattern of breakage?

(1) hardness
(2) streak
(3) cleavage
(4) luster

30 The diagram below shows a meandering stream flowing across nearly flat topography and over loose sediments.

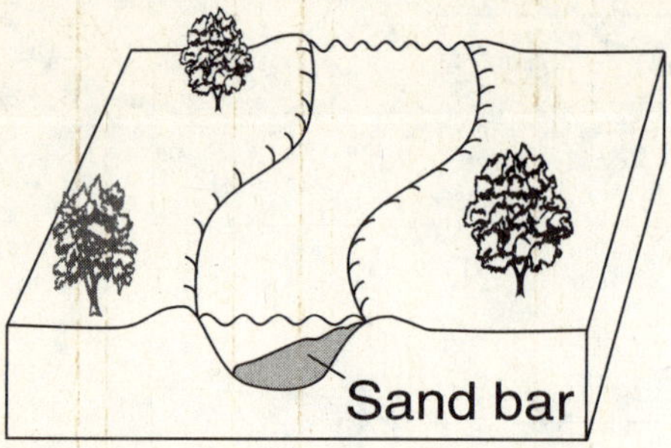

If arrow length represents stream velocity, which diagram best shows the relative stream velocities in this section of the stream?

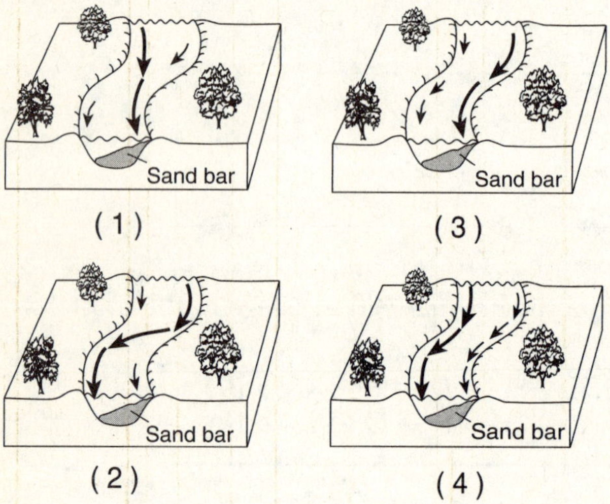

31 Bedrock located near Old Forge, New York, would most likely have which characteristics?

(1) clastic texture consisting of angular sediments of mostly quartz and feldspar cemented together
(2) crystalline texture composed predominantly of gypsum
(3) noncrystalline, glassy texture with a dark color
(4) foliated texture with mica and feldspar separated into bands

32 Bedrock of which four consecutive geologic periods is best preserved in New York State?

(1) Cambrian, Ordovician, Silurian, Devonian
(2) Devonian, Carboniferous, Permian, Triassic
(3) Permian, Triassic, Jurassic, Cretaceous
(4) Jurassic, Cretaceous, Tertiary, Quaternary

33 The diagram below represents bedrock layers found in an outcrop. Three index fossils are found within the bedrock layers.

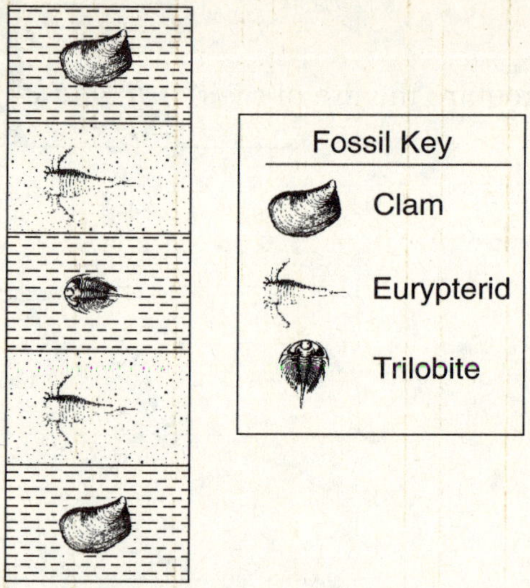

Which evidence best suggests that this outcrop has undergone crustal movement?

(1) The same rock layers appear twice within the outcrop.
(2) The trilobite fossil is not found in all five layers.
(3) The sedimentary layers have the same thickness.
(4) The eurypterid fossil is absent in the middle layer.

34 A student determines the density of a mineral to be 1.5 grams per cubic centimeter. If the accepted value is 2.0 grams per cubic centimeter, what is the student's percent deviation (percent error)?

(1) 25.0%
(2) 33.3%
(3) 40.0%
(4) 50.0%

35 What is the best way to determine if a mineral sample is calcite or quartz?

(1) Observe the color of the mineral.
(2) Place the mineral near a magnet.
(3) Place a drop of acid on the mineral.
(4) Measure the mass of the mineral.

Part B–1

Answer all questions in this part.

Directions (36–50): For *each* statement or question, write on your separate answer sheet the *number* of the word or expression that, of those given, best completes the statement or answers the question. Some questions may require the use of the *Earth Science Reference Tables*.

Base your answers to questions 36 and 37 on the diagram below, which shows the angle of the Sun's noontime rays received at different Earth latitudes on May 1.

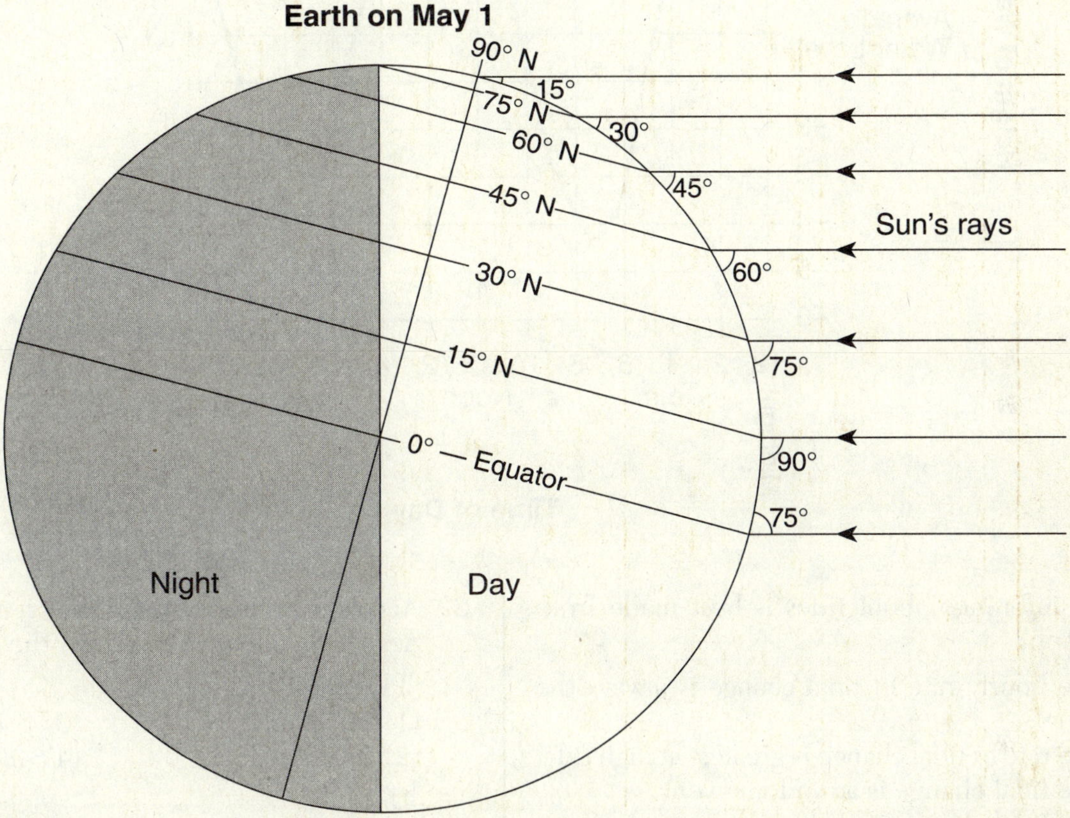

36 Which changes can be expected to occur at 45° N over the next 30 days?

(1) The duration of insolation will decrease and the temperature will decrease.
(2) The duration of insolation will decrease and the temperature will increase.
(3) The duration of insolation will increase and the temperature will decrease.
(4) The duration of insolation will increase and the temperature will increase.

37 At which latitude can the noontime Sun be observed in the northern part of the sky?

(1) 0°
(2) 30° N
(3) 60° N
(4) 90° N

Base your answers to questions 38 and 39 on the graph below. The graph shows the recorded change in water level (ocean tides) at a coastal city in the northeastern United States during 1 day.

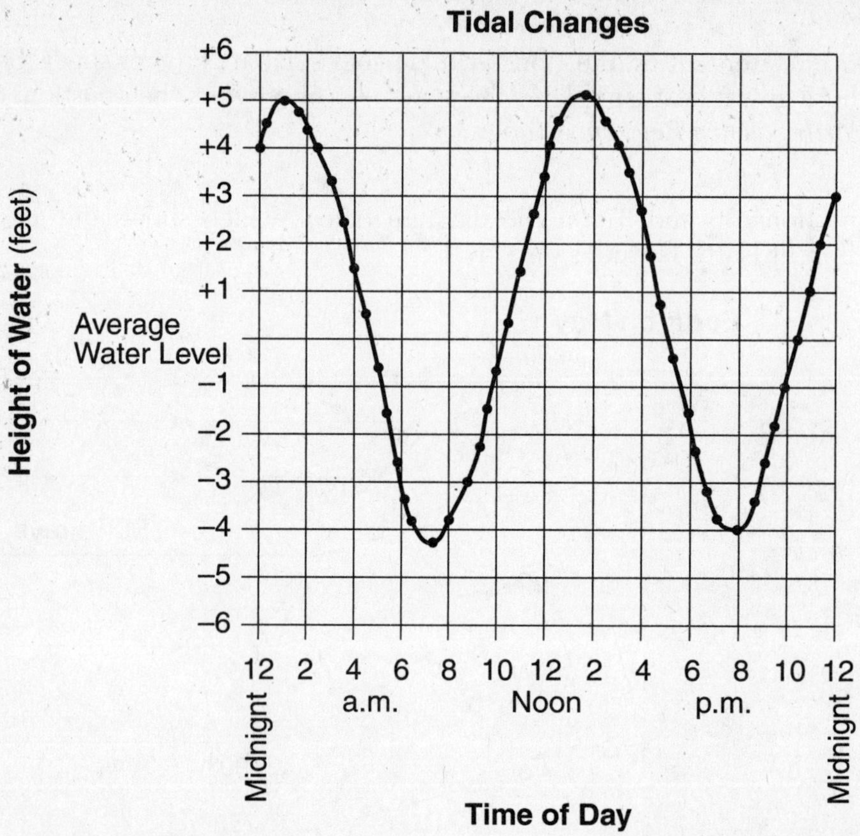

38 Which inference about tides is best made from this graph?

(1) The hourly rate of tidal change is always the same.
(2) The rate of tidal change is greatest at high tide.
(3) The tidal change is a random event.
(4) The tidal change is cyclic.

39 According to the pattern shown on the graph, the next high tide will occur on the following day at approximately

(1) 12:30 a.m.
(2) 2:00 a.m.
(3) 3:15 a.m.
(4) 4:00 a.m.

40 Which model best represents the apparent path of the Sun observed at various times during the year at the Equator?

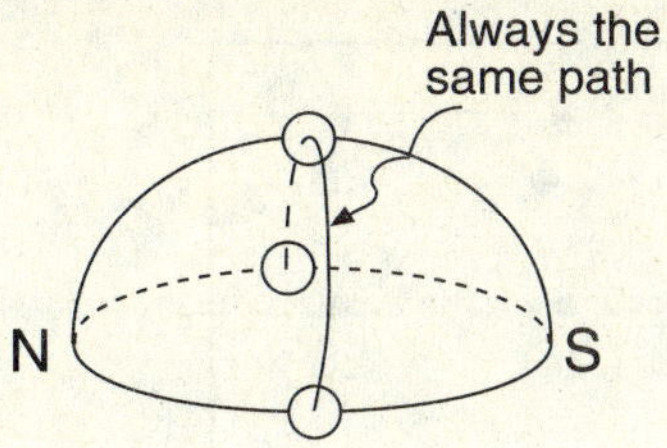

(1)

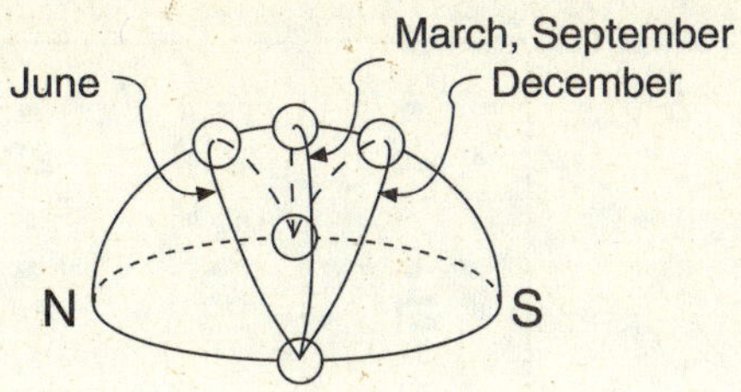

(3)

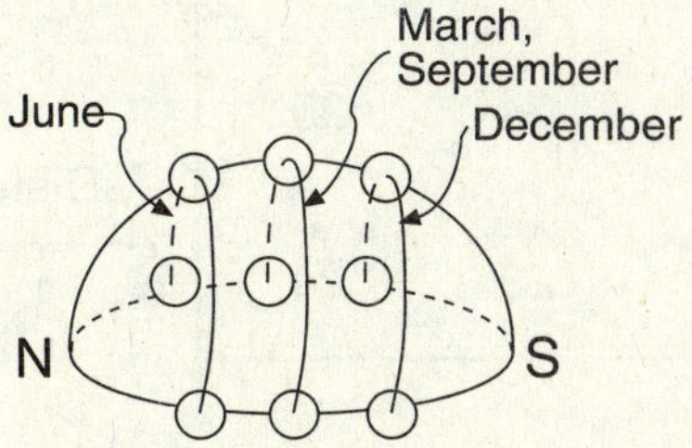

(2)

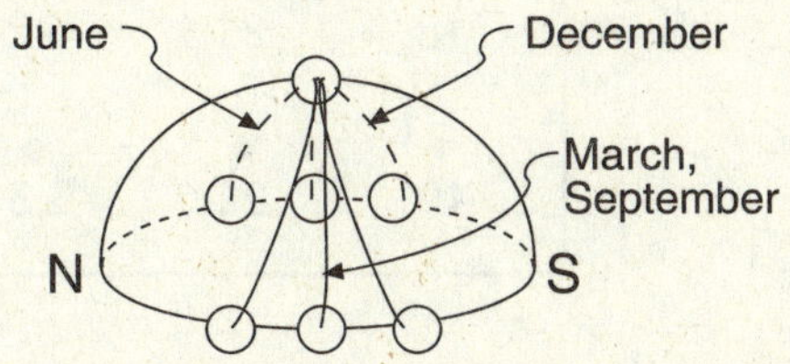

(4)

41 The topographic map below shows a particular landscape.

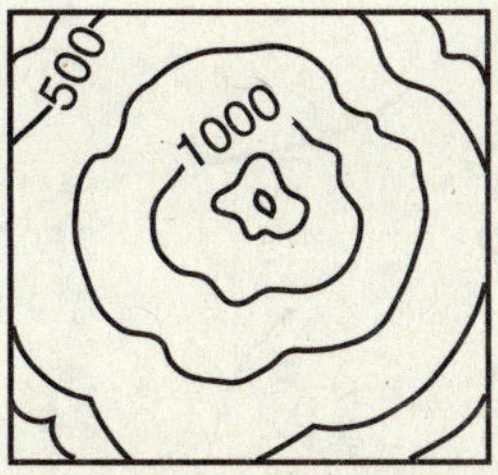

Which map best represents the stream drainage pattern for this landscape?

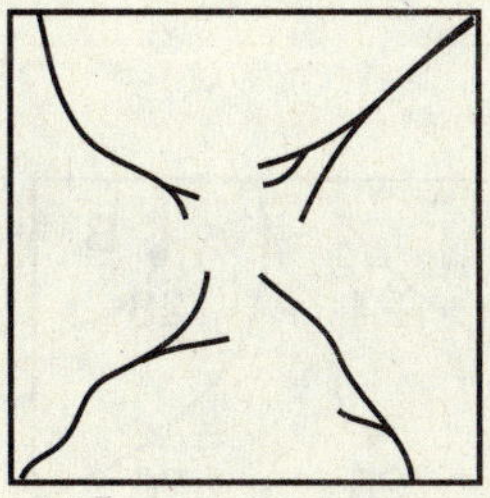

(1)

(3)

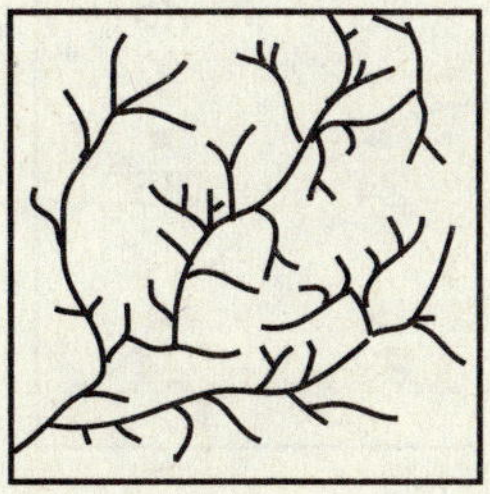

(2)

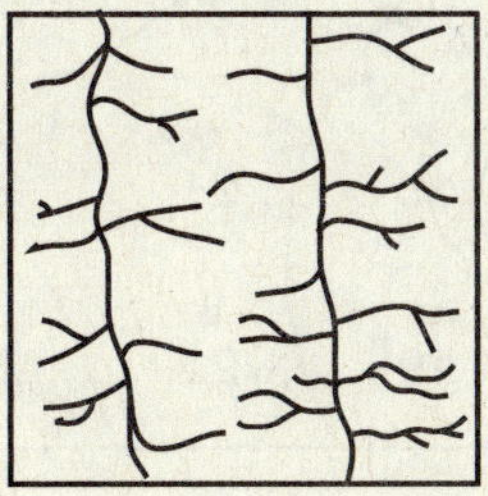

(4)

42 Which diagram correctly shows how mantle convection currents are most likely moving beneath colliding lithospheric plates?

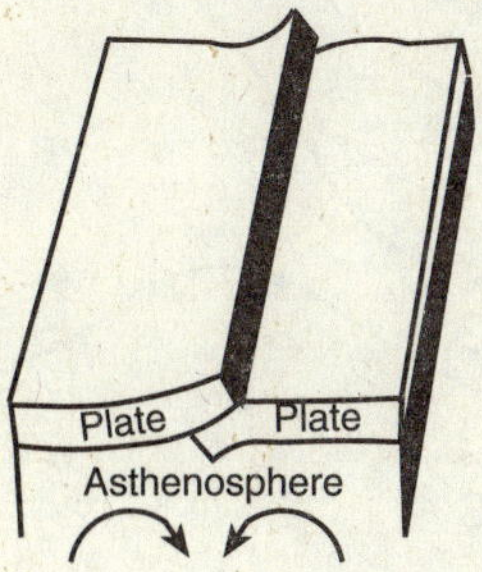

(1)

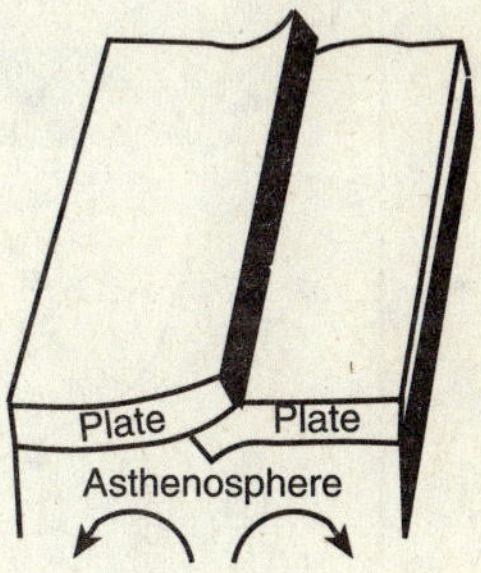

(3)

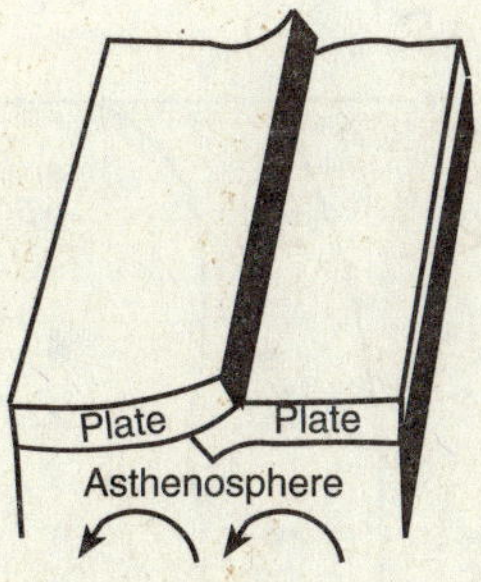

(2)

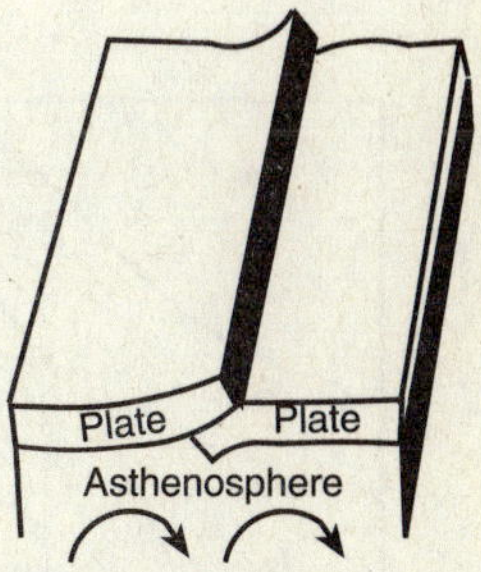

(4)

43 Which two rocks have the most similar mineral composition?

(1) marble and rhyolite
(2) limestone and basalt
(3) quartzite and rock salt
(4) granite and phyllite

44 The field map below shows air temperature measurements, in degrees Celsius, taken at the same elevation within a closed room. Two reference points, *A* and *B*, are shown.

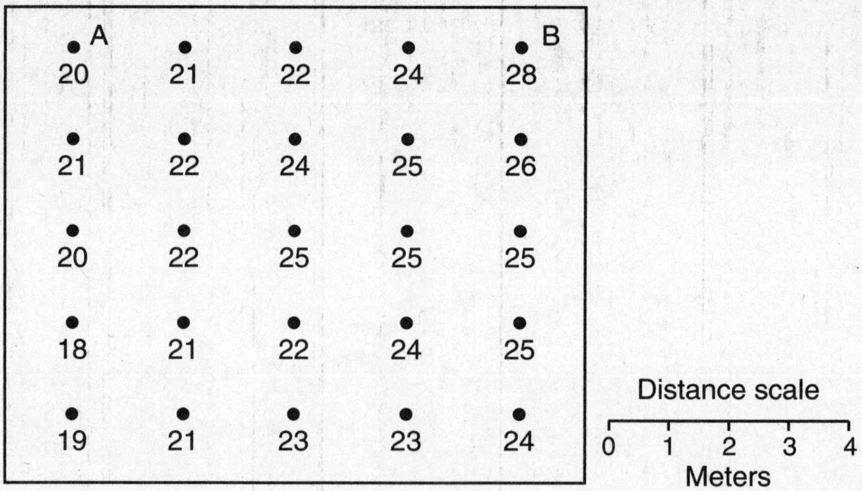

Which temperature field map shows correctly drawn isotherms?

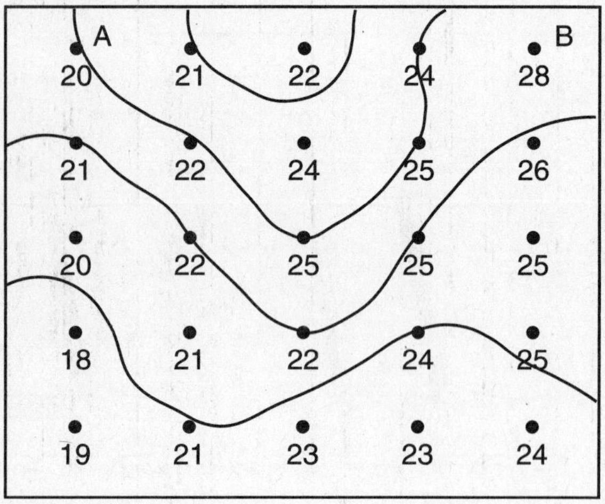

(1)

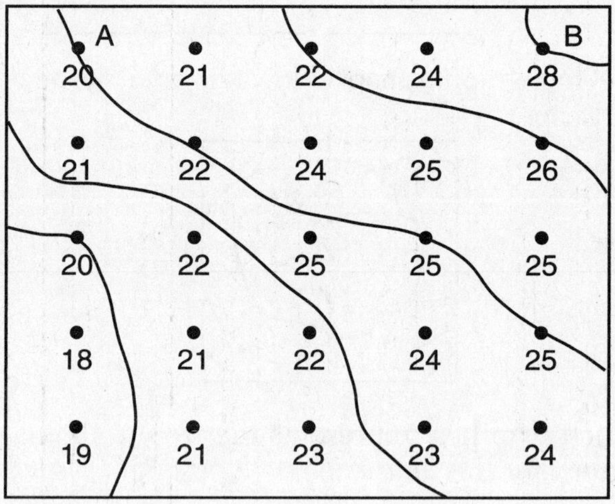

(3)

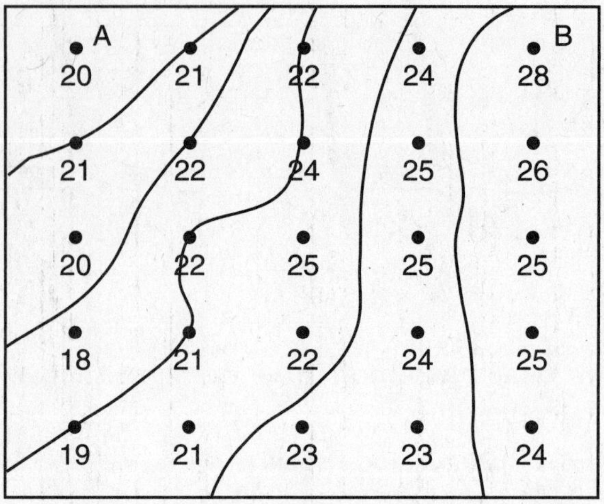

(2)

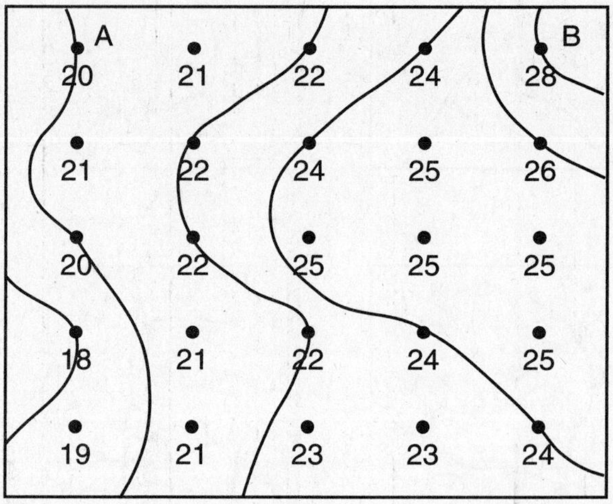

(4)

Base your answers to questions 45 and 46 on the diagram below, which shows models of two types of earthquake waves.

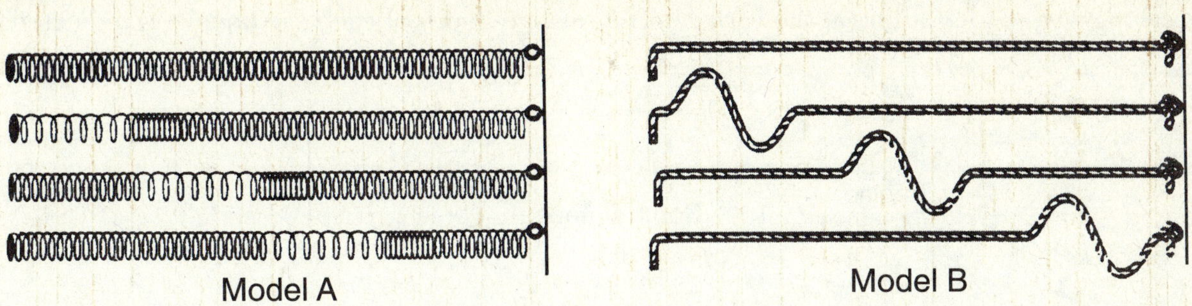

45 Model *A* best represents the motion of earthquake waves called

(1) *P*-waves (compressional waves) that travel faster than *S*-waves (shear waves) shown in model *B*
(2) *P*-waves (compressional waves) that travel slower than *S*-waves (shear waves) shown in model *B*
(3) *S*-waves (shear waves) that travel faster than *P*-waves (compressional waves) shown in model *B*
(4) *S*-waves (shear waves) that travel slower than *P*-waves (compressional waves) shown in model *B*

46 The difference in seismic station arrival times of the two waves represented by the models helps scientists determine the

(1) amount of damage caused by an earthquake
(2) intensity of an earthquake
(3) distance to the epicenter of an earthquake
(4) time of occurrence of the next earthquake

Base your answers to questions 47 and 48 on the cross section and data table shown below. The cross section shows a sediment-laden river flowing into the ocean. The arrows show the direction of river flow. Different zones of sorted sediments, *A*, *B*, *C*, and *D*, have been labeled. Sediments have been taken from these zones and measured. The data table shows the range of sediment sizes in each zone.

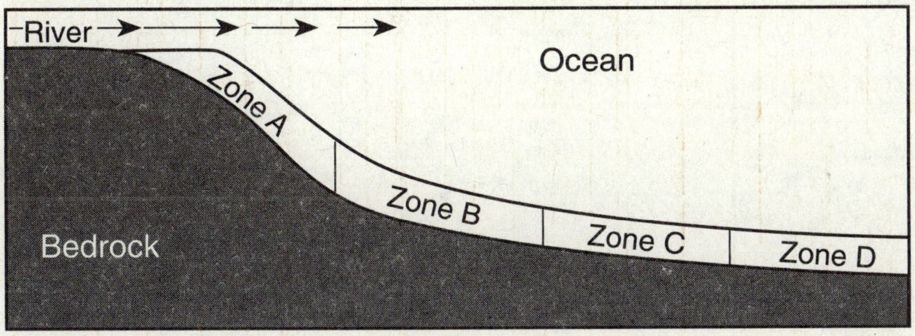

Data Table

Zone	Major Sediment Sizes
A	0.04 cm to 6 cm
B	0.006 cm to 0.1 cm
C	0.0004 cm to 0.006 cm
D	Less than 0.0004 cm

47 How is this pattern of horizontal sorting produced?

(1) High-density materials generally settle more slowly.
(2) Rounded sediments generally settle more slowly.
(3) Dissolved minerals are generally deposited first.
(4) Bigger particles are generally deposited first.

48 The sedimentary rock, siltstone, will most likely form from sediments deposited in zone

(1) *A*
(2) *B*
(3) *C*
(4) *D*

Base your answers to questions 49 and 50 on the map below, which shows the depths of selected earthquakes along the crustal plate boundary near the west coast of South America. Letters *A*, *B*, *C*, and *D* are epicenter locations along a west-to-east line at the surface. The relative depth of each earthquake is indicated.

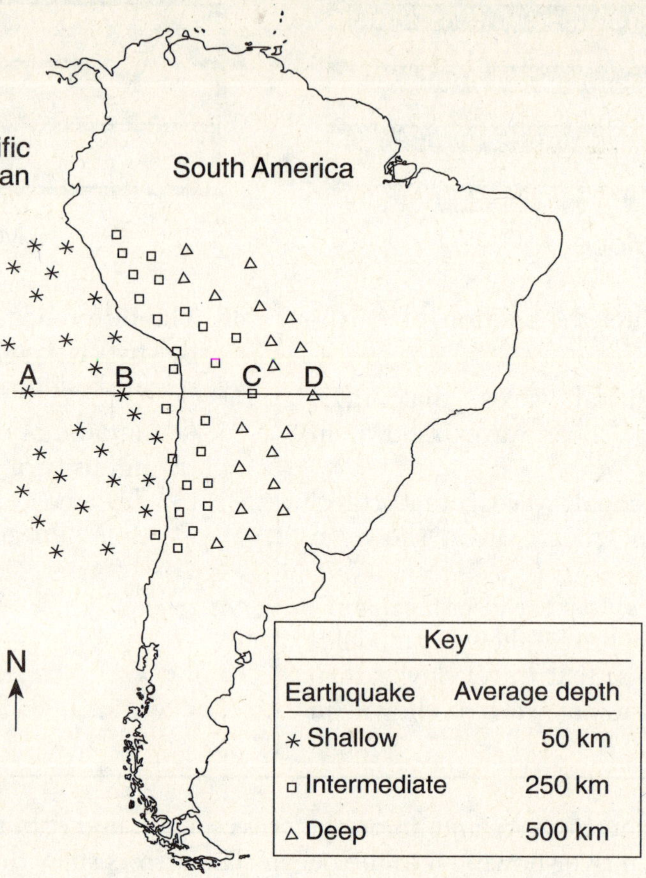

49 Which graph best shows the depth of earthquakes beneath epicenters *A*, *B*, *C*, and *D*?

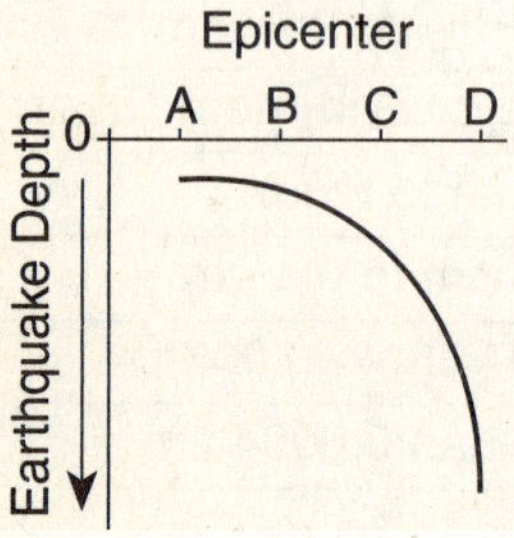

(1)

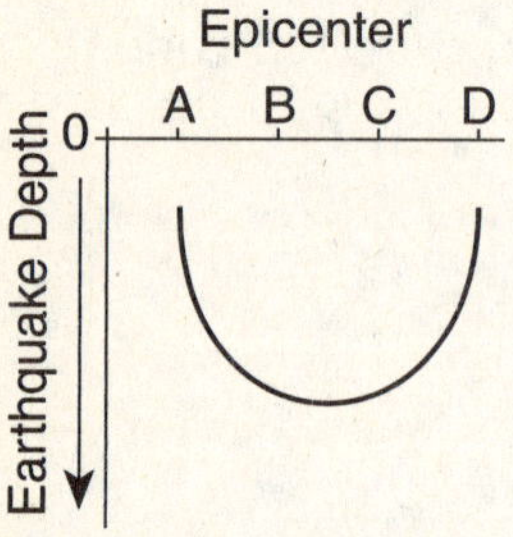

(2)

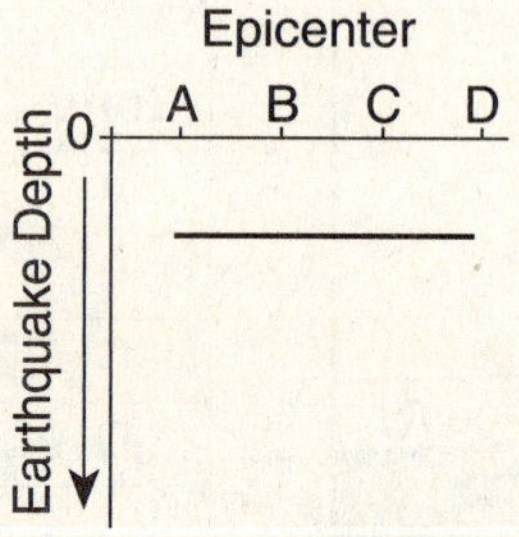

(3)

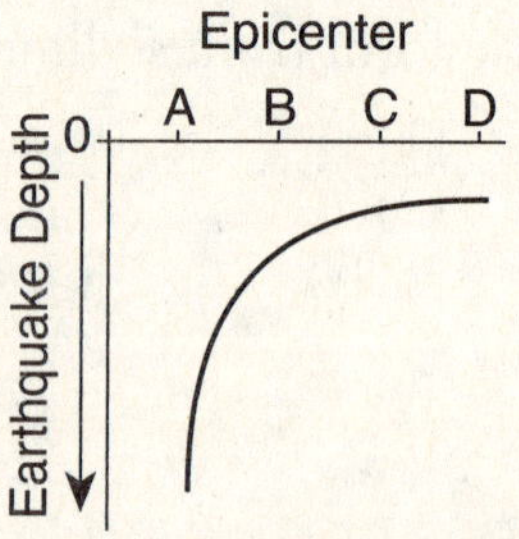

(4)

50 The earthquake beneath epicenter *D* occurred in which part of Earth's interior?

(1) crust
(2) rigid mantle
(3) asthenosphere
(4) stiffer mantle

Part B–2

Answer all questions in this part.

Directions (51–60): Record your answers in the spaces provided in your answer booklet. Some questions may require the use of the *Earth Science Reference Tables.*

Base your answers to questions 51 through 54 on the reading passage below and on your knowledge of Earth science. The reading passage provides some background information about a recent fossil discovery. The map of Canada shows the fossil site. The scale drawing shows the new trilobite fossil compared to other trilobite fossils.

The World's Biggest Trilobite

A team of Canadian paleontologists examining rock units along the shore of Hudson Bay in northern Manitoba has discovered the world's largest recorded complete fossil of a trilobite, a many-legged, sea-dwelling animal inferred to have lived during the late Ordovician Period. The giant creature, measuring 70 centimeters in length, is a new species of the genus *Isotelus*. This remarkable discovery adds to our knowledge of the diversity of life following one of the greatest increases in the number and types of life-forms in history. The new *Isotelus* species existed just before the end of the Ordovician Period.

Map of North America

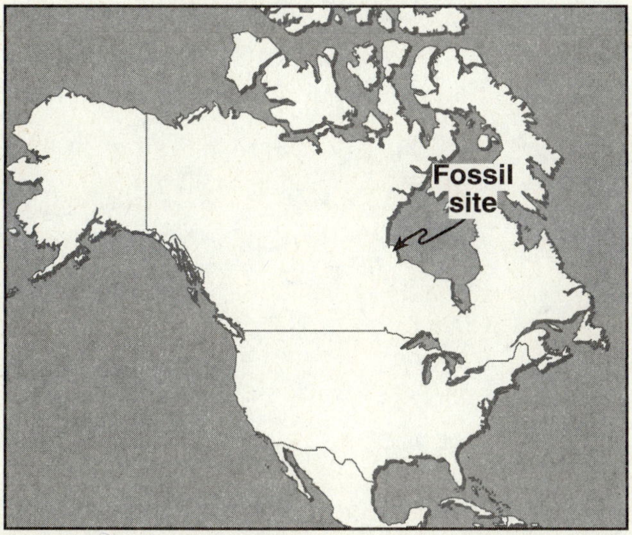

Scale drawings of the new trilobite *Isotelus (A)*, other big species reported from elsewhere *(B,C,D,E)*, and a typical large trilobite *(F)*.

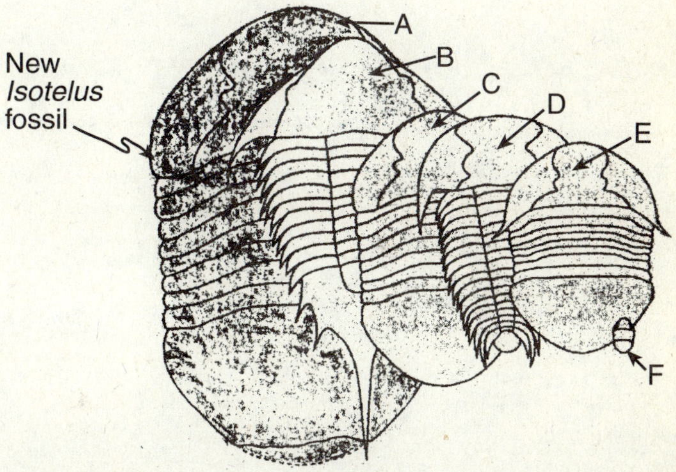

51 In what type of rock was the new *Isotelus* fossil most probably found? [1]

52 At the time the new *Isotelus* fossil lived and died, during the Ordovician Period, what was the approximate latitude of the fossil site according to plate tectonic theory? [1]

53 What New York State nautiloid index fossil would most likely be found in the bedrock just below the new *Isotelus* fossil? [1]

54 The actual new *Isotelus* fossil is approximately how many times larger than scale drawing *A*? [1]

Base your answers to questions 55 through 58 on the weather map provided *in your answer booklet,* which shows a weather system over the northeastern United States and weather data for several locations. Isobars show a low-pressure (L) center. Point ⊗ is a location in Canada.

55 On the weather map provided *in your answer booklet,* draw a curved arrow through point ⊗ to show the general direction of surface winds on that side of the low-pressure center. [2]

56 State the relationship between isobar spacing on the map and wind velocity. [1]

57 Describe the *five* specific weather conditions for Charleston indicated by the station model on the weather map. Complete the chart provided *in your answer booklet* and include appropriate units where necessary. [2]

58 Describe how clouds form when warm, humid air rises along the cold front.

a Include the terms *dewpoint* and either *expansion* or *expands* in your answer. [1]

b State the phase change that occurs at the dewpoint. [1]

Base your answers to questions 59 and 60 on the diagram below, which shows an incomplete concept map identifying the types of plate boundaries. Information in the boxes labeled *A, B, C, D,* and *E* has been deliberately omitted.

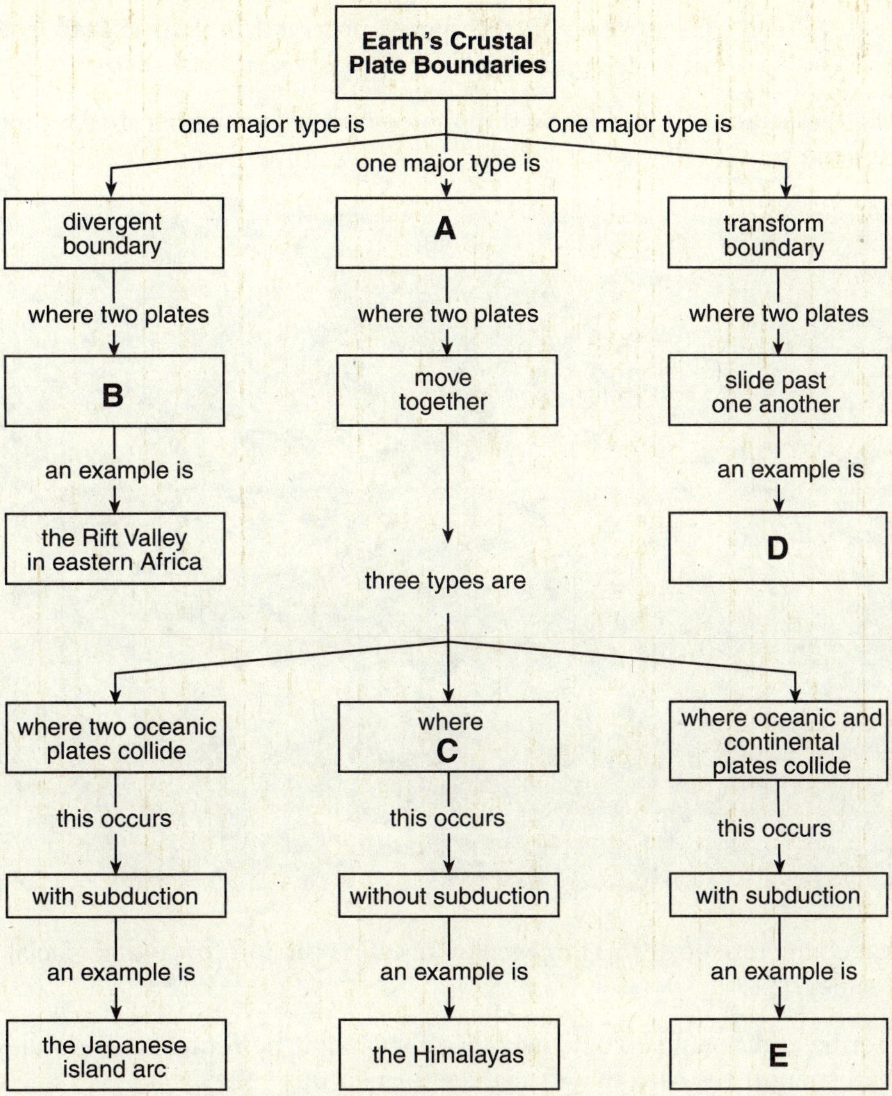

59 On the chart provided *in your answer booklet,* write the information that should be placed in the boxes labeled *A, B,* and *C* that will correctly complete those portions of the concept map. [2]

60 On the geographic map provided *in your answer booklet,* write the letters *D* and *E* on the plate boundary locations where the indicated movements are occurring. Write the letters approximately the same size as shown on the concept map and locate the letters directly on the plate boundary. [2]

Part C

Answer all questions in this part.

Directions (61–74): Record your answers in the spaces provided in your answer booklet. Some questions may require the use of the *Earth Science Reference Tables.*

Base your answers to questions 61 and 62 on the photograph below, which shows a mountainous region cut by a large valley in its center.

61 What characteristic of this large valley supports the inference that glacial ice formed the valley? [1]

62 Describe additional geologic evidence that might be found on the valley floor that would support the idea that glacial ice formed this valley. [1]

Base your answers to questions 63 through 67 on the map and graphs below. The map shows five climate regions of New York State. The bar graphs show average monthly temperatures of four of these climate regions.

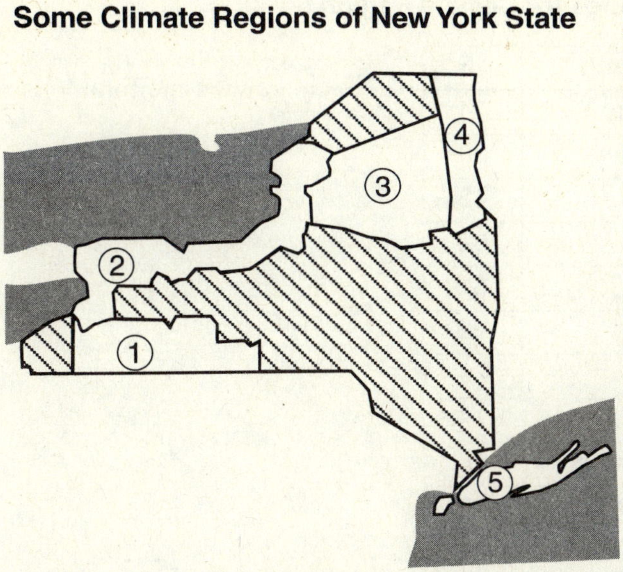

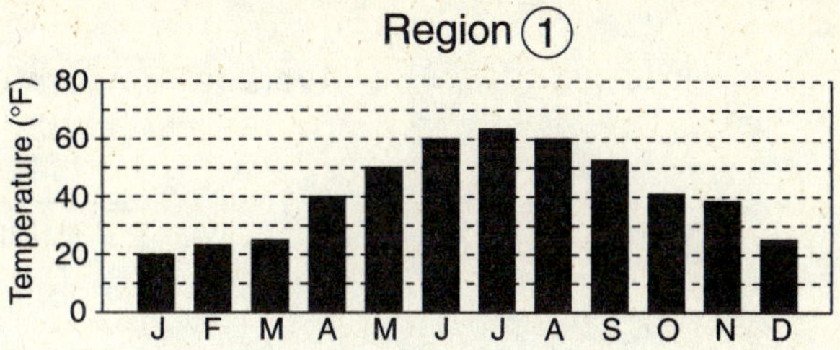

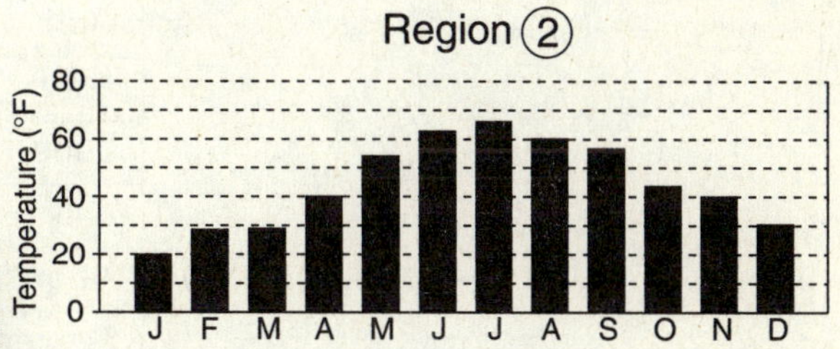

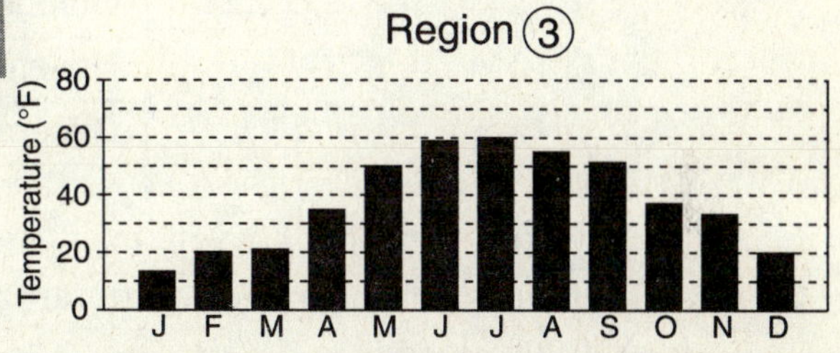

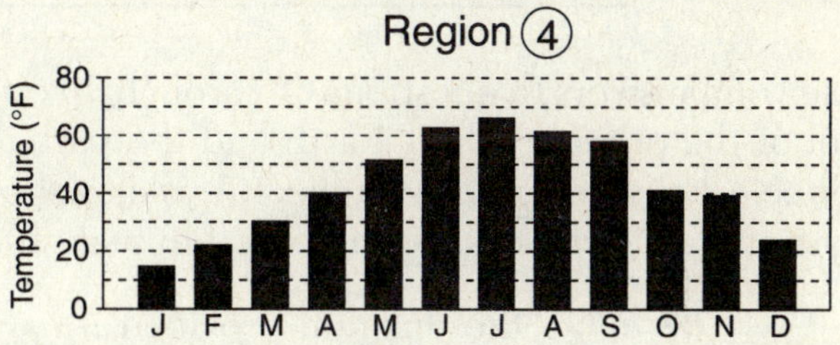

63 The average monthly temperatures for climate regions 1, 2, 3, and 4 show a similar yearly pattern of change. Identify one climate control factor that these four climate regions have in common that most probably causes this similarity in temperature pattern. [1]

64 What climate variable, other than temperature, was also used to identify these areas as four different climate regions? [1]

65 What landscape characteristic of climate region 3 most likely causes it to have both cooler summer temperatures and cooler winter temperatures than climate region 2? [1]

66 On the grid provided *in your answer booklet,* construct a bar graph of the average monthly temperatures provided below for climate region 5. January has been completed for you. [2]

Average Temperatures for Climate Region 5

Month	°F
January	34
February	36
March	42
April	52
May	61
June	72
July	79
August	74
September	68
October	55
November	49
December	39

67 Describe how the Atlantic Ocean surrounding climate region 5 has most probably influenced the average temperatures of this region during January, February, and March. [1]

Base your answers to questions 68 through 70 on the diagram provided *in your answer booklet,* which shows a model of the orbital path of Earth and the partial orbital path of Jupiter around the Sun. A partial orbit of another celestial object, labeled object *A*, is also shown. Celestial object *A* is a natural object that is part of our solar system. [All distances are measured to scale from the center of the Sun in this model.]

68 *a* On the diagram provided *in your answer booklet,* place an **X** to represent the position of Mars at the properly scaled distance from the Sun in this model. [1]

b On the diagram provided *in your answer booklet,* starting at your plotted position of Mars, draw a scale model of Mars' orbital path. Be careful to show the correct shape of the orbit. [1]

69 Identify what type of solar-system object is most probably represented by celestial object *A*. [1]

70 State one reason why determining the exact orbital path and period of revolution of celestial object *A* might be important to the continued existence of life on Earth. [1]

Base your answers to questions 71 through 74 on the topographic map below. The map shows a portion of the Taterskill Creek flowing past the towns of Lawson and Glenton. The shaded area is Taterskill Creek. The arrows in the creek show its direction of flow. Points *A*, *B*, and *C* are locations on the map. Points *A* and *B* are connected with a reference line.

Mercado Dam is located 32 miles upstream from Lawson. In the remote possibility of a failure of the Mercado Dam, the Taterskill Creek is expected to rise to the 600-foot contour line in the vicinity of the two towns.

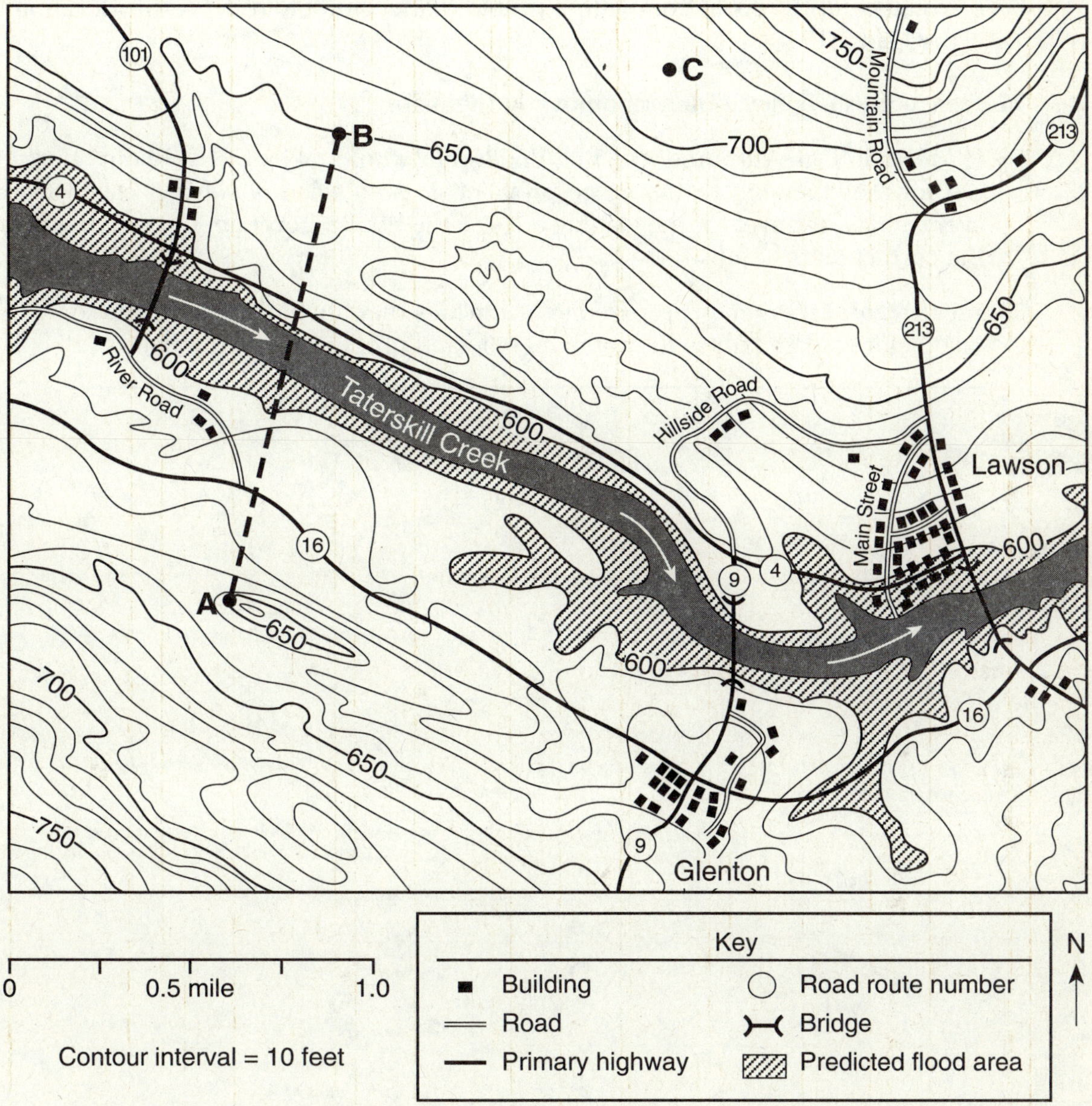

71 On the grid provided *in your answer booklet,* construct a topographic profile from point *A* to point *B*, following the directions below.

a Write numbers along the vertical axis to show an appropriate scale for the elevations crossed by line *AB*. Your number scale should label at least half of the lines along the vertical axis and should not extend beyond the grid provided. [1]

b Plot the elevation along line *AB* by marking an **X** at *each* point where a contour line is crossed. Point *A* and point *B* have been plotted for you. [1]

c Connect all the **X**s to complete a profile that accurately reflects the elevation of the land. [1]

72 State a possible elevation for point *C* on the map. [1]

73 If Mercado Dam ruptured, the first floodwater would take exactly 4 hours to reach the town of Lawson. In the space provided *in your answer booklet,* calculate the average rate of travel for the leading edge of the floodwater. Label your answer with the correct units. [2]

74 Identify *two* emergency preparedness activities that town officials in Lawson could take before a dam failure to protect people and property from the flood. [2]

The University of the State of New York

REGENTS HIGH SCHOOL EXAMINATION

PHYSICAL SETTING
EARTH SCIENCE

Tuesday, January 27, 2004 — 1:15 to 4:15 p.m., only

ANSWER SHEET

Student .... Sex: ☐ Male ☐ Female Grade ....

Teacher .... School ....

Record your answers to Part A and Part B–1 on this answer sheet.

Part A

1	13	25 ....
2	14	26 ....
3	15	27 ....
4	16	28 ....
5	17	29 ....
6	18	30 ....
7	19	31 ....
8	20	32 ....
9	21	33 ....
10	22	34 ....
11	23	35 ....
12	24	

Part A Score ☐

Part B–1

36	44 ....
37	45 ....
38	46 ....
39	47 ....
40	48 ....
41	49 ....
42	50 ....
43	

Part B–1 Score ☐

Write your answers to Part B–2 and Part C in your answer booklet.

The declaration below should be signed when you have completed the examination.

I do hereby affirm, at the close of this examination, that I had no unlawful knowledge of the questions or answers prior to the examination and that I have neither given nor received assistance in answering any of the questions during the examination.

____

Signature

The University of the State of New York

REGENTS HIGH SCHOOL EXAMINATION

PHYSICAL SETTING EARTH SCIENCE

Tuesday, January 27, 2004 — 1:15 to 4:15 p.m., only

ANSWER BOOKLET

Student . . . . Sex: ☐ Male ☐ Female

Teacher . . . .

School . . . . Grade . . . .

Answer all questions in Part B–2 and Part C. Record your answers in this booklet.

Performance Test Score (Maximum Score: 23) ☐

Part	Maximum Score	Student's Score
A	35	
B–1	15	
B–2	15	
C	20	

Total Written Test Score (Maximum Raw Score: 85) ☐

Final Score (from conversion chart) ☐

Raters' Initials:

Rater 1 . . . . **Rater 2** . . . .

Part B–2

For Raters Only

51 ________________________ **51** ☐

52 ________________________ **52** ☐

53 ________________________ **53** ☐

54 ________________________ **54** ☐

[1] [OVER]

For Raters Only

55

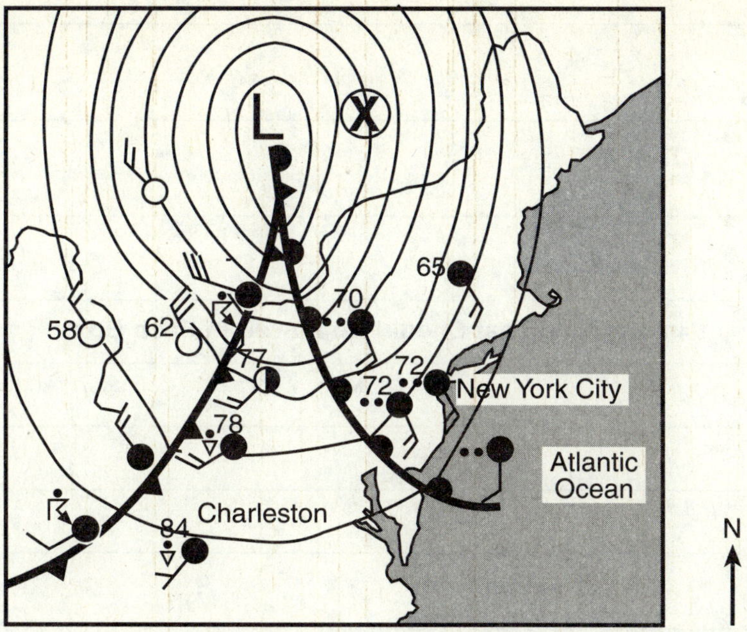

55

56 ____

56

57

Weather Conditions	Description
(1) Air temperature	
(2) Present weather	
(3) Wind speed	
(4) Wind direction	From
(5) Cloud cover	

57

[2]

For Raters Only

58 *a* ____ 58*a* [ ]

____

b ____ *b* [ ]

59

Letter	Information That Should Be Placed in Each Box
A	
B	
C	

59 [ ]

60

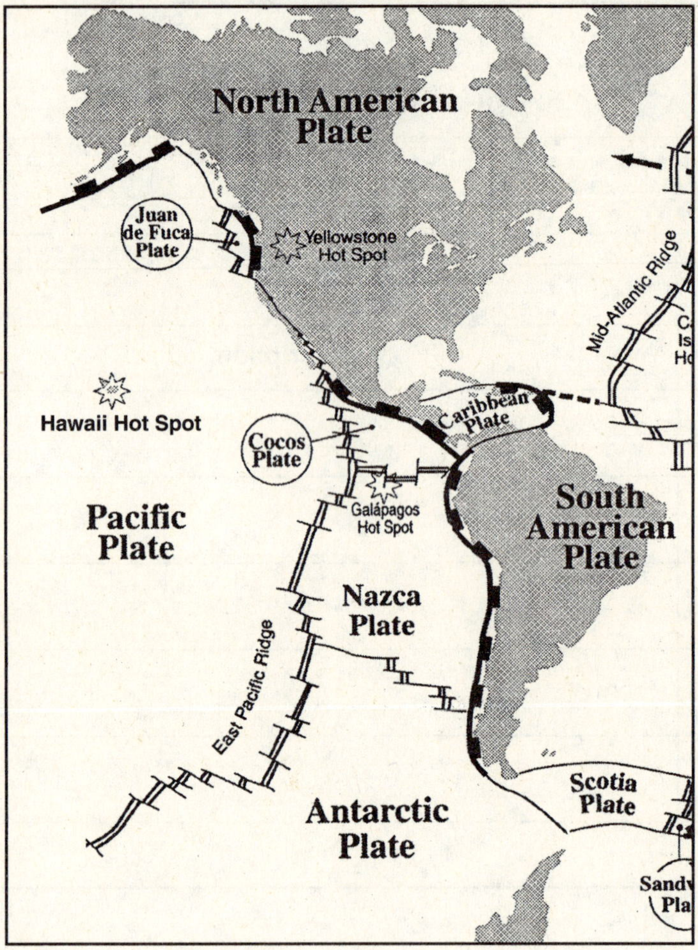

60 [ ]

Total Score for Part B–2

For Raters Only

Part C

61 ______________________________ 61 [ ]

62 ______________________________ 62 [ ]

63 ______________________________ 63 [ ]

64 ______________________________ 64 [ ]

65 ______________________________ 65 [ ]

66 66 [ ]

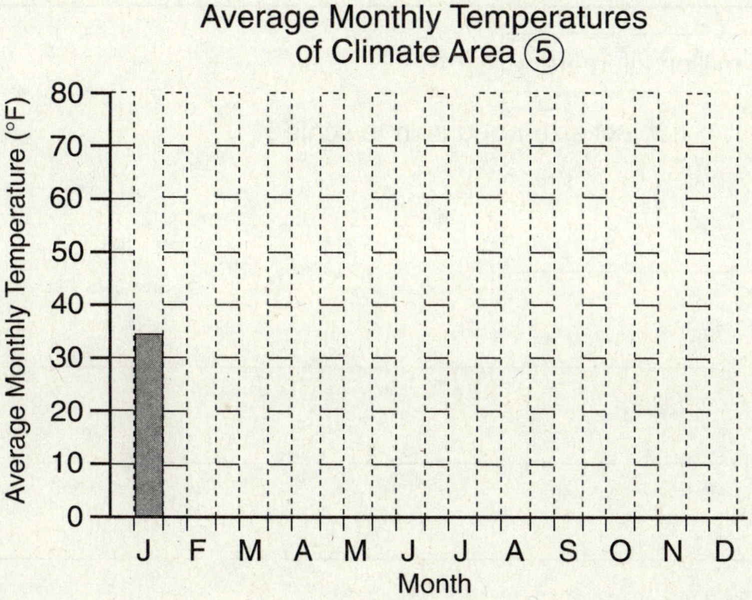

67 ______________________________ 67 [ ]

[4]

For Raters Only

68 *a* and *b*

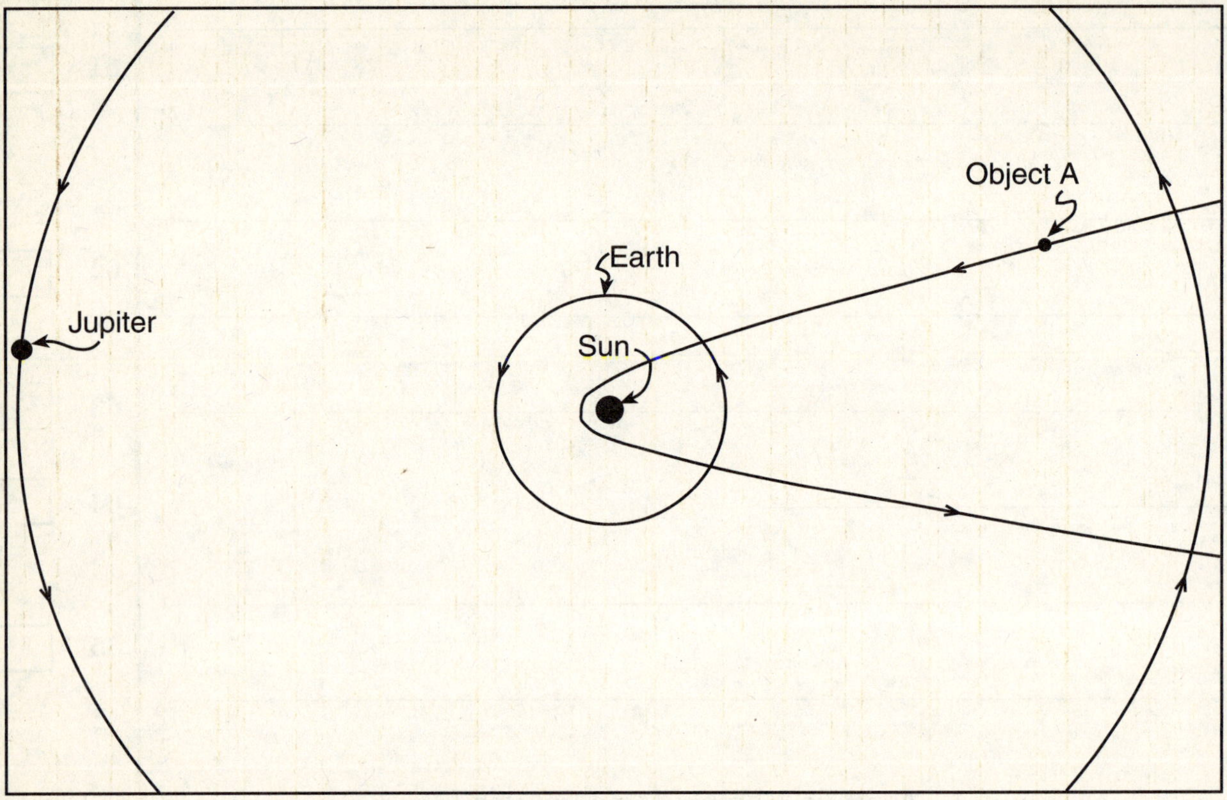

Distance scale 1 cm = 100 million kilometers

(Object size not drawn to scale)

68*a* ☐

b ☐

69 ______________________ 69 ☐

70 ______________________

____

______________________ 70 ☐

For Raters Only

71 *a–c*

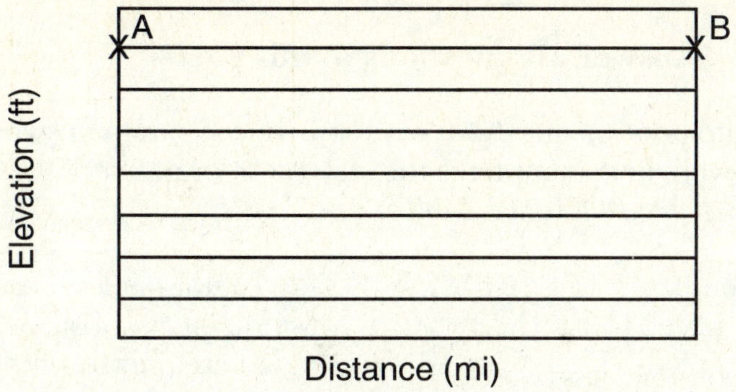

71*a* ☐

b ☐

c ☐

72 ____________________ **feet**

72 ☐

73

Rate = ______________

73 ☐

74 (1) ______________________________

(2) ______________________________

74 ☐

Total Score for Part C ☐

[6]

Physical Setting Earth Science August 13, 2003

Part A

Answer all questions in this part.

Directions (1–35): For *each* statement or question, write on your separate answer sheet the *number* of the word or expression that, of those given, best completes the statement or answers the question. Some questions may require the use of the *Earth Science Reference Tables*.

1 Which statement correctly compares the size, composition, and density of Neptune to Earth?

(1) Neptune is smaller, more gaseous, and less dense.
(2) Neptune is larger, more gaseous, and less dense.
(3) Neptune is smaller, more solid, and more dense.
(4) Neptune is larger, more solid, and more dense.

2 A person in New York State worked outdoors in sunlight for several hours on a day in July. Which type of clothing should the person have worn to absorb the *least* electromagnetic radiation?

(1) dark colored with a rough surface
(2) dark colored with a smooth surface
(3) light colored with a rough surface
(4) light colored with a smooth surface

3 The diagram below shows an observer on Earth measuring the altitude of *Polaris*.

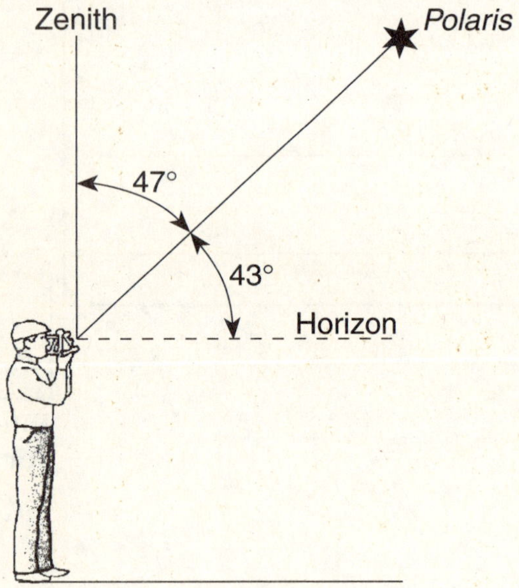

What is the latitude of this observer?

(1) 43° N
(2) 43° S
(3) 47° N
(4) 47° S

4 The diagram below represents Earth at a specific position in its orbit as viewed from space. The shaded area represents nighttime.

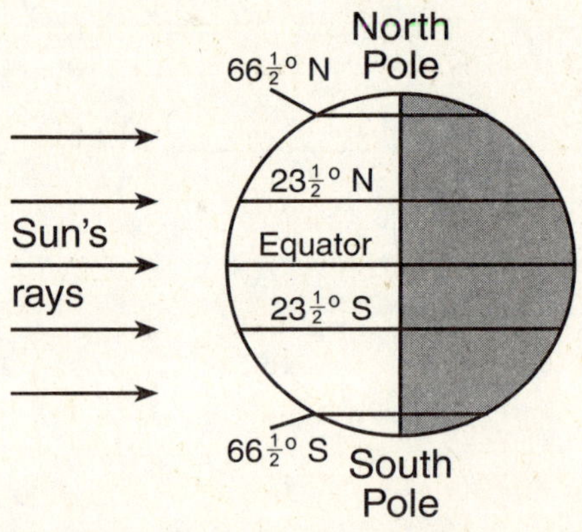

Which Earth latitude receives the greatest intensity of insolation when Earth is at the position shown in the diagram?

(1) 0°
(2) $23\frac{1}{2}$° N
(3) $66\frac{1}{2}$° N
(4) 90° N

5 Surface ocean currents curve to the right in the Northern Hemisphere because

(1) the Moon spins on its axis
(2) the Moon travels in an orbit around Earth
(3) Earth spins on its axis
(4) Earth travels in an orbit around the Sun

6 Which star color indicates the hottest star surface temperature?

(1) blue
(2) white
(3) yellow
(4) red

7 What is the dewpoint temperature when the dry-bulb temperature is 12°C and the wet-bulb temperature is 4°C?

(1) –9°C
(2) 19°C
(3) 8°C
(4) 4°C

8 Letters *A, B, C, D,* and *X* on the map below represent locations on Earth. The map shows the latitude-longitude grid.

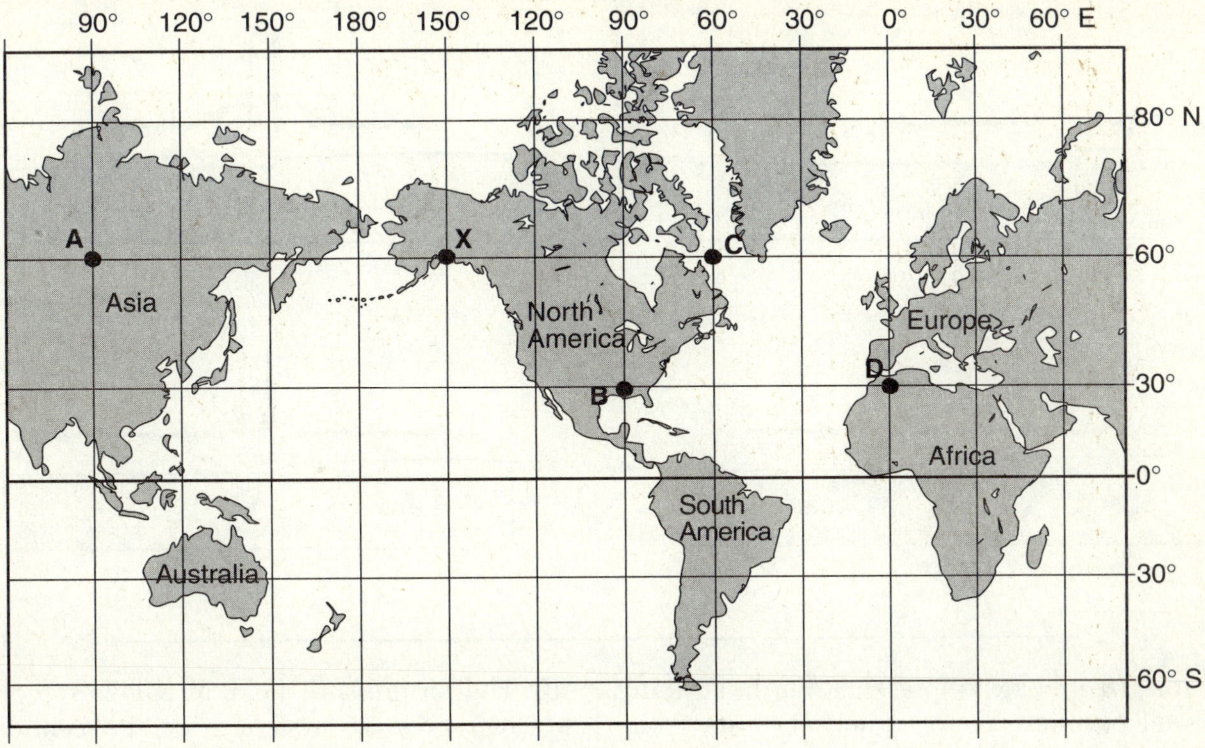

Solar time is based on the position of the Sun. If the solar time is 1 p.m. at location *X*, at which location is the solar time 5 p.m.?

(1) *A*
(2) *B*
(3) *C*
(4) *D*

9 The table below shows the duration of insolation (hours of daylight) measured by four observers, *W, X, Y,* and *Z*, at four different Earth latitudes on both March 21 and June 21. There were clear skies at all four latitudes on both days.

Observer	Duration of Insolation March 21	Duration of Insolation June 21
W	12 hr	0 hr
X	12 hr	12 hr
Y	12 hr	18 hr
Z	12 hr	24 hr

Which observer was located at the Equator?

(1) *W*
(2) *X*
(3) *Y*
(4) *Z*

10 Adjacent land and ocean surfaces have the same temperature at sunrise on a clear, calm, summer day. Then the land and water are heated by the Sun for several hours. Which cross section shows the most likely direction of surface winds that will develop at this ocean shore?

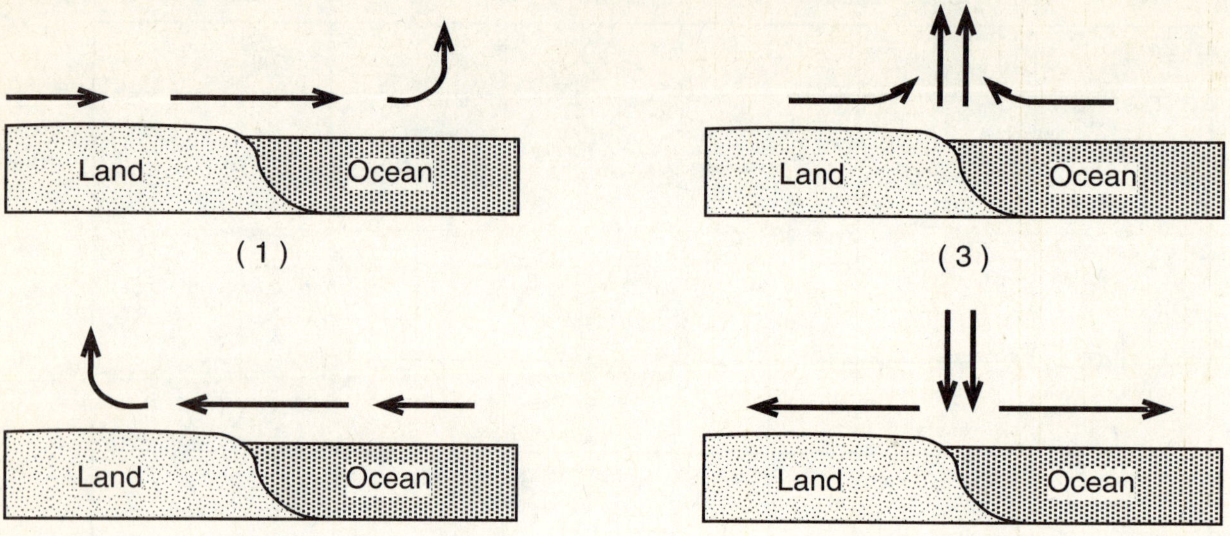

(1) (2) (3) (4)

---

11 The diagram below represents Earth at four different positions, *A*, *B*, *C*, and *D*, in its orbit around the Sun.

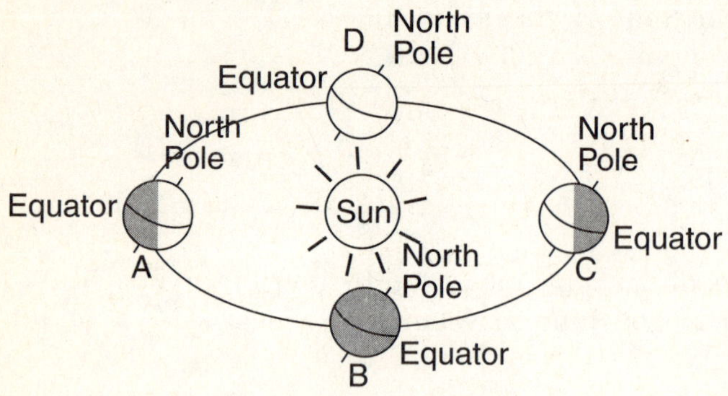

(Not drawn to scale)

Between which positions would New York State be experiencing the summer season?

(1) *A* and *B*
(2) *B* and *C*
(3) *C* and *D*
(4) *D* and *A*

12 An air mass classified as mP usually forms over which type of Earth surface?

(1) warm land
(2) warm ocean
(3) cool land
(4) cool ocean

13 The air-pressure field map below represents a high-pressure system over the central United States. Isobars show the air pressure, in millibars. Letters *A* through *E* represent locations on Earth's surface.

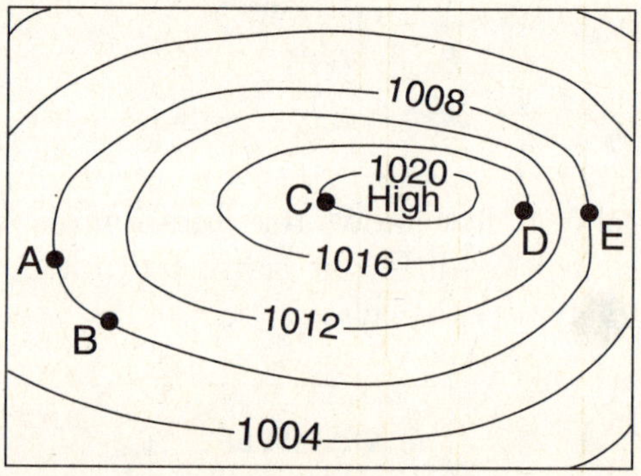

Between which two locations is the wind speed greatest?

(1) *A* and *B*
(2) *B* and *C*
(3) *C* and *D*
(4) *D* and *E*

14 Surface ocean currents located at 40° south latitude, 90° west longitude generally flow toward the

(1) northeast
(2) southeast
(3) southwest
(4) west

15 The arrows on the two maps below show how the monsoon winds over India change direction with the seasons.

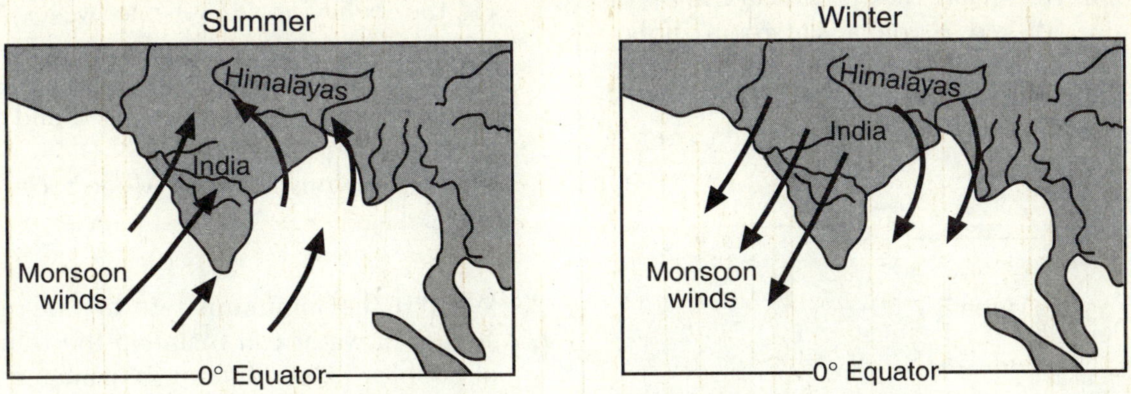

How do these winds affect India's weather in summer and winter?

(1) Summer is cooler and less humid than winter.
(2) Summer is warmer and more humid than winter.
(3) Winter is warmer and less humid than summer.
(4) Winter is cooler and more humid than summer.

---

16 Most water vapor enters Earth's atmosphere by the processes of

(1) condensation and precipitation
(2) radiation and cementation
(3) conduction and convection
(4) evaporation and transpiration

17 At an altitude of 95 miles above Earth's surface, nearly 100% of the incoming energy from the Sun can be detected. At 55 miles above Earth's surface, most incoming x-ray radiation and some incoming ultraviolet radiation can no longer be detected. This missing radiation was most likely

(1) absorbed in the thermosphere
(2) absorbed in the mesosphere
(3) reflected by the stratosphere
(4) reflected by the troposphere

18 Weather along most fronts is usually cloudy with precipitation because the warm air along most fronts is usually

(1) sinking and cooling, causing water to evaporate
(2) sinking and warming, causing water to evaporate
(3) rising and cooling, causing water vapor to condense
(4) rising and warming, causing water vapor to condense

19 Which cross section below best represents the crustal plate motion that is the primary cause of the volcanoes and deep rift valleys found at mid-ocean ridges?

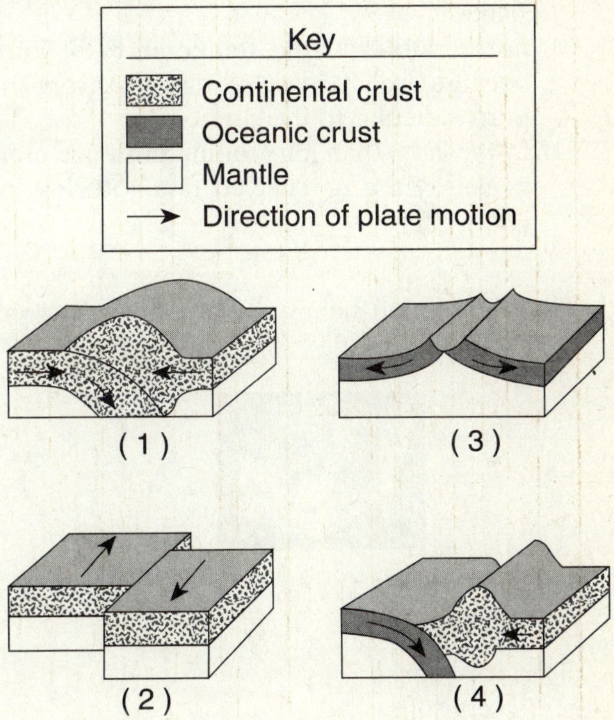

20 The diagram below shows a classroom demonstration. Two identical flashlights were placed in the positions shown and they illuminated areas of varying size, *A* and *B*, on a classroom globe. Thermometers were then placed at the center of each illuminated area to measure the rate of temperature increase. Readings were taken over a period of 30 minutes.

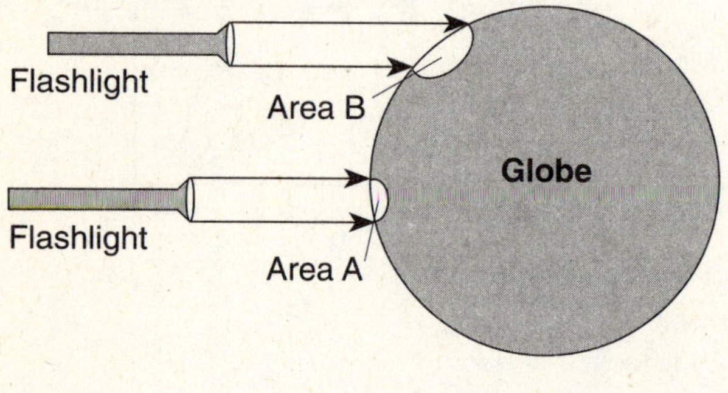

Students most likely observed that the temperature of area *A* increased at a

(1) slower rate than the temperature of area *B* because area *A* received rays that were less concentrated
(2) slower rate than the temperature of area *B* because area *A* received rays that were more slanted
(3) faster rate than the temperature of area *B* because area *A* received rays that were more perpendicular to the surface
(4) faster rate than the temperature of area *B* because area *A* received rays with less total energy

21 The photograph below shows a broken piece of the mineral calcite.

The calcite breaks in smooth, flat surfaces because calcite

(1) is very dense
(2) is very soft
(3) contains certain impurities
(4) has a regular arrangement of atoms

22 Most inferences about the characteristics of Earth's mantle and core are based on

(1) the behavior of seismic waves in Earth's interior
(2) well drillings from Earth's mantle and core
(3) chemical changes in exposed and weathered metamorphic rocks
(4) comparisons between Moon rocks and Earth rocks

23 What is the minimum rate of flow at which a stream of water can maintain the transportation of pebbles 1.0 centimeter in diameter?

(1) 50 cm/sec
(2) 100 cm/sec
(3) 150 cm/sec
(4) 200 cm/sec

24 Which activity demonstrates chemical weathering?

(1) freezing of water in the cracks of a sandstone sidewalk
(2) abrasion of a streambed by tumbling rocks
(3) grinding of talc into a powder
(4) dissolving of limestone by acid rain

25 Unsorted, angular, rough-surfaced cobbles and boulders are found at the base of a cliff. What most likely transported these cobbles and boulders?

(1) running water
(2) wind
(3) gravity
(4) ocean currents

26 The seismogram below shows *P*-wave and *S*-wave arrival times at a seismic station following an earthquake.

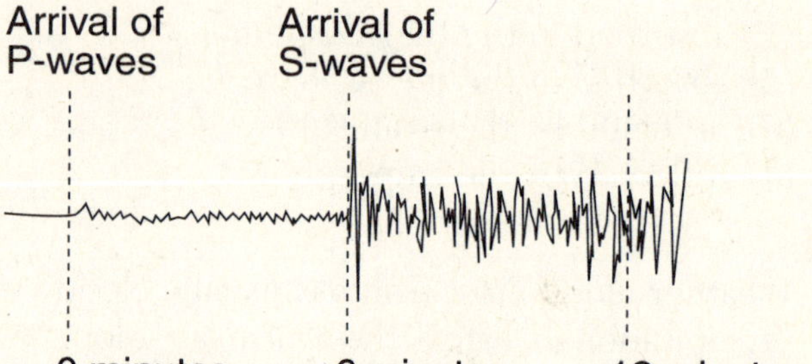

The distance from this seismic station to the epicenter of the earthquake is approximately

(1) 1,600 km
(2) 3,200 km
(3) 4,400 km
(4) 5,600 km

27 The table below gives information about the radioactive decay of carbon-14. Part of the table has been deliberately left blank for student use.

Half-life	Mass of Original Carbon-14 Remaining (grams)	Number of Years
0	1	0
1	$\frac{1}{2}$	5,700
2	$\frac{1}{4}$	11,400
3	$\frac{1}{8}$	17,100
4	$\frac{1}{16}$	
5		
6		
7		

After how many years will $\frac{1}{128}$ gram of the original carbon-14 remain?

(1) 22,800 yr
(2) 28,500 yr
(3) 34,200 yr
(4) 39,900 yr

---

28 In which layer of Earth's interior is the pressure inferred to be 1.0 million atmospheres?

(1) outer core
(2) inner core
(3) rigid mantle
(4) stiffer mantle

29 Which event occurred earliest in geologic history?

(1) appearance of the earliest grasses
(2) appearance of the earliest birds
(3) the Grenville Orogeny
(4) the intrusion of the Palisades Sill

30 A very large, circular, impact crater under the coast of Mexico is believed to be approximately 65 million years old. This impact event is inferred to be related to the

(1) appearance of the earliest trilobites
(2) advance and retreat of the last continental ice sheet
(3) extinction of the dinosaurs
(4) formation of Pangea

31 The greatest amount of rainwater infiltration occurs on the side of a hill if the surface of a permeable soil has

(1) small soil particles and a steep slope
(2) small soil particles and a gentle slope
(3) large soil particles and a steep slope
(4) large soil particles and a gentle slope

32 Which element is most abundant in Earth's lithosphere?

(1) oxygen
(2) silicon
(3) hydrogen
(4) nitrogen

33 A student incorrectly measured the volume of a mineral sample as 83 cubic centimeters when the actual volume was 89 cubic centimeters. What was the student's approximate percent deviation (percentage of error)?

(1) 6.7%
(2) 7.2%
(3) 9.3%
(4) 14.8%

34 The characteristic of the radioactive isotope uranium-238 that makes this isotope useful for accurately dating the age of a rock is the isotope's

(1) organic origin
(2) constant half-life
(3) common occurrence in sediments
(4) resistance to weathering and erosion

35 In which New York State landscape region is Niagara Falls located?

(1) Tug Hill Plateau
(2) St. Lawrence Lowlands
(3) Allegheny Plateau
(4) Erie-Ontario Lowlands

Part B–1

Answer all questions in this part.

Directions (36–50): For *each* statement or question, write on your separate answer sheet the *number* of the word or expression that, of those given, best completes the statement or answers the question. Some questions may require the use of the *Earth Science Reference Tables*.

36 The table below shows gravitational data for a planet traveling in an elliptical orbit around a star. The table shows the relative gravitational force between the star and this planet at eight positions in the orbit (letters *A* through *H*). Higher numbers indicate stronger gravitational attraction.

Planet's Position in the Orbit	*A*	*B*	*C*	*D*	*E*	*F*	*G*	*H*
Relative Gravitational Force Between Star and Planet	52	42	25	12	10	12	25	42

Which diagram best represents the positions of the planet in its orbit that would produce the gravitational forces shown in the data table?

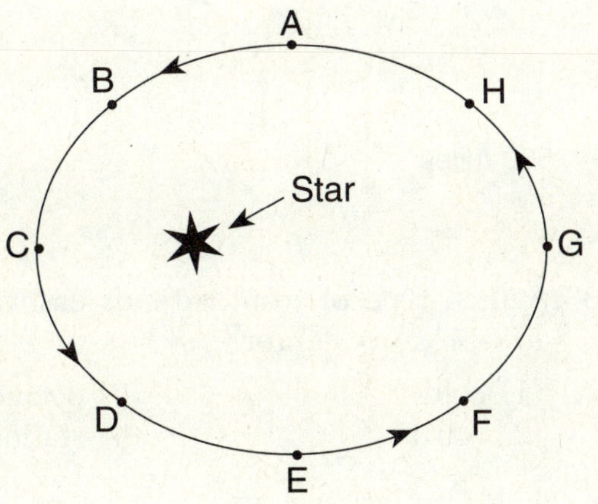

(1)

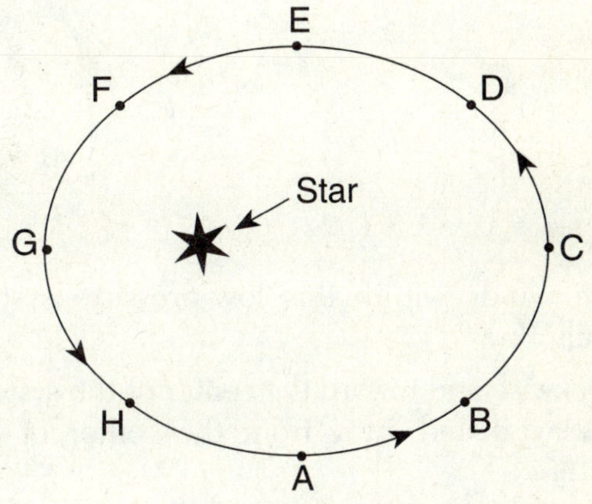

(3)

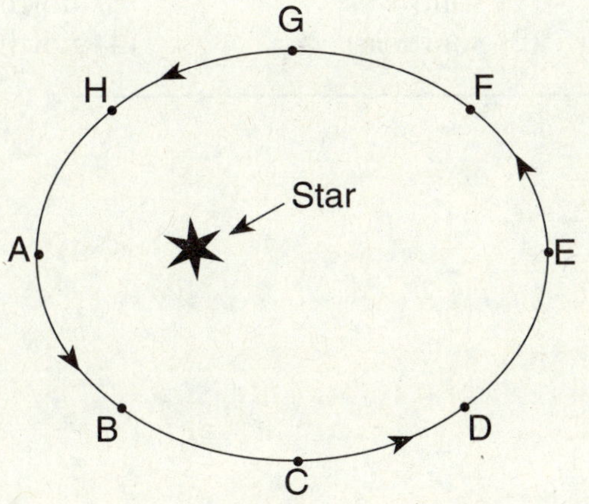

(2)

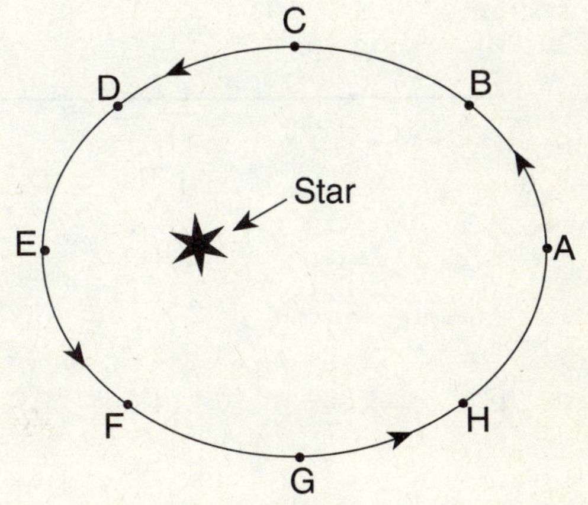

(4)

Base your answers to questions 37 through 39 on the weather map below, which shows air temperature and winds for a few locations in the eastern half of the United States. A large low-pressure system is shown on the map.

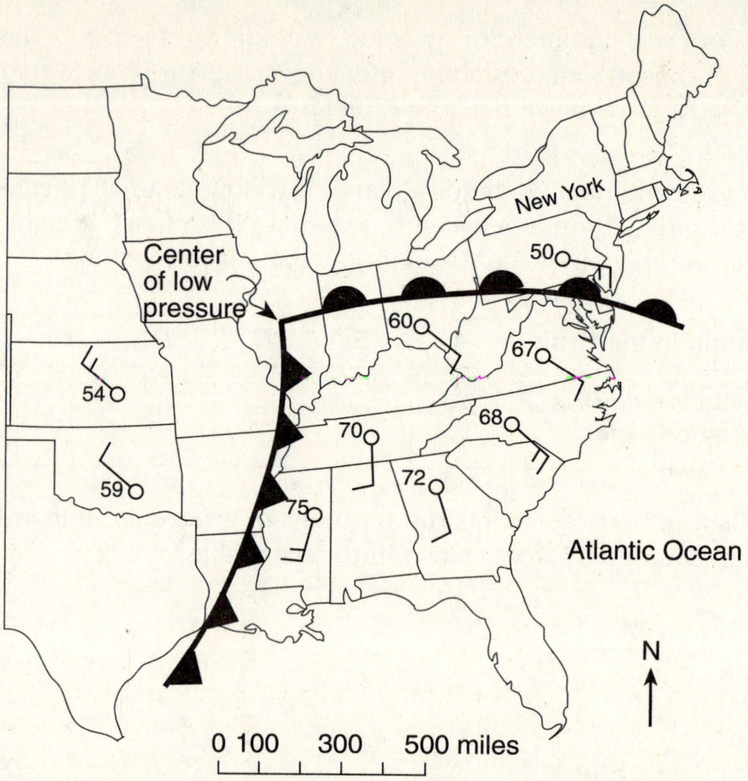

37 Surface winds within this low-pressure system generally flow

(1) clockwise and toward the center of the system
(2) clockwise and away from the center of the system
(3) counterclockwise and toward the center of the system
(4) counterclockwise and away from the center of the system

38 Which type of front extends eastward from the low-pressure center?

(1) cold
(2) warm
(3) occluded
(4) stationary

39 If the low-pressure center follows a typical storm track, it will move toward the

(1) southwest
(2) southeast
(3) northwest
(4) northeast

Base your answers to questions 40 through 42 on the geologic cross section and the table below. The cross section represents the bedrock structure beneath four landscape regions, *A*, *B*, *C*, and *D*.

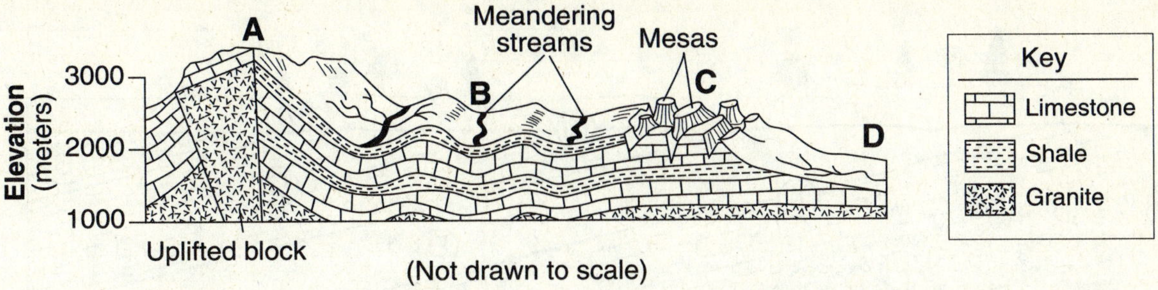

The table below shows characteristics of the four landscape regions *A*, *B*, *C*, and *D*.

Landscape Region	Relief	Bedrock
A	great relief, high peaks, deep valleys	faulted and tilted structure; many bedrock types, including igneous
B	moderate relief, rounded peaks, wide valleys	folded sedimentary bedrock
C	moderate to high relief	horizontal sedimentary bedrock layers
D	very little relief, low elevations	horizontal sedimentary bedrock layers

40 Which terms best describe the surface landscapes of *A*, *B*, *C*, and *D*?

(1) *A*—mountains, *B*—ridges and valleys, *C*—plateau, *D*—plain
(2) *A*—plateau, *B*—plain, *C*—mountains, *D*—ridges and valleys
(3) *A*—plain, *B*—mountains, *C*—plateau, *D*—plain
(4) *A*—ridges and valleys, *B*—plateau, *C*—plain, *D*—mountains

41 The meandering streams shown in landscape region *B* usually form where there are

(1) volcanic cones
(2) gentle gradients
(3) many fractures in the bedrock
(4) numerous escarpments

42 The sharp, angular flat-topped hills (mesas) in landscape region *C* were most likely produced by a climate that was

(1) tropical
(2) humid
(3) dry
(4) polar

Base your answers to questions 43 through 47 on the geologic cross section of bedrock shown below. *A* through *G* identify rock layers and *Q* represents a fault. Lines *W*, *X*, *Y*, and *Z* are locations of unconformities. The rocks have not been overturned.

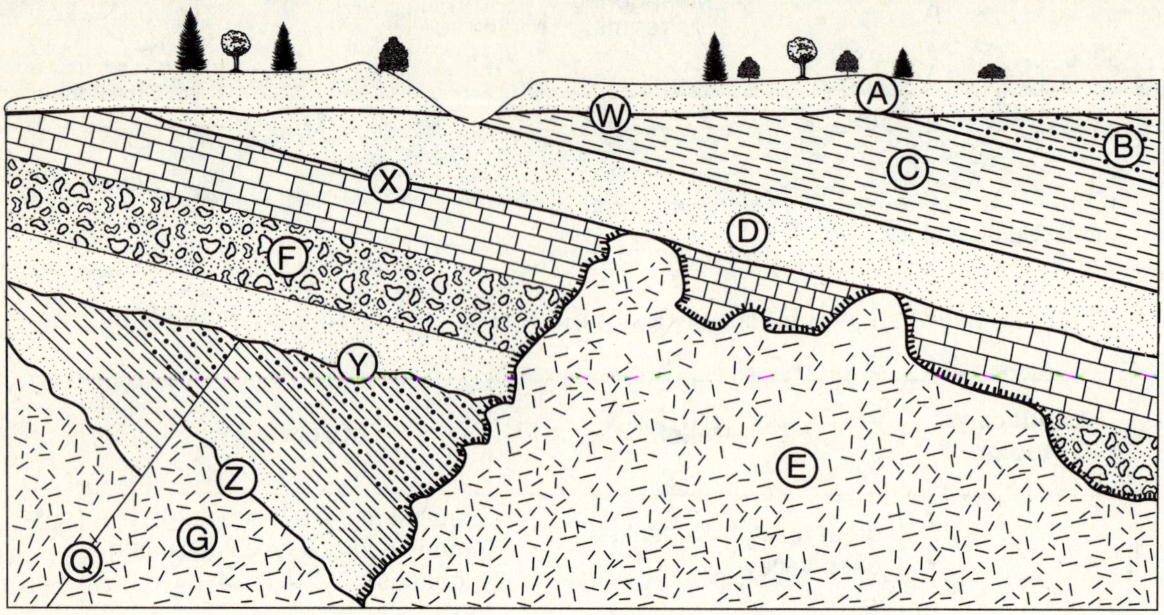

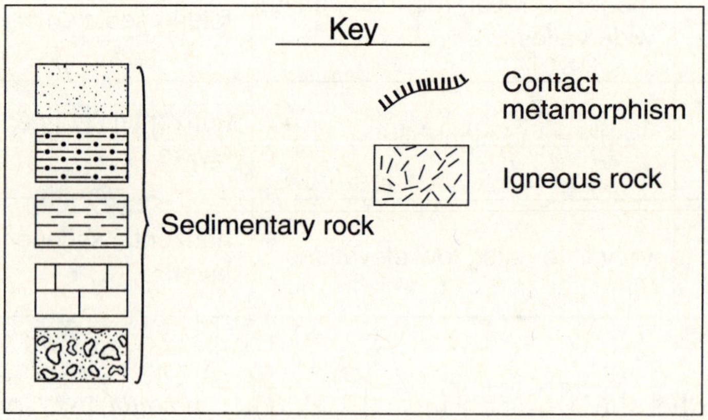

43 Which rock or feature is oldest?

(1) rock *A*
(2) rock *G*
(3) fault *Q*
(4) unconformity *Z*

44 The unconformities shown in the cross section represent

(1) buried erosional surfaces
(2) locations of index fossils
(3) volcanic ash deposits
(4) boundaries between oceanic and continental crust

45 The movement of bedrock along fault *Q* most probably produced

(1) gaps in the rock record
(2) an earthquake
(3) a volcanic lava flow
(4) zones of contact metamorphism

46 Which rock most likely formed in the zone of contact between rock *E* and rock *F*?

(1) obsidian
(2) slate
(3) metaconglomerate
(4) sandstone

47 Rock layers *B*, *C*, and *D* formed during the Devonian Period. Which fossil might be found in these rock layers?

Mastodont
(1)

Elliptocephala
(2)

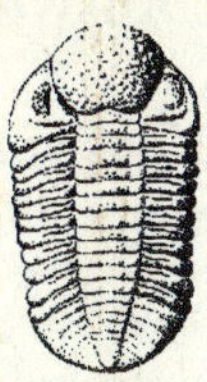

Phacops
(3)

Cystiphyllum
(4)

Base your answers to questions 48 and 49 on the diagram below, which shows numbered positions of the Sun at four different times along the Sun's apparent daily path, as seen by an observer in New York State. Numbers ① through ④ represent apparent positions of the Sun.

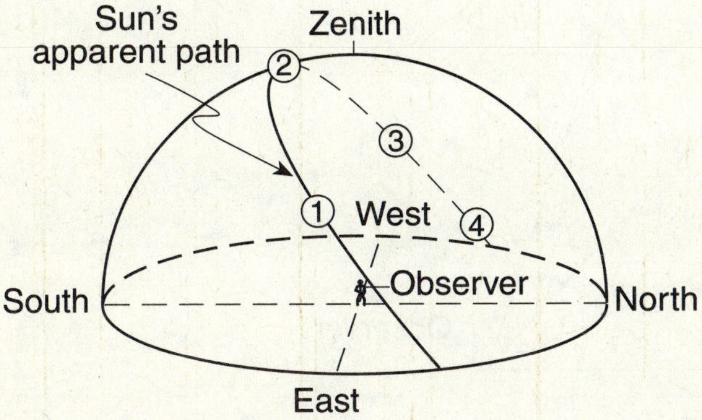

48 The observer had the longest shadow when the Sun was at position

(1) 1
(2) 2
(3) 3
(4) 4

49 During which day of the year is the Sun most likely to follow the apparent path shown?

(1) March 1
(2) July 1
(3) October 1
(4) December 1

50 The two photographs below show dates on tombstones found in a cemetery in St. Remy, New York. The tombstones were 5 meters apart and both faced north. Tombstone *A* had dates cut into the rock in 1922. Tombstone *B* had dates cut into the rock in 1892.

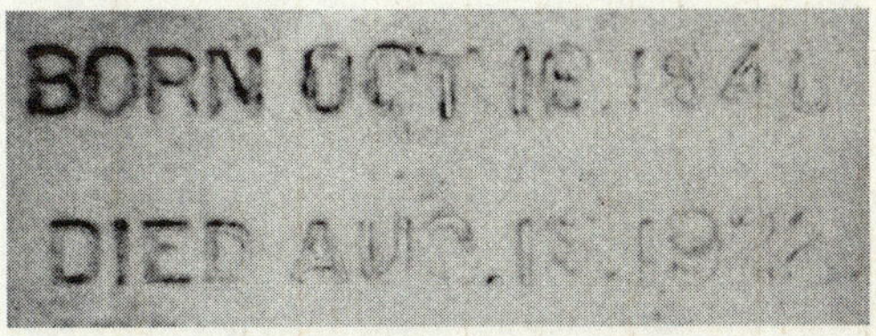

Tombstone A (1922)

Tombstone B (1892)

Which statement best explains why the dates are more difficult to read on tombstone *A* than on tombstone *B*?

(1) Tombstone *A* is composed of minerals less resistant to weathering than tombstone *B*.
(2) Tombstone *A* has undergone a longer period of weathering than tombstone *B*.
(3) Tombstone *A* experienced cooler temperatures than tombstone *B*.
(4) Tombstone *A* was exposed to less acid rain than tombstone *B*.

Part B–2

Answer all questions in this part.

Directions (51–63): Record your answers in the spaces provided in your answer booklet. Some questions may require the use of the *Earth Science Reference Tables.*

51 State the general relationship between a planet's distance from the Sun and the time a planet takes to complete one orbit around the Sun. [1]

Base your answers to questions 52 through 54 on the diagram below, which shows the altitude of the Sun at solar noon on certain dates. The positions of the Sun, labeled *A, B,* and *C,* were measured by an observer at 42° north latitude. The date when the Sun was observed at position *A* has been deliberately left blank.

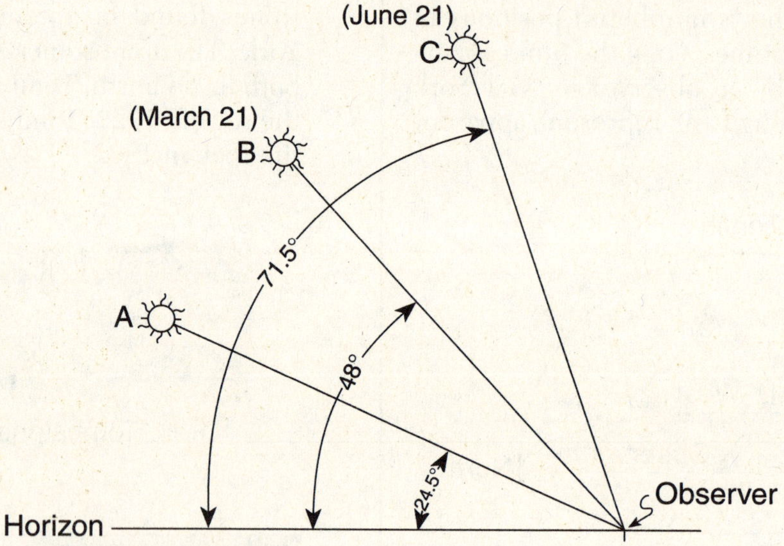

52 Which season begins in New York State when the noontime Sun is observed at position *A*? [1]

53 Position *B* represents the Sun's position at solar noon on March 21. On what other date of the year would the noontime Sun be observed at position *B*? [1]

54 What is the total change in altitude that occurs as the noontime Sun appears to move from position *A* to position *C*? [1]

Base your answers to questions 55 through 57 on the weather station model provided in your answer booklet.

55 On the weather station model provided *in your answer booklet,* draw the proper symbols to indicate a wind of 25 knots blowing from the southeast. [1]

56 What is the actual air pressure shown by this weather station model? [1]

57 *a* What specific type of precipitation is occurring at this weather station? [1]

b State one additional weather condition shown by the station model. Explain how this weather condition provides evidence of high relative humidity. [1]

Base your answers to questions 58 through 61 on the contour map below. Letters *A* through *H* represent locations in the area represented by the map. Contour lines are labeled in feet.

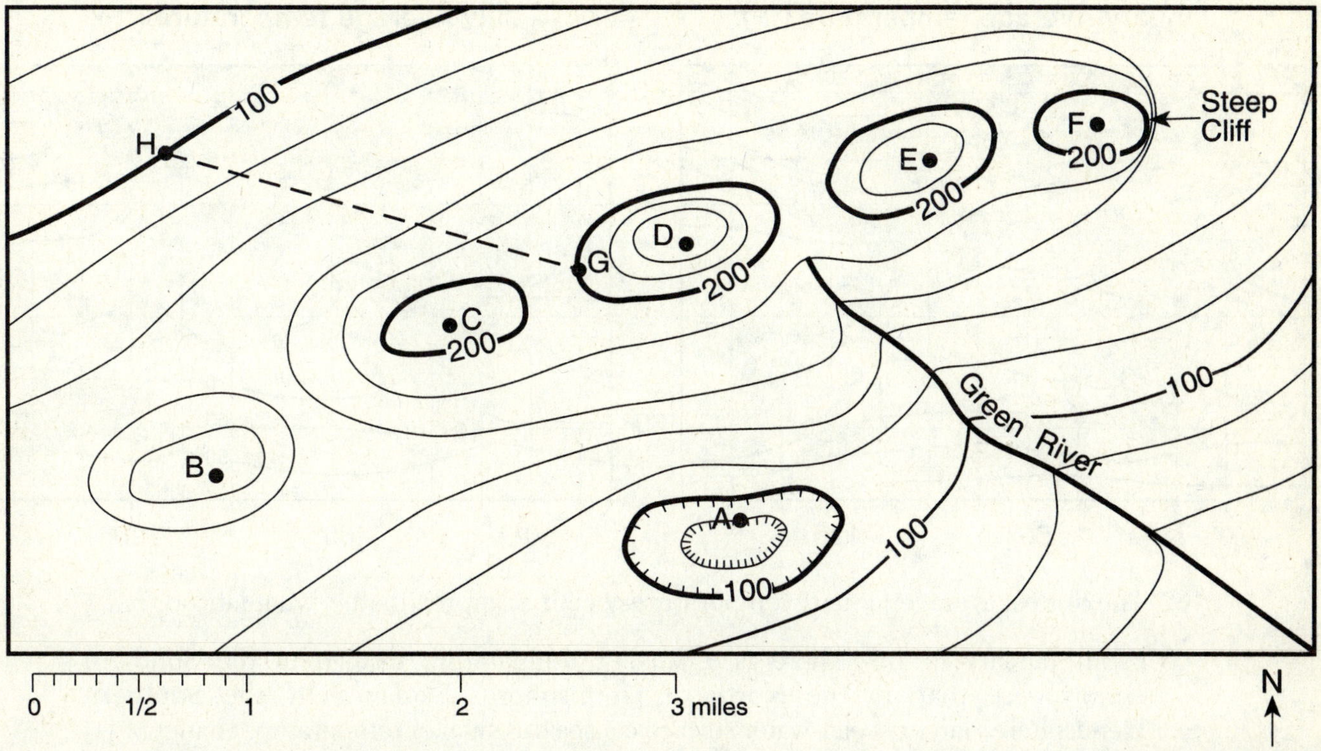

58 Calculate the gradient of the slope along the dashed line between points *G* and *H* on the map. Label the answer with the correct units. [2]

59 State how the shape of the contour lines crossing the Green River indicates that this river flows toward the southeast. [1]

60 Which letter represents the highest elevation? [1]

61 Explain how the contour lines on the map indicate that the location labeled "Steep Cliff" is accurately named. [1]

Base your answers to questions 62 and 63 on the isotherm maps below and on your knowledge of Earth science. The maps show the average monthly air temperatures (°F) over a portion of Earth's surface for January and July.

January Average Temperature (°F)

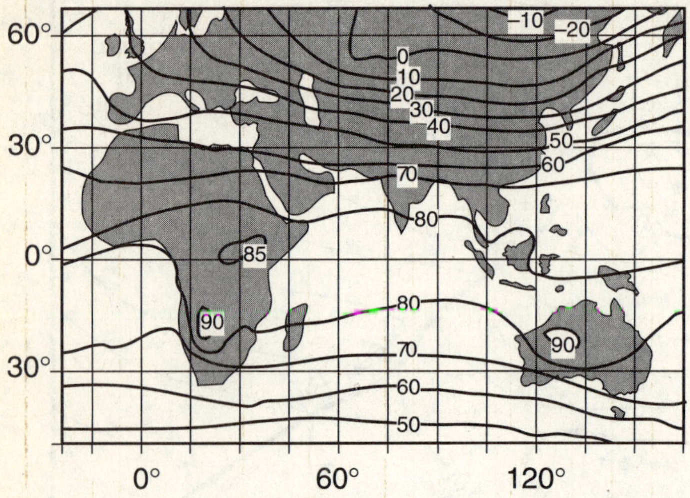

July Average Temperature (°F)

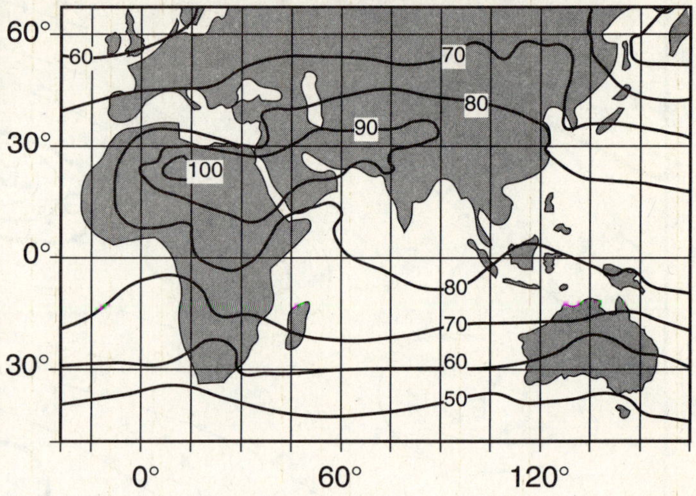

62 The hottest average January temperatures occur at approximately what latitude? [1]

63 From January to July, there is a smaller temperature change in the Southern Hemisphere than in the Northern Hemisphere. Explain why the Southern Hemisphere's larger ocean-water surface causes this smaller temperature change. [1]

____

Part C

Answer all questions in this part.

Directions (64–80): Record your answers in the spaces provided in your answer booklet. Some questions may require the use of the *Earth Science Reference Tables.*

Base your answers to questions 64 through 66 on the diagram provided in your answer booklet and on your knowledge of Earth science. The diagram shows the Sun, Earth, and the Moon's orbit around Earth as viewed from space.

64 On the diagram provided *in your answer booklet,* draw a circle of approximately this size ◯ to represent the Moon's position in its orbit when a solar eclipse is viewed from Earth. [1]

65 Approximately how many complete revolutions does the Moon make around Earth each month? [1]

66 Explain why solar eclipses do not occur every time the Moon revolves around Earth. [1]

Base your answers to questions 67 through 70 on the table below and on your knowledge of Earth science. The table shows air temperatures and air pressures recorded by a weather balloon rising over Buffalo, New York.

Altitude Above Sea Level (m)	**Air Temperature** (°C)	**Air Pressure** (mb)
300	16.0	973
600	16.5	937
900	15.5	904
1,200	13.0	871
1,500	12.0	842
1,800	10.0	809
2,100	7.5	778
2,400	5.0	750
2,700	2.5	721

67 On the grid provided *in your answer booklet,* construct a graph of altitude above sea level and air temperature by following the directions below.

a Plot an **X** for the air temperature recorded at *each* altitude shown on the table. [1]

b Connect the **X**s with a solid line. [1]

68 What weather instrument is usually attached to a weather balloon to measure air pressure? [1]

69 State the relationship shown in the table between altitude above sea level and air pressure recorded by the rising weather balloon. [1]

70 This rising weather balloon also recorded dewpoint temperatures. If the dewpoint at 1,500 meters was 12°C, what was the relative humidity of the air at 1,500 meters above sea level? [1]

Base your answers to questions 71 through 74 on the reading passage below and on your knowledge of Earth science.

Greenhouse Effect

The warming of Earth's surface and lower atmosphere tends to intensify with an increase in atmospheric carbon dioxide. The atmosphere allows a large percentage of the visible light rays from the Sun to reach Earth's surface. Some of this energy is reradiated by Earth's surface in the form of long-wave infrared radiation. Much of this infrared radiation warms the atmosphere when it is absorbed by molecules of carbon dioxide and water vapor. A similar warming effect is produced by the glass of a greenhouse, which allows sunlight in the visible range to enter, but prevents infrared radiation from leaving the greenhouse.

The absorption of infrared radiation causes Earth's surface and the lowest layer of Earth's atmosphere to warm to a higher temperature than would otherwise be the case. Without this "greenhouse" warming, Earth's average surface temperature could be as low as –73°C. The oceans would freeze under such conditions.

Many scientists believe that modern industrialization and the burning of fossil fuels (coal, oil, and natural gas) have increased the amount of atmospheric carbon dioxide. This increase may result in an intensified greenhouse effect on Earth causing significant alterations in climate patterns in the future. Scientists estimate that average global temperatures could increase by as much as 5°C by the middle of the 21st century.

71 The lowest layer of Earth's atmosphere has undergone a large increase in temperature due to the presence of greenhouse gases. State the name of this temperature-zone layer. [1]

72 State a possible wavelength, in centimeters, of infrared radiation. [1]

73 Explain why most scientists believe an increase in the greenhouse effect will cause sea levels to rise. [1]

74 State one possible change humans could make to significantly reduce the amount of greenhouse gases added to the atmosphere each year. [1]

Base your answers to questions 75 through 78 on the Rock Classification flowchart shown below. Letters *A*, *B*, and *C* represent specific rocks in this classification scheme.

Rock Classification Flowchart

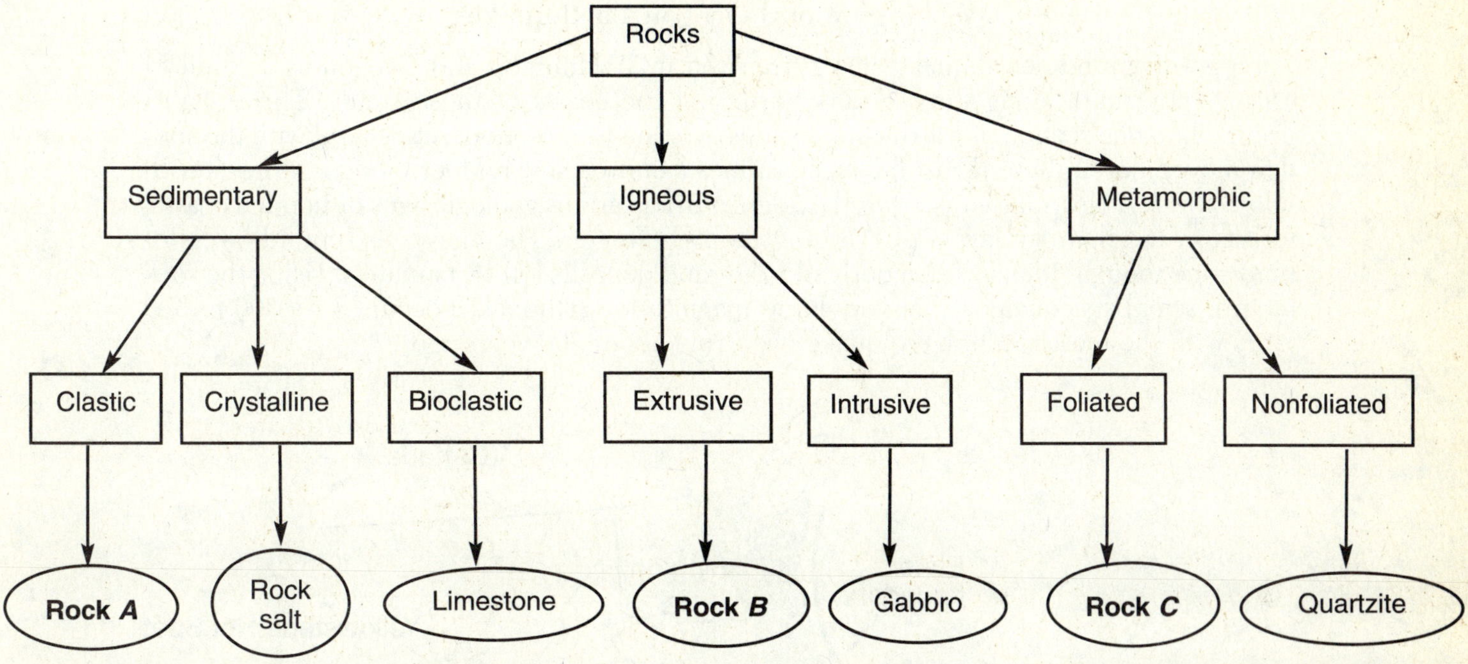

75 Rock *A* is composed of very fine-grained quartz and feldspar particles 0.005 centimeter in diameter. State the name of rock *A*. [1]

76 Rock *B* has a glassy, vesicular texture and is composed mainly of potassium feldspar and quartz. State the name of rock *B*. [1]

77 Granite could be placed in the same position in the flowchart above as gabbro. Describe *two* differences between granite and gabbro. [1]

78 The diagram below represents two magnified views showing the arrangement of minerals before and after metamorphism of rock *C*. State the name of rock *C*. [1]

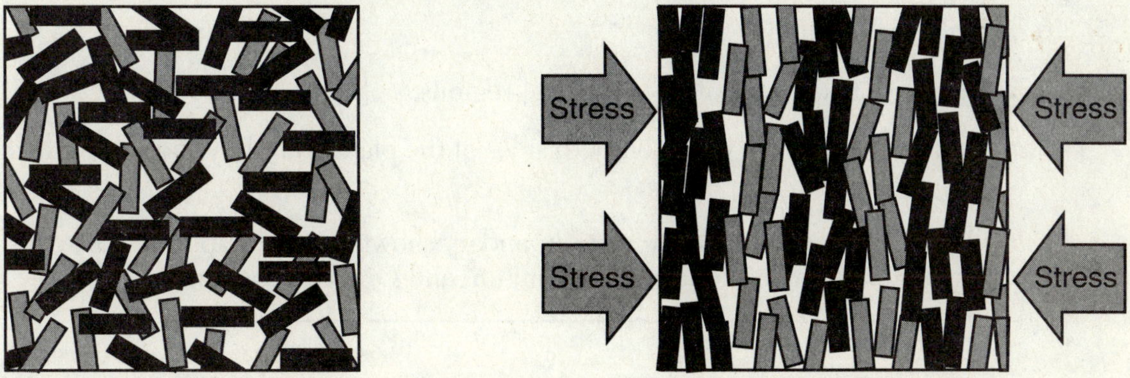

Mineral Arrangement Before Metamorphism

Rock C Showing Banding After Metamorphism

Base your answers to questions 79 and 80 on the reading passage and map of the western United States below and on your knowledge of Earth science. The states of Washington and Oregon have been labeled on the map. The plate boundary shown on the map is the source area for high-magnitude earthquakes in Washington and Oregon. Two hazardous zones associated with these earthquakes are also shown.

Washington and Oregon Earthquakes

Large-magnitude earthquakes have occurred in Washington and Oregon as a result of crustal movement along thrust faults bordering the coasts of these states. Thrust faults occur when one section of Earth's crust slides over another section. Associated with the sudden movement of these thrust faults, coastlines can drop several feet, flooding forests with saltwater. Geologists have discovered evidence from various geologic ages of flooded coastal forests in the bedrock layers of Washington and Oregon. They have also found layers of sandstone thought to have been derived from sand deposits left by tsunamis. Using the rock record, scientists conclude that very large magnitude earthquakes occur every 300 to 500 years with the most recent large quake occurring about 200 years ago.

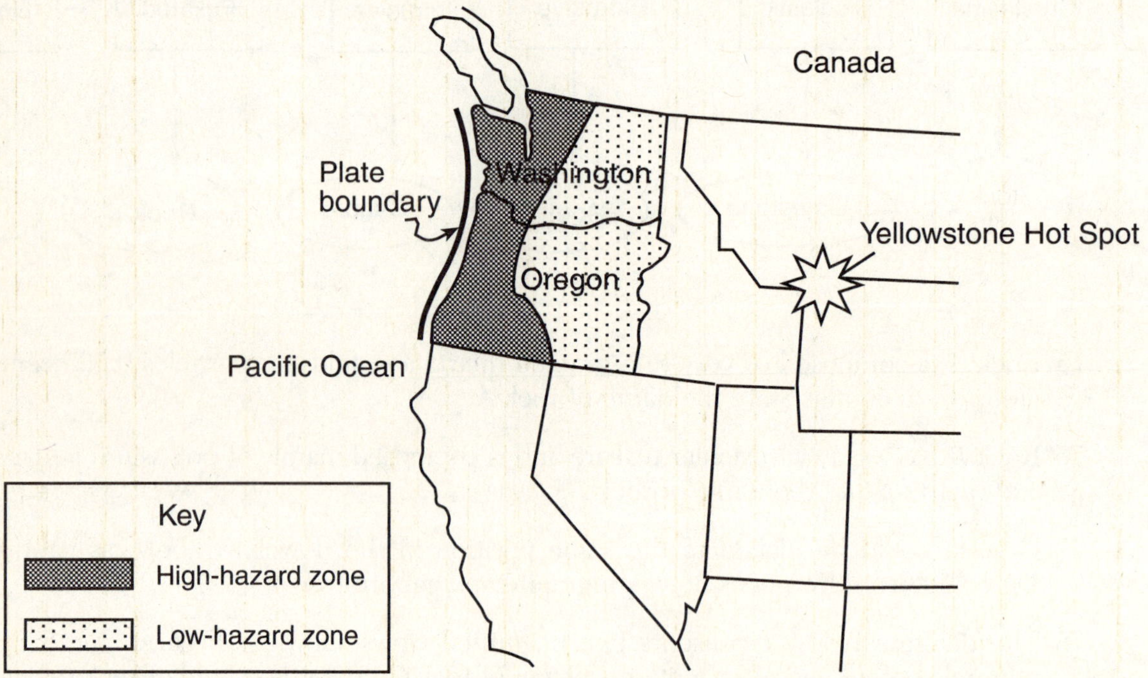

79 *a* What is a tsunami? [1]

b State how tsunamis can affect coastal regions. [1]

80 *a* Identify the tectonic plates on both sides of the plate boundary shown on the map. [1]

b Identify the type of tectonic plate boundary shown on the map that is responsible for the thrust faults along the Washington and Oregon coastline. [1]

The University of the State of New York

REGENTS HIGH SCHOOL EXAMINATION

PHYSICAL SETTING
EARTH SCIENCE

Wednesday, August 13, 2003 — 12:30 to 3:30 p.m., only

ANSWER SHEET

Student .... Sex: ☐ Male ☐ Female Grade ....

Teacher .... School ....

Record your answers to Part A and Part B–1 on this answer sheet.

Part A

1	13	25 ....
2	14	26 ....
3	15	27 ....
4	16	28 ....
5	17	29 ....
6	18	30 ....
7	19	31 ....
8	20	32 ....
9	21	33 ....
10	22	34 ....
11	23	35 ....
12	24	

Part A Score ☐

Part B–1

36	44 ....
37	45 ....
38	46 ....
39	47 ....
40	48 ....
41	49 ....
42	50 ....
43	

Part B–1 Score ☐

Write your answers to Part B–2 and Part C in your answer booklet.

The declaration below should be signed when you have completed the examination.

I do hereby affirm, at the close of this examination, that I had no unlawful knowledge of the questions or answers prior to the examination and that I have neither given nor received assistance in answering any of the questions during the examination.

____

Signature

The University of the State of New York

REGENTS HIGH SCHOOL EXAMINATION

PHYSICAL SETTING EARTH SCIENCE

Wednesday, August 13, 2003 — 12:30 to 3:30 p.m., only

ANSWER BOOKLET

Student .... Sex: ☐ Male ☐ Female

Teacher ....

School .... Grade ....

Answer all questions in Part B–2 and Part C. Record your answers in this booklet.

Performance Test Score (Maximum Score: 23) ☐

Part	Maximum Score	Student's Score
A	35	
B–1	15	
B–2	15	
C	20	

Total Written Test Score (Maximum Raw Score: 85) ☐

Final Score (from conversion chart) ☐

Raters' Initials:

Rater 1 .... **Rater 2** ....

Part B–2

For Raters Only

51 ______________________________ **51** ☐

____

____

52 ______________________________ **52** ☐

53 ______________________________ **53** ☐

54 ______________________________ **degrees** **54** ☐

For Raters Only

55

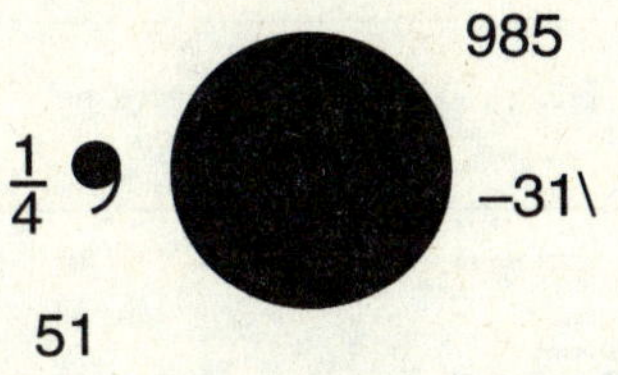

55 ☐

56 ______________ **millibars** — **56** ☐

57 *a* ______________________ — **57*a*** ☐

b Weather condition: ______________________ — ***b*** ☐

Explanation: ______________________

____

____

[b]

For Raters Only

58 Gradient = ________

59 ________

____

____

60 ________

61 ________

____

____

62 ________

63 ________

____

____

For Raters Only	
58	
59	
60	
61	
62	
63	

Total Score for Part B–2

[c]

[OVER]

For Raters Only

Part C

64

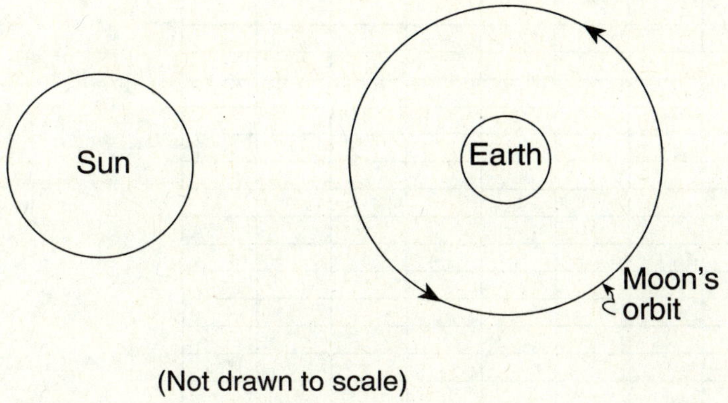

64 [ ]

65 ____________________ revolution(s)

65 [ ]

66 ____

____

____

____

66 [ ]

[d]

For Raters Only

67

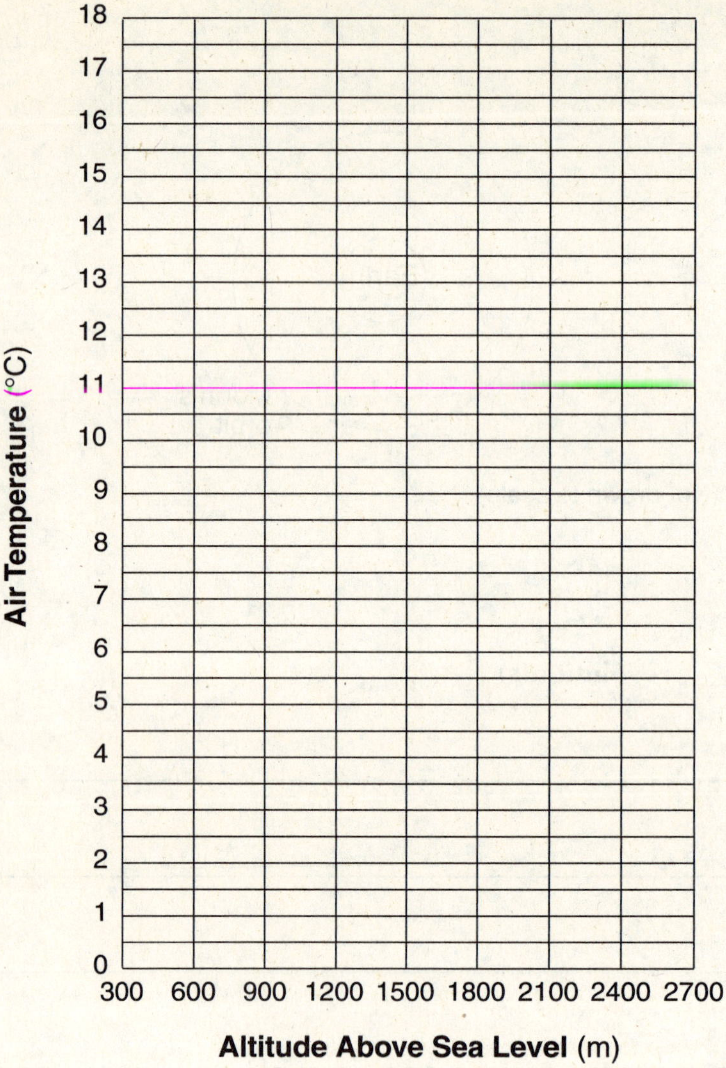

67 ☐

68 ____________________ 68 ☐

69 ____

____ 69 ☐

70 ______________ % relative humidity 70 ☐

[e] [OVER]

For Raters Only

71 ______________________________ 71 [ ]

72 ____________________ cm 72 [ ]

73 ____ 73 [ ]

____

74 ____ 74 [ ]

____

75 ______________________________ 75 [ ]

76 ______________________________ 76 [ ]

77 (1) ____ 77 [ ]

(2) ____

78 ______________________________ 78 [ ]

79 *a* ____ 79*a* [ ]

____

b ____ *b* [ ]

____

80 *a* __________________________ **Plate and** __________________________ **Plate** 80*a* [ ]

b ______________________________ *b* [ ]

[ ]

Total Score for Part C

[f]

Glossary

**Glossary items with asteric are Topic Vocabulary Terms.*

***abrasion** the physical action of scraping, rubbing, grinding, or wearing away of rock surfaces and sediments due to the movement of solid sediments in an erosional system such as a stream or wind

***absolute age** the actual age, or date, in years when a geological event occurred or a rock was formed

absolute humidity the amount (mass) of water vapor in a unit volume of air, such as in grams/cubic meter

absolute zero theoretically, the lowest possible temperature; no heat energy can be extracted at this temperature

absorbed taken into a material

aerosol small solid or liquid particles suspended in a gas; suspended solid or liquid water are the aerosols that compose fog and clouds

***air mass** a large body of air in the troposphere with similar characteristics of pressure, moisture, and temperature

***air pressure** see **atmospheric pressure**

***air pressure gradient** see **pressure gradient**

altitude (1) the vertical distance (elevation) between a point and sea level or Earth's surface (2) the angle of a celestial object above the horizon; usually expressed in degrees

***anemometer** a weather instrument used to measure wind speed

***angle of incidence** the angle at which the sun's rays hit Earth's surface; also called angle of insolation

anticyclone a high-pressure mass of air within the troposphere in which air moves out from the center; rotating clockwise in the Northern Hemisphere and counterclockwise in the Southern Hemisphere; also called a **high**

aphelion the point in a planet's orbit when it is farthest from the sun

apparent diameter the diameter a celestial object appears to have, depending on its distance from an observer; not the actual diameter

apparent motion a motion of an object that is not real but appears to be real, such as the daily motion of the stars in the sky caused by the real motion of Earth's rotation

apparent solar day a day of varying length determined by the time it takes for the sun to arrive at its highest point in the sky on two consecutive days at the same location, often measured by a sundial; see also **mean solar day**

arc a curved line that is part of a circle; the shape of the path of most celestial objects, such as the sun, in their daily motion paths through Earth's sky

arid climate a dry climate where the precipitation is less than the potential evapotranspiraton for a large part of the year, producing a deficit of moisture and a drought much of the time

***asteroid** a solid, rocky and/or metallic body that independently orbits the sun; large, irregularly shaped, except for the few larger spherical ones; mostly located between the orbits of Mars and Jupiter

***asthenosphere** the plastic, partly solid, partly liquid layer of Earth's mantle just below the lithosphere that allows plate movements

***atmosphere** the layers of gases surrounding Earth or other celestial object; Earth's atmosphere is divided into layers according to differences in chemical and physical properties

***atmospheric pressure** the weight of the overlying atmosphere pushing down on a given unit of area; affected by changes in temperature, water vapor, and altitude; also called **air pressure** or **barometric pressure**

***atmospheric transparency** how transparent the atmosphere is to insolation; how easily insolation can pass through the air

atom the smallest part of matter or an element which can't be separated by most chemical or physical processes

atomic structure see **crystal structure**

***axis (of rotation)** an imaginary line through Earth from the north to the south geographic poles, about which Earth rotates; all rotating celestial bodies have an axis

banding the layered arrangement of mineral crystals in some medium to coarse metamorphic rocks; a special type of foliation caused by layering and separation of minerals

***barometer** an instrument used to measure air pressure

***barometric pressure** see **atmospheric pressure**

***barrier island** a long narrow island, parallel to the shore, built of sand deposited by ocean waves, ocean currents, and wind; separated from the coast by a lagoon

beach the narrow portion of the shore or coastline between the low and high tide lines, usually covered with loose sediments

***bedrock** an area's mostly unweathered rock beneath vegetation, soil, other loose materials, and human-built structures; also called local rock

bench mark a permanent marker, usually metal, at a specific location on Earth's surface, that indicates an exact elevation or altitude at the time of installation

***Big Bang theory** states that the present universe started as a big explosion 10 to 17 billion years ago and has been expanding ever since

***bioclastic sedimentary rock** any rock made by living organisms or mostly composed of materials from life forms; also called **organic sedimentary rock**

blizzard a storm with winds of 35 miles an hour or greater with considerable falling or blowing snow causing low visibility

***breaking waves** the result of waves dragging on the ocean bottom causing the water in the waves to fall forward as the waves bunch together, rise up, and break against the shore

***calorie** a unit of heat energy defined as the amount of heat needed to raise the temperature of one gram of water one degree Celsius

calorimeter an instrument used in studies of heat

***capillarity** the process by which water is drawn into openings due to the attractive force between water molecules and the surrounding Earth materials

capillary migration the upward movement of water, against gravity, in part of the soil, loose materials, or bedrock due to capillarity; also called capillary action

capillary water the water held in soil, loose materials, and rocks in the zone of aeration as a result of the process of capillarity

carbon-14 a radioactive isotope of carbon with a short half-life (5,700 years); used to date recent (up to 70,000-year old) remains of organic material

***carbon-14 dating** the use of carbon-14 in dating rocks and organic remains of relatively recent origin

carbon dioxide a colorless, odorless gas, CO_2, an important greenhouse gas present in the atmosphere; given off when a fuel containing carbon is burned

***celestial object** any object in the universe outside of Earth's atmosphere, including moons, comets, planets, stars, and galaxies

cementation the process by which solid sediments or clasts are "glued" together by precipitated minerals, forming a sedimentary rock

change the alteration or modification of the characteristics of a part of the environment

change of state see **phase change**

channel (stream) see **stream channel shape**

***chemical sedimentary rock** a rock composed of interconnected crystals of just one mineral that form by evaporation and/or precipitation of dissolved minerals; include rock salt and rock gypsum; also called an evaporite

***chemical weathering** the processes by which chemicals, such as oxygen, acids, and water, break down rocks and other Earth materials, resulting in more stable new minerals (chemicals); example—rusting

***classification** the grouping together of similar observations and inferences to make the study of objects and events in the environment more meaningful or easier to understand

***clastic sedimentary rock** rock that is largely composed of solid sediments, such as the sand in sandstone

clasts the solid sediments, fragments, or grains in a clastic sedimentary rock such as the sand grains of sandstone

clay (1) a chemical group of minerals (2) very small solid sediments or clasts (less than 0.0004 cm in diameter) that often remain suspended in water for long periods of time

***cleavage** the tendency of a mineral to break along the zones of weakness and form smooth to semi-smooth parallel surfaces

***climate** the overall view of a region's weather conditions over long periods of time; includes not only averages, but extremes

cloud a visible mass of suspended liquid water droplets and/or ice crystals in the atmosphere

***cloud cover** the fraction or percent of the total sky at a location that is covered by clouds; usually expressed in tenths

***cold front** the boundary of an advancing cold air mass and a warmer air mass, where the underlying cold air pushes forward like a wedge; characterized by a steep slope, rapid changes in weather, thunderstorms, and sometimes hail and tornadoes

***comet** a low density object composed of materials (ices) that easily vaporize and some other solids; independently orbits the sun or other stars; usually has highly eccentric orbit; partly vaporizes forming a visible tail when near the sun

compaction the reduction in volume of sediments in the formation of sedimentary rocks, usually caused by the weight of overlying sediments and water causing a reduction of pore space and liquid water

compound a substance made of two or more elements chemically combined in a specific proportion, such as most minerals

***condensation** the change in state or phase from a gas to a liquid, such as when water vapor changes to liquid water droplets as clouds form

***conduction** the transfer of heat energy from atom to atom or molecule to molecule, in any state of matter, when vibrating atoms or molecules collide

***constellation** a group of stars that form a pattern and are used to help people locate celestial objects

contact metamorphic zone a type of interface or transition zone between rock types caused by the baking or altering of older bedrock by contact with molten rock (lava or magma); much of the older rock is changed into metamorphic rock

***contact metamorphism** process in which older rocks come in contact with the magma of an intrusion or lava of an extrusion and the heat and mineral fluids of the liquid rock alter the older rock by recrystallization

continental climate the climate of inland areas not moderated by a large body of water; characterized by hot summers and cold winters, and thus having a wide annual temperature range

***continental crust** the part of Earth's crust (upper lithosphere) that makes up the continents and larger islands; thicker and lower in density than the oceanic crust, and granitic rather than basaltic in composition

continental glacier very large glaciers (like those of Antarctica and Greenland), so thick that they cover all landscape features except the highest mountains; tend to create a smooth and low landscape by erosion and deposition; see **mountain glacier**

continental arctic air mass (cA) the coldest and driest air masses that only invade the contiguous United States from Arctic regions at the coldest times of the year

continental polar air mass (cP) cold and dry air masses that invade the contiguous United States from Canada

continental tropical air mass (cT) relatively rare hot dry air masses that form in the southwestern United States or northern Mexico, that may cause very hot and dry weather for the contiguous United States in the summer

contour interval the difference in elevation indicated by two consecutive contour lines (of different value) on contour or topographic maps

***contour line** an isoline on a topographic, or contour, map that connects points of equal elevation on the surface of any solid celestial body, such as Earth

contour map see **topographic map**

***convection** the transfer of heat energy by circulatory movements in a fluid (usually liquids or gases) that results from differences in density within the fluid

convection current a circulatory motion in a fluid due to convection; also called a convection cell

convergence (1) the coming together of air currents at Earth's surface and at the top of the troposphere (2) the direct collision of lithospheric plates in the **plate tectonic theory**

***convergent plate boundary** the boundary between two colliding plates; often associated with mountain building, ocean trenches, and volcanic island arcs

***coordinate system** a grid or a system of lines for determining location of a point on a surface, such as latitude and longitude of a point on Earth

core the center part of Earth below the mantle thought to be composed of iron and nickel; inner part is a solid and outer part a liquid

***Coriolis effect** the deflection of all moving particles of matter (such as winds and surface ocean currents) at Earth's surface to the right in the Northern Hemisphere and to the left in the Southern Hemisphere, which provides evidence for Earth's rotation

***correlation** in geology, the process of showing that rocks or geologic events from different places are the same or similar in age

crater see **impact crater**

***crust** the outermost portion of Earth's solid lithosphere; separated from the uppermost mantle and lower lithosphere by the thin Moho interface

crystal (1) the individual mineral grains of many rocks (2) a solid with a definite internal structure of atoms arranged in a characteristic, regular, repeating pattern

crystalline composed of intergrown or interconnected mineral crystals; having a specific arrangement of atoms

***crystallization** a type of solidification in which molten rock (magma or lava) cools to form igneous rocks composed of mineral crystals; also see solidification

***crystal shape** the outward 3-D shape or geometric shape of a mineral specimen that reflects the internal atomic structure; also called crystal form

***crystal structure** the pattern or arrangement of atoms that characterizes each mineral; also called atomic structure

***cyclic change** an orderly change in the environment in which an event repeats itself with reference to time and space; the opposite of random change

***cyclone** a low-pressure portion of the troposphere that has air moving towards its center; usually rotates counterclockwise in the Northern Hemisphere and clockwise in the Southern Hemisphere; includes hurricanes, tornadoes, and mid-latitude cyclones, also called a **low**

***cyclonic storm** a large type of low-pressure storm system formed in the mid–latitudes; also called mid-latitude cyclones

daily motion the apparent, usually east to west movement of celestial objects in the sky caused by Earth's west to east rotation; objects appear to move in circular or arc-shaped paths

day the amount of time it takes a planet or other celestial object to make one rotation

***deforestation** the cutting down of trees and other plants of a forest

***delta** the deposit of sediment at the mouth or end of a river or stream where it flows into a quiet body of water such as a lake or ocean

***density** the concentration of matter in an object; the ratio of the mass of an object to its volume—the mass per each unit of volume

***deposition** the process by which sediments are released, dropped, or settled from erosional systems; includes the precipitation of dissolved minerals in the formation of chemical sedimentary rocks; also called sedimentation

desert a region with an arid climate where the average yearly precipitation is much smaller than potential evapotranspiration

dew droplets of liquid water that form on Earth's solid surface by condensation; NOT a type of precipitation

***dew point** the temperature at which the air becomes saturated with water vapor and the relative humidity is 100%; at temperatures below the dew point, condensation or sublimation of water vapor occurs

direct rays rays of sunlight (insolation) that strike Earth at an angle of 90°; also called vertical rays or perpendicular insolation

discharge the amount of water that passes a certain spot in a stream in a specific amount of time, such as liters per minute

distorted structure the curving and folding of the foliations (mineral layers) and grains or crystals in metamorphic rocks caused by heat and pressure

divergence (1) the spreading out of air from rising or falling currents of air in the troposphere (2) the type of plate movement in which Earth's plates spread or rift apart from each other, according to the **plate tectonic theory**

***divergent plate boundary** the boundary between two plates that are spreading apart at a mid-ocean ridge or at a continental rift zone

***Doppler effect** the apparent wavelength shifting of electromagnetic energy (such as visible light) caused by the relative motion between the energy source and the observer; also see **redshift**

down cutting the erosional process whereby a stream "digs" deeper into Earth's surface

down-warped ocean basin an ocean basin, at the margin of a continent, formed by the bending down of the ocean lithosphere by plate movements

drizzle liquid precipitation with drops smaller than raindrops

drought a time of abnormal dry weather with a large enough deficit of water to cause crops and local water supplies to fail

***drumlin** a low, long, narrow, streamlined oval mound of unsorted sediment that is formed at the bottom of glaciers, usually continental ones

dry-bulb thermometer the thermometer of a sling psychrometer without a wick around its bulb, providing normal temperature values

dune see **sand dune**

duration of insolation the length of time sunlight is received at a location in a day, or the amount of time between sunrise and sunset

***dynamic equilibrium** a condition of the balancing out of opposing forces or actions, such as evaporation and condensation or erosion and deposition

***earthquake** a natural, rapid shaking of the lithosphere caused when rocks are displaced due to the release of energy stored in rocks; most caused by rapid movement along faults, but also associated with other events, such as volcanic eruptions

earthquake magnitude scale a numbering system that ranks earthquakes according to the total energy they emit

***Earth's interior** the region extending from the rocky part of Earth's surface to Earth's center

Earth materials water, gases, soils, minerals, rocks, mineral resources, and other materials and energy sources that are of value to people

***eccentricity** the degree of ovalness of an ellipse, or how far an ellipse is from being a circle; computed using the following formula

$$\text{eccentricity} = \frac{\text{distance between foci}}{\text{length of the major axis}}$$

***eclipse** the complete or partial blocking of light when one celestial body moves into the shadow of another celestial body; see **lunar eclipse** and **solar eclipse**

eclipse of the moon see **lunar eclipse**

***electromagnetic energy** energy that is radiated (given off) from all objects not at a temperature of absolute zero, in the form of transverse waves from vibrating matter into any part of the universe; examples—visible light, radio waves, infrared radiation, and ultraviolet radiation; often called light or radiant energy

***electromagnetic spectrum** a model, such as a chart, that shows the full range of types of electromagnetic energy, usually in order of wavelengths

element any of the different types of atoms such as oxygen, iron, and mercury; 90 naturally occurring elements exist on Earth

***elevation** the vertical distance or height above or below sea level

***ellipse** a closed curve around two fixed points, called foci, in which the sum of the distances between any point of the curve and the foci is a constant; example—the shape of all planetary orbits

***El Niño** a series of weather changes on Earth caused by a change from cold surface ocean water to warm surface ocean water in the eastern Pacific Ocean off western South America; occurs every two to ten years

emergency preparedness the steps or plans society, government agencies, organizations, and individuals can do to get ready to cope with disasters; planning an escape route in case of a hurricane or flooding, or storing food at home for a blizzard

***energy** the ability to do work

environmental equilibrium the balance that exists among the natural parts of the environment even though all parts of the environment are constantly changing

eon the largest division of geologic time, which is divided into eras

***epicenter** the place on Earth's surface lying directly above the focus, or the origin of an earthquake

epoch one of the small divisions of geologic time that are combined into periods

***equator** the parallel on Earth midway between the geographic North and South poles with a latitude of 0°

equinox a time when the sun is directly overhead at noon at the equator, and there are 12 hours of daylight and 12 hours of darkness over all of Earth

era the second longest division of geologic time; combined into eons and divided into periods

***erosion** the carrying away of sediment by wind, water, ice, and other agents; the process by which sediments are obtained and transported; the wearing away and lowering of Earth's land surface features

erosional-depositional system the various agents such as streams, glaciers, wind, ocean currents that pick up, transport, and deposit sediments at or near Earth's (or similar celestial object's) surface

error the amount of deviation or incorrectness in a measurement; see also **percent error**

***escarpment** a steep slope or cliff in layered rocks; often formed from certain rock layers that are resistant to weathering and erosion

evaporation the change in state from liquid to a gas, such as liquid water into water vapor; also called vaporization

evapotranspiration the combination of the processes of **evaporation** and **transpiration**

evaporite see **chemical sedimentary rock**

event the name used to describe the occurrence of a change in the environment

evolution see **organic evolution**

***extrusion** a mass of igneous rock formed by the cooling and solidification of molten rock (lava) on Earth's surface; examples—a lava flow and a volcanic mountain

***extrusive igneous rock** a rock formed by solidification of lava at or above Earth's surface; also called volcanic igneous rock

fault a crack in a mass of rock or soil along which there has been displacement, shifting, or movement of the rock or soil on each side of the crack

***faulted (rock)** rock layers that are offset or displaced along a type of crack called a fault

faulting rapid movements along faults

***field** any part of the universe that has some measurable value of a given quantity at every point, such as Earth's magnetic or Sun's gravitational fields

***finger lake** a body of water that forms in a long, narrow U-shaped glacial valley, often partly dammed at one end by a mound of glacial moraine sediment; example—the Finger Lakes of central New York

flood condition in which so much water flows into a stream that the water pours over the stream's confining banks onto areas not normally covered with water

***flood plain** a nearly level plain that borders a stream that is subject to flooding unless protected artificially; usually composed of layers of river-deposited sediment

***focus** (plural **foci**) (1) in an ellipse, either of two fixed points located so that the sum of their distances to any point on the ellipse is constant; example—the sun at one of the two foci of the orbit of each solar system planet (2) the place where an earthquake originates

fog a cloud that is on, or just above, Earth's surface

***folded (rock)** the bends in layered rock due to movement in the lithosphere; a type of deformed rock

***foliation** texture of metamorphic rocks caused by the layering of mineral crystals

***fossil** any physical evidence of former life, either direct or indirect

fossil fuel the organic fuels found within Earth's crust; includes oil (petroleum), natural gas, and coal

***Foucault pendulum** a freely swinging pendulum whose path appears to change in a predictable way, thus providing evidence for Earth's rotation

***fracture** the way in which a mineral breaks producing an uneven breaking surface; types of fracture—earthy, splintery, and curved (conchoidal)

freezing the change in state of a liquid to a solid by the removal of heat; also see **solidification**

freezing rain rain that freezes as it hits Earth's surface

***front** the interface, or boundary, between two air masses of different characteristics

fusion the change of state from a solid to a liquid; also called melting; also see **nuclear fusion**

***galaxy** the large groupings of millions or billions of stars and other forms of mass held together by gravitation; our galaxy is called the Milky Way

***geocentric model** an early concept of celestial objects and their motions in which all celestial objects revolved around Earth, which was stationary and was the center of the universe

geographic poles the North and South poles of Earth, with a latitude of 90°; located at opposite ends of Earth's axis of rotation

***geologic time scale** a chronological model of the geologic history of Earth using the divisions of **eons, eras, periods,** and **epochs;** see History of New York State in the *Earth Science Reference Tables*

***glacial groove** a long narrow channel or furrow on bedrock formed by the gouging and sanding actions of rocks and sediments frozen to the bottom of a glacier; show the direction of former glacial movement

***glacial parallel scratches** parallel cuts in bedrock formed by abrasion (gouging and sanding) of rocks and sediments frozen to the bottom of a glacier; show the direction of former glacial movement

***glacier** a large mass of naturally formed ice on land that moves downhill due to gravity; also see **continental glacier** and **mountain glacier**

***global warming** a recent trend towards a warming of Earth's surface and lower atmosphere (troposphere) possibly caused by human pollution

graded bedding a layering of sediment or sedimentary rock that shows a gradual change in particle size, with the largest particles on the bottom and the smallest ones on top

***gradient** the rate of change from place to place within a field; also called slope

$$\text{gradient} = \frac{\text{amount of change in field}}{\text{distance through which change occurs}}$$

***gravitation** the attractive force that exists between any two objects in the universe; proportional to the product of the masses of the objects and inversely proportional to the square of the distance between their centers; also called gravitational force

$$\text{force} \propto \frac{\text{mass}_1 \times \text{mass}_2}{(\text{distance between their centers})^2}$$

gravity the force that pulls objects toward the center of Earth

***greenhouse gases** gases such as carbon dioxide, water vapor, and methane, present in the atmosphere, which absorb long-wave infrared radiation

greenhouse effect the process by which the atmosphere transmits short-wave radiation from insolation and absorbs long-wave radiation emitted by Earth's surface; warms the atmosphere and reduces heat loss by radiation from Earth's surface

***groundwater** the subsurface water found beneath the water table in the zone of saturation; sometimes used to mean all subsurface water.

hail a round solid form of precipitation composed of concentric layers of ice and snow; only forms from cumulonimbus (thunderhead) clouds

***half-life** the time required for half of the atoms in a given mass of a radioactive isotope to decay, or change, to a different isotope

***hardness** the resistance a mineral offers to being scratched or dented; usually measured by comparison to the Mohs hardness scale of minerals

hazard an object, process, or situation that holds the possibility of injury or death to humans or damage to property

haze a condition of the atmosphere in which the aerosol content is so high that distant images are blurred, and a cloudless sky does not appear blue

***heat budget** the result of the balance between the total amount of energy an object receives and the total energy it emits, or loses; measured as the average temperature of an object

***heat energy** energy that is transferred from one body to another as a result of a difference in temperature or thermal energy of two bodies; also called heat; also see **thermal energy**

***heliocentric model** the modern concept of celestial objects and their motions, in which a rotating Earth and other planets revolve around the sun

high see **anticyclone**

high tide the bulge of ocean water directly under the moon's position and on the opposite side of Earth

***hot spot** major regions of volcanic activity in the interior parts of plates away from plate boundaries; may be the cause of chains of volcanic activity within moving plates, such as the Hawaiian Islands

humid climate a moist or wet climate where the precipitation is greater than potential evapotranspiration on a yearly average

***humidity** the amount of water vapor (gaseous water) in the atmosphere; also see **relative humidity** and **absolute humidity**

hurricane a large, strong cyclonic storm that forms over tropical ocean waters with sustained winds of 74 miles per hour or greater

***hydrologic cycle** see **water cycle**

***hydrosphere** the liquid water (mostly the oceans) that rests on much of the Earth's solid or rocky surface; included by some scientists—the subsurface water, water in the atmosphere, sea ice, glaciers, and water in life forms

***ice ages** times of widespread glaciation outside of polar areas (see the Pleistocene epoch in Geologic History Of New York State At A Glance in the *Earth Science Reference Tables*)

***igneous rock** a rock formed when natural, molten rock-forming material (magma or lava) cools and turns into a solid—above, below, or on Earth's surface

***impact crater** an oval-shaped depression with a raised rim formed by a meteorite, asteroid, or comet colliding with Earth's, or any other solid celestial object's, solid surface

***impact event** the colliding of comets, asteroids, and meteoroids or any other type of celestial body

***inclusion** a body of older rock within an igneous rock formed when pieces of rock surrounding liquid rock fall into the magma or lava, but don't melt before the liquid rock solidifies

***index fossil** a fossil used in correlation and relative dating of rocks; must have lived for a short time and have been distributed over a large geographic area

***inertia** the concept that an object at rest will tend to remain at rest and that an object in motion will maintain the direction and speed of that motion unless an opposing force affects it

***inference** an interpretation of an observation; a mental process that proposes causes, conclusions, or explanations for what has been observed

***infiltrate** water entering or sinking into, under the influence of gravity, the upper parts of Earth's lithosphere where the water becomes subsurface water; also called seep

infrared a type of long-wave electromagnetic radiation

***inner core** the innermost zone of Earth's core, which is thought to be composed of iron and nickel in a solid state

***inorganic** not organic, thus not part of a life form or made by living or former life forms

***insolation (INcoming SOLar radiATION)** the part of the sun's radiation that is received by Earth; also called solar energy

***instrument** a device invented by people to extend the senses beyond their normal limits, thus enabling them to make observations that would otherwise be impossible or highly inaccurate

intensity of insolation the relative strength of the sun's radiation intersecting a specific area of Earth in a specific amount of time, such as calories per square meter per minute

***interface** the boundary between regions with different properties; the usual place for energy exchange

interpretation see **inference**

***intrusion** a mass of igneous rock formed when molten rock (magma) squeezes or melts into preexisting rocks and crystallizes; examples—sills and dykes

***intrusive igneous rock** a rock that forms by the cooling and solidification of magma beneath Earth's solid surface; examples—granite or gabbro; also called plutonic igneous rock

***island arc** a curved series of volcanoes and volcanic islands that stretch hundreds of miles; form at subduction zones where plates converge and magma rises to form volcanoes; also called a **volcanic island arc**

***isobar** an isoline used on weather and climate maps that connects points of equal air pressure

***isoline** a line used on a model of a field, such as a map, which connects points of equal value of a field quantity; examples—isotherms, isobars, and contour lines

isotherm an isoline used on weather and climatic maps to connect points of equal air temperature

***isotope** one of the varieties of an element, which all have the same atomic number and chemical properties, but differ in their atomic masses and physical properties; examples—carbon isotopes carbon-12 and carbon-14

***jet stream** a concentrated curving band of high speed, easterly moving winds usually at the top of Earth's troposphere

joint a crack in rocks along which there has been no relative movement or displacement, such as there is with a fault

***Jovian planets** planets that are far from the sun, largely gaseous, and have relatively large diameters, many moons, rings, and low densities; Jupiter, Saturn, Uranus, and Neptune

***kettle lake** a lake formed when a large block of ice buried in glacial sediments melts, leaving an oval depression which becomes filled with water; very common in New York State

kinetic energy the energy of movement of any object; the greater the speed and mass of an object, the greater the kinetic energy

knots nautical miles per hour; unit of wind speed

landform individual features of Earth's (or any other solid celestial object's) surface from mountain ranges to a mud crack in a puddle

***landscape** the features of Earth's surface at the interfaces between the atmosphere or the hydrosphere and the top of the lithosphere; also on all other solid celestial bodies

***landscape region** a portion of Earth's surface with landscape (topographic) characteristics that distinguish it from other areas; distinguishing characteristics—rock structure, elevation, degree of slope, and stream drainage pattern

***latitude** the angular distance north or south of the equator; usually expressed in units of angular measurement such as degrees; minimum latitude 0° at the equator and maximum 90° N or 90° S at the geographic poles

latitudinal climatic pattern east-west belts, or zones, of climate types on Earth caused by latitudinal changes in climate factors, such as temperature, precipitation amount, wind, and ocean currents

lava (1) liquid rock material at or above Earth's surface (2) the extrusive or volcanic igneous rock that forms from solidification of liquid lava

leveling forces forces that operate at or near Earth's surface and break down rocks, transport material from higher to lower elevations, and tend to level off and lower the land; examples—weathering, erosion, deposition, and subsidence

light see **electromagnetic energy**

***lithosphere** the whole crust and the uppermost portion of the mantle (layer of rock that forms the solid outer shell at the top of Earth's interior)

***lithospheric plate** see **plate**

***local time** time based on the rotation of Earth as reflected in motions of the sun in the sky

***longitude** an angular distance east or west of the prime meridian; usually expressed in degrees; minimum longitude 0° at the prime meridian, which runs through Greenwich, England, and maximum longitude 180° E or 180° W

longshore currents ocean or lake currents that move parallel and close to shore due to the angled advance of waves from one direction; a major agent of erosion and deposition

long-wave electromagnetic energy electromagnetic energy with a wavelength longer than visible light, especially infrared energy

low see **cyclone**

low tide the level of low ocean water, occurring at right angles to the positions of high tide

***luminosity (of a star)** measures how bright a star would be in relation to the sun if all stars were the same distance from the observer

lunar eclipse the darkening of the moon caused by Earth's shadow

***luster** the way a mineral looks or shines in reflected light

***magma** liquid rock material beneath Earth's solid surface

magnitude see **earthquake magnitude scale**

***mantle** the mostly solid part of Earth between the crust and the outer core

marine ocean or sea-like

marine climate a coastal climate moderated by the effects of a large body of water (ocean, sea, or lake); having warmer winters and colder summers than areas of similar latitude not near a large body of water; have a smaller annual temperature range than inland areas

maritime polar air mass (mP) cool and humid air masses that invade the contiguous United States from the oceans to the northeast and northwest

maritime tropical air mass (mT) a very common warm and humid air mass that invades the contiguous United States from the oceans to the south, east, and west

***mass** the amount of matter in an object; unlike weight, not affected by location

***mass movement** any variety of erosion and deposition done directly by gravity; examples—soil creep, earthflows, mudslides, avalanches, slumping, and landslides; also called gravity erosion

mean solar day the 24-hour day established for convenience in time-keeping; derived by averaging the lengths of the apparent solar days in a year

mean solar time a type of local time divided into exactly 24 hours for the convenience of timekeeping

***meander** a curve or bend in a stream or river

***measurement** a means of expressing an observation with greater accuracy or precision; provides a numerical value for an aspect of the object or event being observed by comparision with a standard; example—measuring the length of this page with a ruler

***mechanical energy** all the energy of an object or system not related to the individual motions of atoms and molecules; the total of the potential and kinetic energy of an object or system

melting the change in state or phase from a solid to a liquid

***meridian of longitude** any north–south semicircle of constant longitude on maps and globes connecting the north and south geographic poles: also called a meridian

***metamorphic rock** a rock that forms from changes in previously existing rocks (igneous, sedimentary, or other metamorphic rocks) due to heat, pressure, and/or chemical fluids (not weathering or melting)

***metamorphism** the process by which heat, pressure, and/or exposure to chemical fluids can change previously existing rock into metamorphic rock

***meteor** path of light from a meteoroid burning as it passes through Earth's atmosphere; also called a shooting star

meteorite a meteoroid that has landed on Earth's surface

meteoroid solid particles smaller than asteroids and comets that orbit the sun

methane the organic compound CH_4, often called swamp gas, that is an important greenhouse gas; major part of the fossil fuel called natural gas

mid-latitude cyclone see **cyclonic storm**

***mid-ocean ridge** a mountain range at the bottom of the ocean, composed mostly of volcanoes and lava flows; forms at zones of diverging plates

***Milky Way Galaxy** the spiral-shaped galaxy that Earth and our solar system are part of

***mineral** a naturally occurring, crystalline solid having a definite chemical composition and physical and chemical properties that vary within specified limits; has a unique crystal (atomic) structure

***mineral crystal** an individual grain of a mineral

***mineral resources** Earth materials that people need, including minerals, rocks, and fossils fuels

misconception an idea, notion, or concept believed to be correct, but which is NOT true or correct

***model** any way of representing the properties of an object, event, or system; includes graphs, drawings, charts, mental pictures, numerical data, or scaled physical objects

***Moho** the interface, or boundary zone, between Earth's crust and mantle; short for Mohorovicic discontinuity

Mohs hardness scale a scale used to measure the relative hardness of minerals; ranges from 1 to 10 with talc assigned a hardness of 1 and diamond a hardness of 10

moisture a somewhat vague term used to mean the liquid or gaseous water in the atmosphere or in the ground (soil, loose materials, or bedrock)

moisture capacity a measure of the total amount of water vapor the air can hold at a particular temperature; the maximum absolute humidity of a parcel of air at a particular temperature

***monsoons** cyclic and extreme weather changes caused by the shifting wind and pressure belts; especially strong in southeast Asia where summer brings wet weather from the ocean and winter brings dry weather from the continental interior

***moon** (1) the one natural satellite of Earth (2) a body that orbits a planet or an asteroid as those objects orbit the sun

***moraine** a mound, ridge, or sheet of unsorted, unlayered sediment deposited directly from an edge or bottom of a glacier; types—lateral, medial, and ground

***mountain** usually an area of high elevation, compared to the surrounding area or sea level, that usually has regions of steep gradient and many changes in slope; internally, mountains are characterized by distorted rock structures

mountain glacier a long narrow glacier confined to valleys in mountains; results in sharp angular landscape features; see **continental glacier**

***natural hazard** a non-human-related object, process, or situation that has the possibility of causing loss of life, personal injury, or loss of property; includes volcanic eruptions, earthquakes, landslides, floods, storms, and asteroid impacts; also called a natural disaster

***natural resources** the materials and energy sources found in the environment that humans use in their daily lives

North Pole the location on Earth's surface at the north end of the axis of rotation with a 90° N latitude; also called the geographic north pole

North Star see **Polaris**

***nuclear decay** see **radioactive decay**

***nuclear fusion** the combining of the nuclei of smaller elements to form the nuclei of larger elements with some mass being converted into energy; example—the sun produces energy in this way

***observation** the perception of some aspect of the environment by one or more human senses, with or without the aid of instruments

***occluded front** the boundary of opposing wedges of cold air masses formed when a cold front overtakes a warm front, lifting the warm air mass off Earth's surface, forming mid-latitude cyclones (lows)

ocean the continuous salty water body that covers 70% of Earth's surface or any one of its major parts such as the Atlantic Ocean; the major part of the **hydrosphere**

ocean currents see **surface ocean currents**

ocean floor spreading see **sea floor spreading**

***oceanic crust** the portion of Earth's crust that is usually below the oceans and not associated with continental areas; thinner and higher in density than continental crust and basaltic rather than granitic in composition

***ocean trench** the long, steep, and narrow depression produced by the bending down of subducting plates, which warps the crust

orbit the path of an object revolving around another object, such as the path of Earth around the sun

orbital speed the speed of an orbiting body along its orbit at any given time similar to orbital velocity

ore a rock or mineral deposit that can supply a mineral resource that is worthwhile to mine or drill

***organic** an Earth material that is composed of and/or was formed by life forms

***organic evolution (theory of)** the theory stating that life forms change through time; new species of organisms arise by gradual transitional changes from existing species

organic sedimentary rocks see **bioclastic sedimentary rocks**

***original horizontality** a concept that states that sedimentary rocks and some extrusive igneous rocks, such as lava flows, form in horizontal layers parallel to Earth's surface

orogeny (1) the process of mountain building (2) a portion of geologic time when there is much uplift and mountain building such as the Grenville Orogeny in Geologic History Of New York State in the *Earth Science Reference Tables*; caused by plate convergence or collision

outcrop location at Earth's surface where bedrock is exposed without a cover of soil or other materials

***outer core** the zone of Earth between the mantle and the inner core; thought to be a liquid because earthquake S-waves do not go through it; believed to be composed of iron and nickel

***outgassing** the seeping out of gases from Earth's interior through cracks and volcanic eruptions to Earth's surface

***outwash plain** the landform feature composed of sorted and layered sediments deposited in front of a glacier by running water from the melting ice; example—southern Long Island

oxidation a process where oxygen is added to other elements in chemical weathering; such as in the rusting of the black mineral magnetite to the red mineral hematite

***ozone** a three-atom molecule of oxygen (compared to the more common two-atom molecule of oxygen) that is a pollutant in the troposphere, but is important in the stratosphere where it absorbs much of the ultraviolet insolation from the sun

parallels of latitude east-west circles on maps or globes that are equidistant from the equator at all points; and thus maintain the same value of latitude; also called parallels

passive margin basin a broad and long depression of the crust at the border of a continent that is not the site of a plate boundary; usually the site of much deposition of sediments

***pauses (of atmosphere)** the interfaces, or boundaries, of the layers of Earth's atmosphere

***percent deviation,** or **percent error** the numerical amount, expressed as a percent, by which a measurement differs from a given standard or accepted value

$$\text{deviation (\%)} = \frac{\text{difference from accepted value}}{\text{accepted value}} \times 100\%$$

perihelion the point in a planet's orbit when it is closest to the sun; occurs for Earth about January 3, when Earth is about 147,000,000 kilometers from the sun

period (1) the amount of time it takes a planet to make one orbit, or revolution, around the sun; called a year for that planet (2) in geology, a part of the geologic time scale smaller than an era but larger than an epoch

***permeability** the degree to which a porous material (such as rock or soil) will allow fluids, such as water and oil, to pass through it; commonly measured in centimeters per hour

permeability rate the speed at which a certain amount of fluid, such as water or oil, can pass through a porous material; the speed at which water moves from above to below Earth's surface, becoming subsurface water—a special permeability rate called infiltration rate; commonly measured in liters per minute

perpendicular insolation see **direct rays**

***phase** (1) one of the three main forms of matter—liquid, solid, or gas; also called state of matter (2) the varying portion of the lighted part of the moon, Venus, or Mercury visible from Earth

phase change the change of a substance from one phase or state to another, such as liquid water to gaseous water

***physical weathering** the mechanical or physical breakdown of rock and other Earth materials at or near Earth's surface into smaller pieces (sediments) without a change in the mineral or chemical composition; example—frost action splitting rock

***plain** a landscape of low elevation and generally level surface with little change in slope; usually characterized by horizontal rock structure

planet largest of non-star celestial objects that revolve around a star; Earth and eight other planets revolve around our sun

***planetary wind belts** east-west zones on Earth where the wind blows from one direction much of the time; also see **prevailing winds;** example—the prevailing southwest winds that blow over the contiguous United States

***plate** section of the lithosphere that moves around Earth's solid surface; also called a **lithospheric plate** or **tectonic plate**

***plateau** a landscape of relatively high elevation composed of undistorted horizontal rock structure and often a more level slope or gradient than that of most mountains

***plate tectonic theory** states that Earth's lithosphere is divided into sections called plates that can move up and down or sideways on the plastic part of the upper mantle—diverging, converging, and sliding past each other, which results in many of Earth's major physical features and events, including continent and mountain formation, volcanoes, and earthquakes

plutonic igneous rocks see **intrusive igneous rocks**

***polar front** an ever-changing boundary between the colder air masses toward the poles and the warmer air masses toward the middle latitudes; storm systems (cyclones) of the mid-latitudes are formed at the polar front

Polaris the star that is presently almost directly over the geographic North Pole of Earth; also called the North Star

pollutants substances or forms of energy that pollute the environment; they include solids, liquids, gases, life forms, and forms of energy such as heat, visible light, and sound

***pollution** the occurrence in the environment of a substance or form of energy in concentrations large enough to have an adverse effect on people, their property, or plant or animal life

***porosity** percentage of open space (pores and cracks) in a material compared to its total volume

potential energy the energy possessed by an object as a result of its position or location, chemical conditions, or phase (state) of matter

potential evapotranspiration the amount of water that would be lost from a portion of Earth's surface through evaporation and transpiration over a given time IF the water were available

***precipitation** (1) falling liquid or solid water from clouds toward Earth's surface; (2) a type of sediment deposition in which dissolved minerals come out of solution to form solids, as in the formation of chemical sedimentary rocks such as rock salt

precipitation gauge any instrument used to measure the amount of atmospheric precipitation, such as a rain gauge

***prediction** a type of inference about the conditions and behavior of the environment in the future

present is the key to the past concept that the physical, chemical, biological, and geological events today are similar to those in the past; thus we can interpret the past by understanding the present

present weather the conditions or state of the atmosphere for a short period of time at a location determined by comparison with a standard list produced by the United States Weather Service; partial abbreviated list is found on the sample station model in Weather Map Information in the *Earth Science Reference Tables*

pressure gradient the amount of difference in air pressure over a specific distance; the greater the pressure gradient, the greater the speed of the wind; also called air pressure gradient

***prevailing winds** winds that blow from a certain compass direction for a large part of a year; usually associated with the **planetary wind belts**

primary waves see **P-waves**

***prime meridian** the meridian of 0° longitude; also called the Meridian of Greenwich

***principle of superposition** see **superposition**

***probability** the chance of some environmental event, such as rain or an earthquake, taking place

***profile** a model, usually drawn from a topographic map, that uses upward and downward changes of a line to show changes in elevation and slope; a side view of an area's topography, or landscape

***psychrometer** an instrument that is used to indirectly measure the amount of water vapor in air, such as a sling psychrometer; with the use of data tables, temperature readings from a psychrometer can be used to determine relative humidity and dew point

***P-waves** the waves in earthquakes that travel through Earth's interior and cause particles through which they travel to vibrate in the direction the waves are moving; the fastest-moving of all earthquake waves; also called primary waves

***radar (RAdio Detection And Ranging instrument)** an instrument that uses radio/microwave electromagnetic radiation to observe many weather features such as precipitation, tornadoes, and hurricanes; wind speeds calculated using Doppler radar have greatly aided tornado predictions

***radiation** (1) the emission or giving off of energy in the form of electromagnetic energy; (2) the method by which electromagnetic energy moves from place to place by way of transverse waves

***radioactive dating** the use of radioactive isotopes to determine the absolute age of rocks and geologic events

***radioactive decay** the natural spontaneous breakdown of the nucleus of unstable atoms into more stable atoms of the same or different elements; releasing energy and/or small subatomic particles, also called **nuclear decay**

radioactive material any rock or mineral that contains radioactive atoms

rain liquid precipitation larger than drizzle or 0.5mm

rain gauge a type of precipitation gauge that measures liquid precipitation

rain shower brief, rapidly forming, and rapidly ending liquid precipitation; often heavy rain associated with thunderstorm-type clouds

***rate of change** how much a measurable aspect of the environment, called a field, is altered over a given time—years, hours, or seconds

$$\text{rate of change} = \frac{\text{change in field value}}{\text{change in time}}$$

recrystallization a process in the formation of metamorphic rocks by which mineral crystals grow in size at the expense of older crystals or sediments without true melting

***red shift** the type of Doppler effect caused by an increase in distance between the observer and the source of the electromagnetic radiation; a displacement towards longer wavelengths of electromagnetic energy; used as evidence for the Big Bang theory of the origin of the universe

reflected turned back electromagnetic energy or other type of waves; waves bounced off a material

reflection a change in direction of waves when the waves strike the surface of a material, in which the waves leave the surface at the same angle at which they arrived

refracted waves bent during passage through materials of varying density so that the direction is changed; example—light waves refracted by water

refraction a change in direction and velocity of waves when they pass from one type of material into another with a different density

***regional metamorphism** process of formation of metamorphic rock over large areas due to an increase in temperature and pressure, usually as the result of mountain building associated with plate collision or convergence

relative dating the determination of the age of a rock or event in relation to the age of other rocks or events

***relative humidity** the ratio of the amount of water vapor in the air to the maximum amount it can hold; often expressed as a percent; can be calculated by obtaining temperature readings from a psychrometer and using tables such as Relative Humidity (%) in the *Earth Science Reference Tables*

resources anything that may be useful to humans in the future; see **natural resources**

reversal of Earth's magnetic polarity the fact that Earth's magnetic field and magnetic poles switch polarity (north for south and south for north) in intervals of thousands of years in no known cycle

***revolution** a planet's movement around the sun in a path called an orbit; the movement of one body around another body in a path called an orbit

***ridge** a long, narrow, high, steep-sloped part of the landscape that is usually part of a mountain range

rift a long narrow valley in a continental area bounded by faults caused by rifting or splitting of the lithosphere; usually the site of plate divergence such as the East African Rift or Grand Canyon Rift; also used to mean the valley in the middle of mid-ocean ridges

rock any naturally formed solid that is part of Earth or any other celestial body; usually composed of one or more minerals, except for a few rocks, such as coal, obsidian, and coral, which don't have any minerals

***rock cycle** a model of the interrelationships of the different rock types (sedimentary, igneous, and metamorphic), the materials they form from and the processes that produce the rocks

rock-forming mineral any one of a small number of minerals (20-30) that are commonly found in rocks; mostly silicates

rock structure the features of rock that can be observed in an outcrop; including folds, faults, joints, tilting, and thickness of strata

***rotation** the spinning of an object, such as a planet, moon, or star, on an imaginary axis—like a top

running water any erosional-depositional agent caused by the flowing of liquid water on land including rain splashing, sheet wash, and stream or river flow; called a stream or river when confined to a channel

***runoff** all the natural liquid water flowing at Earth's solid surface, including stream flow

***sandbar** a pile or low ridge of sand, often just above or just below water level, deposited by waves and currents; found near ocean shores or within streams and rivers

***sandblasting** an erosional process by which blowing sand impacts and abrades rock surfaces resulting in shaped landscape features and new sediments

***sand dune** a pile, hill, or mound of sand deposited by wind on a land surface

satellite any celestial object, either natural or humanmade, that revolves around another celestial object

saturation the condition of being filled to capacity; example—at saturation, the relative humidity of the atmosphere is 100%

scattered waves that are refracted and/or reflected in various directions

sea-floor spreading the principle that the oceanic lithosphere spreads outward (plate divergence or rifting) at mid-ocean ridges; also called ocean-floor spreading

seasons the divisions of the year with characteristic weather conditions

secondary waves see **S-waves**

***sediment** particles or materials formed by the weathering and erosion of rocks or organic materials; material transported by erosional systems

***sedimentary rock** rock that forms directly from sediments deposited near Earth's surface by the processes of cementation, precipitation of minerals, loss of water, and compaction

***seep** see **infiltrate**

***seismic waves** the energy waves given off by an earthquake; also called earthquake waves

seismogram the chart or record of earthquake movement (wiggles) recorded by a seismograph

seismograph an instrument used to detect and record seismic or earthquake waves

senses the five human abilities or faculties—sight, touch, hearing, taste, and smell—by which one observes

shore, or **shoreline** the interface between land and a water body such as a lake or ocean; the narrow strip of land at the coast between high and low tide

shower see **rain shower** or **snow shower**

sink in an energy system, a region that has a lower energy concentration than its surroundings

sleet precipitation composed of transparent ice pellets, less than 5 mm in size; forms by freezing of falling rain

slope see **gradient**

smog a haze or fog, usually brownish, which is highly polluted

snow precipitation that is composed of one or more solid six-sided ice crystals that form in the clouds directly by sublimation of water vapor to ice

snow shower a brief, heavy snowfall

soil the part of the ground that will support rooted plants; the product of weathering of rock and the actions of living organisms

solar eclipse an eclipse caused by Earth's moon blocking out part or all of the sun from view, and the moon's shadow passing over part of the day-time side of Earth

solar energy see **insolation**

solar day the time it takes for Earth to rotate from solar noon to solar noon on two successive days at any fixed location on Earth

solar noon the time at which the sun reaches its highest point in the sky at any fixed location

***solar system** (1) the sun and all objects that orbit the sun under its gravitational influence (in this book) (2) any star or a small group of orbiting stars and all the other objects that revolve around the star (s)

solar time the local time based on the actual motions of the sun in the sky; sundial time

***solidification** the processes by which a liquid changes to a solid, such as when molten rock (lava and magma) changes into igneous rock; also see **crystallization**

solstices the two times of the year when the vertical rays of the sun fall the farthest from the equator; see **summer solstice** and **winter solstice**

***sorted** a condition in which all the particles in a material are the same or similar in size

***sorted sediments** a deposit of sediments that are similar in size (or shape, or density); the greater the similarity of particles the more sorted the sediments

source in an energy system, a region that has a higher energy concentration than its surroundings; energy flows from the source

source region the area of Earth's surface over which an air mass forms and acquires its characteristics

space the portion of the universe above Earth's atmosphere; also called outer space

***species** a group of organisms which are similar enough to be able to interbreed and produce fertile young

***specific heat** the quantity of heat, in calories, needed to raise the temperature of one gram of a substance one degree Celsius; the degree of difficulty a material offers to heating up or cooling off

speed how much distance is covered in a specific amount of time, such as kilometers per second

standard time a worldwide system of 24 north–south zones, that are 15 degrees of longitude wide, which keep the same local time

stages of landscape development stages (including youth, maturity, and old age) in the evolution of a landscape feature or region; stages characterized by certain features, including types of dominant forces, the amount of slope, elevation, and the amount of change in slope

***star** usually a large ball of gas held together by gravity that produces tremendous amounts of energy and shines; also called a sun

state of matter see **phase**

***stationary front** a weather condition in which the boundary between two air masses remains in the same position

***station model** cities and other weather station sites on weather maps represented by circles, with symbols in and around each circle that indicate the many weather variables

storm a violent disturbance in the atmosphere that usually creates dangerous, destructive, and/or unpleasant conditions at Earth's surface

strata the layers or beds of sedimentary rock and extrusive igneous rock

***streak** the color of the powder of a mineral which is usually more consistent than normal mineral color

***stream** water flowing through a channel on land; examples—a large river or a narrow creek

***stream abrasion** the rounding, smoothing, and size reduction of sediments resulting from the rolling, sliding, or bouncing of solid sediments along a stream bottom; also the erosion (lowering) of the stream bed by impact of solid sediments

stream bed the bottom or floor of a stream

***stream channel shape** the shape of the body of rock or loose materials that confine the stream

***stream discharge** see **discharge**

***stream drainage pattern** the shape of the stream courses in an area as viewed from the sky

structure see **rock structure**

***subduction** the plate tectonic process in which one of the plates at a convergent boundary sinks under the other plate and eventually melts into the asthenosphere

subduction zone the portion of a plate that sinks down into Earth's interior (asthenosphere) where converging plates cause subduction to take place

sublimation the change of state from a solid directly to a gas or from a gas to a solid with no intermediate liquid state; example—the forming of frost by sublimation of water vapor into ice

subsidence the sinking or depression of part of Earth's surface

subsurface water all liquid water found in soil, sediment, and bedrock beneath Earth's surface; see also **groundwater**

summer solstice on about June 21, the vertical rays of the sun fall on $23\frac{1}{2}°$ N latitude and the duration and angle of incidence are greatest for most of the Northern Hemisphere and least in the Southern Hemisphere; see **winter solstice**

sun (1) star at the center of our solar system (2) another name for any star

sundial an ancient time-keeping device that uses the position of the sun to determine apparent solar, local, or sundial time

***sunspot** a darker region of the sun's visible surface; increased numbers of sunspots are associated with increased electromagnetic energy emitted from the sun

sun's vertical rays see **direct waves**

superposition a principle applied in the relative dating of layered sedimentary and some extrusive igneous lava flow rocks stating that the youngest rock layer is found on top and that rock age increases with depth; exceptions—in deformed rocks, and where there are igneous intrusions; also called **principle of superposition**

surface ocean currents a continuous horizontal flow of water at or near the ocean's surface driven by the prevailing winds; also called **ocean currents**

synoptic method a method of weather prediction that uses a synopsis or summary of the total weather picture (often using a weather map) to predict future weather

***S-waves** earthquake waves that move through Earth's interior and cause the particles through which they travel to vibrate at right angles to the direction of the wave motion; will only travel through solids, not through liquids or gases; also called **secondary waves**

technology the use of scientific information to serve human needs; the means by which a society provides objects required for human subsistence, comfort, and pleasure

tectonic forces the forces that create the large-scale structural features of Earth's lithosphere, such as mountains and continents, caused by lithospheric movements; energy comes from Earth's interior

***tectonic plate** see **plate**

***temperature** a measure of the average kinetic energy of the particles in a body of matter; a measure of how hot or cold a substance is

***terrestrial planets** the planets (Mercury, Venus, Earth, and Mars) that are Earth-like in being relatively close to the sun, mostly solid, and having relatively small diameters and high densities

***texture** (1) the roughness or smoothness of a surface; (2) the size, shape and arrangement of the mineral crystals, sediments, clasts, and glass that make up a rock

thermal energy the energy of motion of atoms and molecules; see **heat energy**

thermometer an instrument used to measure temperature that usually consists of a confined fluid (alcohol or mercury) that expands and contracts with temperature changes

thunderstorm a local, short-lived storm from cumulonimbus (thunderhead) clouds that always has lightning and thunder; often associated with heavy rain, high winds, hail, and less commonly, tornadoes and flooding

***tides** the cyclic rise and fall of ocean water on Earth caused by the gravitational attraction of the moon and to a lesser degree the sun

tilted strata a type of deformed rock in which the strata, or layers, have been forced out of a horizontal position by tectonic movement

time the sense of things happening one after another or the duration of an event

time zone One of the 24 north-south trending zones, that are 15 degrees of longitude wide, which keep the same standard time as the central meridian of the zone. An example is the Eastern Standard Time Zone of the United States which generally extends between 82° and 67° west longitude with the central meridian at 75° west longitude

***topographic map** a model of the elevation field of a solid celestial object, such as Earth, using contour lines and other symbols; also called a contour map

topography see **landscape**

tornado a narrow cyclonic storm with very high wind speeds characterized by a very low air pressure and a twisting funnel that touches the ground; formation often associated with cumulonimbus (thunderhead) clouds and can result in much damage and loss of life; commonly called a twister

track the path of movement of an air mass, front, or storm; weather tracks are often predictable, which helps in forecasting

trade-off an outcome that results when choices are made; example—the choice to build a dam for a local water supply as compared to using the land that will be flooded by building the dam for farming and homes

***transform plate boundary** a boundary at which plates slide by each other, such as the San Andreas fault that separates the North American and Pacific Plates; also called a lateral fault plate boundary

***transpiration** the process by which plants release water vapor into the atmosphere as part of their life functions

transported sediment weathered or eroded rock and organic materials that have been moved by an erosional system from their place of origin

transported soil soil that has been moved from its place of original formation; examples—glacial soils common in the northern United States and the river-deposited soils found in flood plains and deltas

transverse wave a wave that vibrates at right angles to its direction of motion; examples—electromagnetic energy and earthquake S–waves

***tributary** small stream or river that flow into a larger stream or river

***troposphere** the part of the atmosphere immediately above Earth's surface; where most weather changes occur

***tsunami** very fast large wavelength ocean wave or waves produced by disturbances of the ocean floor caused by earthquakes, volcanic eruptions, or landslides; Tsunami can become very high and cause much wave damage near ocean shores

ultraviolet radiation a form of electromagnetic radiation of shorter wavelength than visible light; mostly absorbed by gases of the atmosphere, especially ozone in the stratosphere, before reaching Earth's surface

***unconformity** a buried eroded surface causing a break, or gap, in the rock record

***universe** the totality of all things that exist—all matter, time, energy, and space

***unsorted** a condition in which the particles in a material are of mixed sizes (or shape, or density)

***unsorted sediments** a deposit of sediments that are mixed in size (or shape, or density); the greater the difference in the size of the particles the more unsorted the sediments

***uplifted** raised up; as in mountain building

***uplifting forces** forces that originate beneath or within Earth's lithosphere that raise the land, build mountains, and cause continental growth; examples—volcanic action, earthquakes, and plate movements

***uranium-238** a radioactive isotope of uranium with an atomic mass of 238 units and a half-life of 4.5 billion years; decays to lead-206

***urbanization** the growth of a city environment (an area of high human-population density)

***U-shaped valley** the characteristic shape of a valley after it has been eroded by glaciers, especially mountain glaciers, as compared to the V-shaped valley eroded by streams

***vaporization** see **evaporation**

vein a sheetlike shaped mineral deposit formed from a solution by precipitation of dissolved minerals that has filled a crack or permeable zone in previously formed rocks; younger than the rocks into which it intrudes

vertical rays see **direct rays**

***visibility** the farthest distance at which one can see a prominent object at Earth's surface with the naked eye; decreased by fog, air pollution, and precipitation

***volcanic ash** small pieces of extrusive igneous rock that are shot into the air during a volcanic eruption

***volcanic eruption** the giving off of gases, lava, and/or lava rock onto Earth's surface or into the atmosphere through the opening or vent of a volcano

volcanic igneous rock see **extrusive igneous rock**

volcanic island arc see **island arc**

***volcano** a mound or mountain composed of extrusive (volcanic) igneous rock

***volume** the amount of space that an object occupies

***V-shaped valley** the characteristic shape of a valley eroded by a stream or river, as compared to a U-shaped glacial valley

***warm front** the boundary of an advancing warm air mass and a retreating wedge of a cooler air mass; characterized by a gentle slope, long periods of precipitation, and strato (layered) clouds

***water cycle** a model, often a diagram, used to illustrate the movement and changes of state (phase) of water at and near Earth's surface; also called the hydrologic cycle

***water retention** storing or retaining of precipitation on Earth's land surfaces as ice or snow, or as liquid water on trees and other plants

***watershed** the area of land drained by any one stream; also called a drainage area

***water table** the interface between the zone of saturation and the zone of aeration

***water vapor** gaseous water in the atmosphere; also called moisture or steam

wave the up and down motion of water in an ocean or lake; usually caused by wind

wave action the erosion and deposition of landforms near coastlines caused by ocean or lake waves

***wavelength** the distance between a point on a wave and the corresponding point on the next wave, such as the distance between two successive crests in an electromagnetic wave or an ocean wave

weather the condition of the atmospheric variables, such as temperature, air pressure, wind, and water vapor, at a particular location for a relatively short period of time

***weathering** the chemical and physical alterations of rock and other Earth materials at or near Earth's surface, through the action of temperature changes, water, chemical agents, atmospheric gases, and organic materials

***weather variable** condition of the atmosphere such as temperature, air pressure, wind, moisture conditions, cloud cover, precipitation, and storms

weight the effect of gravity on mass; unlike mass, weight varies with location

wet-bulb thermometer the thermometer of a psychrometer with a wick soaked in water around its bulb

wind the horizontal movement (compared with Earth's surface) of air in the atmosphere

wind direction the compass direction wind is blowing from; often indicated by a wind vane

wind erosion an erosional and depositional agent that is most common in desert and shoreline areas which lack much vegetation

wind speed how fast the wind is; see **wind velocity**

wind vane an instrument that spins in the wind and points to the direction the wind is blowing from, which is the wind direction name

wind velocity wind speed with a directional aspect

winter solstice on about December 21 the vertical rays of the sun fall on $23\frac{1}{2}°$ S latitude and the duration, insolation, and angle of incidence are greatest for most of the Southern Hemisphere and least in the Northern Hemisphere; see **summer solstice**

year the time it takes for a planet to make one revolution around the sun; also see **period**

***young mountains** mountains that are presently rising because they are at the location of converging plates or at the site of a hot spot

zone of aeration the soil, loose materials, or bedrock from Earth's surface down to the water table, where the pores are only partly filled with liquid water; air fills the rest of the pores

zone of saturation the portion of the soil, loose materials, or bedrock that is below the water table, where the pores are filled with ground water

Index

Landscape Regions of New York State and Their Characteristics

New York has a greater number of different landscape regions than any other state. The chief reason for this variety of landscape regions is the great variation in age, structure, and resistance of the bedrock found within the state. The present climate of the state is uniformly humid and has not been a major factor in producing variations in landscape development.

Almost all of New York was affected by the Pleistocene glaciation that ended in New York about 10,000 years ago. As a result, glacial depositional features are observed throughout the state, superimposed on the other characteristics of the different landscape regions. One of the most important features left by the glaciers is the transported soils, which are young and have small and incomplete profiles.

The landscape regions that are shown on the facing map have the following distinguishing characteristics.

ATLANTIC COASTAL PLAIN The bedrock consists of horizontal sedimentary rocks of late Mesozoic and Cenozoic age, with elevations near sea level, covered by thick glacial deposits reaching elevations of about 125 meters above sea level in some areas. The features of glacial deposition result in minor changes of slope on a generally smooth plain. Waves and ocean currents have created many typical shore features, such as bars, beaches, and lagoons.

NEWARK LOWLANDS This region is composed of weak sedimentary rocks of early Mesozoic age, which have been leveled to lower elevations than the surrounding landscape regions. The generally smooth landscape has moderate changes in slope caused by faults. A volcanic intrusion called the Palisades Sill borders the Hudson River. Because of its greater resistance, it forms a cliff that ranges up to more than 150 meters above sea level.

HUDSON HIGHLANDS—MANHATTAN PRONG These two regions are extensions of the New England Highlands to the east. These mountains in the mature and old-age stages are composed of highly distorted metamorphic rocks, mostly early Paleozoic and Precambrian in age and more resistant that the rocks of surrounding regions. In the northern parts of the Hudson Highlands there are elevations up to 500 meters with steep slopes; in the southern parts of the Manhattan Prong, the rocks have been eroded to low elevations and little slope typical of the plainlike topography of mountains in old age.

TACONIC MOUNTAINS These greatly eroded mountains were originally uplifted in the Paleozoic Era. They consist mostly of metamorphic rocks that are highly folded and faulted. Today they have moderate elevations (up to about 600 meters) and gradual changes in slope, so that the topography has the form of rolling hills.

HUDSON-MOHAWK LOWLANDS The rocks of this region are distorted sedimentary rocks of Paleozoic age that have lower resistance than the surrounding rocks. They have therefore been leveled by weathering and the erosion of the Mohawk and Hudson Rivers to generally low elevations. Variations in slope are associated with escarpments that have been formed by resistant layers, and with the valleys carved by the rivers. South of Albany the Hudson River has an elevation of sea level and may be considered an inlet of the ocean.

ALLEGHENY PLATEAU This part of the Appalachian Plateau is composed of horizontal sedimentary rocks, mostly of Paleozoic age. Much erosion by streams and glaciers has resulted in steep slopes and many changes in elevation, so that some sections, such as the Catskills, resemble mountains. Elevations range from about 500 meters in the west to over 1250 meters in the Catskills. The region contains many lakes, such as the Finger Lakes, which are the result of glacial erosion and deposition.

ERIE-ONTARIO LOWLANDS This region is a plain of horizontal sedimentary rocks of Paleozoic age, covered by much glacial transported sediment. The northern section is especially smooth because of deposition of sediments from glacial lakes that were the ancestors of Lakes Erie and Ontario. In some places, resistant rocks have been

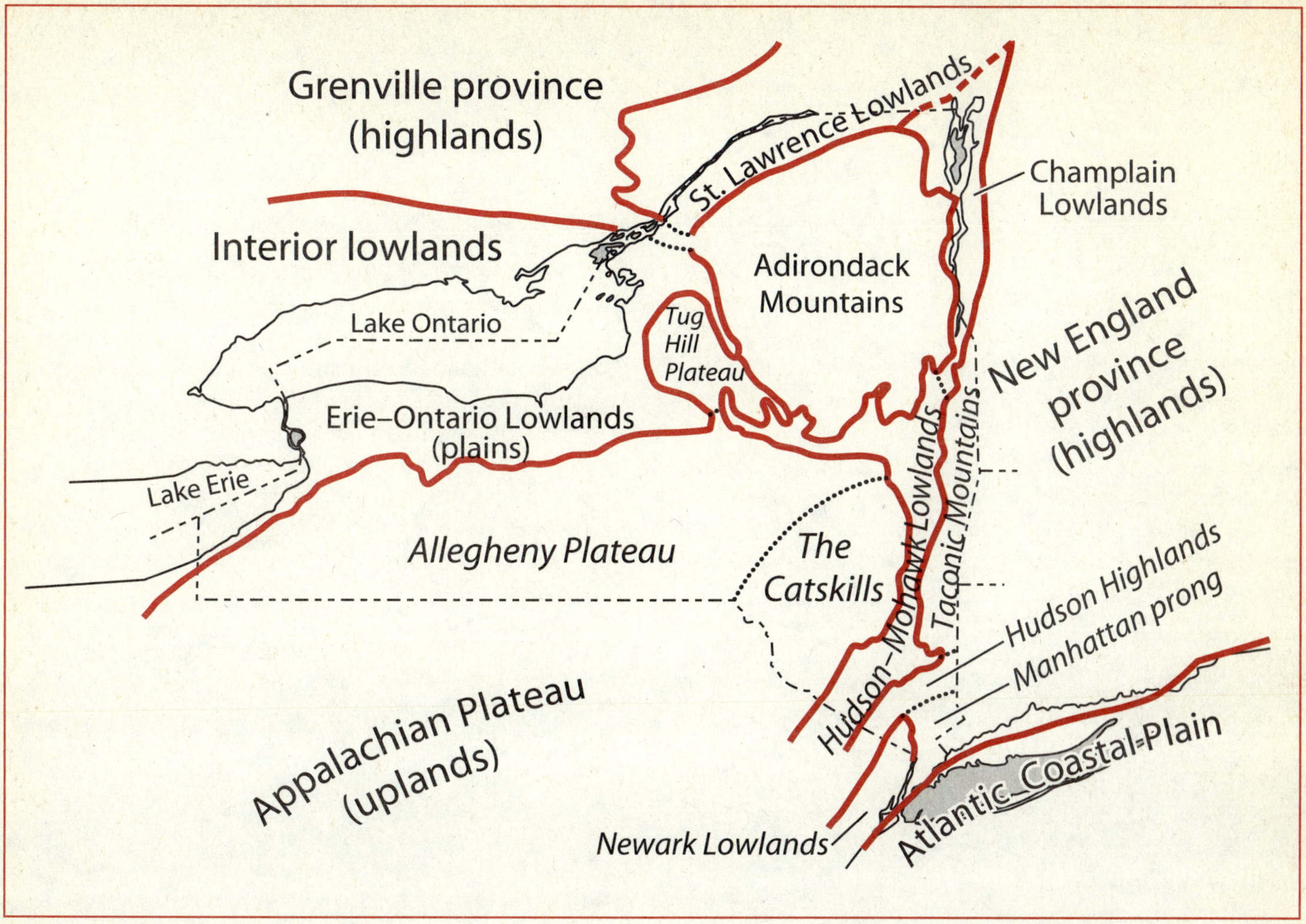

eroded to escarpments up to 500 meters in elevation, but elevations and slopes are generally small throughout the region.

TUG HILL PLATEAU This plateau of horizontal sedimentary rocks of Paleozoic age has a resistant surface layer that has resulted in fairly uniform elevations of about 600 to 700 meters. However, steep slopes occur where rivers have cut valleys, such as the Black River Valley that separates this region from the Adirondacks.

ADIRONDACK MOUNTAINS These are mature mountains composed mostly of metamorphic rocks of Precambrian age. There are moderate elevations in the western section, but the highest elevations in the state (over 1600 meters at mountain peaks) occur in the eastern section. The landscape is rugged, with much change in elevation and steep slopes, and with valleys related to faults and rocks of lesser resistance. Stream drainage patterns are of radial and trellis types except where glacial deposition has blocked the former drainage paths, resulting in many swamps and lakes.

ST. LAWRENCE-CHAMPLAIN LOWLANDS This plain is composed generally of horizontal sedimentary rocks of lower resistance than rocks of surrounding regions, and therefore has lower elevations. The St. Lawrence section is mostly flat, with changes in elevation of usually no more than 30 meters. Some steep slopes exist in the Champlain section as the result of uplift and subsidence caused by faults. Lake Champlain is the site of a subsided block.